Calculus
With Analytic Geometry

RUSSELL V. PERSON
Capitol Institute of Technology

RINEHART PRESS | San Francisco

Copyright © 1970 by Holt, Rinehart and Winston, Inc.
All Rights Reserved
Library of Congress Catalog Card Number: 69-19016
ISBN: 03-075430-5
Printed in the United States of America

   1 2 3   038   15 14 13 12

Preface

This book is an outgrowth of courses in calculus I have taught over a period of several years in a technical institute and in a university. The material and the general approach are intended for use chiefly in technical institutes but also for first year college students who need calculus for early use in their science courses. The book covers the material of an elementary course in differential and integral calculus together with some necessary analytic geometry and an introduction to differential equations.

The subject matter is meant to be practical enough for students in technology, yet theoretical enough to give students an understanding of the underlying principles of calculus as a basis for further study in mathematics. In technical work, students must have mathematics that is practical and useful to them. They want to know not only *what* to do but also *how* to do it, and something about *why*. This book attempts to give them that knowledge and understanding.

As to the organization of the material, some basic analytic geometry is presented at the beginning of the book. Needless to say, this is far from a complete treatment of that subject. However, it is intended as a necessary background for calculus. It is not interspersed at various places throughout the text. Many schools and colleges still offer a separate course in analytic geometry prior to their regular calculus course. Classes that have previously had a full course in analytic geometry can omit this section or portions of it unless they wish to use it for review.

In beginning calculus, the student is introduced to the meaning of limits and the derivative. He gets not only an understanding of the theory but also the beginning of the techniques or "tricks" of the subject. He acquires some insight, it is hoped, into the power of the derivative as something beyond his former understanding of mathematics. Applications of the derivative are given in connection with maxima, minima, linear motion, and related rates. He also discovers the technique of implicit differentiation as a powerful method of finding the derivative.

The integration of algebraic functions is taken up before going on to the study of the transcendental functions. The integral is introduced as the antiderivative, but the term *integral* is also immediately used. Then integration is easily seen as the inverse of differentiation. The reversing of the steps is carefully pointed out, so that the student knows exactly what he is doing and why, before he ever memorizes a formula. Also introduced at this point is the idea of a simple differential equation. The concept *particular integral*, which corresponds to the particular solution of a differential equation, is introduced as a link between the indefinite and the definite integral. Practical applications of the definite integral are shown in connection with areas, volumes, and linear motion.

In the differentiation of transcendental functions, careful attention is given to the application of the power rule, the product rule, and the quotient rule. After the study of the integration of these functions, the usual special techniques of integration are explained. Partial fractions and integration by parts are taken up prior to algebraic and trigonometric substitutions, since the substitution methods often result in expressions that must be integrated by the use of partial fractions or parts. Following these techniques, more complicated applications of the definite integral are shown, such as moments, centroids, length of arc, and area of surface of revolution. Separate chapters cover the hyperbolic functions, parametric equations, polar coordinates, infinite series, and the differentiation and integration involving more than one independent variable.

The final section of the book is given to the introduction of differential equations of simple types, their solution by elementary methods, and their application to specific topics in science, such as electric circuits and harmonic motion. This section is very limited in scope and is intended only to indicate to the student the power of the differential equation in solving some physical problems.

The chapters are shorter, as a rule, than those usually found in books on calculus. In fact, in some instances a chapter can be the subject of a single lecture or two. Yet I believe instructors will find there is a smooth continuity between chapters. I have found this arrangement to be most satisfactory for classroom work. Some topics often included in textbooks on calculus are omitted, not because they are considered unimportant but because they were crowded out by others that seemed more vital for an elementary course.

What is new in this book? What is different? Some special features I feel unusual are:

1. At the beginning of each topic the attempt is made to show the student a reason for what is to come. A student will learn better when he feels a need for what he is about to learn. Motivation is established in terms of the student's interest. Thereafter motivation is maintained by showing practical applications of the knowledge gained. Emphasis is upon the use of mathematics in the student's field of study.

2. The book is written for the student and in his language. Concepts are first explained in ordinary everyday language whenever possible. The student is then led to the more sophisticated symbols and formulas of calculus. The text material has been tried out in the classroom over a period of many years, not only for concepts but also for actual language used in explanations. If students at any time could not understand the

text without elaborate explanation, the language was changed for clearer understanding. The book is based on the theory that understanding must come before memorizing. It is not written as a textbook that is simply a book of texts.

3. Explanations begin where the student is, not where we would like him to be. They begin with what he already knows, not with what we wish he knew or with what we think he ought to know. The book follows a basic principle of teaching: *lead from the known to the unknown.* It contains more detailed explanation than that usually found in a calculus book. The word "obvious" is not used. Nothing is obvious to one who does not understand.

4. Numerous worked-out examples show the student exactly what to do and how to do it. Explanations in the text provide the necessary understanding, and then the worked-out examples illustrate each step in a process.

5. The entire approach is through inductive reasoning. Ideas and concepts are first presented in specific terms, then in general, not the other way around. This order of presentation follows a natural law of learning: *begin with the specific and lead to the general.*

I wish to express my appreciation to my many colleagues who have used and tested many of the techniques of presentation shown in the book, and to the staff members of Holt, Rinehart and Winston whose many valuable suggestions have assisted immeasurably in writing a textbook that will be helpful to students and instructors alike.

Russell V. Person

Washington, D.C.
September 1969

Contents

	Preface	v
1	Coordinate Systems	1
2	Equations and Graphs	29
3	The Straight Line	42
4	The Circle	57
5	The Parabola	72
6	The Ellipse	87
7	The Hyperbola	100
8	Higher Degree and Other Curves	116
9	Functions and Functional Notation	125
10	Increments: Average Rates	134
11	Limits	146
12	The Derivative: Delta Method	164
13	The Derivative by Formula	174
14	Differentiation of Algebraic Functions	183
15	Linear Motion; Distance; Velocity; Acceleration	195
16	The Derivative as the Slope of a Curve	203
17	Maxima and Minima; Inflection Points	211
18	Problems in Maxima and Minima	225
19	Implicit Differentiation	234
20	Related Rates	241
21	The Differential	251
22	Integral as Antiderivative	258
23	Integrals: Particular and Definite	274
24	The Definite Integral as an Area	287
25	Volume of a Solid	299
26	Linear Motion: Integration	307
27	Approximate Integration	326

28	Trigonometric Functions: Differentiation	334
29	Inverse Trigonometric Functions: Differentiation	352
30	Exponential and Logarithmic Functions	365
31	Integration: Power, Logarithmic, and Exponential Forms	384
32	Integration of Trigonometric and Inverse Trigonometric Functions	402
33	Partial Fractions	427
34	Special Techniques of Integration	445
35	Some Applications: Liquid Pressure; Work	464
36	Centroids	480
37	Moment of Inertia	498
38	Length of Arc and Area of Surface of Revolution	515
39	Parametric Equations	527
40	Polar Coordinates	545
41	Hyperbolic Functions	574
42	Partial Differentiation	587
43	Infinite Series	609
44	Differential Equations	626
45	Perfect Differentials	641
46	Homogeneous Equations; Constant Coefficients	649
47	Imaginary and Complex Roots	662
48	Nonhomogeneous Differential Equations	672
	Table I Natural Logarithms of Numbers	681
	Table II Values of e^x and e^{-x}	682
	Answers to Selected Odd-Numbered Problems	683
	Index	697

Calculus
With Analytic Geometry

chapter

1

Coordinate Systems

1.1 THE LOCATION OF POINTS

Whenever we ask, "Where is it?" we are concerned with location. Perhaps hundreds of times a day we ask or are asked this question. The question involves the location of a particular point or place. There are several ways or systems for locating points. These systems may be called one-dimensional, two-dimensional, or three-dimensional. We shall be concerned chiefly with the two-dimensional system, but shall describe briefly the other two.

By means of the one-dimensional system we locate points along a *line*. In the two-dimensional system we locate a point in a *plane*. The three-dimensional system is a means of locating a point in *space*.

1.2 A ONE-DIMENSIONAL SYSTEM

Suppose everyone in a particular area lived along a single straight road, and along this road there is a milepost at every mile, numbered from some point we call *zero*. Then all we should need to do to tell the location of a house along the road would be to give the number of the milepost at the house. For example, we could say Mr. X lives at milepost No. 3. Of course, we might have fractions. For example, we could say that a particular house is located at the point that corresponds to the number 7.3, which is three-tenths of a mile past the 7-mile post.

This is essentially what is done on the so-called *number line*. We draw a horizontal line and call it the *x-line* (Fig. 1.1). We mark a point and call it *zero*, as the *origin* or beginning. Then, beginning at zero, we mark off equal units of length to the right and left of zero. At the right we mark points corresponding to the positive numbers 1, 2, 3, and so on. Toward the left, starting at zero, we mark off units of distance and mark the points with the negative numbers -1, -2, -3, and so on. We can also show

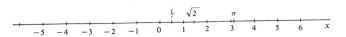

Fig. 1.1

points for fractions and for the irrational numbers, such as $\sqrt{2}$, $\sqrt{3}$, π, and so on. This line then we call the *Real Number Line*. For every real number there is a corresponding point on the line, and for every point on the line there is a corresponding real number. That is, there is a *one-to-one correspondence* between the real numbers and the points on the line.

Specific or definite points on the line are numbered by some real number. For example, the point 5 is a definite point. We sometimes denote the point by enclosing its number in parentheses. Then (5) denotes the point itself, whereas the number 5 indicates its directed distance from zero.

If we wish to refer to the general location of a point along the line, we call it the point x. This is a variable point on the line, like a car driven along our single straight road. By the x we mean the *distance* of the point from zero.

If we wish to denote a point that is understood to be fixed but may be anywhere, we may call it point A, point B, and so on. Or we sometimes use subscripts and say the point x_1, point x_2, and so on. Such a point is understood to be a *specific point that may be anywhere*. Again, the notations x_1 and x_2 indicate the respective distance from zero.

1.3 GREATER THAN AND LESS THAN

If we wish to express the fact that one number is *greater than* another, we use the symbol $>$. For the term *less than*, we use the symbol $<$. For example, $5 > 3$, and $3 < 5$. If one number is greater than another, its corresponding point lies farther toward the right on the number line. Then the point (5) is at the right of the point (3), since $5 > 3$. Conversely, if any point on the line lies at the right of another point, we say the number of the first point is greater than the number of the second. Then

$$1 > -3 \quad \text{or} \quad -3 < 1$$

If we denote any two specific points on the line by x_1 and x_2, such that x_2 is at the right of x_1, then we can say

$$x_2 > x_1; \text{ or, subtracting } x_1 \text{ from both sides, } x_2 - x_1 > 0$$

The *absolute value* of a number means the value of the number without regard to its sign. The absolute value of (-3) is $+3$. That is, for any number, its absolute value is *positive*. The absolute value is indicated by a pair of vertical lines. Then $|-3| = |+3|$. As a result, $1 < |-3|$. In general,

if x is positive, then $|-x| = x$; if x is negative, then $|x| = -x$

The expression $x > 2$ means that x may have any value greater than 2, such as 2.0001, 3, 7, 1000, and so on, but it is not equal to 2. The symbol $x \geq 2$ is read "x is greater than or equal to 2." In this case the values of x include the value 2, as well as all values greater than 2.

Inequalities involving the unknown x can be solved for the values of x, just as we solve equalities (equations). There is a difference in answers, however. When we solve an equation, such as $2x + 5 = 11$, we get a specific value, in this case, $x = 3$. Then x represents one specific point on the number line.

However, suppose we have the inequality $2x + 5 > 11$
Subtracting 5 from both sides we get $2x > 6$
Dividing both sides by 2, $x > 3$

The result means that x represents more than a single value. In fact, it represents an entire domain of positive values beginning with but not including 3 and going on to positive infinity.

In solving inequalities one point must be remembered. Multiplying or dividing both sides of an inequality by a negative number reverses the sign of inequality. For example, if $-x > 2$, then $x < -2$.

1.4 DISTANCE BETWEEN TWO POINTS ON THE LINE

Let A and B represent two points on the number line (Fig. 1.2). The line segment AB is taken as a *directed* distance starting at A and going to B. If the segment is read BA, we mean the distance from B going to A, which is negative. Then $AB = -BA$. For absolute values we have $|AB| = |-AB|$.

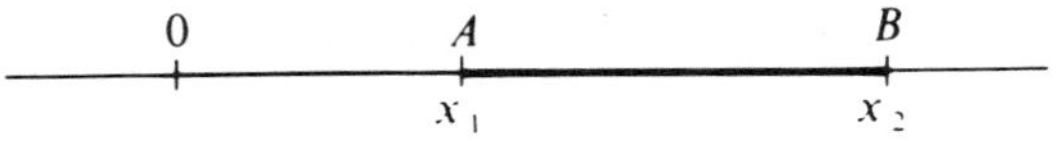

Fig. 1.2

Let us suppose that x_1 represents the distance from the origin O to A and x_2 represents the distance from O to B. To find the distance from A to B, we take

$$x_2 - x_1$$

For example, if A is at the point (4) and B is at the point (10), then the distance from A to B is

$$10 - 4 = 6$$

If we subtract in the opposite way, we get $4 - 10 = -6$. The result represents a negative distance. If we are interested only in the absolute value of the distance, we have

$$d = |4 - 10| = 6$$

That is,

$$x_2 - x_1 = -(x_1 - x_2); \quad \text{but} \quad |x_2 - x_1| = |x_1 - x_2|$$

In most instances we are concerned only with the absolute values of distances. However, sometimes it is necessary to hold to a signed distance.

1.5 MIDPOINT OF A SEGMENT OF THE NUMBER LINE

Suppose we wish to find the midpoint of the line segment from the point (4) to the point (10) on the number line (Fig. 1.3). To find the midpoint we can of course first find the entire length of the segment $10 - 4 = 6$ units. Then we divide the length by 2 and get the length of *half* the segment; that is, $\frac{6}{2} = 3$. Then we add this half to the first value 4 and get $4 + 3 = 7$. Therefore, the point (7) is the midpoint.

Fig. 1.3

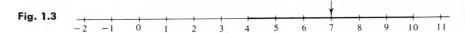

We might get the midpoint in a somewhat simpler way. Let us suppose the end points of the segment are x_1 and x_2. If we work this out in the same way as above we get, $x_2 - x_1$, the length of the segment. Then one-half the segment is given by

$$\frac{x_2 - x_1}{2}$$

Adding this quantity to x_1, we get

$$x_1 + \frac{x_2 - x_1}{2}$$

which equals the distance to the midpoint of the segment.
Combining the terms we get

$$\frac{2x_1 + x_2 - x_1}{2} = \frac{x_1 + x_2}{2}$$

That is, to find the midpoint of a line segment, we *add* the values of the end points and divide by 2. This is the same as saying the *average* of the two x-values. To find the midpoint of the segment from (4) to (10), we have

$$\frac{4 + 10}{2} = 7$$

1.6 INCREMENT IN x

There is one other notation we introduce here. Suppose P is a variable point that moves along the line. Let us assume some particular x-value for P, and call it x_1 (Fig. 1.4). Now suppose the point P moves to some new x-value, x_2. The change in the x-value is often called an *increment in x*. An increment means a change in the value of x, positive or negative. This increment is called *delta x*, and is written with capital Greek letter:

Fig. 1.4

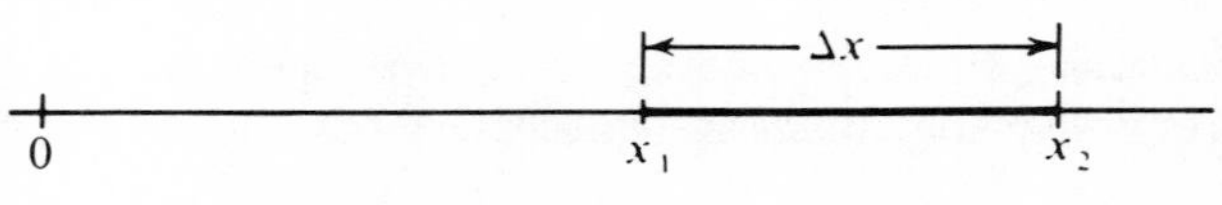

Δx. This notation does not mean a product, but simply a change in x. Then we can say that the change in x can be stated as

$$\Delta x = \text{second value} - \text{first value}$$

or $\quad \Delta x = x_2 - x_1$

Any increment in x from one value to a second value will always have the correct sign if we take the values in this order:

second value $-$ *first value*

For example, if P is first at the point (7) and later at the point (4), then $x_1 = 7$ and $x_2 = 4$. Then the change is given by

$$\Delta x = 4 - 7 = -3$$

This means that x changed by a negative 3 units.

As another example, if $x_1 = -2$ and $x_2 = 4$, then

$$\Delta x = x_2 - x_1 = 4 - (-2) = +6$$

Note that the change from (-2) to $(+4)$ is $+6$ units.

EXERCISE 1.1

1. Arrange the following numbers in the order of magnitude:
   $5, -2, -5, 3, -1, 0, 2, -4, 1, -7$
2. Arrange in the order of magnitude:
   $0, 3.2, -1.3, -4.7, 0.3, 6.1, \sqrt{3}, -\sqrt{5}, 1.5, -\sqrt{2}, -\sqrt{6}$
3. Place the proper equality or inequality sign between the two numbers in each of the following sets:
   (a) $-3;5$ (b) $4;-1$ (c) $2;-7$ (d) $-5;3$ (e) $6;6$
   (f) $0;2$ (g) $0;-3$ (h) $-2;2$ (i) $\sqrt{2};\sqrt{3}$ (j) $2\sqrt{3};\sqrt{8}$
4. In each of the following, find the distance between the two points indicated by the two values of x:
   (a) -3 and 4 (b) 2 and 7 (c) 1 and -4 (d) -3 and -7
   (e) -5 and -2 (f) -1.3 and -4.1 (g) 5.7 and 2.3
5. Find the midpoint of the line segment indicated by the following intervals (the numbers in brackets denote the end points of the segment):
   (a) $[7,13]$ (b) $[-5,3]$ (c) $[-7,-1]$ (d) $[4,7]$ (e) $[3,8]$
   (f) $[-3,10]$ (g) $[-4,4]$ (h) $[-1,6]$ (i) $[-9,8]$ (j) $[-3,-4]$
6. Find the value of Δx as x changes from the first value to the second:
   (a) $x_1 = 4, x_2 = 9$ (b) $x_1 = 3, x_2 = -5$ (c) $x_1 = -2, x_2 = 4$
   (d) $x_1 = -1, x_2 = -6$ (e) $x_1 = 3, x_2 = 0$ (f) $x_1 = 0, x_2 = -4$
   (g) $x_1 = -2, x_2 = 0$ (h) $x_1 = 0, x_2 = -5$ (i) $x_1 = -3, x_2 = -7$
7. Find the set of all values of x that satisfy each of the following:
   (a) $|x| = 4$ (b) $x = |-3|$ (c) $|x - 2| = 5$ (d) $|x| = -3$
   (e) $x = 2$ (f) $x > 2$ (g) $x < -4$ (h) $x > -3$
   (i) $x \leq 3$ (j) $x \geq 4$ (k) $3x - 2 > 13$ (l) $2x + 3 < 8$
   (m) $4 - x > 7$ (n) $\dfrac{3x + 2}{5} < 4$ (o) $\dfrac{4 - 3x}{2} < 5$

1.7 POINTS IN A PLANE

To locate points in a plane (a flat surface) we need a *two-dimensional* system. The location of points can then be described in two ways.

Suppose we stand at point O, the intersection of two city streets laid out at right angles to each other (Fig. 1.5). A man comes up to us and asks, "How can I get to point P, a point diagonally across the city block?"

Fig. 1.5

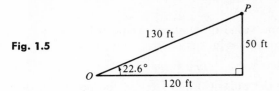

If it is necessary to follow the walks along the streets, we might say, "Walk 120 feet east and then 50 feet north." In this case we make use of the two distances, one east, the other north. The two numbers, 120 feet and 50 feet, are called the *rectangular coordinates* of point P with reference to point O as a starting point. This system of showing the location of points in a plane is called the *rectangular* system, often called the *Cartesian* system in honor of René Descartes, the Frenchman who originated the system. The two numbers used to locate a point are sometimes called *Cartesian coordinates*. The two distances are measured at right angles, hence the term *rectangular*.

On the other hand, if there is a vacant lot between the two points O and P and no obstruction, we might tell the man, "If you are in a hurry, you can turn your steps in a direction 22.6° north of east and walk a distance of 130 feet directly to your destination" (Fig. 1.5). The two numbers 22.6° and 130 feet are the two numbers that designate the point P with reference to point O. These two numbers are called *polar coordinates*. Polar coordinates consist of the two numbers denoting an *angle* and a *distance*. The angle is measured with reference to a ray called the *polar axis*, and the distance is measured from a point called the *pole*.

The system of polar coordinates is studied in detail in Chapter 40. In this chapter we study the rectangular system, since it is used in connection with most work in beginning calculus. However, in much of our moving about in our everyday activities, we make use of the principle of polar coordinates. We simply set our direction and then walk directly toward our objective. The main purpose is to save time. Whenever you see a diagonal path that hurrying feet have made across a nice green lawn, you know someone has misused polar coordinates.

1.8 THE RECTANGULAR COORDINATE SYSTEM

To locate a point in a plane by means of rectangular coordinates, we draw two straight lines, one horizontal which we call the *x-axis*, the other vertical which we call the *y-axis* (Fig. 1.6). These two lines, or axes, may be compared to two main streets of a town, one having an *east-west* direction, the other a *north-south* direction. The intersection O of the *x*-axis and the *y*-axis is called the *origin*. The two axes divide the entire plane

1.8 The Rectangular Coordinate System

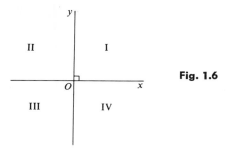

Fig. 1.6

into four *quadrants*, numbered I, II, III, IV, in a counterclockwise direction, as shown in Figure 1.6.

Any point in the plane can now be located by stating its distance from the two axes. The distance in the *x*-direction, right or left, is called the *abscissa* of the point. The distance in the *y*-direction, up or down, is called the *ordinate* of the point. The abscissa and the ordinate are called the *coordinates* of the point. (When we use the terms *up* and *down*, it must be understood that one point is not actually *above* another in space, since we are here dealing with points on a flat surface. By the term *up* we mean toward the top of the paper; *down* means toward the bottom of the paper.)

Distances measured to the *right* and *up* are called *positive;* distances to the left and downward are called *negative*. Note that the abscissa, or *x*-distance, is actually measured from the *y*-axis; the ordinate, or *y*-distance, is actually measured from the *x*-axis. For example, the point *P* (Fig. 1.7) is located 8 units in the *x*-direction and 3 units in the *y*-direction. The two numbers showing the location of the point are written in parentheses and separated by a comma, (8,3). *The abscissa is always written first*, the ordinate second. Then we can say that every point in the plane can be represented by a pair of numbers in a definite order, and every ordered pair of numbers represents a point in the plane. That is, there is *one-to-one correspondence* between ordered pairs of numbers and the points in the plane. Note that the pair is *ordered*. That is, the point (8,3) is not the same as the point (3,8).

Figure 1.8 shows the location of a number of points with the coordinates of each point given. Note especially the sign of each coordinate of each

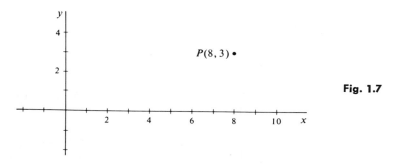

Fig. 1.7

8   Coordinate Systems

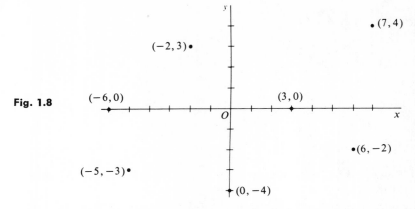

Fig. 1.8

point. In the first quadrant both coordinates are positive. In the second quadrant the abscissa is negative. In the third quadrant both coordinates are negative, and in the fourth quadrant the ordinate is negative. If a point lies on one of the coordinate axes, then one of the coordinates is zero. The use of rectangular coordinate paper (marked off into squares) makes it fairly easy to locate any point in the plane. Fractions can also be estimated. Remember, the plane is not limited to the size of the paper but extends infinitely far in positive and negative directions.

1.9 NOTATION FOR POINTS

It is well at this time to identify three kinds of points and their notations.

(1) *Specific* (or *definite*) points. The number pair (7,4) represents a specific, definite point in the first quadrant. It is fixed and is not found in any other position (Fig. 1.9). The same is true with respect to the point (−5,−3), a specific point in the third quadrant. A specific or definite point is any point whose position is definitely fixed.

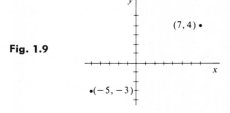

Fig. 1.9

(2) *General* (or *variable*, or *moving*, or *tracing*) point. If we wish to denote a point that may vary and be anywhere, we denote it by (x,y) (Fig. 1.10). This general point may be considered as representing all points in the plane, or it may be considered as moving about and tracing out a curve of some kind. The notation (x,y) is used to denote the general point in *all* quadrants. The student is warned against a common error of writing $-x$ for the abscissa of the general point in the second and third

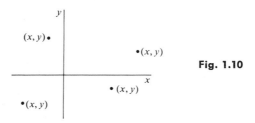

Fig. 1.10

quadrants, or $-y$ for the ordinate in the third and fourth quadrants. It is true, the numerical values are negative for some quadrants, but the general point should always be denoted (x,y). For example, in the second quadrant, we may have $x = -4$, but do not say $-x = -4$. To do so would make $x = +4$.

(3) A *"hybrid"* point. This kind of point is mentioned here to clarify a trouble some students have with regard to notation. Often, especially in deriving formulas, we wish to indicate a particular point as fixed, but we are not concerned with just where it is fixed. It may be anywhere, but it is understood to be a definite point. It does not vary. Such a point is usually denoted by numbers that are called arbitrary constants, such as (h,k), (a,b), and so on. Often subscripts are used, such as in $P_1(x_1,y_1)$, $P_2(x_2,y_2)$, $P_n(x_n,y_n)$. These points (Fig. 1.11) may look like variable points, but they are specific points. Then a "hybrid" point may be defined as a *definite point that may be anywhere.*

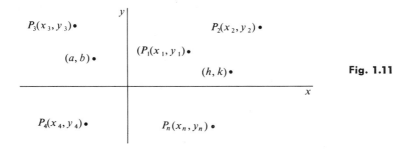

Fig. 1.11

1.10 PROJECTIONS

The projection of a point on a line is the foot of the perpendicular drawn from the point to the line. Thus the projection of point P on line l is the point A (Fig. 1.12).

Fig. 1.12

Suppose now we have a line segment from point $P_1(x_1,y_1)$ to $P_2(x_2,y_2)$. The projection of P_1 on the x-axis is point A (Fig. 1.13), and the projection of the same point on the y-axis is point C. The projection of P_2 on the x-axis is point B, and on the y-axis point D. Then the projection of the line segment P_1P_2 on the x-axis is AB; its projection on the y-axis is CD. It is well to consider these segments as directed distances. In going from one point to another, we read the projection in the same order. If we wish to find the length of the projection, we subtract in the following order:

second value − first value

The projection of the segment P_1P_2 (in that order) on the x-axis is $x_2 - x_1$. Its projection on the y-axis is $y_2 - y_1$. These formulas hold even when the two points P_1 and P_2 are on opposite sides of an axis.

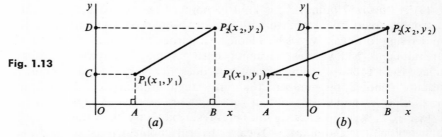

Fig. 1.13

1.11 DISTANCE BETWEEN TWO POINTS IN A PLANE

One important problem in analytic geometry is that of finding the distance between two given points. There is a simple formula for this distance in terms of the coordinates of the points.

To see how the formula comes about, let us first take two specific points, say, $A(2,5)$ and $B(10,11)$ (Fig. 1.14). We denote the distance by d. In this case we are not concerned with direction, only with the absolute value of the distance, the length of the line segment AB. We draw a line through A parallel to the x-axis and a line through B parallel to the y-axis. These

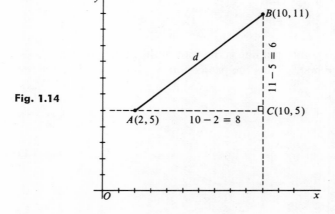

Fig. 1.14

lines intersect at some point C, forming a right triangle. The hypotenuse of this right triangle is the required distance d. Note that the point C has the coordinates $(10,5)$.

To find the length of the hypotenuse we apply the Pythagorean Theorem. Note that one leg of the triangle, the base, has a length equal to

$$10 - 2 = 8 \text{ units}$$

The difference of the abscissas is actually the projection on the x-axis. The other leg of the right triangle is found as the difference between the ordinates:

$$11 - 5 = 6 \text{ units}$$

The difference between the ordinates is actually the projection on the y-axis. For the hypotenuse we square the lengths of the legs, 8 and 6, and find the square root of the sum, that is,

$$8^2 + 6^2 = 100$$

then

$$d = \sqrt{100} = 10$$

The formula itself for the distance between any two points is derived in the same manner. We denote the two given definite points by $P_1(x_1,y_1)$ and $P_2(x_2,y_2)$ (Fig. 1.15). Again we draw a line through P_1 parallel to the x-axis and a line through P_2 parallel to the y-axis, intersecting at another point $P_3(x_2,y_1)$, forming a right triangle. The hypotenuse of the right triangle is the required distance d. One leg of the right triangle is the length, $x_2 - x_1$, the projection on the x-axis. The other leg is $y_2 - y_1$, the projection on the y-axis.

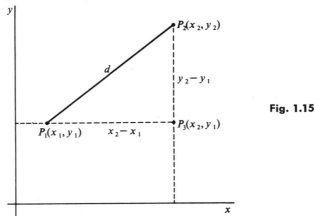

Fig. 1.15

We first square the lengths of the two legs and express their sum:

$$(x_2 - x_1)^2 + (y_2 - y_1)^2$$

The sum is the square of the hypotenuse d, which is then d^2. Taking the square root of the sum, we get the formula for the distance d:

$$d = \sqrt{(x_2 - x_1)^2 + (y_2 - y_1)^2}$$

12   Coordinate Systems

Since we are concerned only with the magnitude of the distance as a non-directed quantity, we take the positive form of the radical.

Stated in words, *the distance between any two points can be found by taking the difference between the abscissas, and the difference between the ordinates. The differences are squared, and these squares are then added. Finally, we take the square root of the sum.*

Actually it makes no difference which point is called P_1 or P_2. The formula applies to all situations in which we wish to find the distance between *any* two given points.

In some formulas involving distances it is important that we take points in a certain order. However, for the distance between two points, the difference values are squared, and the result is always positive.

EXAMPLE 1. Find the distance between the two points, (3,8) and (5,13).

Solution. If we call the first point P_1 and the second point P_2, we have

$$d = \sqrt{(5-3)^2 + (13-8)^2} = \sqrt{2^2 + 5^2} = \sqrt{4 + 25} = \sqrt{29} = 5.385 \text{ (approx.)}$$

If we reverse the points, we get the same answer:

$$d = \sqrt{(3-5)^2 + (8-13)^2} = \sqrt{(-2)^2 + (-5)^2} = \sqrt{4 + 25} = \sqrt{29}$$

EXAMPLE 2. A triangle has two of its vertices at the points $B(-4,0)$ and $C(3,-5)$, respectively. Find the length of this side of the triangle.

Solution. $d = \sqrt{(-4-3)^2 + (0+5)^2} = \sqrt{49 + 25} = \sqrt{74}$

Answers are often left in radical form. If the points are reversed, we get the same answer.

1.12   MIDPOINT OF A LINE SEGMENT

The problem of finding the midpoint of a line segment shows the difference between analytic geometry and Euclidean high school geometry. In high school geometry, to find the midpoint of a line segment, we swing a couple of pairs of intersecting arcs with a compass, using the end points of the segment as centers. We line up a straightedge with the two points of intersection of the arcs. Then we make a mark on the line segment and say, "There is the midpoint."

In analytic geometry the approach is entirely different. A midpoint of a line segment is not necessarily a mark on a piece of paper. It is a *pair of coordinates* that name the point. In analytic geometry, points are denoted by pairs of numbers, not by pencil marks on paper.

For example, suppose we have the line segment whose end points are (2,3) and (14,9) (Fig. 1.16). Now we wish to find the midpoint of the segment. We denote the midpoint by $P_0(x_0, y_0)$. The abscissa of the midpoint is the *average* of the abscissas of the end points. That is,

$$x_0 = \frac{2 + 14}{2} = \frac{16}{2} = 8$$

1.12 Midpoint of a Line Segment   13

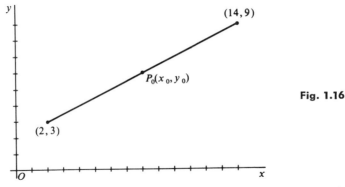

Fig. 1.16

In like manner, the ordinate of the midpoint is the average of the ordinates of the end points. That is,

$$y_0 = \frac{3+9}{2} = \frac{12}{2} = 6; \quad \text{midpoint } (8,6)$$

Using more general terms, we can derive the formulas for the midpoint of a line segment. Call the endpoints $P_1(x_1,y_1)$ and $P_2(x_2,y_2)$, two given points, and let us call the midpoint $P_0(x_0,y_0)$ (Fig. 1.17). To find x_0, the abscissa of the midpoint, we note that points A and B are the projections of P_1 and P_2, respectively, on the x-axis. Also point M is the projection of the midpoint P_0. Now, since AP_1 and BP_2 are both perpendicular to the x-axis, they are parallel to each other and also to MP_0. Therefore, M is the midpoint of AB. Then we have the following:

$$AB = OB - OA; \quad OB = x_2 \quad \text{and} \quad OA = x_1; \quad AM = \frac{1}{2} AB;$$

then

$$AM = \frac{1}{2}(x_2 - x_1)$$

Since $OM = OA + AM$, we have

$$x_0 = x_1 + \frac{1}{2}(x_2 - x_1)$$

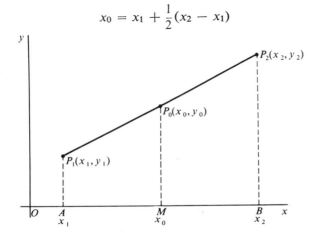

Fig. 1.17

Simplifying, $$x_0 = \frac{x_1 + x_2}{2}$$

In like manner, $$y_0 = \frac{y_1 + y_2}{2}$$

EXAMPLE 3. A triangle has its vertices at $A(8,7)$; $B(-4,1)$; $C(0,-7)$. Find the midpoints of the sides.

Solution. Note that we first add the abscissas of the two points and then take one-half of the sum. We do the same for the ordinates. For the midpoint of AB we have

$$x_0 = \frac{8-4}{2} = \frac{4}{2} = 2; \quad y_0 = \frac{7+1}{2} = \frac{8}{2} = 4; \text{ midpoint is } (2,4)$$

For side BC:

$$x_0 = \frac{-4+0}{2} = -2; \quad y_0 = \frac{1-7}{2} = -3; \text{ midpoint is } (-2,-3)$$

For side AC:

$$x_0 = \frac{8+0}{2} = 4; \quad y_0 = \frac{7-7}{2} = 0; \text{ midpoint is } (4,0)$$

The student should sketch the triangle showing vertices and midpoints of the sides.

EXERCISE 1.2

1. Find (a) the distance between the two points indicated in each of the following sets of coordinates; (b) the midpoint of the line segment joining the two points.
   (a) $A(-4,7); B(3,-2)$ (b) $C(3,-5); D(3,2)$ (c) $E(7,0); F(0,-5)$
   (d) $G(4,0); H(-2,4)$ (e) $I(-3,-1); J(-4,6)$ (f) $K(6,-2); L(-2,-8)$
2. Find (a) the distance between the two points indicated in each of the following sets of coordinates; (b) the midpoint of the line segment joining the two points.
   (a) $A(7,-4); B(-5,4)$ (b) $C(4,-4); D(0,-2)$ (c) $E(-3,-3); F(6,-3)$
   (d) $G(-1,-5); H(-6,0)$ (e) $I(5,-5); J(-5,5)$ (f) $K(-1,2); L(0,0)$
3. The following points are vertices of a triangle: $A(8,6); B(-4,2); C(-2,-6)$. Find (a) the length of each side of the triangle; (b) the midpoint of each side; (c) the length of each median; (d) the perimeter of the triangle.
4. The following points are vertices of a triangle: $D(1,6); E(-7,2); F(5,-4)$. Find (a) the length of each side of the triangle; (b) the midpoint of each side; (c) the length of each median; (d) the perimeter of the triangle.
5. The points in each of the following sets are vertices of a triangle. By use of the distance formula show whether each triangle is equilateral, isosceles, or scalene:
   (a) $A(1,5); B(-4,0); C(2,-2);$ (b) $D(2,7); E(-3,-7); F(5,-2);$ (c) $R(0,6); S(-9,-1); T(3,-5)$.
6. Show by the distance formula and the Pythagorean Theorem that the following points are vertices of a right triangle: $A(6,1); B(2,7); C(-6,-7)$.
7. Find the radius of each of the following circles: (a) center at $(5,-2)$ and the point $(1,6)$ on the circumference; (b) center at $(7,-3)$ with the x-axis tangent to the circle; (c) center $(-4,-6)$ with the y-axis tangent to the circle.
8. A circle has its center at $(-1,2)$. How many of the following points could lie on its circumference: $A(5,0); B(-3,8); C(-5,-3); D(1,-4)$?
9. How many of the following points could lie on the same circle with center at the point $(2,-1)$: $A(1,5); B(-4,0); C(-1,-6); D(8,-2)$.

10. A circle has the ends of a diameter at the points $(7,-3)$ and $(-1,7)$. Find the center of the circle, and then find the radius of the circle.
11. A circle has the ends of a diameter at the points $(-5,-2)$ and $(5,-4)$. Find the center of the circle and then find the radius of the circle.
12. A circle has the ends of a diameter at the points $(-6,-4)$ and $(-9,4)$. Find the center of the circle and then find the radius of the circle.
13. Show by the distance formula that the following points are vertices of a parallelogram: $A(8,2)$; $B(1,6)$; $C(-10,0)$; $D(-3,-4)$.
14. Show by the distance formula that the following points are vertices of a parallelogram: $E(5,2)$; $F(-2,3)$; $G(-3,-4)$; $H(4,-5)$.
15. Show that the following points are the vertices of a square: $A(6,0)$; $B(3,5)$; $C(-2,2)$; $D(1,-3)$.
16. Show that the following points are the vertices of a square: $E(5,2)$; $F(-2,3)$; $G(-3,-4)$; $H(4,-5)$.
17. Show that these points are collinear: $A(2,1)$; $B(5,3)$; $C(-4,-3)$.
18. Show that the point $(2,1)$ is the midpoint of the line segment joining the points $A(-7,6)$ and $B(11,-4)$.
19. A quadrilateral has its vertices at the points $A(9,3)$; $B(-7,7)$; $C(-3,-7)$; $D(5,-5)$. Show by the distance formula that the figure is not a parallelogram.
20. Find the midpoints of the sides of the quadrilateral in No. 19. Then show that the figure formed by joining the midpoints consecutively is a parallelogram.

1.13 SLOPE OF A LINE

The word *line* is here understood to mean *straight line*. The *slope* of a line refers to its steepness or, in other words, to its rate of rising or falling. In everyday language we often describe the slope with such words as "steep," "very steep," "not very steep," "extremely steep," and so on. (One student, when asked to describe the slope of a very steep line, called it "treacherous.")

However, in mathematics the slope must be described by the use of numbers, just as in all science and engineering. By using numerical values we can describe the slope precisely whether we refer to a roadway, the roof of a house, a railroad grade, or a church steeple.

Suppose a line passes through the two points A and B (Fig. 1.18). If we follow along the line in the direction from A to B, our elevation changes as our horizontal distance changes. To determine the amount of these changes, we draw a horizontal line through A and a vertical line through B, these two lines intersecting at C. Then, as we move from A to B, our change in elevation, called the *rise*, is the distance CB. The horizontal change, called the *run*, is the distance AC. Then we define the slope of the line by the ratio

$$\text{slope} = \frac{\text{rise}}{\text{run}}$$

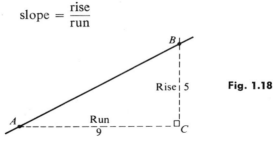

Fig. 1.18

The changes should be considered as directed distances in going *from* a first point *to* a second point. In going from A to B, let us suppose the rise is 5 units and the run is 9 units. Then the slope of the line is given by the fraction (or ratio)

$$\text{slope} = \frac{\text{rise}}{\text{run}} = \frac{5}{9}$$

It should be noted that the slope is an abstract number. It does not signify any kind of units. The rise and the run are measured in linear units, such as inches, feet, and so on. But the slope has no denomination. For example,

$$\text{slope} = \frac{\text{rise}}{\text{run}} = \frac{5 \text{ inches}}{9 \text{ inches}} = \frac{5}{9}$$

1.14 FORMULA FOR SLOPE

To get a formula for the slope of a given line l, we take any two points, $P_1(x_1,y_1)$ and $P_2(x_2,y_2)$ on the line (Fig. 1.19). We draw a horizontal line through P_1 and a vertical line through P_2, these two lines intersecting at C. Then as a point moves along the line l in the direction P_1 to P_2, the rise is the directed distance CP_2 and the run is the directed distance P_1C. The *run* represents a change in the abscissa of the moving point from x_1 to x_2. The rise represents a change in the ordinate from y_1 to y_2.

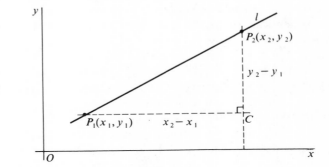

Fig. 1.19

As we have stated (Sec. 1.6) changes in variables are called *increments*. A change in x is called an increment in x, denoted by Δx. The increment in y is called *delta y* and is denoted by Δy. To find the amount of an increment or change in moving from one point to another, we subtract the first value from the second; that is,

$$\text{second value} - \text{first value}$$

The *run*, or the increment in x, is given by

$$\Delta x = x_2 - x_1$$

The *rise*, or increment in y, is given by

$$\Delta y = y_2 - y_1$$

1.14 Formula for Slope

The slope of a line is therefore given by the formula

$$\text{slope} = \frac{\text{rise}}{\text{run}} = \frac{\Delta y}{\Delta x} = \frac{y_2 - y_1}{x_2 - x_1}$$

The slope of a line is often denoted by m. Then, as a general point (x,y) moves along the line from P_1 to P_2, the slope of the line is given by

$$m = \frac{y_2 - y_1}{x_2 - x_1}$$

The order here is sometimes important since increments are directed distances. The directed line segment from the first point to the second is often called a *vector*. Then Δx and Δy are called the *scalar components* of the vector. If the two points are reversed, the slope of the line will turn out to be unchanged, but the increments (the scalar components) will have reversed signs. In moving from one point to another, we shall always get the correct sign for the increments if we subtract in this order:

second value — first value

EXAMPLE 4. Find the slope of the line passing through the points $P_1(1,4)$ and $P_2(7,8)$. What are the increments Δy and Δx?

Solution. Applying the formula

$$m = \frac{y_2 - y_1}{x_2 - x_1} = \frac{8 - 4}{7 - 1} = \frac{4}{6} = \frac{2}{3}$$

The fraction $\frac{2}{3}$ represents the numerical value of the slope. However, the increments are: $\Delta y = 4$; $\Delta x = 6$. If the points are reversed and we assume moving from P_2 to P_1, we get the same value for the slope:

$$m = \frac{4 - 8}{1 - 7} = \frac{-4}{-6} = +\frac{2}{3}$$

However, in this case, the increments. Δx and Δy, being directed distances, become

$$\Delta y = 4 - 8 = -4; \Delta x = 1 - 7 = -6$$

If a line falls as it moves from left to right, then the slope is negative, as in the following example:

EXAMPLE 5. Find the slope of the line passing through the points $P_1(-2,6)$ and $P_2(4,1)$. Sketch the line and find the increments in going from P_1 to P_2.

Solution. Note that in this example the *rise is negative* (Fig. 1.20). That is $\Delta y = 1 - 6 = -5$. The slope is

$$m = \frac{\Delta y}{\Delta x} = \frac{1 - 6}{4 + 2} = \frac{-5}{6} = -\frac{5}{6}$$

In this case, $\Delta y = -5$; $\Delta x = +6$. If we reverse the points, the numerical value of the slope will be the same as before; however, the increments will have different signs. For the slope we have

$$m = \frac{6 - 1}{-2 - 4} = \frac{+5}{-6} = -\frac{5}{6}$$

however,

$$\Delta y = +5; \Delta x = -6$$

18   Coordinate Systems

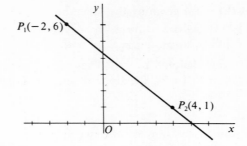

Fig. 1.20

If the two given points on the line are such that $y_2 = y_1$, then the numerator for the slope becomes zero and the slope is zero. For example, the slope of a line through $P_1(-4,3)$ and $P_2(5,3)$ is

$$m = \frac{3-3}{5+4} = \frac{0}{9} = 0$$

That is, a horizontal line has a *zero slope*.

If the two points on the line are such that $x_2 = x_1$, then the denominator of the slope formula becomes zero; that is, $x_2 - x_1 = 0$. The line can be said to have an *infinite* slope. For example, if a line passes through $P_1(5,-3)$ and $P_2(5,+7)$, the slope is

$$m = \frac{7+3}{5-5} = \frac{10}{0} = \text{---}$$

Since we cannot divide by zero, the result has no meaning. In this case, the line is vertical and is said to have an infinite slope.

1.15   INCLINATION OF A LINE

The *inclination* of a line is the *positive angle* the line makes with the *positive* direction of the *x*-axis. The inclination is often called *alpha* (α) (Fig. 1.21). Note that the slope of the line is equal to the tangent of α; that is,

$$m = \tan \alpha = \frac{\Delta y}{\Delta x}$$

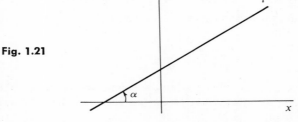

Fig. 1.21

When the slope is known, it is possible to find α. In Figure 1.21 the slope of the line l is $\frac{5}{8}$. Then $\tan \alpha = \frac{5}{8}$. Then $\alpha = \arctan (\frac{5}{8}) = \arctan 0.6250 = 32°$ (approx.).

1.15 Inclination of a Line

If a line slopes upward to the right it has a positive slope, and α is an acute angle (Fig. 1.22). Then tan α is also positive. If a line slopes downward to the right, it has a negative slope, and α is an obtuse angle. This agrees with the fact that the tangent of an obtuse angle is negative.

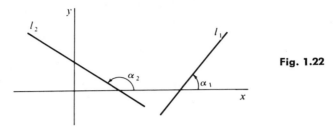

Fig. 1.22

For a horizontal line, $\alpha = 0$, because the line does not intersect the x-axis. Then tan $\alpha = 0$, and the line has a zero slope. A vertical line intersects the x-axis at right angles, and $\alpha = 90°$. Then tan α is infinite. Therefore, as we have already stated, the slope of a vertical line is infinite.

EXAMPLE 6. Find the slope and the inclination of a line passing through the points $(-1,5)$ and $(5,-7)$. Sketch the line and show the inclination.

Solution. Note that the line slopes downward to the right, indicating a negative slope. Applying the formula we get

$$m = \frac{-7-5}{5+1} = \frac{-12}{6} = -2$$

As we move from left to right we have the increments

$$\Delta x = +6; \; \Delta y = -12$$

Note that the line makes an obtuse angle with the positive direction of the x-axis (Fig. 1.23). Since the slope is equal to -2, we have

$$\tan \alpha = -2; \text{ then } \alpha = \arctan(-2) = 116°34' \text{ (approx.)}$$

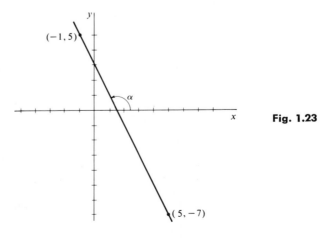

Fig. 1.23

1.16 PARALLEL AND PERPENDICULAR LINES

If two lines l_1 and l_2 are parallel, they have the same slope. In that case the inclinations, α_1 and α_2, are equal; $\tan \alpha_1 = \tan \alpha_2$; and $m_1 = m_2$. Conversely, if two lines have the same slope, they are parallel or they coincide. If the slope of one line is $\frac{2}{5}$, then the slope of a line parallel to this also has a slope of $\frac{2}{5}$ (Fig. 1.24a).

If two lines are perpendicular, their slopes are related by the equation: $m_1 = -1/m_2$. That is, one slope is the negative reciprocal of the other. If one of two perpendicular lines has a slope of $\frac{2}{5}$, then the slope of the other is $-\frac{5}{2}$. This relation may be seen from the fact that $\tan \alpha = -\cot(\alpha + 90°)$ (Fig. 1.24b).

Fig. 1.24

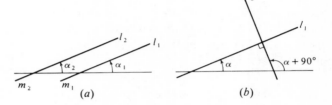

EXERCISE 1.3

1. Find the slopes and inclinations of the lines determined by the following pairs of points. Sketch each line in position, showing the two given points on the line. Which lines are perpendicular and which are parallel?
   (a) $(-2,-1); (3,4)$ (b) $(6,5); (3,-1)$ (c) $(-4,6); (-1,-3)$
   (d) $(5,3); (-4,2)$ (e) $(-3,-2); (5,-2)$ (f) $(-2,4); (-2,-5)$
   (g) $(4,0); (-5,-3)$ (h) $(3,-2); (2,7)$ (i) $(0,-4); (5,6)$
2. Find the slopes and inclinations of the lines determined by the following pairs of points. Sketch each line in position, showing the two given points on the line. Which lines are perpendicular and which are parallel?
   (a) $(-1,0); (6,0)$ (b) $(-5,2); (9,0)$ (c) $(-3,5); (5,-1)$
   (d) $(-5,2); (-4,-5)$ (e) $(-1,-3); (1,5)$ (f) $(2,-3); (-5,4)$
   (g) $(-2,7); (1,-5)$ (h) $(4,-5); (-2,3)$ (i) $(-2,-7); (7,5)$
3. A triangle has its vertices at the points: $A(8,6); B(-4,2); C(-2,-6)$. Find (a) the slope of each side of the triangle; (b) the slope of each median; (c) the slope of each altitude.
4. A triangle has its vertices at the points: $D(1,6); E(-7,2); F(5,-4)$. Find (a) the slope of each side of the triangle; (b) the slope of each median; (c) the slope of each altitude.
5. By use of the slope formula, show that the triangle with its vertices at $A(6,1), B(2,7)$, and $C(-6,-7)$ is a right triangle.
6. A circle has its center at the point $(-1,3)$, and the circle passes through the point $(5,7)$. What is the slope of the radius joining this point to the center of the circle? What is the slope of the line tangent to the circle at the point $(5,7)$?
7. Show by slopes that these points are collinear: $A(2,1); B(5,3); C(-4,-3)$.
8. The following points, taken in order, are the vertices of a quadrilateral: $A(8,2); B(1,6); C(-10,0); D(-3,-4)$. (a) Show by use of the slope formula that the figure is a parallelogram. (b) Find the lengths and the slopes of the diagonals of the figure.

9. A quadrilateral has its vertices at the following points: $A(5,2)$; $B(-2,3)$; $C(-3,-4)$; $D(4,-5)$. (a) Show by use of the slope formula that the figure is a parallelogram. (b) Find the lengths and the slopes of the diagonals.
10. A quadrilateral has its vertices at the points: $A(6,0)$; $B(3,5)$; $C(-2,2)$; $D(1,-3)$. (a) Show by use of the slope formula that the figure is a square. (b) Find the lengths and the slopes of the diagonals of the figure.
11. A quadrilateral has its vertices at the following points: $A(9,3)$; $B(-7,7)$; $C(-3,-7)$; $D(5,-5)$. Determine by the slope formula whether or not this figure is a parallelogram.
12. Find the midpoints of the sides of the quadrilateral in No. 11, and then show by the slope formula whether or not the figure formed by joining these midpoints is a parallelogram.
13. A quadrilateral has its vertices at the following points: $A(-1,1)$; $B(-6,-4)$; $C(1,-5)$; $D(6,0)$. (a) Determine whether or not the figure is a parallelogram. (b) Determine whether or not it is a square. (c) Compare the slopes of the diagonals.

1.17 ANGLE BETWEEN TWO LINES

When two straight lines intersect, it is sometimes necessary to find the angle between the lines. Of course, four angles are formed consisting of two pairs of vertical angles. Then we need consider only the two supplementary angles, θ and ϕ (Fig. 1.25). Now suppose we wish to find the size of θ, the acute angle. Then we make use of the formula for the tangent of the difference between two angles.

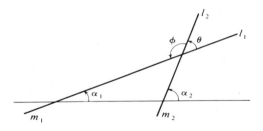

Fig. 1.25

Let us denote the two lines by l_1 and l_2 and corresponding inclinations and slopes by α_1, α_2, m_1, and m_2, respectively. Then

$$\tan \alpha_1 = m_1 \quad \text{and} \quad \tan \alpha_2 = m_2$$

Now note that the angle θ between the two lines has the following relation:

$$\theta + \alpha_1 = \alpha_2 \quad \text{(from geometry)}$$

Then
$$\theta = \alpha_2 - \alpha_1$$
$$\tan \theta = \tan(\alpha_2 - \alpha_1)$$

Now we make use of the formula for the tangent of the difference between two angles:

$$\tan \theta = \tan(\alpha_2 - \alpha_1) = \frac{\tan \alpha_2 - \tan \alpha_1}{1 + (\tan \alpha_2)(\tan \alpha_1)}$$

Substituting slope values we get

$$\tan \theta = \frac{m_2 - m_1}{1 + m_2 m_1}$$

This formula gives us the value of the tangent of the angle between the two lines. Once we know the tangent value we can find the angle itself.

EXAMPLE 7. Find the angle between two lines having the following slopes, respectively:

$$m_1 = \frac{2}{5}; \quad m_2 = \frac{3}{2}$$

Solution. If we construct the two lines rather carefully, we can estimate the angle between the lines (Fig. 1.26). Inserting the values in the formula we have

$$\tan \theta = \frac{3/2 - 2/5}{1 + (3/2)(2/5)} = \frac{15 - 4}{10 + 6}$$
$$= 11/16 = 0.6875$$

Then $\theta = \arctan 0.6875 = 34.5°$ (approx.).

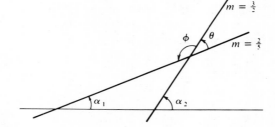

Fig. 1.26

In this example, if we reverse the order and take $m_1 - m_2$, we get

$$\tan \phi = \frac{2/5 - 3/2}{1 + (2/5)(3/2)} = \frac{4 - 15}{10 + 6} = -\frac{11}{16} = -0.6875$$

In this case, $\phi = \arctan(-0.6875) = 145.5°$, the supplement of $34.5°$.

The question arises: which of the following forms shall we use?

$$m_2 - m_1 \quad \text{or} \quad m_1 - m_2$$

In other words, which line shall we call l_1 and which l_2? In one case we get the tangent of the acute angle, in the other case the tangent of the obtuse angle. The numerical value of the tangent is the same for both angles. A positive value indicates the acute angle; a negative value indicates the obtuse angle.

There is a simple way to determine which slope should be taken first in the formula. Consider the point P of intersection of the lines as a point of rotation (Fig. 1.27). Then, if we draw a curved arrow counterclockwise showing the angle to be considered between the lines, then the arrowhead points to l_2 and therefore this slope is to be taken as m_2, which comes first

1.17 Angle between Two Lines

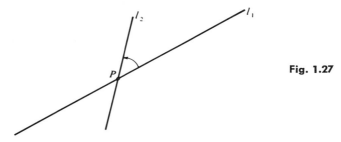

Fig. 1.27

in the formula. If we assume that the two lines have rotated in a positive direction about the point of intersection P, then we can consider l_2 as the line having the *greater* amount of rotation.

EXAMPLE 8. Two lines have the slopes $\frac{5}{2}$ and $-\frac{7}{3}$, respectively. Find the obtuse angle between the lines (Fig. 1.28).

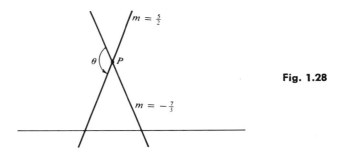

Fig. 1.28

Solution. We wish to find the size of θ, the angle indicated by the arrow. Note that the line at the head of the arrow has a slope of $\frac{5}{2}$. Then we consider that this line represents the greater amount of rotation about point P, the point of intersection. This slope, $\frac{5}{2}$, is then considered as m_2 and is used first in the formula. We get

$$\tan\theta = \frac{5/2 - (-7/3)}{1 + (5/2)(-7/3)} = \frac{15 + 14}{6 - 35} = \frac{29}{-29} = -1$$

Then $\theta = \arctan(-1) = 135°$.

The formula fails when the denominator becomes zero. This happens when $(m_1)(m_2) = -1$. But then we know that the lines are perpendicular and the angle between them is a right angle. This agrees with the fact that the tangent of 90° is infinite.

EXAMPLE 9. Find the three angles of the triangle having vertices at $A(-2,11)$, $B(-6,-3)$, and $C(4,-9)$ (Fig. 1.29).

Solution. Let us denote the sides of the triangle by a, b, and c, corresponding to the opposite angles. Then we shall denote the slopes of these sides by m_a, m_b, and m_c, respectively.

24 Coordinate Systems

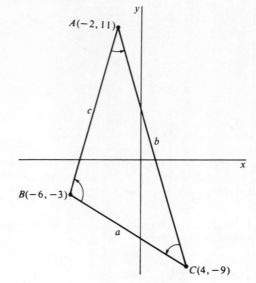

Fig. 1.29

To find m_a we use the two points B and C; to find m_b we use the two points C and A; to find m_c we use the two points A and B. Then by the slope formula we get

$$m_a = \frac{-9+3}{4+6} = -\frac{3}{5}; \quad m_b = \frac{11+9}{-2-4} = -\frac{10}{3}; \quad m_c = \frac{11+3}{-2+6} = \frac{7}{2}$$

Now note that to find the angles A, B, and C we use the slopes in this order:

for angle A, we use $m_b - m_c$, in that order

for angle B, we use $m_c - m_a$, in that order

for angle C, we use $m_a - m_b$, in that order

Then

$$\tan A = \frac{-10/3 - 7/2}{1+(-10/3)(7/2)} = \frac{-20-21}{6-70} = \frac{-41}{-66} = 0.6212; \; A = 31°51'$$

$$\tan B = \frac{7/2 + 3/5}{1+(7/2)(-3/5)} = \frac{35+6}{10-21} = -\frac{41}{11} = -3.7273; \; B = 105°01'$$

$$\tan C = \frac{-3/5 + 10/3}{1+(-3/5)(-10/3)} = \frac{-9+50}{15+30} = \frac{41}{45} = 0.9111; \; C = 42°20'$$

EXERCISE 1.4

1. One line passes through the points (8,4) and (−2,−1). A second line passes through the points (11,12) and (8,7). Find the slope of each line, and then find the acute angle between the two lines.
2. One line passes through the points (−7,4) and (3,−2). A second line passes through the points (−3,−1) and (0,4). Find the angles between them.
3. One line passes through the points (1,−7) and (6,−4). A second line passes through the points (−1,2) and (−6,−1). Find the angles between them.

Find the interior angles of the triangle whose vertices are indicated by each of the following sets of coordinates:

4. $A(-6,6)$; $B(0,-8)$; $C(9,-4)$
5. $D(-5,4)$; $E(-2,-1)$; $F(7,-5)$
6. $G(6,1)$; $H(2,7)$; $K(-6,-7)$
7. $R(3,0)$; $S(-7,4)$; $T(-3,-2)$
8. $L(1,5)$; $M(-4,0)$; $N(2,-2)$
9. $P(2,7)$; $Q(-3,-7)$; $R(5,-2)$
10. $A(0,6)$; $B(-9,-1)$; $C(3,-5)$
11. $D(6,1)$; $E(2,7)$; $F(-6,-7)$
12. A quadrilateral has its vertices at the points $A(5,2)$; $B(-2,3)$; $C(-3,-4)$; and $D(4,-5)$. Find the interior angles of the quadrilateral.
13. A quadrilateral has its vertices at $A(9,3)$; $B(-7,7)$; $C(-3,-7)$; and $D(5,-5)$. Find the interior angles. What conclusion can you draw concerning the quadrilateral? Is it a parallelogram?
14. Find the interior angles of the polygon whose vertices are at $A(8,2)$; $B(1,6)$; $C(-10,0)$; and $D(-3,-4)$.
15. A polygon has its vertices at $A(6,0)$; $B(3,5)$; $C(-2,2)$, and $D(1,-3)$. Find the interior angles of the polygon. What conclusion can you draw?
16. One line passes through the points $(4,-3)$ and $(-2,0)$. A second line passes through the points $(-6,2)$ and $(-2,0)$. Find the angle between the lines.

1.18 SPACE COORDINATES

There are several systems that are used to describe the location of a point in space. One of the most common is the rectangular coordinate system. In this system we set up three axes concurrent at some point. Each axis is assumed to be drawn at right angles to the other two. We attempt in this way to represent three dimensions.

There is always some difficulty in showing three dimensions on a flat sheet of paper, a plane. For example, if we wish to draw a figure to represent a rectangular box, we make a drawing that only *looks* like a box (Fig. 1.30). Let us suppose the box is 8 inches long, 5 inches wide, and 5 inches high. The length and the height can be drawn to scale with units of the same length. However, the width represented by the line segment CF extends out of the paper and is therefore foreshortened. For the most realistic appearance, a unit along such a line should be approximately 0.7 as long as the units along the other sides.

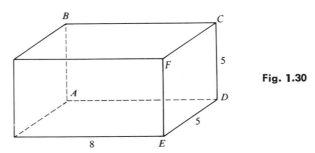

Fig. 1.30

To locate a point in space by the rectangular system, we set up three axes — called the x-axis, the y-axis, and the z-axis — concurrent at some point O called the *origin* (Fig. 1.31). We take the direction from O shown by arrows as positive. The opposite directions, shown by dotted lines, are negative. Note that any set of two axes determine a plane. The x-axis and the y-axis determine the plane xy. Try to imagine that the xy-plane is

Fig. 1.31

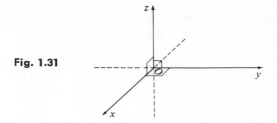

this sheet of paper, and that you are at z looking down on the paper, the plane xy. This xy-plane is the horizontal plane that we have been using to show the location of points in a plane. The z-axis then extends upward out of the paper. (You may instead think of the yz-plane as the paper and the x-axis extending outward toward you.) The xy-plane, the yz-plane, and the xz-plane divide the entire space into eight octants. Octant I is the octant in which all space coordinates are positive.

A point is located in space by stating its distance in each of the three directions, the x-distance, the y-distance, and the z-distance. These distances represent the respective distances from the three planes. The coordinates of a point are always stated in the order (x,y,z). For example, to locate the point (3,7,4) (Fig. 1.32), we count 3 units in the x-direction, 7 units in the y-direction, and 4 units in the z-direction. Again, the units in the x-direction are foreshortened. If a coordinate is negative, we take the units in the negative direction.

Fig. 1.32

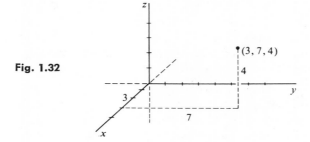

1.19 DISTANCE FROM ORIGIN TO A POINT IN SPACE

To find the distance from the origin $O(0,0,0)$ to the point (3,7,4), let us sketch the complete rectangular solid (Fig. 1.33). Now note that the base $OADB$ is a rectangle, and OD is the diagonal of the rectangle. Then $OD = \sqrt{7^2 + 3^2} = \sqrt{58}$. Now note that ODG is a right triangle of which OG is the hypotenuse. The line OG is the distance we wish to find. In the right triangle ODG one side is the z-distance, 4. One side, OD, we have seen is equal to $\sqrt{58}$ or $\sqrt{7^2 + 3^2}$. Squaring the two sides of triangle ODG, the sum is equal to the square of the hypotenuse, OG. Then

$$OG = \sqrt{(OD)^2 + (DG)^2}$$

1.20 Distance between Two Points in Space

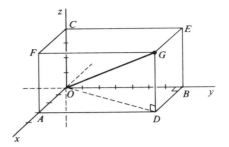

Fig. 1.33

That is, $OG = \sqrt{7^2 + 3^2 + 4^2} = \sqrt{74}$

Then the diagonal of a rectangular solid is $d = \sqrt{l^2 + w^2 + h^2}$. The distance of a point in space from the origin is then given by the formula

$$d = \sqrt{x^2 + y^2 + z^2}$$

1.20 DISTANCE BETWEEN TWO POINTS IN SPACE

For the distance between two points $P_1(x_1,y_1,z_1)$ and $P_2(x_2,y_2,z_2)$ in space, we first find the difference in the x-direction, the difference in the y-direction, and the difference in the z-direction. If we think of going from P_1 to P_2 as a directed distance, then we have the following increments:

$$\Delta x = x_2 - x_1; \quad \Delta y = y_2 - y_1; \quad \Delta z = z_2 - z_1$$

These increments represent the dimensions of a rectangular solid, and we have for the distance between the points

$$d = \sqrt{(\Delta x)^2 + (\Delta y)^2 + (\Delta z)^2}$$

or

$$d = \sqrt{(x_2 - x_1)^2 + (y_2 - y_1)^2 + (z_2 - z_1)^2}$$

Note that this is simply the extension of the formula for the distance between two points in a plane. The student might like to think of what might be the formula for the distance if we have a fourth dimension, which of course cannot be pictured. Could the formula be extended to cover any number of dimensions n?

EXERCISE 1.5

1. Graph the following points in space: $A(5,-2,3)$; $B(6,4,-5)$; $C(-2,3,-1)$; $D(-5,-3,-2)$; $E(7,0,2)$; $F(-3,5,0)$.
2. Graph the following points and then outline the rectangular solid in the first octant determined by these three points: $A(4,0,0)$; $B(0,6,0)$; $C(0,0,3)$. Write the coordinates of the remaining five vertices of the rectangular solid. What is the length of its diagonal?
3. Outline the rectangular solid in the first octant determined by the three points: $A(1,0,0)$; $B(0,1,0)$; $C(0,0,1)$. Find the length of the diagonal of the solid. How far is the point $(1,1,1)$ from the origin?
4. Find the distance of each of the points in No. 1 from the origin.

5. Find the distance between the two points $A(-3,-4,-5)$ and $B(2,3,-1)$.
6. Plot the two points $C(-2,4,-6)$ and $D(2,-4,6)$. Then find the distance of each point from the origin. Find the distance between the two points and show that this distance is twice the distance of each from the origin.
7. In No. 1 find the distance from A to each of the other points, that is, find AB, AC, AD, AE, and AF.
8. A polyhedron of 6 faces, 8 vertices, and 12 edges has one of its faces determined by the following vertices: $A(-5,4,-1)$; $B(-5,-2,2)$; $C(2,-1,4)$; $D(1,6,1)$. Four of the edges are formed by joining A, B, C, and D with the following vertices, respectively: $E(-3,5,-4)$; $F(0,0,0)$; $G(3,-3,-2)$; $H(2,4,-5)$. Find the lengths of all 12 edges of the polyhedron.

chapter

2

Equations and Graphs

2.1 THE SOLUTION OF AN EQUATION

In algebra we are concerned with solutions of equations. If an equation contains two unknowns, such as x and y, we attempt to find corresponding values of x and y that will satisfy the equation. Any *ordered pair* of numbers that will make the equation true is called a *solution* of the equation.

Suppose we have the equation $x + y = 10$. In this equation, if $x = 7$, then $y = 3$. This pair of numbers is a solution. However, there are other solutions of this equation. Some of them are shown in the following table:

If $x =$	8	6	10	1	-3	7.5	0
Then $y =$	2	4	0	9	13	2.5	10

An equation in x and y may have an infinite number of solutions, or the number of solutions may be limited. For example, in the equation $y^2 = x$, if x is negative, then y becomes imaginary. In such equations, if we wish only real values of x and y, then we must exclude any values that will make one or both unknowns imaginary. As another example, in the equation $y = x^2$, the value of y is nonnegative for all real values of x; that is, y is either zero or a positive number. In the equation $x^2 + y^2 = 0$, the only pair of real values of x and y is the pair (0,0).

In any equation, if a symbol, such as x, may take on a number of values, it is called a *variable*. In the equation $x + y = 10$, x and y are both variables. Since the value of one variable depends on the value assigned to the other, one variable is called the *independent variable* and the other the *dependent variable*. The independent variable is usually taken to be the one whose value is assumed first. This is often x.

For example, suppose we have the equation

$$2x + 3y = 18$$

In this equation, if we first give x the value 6, then y must be 2. Then we call x the *independent* variable and y the *dependent* variable. As an ordered pair the solution is written (6,2). In writing the solution as an ordered pair, we always write the *x*-value first.

Sometimes, for convenience, we first assign a value to *y* and then compute the corresponding value of *x*. For example, in the equation

$$y^2 - 2y - 3 = x$$

it is simpler to assign a value to *y* first: if $y = 4$, then $x = 5$. As an ordered pair, the solution is written (5,4).

2.2 GRAPH OF AN EQUATION

An equation in *x* and *y* can usually be shown by a picture called a *graph*. Let us see what is involved in the process of graphing an equation.

In Chapter 1 we have been concerned with points, lines, and angles. These are *geometric* concepts. On the other hand, in algebra we are concerned with equations and their solutions. When we find numbers that satisfy equations, we are engaged in *algebra*. When we locate points, draw lines, and show angles, we are engaged in *geometry*. For a long time no one saw any relation between these two kinds of activities.

Let us see how we may bring about a union of algebra and geometry. Suppose we have the equation $x + y = 10$. We find a pair of numbers, say, $x = 7$ and $y = 3$, that satisfies the equation. This operation involves *algebra*. It has essentially nothing to do with geometry. However, now we say that this ordered pair (7,3) shall be the coordinates of a *geometric* point on a graph. This was the revolutionary idea developed by Descartes in the early 1600's, an idea that made possible the analytic study of geometric concepts by means of algebra.

In general, to graph an equation in *x* and *y* we find ordered pairs of values of *x* and *y* that satisfy the equation. Then we say that each *ordered pair* of corresponding values shall represent a geometric point in the *XY*-plane. We find several pairs of values and then plot the corresponding points. In most instances we connect the points with lines, straight or curved, and say, "There is a picture of the equation."

If a figure is to represent the true and complete graph of an equation, it must include all the points and only those points that satisfy the equation. That is,

(a) each ordered pair that satisfies the equation will represent a point on the graph, and, conversely,

(b) each point on the graph will have an ordered pair of coordinates that satisfies the equation.

A few facts should be mentioned concerning the graphing of an equation. In analytic geometry we often call all graphs *curves*, even straight lines. If we know the graph of an equation is a straight line, then we need only two points to determine the graph. However, it is best to plot one or two extra points as a check on the work.

If the graph is not a straight line, it is often necessary to plot a few points rather near together in some places so as to determine correctly the shape of the curve. This is especially necessary where curves have rather sharp changes in direction.

Some curves are discontinuous; that is, they break off at some point and appear later at another place. In some instances they approach but do not touch straight lines called *asymptotes*. It may happen that some values of one variable will produce imaginary values of the other, which cannot be shown on the graph.

For the foregoing reasons it is important to try to judge the shape and extent of a curve from only a few plotted points. Do not draw a curve or extend a curve unless you are sure the result will truly represent the equation. Some of the factors to be considered will be pointed out in connection with various kinds of equations and their graphs.

2.3 THE LINEAR EQUATION

A *first-degree equation* is an equation containing one or two variables, such as x and y, each raised to a power no higher than the first and no term containing the product xy. An example is the equation $2x - y = 5$. Note that the equation does not contain a term in x^2, y^2, or xy. A first-degree equation will always result in a straight-line graph. For this reason the equation is called a *linear* equation.

To graph the equation $2x - y = 5$, we may use the following pairs of values:

If $x =$	0	1	2	4	6	-2	2.5
Then $y =$	-5	-3	-1	3	7	-9	0

Since the equation is of the first degree, we know the graph is a straight line. Then we need only two points, but it is advisable to take a few more.

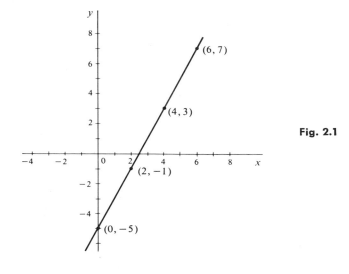

Fig. 2.1

When the points represented by the pairs of numbers are plotted and connected, the result is a straight line (Fig. 2.1).

However, we should ask: Can we be sure that any point on the line besides those plotted will also have a pair of coordinates satisfying the equation? To answer this question we might note that in the equation, $2x - y = 5$, the x-values can be taken so closely together that they can be assumed to form a continuum, and there would always be a corresponding value of y. The same would be true if points were taken beyond those plotted. Then the line should be drawn, not only *to* the extreme points shown, but *through* those points to indicate that the line extends further.

2.4 INTERCEPTS

In graphing an equation it often happens that the easiest points to locate are those where the curve intersects the x-axis and the y-axis. Where a curve cuts the x-axis, the abscissa of the point is called the *x-intercept*. Where the curve cuts the y-axis, the ordinate of the point is called the *y-intercept*. For example, if a curve passes through the point $(5,0)$, the x-intercept is 5. If it passes through the point $(0,-2)$, the y-intercept is -2.

For any x-intercept, $y = 0$; and for any y-intercept, $x = 0$. Therefore, to find the intercepts, we set each variable in turn equal to zero in the equation and then solve the equation for the value of the other variable. As a rule, if there are intercepts, these points are often the first to be found and plotted. Of course, some curves do not have intercepts.

If a line is parallel to one of the axes, then it does not intersect that axis. For example, take the equation

$$x = 4$$

The equation really says that for any value of y whatever, the value of x is 4. For example, we might take the values shown here:

$x =$	4	4	4	4	4
$y =$	0	1	7	-3	-8

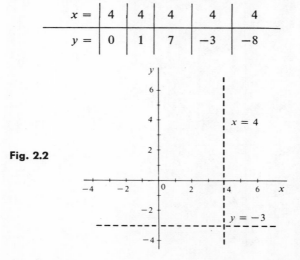

Fig. 2.2

When the points represented by these pairs of numbers are plotted, they determine a straight line parallel to the y-axis and intersecting the x-axis at the point (4,0), (Fig. 2.2). In a similar manner we shall find that the equation, $y = -3$, is satisfied by any and all points on a line parallel to the x-axis and intersecting the y-axis at the point $(0, -3)$. The y-axis is often called the line $x = 0$. The x-axis is often called the line $y = 0$.

2.5 EXTENT OF A CURVE: DOMAIN AND RANGE

The graph of the equation, $2x - y = 5$, which we have just shown, extends infinitely far in the x-direction and in the y-direction, positively and negatively. The extent of a curve in the x-direction is determined by the totality of all permissible values of x in the equation. This totality of values is called the *domain* of the variable x. From the equation $2x - y = 5$ we can tell that x may take on any value in the domain $-\infty < x < +\infty$. The extent of the curve in the y-direction is determined by the totality of all permissible values of y in the equation. The totality of y-values is called the *range*. From this equation note that y may take on any values in the range $-\infty < y < +\infty$. The domain and range should be found by inspection of the equation itself, so as to determine all possible real values of x and y and to omit any imaginary values.

2.6 GRAPHS INVOLVING ABSOLUTE VALUES

In graphing an equation involving absolute values, we must note carefully which values cannot be negative. Consider, for example, the equation

$$y = |x|$$

Note that since y is equal to an absolute value, then y cannot be negative. We might then have corresponding values such as these:

If $x =$	0	1	-1	2	-2	$+5$	-5
Then $y =$	0	1	1	2	2	5	5

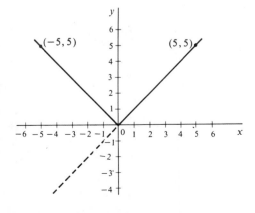

Fig. 2.3

The resulting graph (Fig. 2.3) consists of two rays extending diagonally upward and outward from the origin, one as the 45° line and the other as the 135° line. Compare the result with the graph of the line $y = x$, which extends along the broken line.

EXAMPLE 1. Find the graph of the equation: $y = |2x - 3|$.

Solution. It is sometimes advisable to set up a table showing the values of the function, $2x - 3$, as well as the values of y, such as the following:

If $x =$	0	1	1.5	2	3	4	-1	-2	-3
Then $2x - 3 =$	-3	-1	0	1	3	5	-5	-7	-9
and $y =$	3	1	0	1	3	5	5	7	9

The graph (Fig. 2.4) cannot extend below the x-axis, since y cannot be negative. Note again that the graph consists of two rays extending upward from point (1.5,0). Note the difference between this graph and the graph of the equation

$$y = 2x - 3$$

a portion of which extends below the x-axis.

Fig. 2.4

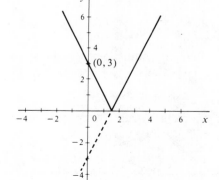

EXERCISE 2.1

Sketch the graph of each of the following equations. Note the intercepts and the extent of the curve. (The word *curve* is here used in its general sense to refer to all graphs.)

1. $y = x - 4$
2. $y = 3 - 2x$
3. $y = 4x + 1$
4. $y = 2 - 5x$
5. $x = 2y + 3$
6. $x = 1 - 3y$
7. $4x - 3y = 24$
8. $2x + 5y = 20$
9. $2y - 3x = 9$
10. $3x - 2y = 0$
11. $2y - 3x = 0$
12. $x + 5y = 0$
13. $x = -6$
14. $y = 5$
15. $x - 3 = 0$
16. $3y = 2x - 7$
17. $4x = 5y + 2$
18. $2x = 7y + 1$
19. $y = \dfrac{x + 3}{5}$
20. $x = \dfrac{3y + 4}{7}$
21. $y = \dfrac{2}{5}x - 4$
22. $\dfrac{x}{3} + \dfrac{y}{5} = 1$
23. $\dfrac{x}{-4} + \dfrac{y}{6} = 1$
24. $y = -\dfrac{3}{8}x + 2$

25. $x = 0$
26. $y = 0$
27. $y + 3 = 0$
28. $y = |x - 1|$
29. $x = |y + 3|$
30. $y = 3 - |x|$
31. $y = x + |x|$
32. $x = |y| - y$
33. $y = 2x + |3x|$
34. $x = 2 - |3y|$
35. $x = |2 - 3y|$
36. $y = \left|\dfrac{x}{2}\right|$

2.7 THE SECOND-DEGREE EQUATION

A *second-degree equation* is an equation containing at least one term of the second degree, but no higher degree. A *second-degree term* may be defined as a term containing the product of two variables. The variables may be different, as in the term xy, or they may be the same, as in the term x^2, which means $(x)(x)$. The following are second-degree terms: $3x^2$; $4y^2$; $5xy$. The following are second-degree equations:

$$y = x^2 - 4; \quad x^2 + y^2 = 16; \quad xy = -12; \quad x^2 - 3x - 4y = 7$$

The graphs of most second-degree equations are not straight lines. Therefore, we must find points rather close together at some places to determine the correct shape of the curve. Here, too, we should note the intercepts, if any, and the extent, including domain and range. In addition, a few other facts should be noted, as we shall point out in the examples shown.

EXAMPLE 2. Graph the equation $y = x^2 - 4$.

Solution. For the intercepts we have, if $x = 0$, $y = -4$, the y-intercept; if $y = 0$, $x = +2$ and -2, the x-intercepts. For pairs of values we might use the following:

If $x =$	0	1	2	3	−1	−2	−3
Then $y =$	−4	−3	0	5	−3	0	5

When we have plotted these points and connected them with a smooth curve, we can tell approximately where other points on the curve would lie. From the equation itself we can tell that for more values of x beyond ± 4, the curve will continue to rise (Fig. 2.5).

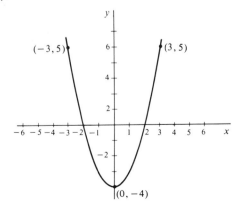

Fig. 2.5

This curve is called a *parabola*. The point of sharpest turning is called the *vertex* of the parabola. In this example the vertex is at $(0,-4)$.

As to the domain, the graph as well as the equation itself indicates that the value of x is not limited in either the positive or negative direction. Then for the *domain* we have $-\infty < x < +\infty$. However, the *range* is limited in a negative direction. Whatever value we give x, we shall never have x^2 negative. Therefore the value of y can never be less than -4. However, y is unlimited in a positive direction. Then for the range we have $-4 \leq y < +\infty$.

If a second-degree equation contains one variable raised to the second power and the other variable raised to the first power only, then the graph will always be the curve called a parabola. Parabolas also occur in some equations containing the product $(x)(y)$. However, not all second-degree equations represent parabolas. The graphs of other second-degree equations are circles, ellipses, hyperbolas, and in some special cases straight lines.

2.8 SYMMETRY

Note that in the graph of the equation, $y = x^2 - 4$, the line $x = 0$ (the y-axis) divides the graph into two congruent halves. If the right half is folded over along the y-axis, it will fit on the left half. We say then that the curve is *symmetric with respect to the line*, the y-axis.

Two points are *symmetric* with respect to a line if the line is the perpendicular bisector of the line segment connecting the two points. In Figure 2.6 points A_1, B_1, and C_1 are symmetric to the points A_2, B_2, and C_2, respectively, with respect to the line MN. A curve is symmetric with respect to a line if every point on the curve has its symmetric point also on the curve. In other words, a figure is symmetric with respect to a line if one half of the curve is a reflection of the other half in the line. The line that divides a figure into two symmetric halves is called the *axis of symmetry*. If it is known that a curve is symmetric with respect to a line, then one half of the curve may be drawn from plotted points and the other half sketched as a reflection of the first half.

Fig. 2.6

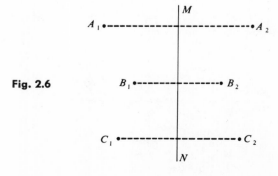

2.8 Symmetry

As tests for symmetry, we have the following:

(1) If an equation remains unchanged when every x is replaced by $(-x)$, then the figure is symmetric with respect to the y-axis, because for every positive x there is a corresponding negative x.
(2) If the equation remains unchanged when every y is replaced by $(-y)$, then the figure is symmetric with respect to the x-axis, because for every positive y there is a corresponding negative y.
(3) If an equation remains unchanged when x is replaced by $(-x)$ and y is replaced by $(-y)$ at the same time, then the figure is symmetric with respect to the origin.

EXAMPLE 3. Graph the equation $x^2 + y^2 = 16$. Find the intercepts, domain and range, and any symmetry of the curve.

Solution. In this equation note that there is more than one value for each value of x, except for $x = \pm 4$. We can solve the equation for y and get two equations:

$$y = +\sqrt{16 - x^2} \quad \text{and} \quad y = -\sqrt{16 - x^2}$$

To plot the graph of the first equation, we may use the following values:

If $x =$	0	1	2	3	4	-1	-2	-3	-4
Then $y =$	$+4$	$\sqrt{15}$	$\sqrt{12}$	$\sqrt{7}$	0	$\sqrt{15}$	$\sqrt{12}$	$\sqrt{7}$	0

For the first equation note that all values of y are positive or zero. For these pairs of values the curve is a semicircle, the upper half of Figure 2.7. Now, if y is equal to the negative radical, this is equivalent to replacing y with $(-y)$. Then we get the lower half of the circle. By the three tests for symmetry, the curve is symmetric with respect to the x-axis, to the y-axis, and to the origin. Note that the values of both x and y are confined within the limits, $+4$ and -4. For any value of x such that $|x| > 4$, the value of y becomes imaginary. Also, if $|y| > 4$, then x becomes imaginary. For the domain we have $-4 \leq x \leq +4$; for the range, we have $-4 \leq y \leq +4$.

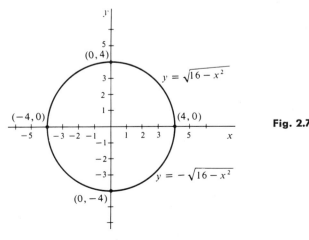

Fig. 2.7

38   Equations and Graphs

EXAMPLE 4. Graph the equation $xy = -12$.

Solution. Let us first note that neither x nor y can be zero. Therefore, there are no intercepts. Note also from the equation that either x or y can extend infinitely far positively or negatively. Since the product xy is negative, then the signs of x and y must be *opposite*. For graphing the equation, we might use the following values:

If $x =$	1	2	3	4	6	12	24
Then $y =$	−12	−6	−4	−3	−2	−1	$-\frac{1}{2}$
If $x =$	−1	−2	−3	−4	−6	−12	−24
Then $y =$	12	6	4	3	2	1	$\frac{1}{2}$

Note that x may increase numerically without limit. The absolute value of y continues to decrease toward zero but can never reach zero. As the numerical value of one variable becomes larger and larger, the numerical value of the other becomes smaller and smaller. The curve therefore approaches nearer and nearer the axes (Fig. 2.8). For example, if $x = 12{,}000$, $y = -0.001$.

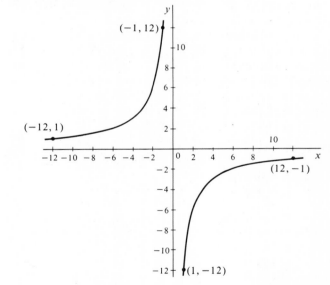

Fig. 2.8

This particular kind of curve is called a *hyperbola*. In general, a hyperbola has two branches or separate disconnected portions of the curve. Such a curve is called *discontinuous*. This particular hyperbola is discontinuous for the values $x = 0$ and $y = 0$. These values are then excluded from the domain and range.

2.9 ASYMPTOTES

We have seen in Example 4 in the foregoing section how a curve approaches a straight line but never touches it. In the graph of the equation

$xy = -12$, the curve approaches but never touches the x-axis and the y-axis. Whenever a curve approaches a straight line in this manner, the straight line is called an *asymptote* of the curve. Then, by definition, *an asymptote is a straight line that is approached by a curve in such a way that the curve moves closer and closer to the line but never touches the line.* The asymptote is often called the *limiting position* of the curve because the curve may be drawn as close to the line as we please by taking the proper corresponding values of the variables.

An asymptote can be a great help in sketching a curve. If we know that a particular line is an asymptote, then we know the curve comes nearer and nearer to the line. An asymptote represents a break in the curve, and therefore the curve does not touch the asymptote. However, there are some curves that approach a straight line asymptotically in one direction while at another point the same curve may cross the same straight line.

In some cases a curve approaches an asymptote in one direction and may at some other place appear again on the opposite side of the asymptote. The separate portions of the curve are called separate branches of the same curve. This is true for the equation in Example 4: $xy = -12$ (Fig. 2.8). Note that in the graph, as x begins at $-\infty$ in the second quadrant and approaches zero, the curve approaches the y-axis from the left but does not touch it. Then just to the right of the y-axis, the curve appears again in the fourth quadrant and moves away from the y-axis, again approaching the x-axis. To get across the y-axis, we might almost imagine the first branch went around to infinity to get to the other portion of the curve.

Since an asymptote can show much about the nature of a curve and aid in sketching the graph, let us see then how asymptotes can be determined. Any known asymptotes should be sketched first. Let us again take the equation $xy = -12$. Solving this equation for y we get

$$y = \frac{-12}{x}$$

Note especially that if x becomes very small, then y becomes very large negatively, as shown by the graph. However, x cannot be equal to zero. We can make x as small as we please but not zero. This value then, $x = 0$ (or the y-axis), is the asymptote approached by the curve.

The foregoing problem then gives us a clue toward finding asymptotes. An asymptote will be indicated by a *zero denominator*. Therefore, *if a variable appears in a denominator, we set the denominator equal to zero and solve for the value of the variable.* The result will be the equation of the asymptotes.

Most asymptotes that we shall encounter are either vertical or horizontal. To find vertical asymptotes, we solve an equation for y. Then if x appears in the denominator of a fraction, we set the denominator equal to zero and solve for x. The result is the value of x for a vertical asymptote. To find horizontal asymptotes, we solve the equation for x. Then if y appears in the denominator of a fraction, we set the denominator equal to zero and solve for y. The result is the value of y for a horizontal asymptote.

40 Equations and Graphs

EXAMPLE 5. Find any asymptotes of the following curve and sketch the curve. Also note intercepts and any symmetry of the curve.

$$(y - 2)(x + 1) = 4$$

Solution. Solving for y we get

$$y = 2 + \frac{4}{x + 1}$$

Since x appears in the denominator of the fraction, we set the denominator equal to zero and solve for x:

$$x + 1 = 0; \quad \text{then} \quad x = -1$$

The line $x = -1$ is then a vertical asymptote. To find any horizontal asymptotes, we solve the equation for x and get

$$x = \frac{4}{y - 2} - 1$$

Now we set the denominator equal to zero and get

$$y - 2 = 0; \quad \text{then} \quad y = 2$$

The line $y = 2$ is a horizontal asymptote. The curve approaches but never touches these lines (Fig. 2.9). Now, with the intercepts and a few other corresponding values of x and y, we can get a good graph showing the general shape as well as the extent of the curve.

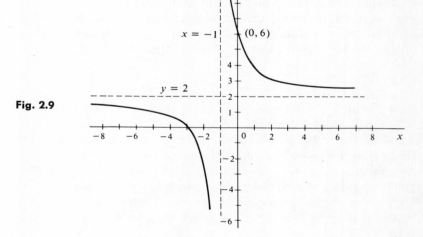

Fig. 2.9

EXAMPLE 6. What are the asymptotes, if any, of the curve

$$(y - 1)(x^2 - x - 2) = 4$$

Solution. Dividing both sides of the equation by the quantity $(y - 1)$ we get

$$x^2 - x - 2 = \frac{4}{y - 1}$$

2.9 Asymptotes

Now we set the denominator $y - 1$ equal to zero and solve for y:

$$y - 1 = 0; \quad \text{then} \quad y = 1, \text{ a horizontal asymptote.}$$

Dividing the original equation by the quantity $x^2 - x - 2$ we get

$$y - 1 = \frac{4}{x^2 - x - 2}$$

Setting the denominator $x^2 - x - 2$ equal to zero and solving for x we get

$$x^2 - x - 2 = 0; \quad \text{then} \quad x = 2 \quad \text{and} \quad x = -1$$

The curve therefore has two vertical asymptotes.

EXERCISE 2.2

Graph the following functions and equations. Recall that in graphing a function we first set the function equal to y and then graph the resulting equation.

1. $x^2 - 3x - 4$
2. $x^2 + 2$
3. $x^2 - 4x$
4. $x - x^2$
5. $3 - 2x - x^2$
6. $4 - x^2$
7. $x^2 + 2x + 4$
8. $x^2 - 2x + 10$
9. $x^2 - 6x + 9$
10. Compare the graphs of the two equations $y = \sqrt{4x}$ and $y^2 = 4x$
11. Compare the graphs of the two equations $y = \sqrt{25 - x^2}$ and $y^2 = 25 - x^2$
12. Compare the graphs of the equations $y = \sqrt{9 + x^2}$ and $y^2 = 9 + x^2$
13. Compare the graphs of the two equations $y = \sqrt{x + 3}$ and $y^2 = x + 3$

Graph the following equations:

14. $x^2 + y^2 = 16$
15. $x^2 - y^2 = 4$
16. $xy = 6$
17. $y = \dfrac{4}{x - 3}$
18. $y = \dfrac{12}{x^2 - 4}$
19. $y - 1 = \dfrac{6}{x + 2}$
20. $y = \dfrac{6}{x^2}$
21. $y^2 = \dfrac{12}{x^2 - 9}$
22. $y^2 = \dfrac{x^2}{x^2 - 1}$
23. $(y - 2)(x^2 - x - 6) = 12$
24. $(y + 1)^2(x - 3)^2 = 24$

chapter
3

The Straight Line

3.1 THE TWO PROBLEMS OF ANALYTIC GEOMETRY

Analytic geometry is concerned with two chief types of problem: (1) A relation between x and y may be given by an equation, such as $3x + 2y = 12$. Then our problem is to find the graph of the equation. (2) If a particular curve or graph is given or described, then our problem is to find the equation whose graph is given.

We have already had experience with the first type of problem. In graphing an equation, we find pairs of values that satisfy the equation. These pairs of numbers are then plotted as points in the plane. From the equation we determine such facts as intercepts, domain, range, and asymptotes. Finally we connect the points with a smooth curve.

The second type of problem is more difficult. In this case, a curve of some kind is given or described, perhaps a straight line. Then we try to find the equation represented by the curve. When we say a curve is given, we mean that enough information is given so that we can draw the curve.

For example, we may know that a given straight line passes through the two points (2,3) and (5,7). Or a circle may be given by simply stating that its center is at the point (3,1) and its radius is 5 units. Our problem then is to find the equation of the straight line or circle or other given curve.

There are three steps that might be followed here:

1. From the information given, sketch the curve as you think it looks.
2. Take the *general point* (x,y) in any general position on the curve. This is the moving point that traces out the curve.
3. Set up an equation from the given information.

3.2 THE STRAIGHT LINE: POINT-SLOPE FORM

To write the equation of a given straight line we begin with a line that passes through a given point and has a given slope. Suppose a line passes through the point (5,2) and has a slope of $\frac{3}{4}$ (Fig. 3.1).

3.2 The Straight Line: Point-Slope Form

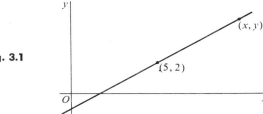

Fig. 3.1

Step 1. First we draw the line in the position as described.

Step 2. Next we take the general point (x,y) in any general position on the curve.

Step 3. Now we use the two points (x,y) and $(5,2)$ and set up the expression for the slope. We get

$$\frac{y-2}{x-5} = m$$

Now we simply state the fact that this expression is equal to the given slope:

$$\frac{y-2}{x-5} = \frac{3}{4}$$

This equation is the equation of the line because it meets all the requirements. However, it is simplified as follows:

Clearing of fractions, $\quad\quad\quad\quad\quad\quad 4(y-2) = 3(x-5)$
Simplified, the equation becomes $\quad\quad 3x - 4y = 7$

This final form, $3x - 4y = 7$, is called the *general form* of the equation of a straight line. It is usually stated in general terms as

$$Ax + By = C$$

where A, B, and C are constants. We shall look at the general form a little more in detail in Sec. 3.7.

We can now develop a general rule for writing the equation of any line if the slope and a point on the line are given. Let us call the given point $P_1(x_1,y_1)$ and the given slope m_1. Then we follow the same steps as in the foregoing example.

1. Again we draw the given line l, through the given point P_1 (Fig. 3.2).
2. We take the general point $P(x,y)$, in any general position on the line.

Fig. 3.2

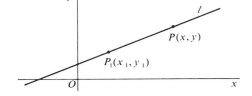

3. Using the two points $P_1(x_1,y_1)$ and $P(x,y)$, we set up the expression for the slope and say that it is equal to m_1, the given slope.

$$\frac{y - y_1}{x - x_1} = m_1$$

The equation is simplified to the form:

$$y - y_1 = m_1(x - x_1)$$

This final form is called the *point-slope* form of the equation of a straight line. It is the most useful of all forms and should be thoroughly memorized. It can be used to find the equation of any line when the slope and a point on the line are given. It can be used when all other forms are forgotten. The only exception is for a *vertical* line. In that case we simply write $x = x_1$, where x_1 is the abscissa of the given point.

In the *point-slope* form, the subscript for m may be omitted. The coordinates x_1 and y_1 represent the given point, m represents the given slope, and the point (x,y) is the variable tracing point.

EXAMPLE 1. Write the equation of the line having a slope of $-\frac{3}{8}$ and passing through the point $(-7,2)$.

Solution. Since a point and the slope are given we use the point-slope form:

$$y - y_1 = m(x - x_1)$$

First, we must recognize that y_1, the ordinate of the given point, is 2; and x_1, the abscissa of the given point, is -7. Substituting the given values in the formula, we get

$$y - 2 = -\frac{3}{8}(x + 7)$$

Simplifying the equation, we get the general form,

$$3x + 8y = -5$$

Sometimes the constant is written at the left:

$$3x + 8y + 5 = 0$$

3.3 A LINE THROUGH TWO GIVEN POINTS

When two points on a line are given, the best way to write the equation is first to find the slope by the slope formula, and then use the point-slope form, which we have just shown.

EXAMPLE 2. Write the equation of the line passing through $(-5,-3)$ and $(-2,1)$.

Solution. First we find the slope:

$$m = \frac{1 + 3}{-2 + 5} = \frac{4}{3}$$

Now we use the slope with either of the points. Let us use the first point. Then
$$y + 3 = \frac{4}{3}(x + 5)$$
Simplifying, we get
$$3y + 9 = 4x + 20$$
Transposing and combining,
$$4x - 3y = -11$$
Since we have used the first point to get the equation of the line, we might substitute the second point coordinates in the equation as a check.

3.4 INTERCEPT FORM

To derive this form we let a represent the x-intercept and b represent the y-intercept. (Note that lowercase letters are used to represent intercepts.) Then the intercepts represent the points $(a,0)$ and $(0,b)$, respectively. If the intercepts are given, the equation can be written directly from the intercepts. For the slope, we get
$$m = \frac{b - 0}{0 - a} = -\frac{b}{a}$$
Now, using this slope and one point, say, $(a,0)$, by the point-slope form we get
$$y - 0 = -\frac{b}{a}(x - a)$$

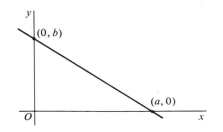

Fig. 3.3

Multiplying both sides of the equation by a we get
$$ay = -bx + ab$$
This can be written
$$bx + ay = ab$$
Finally, dividing both sides of the equation by the quantity ab, we get the so-called *intercept* form:
$$\frac{x}{a} + \frac{y}{b} = 1$$

This form is convenient when the two intercepts are given.
Note the following facts about the intercept form:

46 The Straight Line

(1) Each intercept is directly below the corresponding variable.
(2) The coefficients of x and y are 1.
(3) Both fractions are positive.
(4) The right side of the equation is $+1$.

EXAMPLE 3. Write the equation of the line having intercepts: $a = 6; b = -4$.

Solution. Substituting numerical values in the intercept form, we get

$$\frac{x}{6} + \frac{y}{-4} = 1$$

The equation is simplified to the general form: $2x - 3y = 12$.

EXAMPLE 4. Change the following equation to the intercept form:

$$5x - 3y + 13 = 0$$

Solution. When we have changed this equation to the intercept form, we can see at a glance the two intercepts. Transposing the constant term,

$$5x - 3y = -13$$

Now we divide both sides by -13, so that a $+1$ appears on the right side.

$$\frac{5x}{-13} - \frac{3y}{-13} = +1$$

We change this form slightly so that we have $+1$ for the coefficients of x and y and both fractions positive, and get

$$\frac{x}{-\frac{13}{5}} + \frac{y}{+\frac{13}{3}} = 1$$

Now we recognize the intercepts: $a = -\frac{13}{5}; b = \frac{13}{3}$.

Of course, another way to write the intercept form is to solve the equation for each intercept and then insert the intercepts in the formula. Given the equation

$$5x - 3y + 13 = 0$$

if $y = 0$, then $x = -\frac{13}{5}$, the x-intercept

if $x = 0$, then $y = \frac{13}{3}$, the y-intercept

Now we simply insert these intercepts in the intercept form.

3.5 SLOPE-INTERCEPT FORM

The slope-intercept form can be used directly when the slope m and the y-intercept b are known. This is a special case of the point-slope form. Here the intercept refers only to the y-intercept b.

Taking the y-intercept b as representing the point $(0,b)$ and using the point-slope form we get

$$y - b = m(x - 0)$$

Simplifying, $y - b = mx$

Now if we solve this equation for y, we get the *slope-intercept* form:

$$y = mx + b$$

Note especially that in this form *the equation is solved for y.*

3.6 Horizontal and Vertical Lines 47

EXAMPLE 5. Write the equation of a line having a slope of $\frac{5}{2}$ and cutting the y-axis at the point $(0,-4)$.

Solution. We use the form
$$y = mx + b$$
Substituting given values,
$$y = \frac{5}{2}x - 4$$
Simplifying the equation we get
$$5x - 2y = 4$$

The slope-intercept form is especially valuable because it shows at a glance the slope and the y-intercept. If an equation is solved for y, the *coefficient of x will show the slope*, and the *constant term will show the y-intercept*.

EXAMPLE 6. Change the following equation to the slope-intercept form, and from the result identify the slope and the y-intercept:
$$7x + 3y = -25$$

Solution. Solving for y,
$$3y = -7x - 25$$
$$y = -\frac{7}{3}x - \frac{25}{3}$$
Now we see that the slope is $-\frac{7}{3}$ and the y-intercept is $-\frac{25}{3}$.

Note: There is no corresponding simple form for a given slope and a given x-intercept. If the x-intercept is given, we go back to the point-slope form.

3.6 HORIZONTAL AND VERTICAL LINES

There is a simple way to write the equations of lines parallel to the coordinate axes. Of course, since the slope of a horizontal line is zero, a value that can be used in a formula except in the denominator, then the point-slope form could be used for such a line. On the other hand, the slope of a vertical line is infinite, something that cannot be used in a formula. However, there is a simple way to write the equations of such lines.

Suppose a line is vertical and passes through the point $(5,3)$ (Fig. 3.4). Then every point on the line has its abscissa equal to 5. The equation of the line is the simple statement $x = 5$. In like manner, if a horizontal line passes through the point $(3,-4)$, then every point on the line has its ordinate

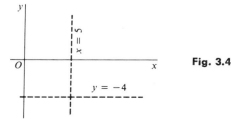

Fig. 3.4

48 The Straight Line

equal to -4. The equation of the line is $y = -4$. In general, for vertical and horizontal lines we have the equations:

$$x = a \quad \text{and} \quad y = b$$

EXERCISE 3.1

Write the equations of lines indicated by the following conditions:
1. Through $(4,-2)$; $m = \frac{3}{5}$.
2. Through $(5,-3)$; $m = -\frac{7}{2}$.
3. Through $(-7,-2)$; $m = \frac{4}{3}$.
4. Through $(-1,3)$; $m = -\frac{3}{5}$.
5. Through $(0,-4)$; $m = \frac{3}{8}$.
6. Through $(5,-1)$; $m = -\frac{4}{7}$.
7. Through $(-5,-3)$; $m = 3$.
8. Through $(-4,-1)$; $m = -3$.
9. Through $(0,3)$; $m = -\frac{1}{3}$.
10. Through $(7,-3)$; $m = -\frac{3}{7}$.
11. Through $(-3,4)$ and $(5,-2)$.
12. Through $(6,-3)$ and $(-4,-1)$.
13. Through $(0,0)$ and $(-2,-7)$.
14. Through $(7,2)$ and $(4,-5)$.
15. Through $(5,-1)$ and $(0,-3)$.
16. Through $(0,5)$ and $(4,0)$.
17. Through $(4,-2)$ and $(4,6)$.
18. Through $(3,-5)$ and $(-4,-5)$.
19. Through $(-3,2)$ and parallel to the y-axis.
20. Through $(-3,2)$ and parallel to the x-axis.
21. $b = 5; m = \frac{3}{4}$.
22. $m = -\frac{2}{5}; b = 7$.
23. $m = \frac{4}{7}; b = -3$.
24. $m = \frac{2}{3}; a = 5$.
25. $a = -4; m = -\frac{3}{2}$.
26. $a = 4; b = 7$.
27. $a = 6; b = 8$.
28. $a = -4; b = 3$.
29. $a = 6; b = -1$.
30. $a = 5; b = \frac{2}{3}$.
31. $a = -\frac{3}{4}; b = \frac{2}{3}$.
32. $a = -\frac{5}{3}; b = \frac{1}{4}$.
33. Write the equation of the line parallel to the given line and through the given point: (a) line, $3x + 5y = 8$; $(4,1)$ (b) line, $2x - 3y = 12$; $(-5,2)$.
34. Write the equation of the line parallel to the given line and through the given point: (a) $4x - 7y = 13$; $(-1,-5)$ (b) $5x + 2y = 0$; $(-4,3)$.
35. Write the equations of the lines perpendicular to the given lines in No. 33 and passing through the given points.
36. Write the equations of the lines perpendicular to the given lines in No. 34, and passing through the given points.
37. Write the following equations in slope-intercept form and tell the slope of each:
    (a) $3x + 7y = 25$; (b) $4y - 3x = 7$; (c) $6y - 9x = -40$.
38. Write the following equations in slope-intercept form and tell the slope of each:
    (a) $3y - 2x = 26$; (b) $2x + 5y = -20$; (c) $4y + x = -13$.

3.7 THE GENERAL FORM

The general form of the equation of a straight line is usually shown as

$$Ax + By = C$$

Note especially that the coefficients are denoted by *capital* letters. (The constant C is sometimes shown on the left side of the equal sign, with a zero on the right. Usually we shall use the form shown here.)

We have seen that if we solve the general form for y we get the slope-intercept form. For example, consider the equation

$$2x + 3y = 7$$

Solving for y we get

$$y = -\frac{2}{3}x + \frac{7}{3}$$

3.7 The General Form

In this form, the slope of the line is seen as $-\frac{2}{3}$, the coefficient of x; and the y-intercept is the constant term $\frac{7}{3}$. This is the meaning of the form, $y = mx + b$.

Let us solve a few equations for y and *from the result note especially the slope of each line:*

(a) $4x + 7y = 31$ $\quad y = -\frac{4}{7}x + \frac{31}{7};\quad m = -\frac{4}{7}$

(b) $3x + 8y = 17$ $\quad m = -\frac{3}{8}$

(c) $5x + 2y = 23$ $\quad m = -\frac{5}{2}$

Our question now is this: Can the slope of a line be seen directly from the general form $Ax + By = C$? Compare the slope of each line with the coefficients A and B. We might guess that the slope is always $-A/B$. To see whether this is true, we solve the general form for y:

$$Ax + By = C$$

Solving for y we get
$$By = -Ax + C$$
$$y = -\frac{A}{B}x + \frac{C}{B}$$

It is indeed true that the slope can be determined by $-A/B$.

Assuming we always take A as positive, then the slope will always have a sign that is the negative of the sign of B. Taking $m = -A/B$, we can immediately tell the slope of the following equations:

(a) $5x + 3y = 13;\ m = -\frac{5}{3}$ (b) $3x - 7y = 4;\ m = -\frac{3}{-7} = \frac{3}{7}$

(c) $x - 2y = 5;\ m = \frac{1}{2}$ (d) $4x + y = 9;\ m = -4$

A fast way of writing the equation of a line.

Let us reverse the preceding process. We have said that the slope can be seen directly from the general form: $Ax + By = C$. For example, in the equation,

$$3x + 7y = 19,\quad m = -\frac{3}{7}$$

Then it is reasonable to assume that if a line has a slope of $-\frac{3}{7}$, the equation starts out

$$3x + 7y = \underline{\qquad}$$

Of course, the coefficients might be other numbers, but their ratio would be the same, and we may as well use the numbers indicating the slope.

50   The Straight Line

As another example, if the slope of a line is $\frac{5}{8}$, it is reasonable to assume that the equation starts out
$$5x - 8y = \underline{}$$
Then we have a quick way of writing the equation of a line if the slope and a point on the line are known.

EXAMPLE 7.   Write the equation of the line having a slope of $\frac{3}{5}$ and passing through the point (6,2).

Solution.   We could use the point-slope form, but let us say instead that since the slope is $\frac{3}{5}$, the equation starts out
$$3x - 5y = \underline{}$$
All we need now is the constant.   Since the equation must be satisfied by the point (6,2), we substitute these coordinates for x and y, and we find the constant must be
$$3(6) - 5(2) = 8$$
Since 8 is the constant, we have the equation: $3x - 5y = 8$

EXAMPLE 8.   Write the equation of the line having a slope of $-\frac{7}{4}$ and passing through the point $(-2,5)$.

Solution.   The equation starts out $7x + 4y = \underline{}$.   When the coordinates of the point $(-2,5)$ are substituted for x and y, respectively, the value is 6.   Then the constant is 6, and the equation is
$$7x + 4y = 6$$

EXAMPLE 9.   Write the equation of the line through $(5,-2)$ and parallel to the line $4x - 7y = 9$.

Solution.   Since the lines are parallel they have the same slope.   Therefore, the second-line equation also starts out
$$4x - 7y = \underline{}$$
The substitution of the given point gives the constant a value of 34.   Then the equation of the second line is
$$4x - 7y = 34$$

EXAMPLE 10.   Write the equation of the line through $(4,-3)$ and perpendicular to the line $2x - 5y = 7$.

Solution.   Since the second line is perpendicular to the first, its slope is the negative reciprocal of the slope of the given line.   Therefore, the coefficients of x and y are reversed and one sign is changed.   Then the equation of the perpendicular starts out
$$5x + 2y = \underline{}$$
Substituting the coordinates of the point $(4,-3)$ we get a constant value of 14. The equation of the perpendicular is
$$5x + 2y = 14$$

3.8 The Normal Form of the Equation of a Straight Line

EXERCISE 3.2

By inspection (using the fast method described) write the general form of the equations of the lines indicated in Exercise 3.1, No. 1–25 and No. 33–36.

3.8 THE NORMAL FORM OF THE EQUATION OF A STRAIGHT LINE

The normal form of the straight-line equation has some important uses. We have seen that, in general, a straight line is determined by two conditions. These two conditions may be two points, a point and a slope, two intercepts, or the slope and an intercept.

There are two other conditions that determine a straight line:

(a) The perpendicular distance p of the line from the origin.
(b) The positive angle *omega* (ω) that this perpendicular makes with the positive x-axis.

Suppose we draw a line OE from the origin forming some positive angle ω with the positive x-axis (Fig. 3.5). Then we lay off on this line some distance OD from the origin. Call this distance p. Then the line through point D perpendicular to OE is completely determined by the angle ω and the distance p. The equation of this line can be written in terms of the distance p and the angle ω.

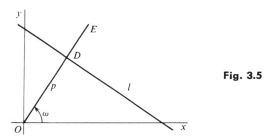

Fig. 3.5

If we take $(\cos \omega)$ as the coefficient of x and $(\sin \omega)$ as the coefficient of y, then we shall have, for the equation of the line,

$$(\cos \omega)x + (\sin \omega)y = p$$

This is the normal form of the equation of a line. This means that in this form, the coefficient of x and the coefficient of y must be the *cosine* and the *sine*, respectively, of some angle. This turns out to be the angle ω, and the constant term is then equal to p, the distance of the line from the origin. The distance p is often called the *normal intercept*.

The normal form can be derived in several ways. Perhaps the following is as good as any. We begin with the slope-intercept form of the equation of the given line (Fig. 3.6):

$$y = mx + b$$

In Figure 3.6, since the inclination is the angle alpha (α), we have

$$m = \tan \alpha$$

The Straight Line

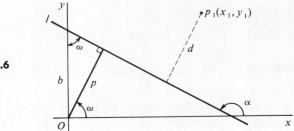

Fig. 3.6

Then
$$y = x \tan \alpha + b$$

Now we note that
$$\alpha = 90° + \omega$$

Then
$$\tan \alpha = \tan(90° + \omega) = -\cot \omega$$

Substituting,
$$y = -x \cot \omega + b$$

Now we seek equivalents for b and $\cot \omega$. Note that

$$\frac{p}{b} = \sin \omega \text{ or } b = \frac{p}{\sin \omega}; \text{ also } \cot \omega = \frac{\cos \omega}{\sin \omega}$$

Substituting we get
$$y = -x\left(\frac{\cos \omega}{\sin \omega}\right) + \frac{p}{\sin \omega}$$

Multiplying both sides of this equation by $(\sin \omega)$ and transposing, we get the normal form

$$x \cos \omega + y \sin \omega = p$$

This is called the *normal form* because it is stated in terms of the angle ω formed by the normal (perpendicular) to the line, and the distance p, the *normal intercept*.

One use of the normal form is to find the distance from a line to any external point. To find this distance we first convert an equation to normal form and then substitute the coordinates of the external point in the normal form. The result will show the distance from the line to the point.

We first write the normal form as

$$x \cos \omega + y \sin \omega - p = 0$$

Now we substitute the coordinates of the external point (x_1, y_1) in the left side of the equation. The result will show the distance d from the line to the point. That is,

$$x_1 \cos \omega + y_1 \sin \omega - p = d$$

To get the normal form of the equation is not so difficult as it might at first seem. Suppose we have the general form of the equation of a line

$$4x + 3y = 30$$

Now we have a right to divide both sides of the equation by any number we wish (except, of course, zero). We must divide by a number so that the

3.8 The Normal Form of the Equation of a Straight Line

coefficients of x and y, respectively, will be the cosine and sine of the same angle. Note that the coefficients of x and y are 4 and 3, respectively. If we square these coefficients, add these squares, and then take the square root of the sum, we shall have the proper divisor. In this case we divide by $\sqrt{4^2 + 3^2} = 5$. Then we get

$$\frac{4}{5}x + \frac{3}{5}y = 6$$

Now the coefficients $\frac{4}{5}$ and $\frac{3}{5}$ are the cosine and sine, respectively, of the same angle, and this angle is ω. The constant 6 represents the perpendicular distance of the line from the origin. If we sketch the graph from the general form (Fig. 3.7), we see that the intercepts are (7.5,0) and (0,10). Note that the distance $OD = p = 6$. From the value $\sin \omega = \frac{3}{5}$, we can find $\omega = 36.9°$ (approximately).

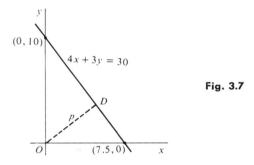

Fig. 3.7

Now suppose we wish to find the distance from the line $4x + 3y = 30$ to the point (8,12). We write the equation in normal form with the constant on the left side:

$$\frac{4}{5}x + \frac{3}{5}y - 6 = 0$$

Now we substitute the coordinates of the point in the equation and get

$$\frac{4}{5}(8) + \frac{3}{5}(12) - 6 = d$$

and, therefore

$$d = 7.6$$

This means that the point is 7.6 units from the line.

The normal form is especially useful in finding the altitude of a triangle. If d turns out to be negative, then the point is on the same side of the line as the origin. If d is positive, the point and the origin are on opposite sides of the line. If $d = 0$, the point is on the line because the equation is satisfied by the coordinates of the point. Of course, in finding the altitude of a triangle, if the distance d turns out to be negative, we simply call it positive, since the altitude refers to the absolute value of the distance.

54   The Straight Line

In the general form $Ax + By = C$, we shall always get the normal form if we divide both sides of the equation by the quantity

$$\pm \sqrt{A^2 + B^2}$$

In choosing the sign of the radical, we shall always choose the radical with the same sign as the sign of C. Then the distance p will always be the *positive* distance from the origin to the line. In general, the normal form may be stated as

$$\frac{A}{\pm\sqrt{A^2 + B^2}} x + \frac{B}{\pm\sqrt{A^2 + B^2}} y = \frac{C}{\pm\sqrt{A^2 + B^2}}$$

The form may be written with a single radical:

$$\frac{Ax + By = C}{\pm\sqrt{A^2 + B^2}}$$

EXAMPLE 11. Write the following equation in normal form: $2x + 5y = -15$. Then find the distance of the line from the origin. Also find the distance from the line to the points $(4,-7)$, $(-1,3)$, and $(-5,-1)$. Sketch the line.

Solution. We graph the line from the general form (Fig. 3.8). To write the normal form, we divide both sides of the equation by $-\sqrt{29}$, since $2^2 + 5^2 = 29$.

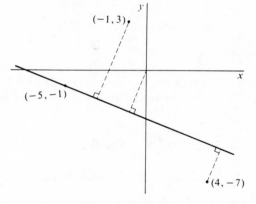

Fig. 3.8

We choose the negative sign of the radical because the constant C is negative. We get

$$-\frac{2}{\sqrt{29}} x - \frac{5}{\sqrt{29}} y = +\frac{15}{\sqrt{29}}$$

The normal form now shows that the line is $+\dfrac{15}{\sqrt{29}}$ (or approximately 2.78) units from the origin.

To find the distance from the line to each of the points, we substitute the coordinates of each point in turn. For the point $(4,-7)$ we have

$$-\frac{2}{\sqrt{29}}(4) - \frac{5}{\sqrt{29}}(-7) - \frac{15}{\sqrt{29}} = d; \quad \text{or} \quad d = \frac{-8 + 35 - 15}{\sqrt{29}} = \frac{12}{\sqrt{29}} = 2.23$$

3.8 The Normal Form of the Equation of a Straight Line

For the point $(-1,3)$ we can write
$$\frac{(-2)(-1) - (5)(3) - 15}{\sqrt{29}} = d; \quad \text{then} \quad d = \frac{2 - 15 - 15}{\sqrt{29}} = \frac{-28}{\sqrt{29}} = -5.20$$

Note that for the point $(-1,3)$, the distance is negative. This means that the point $(-1,3)$ and the origin are on the *same side* of the line. For the point $(-5,-1)$ we have
$$d = \frac{(-2)(-5) - (5)(-1) - 15}{\sqrt{29}} = \frac{10 + 5 - 15}{\sqrt{29}} = 0$$

Note that for the point $(-5,-1)$ the distance is zero. This means the point lies on the line and its coordinates satisfy the equation.

It might be pointed out that the angle ω has its terminal side in the third quadrant. This is also shown by the fact that the coefficients of x and y in the normal form are both negative, since they represent sine and cosine of ω, respectively.

EXAMPLE 12. A triangle has its vertices at $A(4,5); B(-6,-1); C(1,-5)$. Find the length of the altitude from side c to the vertex C by writing the equation of side c and then changing the equation to normal form. Find the area of the triangle by use of the formula $A = (\frac{1}{2})(b)(h)$. Finally, write the equation of this altitude.

Solution. First, let us sketch the triangle, showing the vertices (Fig. 3.9). By the slope formula we find the slope of side c is $\frac{3}{5}$. (This is the side AB.) Then the equation of this side starts out
$$3x - 5y = \underline{\qquad}$$

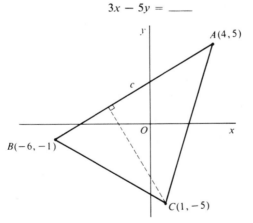

Fig. 3.9

Taking either point on the line, we find the constant is -13. Then the general form of the equation of c is $3x - 5y = -13$. To change to the normal form, we divide both sides of the equation by $-\sqrt{34}$ and get
$$\frac{3x - 5y}{-\sqrt{34}} = \frac{-13}{-\sqrt{34}}$$

To find the distance of point C from this line, we transpose the constant term and insert the coordinates of point C. We get
$$\frac{3 + 25 + 13}{-\sqrt{34}} = d = \text{altitude from side } c \text{ to vertex } C$$

56 The Straight Line

Since the altitude must be an absolute value, we disregard the negative sign. For the area of the triangle we must have also the length of side c, which is the base of the triangle if we call C the vertex angle. For the length of side c, by use of the distance formula, we get $c = 2\sqrt{34}$. Then area is given by

$$A = \left(\frac{1}{2}\right)(2\sqrt{34})\left(\frac{41}{\sqrt{34}}\right) = 41 \text{ (square units)}$$

For the equation of the altitude we first find its slope, which is the negative reciprocal of the slope of side c, or $-\frac{5}{3}$. The constant is determined by the point $C(1,-5)$ on the altitude. The equation of the altitude is then $5x + 3y = -10$.

EXERCISE 3.3

1. Write each of the following equations in normal form and tell how far the line is from the origin: (a) $3x + 7y = 25$; (b) $4y - 3x = 7$; (c) $6y - 9x = -40$.
2. Find the distance of each of the following lines from the origin:
   (a) $3y - 2x = 26$; (b) $2x + 5y = -20$; (c) $4y + x = -13$
3. Show that the following lines are parallel. Then find the distance of each line from the origin and the distance between the lines:
   (a) $3x + 4y = 15$ and (b) $3x + 4y = 45$
4. Find the distance between the lines: (a) $5x - 2y = 10$ and (b) $5x - 2y = 40$.

Find the distance from each of the following lines to the given point:

5. $3x + 4y = 20$; point $(6,3)$.
6. $5x - 2y = 20$; point $(2,4)$.
7. $2x + 3y = 12$; point $(3,-4)$.
8. $x + 4y = 17$; point $(-1,2)$.
9. $3x + y = 20$; point $(1,1)$.
10. $5x - 2y = 10$; point $(6,-1)$.
11. $2x - 3y = 7$; point $(2,-1)$.
12. $2x + 3y = -12$; point $(-3,-2)$.
13. Two vertices of a triangle are at the points $A(-5,1)$ and $B(-1,-4)$. What is the equation of this side? If the third vertex of the triangle is at $(3,5)$, what is the altitude of the triangle? Find the area of the triangle.
14. A triangle has its vertices at $A(8,6)$; $B(-4,2)$; $C(-2,-6)$. Find the following:
    (a) The slopes of the sides and the equations of the lines forming the sides
    (b) The slopes of the medians and the equations of the medians
    (c) The slopes of the altitudes and the equations of the altitudes
    (d) The length of each altitude, and the area of the triangle in three ways
    (e) The slopes and equations of the perpendicular bisectors of the sides
    (f) The point of intersection of the medians, of the altitudes, and of the perpendicular bisectors of the sides of the triangle. Show that these three points of intersection are in a line. Show that one point is $\frac{2}{3}$ of the distance between the other two.
15. A triangle has its vertices at $A(1,6)$; $B(-7,2)$; and $C(5,-4)$. Find the facts asked for in No. 14.

chapter
4

The Circle

4.1 DEFINITIONS

The circle may be defined as the result of a motion, or it may be defined as something static. From the dynamic standpoint, a circle is the path of a point that moves (in a plane) so that its distance from a fixed point is a constant. If the idea of motion is omitted, the circle is defined as the locus of all points in a plane that are equally distant from a fixed point called the *center*.

It is well to distinguish between the word *circle* and the word *circumference*. Usually the word *circle* refers to the curved line forming the locus whereas the *circumference* is the *length* of this curved line. The *radius* of a circle is the *distance* from the center of the circle to any point on the circle. For any given circle this distance is the same for any point on the circle. A circle therefore has only one radius, the distance from the center to a point on the circle. When we say "all radii of the same circle are equal," we are thinking of the radius as a line segment joining the center with a point on the circle.

4.2 EQUATION OF A CIRCLE

One circle, and only one, can be drawn with a given center and a given radius. Therefore, the equation of any circle can be written when its center and its radius are known.

As an example, consider the circle whose center is the point (7,4) and whose radius is equal to 5 units (Fig. 4.1). Since a circle is completely determined by these two conditions, its equation can be written from them.

To write the equation of the circle we take a general point (x,y) at any general position on the locus. For any position of point (x,y), its distance from (7,4) must always be equal to 5 units. This requirement is stated by the formula for the distance between two points:

$$d = \sqrt{(x_1 - x_2)^2 + (y_1 - y_2)^2}$$

58   The Circle

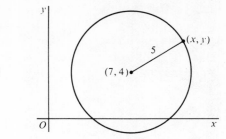

Fig. 4.1

In the formula, we take the variable point (x,y) as one of the two points, and the center $(7,4)$ as the other point, and get

$$d = \sqrt{(x-7)^2 + (y-4)^2}$$

The expression is now set equal to 5, the required condition:

$$\sqrt{(x-7)^2 + (y-4)^2} = 5$$

This equation states the simple fact that the distance from the center to any point on the circle is always equal to 5. It is therefore the required equation. Of course, the equation is usually simplified. Squaring both sides:

$$(x-7)^2 + (y-4)^2 = 25$$

Expanding squares, transposing, and combining, we get the so-called *general* form of the equation of a circle:

$$x^2 + y^2 - 14x - 8y + 40 = 0$$

4.3 THE STANDARD FORM OF THE EQUATION OF A CIRCLE

To obtain a formula for the equation of any circle we begin with the more general center (h,k) and the more general radius r (Fig. 4.2). Using the distance formula again, we get

$$\sqrt{(x-h)^2 + (y-k)^2} = r$$

Squaring both sides, we get

$$(x-h)^2 + (y-k)^2 = r^2$$

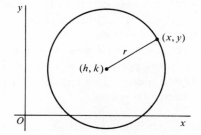

Fig. 4.2

This equation is the *standard form* of the equation of any circle. It can be used to write the equation of any circle when its center and its radius are known. If the center represented by (h,k) and the radius represented by r are known, they are inserted in the standard form to obtain the equation of the specific circle. The result is then simplified.

EXAMPLE 1. Write the equation of the circle whose center is $(5,-2)$ and whose radius is 6 units.

Solution. In this example $h = 5$, $k = -2$, and $r = 6$. Substituting numerical values,

$$(x - 5)^2 + (y + 2)^2 = 36$$

Expanding, $\quad x^2 - 10x + 25 + y^2 + 4y + 4 = 36$

Rearranging terms, $\quad x^2 + y^2 - 10x + 4y - 7 = 0$

EXAMPLE 2. Write the equation of the circle whose center is at the origin and whose radius is equal to 8 units.

Solution. Using the point $(0,0)$ as the center in the formula, we get $(x - 0)^2 + (y - 0)^2 = 64$, which reduces to $x^2 + y^2 = 64$

A circle with its center at the origin is said to be in *standard position*. Then the equation becomes

$$(x - 0)^2 + (y - 0)^2 = r^2$$

or simply $\quad x^2 + y^2 = r^2$

Moreover, any equation of the form $x^2 + y^2 = K$ represents a circle with its center at the origin and with radius equal to $\sqrt{K}$. If the center of a circle is not at the origin, the circle is said to be *translated*.

EXAMPLE 3. If a circle has the equation, $x^2 + y^2 = 20$, its center is at the origin, and its radius is equal to $\sqrt{20}$.

Two special cases called *degenerate* forms of the circle should be mentioned. Whatever the center of a circle, if $r = 0$, the circle is called a *point-circle*, since it has a center but the radius is equal to zero. The only pair of numbers that satisfy such an equation is the pair representing the center of the circle.

As another degenerate form, if r^2 is negative, the radius is imaginary, as in $x^2 + y^2 = -9$, and the circle is called *imaginary*. In practical work, the two degenerate forms of the circle have little significance.

4.4 GENERAL FORM OF THE EQUATION OF A CIRCLE

Consider again the equation of the circle in Example 1, with center $(5,-2)$ and radius 6. For the *standard form* we write

$$(x - 5)^2 + (y + 2)^2 = 36$$

When this equation is expanded and simplified, it takes the form

$$x^2 + y^2 - 10x + 4y - 7 = 0$$

This last result is called the *general form* of the equation of the circle.

The general form of the equation of any circle is

$$x^2 + y^2 + Dx + Ey + F = 0$$

Note that this equation shows the sum of x^2 and y^2, each with a coefficient of 1. The equation also contains an x term and a y term, with coefficients D and E, respectively. It also contains a constant term F. Either or all of the constants, D, E, and F, may be positive, negative, or zero. Note especially that the equation contains *no* term involving the product (xy).

It can be proved that every equation of the second degree in which the coefficients of x^2 and y^2 are equal represents a circle provided the equation contains no term involving the product xy. Such an equation can be reduced to the general form by dividing through by the coefficient of x^2 and y^2.

Let us note the relation between the standard form and the general form of the equation of the circle. The standard form is

$$(x - h)^2 + (y - k)^2 = r^2$$

Expanding, we get

$$x^2 - 2hx + h^2 + y^2 - 2kx + k^2 = r^2$$

Transposing and rearranging,

$$x^2 + y^2 - 2hx - 2ky + h^2 + k^2 - r^2 = 0$$

The general form is

$$x^2 + y^2 + Dx + Ey + F = 0$$

Comparing the two we see that, as coefficients of x, $D = -2h$; as coefficients of y, $E = -2k$. The constant is $h^2 + k^2 - r^2 = F$.

Solving for h, for k, and for r^2,

$$h = -\frac{D}{2}; \qquad d = -\frac{E}{2}; \qquad r^2 = h^2 + k^2 - F$$

By means of these relations, the center and radius of a circle may be determined from the general form.

4.5 GRAPHING A CIRCLE FROM A GIVEN EQUATION

If the equation of a circle indicates that the center is at the origin, the radius can then be easily determined and the graph constructed by use of a compass. For example, the equation

$$x^2 + y^2 = 49$$

indicates a circle with center at the origin and radius of 7 units. The circle can now be constructed with a compass.

The following circle is a little more difficult:

$$x^2 + y^2 - 8x + 12y + 27 = 0$$

4.6 Intercepts of a Circle

The circle could be graphed by finding several pairs of corresponding values for x and y and then plotting them as points on the graph. The points could then be connected by a smooth curve.

However, the above procedure would be very slow and laborious. Instead, if the center and the radius are first determined, the circle can be drawn by use of a compass.

The center and the radius can be found by reducing the equation to the standard form, $(x - h)^2 + (y - k)^2 = r^2$. This is done by completing the squares in x and y. As an example, consider the equation

$$x^2 + y^2 - 8x + 12y + 27 = 0$$

The terms are first rearranged and the constant, 27, is written at the right:

$$x^2 - 8x \ldots + y^2 + 12y \ldots = -27$$

Now we complete the squares in x and in y by adding 16 and 36 to both sides of the equation to obtain

$$x^2 - 8x + 16 + y^2 + 12y + 36 = -27 + 16 + 36$$

The equation is changed to the standard form by expressing the squares in x and y:

$$(x - 4)^2 + (y + 6)^2 = 25$$

In this form the equation shows at a glance that the center is at $(4, -6)$; that is, $h = 4$ and $k = -6$, where (h,k) is the center. Since the square of the radius is 25, the radius is equal to 5. Now that we have determined the center and the radius, the circle can be constructed with a compass.

Another method is to make use of the relation between the standard form and the general form. We have seen that

$$h = -\frac{D}{2} \qquad k = -\frac{E}{2} \qquad r^2 = h^2 + k^2 - F$$

In the given equation, $D = -8$; $E = +12$; $F = 27$. Then

$$h = -\frac{-8}{2} = 4, \quad k = -\frac{12}{2} = -6, \quad r^2 = 16 + 36 - 27 = 25, \text{ then } r = 5$$

Now that we have the center, $(4, -6)$, and the radius, 5, we can use a compass to construct the circle.

4.6 INTERCEPTS OF A CIRCLE

The intercepts of a circle can be found by setting each variable, in turn, equal to zero and then solving for the other variable. As an example, consider the following:

EXAMPLE 4. Find the intercepts of the circle $x^2 + y^2 + 10x - 8y + 16 = 0$.

Solution. To find the x-intercepts, we set $y = 0$ and solve for x. If $y = 0$ we have

$$x^2 + 10x + 16 = 0$$

Solving this quadratic, $\qquad x = -2 \quad$ and $\quad x = -8$

These values represent the points $(-2,0)$ and $(-8,0)$. To find the y-intercepts, we set $x = 0$ and get
$$y^2 - 8y + 16 = 0$$
Solving, we get $\qquad\qquad\qquad y = 4 \quad$ and $\quad y = 4$

The two equal roots of the quadratic equation represent the double point $(0,4)$. The graph of this circle is shown in Fig. 4.3. The double point $(0,4)$ is indicated by the fact that the circle is *tangent* to the y-axis at the point $(0,4)$. The circle can be constructed by compass after the center is found to be $(-5,4)$ and the radius is equal to 5.

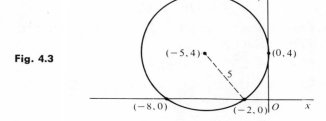

Fig. 4.3

EXAMPLE 5. Find the intercepts of the following circle and sketch the graph:
$$x^2 + y^2 - 10x - 4y + 13 = 0$$

Solution. First we determine the center $(5,2)$ and the radius 4. Then the circle is easily sketched (Fig. 4.4). To find the x-intercepts we have
$$x^2 - 10x + 13 = 0$$
Solving by formula we get $\qquad x = 5 \pm 2\sqrt{3}$

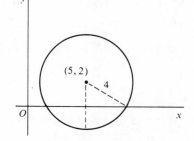

Fig. 4.4

Note that the x-intercepts are real but irrational. To find the y-intercepts, we have
$$y^2 - 4y + 13 = 0$$
Solving by formula we get $\qquad y = 2 \pm 3i$

The imaginary roots of the quadratic indicate that the circle does not touch or intersect the y-axis.

4.7 INTERSECTION OF THE CIRCLE AND STRAIGHT LINES

The points of intersection of any two curves can be found by solving their equations as a system. The points can often be estimated by graphing the equations and noting their points of intersection.

Since the equation of a straight line is of the first degree and the equation of a circle is of the second degree, solving for their points of intersection is simply a matter of solving a linear equation with a quadratic. Such a system can be solved as follows:

(1) Solve the linear equation for one letter, x or y, in terms of the other.
(2) Substitute the result in the quadratic equation, obtaining an equation in one unknown.
(3) Solve the quadratic for the unknown. The result, in general, will be two roots or values.
(4) Substitute these two roots in turn in the *linear* equation to find the value of the second variable.

The intersection points of a straight line and a circle will correspond to one of the three following conditions (Fig. 4.5):

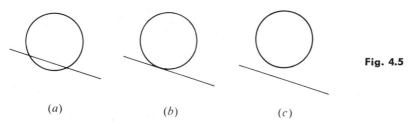

Fig. 4.5

(a)   (b)   (c)

(1) Two real and distinct points if the straight line passes through the circle (Fig. 4.5a).
(2) Two real but equal points (called a double point) if the line is tangent to the circle (Fig. 4.5b).
(3) Two imaginary points if the line does not intersect or touch the circle (Fig. 4.5c).

Note that in each instance, two points are indicated.

It may be noted here that the number of points of intersection of any two curves is equal to the maximum number it is possible to produce by letting the curves intersect in any way at all.

EXAMPLE 6. Find the points of intersection of the circle (Fig. 4.6):

$$x^2 + y^2 + 2x - 6y - 19 = 0$$

with each of the lines: (a) $3x - 7y = 5$, (b) $2x - 5y = 12$, (c) $x - 2y = 6$.

Solution. We use the method of substitution to solve each linear equation with the quadratic as a system.

(a) From linear equation (a)

$$y = \frac{3x - 5}{7}$$

64   The Circle

Fig. 4.6

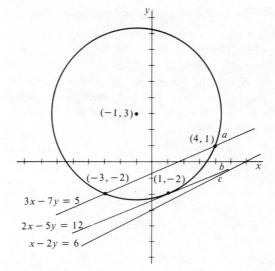

Substituting in the quadratic we get

$$x^2 + \left(\frac{3x-5}{7}\right)^2 + 2x - 6\left(\frac{3x-5}{7}\right) - 19 = 0$$

Expanding,

$$x^2 + \frac{9x^2 - 30x + 25}{49} + 2x - \frac{18x - 30}{7} - 19 = 0$$

Multiplying by 49,

$$49x^2 + 9x^2 - 30x + 25 + 98x - 126x + 210 - 931 = 0$$

This reduces to

$$58x^2 - 58x - 696 = 0, \quad \text{or} \quad x^2 - x - 12 = 0$$

Now we solve the quadratic for x and find corresponding values of y:

$$x = 4 \text{ and } -3. \quad \text{Corresponding values of } y \text{ are } 1 \text{ and } -2$$

The points of intersection are $(4,1)$ and $(-3,-2)$.

(b) When the line (b) is solved for y and the result substituted in the quadratic, the equation reduces to $x^2 - 2x + 1 = 0$, an equation having the double root, $x = 1$. The line is tangent to the circle at $(1,-2)$.

(c) When the line (c) is solved for y and the result substituted in the quadratic, the equation reduces to

$$5x^2 - 16x + 32 = 0$$

The roots of this equation are imaginary, and the line therefore does not intersect the circle in any real point.

EXERCISE 4.1

Write the equation of each of the following circles and find the intercepts:
1. Center $(5,1)$; radius $= 6$.
2. Center $(-4,3)$; radius $= 5$.
3. Center $(-7,-4)$; radius $= 7$.
4. Center $(6,-1)$; radius $= 3$.

5. Center $(3,-5)$; radius $= \sqrt{34}$.
6. Center $(-2,6)$; radius $= 2\sqrt{10}$.
7. Center $(-5,-2)$; radius $= 2\sqrt{5}$.
8. Center $(4,-7)$; radius $= 2\sqrt{13}$.
9. Center $(3,7)$; tangent to y-axis.
10. Center $(-6,-4)$; tangent to x-axis.

Write the equation of each of the following six circles and then write the equation of the tangent line at the given point, which lies on the circle.
11. Center $(3,-6)$; point $(-1,-3)$.
12. Center $(-1,-3)$; point $(3,-2)$.
13. Center $(2,-4)$; point $(0,-1)$.
14. Center $(-3,4)$; point $(4,3)$.
15. Center $(1,-2)$; point $(-3,-2)$.
16. Center $(2,-3)$; point $(2,3)$.

Find the center and the radius of each of the following circles:
17. $x^2 + y^2 - 8x + 10y + 5 = 0$
18. $x^2 + y^2 + 4x - 12y + 4 = 0$
19. $x^2 + y^2 + 2x - 4y - 8 = 0$
20. $x^2 + y^2 - 6x - 10y + 9 = 0$
21. $x^2 + y^2 - 8x - 7y + 12 = 0$
22. $2x^2 + 2y^2 - 5x - 3y + 8 = 0$
23. $3x^2 + 3y^2 + 10x - 2y - 8 = 0$
24. $5x^2 + 5y^2 - 12x + 8y + 10 = 0$
25. $x^2 + y^2 - 8x + 4y + 20 = 0$
26. $x^2 + y^2 - 4x + 6y + 29 = 0$

Find the points of intersection of the following pairs of curves:
27. $x^2 + y^2 = 20$
    $2x + y = 10$
28. $x^2 + y^2 = 16$
    $x - y = 6$
29. $x^2 + y^2 = 25$
    $4x + 3y = 0$
30. $x^2 + y^2 = 25$
    $4x - 3y = 25$
31. $x^2 + y^2 = 9$
    $x + y = 6$
32. $x^2 + y^2 + 2x = 24$
    $2x - 3y = -8$
33. $x^2 + y^2 + 6x = 7$
    $x^2 + y^2 = 25$
34. $x^2 + y^2 + 2x - 6y - 15 = 0$
    $x^2 + y^2 - 6x + 10y + 9 = 0$

4.8 A CIRCLE DETERMINED BY THREE CONDITIONS

A circle is completely determined by any of the following conditions:

1. If the circle passes through three points, not in a straight line.
2. If the circle passes through two points and is tangent to a straight line not on either of the two points. In this case there may be two circles.
3. If the circle passes through one point and is tangent to two straight lines. In this case there are two circles.
4. If the circle is tangent to three straight lines. In this case there may be four circles.

If three points on a circle are given, then the equation of the circle can be written in at least three different ways.

First Method. We assume the center is at the point which we denote by (h,k). Using the distance formula, we write the expression for the distance from the center (h,k) to each of the three given points. These distances are radii of the circle and are therefore equal. We then have three equations in three unknowns, h, k, and d. (We may use r for d to represent the radius.) The three equations are then solved, as shown in the following example:

EXAMPLE 7. Find the equation of the circle passing through the three points:
$$A(4,7); \; B(-8,3); \; C(0,-5)$$

Solution. Now if we knew the center, we could find the radius and then write the equation of the circle. Since we do not know the center, we assume we already know it and call it (h,k). Now all we need to do is to find h and k. We write the equations for the radius of the circle from the center (h,k) to each of the points. Using d_1, d_2, and d_3 as the distances, respectively, from (h,k) to each of the three points on the circle, we have

$$d_1 = \sqrt{(h-4)^2 + (k-7)^2}$$
$$d_2 = \sqrt{(h+8)^2 + (k-3)^2}$$
$$d_3 = \sqrt{(h-0)^2 + (k+5)^2}$$

Since these three distances are all equal to the radius of the circle, they are equal to each other. Setting $d_1 = d_2$ we get

$$\sqrt{(h-4)^2 + (k-7)^2} = \sqrt{(h+8)^2 + (k-3)^2}$$

Squaring and expanding both sides of the equation we get

$$h^2 - 8h + 16 + k^2 - 14k + 49 = h^2 + 16h + 64 + k^2 - 6k + 9$$

This equation simplifies to

$$3h + k = -1$$

In like manner, setting $d_1 = d_3$ and simplifying the equation we get

$$h + 3k = 5$$

Solving this system of two linear equations we get $h = -1$; $k = 2$. The center of the circle is therefore at the point $(-1,2)$. Now the radius can be found as the distance from the center to any of the given points. For the radius we get

$$r = \sqrt{50}$$

The standard form of the circle becomes

$$(x+1)^2 + (y-2)^2 = 50$$

This simplifies to the general form:

$$x^2 + y^2 + 2x - 4y - 45 = 0$$

Second Method. As another method we recall from plane geometry that the perpendicular bisector of a chord passes through the center of a circle. We can consider the slopes and midpoints of any two chords between the given points. The chord AB has a slope of $\frac{1}{3}$. Its midpoint is $(-2,5)$. The perpendicular bisector of the chord has a slope of -3, the negative reciprocal of the slope of the chord. Then the equation of the perpendicular bisector of the chord AB is $3x + y = -1$. In like manner, the chord AC has a slope of 3, and its midpoint is $(2,1)$. Its perpendicular bisector has a slope of $-\frac{1}{3}$ and passes through the point $(2,1)$. Its equation is therefore $x + 3y = 5$. The two perpendicular bisectors of the chords pass through the center of the circle and their point of intersection is therefore the center. Solving the two equations, we get: $x = -1$; $y = 2$.

Third Method. We have seen that the general equation of a circle has the form

$$x^2 + y^2 + Dx + Ey + F = 0$$

Since the three given points all lie on the circle, they all satisfy the general equation. Substituting the three points in turn in the general equation, we get three equations in D, E, and F:

$$16 + 49 + 4D + 7E + F = 0$$
$$64 + 9 - 8D + 3E + F = 0$$
$$0 + 25 + 0D - 5E + F = 0$$

Solving these three equations we get

$$D = 2; \quad E = -4; \quad F = -45$$

Substituting in the general equation we get the equation of this circle:

$$x^2 + y^2 + 2x - 4y - 45 = 0$$

This method does give the equation of the circle, but it does not show directly the center of the circle. However, the center can easily be seen from the result.

EXAMPLE 8. Find the equation of the circle passing through the two points $A(1,4)$ and $B(-1,8)$ and tangent to the line $x + 3y = 3$ (Fig. 4.7).

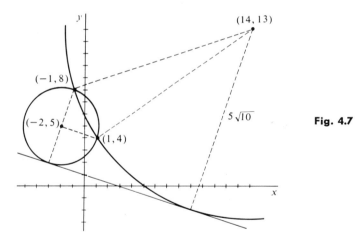

Fig. 4.7

Solution. First we assume the center is at the point (h,k). We write the expression for the distance from the center (h,k) to each of the given points:

$$d_1 = \sqrt{(h-1)^2 + (k-4)^2}$$
$$d_2 = \sqrt{(h+1)^2 + (k-8)^2}$$

We set these two expressions equal to each other and simplify:

$$\sqrt{(h-1)^2 + (k-4)^2} = \sqrt{(h+1)^2 + (k-8)^2}$$

Squaring both sides of this equation and simplifying we get the equation

$$h - 2k = -12, \quad \text{or} \quad h = 2k - 12$$

Now we must write the expression for the distance from the center (h,k) to the line $x + 3y = 3$. We do this by means of the normal form of the equation. Writing the equation in normal form and inserting the coordinates of the center (h,k) we get the distance from the point to the line:

$$\frac{h + 3k - 3}{\sqrt{10}} = d_3$$

Now we equate this distance (again the radius of the circle) to one of the other distances:

$$\sqrt{(h-1)^2 + (k-4)^2} = \frac{h + 3k - 3}{\sqrt{10}}$$

Squaring both sides of this equation and simplifying we get

$$9h^2 - 6hk + k^2 - 14h - 62k + 161 = 0$$

Now we solve this quadratic with the linear equation, $h = 2k - 12$, above, as a system by substitution. Substituting $(2k - 12)$ for h in the quadratic, we get the simplified form:

$$k^2 - 18k + 65 = 0$$

Solving for k, $\qquad k = 5 \quad \text{and} \quad k = 13$

For corresponding values of h we have $h = -2$ and $h = 14$. The result shows that there are actually two circles that satisfy the given conditions having the centers, respectively, at $(-2,5)$ and $(14,13)$. The radius of the first circle is $\sqrt{10}$; the radius of the second is $5\sqrt{10}$. The equations of the circles can now be written from the known centers and radii:

$$(x + 2)^2 + (y - 5)^2 = 10 \quad \text{or} \quad x^2 + y^2 + 4x - 10y + 19 = 0$$
$$(x - 14)^2 + (y - 13)^2 = 250 \quad \text{or} \quad x^2 + y^2 - 28x - 26y + 115 = 0$$

4.9 TANGENTS TO A CIRCLE

From plane geometry we know that a tangent to a circle is perpendicular to the radius drawn to the point of tangency. This theorem is used in finding the equation of a tangent to a circle at a given point on the circle. The method is shown by an example.

EXAMPLE 9. Find the equation of the tangent to the following circle (Fig. 4.8) at the point $(2,-3)$:

$$x^2 + y^2 + 10x - 2y - 39 = 0$$

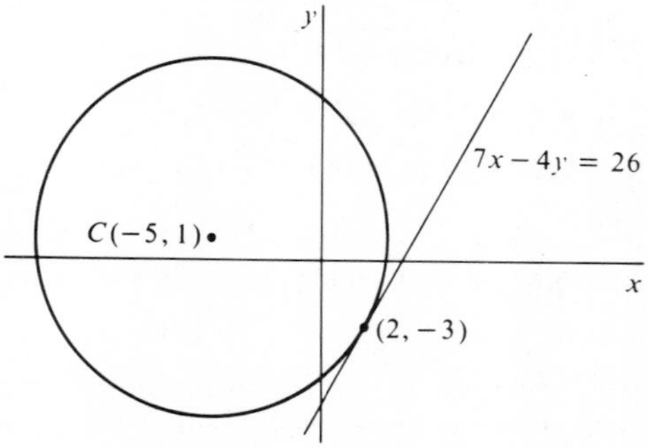

Fig. 4.8

Solution. First, we verify that the point $(2,-3)$ satisfies the equation: $4 + 9 + 20 + 6 - 39 = 0$. To find the slope of the radius to the point $(2,-3)$ we must know the center of the circle. From the equation we identify the center $C(-5,1)$. Then the slope of the radius to point $(2,-3)$ is $-\frac{4}{7}$. Since the tangent is perpendicular to the radius at the point $(2,-3)$, its slope is $+\frac{7}{4}$. Then for the tangent we have the following facts:

$$m = +\tfrac{7}{4}; \quad \text{point } (2,-3); \quad \text{equation: } 7x - 4y = 26$$

4.10 LENGTH OF A TANGENT TO A CIRCLE

By the length of a tangent from an external point to a circle we mean the length of the line segment joining the external point to the point of tangency. It is sometimes necessary to find the length of such a line segment. There is a simple formula for the length of a tangent.

Suppose we have the general circle with center $C(h,k)$ and radius r (Fig. 4.9). Now we wish to find the length of the tangent from the external point $P_1(x_1,y_1)$ to the point of tangency A. Note that the radius CA is perpendicular to the tangent PA. In the right triangle PCA the hypotenuse is PC, which we shall denote by d. The leg PA is the line segment whose length we wish to find. We denote this length by t for tangent.

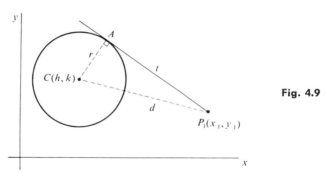

Fig. 4.9

For the distance d from the external point (x_1,y_1) to the center of the circle we have

$$(x_1 - h)^2 + (y_1 - k)^2 = d^2$$

Also note that in the right triangle we have

$$d^2 = r^2 + t^2$$

Substituting we get

$$(x_1 - h)^2 + (y_1 - k)^2 = r^2 + t^2$$

or

$$(x_1 - h)^2 + (y_1 - k)^2 - r^2 = t^2$$

Here we have the formula for the square of the tangent t^2. Note the similarity of the formula to the form of the equation of the circle itself:

$$(x - h)^2 + (y - k)^2 - r^2 = 0$$

70   The Circle

This means that whatever the form of the equation of a circle, if the right side is zero, then the coordinates of the external point may be inserted in the formula for the circle and the result will show the square of the tangent.

EXAMPLE 10. Find the lengths of the tangents to the following circle from the points $A(8,-7)$; $B(-4,7)$; $C(-5,4)$: (Fig. 4.10)

$$x^2 + y^2 + 4x - 6y - 7 = 0$$

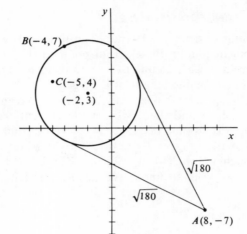

Fig. 4.10

Solution. To find the length of the tangent from any point, we insert the coordinates of the point into the equation of the circle. The result is the square of the tangent. For the point $A(8,-7)$ we get

$$64 + 49 + 32 + 42 - 7 = t^2 = 180$$

Then the length of the tangent is $t = \sqrt{180} = 6\sqrt{5} = 13.42$ (approx.).

For the length of the tangent from point $B(-4,7)$ we have

$$16 + 49 - 16 - 42 - 7 = t^2 = 0; \quad \text{or} \quad t = 0$$

Since the length of the tangent from point $B(-4,7)$ is zero, the point lies on the circle, and the coordinates of point B satisfy the equation.

To find the length of the tangent from point $C(-5,4)$ we insert the coordinates and get

$$25 + 16 - 20 - 24 - 7 = t^2 = -10; \quad \text{then} \quad t = \sqrt{-10}, \text{ imaginary}$$

The fact that the length of the tangent turns out to be imaginary means that the point C lies inside the circle.

EXERCISE 4.2

Write the equation of the circle that passes through each of the following sets of points:
1. $A(4,5)$; $B(-4,-3)$; $C(8,-3)$
2. $A(7,1)$; $B(-6,0)$; $C(-2,-8)$
3. $A(4,7)$; $B(-8,1)$; $C(0,-5)$
4. $A(6,3)$; $B(-5,5)$; $C(3,-3)$
5. $A(-4,2)$; $B(-2,-4)$; $C(2,-1)$
6. $A(5,6)$; $B(-3,2)$; $C(3,-4)$

4.10 Length of a Tangent to a Circle

Write the equation of the tangent to each of the following circles at the given point. Check each point to be sure it lies on the circle. Sketch the graph of each circle.
7. $x^2 + y^2 - 6x - 4y + 3 = 0$, at point $(4,-1)$.
8. $x^2 + y^2 + 6x - 4y + 8 = 0$, at point $(-1,1)$.
9. $x^2 + y^2 - 2x + 4y - 24 = 0$, at point $(-4,0)$.
10. $x^2 + y^2 + 4x - 8y + 7 = 0$, at point $(-4,1)$.

Find the length of the tangents to each of the following circles from each of the given points. Sketch each circle and show the tangents.
11. $x^2 + y^2 + 4x - 8y + 4 = 0$; points: $A(2,-1)$; $B(4,2)$.
12. $x^2 + y^2 - 6x - 4y + 3 = 0$; points: $A(8,3)$; $B(4,1)$; $C(4,-1)$.
13. $x^2 + y^2 + 2x + 6y - 6 = 0$; points: $A(7,-2)$; $B(-2,1)$.
14. $x^2 + y^2 - 4x + 10y = 20$; points: $A(-5,2)$; $B(7,5)$.
15. $x^2 + y^2 + 8x - 2y + 16 = 0$; points: $A(4,0)$; $B(-3,1)$; $C(-4,1)$.
16. $x^2 + y^2 - 6x + 2y - 5 = 0$; points: $A(6,3)$; $B(-4,3)$.
17. $x^2 + y^2 - 4x + 10y - 16 = 0$; points: $A(5,1)$; $B(-7,3)$; $C(-3,-1)$.
18. $2x^2 + 2y^2 + 10x - 6y - 7 = 0$; points: $A(-3,0)$; $B(-4,6)$.

chapter

5

The Parabola

INTRODUCTION

A ship captain finds it necessary to sail his ship between a straight shoreline and a small island 2000 feet from the shore. Not knowing the depth of the sea at any point in the locality, he assumes the deepest part of the channel is midway between the island and the shore. He decides to sail the ship so that it is always the same distance from the island as it is from the shore. What is his path of travel?

5.1 THE PARABOLA: DEFINITION

A *parabola* is a curve traced by a point that moves, in a plane, so that its distance from a fixed point is always equal to its perpendicular distance from a fixed straight line.

In Figure 5.1, let F be the fixed point, called the *focus*, and let the line DD' be the fixed straight line, called the *directrix*. Let P be the moving point whose distance from the focus F is denoted by d_1, and whose distance from the directrix DD' is denoted by d_2. Then, if d_1 is always equal to d_2, the path of the moving point is a parabola. That is, for a parabola $d_1 = d_2$, or as a ratio, $d_1/d_2 = 1$.

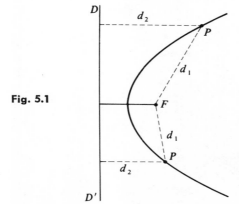

Fig. 5.1

72

It might be noted here that other curves will be formed if the ratio d_1/d_2 has some value greater than or less than 1. In any case, the ratio d_1/d_2 is called the *eccentricity* of the curve, denoted by e. If the variable point moves so that the eccentricity is equal to some fraction between zero and 1, exclusive, the figure is an ellipse. If the value of the eccentricity is equal to some constant greater than 1, the curve is a hyperbola. For the parabola, the eccentricity e is equal to 1. In all instances, the two distances, d_1 and d_2 must be taken in the correct order.

The line through the focus perpendicular to the directrix is called the *principal axis* and divides the parabola into two symmetric halves (Fig. 5.2). This line passes through the parabola at the *vertex V*, the point of greatest curvature. Line segments, such as AB and MN, through the focus and terminating on the parabola are called *focal chords*. The focal chord MN, called the *latus rectum*, is perpendicular to the principal axis and parallel to the directrix. The vertex is equally distant from the focus and the directrix, so that in the figure, $CV = VF$. Then we have the following:

$$FM = CF = 2VF \quad \text{and} \quad MN = 4VF$$

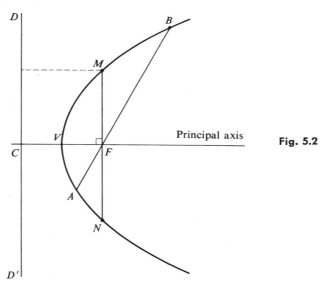

Fig. 5.2

That is, the latus rectum is four times the *focal length* of the parabola, where the focal length is the distance from the vertex to the focus.

5.2 THE EQUATION OF THE PARABOLA

The simplest form of the equation of a parabola is obtained if the parabola is placed with its vertex at the origin (0,0) with the principal axis along one of the coordinate axes. In the parabola shown (Fig. 5.3), the parabola opens to the right with the principal axis along the x-axis. Note that the directrix is not along the y-axis. We denote the focal length by a, so that the coordinates of the focus are $(a,0)$. Then the distance $CF = 2a$.

74   The Parabola

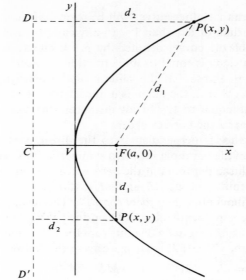

Fig. 5.3

Now we take the general point $P(x,y)$ in any general position and denote its distance from the focus by d_1 and its distance from the directrix by d_2. Then

$$d_1 = \sqrt{(x-a)^2 + (y-0)^2} \quad \text{and} \quad d_2 = x + a$$

Now we set these two quantities equal to each other as required for a parabola:

$$\sqrt{(x-a)^2 + (y-0)^2} = x + a$$

Squaring each side, expanding, and simplifying, we get

$$y^2 = 4ax$$

This is the form of the equation of a parabola in standard position with its principal axis along the x-axis. Note that the equation contains one variable to the *first degree* and the other to the *second degree*. With the parabola in this position, there is no constant term. The coefficient of x is $4a$, which represents 4 times the focal length a. The directrix is located a units to the left of the origin so that its equation is $x = -a$.

The ends of the latus rectum are at the points $(a, \pm 2a)$, and its length is therefore $4a$, the same as the coefficient of x.

In the equation of the parabola,

$$y^2 = 4ax$$

if a is negative the parabola is still considered to be in standard position, but it opens toward the left and the directrix is at the right of the origin. If the variables are reversed, we get the standard x^2 parabola:

$$x^2 = 4ay$$

This equation represents two other standard positions of parabolas, both with vertex at (0,0), principal axis along the y-axis, and focus at $(0,a)$. If a is positive, the parabola opens upward; if a is negative, it opens downward.

5.3 GRAPHING THE PARABOLA

We have seen that the equation of a parabola in standard position, $y^2 = 4ax$, represents a parabola opening to the right or left, in which the constant a represents the focal length, provided that the coefficient of y^2 is 1. Therefore, whatever the coefficient of x, the focal length is one fourth of this coefficient. This fact enables us to locate the focus. The graph can then be sketched fairly accurately by locating a few particular points on the curve. The procedure is shown by examples.

EXAMPLE 1. Graph the equation $y^2 = 20x$, and state important facts about the curve.

Solution. Since the coefficient of x is positive, the parabola opens to the right, with vertex at (0,0). To locate the focus, we take $(\frac{1}{4})(20) = 5$. The focal length is 5, and the focus is at (5,0). To graph the parabola, we locate four more points. The ends of the latus rectum are at $(5,\pm 10)$. Two other points are found by setting x equal to its coefficient 20. Then $y^2 = 400$, and $y = \pm 20$. We now have the points:

$x =$	0	5	10
$y =$	0	± 10	± 20

These five points will give a good idea of the shape of the curve (Fig. 5.4). The directrix is 5 units to the left of the vertex. Its equation is $x = -5$. The length of the latus rectum is 20. Note that if x is assigned any negative value, then y^2 is negative, and therefore y is imaginary. Then no portion of the curve extends to the left of $x = 0$. The domain is $0 \leq x < +\infty$. The range of the function y is $-\infty < y < +\infty$.

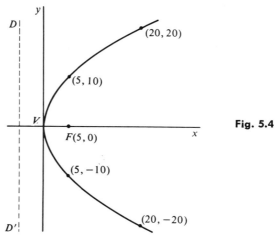

Fig. 5.4

The Parabola

EXAMPLE 2. Sketch the curve $3x^2 + 28y = 0$.

Solution. We first write the equation in standard form:

$$x^2 = -\frac{28}{3}y$$

The equation represents a parabola with vertex at (0,0) and with its principal axis along the y-axis. Since the coefficient of y is negative, the parabola opens downward. Note that all positive values of y yield imaginary values of x. To determine the focus, we first take $(\frac{1}{4})(-\frac{28}{3}) = -\frac{7}{3}$. The focus is therefore at the point $(0, -\frac{7}{3})$. The ends of the latus rectum are at $(\pm\frac{14}{3}, -\frac{7}{3})$. For two more points, when $y = -\frac{28}{3}$, $x = \pm\frac{28}{3}$. We now have five important points on the curve, and the curve can then be sketched fairly accurately (Fig. 5.5). The length of the latus rectum is $\frac{28}{3}$. The directrix is $\frac{7}{3}$ units above the x-axis. Its equation is therefore $y = \frac{7}{3}$.

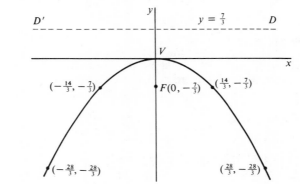

Fig. 5.5

5.4 WRITING THE EQUATION OF A GIVEN PARABOLA

The equation of a parabola can often be written directly from certain known facts. Recall the standard forms

$$y^2 = (4a)x \quad \text{and} \quad x^2 = (4a)y$$

In these forms the focal length is a. If a parabola is in standard position, and the coordinates of the focus are known, the equation can be written at once.

EXAMPLE 3. Write the equation of the parabola with its focus at $(-3,0)$ and its vertex at the point $(0,0)$.

Solution. Here we have $a = -3$. Since the parabola opens to the left, it is a y^2 parabola. The ends of the latus rectum are at $(-3, \pm 6)$. The parabola is shown in Figure 5.6. The equation is

$$y^2 = 4(-3)x, \quad \text{or} \quad y^2 = -12x$$

The directrix is the line $x = 3$.

5.4 Writing the Equation of a Given Parabola 77

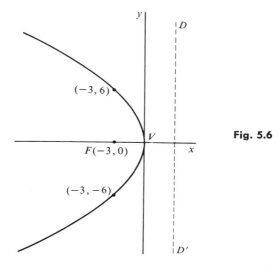

Fig. 5.6

The equation for this parabola could also be found by setting up the expressions for d_1 and d_2 and then equating the two:

$$d_1 = \sqrt{(x+3)^2 + (y-0)^2} \qquad d_2 = 3 - x$$

Then
$$\sqrt{(x+3)^2 + y^2} = 3 - x$$

Squaring and simplifying,

$$x^2 + 6x + 9 + y^2 = 9 - 6x + x^2$$

or
$$y^2 = -12x$$

EXAMPLE 4. Sketch the parabola with focus at $(0,5)$ and directrix, $y = -5$.

Solution. We first locate the focus and the directrix (Fig. 5.7). Now we shall work the problem first by taking a general point (x,y) on the curve and then setting up the two distances d_1 and d_2. Then we have

$$d_1 = \sqrt{(x-0)^2 + (y-5)^2}$$

and
$$d_2 = y - (-5), \quad \text{or} \quad d_2 = y + 5$$

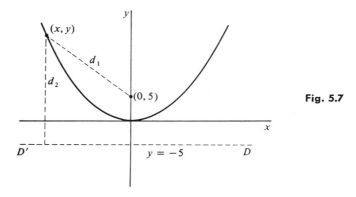

Fig. 5.7

78 The Parabola

Since d_1 must equal d_2 for a parabola, we get

$$\sqrt{x^2 + (y-5)^2} = y + 5$$

Squaring both sides and simplifying we have

$$x^2 = 20y$$

The same equation of the curve could be found more easily by noting that the equation has the form $x^2 = (4a)y$. Since $a = 5$, we have $x^2 = 20y$.

EXAMPLE 5. Write the equation of the particular parabola in standard position, opening to the left and passing through the point $(-7,3)$.

Solution. Substitute the coordinates $(-7,3)$ for x and y, respectively, in the general equation, $y^2 = Kx$, and solve for the value of K. We get: $7y^2 = -9x$.

EXERCISE 5.1

Sketch each of the following parabolas; then (a) locate the vertex and the focus of each, (b) write the equation of the directrix, (c) find the length of the latus rectum:

1. $y^2 = 14x$
2. $x^2 = 10y$
3. $x^2 + 6y = 0$
4. $y^2 + 9y = 0$
5. $3y^2 - 28x = 0$
6. $5y^2 + 576x = 0$
7. $7x^2 + 32y = 0$
8. $2x^2 - 45y = 0$
9. $y^2 - x = 0$
10. $y = 8x^2$
11. $3x^2 + y = 0$
12. $x = y^2$

Write the equation of each of the following parabolas. All are in standard position with the vertex at (0,0).

13. Focus at (6,0).
14. Focus at (0,-7.5).
15. Focus at (-24.5,0).
16. Directrix: $y = 2$.
17. Directrix: $x = -4.5$.
18. Directrix: $y + 0.5 = 0$.
19. Passing through the point (3,5), (two parabolas).
20. Passing through the point (-7,4), (two parabolas).
21. Passing through the point (-2,-3), (two parabolas).
22. Passing through the point (4,-2), (opening downward).
23. Passing through the point (-5,1), (opening toward the left).
24. Latus rectum equal to 10 units, (parabola opening toward the left).
25. Latus rectum equal to 7 units, (parabola opening upward).
26. Find the angle between the two chords drawn from the vertex of a parabola to the ends of the latus rectum.
27. Write the equation of the locus of the midpoints of all chords from the vertex of the parabola $y^2 = 12x$.
28. Find the area of an isosceles right triangle inscribed in the parabola $y^2 = 12x$, if the right angle is at the vertex of the parabola.
29. Find the area of an equilateral triangle inscribed in the parabola $x^2 = 8y$ if one vertex is at the origin.
30. Write the equation of the circle whose diameter is the latus rectum of the parabola $y^2 = 8x$. Where does the circle intersect the principal axis?

5.5 TRANSLATION

If a parabola is not in standard position, its equation is somewhat more complicated. If the vertex is not at the origin, $O(0,0)$, but the principal axis is still horizontal or vertical, the curve is said to be *translated*. *Translation* of any curve may be considered a dislocation without any rotation.

5.5 Translation

If any figure is translated to a new position, we imagine a new set of axes, the x'-axis and the y'-axis, drawn so that the figure will be in standard position with reference to the new axes. We call the new origin O'. Let us suppose the new origin is at the point (h,k) with reference to the old axes. Now let us take a general point $P(x,y)$ on the curve (Fig. 5.8). These coordinates (x,y) refer to the original axes. Note from the figure that the original abscissa x of the point P means the distance AC and the original ordinate y means the distance DC.

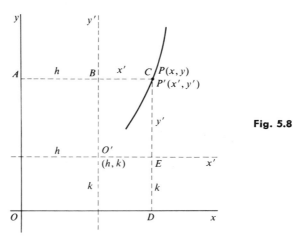

Fig. 5.8

With reference to the new set of axes, we call the point $P'(x',y')$. Then the new abscissa x' means the distance BC. The new ordinate y' means the distance EC. Now, since the new origin O' is at the point (h,k) with reference to the old set of axes, then from the figure we can see that the following relations are true:

$$x' = x - h \quad \text{and} \quad y' = y - k$$

These are the formulas we use for translating any curve to a new position. Then we can say the new origin (h,k) represents the amount of translation of the curve, that is, h represents the translation in the x direction and k represents the translation in the y direction. For translated parabolas we then have the following standard forms of equations:

$(y - k)^2 = 4a(x - h)$ for parabolas opening to the right or left

$(x - h)^2 = 4a(y - k)$ for parabolas opening upward or downward

EXAMPLE 6. Write the equation of the parabola whose focus is at (8,4) and whose directrix is the line $x = 2$.

Solution. First, we show on the graph the focus and the directrix. The parabola may then be sketched roughly (Fig. 5.9). Since the vertex is midway between the focus and the directrix, it is at the point (5,4). The focal length is 3 units.

The Parabola

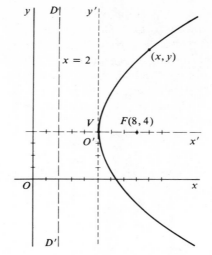

Fig. 5.9

Now we tentatively assume another set of coordinate axes, x' and y', having the new origin at the point $O'(0',0')$, such that the parabola is in standard position with reference to this new set of axes. Then we can write its equation in terms of the new set of axes. With a focal length of 3, the equation is

$$(y')^2 = 12(x')$$

However, the parabola is translated 5 units in the x-direction and 4 units in the y-direction. For the correct equation for the translated parabola, we make the following substitutions in the tentative equation we have set up:

x' is replaced by $x - 5$; y' is replaced by $y - 4$

We get $(y - 4)^2 = 12(x - 5)$. Simplifying, $y^2 - 12x - 8y + 76 = 0$.

The equation for the foregoing translated parabola may be checked by setting up the expressions for d_1 and d_2 and equating the results. We have

$$d_1 = \sqrt{(x - 8)^2 + (y - 4)^2} \quad \text{and} \quad d_2 = x - 2$$

for point (x,y) on the curve. Then

$$\sqrt{(x - 8)^2 + (y - 4)^2} = x - 2$$

Squaring and expanding,

$$x^2 - 16x + 64 + y^2 - 8y + 16 = x^2 - 4x + 4$$

or

$$y^2 - 12x - 8y + 76 = 0$$

EXAMPLE 7. Graph the equation of the translated parabola

$$x^2 - 8x + 6y + 28 = 0$$

Solution. To graph this equation we could find pairs of values of x and y that satisfy the equation and then plot the corresponding points. Instead of laboriously computing values and plotting points, we attempt to discover the location and general shape of the parabola. If we can locate the vertex and the focus, we can sketch the parabola in its translated position. To do so we put the equation in the standard form for a translated parabola. First we note that the equation contains the squared term x^2. Then we conclude that the parabola opens upward or downward and is of the standard form

$$(x - h)^2 = 4a(y - k)$$

Now, in the given equation, we complete the square in x. First, isolating the x terms we have

$$x^2 - 8x = -6y - 28$$

Adding 16 to both sides,

$$x^2 - 8x + 16 = -6y - 12$$

In standard form, this is

$$(x - 4)^2 = -6(y + 2)$$

Now we can recognize that the vertex of the parabola is at the point $(4, -2)$. The parabola opens downward, and $a = -1.5$. This means that the focus is 1.5 units below the vertex. The parabola can now be sketched in its position away from the origin (Fig. 5.10).

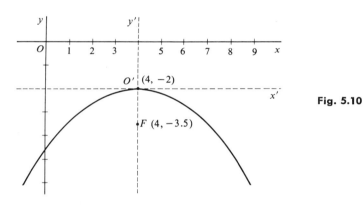

Fig. 5.10

EXERCISE 5.2

Write the equation of each of the following parabolas:
1. Vertex at (5,1); focus at (5,3).
2. Vertex at $(-2,-4)$; focus at $(0,-4)$.
3. Vertex at $(-2,3)$; focus at $(-2,-1)$.
4. Vertex at (2,1); focus at $(-1,1)$.
5. Vertex at (2,1); directrix: $y = 3$.
6. Vertex at $(-2,-2)$; directrix: $x = 4$.
7. Focus at $(-1,-1)$; directrix: $y = -3$.
8. Focus at $(-1,-3)$; directrix: $x = -5$.
9. Vertex at $(-1,4)$; focus at (2,4).
10. Focus at $(-1,3)$; directrix: $x = -3$.
11. Focus at $(-2,3)$; vertex at $(-2,0)$.
12. Focus at $(2,-3)$; directrix: $y = 5$.

Locate the vertex and focus of each of the following translated parabolas and sketch each parabola in its translated position:
13. $y^2 - 6y - 8x + 41 = 0$
14. $x^2 - 6x - 14y - 5 = 0$
15. $x^2 + 2x - 4y + 13 = 0$
16. $y^2 - 2y - 12x + 25 = 0$
17. $y^2 + 6y - 4x + 17 = 0$
18. $x^2 - 6x - 8y - 7 = 0$
19. $x^2 - 6x + 10y + 9 = 0$
20. $y^2 - 4y + 12x + 16 = 0$
21. $x^2 + 8x - 4y + 8 = 0$
22. $y^2 - 2y + 8x - 23 = 0$
23. Write the equation of the path of a point that moves so that it is always the same distance from the point $(5,-2)$ as it is from the line $x = -4$.

24. Write the equation of the path of a point that moves so that it is always the same distance from the point $(-3,7)$ as it is from the line $y = 2$.

5.6 INTERSECTION OF A PARABOLA WITH OTHER CURVES

Knowing the points of intersection of two curves is useful when the area bounded by the two curves is to be found. This is a common problem in calculus.

A straight line and a parabola can intersect in at most two points. In solving such a system of equations, the same procedure is followed as in the case of a straight line and a circle. The linear equation is first solved for one of the variables, and the result is substituted in the quadratic. After one variable has been determined, the linear equation is used to find the values of the other variable. For the points of intersection of a parabola and a straight line, we may get

1. two real and distinct points
2. two real but equal values, a double point
3. two imaginary points.

In the case of a parabola and a circle, it is possible to get a maximum of four points of intersection. Therefore the solution of such a set of equations will yield four possible types of answers:

1. four real and distinct points
2. two real and distinct points and one real *double* point
3. two real, distinct or double, and two imaginary points
4. four imaginary points.

In the algebraic solution of such a system of equations, all four points should be identified. For imaginary points, there is no real intersection of the curves. For a double point, the curves are tangent to each other.

EXAMPLE 8. Find the points of intersection of the parabola $y^2 - 6x - 24 = 0$ and the circle $x^2 + y^2 = 40$. Sketch the figure.

Solution. The circle has its center at $(0,0)$, and its radius is equal to $\sqrt{40}$. To construct the parabola, we rewrite its equation $y^2 = 6x + 24$. This is changed to $y^2 = 6(x + 4)$. The quantity $(x + 4)$ represents a translation of 4 units in the negative x-direction. The vertex is therefore at the point $(-4,0)$. We now solve the equations simultaneously. From the first equation we have $y^2 = 6x + 24$. Then the term y^2 in the circle equation is replaced by $(6x + 24)$. It becomes $x^2 + 6x + 24 = 40$, or $x^2 + 6x - 16 = 0$. The solutions of this quadratic are $x = 2$ and $x = -8$. The values of x are now used to compute the values of y from $y^2 = 6x + 24$:

when $x = 2$, $y^2 = 12 + 24$; $y^2 = 36$; or $y = \pm 6$
when $x = -8$, $y^2 = -48 + 24$; $y^2 = -24$; or $y = \pm\sqrt{-24}$, imaginary

The points of intersection are therefore the following: the real points $(2,6)$ and $(2,-6)$; the imaginary points $(-8, +\sqrt{-24})$ and $(-8, -\sqrt{-24})$. The real points of intersection are shown in Figure 5.11.

5.7 The Reflecting Principle of the Parabola

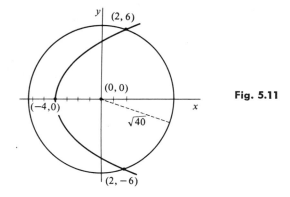

Fig. 5.11

5.7 THE REFLECTING PRINCIPLE OF THE PARABOLA

The parabola has an important reflecting principle that has many applications. To understand this principle, first let us recall a fact from physics:

If a light ray strikes a reflecting surface at any angle, the reflected ray will form the same angle as the incident ray with the reflecting surface. In Figure 5.12 the ray from A to B is the *incident ray i* and the ray from B to C is the *reflected ray r*. Point B is the point of *incidence*. If the normal (the perpendicular) is drawn to the reflecting surface MN at point B, it will bisect angle ABC. Then $\theta_1 = \theta_2$. Angle θ_1 between the incident ray and the normal is called the angle of *incidence*. Angle θ_2 between the normal and the reflected ray is called the angle of *reflection*. Then *the angle of incidence is equal to the angle of reflection*. Note that angles ϕ_1 and ϕ_2, formed by the two rays with the reflecting surface, are also equal since they are complements of equal angles.

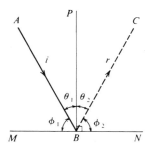

Fig. 5.12

If parallel rays strike a reflecting plane surface, then the reflected rays also will be parallel (Fig. 5.13). If the reflecting surface is curved, then parallel incident rays will be reflected in different directions. The reflected

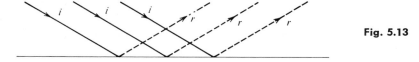

Fig. 5.13

84 The Parabola

rays will not be parallel. In this case we consider the incident ray *i* and the reflected ray *r* with reference to the tangent to the curve at the point of incidence *P* (Fig. 5.14).

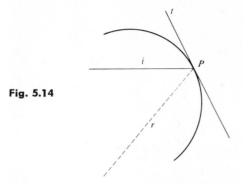

Fig. 5.14

Let us see what happens when parallel rays strike a concave, spherical surface. Figure 5.15 shows parallel incident rays striking the surface at points *A*, *B*, *C*, *D*, and *E*, and the direction of the reflected rays. The rays will be reflected in such a way that they do not come to a focus but instead intersect along a curved line. This result is called *spherical aberration*.

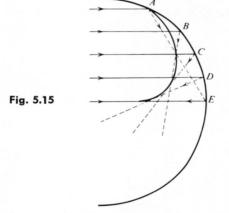

Fig. 5.15

Now suppose the concave surface is parabolic. When rays parallel to the principal axis strike the concave, parabolic surface, all reflected rays pass through the focus (Fig. 5.16). If a parabolic mirror such as the reflector of a headlight, searchlight, or flashlight is held toward the sun, the incident rays may be considered as being parallel. Then all the heat rays as well as light rays are reflected to the focus, which becomes a point of concentration of heat and light. This point, although not visible, will be a "hot spot," hot enough to ignite a piece of paper placed at the focus. This reflecting principle is also utilized in radar and radio reception and in reflecting telescopes.

5.7 The Reflecting Principle of the Parabola

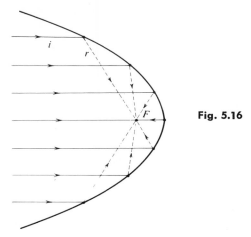

Fig. 5.16

Now let us reverse the process. If the light rays originate at the focus of a parabolic reflector, all reflected rays will be parallel to the principal axis (Fig. 5.17). The light beam of a car headlight will consist of parallel rays. If the source of light is not exactly at the focus, the reflected rays will not be parallel but will be diffused. The reflecting principle of the parabola is proved in calculus.

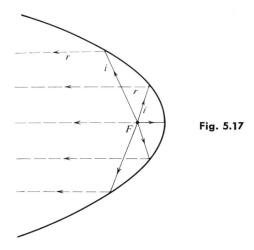

Fig. 5.17

EXERCISE 5.3

Find the points of intersection of the following pairs of curves. Sketch each.
1. $y^2 = 4x$ and $2x - y = 4$.
2. $x^2 = 4y$ and $x - 2y = 4$.
3. $x^2 = 4y$ and $x - y - 2 = 0$.
4. $y^2 = 4x$ and $x - y + 1 = 0$.
5. $y^2 = 12x$ and $y - x = 3$.
6. $y^2 = 12x$ and $x^2 = 6y$.
7. $x^2 + y^2 = 10y$ and $x^2 = 2y$.
8. $x^2 + y^2 + 2x - 14y = -32$ and $y^2 = 8x$.
9. $y^2 - 2y - 4x - 11 = 0$ and $y^2 - 2y + 8x - 23 = 0$.
10. A circle with a radius of 6 inches has its center at the focus of the parabola $y^2 + 12x = 0$ (units in inches). Where do the two graphs intersect?

The Parabola

11. Find the points of intersection of the parabola $x^2 + 8y + 8 = 0$, and the circle of radius 6 with the center of the circle at the focus of the parabola.
12. A parabolic mirror, 6 inches in diameter, has a maximum depression of $\frac{1}{8}$ inch. What is the focal length of the mirror?
13. If a parabola is rotated about its principal axis, it generates a geometric solid called a *paraboloid*. A tank in the shape of a paraboloid with the vertex at the bottom has a circular top 8 feet across and a depth of 10 feet. What is the equation of the parabolic vertical section through the vertex? [Take the vertex at (0,0) and the principal axis along the y-axis.]
14. A bridge with a parabolic arch 36 feet high at the vertex spans a roadway. The center portion of the roadway 48 feet wide has a minimum clearance of 20 feet. Find the width of the arch.
15. A 6-inch telescopic concave mirror has a focal length of 54 inches. What is the equation of the parabola formed by a longitudinal section through the vertex? What is the greatest depression of the mirror?
16. If you wished to adjust the beam of your car's headlight so that the light would cover a wider spread, would you move the light bulb forward or backward in the headlight?

chapter

6

The Ellipse

6.1 DEFINITION

The ellipse can be defined in either one of two ways. As a first definition we begin with a fixed straight line, the directrix, and a fixed point, the focus, just as we did with the parabola. Again, let us consider the variable point $P(x,y)$, and as it moves, we take the following definitions:

Let d_1 = the distance of the variable point from the focus
Let d_2 = the distance of the variable point from the directrix

We have already seen that if the two distances are equal, the figure formed is a parabola. In this case the eccentricity of the curve is equal to 1. That is, for a parabola, $d_1/d_2 = e = 1$.

Now let us suppose that the variable point moves so that the eccentricity is equal to some constant proper fraction, that is, $e < 1$. In this case the curve formed will be an ellipse. We can now state the definition:

Definition:

The ellipse is the path of a moving point that moves so that the ratio of its distance from the focus to its distance from the directrix is a constant proper fraction. Stated briefly, the ellipse is a curve whose eccentricity is greater than zero but less than 1.

As an example, consider the ellipse shown in Figure 6.1. Let us suppose that the eccentricity in this ellipse is equal to the fraction $\tfrac{2}{3}$. This means

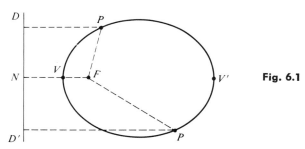

Fig. 6.1

88 The Ellipse

that the variable point P is always $\frac{2}{3}$ as far from the focus F as it is from the fixed line, the directrix DD'. This ratio can be checked roughly by actual measurement for any position of the variable point. When the variable point is at the position V, we have $\dfrac{VF}{VN} = \frac{2}{3}$. The same ratio holds for position V'.

Now we shall find that the same ellipse would be formed if the focus were at the position F_2 and another directrix at the position $D_2 D_2'$ (Fig. 6.2). An ellipse can therefore be considered as having two foci and two directrices.

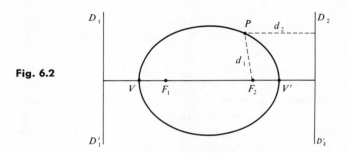

Fig. 6.2

6.2 SECOND DEFINITION OF THE ELLIPSE

The ellipse may be defined in a second way. In this definition we begin with the two fixed points, or foci, F_1 and F_2 (Fig. 6.3). Let point P be the moving point and let its distances from the two foci be represented by d_3 and d_4, respectively. (The notations d_3 and d_4 are used here to avoid any possible confusion with the notations used in the first definition of the ellipse.)

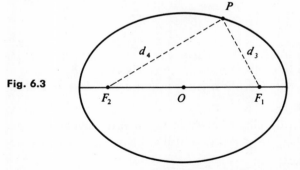

Fig. 6.3

Now, for the positions of point P, if the sum of the two distances, d_3 and d_4, is some constant, an ellipse is formed. We now have the second definition.

Definition:

An ellipse is the path of a moving point that moves so that the sum of its distances from two fixed points is a constant.

Note that in this definition no mention is made of a directrix. Yet we shall see the directrices exist and can be located.

The second definition suggests a method of constructing an ellipse mechanically. A string whose length is greater than the distance F_1F_2 is fastened with one end at F_1 and the other end at F_2. If a pencil is placed inside the loop F_1PF_2 and moved along while the string is kept taut, an ellipse will be traced. Whatever the position of point P, the sum, $d_3 + d_4$, will be equal to the length of the string. The distances, d_3 and d_4, are called *focal radii* from the point P. The point midway between the two foci is called the *center* of the ellipse.

A *diameter* of an ellipse is a line segment through the center with its end points on the ellipse (Fig. 6.4). The greatest diameter, V_1V_2, is called the *major axis*, and the shortest diameter, GH, is called the *minor axis*. The major and the minor axes are perpendicular and bisect each other. The end points of the major axis are called the *vertices* and represent the points of sharpest turning. From the figure note especially:

$$PF_1 + PF_2 = d_3 + d_4 = V_1V_2, \text{ the major axis}$$

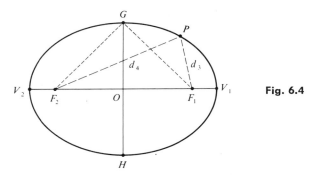

Fig. 6.4

Now, let $a = OV_1$, the semimajor axis. Then we have

$$PF_1 + PF_2 = 2a$$

6.3 EQUATION OF THE ELLIPSE

For a simple form of the equation we place the ellipse on the xy-coordinate system with its center at the origin (0,0) and its major axis along one of the coordinate axes. Let us first place the major axis along the x-axis (Fig. 6.5). Then the foci, F and F', and the vertices, V_1 and V_2, also lie on the x-axis.

Let a represent OV_1, the length of the semimajor axis.
Let c represent OF, the distance from the center to a focus.
Let b represent OG, the semiminor axis.

The two vertices are at the points $(a,0)$ and $(-a,0)$, respectively. The two foci are at the points $(c,0)$ and $(-c,0)$, respectively.

The Ellipse

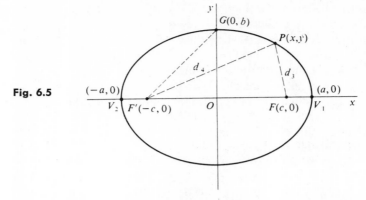

Fig. 6.5

Now we take the variable point $P(x,y)$ in any general position on the curve. Let d_3 and d_4 represent the distances from point P to the foci, F and F', respectively. Then, by definition of the ellipse,

$$d_3 + d_4 = 2a, \text{ a constant}$$

At one time when point $P(x,y)$ is at V_1, we have

$$V_1F + V_1F' = d_3 + d_4 = 2a, \text{ the major axis}$$

We are now prepared to derive the equation of the ellipse. We express the two distances d_3 and d_4 by means of the distance formula, set the sum equal to the quantity $2a$, and simplify the result:

$$d_3 = \sqrt{(x-c)^2 + (y-0)^2} \qquad d_4 = \sqrt{(x+c)^2 + (y-0)^2}$$

Then

$$\sqrt{(x-c)^2 + (y-0)^2} + \sqrt{(x+c)^2 + (y-0)^2} = 2a$$

Now we transpose one radical, square both sides, and simplify the result:

$$\sqrt{(x-c)^2 + y^2} = 2a - \sqrt{(x+c)^2 + y^2}$$

$$x^2 - 2cx + c^2 + y^2 = 4a^2 - 4a\sqrt{(x+c)^2 + y^2} + x^2 + 2cx + c^2 + y^2$$

Transposing the radical and simplifying,

$$4a\sqrt{(x+c)^2 + y^2} = 4a^2 + 4cx$$

Dividing both sides of $4a$,

$$\sqrt{(x+c)^2 + y^2} = a + \frac{cx}{a}$$

Squaring again and expanding,

$$x^2 + 2cx + c^2 + y^2 = a^2 + 2cx + \frac{c^2 x^2}{a^2}$$

This simplifies to

$$\frac{a^2x^2 - c^2x^2}{a^2} + y^2 = a^2 - c^2$$

Dividing both sides by the quantity $(a^2 - c^2)$,

$$\frac{x^2}{a^2} + \frac{y^2}{a^2 - c^2} = 1$$

Now we make one more simplification. Note that in Fig. 6.5, $GF' = a$. Then in the right triangle OGF', $a^2 = c^2 + b^2$, or $a^2 - c^2 = b^2$. Replacing $(a^2 - c^2)$ with b^2 we get the simple form of the equation of the ellipse:

$$\frac{x^2}{a^2} + \frac{y^2}{b^2} = 1 \tag{1}$$

This is the standard form of the equation of an ellipse in standard position with major axis along the x-axis. In the equation a represents the semimajor axis; b represents the semiminor axis. Then the x and y intercepts are $\pm a$ and $\pm b$, respectively. The distance c from the center to the foci can be determined by the formula

$$c^2 = a^2 - b^2 \tag{2}$$

For an ellipse in standard position with the foci on the y-axis, we have the corresponding standard equation:

$$\frac{x^2}{b^2} + \frac{y^2}{a^2} = 1 \tag{3}$$

Although the derivation of this equation does not involve the distance to a directrix, it can be shown that the eccentricity of the ellipse is given by the formula

$$e = c/a \tag{4}$$

Note that $c < a$, so that the ratio $c/a < 1$.

The directrices can be located by the fact that their distance from the center of the ellipse is equal to the fraction a/e or its equivalent, a^2/c. Their position can be estimated to be some distance from each of the vertices, outside the ellipse and perpendicular to the major axis (Fig. 6.6).

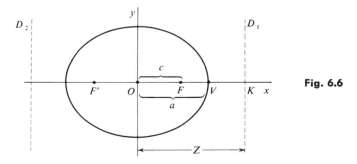

Fig. 6.6

The Ellipse

The distance from a vertex V to its corresponding directrix VK must be greater than the distance from vertex to focus VF. In fact, the ratio $\dfrac{VF}{VK}$ must equal c/a, and $Z = a/e = a^2/c$. Then the equations of the directrices are: $x = \pm a/e$, or $x = \pm a^2/c$.

In the equation of the ellipse, if a approaches b in value so that the major and minor axes become more nearly equal, the ellipse becomes more nearly the shape of a circle, and the foci approach each other at the center. Therefore, a circle may be considered as a special ellipse in which the minor and the major axes are equal, and $a = b$. The standard equation then becomes

$$\frac{x^2}{a^2} + \frac{y^2}{a^2} = 1 \quad \text{or} \quad x^2 + y^2 = a^2$$

which is the equation of a circle. The two foci then coincide, and $c = 0$. The eccentricity of a circle is therefore zero.

6.4 GRAPHING THE ELLIPSE FROM A GIVEN EQUATION

To graph the equation of an ellipse we first write the equation in standard form:

$$\frac{x^2}{a^2} + \frac{y^2}{b^2} = 1 \quad \text{or} \quad \frac{x^2}{b^2} + \frac{y^2}{a^2} = 1$$

The greater denominator is called a^2; the smaller is called b^2. We then identify a and b as the semimajor and semiminor axes, respectively.

EXAMPLE 1. Graph the following ellipse; locate its foci and its vertices; state its eccentricity and the lengths of the major and the minor axes:

$$16x^2 + 25y^2 = 400$$

Solution. Dividing both sides by 400 we get $\dfrac{x^2}{25} + \dfrac{y^2}{16} = 1$

Now we identify the following:

$$a^2 = 25, \text{ from which } a = 5$$
$$b^2 = 16, \text{ from which } b = 4$$

The x-intercepts are $(\pm 5, 0)$; the y-intercepts are at $(0, \pm 4)$. These points represent the ends of the major and the minor axes, respectively, and they afford four vital points in the sketching of the graph (Fig. 6.7). The vertices are at the points $(\pm 5, 0)$. The foci are located by finding the value of c:

$$c^2 = a^2 - b^2 \qquad c^2 = 25 - 16 = 9$$

Then $c = \pm 3$. The foci are at $(\pm 3, 0)$. Knowing a and c, we can find $e = c/a = \tfrac{3}{5}$. The major axis $= 2a = 10$. The minor axis $= 2b = 8$. The graph may be drawn more accurately by considering a *latus rectum*. A *latus rectum* is a chord of the ellipse through the focus and perpendicular to the major axis. Its end points may be determined by computing the value of the ordinate when $x = c$. Then $y = b^2/a$. In the example shown, when $x = \pm 3$, $y = \pm \tfrac{16}{5}$. These four points represent the end points of the *latera recta*.

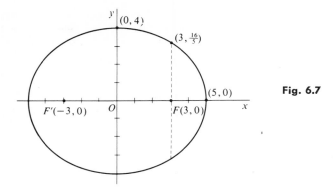

Fig. 6.7

6.5 WRITING THE EQUATION OF A GIVEN ELLIPSE

The equation of a particular ellipse can be written from certain known facts. If the intercepts of an ellipse in standard position are known, then the semimajor axis a and the semiminor axis b are also known. These values are simply inserted in the standard equation, which can then be cleared of fractions.

EXAMPLE 2. Write the equation of the ellipse in standard position whose intercepts are at $(\pm 4, 0)$ and $(0, \pm 6)$; sketch the figure, and state important facts.

Solution. In this ellipse, $a = 6$, $b = 4$. The major axis is along the y-axis. The standard equation is

$$\frac{x^2}{16} + \frac{y^2}{36} = 1$$

Clearing of fractions, $\quad 9x^2 + 4y^2 = 144$

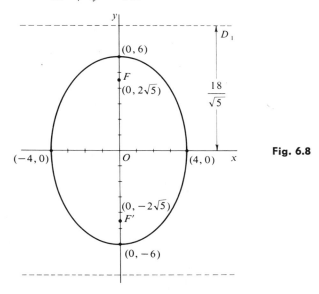

Fig. 6.8

94   The Ellipse

The ellipse is shown in Fig. 6.8. For the foci, we have $c^2 = a^2 - b^2$, $c^2 = 36 - 16 = 20$; $c = \pm 2\sqrt{5}$. Since the foci are always located on the major axis, their coordinates are $(0, \pm 2\sqrt{5})$. The eccentricity of the ellipse is given by

$$e = \frac{c}{a} = \frac{2\sqrt{5}}{6} = \frac{\sqrt{5}}{3} = 0.745 \text{ (approx.)}$$

The major axis $= 2a = (2)(6) = 12$. The minor axis $= 2b = (2)(4) = 8$.
The directrices being perpendicular to the y-axis, their equations are given by

$$y = \pm a/e = a^2/c, \quad \text{or} \quad y = \pm \frac{18}{\sqrt{5}}$$

EXAMPLE 3. Write the equation of the ellipse in standard position whose minor axis is equal to 6, and lies along the y-axis; and $e = \sqrt{3}/2$. Sketch the graph.

Solution. Since the minor axis $= 2b = 6$, we have $b = 3$. Then the y-intercepts are $(0, \pm 3)$ and $b^2 = 9$. To find a, the semimajor axis, we use the following formulas:

$$e = c/a \quad \text{and} \quad a^2 - b^2 = c^2$$

Supplying known information we get

$$\frac{c}{a} = \frac{\sqrt{3}}{2} \quad \text{and} \quad a^2 - 9 = c^2$$

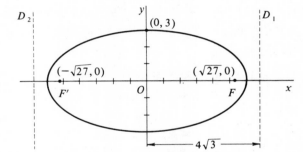

Fig. 6.9

Solving the system we get $a = 6$ and $c = 3\sqrt{3}$. Since $a = 6$, then $a^2 = 36$ and the major axis lies along the x-axis. Then the standard equation is

$$\frac{x^2}{36} + \frac{y^2}{9} = 1$$

The general form is $x^2 + 4y^2 = 36$

EXERCISE 6.1

Sketch the graph of each of the following ellipses and state the following: (a) length of major axis, (b) length of minor axis, (c) coordinates of foci, (d) coordinates of vertices, (e) eccentricity, (f) equations of directrices, (g) length of a latus rectum.

1. $4x^2 + 9y^2 = 36$ 2. $3x^2 + 5y^2 = 60$ 3. $25x^2 + 9y^2 = 225$
4. $5x^2 + 4y^2 = 80$ 5. $3x^2 + y^2 = 24$ 6. $2x^2 + 3y^2 = 54$
7. $5x^2 + 2y^2 = 40$ 8. $16y^2 + x^2 = 64$ 9. $9y^2 + 8x^2 = 144$
10. $y^2 + 9x^2 = 144$ 11. $11x^2 + 12y^2 = 396$ 12. $12x^2 + 13y^2 = 3120$

Each of the following ellipses is in standard position. Write the equation of each and sketch the curve:
13. Vertex at (5,0); focus at (4,0)
14. Vertex at (0,−3); focus at (0,−2)
15. Vertex at (−4,0); $e = \frac{3}{4}$
16. Vertex at (0,6); $e = \frac{2}{3}$
17. Focus at (8,0); $e = \frac{4}{5}$
18. Focus at (−6,0); $e = \frac{2}{3}$
19. Focus at (2,0); major axis = 8
20. Focus at (0,−3); major axis = 12
21. Focus at (−4,0); minor axis = 6
22. Focus at (0,5); minor axis = 4
23. Major axis = 8; along the x-axis; $e = \frac{2}{3}$
24. Major axis = 12; along the y-axis; $e = \frac{1}{4}$
25. Minor axis = 4; along the x-axis; $e = \frac{2}{5}$
26. Minor axis = 6, along the y-axis; $e = \frac{4}{5}$

6.6 TRANSLATION OF THE ELLIPSE

If an ellipse is translated to a new position on the coordinate axes (without any rotation), the amount of translation is determined by the new position of the center. In translation, the axes of the ellipse are still parallel to the coordinate axes.

The ellipse in *standard position* has its center at the point (0,0). Now if the ellipse is translated to a new position, the center is moved to a new position, a point which we denote by the coordinates (h,k). Then the equation of the translated ellipse can be written as follows:

First, write the equation of the ellipse as though it were in standard position with a temporary set of axes through its center. These axes may be called x' and y' if desired. Then make these substitutions:

x' is replaced by $x - h$; $\quad y'$ is replaced by $y - k$

We have seen that the standard equation of the ellipse is

$$\frac{x^2}{a^2} + \frac{y^2}{b^2} = 1$$

If this ellipse is translated to a new position with center at (h,k), its equation becomes

$$\frac{(x-h)^2}{a^2} + \frac{(y-k)^2}{b^2} = 1 \tag{5}$$

As an example, consider the following ellipse, with center at (0,0):

$$\frac{x^2}{16} + \frac{y^2}{9} = 1$$

If this ellipse is translated to a new position with its center at $(2,-1)$, without change of shape or size, its equation becomes (Fig. 6.10)

$$\frac{(x-2)^2}{16} + \frac{(y+1)^2}{9} = 1$$

Expanding and simplifying this equation we get

$$9x^2 + 16y^2 - 36x + 32y - 92 = 0$$

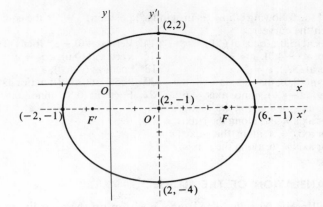

Fig. 6.10

Note in Figure 6.10 that translation does not alter the size or shape of an ellipse. In the foregoing example, the major axis is still 8 and the minor axis is still 6. The distance c from the center of the ellipse to the focus is still equal to $c = \sqrt{a^2 - b^2} = \sqrt{7}$. The eccentricity is still equal to $\frac{\sqrt{7}}{4}$.

However, the intercepts are no longer the values $(+a,0)$ and $(0,\pm b)$. Moreover, the coordinates of particular points on the ellipse, such as the ends of the minor axis, have been changed since the entire ellipse has been moved to a new position on the main coordinate system.

6.7 GRAPHING A TRANSLATED ELLIPSE

It would be a long and tedious process to graph the following equation by finding corresponding values of x and y:

$$16x^2 + 9y^2 + 64x - 54y + 1 = 0$$

For example, when $x = 3$, what is the value of y? When $x = -4$, what is y? Needless to say, this procedure would involve much time and calculation. There is a better way.

This is a translated ellipse. Now if the origin $(0,0)$ is translated to the center of the ellipse, the result is equivalent to translating the ellipse back to standard position. If we can locate the center of the ellipse and translate the origin to this point, then we shall be able to graph the ellipse by inspection of its major and minor axes. The method involves completing the squares in x and y, as shown by the following example.

EXAMPLE 4. Find the center of the following ellipse and graph the figure by inspection:

$$16x^2 + 9y^2 + 64x - 54y + 1 = 0$$

Solution. First we complete the squares in x and y. Rewriting the equation,

$$16x^2 + 64x + \underline{} + 9y^2 - 54y + \underline{} = -1$$

6.7 Graphing a Translated Ellipse

Now we add 64 and 81 to both sides of the equation to complete the squares. We get
$$16x^2 + 64x + 64 + 9y^2 + 54y + 81 = -1 + 64 + 81$$
The equation can be written
$$16(x^2 + 4x + 4) + 9(y^2 - 6y + 9) = 144$$
We divide both sides of the equation by 144 to get the standard form:
$$\frac{x^2 + 4x + 4}{9} + \frac{y^2 - 6y + 9}{16} = 1$$
Rewriting to show the squares,
$$\frac{(x+2)^2}{9} + \frac{(y-3)^2}{16} = 1$$
Comparing this equation with the standard form (5), we recognize that this ellipse has its center at $(-2,3)$ and that the denominators 9 and 16 represent the squares of the semiminor and semimajor axes, respectively. Then the semiminor axis is 3 and the semimajor axis is 4 (Fig. 6.11). That is,

$a = 4$ along the vertical, that is, 4 units from the center $(-2,3)$;
$b = 3$ along the horizontal; the foci are $\sqrt{7}$ units from the center.

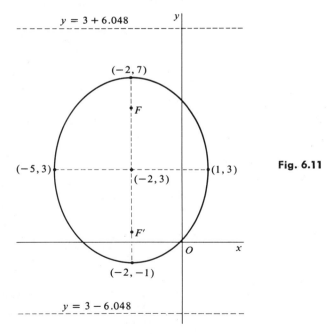

Fig. 6.11

Note the following: The *vertices* are at $(-2,-1)$ and $(-2,7)$. The *ends* of the *minor axis* are at $(1,3)$ and $(-5,3)$. The *foci* are at $(-2,3 \pm \sqrt{7})$. The *eccentricity* is $\sqrt{7}/4$, or 0.661 (approx.). The *intercepts* are most easily found by substituting zero for x and y in turn in the original equation and then solving for the other unknown. For example, if $x = 0$, we have $9y^2 - 54y + 1 = 0$. By use of the quadratic formula, we get $y = \frac{1}{3}(9 \pm 4\sqrt{5})$, or 3 ± 2.981.

The directrices can be located by the value a/e, or a^2/c. This value turns out to be $16/\sqrt{7}$, which is approximately 6.048. If the ellipse were in standard position, the equations of the directrices would be $y = \pm 6.048$. However, in the translated position, this value indicates the distance from the *center* to the directrices. They are then at $y = 3 \pm 6.048$. This value can also be obtained by substituting $(y - 3)$ for y, in which case we get $y - 3 = \pm 6.048$, or $y = 3 \pm 6.048$. Their approximate location is shown by dashed lines.

6.8 THE REFLECTING PRINCIPLE OF THE ELLIPSE

If an ellipse is rotated about its major axis, a solid called an *ellipsoid* is generated. Now, if the inner surface of an ellipsoid is a reflector, then any ray of any kind originating at one focus will be reflected to the other focus. This is the principle involved in the so-called "whispering galleries" (Fig. 6.12). A whisper emanating at F_1 may be easily heard at F_2 but may be entirely inaudible at a point such as P much nearer the source. If a billiard table were made in the shape of an ellipse, then any ball propelled in any direction from one focus F_1 would be reflected to the other focus F_2, provided there were no lateral "spin" on the ball. The reflecting principle of the ellipse can be proved by calculus.

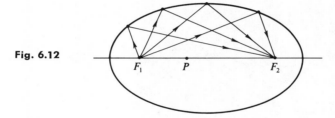

Fig. 6.12

EXERCISE 6.2

Write the equation of each of the first four ellipses below with the center translated to the given point. Expand to the general form:
1. $4x^2 + 9y^2 = 36$, to point $(-2,1)$
2. $3x^2 + 5y^2 = 60$, to point $(2,-1)$
3. $5x^2 + 2y^2 = 40$, to point $(1,-3)$
4. $3x^2 + 2y^2 = 12$, to point $(-2,3)$
5. Write the equation of the curve traced by the point that moves so that the sum of its distances from $(5,3)$ and $(-3,3)$ is always equal to 12.
6. Write the equation of the path of the point that moves so that the sum of its distances from $(4,1)$ and $(4,-7)$ is always equal to 10.

Write the equations of the ellipses satisfying the following conditions:
7. One focus at $(5,2)$, one vertex at $(7,2)$, major axis $= 10$.
8. One focus at $(-4,-1)$, one vertex at $(-4,-2)$, $e = \frac{4}{5}$.
9. Foci at $(0,-1)$ and $(6,-1)$, major axis $= 8$.
10. One focus at $(-2,0)$, minor axis $= 6$, $e = \frac{1}{2}$.

Locate the center of each of the following ellipses, translate the origin to that point, and write the equation in general form in terms of the new set of axes. Sketch the curve:
11. $9x^2 + 16y^2 + 36x - 32y - 92 = 0$.
12. $4x^2 + 9y^2 - 24x - 36y + 36 = 0$.
13. $x^2 + 4y^2 - 12x + 16y + 48 = 0$.
14. $4x^2 + y^2 + 24x - 10y + 35 = 0$.

6.8 The Reflecting Principle of the Ellipse

15. $25x^2 + 16y^2 - 50x + 64y = 311$. 16. $x^2 + 4y^2 + 12x - 32y + 96 = 0$.
17. The dome of a "whispering gallery" has the shape of a semi-ellipsoid with the foci on the floor. If the greatest height of the dome above the floor is 50 feet and the distance between the foci is 75 feet, how far does the sound travel from one focus to the other by being reflected from the ceiling?
18. Suppose a satellite travels about the earth in an elliptical orbit so that the perigee (least distance from the earth) of the satellite is 100 mi. and the apogee (greatest distance from the earth) is 300 mi. Taking the radius of the earth as 4000 mi., find the eccentricity of the orbit of the satellite and write the equation of its path.

Find the points of intersection of each of the following sets of curves and sketch the curves:

19. $x^2 + 4y^2 = 25$
    $x + 2y = 7$
20. $2x^2 + y^2 = 18$
    $2x - y = 2$
21. $2x^2 + 3y^2 = 29$
    $2x - y = 1$
22. $3x^2 + y^2 = 12$
    $x + y = 2$
23. $3x^2 + 4y^2 = 36$
    $x - 2y = 6$
24. $3x^2 + y^2 = 12$
    $y^2 + 2x = 4$
25. $4x^2 + y^2 = 12$
    $y^2 = 2x$
26. $4x^2 + y^2 = 36$
    $x^2 + y^2 = 24$
27. $x^2 + y^2 = 13$
    $x^2 + 4y^2 = 25$
28. $x^2 + 4y^2 = 25$
    $x - 4y = -5$
29. $4x^2 + y^2 = 16$
    $x^2 + y^2 + 2y = 8$
30. $6x^2 + y^2 = 36$
    $y^2 + 2x = 8$

chapter

7

The Hyperbola

7.1 DEFINITION

The hyperbola, like the ellipse, can be defined in two different ways. Again, we begin with a fixed line, the directrix, and a fixed point, the focus. As the variable point P moves and traces out a curve, its distance from the focus and its distance from the directrix have a ratio greater than 1. Again, we let these two distances be represented by d_1 and d_2, respectively. Then, for the hyperbola, we have $d_1/d_2 > 1$. In fact, a hyperbola may be defined as a conic section whose eccentricity is greater than 1, or

$$e > 1$$

In the hyperbola in Figure 7.1, let us assume that $e = \frac{3}{2}$, approximately. This ratio can be checked roughly for any position of the point P on the curve. In the case of the hyperbola, point P may be located on both sides of the directrix. Note that for P_2, d_1/d_2 is still equal to $\frac{3}{2}$.

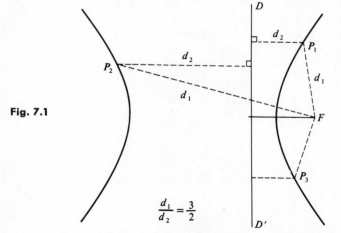

Fig. 7.1

$$\frac{d_1}{d_2} = \frac{3}{2}$$

We see then that the hyperbola is a discontinuous curve with two branches, not connected. Note especially that although the hyperbola is symmetrical, the directrix is *not* the axis of symmetry.

If a second focus F' and a second directrix $D_2 D_2'$ are placed as shown in Figure 7.2, then with the same eccentricity, $\frac{3}{2}$, the same hyperbola with the same two branches will be formed. The hyperbola, therefore, like the ellipse, has two foci and two directrices. The line segment joining the two foci intersects the hyperbola in the two vertices, V and V'. The line segment VV' is called the *transverse* axis. The midpoint of the transverse axis is the *center of symmetry* and is called the *center* of the hyperbola.

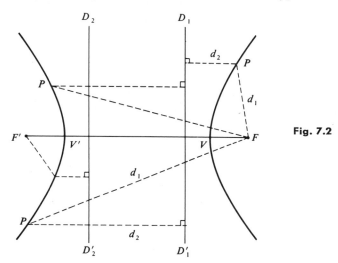

Fig. 7.2

7.2 SECOND DEFINITION OF THE HYPERBOLA

The hyperbola, like the ellipse, may be defined in a second way by starting with the two fixed points, or foci, F and F' (Fig. 7.3). Let P be the moving point, and let the distances from the foci be represented by d_3 and d_4, respectively. Then, if the absolute value $|d_3 - d_4|$ is equal to a constant as the point P traces out the curve, a hyperbola is formed. Note again that this fact holds for any position of point P on either branch of the curve. We therefore have the second definition:

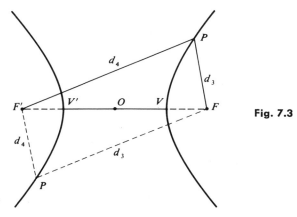

Fig. 7.3

102 The Hyperbola

Definition:

A hyperbola is the locus of a point that moves so that the absolute value of the difference between its distances from two fixed points is a constant.

This definition makes no mention of a directrix. Yet the two directrices exist and can be located, as we shall see. The point O, midway between the foci, is the center of the hyperbola. It should be noted also that the constant difference, $|d_3 - d_4|$, is equal to the length of the transverse axis of the hyperbola, the line segment VV'.

7.3 EQUATION OF THE HYPERBOLA

For the simplest form of the equation, we place the hyperbola in standard position on the coordinate system, with the center O at the origin and with the transverse axis along one of the coordinate axes. First let us take the transverse axis along the x-axis, with the foci F and F' also on the x-axis (Fig. 7.4). Let point $P(x,y)$ be a point on the curve, and let d_3 and d_4 represent the distances to the point from F and F', respectively. Then $|d_3 - d_4|$ = a constant. Let a be the distance OV, or the semitransverse axis. Then the transverse axis = $2a = VV'$. The coordinates of the vertices are $(\pm a, 0)$. (Note the resemblance between the notation here and that used for the ellipse.)

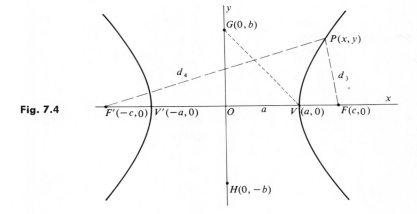

Fig. 7.4

Now we let c represent the distance from the center to a focus. Then the foci are at $(c,0)$ and $(-c,0)$, as in the case of the ellipse. Note $c > a$.

We are now prepared to write the equation of the hyperbola. By definition,

$$|d_4 - d_3| = \text{a constant}$$

We have seen that this constant is equal to $2a$, the transverse axis. The two distances are

$$d_4 = \sqrt{(x+c)^2 + (y-0)^2} \quad \text{and} \quad d_3 = \sqrt{(x-c)^2 + (y-0)^2}$$

7.3 Equation of the Hyperbola

Then
$$\sqrt{(x+c)^2 + y^2} - \sqrt{(x-c)^2 + y^2} = 2a$$

Simplifying this equation in the same way as for the ellipse, we get

$$\frac{x^2}{a^2} - \frac{y^2}{c^2 - a^2} = 1$$

For $c^2 - a^2$ we substitute b^2, whose value will be explained presently, and we get

$$\frac{x^2}{a^2} - \frac{y^2}{b^2} = 1$$

This is the standard form of the equation of a hyperbola in standard position with foci on the x-axis. Note the resemblance to the equation of the ellipse.

If a hyperbola in standard position has its transverse axis along the y-axis, its foci and its vertices also lie on the y-axis. The foci are then the points $(0, \pm c)$, and the vertices are at $(0, \pm a)$. In this case the "x^2" term is negative and the "y^2" term positive, provided the right side of the equation remains positive. Then the equation has the form

$$\frac{y^2}{a^2} - \frac{x^2}{b^2} = 1$$

The hyperbola then opens upward and downward and does not intersect the x-axis.

Let us now see the significance of the value b in the equation of the hyperbola. We have seen that in the case of the ellipse, the values a and b represent the intercepts on the two axes, respectively, when the ellipse is in standard position. In the case of the standard hyperbola, the graph intersects only one axis. However, the value b does represent significant points on the other axis. Referring again to Figure 7.4, the two points $G(0,b)$ and $H(0,-b)$ are located such that the line segment $GV = c$. Then, from the right triangle OGV, we have $b^2 + a^2 = c^2$, or $b^2 = c^2 - a^2$. The two points $(0, \pm b)$ are important guide points for graphing the hyperbola.

Although this formula does not involve distances to the directrices, it can be shown that the eccentricity of the hyperbola is given by the formula

$$e = c/a \quad \text{(the same as for the ellipse)}$$

Note that $\quad c/a > 1$

The distance from the center to each directrix is again given by a/e, or its equivalent, a^2/c.

The following statements hold for both the ellipse and the hyperbola:
1. a = distance from center to a vertex.
2. c = distance from center to a focus.
3. c/a = the eccentricity, e; for the ellipse, $c < a$; for the hyperbola, $c > a$.

104   The Hyperbola

4. a/e or a^2/c represents the distance from center to a directrix.

5. $\dfrac{2b^2}{a}$ = length of the latus rectum, the focal chord perpendicular to the transverse axis.

7.4   GRAPHING THE HYPERBOLA FROM A GIVEN EQUATION

To graph the equation of a hyperbola, we first write the equation in standard form:

$$\frac{x^2}{a^2} - \frac{y^2}{b^2} = 1 \quad \text{or} \quad \frac{y^2}{a^2} - \frac{x^2}{b^2} = 1$$

Note that the denominator a^2 is used with the positive fraction. From the equation, we then identify the values of a and b, respectively. Next, we locate the points $(\pm a, 0)$ and $(0, \pm b)$ on the axes. Through these points we draw lines parallel to the coordinate axes, forming a rectangle (Fig. 7.5). The diagonals of this rectangle are then extended and form the asymptotes of the hyperbola. The vertices are at the points $(\pm a, 0)$, when the "x^2" term is positive.

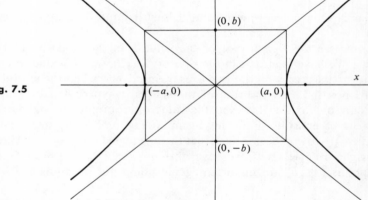

Fig. 7.5

If the "y^2" term is positive, we consider the denominator of this fraction as "a^2". Then the coordinates of the vertices are $(0, \pm a)$, and the distance b is laid off on the x-axis. The two forms of the equation of the hyperbola will now be illustrated by examples.

EXAMPLE 1. Graph the following hyperbola, using the asymptotes as guide lines, and state important facts about the curve: $9x^2 - 16y^2 = 144$.

  Solution.   First we reduce the equation to the standard form by dividing both sides by the constant, 144:

$$\frac{x^2}{16} - \frac{y^2}{9} = 1$$

7.4 Graphing the Hyperbola from a Given Equation

From the standard form we identify the following:
$$a^2 = 16, \quad a = 4, \quad b^2 = 9, \quad b = 3$$
On the x-axis, we mark the points $(\pm 4, 0)$. On the y-axis, we mark the points $(0, \pm 3)$. These points establish the rectangle, the asymptotes, and the vertices of the hyperbola (Fig. 7.6). From the relation $a^2 + b^2 = c^2$, $c = \pm 5$, so that the foci are at the points $(\pm 5, 0)$. The eccentricity is $c/a = \frac{5}{4}$. The equations of the directrices are $x = \pm a/e = \pm a^2/c = \pm \frac{16}{5}$. The length of the transverse axis is $2a = 8$. The length of a latus rectum MN is $2b^2/a = \frac{9}{2}$.

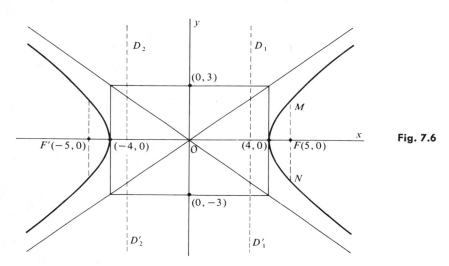

Fig. 7.6

The asymptotes are straight lines and therefore have straight-line equations. To see how we might write these equations, we recall that the asymptotes are the diagonals of a rectangle. Since the lengths a and b are used in constructing this rectangle, note that the slopes of the asymptotes are given by the formula, $m = \pm \frac{b}{a}$. Both asymptotes pass through the center of the hyperbola, in this case, through the point $(0,0)$. Using the slopes $\pm b/a$, and the point $(0,0)$, we get the equations: $bx + ay = 0$ and $bx - ay = 0$. To get the equations directly from the equation of the hyperbola $9x^2 - 16y^2 = 144$, we put zero for 144 and write
$$9x^2 - 16y^2 = 0$$
Factoring, we get the equations of the asymptotes: $3x + 4y = 0$ and $3x - 4y = 0$.

EXAMPLE 2. Graph the equation $16y^2 - 9x^2 = 144$.

Solution. First, we reduce the equation to the form
$$\frac{y^2}{9} - \frac{x^2}{16} = 1$$
Note that this equation is the same as the equation in Example 1 except that the signs of the variables are reversed. Again we mark off the points $(0, \pm 3)$ on the

106   The Hyperbola

y-axis, and the points ($\pm 4,0$) on the x-axis. We construct the same rectangle and the same asymptotes as for Example 1. In this case, however, we say $a^2 = 9$, and $a = \pm 3$. Therefore, the equation represents another hyperbola with the same asymptotes but opening upward and downward, with vertices on the y-axis at points $(0, \pm 3)$, and with foci at points $(0, \pm 5)$, also on the y-axis (Fig. 7.7).

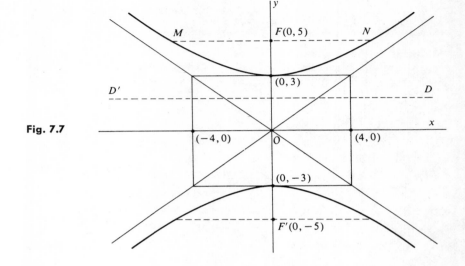

Fig. 7.7

For this hyperbola, $e = c/a = \frac{5}{3}$. The equations of the directrices are $y = \pm a/e = \pm a^2/c = \pm \frac{9}{5}$, or $y = \pm 1.8$. The length of the latus rectum is $2b^2/a$ (MN), which in this case is $\frac{32}{3}$, or $10\frac{2}{3}$.

For this hyperbola, the length of the transverse axis (distance between vertices) is equal to $(2)(3) = 6$ units.

The two hyperbolas shown in Figures 7.6 and 7.7 are called *conjugate hyperbolas*, since they are determined by the same rectangle and the same asymptotes. The transverse axis of one hyperbola is called the *conjugate axis* of the other, and *vice versa*. Note, however, that the eccentricity is different for the two. This will always be true unless $a = b$.

In the case of a hyperbola in which $a = b$, the rectangle formed becomes a square. Then the two conjugate hyperbolas are identical in shape except that one is the *inverse* of the other. A hyperbola in which $a = b$ is called an *equilateral* hyperbola.

7.5 WRITING THE EQUATION OF A GIVEN HYPERBOLA

The equation of a hyperbola can be written if certain facts concerning the figure are known. For a hyperbola in standard position, we usually

7.5 Writing the Equation of a Given Hyperbola

attempt to determine the values of a and b, and then insert these values in the proper standard form:

$\dfrac{x^2}{a^2} - \dfrac{y^2}{b^2} = 1$ for a hyperbola opening toward the right and left, with foci and vertices on the x-axis

$\dfrac{y^2}{a^2} - \dfrac{x^2}{b^2} = 1$ for a hyperbola opening upward and downward, with foci and vertices on the y-axis

EXAMPLE 3. Write the equation of the hyperbola in standard position with one focus at $(0,5)$ and $e = \frac{3}{2}$. Sketch the figure.

Solution. First we note that since the focus is on the y-axis, the hyperbola opens upward and downward. Then the other focus is at $(0,-5)$. The vertices are also on the y-axis at $(0, \pm a)$. Now we refer to the two formulas:

$$e = c/a \quad \text{and} \quad a^2 + b^2 = c^2$$

From the given information we know that

$$c = 5 \quad \text{and} \quad c/a = \tfrac{3}{2}$$

Supplying this information in the formulas we have

$$\frac{c}{a} = \frac{3}{2} \quad \text{or} \quad \frac{5}{a} = \frac{3}{2}$$

Then $a = \tfrac{10}{3}$; $a^2 = \tfrac{100}{9}$. Knowing the value of a^2 we have

$$100/9 + b^2 = 25$$

Then $b^2 = \tfrac{125}{9}$. Since the values of a^2 and b^2 are now known, they are inserted into the formula,

$$\frac{y^2}{100/9} - \frac{x^2}{125/9} = 1 \quad \text{or} \quad 45y^2 - 36x^2 = 500$$

When the equation of the hyperbola is known, the figure can be graphed as explained in the previous section. The graph is shown in Figure 7.8.

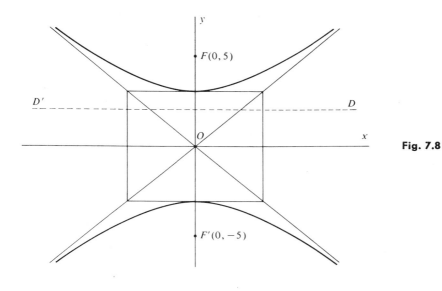

Fig. 7.8

EXERCISE 7.1

Sketch the graph of each of the following hyperbolas and state the following: (a) length of the transverse axis, (b) length of the conjugate axis, (c) coordinates of the foci, (d) coordinates of the vertices, (e) eccentricity, (f) equations of the asymptotes, (g) equations of the directrices, (h) length of a latus rectum:

1. $16x^2 - 9y^2 = 144$.
2. $2x^2 - 5y^2 = 20$.
3. $3y^2 - 5x^2 = 60$.
4. $y^2 - 2x^2 = 16$.
5. $3y^2 - 2x^2 = 24$.
6. $3x^2 - 8y^2 = 48$.
7. $9x^2 - 25y^2 = 225$.
8. $x^2 - y^2 = 16$.
9. $2x^2 - 5y^2 + 60 = 0$.
10. $5y^2 - 12x^2 = 60$.
11. $4x^2 - 9y^2 + 49 = 0$.
12. $3y^2 - x^2 + 36 = 0$.

Exercises 13–22 below represent hyperbolas in standard position. Write the standard form of the equation of each, and sketch the graph. Locate the foci and the directrices, and tell important facts about each—transverse and conjugate axes; eccentricity; equations of the asymptotes, directrices; conjugate hyperbola; coordinates of the foci and vertices:

13. Vertex at (4,0); focus at (5,0).
14. Vertex at (0,−3); focus at (0,4).
15. Vertex at (0,4); $e = \frac{3}{2}$.
16. Focus at (−2,0); $e = 2$.
17. Transverse axis = 6; focus at (0,−4).
18. Conjugate axis = 4; focus at (0,5).
19. Transverse axis = 8, along the x-axis; $e = 2/\sqrt{3}$.
20. Conjugate axis = 4, along the x-axis; $e = \frac{3}{2}$.
21. One asymptote: $2x - 3y = 0$; focus at $(0, \sqrt{13})$.
22. One asymptote: $2x + y = 0$; vertex at (−2,0).
23. Two listening posts hear the sound of an enemy gun. The difference in time is one second. If the listening posts are 1400 feet apart, write the equation of the hyperbola passing through the position of the enemy gun. Take the center of the hyperbola midway between the listening posts. (Sound travels approximately 1080 ft/sec.)

7.6 TRANSLATION OF THE HYPERBOLA

If a hyperbola is translated to a new position with reference to the coordinate axes, the amount of translation is determined by the new location of the center. A hyperbola in standard position, you will recall, has its center at the origin (0,0). The new location of the center is usually denoted by (h,k). This means that the hyperbola has been translated h units in the x-direction and k units in the y-direction.

For example, if we find that the center of a hyperbola is at the point $(5,-2)$, then the figure has been translated 5 units in the x-direction and -2 units in the y-direction. In this case, $h = 5$ and $k = -2$.

To write the equation of a translated hyperbola, we first write it as though it were in standard position with a temporary set of axes through the center. These new axes may be called x' and y'. For example, we may have then

$$\frac{(x')^2}{a^2} - \frac{(y')^2}{b^2} = 1$$

Now we make the following substitutions:

x' is replaced by $(x - h)$; y' is replaced by $(y - k)$

Then we get the equation

$$\frac{(x - h)^2}{a^2} - \frac{(y - k)^2}{b^2} = 1$$

7.7 Graphing a Translated Hyperbola

As an example, suppose we have the following equation representing a hyperbola with center at (0,0):

$$\frac{x^2}{9} - \frac{y^2}{4} = 1$$

If this hyperbola is translated to a new position (without rotation) with its center at $(-4,1)$, its equation becomes

$$\frac{(x+4)^2}{9} - \frac{(y-1)^2}{4} = 1$$

Note in this equation that we have the *squares of binomials*. This fact is important when we come to the problem of trying to discover the center and the amount of translation from a given expanded equation. Expanding and simplifying this equation, we get the general form

$$4x^2 - 9y^2 + 32x + 18y + 19 = 0$$

The hyperbola is graphed with reference to the temporary set of axes x' and y' through the center, which is actually at the point $(-4,1)$. The size or shape of the hyperbola has not been altered (Fig. 7.9). The transverse axis, the conjugate axis, and the eccentricity are the same as before translation. However, particular points, such as the intercepts, the vertices, and the foci, are now changed with reference to the original axes.

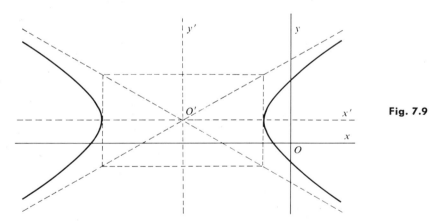

Fig. 7.9

7.7 GRAPHING A TRANSLATED HYPERBOLA

It would be a long process to graph the following equation by finding corresponding values of x and y:

$$5y^2 - 2x^2 + 10y + 12x - 63 = 0$$

Instead, if we can discover the location of the center, and then determine the transverse axis and the conjugate axis, then by the aid of the asymptotes, we can sketch the graph in its translated position. The method involves completing the squares in x and y, as shown in the following example:

EXAMPLE 4. Find the center of the following hyperbola and sketch the graph:
$$5y^2 - 2x^2 + 10y + 12x - 63 = 0$$

Solution. To complete the squares in x and y, we first rewrite the equation
$$5y^2 + 10y + \underline{} - 2x^2 + 12x - \underline{} = 63$$
or
$$5(y^2 + 2y + \underline{}) - 2(x^2 - 6x + \underline{}) = 63$$

For completed squares in x and y, the left side of the equation must be
$$5(y^2 + 2y + 1) - 2(x^2 - 6x + 9)$$

Note that we have added 5 and -18 to the left side. Adding the same to the right side we get
$$5(y^2 + 2y + 1) - 2(x^2 - 6x + 9) = 50$$

Dividing both sides of the equation by 50, we can write it in the form
$$\frac{(y+1)^2}{10} - \frac{(x-3)^2}{25} = 1$$

The equation now represents a hyperbola with center at the point $(3, -1)$. To construct the rectangle we have
$$a = \sqrt{10} \quad \text{and} \quad b = 5$$

Using these values, we can draw the rectangle, show its diagonals as the asymptotes of the hyperbola, and then sketch the figure with reference to its translated location (Fig. 7.10).

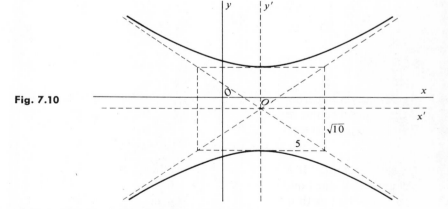

Fig. 7.10

7.8 APPLICATION OF THE HYPERBOLA

The definition of the hyperbola is the basis of its use in warfare in determining the location of enemy guns. The hyperbola is defined as the path of a point that moves so that the difference between its distances from two foci is a constant. If the sound of an enemy gun is heard at two listening posts and the difference in time is calculated, then the gun is known to be located on a particular hyperbola. A third listening post will determine a second hyperbola, and then the gun emplacement can be spotted as the intersection of the two hyperbolas.

7.9 THE EQUILATERAL HYPERBOLA

In the equation of the hyperbola, if $a = b$ the hyperbola is called *equilateral* (sometimes *rectangular*). The standard form can then be written

$$\frac{x^2}{a^2} - \frac{y^2}{a^2} = 1$$

Multiplying both sides by a^2 we get

$$x^2 - y^2 = a^2$$

A specific example is the equation $x^2 - y^2 = 9$. In drawing the auxiliary rectangle to aid in sketching the curve, we get a square; hence, the term "equilateral" (Fig. 7.11). In this case the asymptotes are perpendicular to each other. This form of the hyperbola is important because of its connection with the hyperbolic functions.

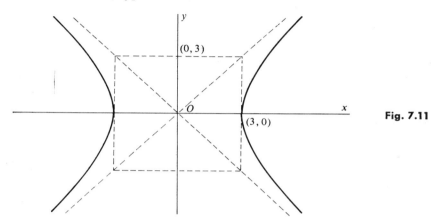

Fig. 7.11

If the equilateral hyperbola is rotated through an angle of 45°, the x and y coordinate axes become asymptotes of the curve (Fig. 7.12). The equation of the hyperbola is then

$$xy = K$$

where K is a constant. If K is positive, the two branches of the hyperbola fall in the first and third quadrants. If K is negative, the hyperbola lies in the second and the fourth quadrants.

EXERCISE 7.2

Write the equation of each of the first four hyperbolas below with the center translated to the given point; expand to the general form:
1. $2x^2 - 5y^2 = 20$, to point $(3, -1)$.
2. $3y^2 - 5x^2 = 60$, to point $(-1, 2)$.
3. $y^2 - 2x^2 = 16$, to point $(2, -5)$.
4. $3x^2 - 2y^2 = 24$, to point $(-2, 4)$.
5. Write the equation of the path of a point that moves so that the difference between its distances from $(3,1)$ and $(-7,1)$ is always equal to 8.
6. Write the equation of the path of a point that moves so that the difference between its distances from $(-2, -3)$ and $(-2, 5)$ is always equal to 4.

The Hyperbola

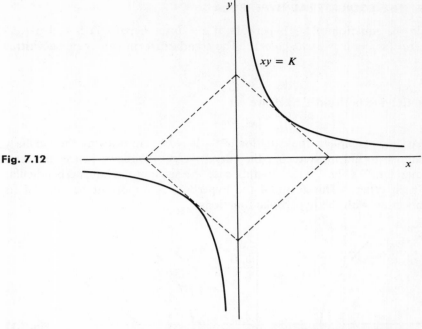

Fig. 7.12

Write the equations of the hyperbolas satisfying the following conditions:
7. One focus at (5,2); one vertex at (3,2); transverse axis = 4.
8. One focus at (−1,5); one vertex at (−1,3); conjugate axis = 6.
9. Foci at (2,3) and (2,−5); $e = \frac{4}{3}$.
10. Foci at (4,−1) and (−4,−1); $e = \frac{3}{2}$.
11. Sketch the graph of each of the following and locate the vertices and the foci:
    (a) $xy = 4$;   (b) $xy = -6$;   (c) $xy = 12$;   (d) $xy = 6\sqrt{2}$.

Locate the center of each of the following hyperbolas, translate the origin to that point, and write the equation in general form in terms of the new set of axes. Sketch the curve:

12. $4x^2 - 9y^2 + 32x + 18y + 19 = 0$. 13. $3x^2 - y^2 - 12x - 6y = 0$.
14. $2y^2 - x^2 + 4y - 8x - 30 = 0$. 15. $5y^2 - 4x^2 - 60y + 40x = 0$.
16. $4x^2 - 5y^2 + 40x + 30y = -135$. 17. $2x^2 - 3y^2 - 16x - 18y - 19 = 0$.
18. $9y^2 - x^2 - 36y - 10x - 25 = 0$.

Find the points of intersection of each of the following sets of curves:

19. $3x^2 - 2y^2 = 30$ 20. $3x^2 - 2y^2 = 12$ 21. $x^2 + y^2 = 36$
    $x - 3y = 5$. $y^2 - x = 2$. $x^2 - 2y^2 = 12$.
22. $x^2 + y^2 = 11$ 23. $xy = 3$ 24. $xy = 6$
    $2x^2 - 5y^2 = 50$. $x^2 - y^2 = 8$. $3x + 2y = 6$.
25. $xy = 6$ 26. $xy = 6$ 27. $xy = 8$
    $3x + 2y = 12$. $x + y = 3$. $x^2 - y^2 = 12$.
28. $x^2 - 3y^2 = 12$ 29. $3x^2 - 4y^2 = 12$ 30. $x^2 + y^2 = 2x$
    $x^2 + 3y^2 = 36$. $3y^2 - 2x^2 = 7$. $4x^2 - y^2 = 16$.

7.10 CONIC SECTIONS

If a right circular cone is cut by a plane, the section is a circle, a parabola, an ellipse, or a hyperbola. Therefore, these curves are called *conic sections*, or sometimes simply *conics*. Let us see what condition determines the type of curve.

First, let us define a right circular cone. Suppose a straight line l_1 passes through a fixed point P on another line l_2 that is fixed. Now, if line l_1 moves so that it always forms a constant angle θ with the fixed line l_2, a *right circular cone* is generated. The moving line l_1 is called the *generator* (Fig. 7.13). Any position of the moving line l_1 is called an *element* of the cone. The fixed line l_2 becomes the axis of the cone. Note that the figure formed consists of two parts, called *nappes*, meeting at point P, the *vertex* or *apex* of the cone.

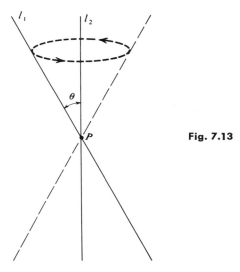

Fig. 7.13

If an intersecting plane is perpendicular to the axis of the cone, the section is a *circle* (Fig. 7.14a). The size of the circle will depend on how near the plane is to the apex of the cone. If the plane passes through the apex, the section is a single point, or a *point-circle*. If the plane forms an acute angle θ with the axis of the cone, the section is an *ellipse* (Fig. 7.14b), which degenerates into a single point if the plane passes through the apex. The particular shape of the ellipse will depend on the size of the acute angle θ. If the intersecting plane is parallel to an element of the cone, the section is a

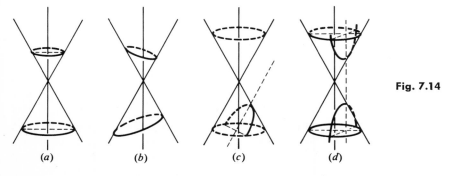

Fig. 7.14

(a)   (b)   (c)   (d)

114   The Hyperbola

parabola (Fig. 7.14c), which degenerates into a single straight line if the plane passes through the apex of the cone. If the plane is parallel to the axis of the cone, the section is a *hyperbola* (Fig. 7.14d), which degenerates into a pair of intersecting straight lines if the plane passes through the apex of the cone.

7.11   THE GENERAL EQUATION OF THE SECOND DEGREE

We have discussed the conic sections chiefly with reference to standard position on the coordinate axes. We have also seen how the equations of these curves are affected when the curves are *translated* to new positions, that is, away from standard position. In translated curves, the axes of the curves are still parallel to one of the coordinate axes.

Now it often happens that second-degree curves are transformed by *rotation* of their axes. A complete discussion of rotation would require an extended course in analytic geometry. However, we shall here simply note a few facts concerning the equations of rotated conic sections.

Any rotation of a second-degree curve can be recognized by the presence of a term containing the product (xy) in the equation. Let us represent any second-degree curve by the general equation

$$Ax^2 + Bxy + Cy^2 + Dx + Ey + F = 0$$

Of course, if A, B, and C are all zero, the equation represents a straight line. If A or B or both are not zero, then we have an equation representing a conic section, or one of the degenerate forms. Note these examples:

$$3x^2 + 4y^2 - 12x - 20y + 10 = 0$$

$$4x^2 - y^2 + 16x - 6y - 6 = 0$$

$$2x^2 - 8y + 6x - 5 = 0$$

$$2x^2 + 7xy + 2y^2 + 3x + 2y + 7 = 0$$

The first three equations can be recognized as being an ellipse, a hyperbola, and a parabola, respectively. In the fourth example, the product, (xy), indicates that the curve is *rotated* so that the axis of the curve is not parallel to one of the coordinate axes. Therefore, the curve cannot be a circle.

To test any second-degree equation to determine the type of curve, we use the coefficients A, B, and C, in the order shown in the general equation. We compute the value of the following expression:

$$B^2 - 4AC$$

This expression is called the *characteristic* of the general second-degree equation. Note that capital letters are used for these coefficients. This quantity is called an *invariant* under rotation as well as under translation because its value does not change.

7.11 The General Equation of the Second Degree

Now we shall find that we can use the following test:
If $B^2 - 4AC = 0$, the equation represents a parabola.
If $B^2 - 4AC < 0$, the equation represents an ellipse.
If $B^2 - 4AC > 0$, the equation represents a hyperbola.

EXAMPLES. Tell what type of curve each of the following represents:

$3x^2 + 4xy + 3y^2 - 6x + 5y = 6.$ $B^2 - 4AC = 16 - 36 = -20$; an ellipse.
$2x^2 + 5xy + 2y^2 + 3y - x = 4.$ $B^2 - 4AC = 25 - 16 = 9$; a hyperbola.
$4x^2 - 12xy + 9y^2 + 7x = 2y.$ $B^2 - 4AC = 144 - 144 = 0$; a parabola.
$x^2 + xy + y^2 + 4x - 8y = 10.$ $B^2 - 4AC = 1 - 4 = -3$; an ellipse.
$x^2 + 2xy + y^2 + 4y - 8 = 0.$ $B^2 - 4AC = 4 = 4 = 0$; a parabola.
$2x^2 - 3xy + 4x - 6y + 8 = 0.$ $B^2 - 4AC = 9 = 0 = 9$; a hyperbola.

EXERCISE 7.3

Identify each of the following conics:
1. $3x^2 - 2xy + 3y^2 - 5x + 2y = 7.$
2. $x^2 + 5xy + 4y^2 + 2x + 9 = 0.$
3. $2x^2 - 4xy + 2y^2 - 7x + 3y = 0.$
4. $5x^2 - 2xy + 8x - 6y + 2 = 0.$
5. $xy + 2y^2 - 3x + 2y + 8 = 0.$
6. $4x^2 - 3xy + y^2 + 4y - 60 = 0.$
7. $x^2 - 6xy + 15y^2 + 3x - y = 0.$
8. $x^2 + xy + y^2 - 5x + 5y = 8.$
9. $4x^2 + 12xy + 9x + 4y + 8 = 0.$
10. $25x^2 - 60xy + 36y^2 + 20x = 0.$
11. $3x^2 + 3y^2 - 12x + 18y = 11.$
12. $2x^2 - 3xy + 4y^2 - 5x = 6y.$
13. $2x^2 - 2y^2 - 8x + 18y = 15.$
14. $x^2 - 8x + 4y - 6 = 0.$
15. $xy = 12.$
16. $xy + 3x - 6y - 10 = 0.$

chapter

8

Higher Degree and Other Curves

8.1 GENERAL CONSIDERATIONS

Higher algebraic curves are of many kinds. They are best studied by the use of calculus. However, at this point we shall mention a few of the more common types of equations and indicate how the graphs may be constructed.

In graphing higher degree equations, we shall find it helpful to note several facts already mentioned. Some points may be overlooked if we are not careful to include all possible pairs of values of x and y that satisfy the equation. On the other hand, since we can find only a limited number of points, we must be careful not to assume a portion of a curve where there is none.

Some of the things to look for have already been mentioned: (1) intercepts; (2) extent, including domain and range; (3) symmetry; and (4) asymptotes. Another point that might be noted about a graph is its *curvature*. We may intuitively define curvature as the *amount of change in direction* of a curve from one point to another. If a curve changes its direction more sharply at one point than at another, we say its curvature is greater at the point of sharper turning. For example, a small circle has a greater curvature than a large circle because its direction changes more over a particular length of the curve. As you drive a car along a curve, the curvature may roughly be called the amount by which the steering

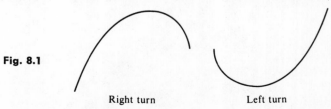

Fig. 8.1

Right turn   Left turn

8.2 THE THIRD-DEGREE EQUATION

Suppose we have the equation $x^3 - 4x^2 - 7x + 10 = 0$. This is a *third-degree equation* because it contains x raised to the third power but no higher. The powers below the third may or may not be present. A third-degree equation is often called a *cubic*. To graph an equation of this kind, we set the expression equal to y and graph the resulting equation in x and y. That is, $y = x^3 - 4x^2 - 7x + 10$.

Before we graph this equation let us consider the general shape of the graph. The graph of a cubic of this kind has, in general, the shape of the letter S (Fig. 8.2). Note that in the graph the curvature changes from *right* to *left*. There is a change in the *sense* of curvature.

Recall that in all the second-degree curves we have studied, the sense of curvature does not change. If we drive a car along a curve that has the shape of a circle, a parabola, an ellipse, or one branch of a hyperbola, the steering wheel of the car is always turned in the same direction. It may be turned more at one time than at another, but it never changes from left to right or from right to left. In Fig. 8.2, as a car goes along the curve from A to B, the steering wheel is turned toward the right. From B to C it is turned toward the left. Therefore there must be an instant when it is pointed straight ahead, since the curve is continuous. This instant is at a single point B, called the point of *inflection*. This reversal of the sense of curvature is characteristic of the graph of a third-degree equation.

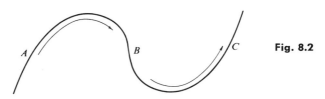

Fig. 8.2

EXAMPLE 1. Graph the equation $y = x^3 - 4x^2 - 7x + 10$. Discuss the curve.

Solution. In graphing any equation it is helpful first to find the intercepts. To find any y-intercept, we let $x = 0$. In this equation, if $x = 0$ $y = 10$. To find the x-intercepts, we set $y = 0$ and solve for x. Then we must find the *roots* of the equation $x^3 - 4x^2 - 7x + 10 = 0$. By synthetic division, we find that all roots of this equation are integral. We have

118 Higher Degree and Other Curves

```
 1  | 1   −4   −7    10
    |      1   −3   −10
    |_____
−2  | 1   −3  −10    0
    |     −2   10
    |_____
 5  | 1   −5    0
    |      5
    |_____
      1    0
```

Then the roots of the equation are $x = 1, -2$, and 5. These values represent *the x-intercepts* and are an aid in sketching the curve (Fig. 8.3). To get a good idea of the shape of the curve, we might also use other values of x and then compute corresponding values of y by synthetic division such as the following: when $x = -3$, $y = -32$; other points are $(-1,12)$, $(3,-20)$, and $(6,40)$. If the y-values are rather large, the vertical scale can be condensed.

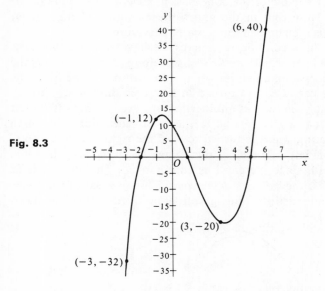

Fig. 8.3

Note that the curve has an unlimited extent in positive and negative x and y directions. Therefore the domain and range are infinite. The curve has no asymptotes. It is not symmetric with respect to either the x-axis or the y-axis, although it appears to be symmetric with respect to the point that has approximately the coordinates $(1.3, -4)$. Note reversal in curvature at this point.

Note that as x moves from left to right the y-value increases, then decreases, and finally increases infinitely. As x moves toward the right, the *final trend* of the y-value to increase or decrease is determined by the sign of the highest power of x, in this case x^3. If the sign of x^3 had been negative, the final trend of the y-value would be to decrease and become infinite negatively.

The *roots of a cubic equation* may be of the following types (Fig. 8.4):

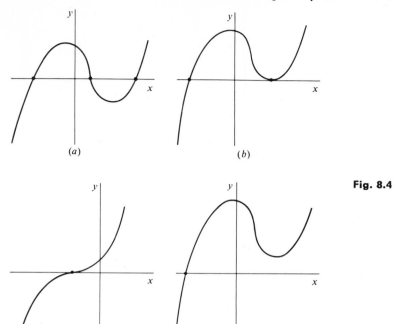

Fig. 8.4

(a) Three real roots, all different.
(b) Three real roots, two of which are equal and the third different.
(c) Three real roots, all equal (called a *triple* root).
(d) One real root and two imaginary roots. Imaginary roots always occur in conjugate *pairs*. If one root is $(3 + 2i)$, then another root is $(3 - 2i)$.

Note that every cubic must have at least one real root because the curve must cross the x-axis at least once. The graph of a cubic equation has no asymptotes. It is not symmetrical with respect to either axis or to any line but may be symmetrical with respect to the origin or to some other point.

EXAMPLE 2. Graph the equation $y = x^3$ and note the change in curvature.

Solution. For the intercepts, we see that if $x = 0$, $y = 0$. In fact, the point $(0,0)$ is the only intercept. The equation $x^3 = 0$ therefore has three equal roots, all $x = 0$. For the extent of the curve we note that x may have any value, positive or negative or zero, so that the domain is unlimited, that is, $-\infty < x < +\infty$. The range is also unlimited (Fig. 8.5).

In the equation, $y = x^3$, if x is replaced with $(-x)$, or if y is replaced with $(-y)$, the equation is changed. Therefore, the figure is not symmetrical with respect to either axis. However, if the two changes are made at once, then the equation is unchanged. The origin, therefore, is a center of symmetry, as is apparent from the graph. Note that the curvature changes from clockwise to counterclockwise as x moves from left to right.

120   Higher Degree and Other Curves

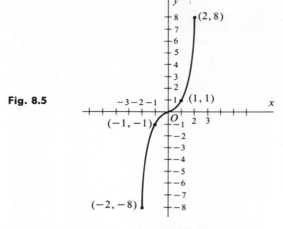

Fig. 8.5

8.3   A FOURTH-DEGREE EQUATION

We have seen that in the graph of a third-degree equation there is a change or *reversal* in the *sense of curvature*. In the graph of a fourth-degree equation, there are, in general, two such reversals. In fact, we might say that theoretically there is one reversal of curvature for each degree above the second. However, in some cases there is no reversal, only a situation in which a section of the curve appears to have the intention of reversing curvature but instead continues on without any reversal.

For an example of a fourth-degree equation, we take one having integral roots so that all are easily found by synthetic division.

EXAMPLE 3.   Graph the equation $y = x^4 - x^3 - 7x^2 + x + 6$.

Solution. For the *y*-intercept, we note that when $x = 0$, $y = 6$. For the *x*-intercepts, we set $y = 0$ and then solve the resulting equation,

$$x^4 - x^3 - 7x^2 + x + 6 = 0$$

First, we note that $x = 1$ is a root. Then, by synthetic division, we get

```
  1 |  1   -1   -7    1    6
    |       1    0   -7   -6
 -1 |  1    0   -7   -6    0
    |      -1    1    6
 -2 |  1   -1   -6    0
    |      -2    6
  3 |  1   -3    0
    |       3
       1    0
```

8.3 A Fourth-Degree Equation

The roots, 1, −1, −2, and 3, are the *x-intercepts*, and are an aid in sketching the graph (Fig. 8.6). To get a good idea of the shape of the curve we might also find by synthetic division the *y*-values for $x = -3, -1.5, -0.5, 2,$ and 4. We must not assume that the lowest point or the highest point of a curve is equally distant from two *x*-intercepts. The exact lowest or highest point can be found by calculus, as we shall see later.

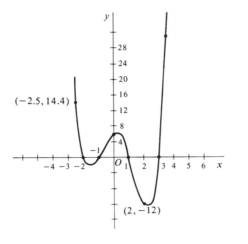

Fig. 8.6

EXAMPLE 4. Graph the equation $y = x^4 - 5x^3 + 6x^2 + 4x - 8$.

Solution. We note that when $x = 0$, $y = -8$, the *y*-intercept. For the *x*-intercepts we solve the equation
$$x^4 - 5x^3 + 6x^2 + 4x - 8 = 0$$
By synthetic division we find the roots are 2, 2, 2, and −1. Note the triple root at $x = 2$. This is a point of inflection. Also note that there are two reversals of curvature. The graph is shown in Figure 8.7.

EXAMPLE 5. Graph the equation $y = x^4 - 4x^3 + 6x^2 - 4x + 1$; note symmetry.

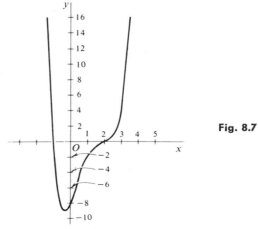

Fig. 8.7

122 Higher Degree and Other Curves

Solution. When $x = 0$, $y = 1$, the y-intercept. For the x-intercepts, we solve the equation

$$x^4 - 4x^3 + 6x^2 - 4x + 1 = 0$$

In this equation we might expect two reversals of curvature. However, solving the equation we find the roots are 1, 1, 1, and 1. The equation can then be expressed as the product of four equal factors set equal to zero, that is, $(x - 1)^4 = 0$. The equation can be written $y = (x - 1)^4$. Note that y can never be negative. Instead, the curve descends to the x-axis at the point (1,0), lingers there for a count of 4, and then rises slowly at first, then very rapidly, without any reversal of curvature (Fig. 8.8). Note that the curve is symmetrical with respect to the line, $x = 1$.

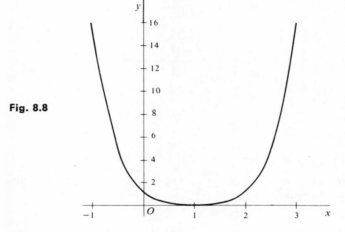

Fig. 8.8

8.4 OTHER ALGEBRAIC CURVES

If an equation contains a variable denominator, then the curve will sometimes be discontinuous for some value of x. In such equations x cannot be equal to any value that makes the denominator equal to zero. However, x may approach such values, thus leading to asymptotes.

EXAMPLE 6. Graph the equation

$$y = \frac{12}{x^2 - 4}$$

Solution. When $x = 0$, $y = -3$, the y-intercept. There is no x-intercept. To find the asymptotes, we set the denominator equal to zero and solve for x:

$$x^2 - 4 = 0; \quad \text{then} \quad x = 2 \quad \text{and} \quad x = -2$$

The two lines $x = 2$ and $x = -2$ are the vertical asymptotes of the curve. If we write the equation

$$x^2 - 4 = \frac{12}{y}$$

we set $y = 0$ and note that this line is a horizontal asymptote. For graphing the curve we make use of the asymptotes and may use the following values:

$x =$	0	1	-1	2	± 3	± 4	± 1.5
$y =$	-3	-4	-4	—	2.4	1	-6.86

The graph is shown in Fig. 8.9. Note that if x is replaced by $(-x)$, the equation is unchanged. Then the y-axis is an axis of symmetry. The values of x are unlimited, except that the values $x = \pm 2$ must be excluded from the domain. In the range, the values $-3 < y \leq 0$ must be excluded, otherwise the range is unlimited positively and negatively. If the equation is solved for x, it may be easier to see what values of y must be excluded:

$$x = \pm\sqrt{1 + (3/y)}$$

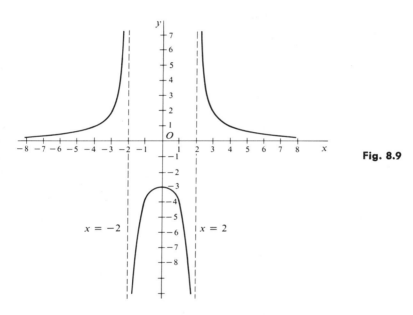

Fig. 8.9

The result shows that y cannot be zero, for then we should have division by zero. Moreover, the radicand must be equal to or greater than zero so that $3/y$ cannot be less than -1.

EXAMPLE 7. Graph the semicubical parabola $y^3 = x^2$.

Solution. In this equation note that y cannot be negative, for then y^3 would also be negative, and this would make x imaginary. Hence, none of the curve can be below the x-axis. The only intercept is the point $(0,0)$. The domain includes all values of x from $-\infty$ to $+\infty$. However, the range (extent of y values) is limited to all values from 0 to $+\infty$. The y-axis is an axis of symmetry since the equation is not changed if x is replaced by $(-x)$. The curve is continuous, since it can be drawn without raising the pencil from the paper. However, there is a sharp point, called a *cusp*, at the point $(0,0)$ (Fig. 8.10).

124 Higher Degree and Other Curves

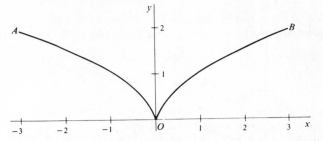

Fig. 8.10

It is interesting to note in this curve that the curvature appears to be clockwise throughout, from A to O and also from O to B. However, we must remember the curve is not really a third-degree polynomial, for if we solve the equation for y, we get $y = x^{2/3}$. Besides, if you drive a car from A to O, the steering wheel is turned toward the right. Then, to continue on from O to B, the wheel must be turned toward the left since you must back up.

EXERCISE 8.1

Sketch the graphs of the following algebraic curves. Note especially any asymptotes, intercepts extent (domain and range) and symmetry of each curve:

1. $y = x^3 - 3x^2 - x + 3$
2. $y = 6 - 11x + 6x^2 - x^3$
3. $y = 2x^3 - 3x^2 - 2x + 3$
4. $y = 2x^3 + 3x^2 - 9x - 10$
5. $y = 4x^3 - 16x^2 + 19x - 6$
6. $y = 3x + 5x^2 - 2x^3$
7. $y = x^3 - 2x - 4$
8. $y = 2x^3 - 9x^2 + 12x - 4$
9. $y = 7 - 12x + 6x^2 - x^3$
10. $y = x^3 + 6x^2 + 12x + 8$
11. $y = 12 - 8x + x^2 + 2x^3 - x^4$
12. $y = x^4 - 7x^2 + 9x^2 + 7x - 10$
13. $y = x^4 - 13x^2 + 36$
14. $y = 3 - 4x - 4x^2 + 4x^3 - x^4$
15. $y = x^4 - 16$
16. $y = x^4 + 4x^2 - 3x + 4$
17. $y = x^4 - 6x^3 + 13x^2 - 12x + 4$
18. $y = 4 - 4x - 3x^2 + 4x^3 - x^4$
19. $y = x^5 - x^3$
20. $y = x^5 - x^3 - 8x^2 + 8$
21. $y = \dfrac{12}{x - 3}$
22. $y = \dfrac{x - 1}{x + 4}$
23. $y = \dfrac{2x}{3x + 4}$
24. $y = \dfrac{6}{x^2 - 1}$
25. $y = \dfrac{x}{9 - x^2}$
26. $y + 2 = \dfrac{x}{x + 3}$
27. $y - 3 = \dfrac{x}{x^2 - 4}$
28. $y = \dfrac{8}{x^2 + 4}$
29. $y^2 = \dfrac{x^2 + x^4}{x - 1}$
30. $y^2 = x^3$
31. $y^3 = 4x^2$
32. $x^3 = 9y^2$
33. $y = x^2 + \dfrac{1}{x}$
34. $x^2 - 1 = \dfrac{6}{y^2}$
35. $x^2 + 1 = \dfrac{6}{y^2}$
36. $x^2y^2 = 16$
37. $y^2 = x^3 - x$
38. $y^2 = x^3 - 5x^2 + 6x$
39. $y = \dfrac{12}{x^2 - 6x + 8}$
40. $y = \dfrac{12}{x^2 - 6x + 9}$
41. $y = \dfrac{12}{x^2 - 6x + 10}$

chapter

9

Functions and Functional Notation

9.1 CALCULUS AS A WAY OF THINKING

Probably the outstanding thing about calculus is the kind of thinking involved. It is true we make use of the mathematics we have already learned, such as arithmetic, algebra, geometry, logarithms, and trigonometry. Calculus does not replace any of these. But calculus involves a kind of thinking that sets it apart from our former mathematics and makes it a most powerful tool in solving problems.

Of course, we do learn certain short-cuts or "tricks" that can be used in arriving at answers. Yet calculus does not consist merely of "tricks" with symbols. It is much more. If you memorize the tricks without understanding the logic behind them, you will reach a point in your study where you will be lost.

The concepts of calculus are not difficult to anyone who desires to learn and who applies himself. It is when the concepts are explained in mathematical terminology that they seem difficult. Yet it is necessary that you become acquainted with the terminology if you are to make progress.

As an example of the kind of thinking involved, let us consider a problem in velocity. Suppose we drive a car for exactly 4 hours and cover 120 miles. To find the average velocity, we divide *distance* by *time*. Dividing 120 by 4, we get 30, which represents a velocity of 30 miles per hour. However, here we must remember that the answer represents *average velocity* over the distance. From a practical standpoint it is impossible to drive for 4 hours at a steady rate of 30 miles per hour during every instant of the time.

Now, we may wish to know our *instantaneous velocity* at some particular point, say, when we passed a certain highway sign. After all, if we happen to hit a tree or another car, the damage done depends on the instantaneous velocity at the time of impact, not on the average velocity for the entire distance. Average velocity never hurt anyone; it is instantaneous velocity that kills.

Whenever we wish to determine velocity, we measure the distance between two points and then divide this distance by the time required to cover the distance. For example, if we wish to determine the velocity of a car, we might make two marks on a road and measure the distance between them. Suppose this distance is 30 feet, and suppose also the measured time required for a car to cover this distance is 0.375 second. Then velocity is given by

$$\frac{\text{distance}}{\text{time}} = \frac{30}{0.375} = 80, \text{ which represents } 80 \text{ ft/sec}$$

However, even in this short distance, the result is only *average* velocity, since the car may not have been traveling at exactly the same velocity during the entire time interval. To arrive at a closer estimate of instantaneous velocity, we might shorten the distance to, say, 10 feet and measure the elapsed time. Yet no matter how short we make the distance between the two marks, the result in every case represents only average velocity. If we wish to find the instantaneous velocity at a single point, we discover that we cannot do so. However, it can be found by calculus.

9.2 VARIABLES AND CONSTANTS

In calculus we deal with two kinds of quantities, *variables and constants*. As we have stated (Chap. 2), a *variable* is a quantity that may take on different values in a problem. Variables are usually represented by letters near the end of the alphabet, such as u, v, w, x, y, and z, but any letter or symbol may be considered a variable. In some problems a variable may be limited to certain particular values. For example, in some applications imaginary values must be excluded.

A *constant* is a quantity that is understood to have the same value throughout a problem. Constants are of two kinds: (1) *absolute* constants, such as 4, 8, $-\frac{2}{3}$, and π, whose values never change; and (2) *arbitrary* constants, represented by symbols whose values are assumed to remain constant throughout a problem. If we use the letter n, for example, in a problem and then assume that n does not change in value in that particular problem, then n also is considered a constant. We often denote arbitrary constants by letters at or near the beginning of the alphabet, but any letter may be considered a constant. Letters with subscripts, such as x_1, x_2, and x_n, usually denote constants.

9.3 FUNCTIONS

Calculus is basically a study of relations between variables. It often happens that two variables, such as x and y, are related in some way. If for each value of the variable x there exists a corresponding value of y, then we say y is a *function* of x. In a similar manner, the temperature T at a particular spot is a function of the hour H on a certain day because for each hour there exists a corresponding temperature.

9.3 Functions

A function is often indicated by an equation, such as

$$y = 3x + 2$$

In this equation, for each value of x there is a corresponding value of y. For example, if $x = 5$, then $y = 17$. The value of y depends on what value is assigned to x. Then we call one variable *independent* and the other *dependent*. In an equation involving x and y, we usually take x to be the independent variable. In all cases it is important to identify the independent variable. In the equation for the area of a square, $A = s^2$, the independent variable is considered to be s; then the dependent variable is A.

A function may also be indicated by a table of values. Consider the following table of temperature readings for various hours as shown by a thermometer at a particular spot on a particular day:

hour (H)	9 AM	10 AM	11 AM	12 noon	1 PM	2 PM	3 PM
temperature (T), deg.	68	70	71	73	73	74	73

The temperature T is a function of the hour H because for each hour there is a corresponding temperature reading. The function may be written as *ordered number pairs:* (9,68), (10,70), (11,71), (12,73), (1,73), (2,74), and (3,73). Note that a change may occur in H without any change in T.

In an equation expressing a relation between x and y, the dependent variable y may have more than one value corresponding to each value of x; as in the example,

$$y^2 = x^2 + 16$$

If $x = 3$, then y has the two values $+5$ and -5. If we solve the equation for y, we get two solutions:

$$y = +\sqrt{x^2 + 16} \quad \text{and} \quad y = -\sqrt{x^2 + 16}$$

Each solution represents a function of x. Then we sometimes say that y is a *multiple-valued function* of x.[1]

A functional relationship can also be shown by a graph. As an example, let us graph the function (Fig. 9.1)

$$y = x^2$$

In this case the function is expressed as an equation. We have already seen how to graph equations (Chap. 2). For this function we may use the following ordered pairs of values:

$x =$	0	1	2	3	-1	-2	-3
$y =$	0	1	4	9	1	4	9

[1] The equation $y^2 = x^2 + 16$ is sometimes called a simple *relation* rather than a function. Then the term *function* is restricted to relations in which there is only one value of the dependent variable for each value of the independent variable.

128   Functions and Functional Notation

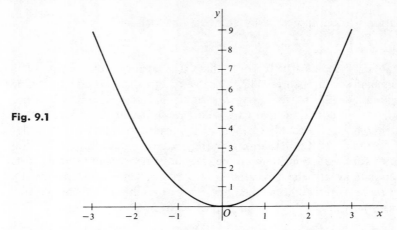

Fig. 9.1

Note in this example there is only one value of y for each value of x. This means that there is only one ordinate for each abscissa.

Functions that arise in connection with practical problems in science are often shown by graphs. For example, the current i in an electric circuit is often a function of time t. Then we usually take time t as the horizontal axis, and current i as the vertical axis. For an object in "free fall" (that is, falling freely under the influence of gravity alone near the earth's surface), the *distance* fallen is a function of *time*, as shown by the formula $s = 16t^2$. Then we take distance s along the vertical axis.

A table of related values showing a function, such as a table of temperature readings, cannot always be expressed by an equation. However, such a function can be shown by a graph. For example, Figure 9.2 shows the relation between the hour H and the temperature reading T, as stated in the table on the previous page.

9.4   A FUNCTION OF x

We have seen that a function may be stated as an equation, such as

$$y = 3x + 5$$

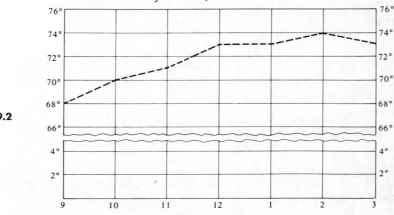

Fig. 9.2

In this equation we say y is a function of x because once the value of x is known, the value of y is determined.

However, there is a way of defining a function of x without using another letter, such as y. The following expression is a function of x since its value is determined when the value of x is given:

$$3x + 5$$

For example, if $x = 4$, then the value of the function is 17.

For an operational definition of a function of x, we can say: *A function of x is any expression that contains x.* This definition includes a constant, such as 4, which may be written $4x^0$. The function, $3x + 5$, may be written as the second element of an ordered pair: $(x, 3x + 5)$. Here are some examples of functions of x:

$$5x - 2; \quad x^3 - 4x; \quad \sqrt{x - 2}; \quad \sin 3x; \quad \log_{10}(4x + 3); \quad 5^{2x}$$

A similar definition holds for functions of other variables:

$(u^2 - 3)$ is a function of u; $\quad$ $(2v^3 + 5v)$ is a function of v

$3t^2$ is a function of t; $\quad$ $(5y^2 + 4y)$ is a function of y

Note: When we use a letter to represent a function of x and write $y = 3x + 5$, we should understand clearly that the function of x is *not* y but the *expression itself:* $3x + 5$. The letter y is used only for convenience to represent the function. Students can avoid a lot of confusion if they will remember this fact.

9.5 FUNCTIONAL NOTATION

Suppose we write: "A certain function of x is $(3x + 5)$." The statement can be expressed in a shortened form:

$$f(x) = 3x + 5$$

The letter f is used instead of the word *function*, and the short form is read "*f* of $x = 3x + 5$." It means that one particular function of x is the expression $3x + 5$.

It is important to remember that when we say, $f(x) = 3x + 5$, the f refers only to the particular function stated. If we have another function of x in the same problem, we must use another letter to represent the second function. Any letter may be used to represent a function. As an example, we may have the two functions:

$$f(x) = 3x + 5 \quad \text{and} \quad g(x) = x^2 + 2$$

The symbol $g(x)$ is read: "The g function of x," or simply "g of x." We sometimes use only the letters, such as f and g, letting the letters represent the respective functions of x: as $f = 3x + 5$ and $g = x^2 + 2x$.

Note: When the symbol $f(x)$ is used to represent a function, as in

$$f(x) = 3x + 5,$$

it should be understood that the function is *not* the symbol $f(x)$. The function of x is the *expression itself*, $3x + 5$. The symbol $f(x)$ is used only for convenience to represent the function.

To represent a general function of x, we often write

$$y = f(x)$$

This statement does not tell us the particular relation between x and y; it says only that y is related to x in some way. A functional relationship between any two variables can be expressed in the same way. For example, the area of a circle is a function of the radius, because if the radius is given, then the area is determined. The function can be written

$$A = f(r)$$

Sometimes the value of a dependent variable is determined by the values of two or more independent variables. For example, the distance s traveled by an object in uniform motion depends on the velocity v and the time t. Then we can say

$$s = f(v,t)$$

That is, the distance is a function of *velocity* and *time*.

9.6 THE FUNCTION VALUE

The symbol $f(x)$ used to represent a function of x affords a convenient way to denote the value of a function for any particular value of x. For example, suppose we have the function

$$2x + 3$$

Now we may wish to know the value of this function for some particular value of x, say, when $x = 5$. Substituting 5 for x in the function, we find that the function value is 13. To denote this substitution we replace x with 5 in the *symbol* as well as in the *function*. That is, we indicate the function value by $f(5)$. Then we can write:

"If $f(x) = 2x + 3$, then $f(5) = (2)(5) + 3 = 13$."

We can also say: $f(0) = 3; f(1) = 5; f(3x + 1) = 2(3x + 1) + 3 = 6x + 5$. The symbol indicates simple *substitution*. However, if a symbol such as $f(5)$ is to have meaning, we must know the function itself. Here are other examples:

if $f(x) = x^2 - 3x + 4$, $f(0) = 4$;   if $g(x) = x^3 + 5x^2$, $g(-1) = 4$

if $h(x) = \log_{10}(x^2 + 19)$, $h(9) = 2$;   if $u(x) = \sin x$, $u(1) = 0.8415$

If we have a function of two or more independent variables, we can evaluate the function in the same way. For example, by substitution,

if $f(x,y,z) = x^2 - 3xy + y^2 - 3z + 2yz$,   then $f(1,-2,4) = -17$

9.7 COMBINATIONS OF FUNCTIONS

Suppose we have the following two functions of x:
$$f(x) = x^2 + 4x + 5 \quad \text{and} \quad g(x) = 2x - 3$$
Then for the *sum* of the functions we have
$$f(x) + g(x) = (x^2 + 4x + 5) + (2x - 3) = x^2 + 6x + 2$$
We may write simply
$$f + g = x^2 + 6x + 2$$
The *difference* between the functions may be expressed as
$$f(x) - g(x) = (x^2 + 4x + 5) - (2x - 3)$$
$$= x^2 + 4x + 5 - 2x + 3 = x^2 + 2x + 8$$
or, simply, $\quad f - g = x^2 + 2x + 8$

For the product of the two functions we have
$$[f(x)][g(x)] = (x^2 + 4x + 5)(2x - 3) = 2x^3 + 5x^2 - 2x - 15$$
This product may be expressed simply as
$$(f)(g) = 2x^3 + 5x^2 - 2x - 15$$
The quotient of the two functions may be expressed in a similar way, but we must be sure the denominator function is not equal to zero. Then
$$\frac{f(x)}{g(x)} = \frac{x^2 + 4x + 5}{2x - 3} = \frac{1}{4}\left(2x + 11 + \frac{53}{2x - 3}\right)$$

There is another combination of two functions that is important, called the *composite* function. This may be called a *function* of a *function*. Let us suppose we substitute the entire g function in place of x in the f function. For example,
$$\text{if } f(x) = x^2 + 4x + 5 \quad \text{and} \quad g(x) = 2x - 3$$
then $\quad f(2x - 3) = (2x - 3)^2 + 4(2x - 3) + 5$

Note that we have taken the entire g function, $2x - 3$, for x in the f function. We are really taking the f function of the g function. Simplifying,
$$f(2x - 3) = 4x^2 - 4x + 2$$
The composite function of two functions, $f(x)$, and $g(x)$ is often shown by
$$f[g(x)] \quad \text{or} \quad f \circ g$$
As an illustration of a composite function, suppose we have
$$f(x) = 2x^2 - 3x - 4 \quad \text{and} \quad g(x) = 3x - 2$$
Then
$$f[g(x)] = f(3x - 2) = 2(3x - 2)^2 - 3(3x - 2) - 4 = 18x^2 - 33x + 10$$

132 Functions and Functional Notation

Let us consider some combinations of functions using specific values. Suppose we have

$$(x) = x^2 - 4x - 5 \quad \text{and} \quad g(x) = x - 3$$

Then
$$f(3) + g(5) = -8 + 2 = -6$$

$$f(-2) - g(2) = 7 - (-1) = +8$$

$$[f(0)] \cdot [g(0)] = (-5)(-3) = +15$$

$$\frac{f(2)}{g(2)} = \frac{-9}{-1} = +9; \qquad \frac{f(1)}{g(3)} = \frac{-8}{0} \text{ (not defined)}$$

$$f[g(-1)] = f[-4] = 27; \qquad f[g(2)] = f[-1] = 0$$

9.8 TWO VIEWS OF A FUNCTION

Consider the function $x^2 + 3x$. If we give x a value of 4, then the value of the function becomes 28. Now, we may consider the function to be the actual expression $x^2 + 3x$, or we may say that the function is the value obtained when we give x some particular value. In most instances we say that the function is the *value* obtained. For example,

$$\text{if } x = 5, \text{ the function} = 40$$

$$\text{if } x = 6, \text{ the function} = 54$$

If the function is understood to be the value obtained, then from this viewpoint, the *function changes* as x changes.

On the other hand, if we consider the function to be the actual expression itself, then the function does not change. The function, $x^2 + 3x$, is simply the rule that states what is to be done with any particular x-value. In this case it is only the function *value* that changes, not the function itself.

From time to time it may be advisable and convenient to change from one viewpoint to the other. As a student you should have no trouble with the two views if you realize there are two. Then you can shift from one to the other whenever you find it convenient to do so.

If we look upon the function as the expression itself, then it may be likened to a machine (Fig. 9.3). Now if we put $x = 4$ into the machine, we can say that 4 is the input. Then 28 is the output. The machine, $x^2 + 3x$ (the function), operates on the input, 4, and produces the output, 28. The function (the machine) does not change, but the output depends on the input and what the function does to it. Even here, we may call the output, 28, the value of the function.

Fig. 9.3

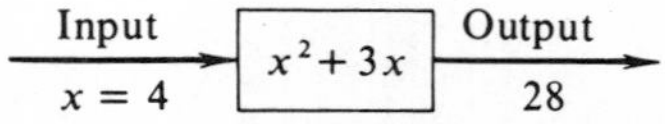

Input $x = 4$ → $x^2 + 3x$ → Output 28

The function machine

9.8 Two Views of a Function

EXERCISE 9.1

1. If $f(x) = x^2 + 3x - 5$, find $f(1)$; $f(-2)$; $f(0)$; $f(a)$; $f(3x)$; $f(x + 2)$.
2. If $g(x) = 3x^2 - 2x - 7$, find $g(1)$; $g(3)$; $g(0)$; $g(\frac{1}{2})$; $g(\frac{1}{3})$; $g(x^2)$.
3. If $h(x) = 9 - x^2$, find $h(0)$; $h(1)$; $h(-1)$; $h(-3)$.
4. If $f(x) = \sqrt{x^2 + 9}$, find $f(0)$; $f(4)$; $f(-4)$; $f(1)$; $f(-1)$.
5. If $g(x) = \sqrt{25 - x^2}$, find $g(3)$; $g(0)$; $g(5)$; $g(-3)$; $g(\sqrt{34})$.
6. If $F(x) = x^3 - 3x^2 - 7x - 4$, find $F(0)$; $F(1)$; $F(-1)$; $F(2x)$; $F(x^3)$.
7. If $k(t) = 3 - 2t - t^2$, find $k(0)$; $k(1)$; $k(-1)$; $k(-3)$; $k(t-2)$.
8. If $f(s) = 2s^4 - 4s^3 + 5s^2 + 3s - 3$, find $f(0)$; $f(1)$; $f(-1)$; $f(-2)$.
9. If $P(u) = 5u^4 - 4u^3 - 3u^2 + 7u - 2$, find $P(0)$; $P(1)$; $P(-1)$; $P(-3)$.
10. If $Q(x) = 6x^3 - 3x^2 + 4x - 5$, find $Q(0)$; $Q(1)$; $Q(3)$; $Q(10)$; $Q(100)$.
11. If $f(t) = 10 - 9t - 3t^2 - t^3$, find $f(1)$; $f(-2)$; $f(b)$; $f(c^2)$; $f(4)$.
12. If $f(x) = x^3 - 2x^2 + 9t - 18$, find $f(2)$; $f(3i)$; $f(-3i)$; $(i = \sqrt{-1})$.
13. If $f(y) = y^4 - y^3 - y^2 - y - 20$, find $f(1)$; $f(-1)$; $f(i)$; $f(-i)$; $f(2i)$.
14. If $f(x) = 2^{x+1}$, find $f(4)$; $f(0)$; $f(-1)$; $f(-2)$; $f(5)$; $f(-5)$; $f(-9)$.
15. If $f(x) = x^x$, find $f(1)$; $f(2)$; $f(-2)$; $f(4)$; $f(5)$; $f(6)$.
16. If $f(x) = 2^{1-2x}$, find $f(0)$; $f(1)$; $f(2)$; $f(-3)$; $f(\frac{1}{2})$; $f(-\frac{1}{2})$.
17. If $g(x) = x^{\sqrt{x}}$, find $g(4)$; $g(9)$; $g(\frac{1}{4})$; $g(\frac{9}{4})$; $g(16)$.
18. If $f(x) = 4^x$, find $f(2)$; $f(-2)$; $f(0)$; $f(3)$; $f(\frac{3}{2})$; $f(-\frac{3}{2})$.
19. If $f(x) = 12/(4 - x^2)$, find $f(0)$; $f(-4)$; $f(1)$; $f(-1)$; $f(2)$.
20. If $f(x) = (x + 3)/(x - 4)$, find $f(0)$; $f(1)$; $f(-1)$; $f(-3)$; $f(4)$.
21. If $f(x) = \log_{10}(x + 7)$, find $f(3)$; $f(93)$; $f(993)$; $f(-6)$.
22. If $f(x) = 3\log_{10}(x + 96)$, find $f(4)$.
23. If $f(\theta) = \sin 4\theta$, find $f(15°)$; $f(\pi/6)$; $f(22.5°)$; $f(\pi/4)$.
24. If $f(\phi) = 1 + \cos \phi$, find $f(\pi)$; $f(0)$; $f(450°)$; $f(-60°)$.
25. If $f(x) = \sin x$, find $f(0)$; $f(1)$; $f(2)$; $f(-1)$; $f(1.5708)$; $f(\cos 60°)$.
26. If $f(x) = \sin x$ and $g(x) = \cos x$, find $f(\pi/2) + g(\pi)$, $[f(x)]^2 + [g(x)]^2$.
27. If $R(x) = 5x^2 - 2x - 3$, find $R(y^2 + 2)$; $R(y^2 + 3y - 1)$.
28. If $f(r) = r^2 - 3r - 4$, find $f(x^2)$; $f(y^3)$; $f(x + 2)$; $f(3) - f(1)$; $f(4) + f(2)$.
29. If $f(x) = x^2 + 3x - 2$ and $g(x) = x - 3$, find $(f + g)(x)$; $f - g$; $(f)(g)$; $f \div g$; $f(2) + g(3)$; $f \circ g$; $f[g(2)]$.
30. If $f(x) = 6x^2 - 7x + 6$ and $g(x) = 3x - 2$, find $f(1) - g(-1)$; $f(x) + g(x)$; $f - g$; $(f)(g)$; $f[g(x)]$; $g[f(x)]$.
31. If $f(x) = x^2$, find $\dfrac{[f(10) - f(2)]}{(10 - 2)}$; $\dfrac{[f(9) - f(6)]}{(9 - 6)}$; $\dfrac{[f(b) - f(a)]}{(b - a)}$.
32. If $f(x) = \dfrac{(x^2 - 9)}{(x - 3)}$ and $g(x) = x + 3$, is $f(x)$ the same as $g(x)$? Formulate your conclusion after you compare the following:
    (a) $f(4)$ and $g(4)$, (b) $f(2)$ and $g(2)$, (c) $f(1)$ and $g(1)$, (d) $f(3)$ and $g(3)$.
33. If $f(x,y) = 3x^2 - 5xy - y^2$, find $f(1,2)$; $f(2,-1)$; $f(-1,0)$; $f(0,-1)$.
34. If $f(m,n) = m^3 - n^3$, find $f(3,1)$; $f(-2,-2)$; $f(3,-3)$; $f(0,-1)$.
35. If $f(a,b) = 4a^2 - 3ab - 2b^2 - 5a - 3b - 4$, find $f(1,2)$; $f(-2,3)$; $f(0,-2)$.
36. If $f(i,s,h) = 3i^2 - 2is^2 - 3h^2s - 4his$, find $f(1,-1,2)$; $f(0,-1,-2)$.
37. If $g(n,u) = n^3 - 2n^2u + 4nu^3$, find $g(2,1)$; $g(-2,-1)$; $g(1,-2)$.
38. If $b(l,a,c) = l^2 - 2la - 3a^2c + 3$, find $b(1,2,3)$; $b(2,3,1)$; $b(-1,-2,-3)$.
39. If $f(x,y,z) = x^2 - 3xy + 2xz - y^2 - z^2$, find $f(1,1,1)$; $f(-1,-1,-1)$.
40. If $F(r,s,t) = r^3 + r^2s - rst - s^2t - t^3$, find $F(2,-1,0)$; $F(0,-1,-2)$.

chapter
10

Increments: Average Rates

10.1 INCREMENTS

We have seen that a variable is a quantity that may change in value in a particular problem. Any change in the value of a variable is called an *increment*. The word *increment* in everyday use usually means an increase.

In calculus we use the term *increment* to represent any change whatever, positive, negative, or zero. If the radius of a circle changes from 6 inches to 6.2 inches, the increment in the radius is 0.2 inch. If the radius changes from 6 inches to 5.8 inches, the increment is -0.2 inch. As we have already stated (Sec. 1.6), an increment, whether positive, negative, or zero, is usually denoted by the Greek letter capital *delta* (Δ) placed before the variable. An increment in the variable x is called "delta x" and is written Δx. This notation is not to be taken as a product. It is *not* "delta times x" but simply represents a change in the value of x. A bar is sometimes placed over the expression, such as $\overline{x\Delta}$, to emphasize that it represents a single quantity.

In like manner, an increment in the variable y is called "*delta y*" and is written Δy. An increment in the radius of a circle is written Δr. An increment in the area is denoted by ΔA and may be called "delta A" or "delta area." An increment in time t is written Δt; an increment in electric current i is written Δi.

It should be noted that as a variable changes from one value to another, the increment is found by subtracting the first value *from* the second value. If x changes from 10 to 12, the increment is found by

$$(\text{second value}) - (\text{first value})$$

This rule holds for all increments, positive, negative, or zero. In the values shown here, the first value of x is 10; the second value is 12. Then the increment is

$$12 - 10 = +2, \text{ the increment in } x$$

If we let x_1 represent the first value and x_2 represent the second value, then the increment Δx is given by

$$\Delta x = x_2 - x_1$$

If an electric current i changes from 10 amperes to 8 amperes, the increment is

$$\Delta i = i_2 - i_1 = 8 - 10 = -2 \text{ amperes}$$

In this instance Δi, the increment, is negative.

When temperature T changes from $-10°$ to $-7°$, the increment ΔT is found by

$$\Delta T = T_2 - T_1 = (-7°) - (-10°) = +3°$$

A change in temperature from $-10°$ to $-7°$ represents an increase, or positive increment, as shown by the $+3°$.

10.2 INCREMENT OF A FUNCTION

Whenever one variable is a function of another, then for any increment in the independent variable there will be some corresponding increment in the dependent variable. For any increment in the radius of a circle there will be corresponding increments in the diameter, the circumference, and the area, since all these are functions of the radius.

Of course, a change in a variable may not bring about an actual change in the function. For example, on any particular day at a particular location the temperature is a function of the hour of day because for each hour there is a corresponding temperature. Yet it may be that the time changes from 3 PM to 4 PM without any change in the temperature. Moreover, it is possible that a positive increment in a variable may correspond to a negative increment in the function. For example, on a trip of, say, 300 miles the time required is a function of the velocity, but if the velocity has a positive increment, the time will have a negative increment.

However, if we take the term *increment* to include *negative* and *zero* values as well as *positive*, then we can say that for *any* increment in a variable there will always be a corresponding increment in the function. If y is a function of x, then as x changes by some increment, there will be a corresponding increment in y, the function, $f(x)$. The increment in the function is called "delta y" or "delta function" and can be denoted by

$$\Delta y \quad \text{or} \quad \Delta f(x)$$

In order to see the corresponding increments in x and its function, consider the graph (Fig. 10.1) of the general function

$$y = f(x)$$

Fig. 10.1

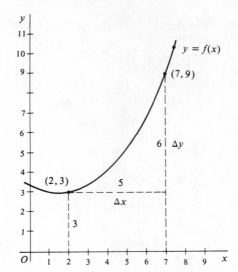

Suppose that when $x = 2$, $y = f(x) = 3$, and when $x = 7$, $y = f(x) = 9$.
Then
$$\Delta x = 7 - 2 = 5$$
$$\Delta y = \Delta f(x) = 9 - 3 = 6$$

Let us consider the same problem in more general terms. Again we take the graph (Fig. 10.2) of the general function

$$y = f(x)$$

Fig. 10.2

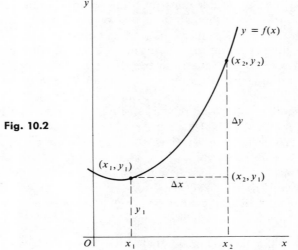

This time we let the two values of x be represented by x_1 and x_2. Then we have the corresponding values of the function, y_1 and y_2, or $f(x_1)$ and $f(x_2)$. As the value of x changes from x_1 to x_2, we have the following increments:

$$\Delta x = x_2 - x_1, \text{ and } \Delta y = y_2 - y_1, \text{ or } \Delta f(x) = f(x_2) - f(x_1)$$

If we always take the two values of x from left to right, that is, in a positive direction, then Δx will always be positive. However, Δy may be negative. This will be true if the function has a smaller value corresponding to x_2 than it has for x_1. If the function has the same value for the two values of x, then Δy will be zero.

Remember: To find the value of any increment, take

$$(\text{second value}) - (\text{first value})$$

10.3 RELATION BETWEEN INCREMENTS

Now we are interested in the relation between the increments of a variable and its function. If two variables x and y are so related that y is a function of x, then their respective increments, Δx and Δy, will also be related in some way. This relation will depend on the particular relation between x and y. The two increments are not usually equal, although they could be equal. It was a careful study of the relation between increments of related variables that led to the invention of calculus, which has turned out to be one of the most powerful tools in mathematics. Our chief concern at this point is to study specifically the ratio of the two increments in the order $\Delta y / \Delta x$. That is, we wish to find the ratio between the increment of the function y and the increment of the independent variable x.

To find the ratio between corresponding increments, consider again the general function

$$y = f(x)$$

Suppose the points $(-1, 2)$ and $(3, 10)$ lie on the curve (Fig. 10.3). Then, as x changes in value from -1 to 3, the function changes from 2 to 10. For the increments we have

$$\Delta x = x_2 - x_1 = 3 - (-1) = 4; \qquad \Delta y = \Delta f(x) = 10 - 2 = 8$$

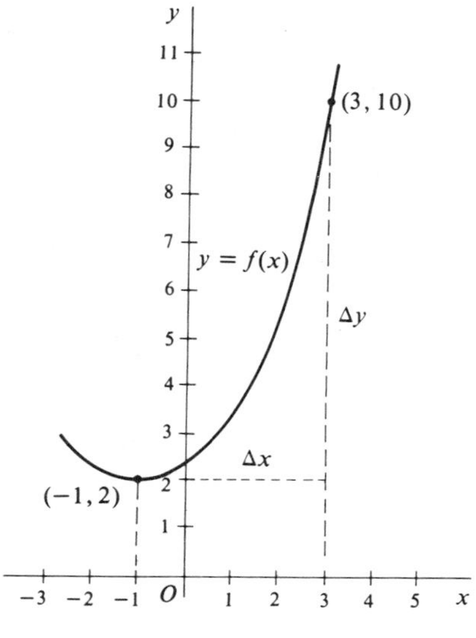

Fig. 10.3

138 Increments: Average Rates

Our concern now is the ratio of the increments:
$$\frac{\Delta y}{\Delta x} = \frac{8}{4} = 2$$

This ratio 2 represents the average rate of change in the function y with respect to x. That is, between the two given values of x, the function changes, *on the average*, 2 times as much as x. In general terms, the ratio of the increments is given by

$$\frac{\Delta y}{\Delta x} = \frac{y_2 - y_1}{x_2 - x_1}$$

10.4 AVERAGE RATE OF CHANGE

Average rate of change is an important concept in many everyday problems. If the temperature rises 20° in 5 hours, the average rate of change in temperature is 4° for each hour of change in time. This does not mean that the temperature changed 4° each hour, or 2° each half-hour, or 1° each 15 minutes. But it means that, on the average, the temperature changed 4° per hour over this particular 5-hour interval.

A rate of change is often considered with reference to *time*. For instance, if the temperature falls 18° in 6 hours, then the average rate of change in temperature is $-3°$ per hour. If a man runs 100 yards in 9.8 seconds, his average velocity is approximately 10.2 yards per second. If the population of a city increased by 50,000 over a period of 10 years, the average rate of change in population was 5000 per year.

If the electric charge on a capacitor changes from 0.00004 to 0.00016 coulomb in 6 seconds, then the average rate of change of charge Q is 0.00002 coulomb per second. In this case,

$$\Delta Q = Q_2 - Q_1 = 0.00016 - 0.00004 = 0.00012, \quad \text{and} \quad \Delta t = 6 \text{ seconds}$$

Then the *average rate of change* of charge is given by

$$\frac{\Delta Q}{\Delta t} = \frac{0.00012}{6} = 0.00002 \text{ coulomb per second}$$

The distance an object falls in a "free fall" is given by the formula

$$s = 16t^2 \text{ (approximately, at sea level)}$$

In the formula, s represents distance in feet and t represents time in seconds. Let us compute the average velocity or average rate of change in distance in the time interval from $t = 3$ seconds to $t = 5$ seconds. At the end of 3 seconds we have

$$s = 16(3)^2 = (16)(9) = 144 \text{ ft}$$

At the end of 5 seconds

$$s = 16(5)^2 = (16)(25) = 400 \text{ ft}$$

10.4 Average Rate of Change

To find the average velocity in the 2-second interval we have,

for increment in distance, $\Delta s = s_2 - s_1 = 400 - 144 = 256$ feet

for increment in time, $\Delta t = t_2 - t_1 = 5 - 3 = 2$ seconds

For average velocity we have

$$\frac{\Delta s}{\Delta t} = \frac{256 \text{ ft}}{2 \text{ sec}} = 128 \text{ ft/sec}$$

In this example, the 128 ft/sec represents the *average rate of change* of s with respect to time t. In Chapter 12 we shall see how to determine the exact velocity *at the instant* when $t = 3$ seconds. This is called *instantaneous velocity* or instantaneous rate of change. However, in this chapter we are interested only in *average rates of change*.

Although rate of change often refers to *time*, it need not do so. We may have any two related variables, neither of which is *time*. Then we find the rate of change in one variable *with reference to the other variable*. For example, in an electric circuit, the power P is given by the formula

$$P = I^2 R$$

If $R = 20$ ohms, we have the equation

$$P = 20 I^2$$

Now when the current changes from 4 amperes to 7 amperes, we wish to know the average rate of change in power with respect to current. To find the answer we compute the power for each value of I. We let I_1 and I_2 represent first and second values of current, and we let P_1 and P_2 represent first and second corresponding values of power. Then,

for $I_1 = 4$ amperes, $P_1 = 20(16) = 320$ watts

for $I_2 = 7$ amperes, $P_2 = 20(49) = 980$ watts

For the increments we have

$$\Delta P = P_2 - P_1 = 980 - 320 = 660 \text{ watts}$$

$$\Delta I = I_2 - I_1 = 7 - 4 = 3 \text{ amperes}$$

Then *average rate* of change of power *with respect to current* is

$$\frac{\Delta P}{\Delta I} = \frac{660 \text{ watts}}{3 \text{ amperes}} = 220 \text{ watts per ampere}$$

That is, the power changes *on the average* 220 watts for each change of 1 ampere of current.

EXAMPLE 1. In an electric circuit, the current I is given by the formula $I = E/R$, where E is the electromotive force (in volts) and R is the resistance (in ohms). If $E = 120$ volts, find the average rate of change of I with respect to R when R changes from 20 ohms to 30 ohms.

140 Increments: Average Rates

Solution. We let first and second values of I be represented by I_1 and I_2 and first and second values of R by R_1 and R_2. Then

for $R_1 = 20$ ohms, $I_1 = 120/20 = 6$ amperes

for $R_2 = 30$ ohms, $I_2 = 120/30 = 4$ amperes

Then
$$\Delta I = I_2 - I_1 = 4 - 6 = -2 \text{ amperes}$$
$$\Delta R = R_2 - R_1 = 30 - 20 = 10 \text{ ohms}$$

For average rate of change

$$\frac{\Delta I}{\Delta R} = \frac{-2 \text{ amperes}}{10 \text{ ohms}} = -0.2 \text{ amp/ohm}$$

That is, the average rate of change of current is -0.2 ampere for each change of 1 ohm in resistance.

10.5 A FORMULA FOR $\Delta y/\Delta x$

Up to this point we have used specific arithmetic numbers for initial values of x and for its increment Δx. Then from these we have obtained specific arithmetic values for the function and its increment Δy. The ratio $\Delta y/\Delta x$ has also been a specific number.

One important point should be noted here. Although we can assign some value to Δx, the value of the increment in y must be found by subtraction. For example, consider the equation $y = x^2$. If we begin with an initial value of $x = 4$, then the initial value of y, the function, is 16. Now if we change x by some increment, say, 3, so that $\Delta x = 3$, then the new value of x is 7. The new value of y in this case is 49. Then we find the increment in y by subtraction:

$$\Delta y = 49 - 16 = 33 \quad \text{and} \quad \frac{\Delta y}{\Delta x} = \frac{33}{3} = 11$$

Note especially that the new value of y, that is, 49, represents

$$y_1 + \Delta y$$

Now, instead of using arithmetic numbers, 4 for x_1 and 3 for Δx, let us use more general numbers. Then we shall get a *formula* or *rule* for finding the ratio $\Delta y/\Delta x$, *whatever the increments*. We begin again with the equation

$$y = x^2$$

Now we take some specific value of x, but in more general terms we call it x_1. Remember a subscript indicates a *specific* value of a letter. When we substitute this value for x, we get a specific value for y, which we call y_1. Then we have, for initial values,

$$y_1 = (x_1)^2$$

Now we change x_1 by some increment. We use the general increment Δx. The new value of x is then

$$x_1 + \Delta x$$

10.5 A Formula for $\Delta y/\Delta x$

Using this new value for x in the equation, we get a new value for y which can be represented by

$$y_1 + \Delta y$$

Then the new equation is

$$y_1 + \Delta y = (x_1 + \Delta x)^2$$

or, expanding,

$$y_1 + \Delta y = (x_1)^2 + (2)(x_1)(\Delta x) + (\Delta x)^2$$

Subtracting the original function to find the expression for Δy we get

$$\Delta y = 2(x_1)(\Delta x) + (\Delta x)^2$$

To get the ratio $\Delta y/\Delta x$, which is the thing we are after, we divide through by Δx, and get

$$\frac{\Delta y}{\Delta x} = 2(x_1) + (\Delta x)$$

This equation shows precisely what we want: the rule for finding the ratio of $\Delta y/\Delta x$. The rule says that for this particular equation, $y = x^2$, the ratio, $\Delta y/\Delta x$, for the average rate of change, will always be equal to *twice the initial value of x plus the increment in x*. We can check the result with our example using the arithmetic numbers, $x = 4$, and $\Delta x = 3$. Then

$$\frac{\Delta y}{\Delta x} = 2(4) + 3 = 11$$

The result is the same as we get for the ratio by using arithmetic numbers.

The rule or formula (for this problem) can now be used to find this ratio for any initial values and for any increments. For example, if we begin with an initial value of $x = 5$, and then change x by an increment from 5 to 5.4, the ratio will be

$$\frac{\Delta y}{\Delta x} = 2(5) + 0.4 = 10.4$$

If we change x from 8 to 7.7, then the average rate of change will be

$$\frac{\Delta y}{\Delta x} = 2(8) + (-0.3) = 15.7$$

Note: If we understand that the x in the final formula represents *any* initial x-value, then it is not necessary to use subscripts in the equation.

Then we follow these steps:

(1) Replace x with $x + \Delta x$ and replace y with $y + \Delta y$, in the original equation. Expand the result if necessary.
(2) Subtract the original equation.

142 Increments: Average Rates

(3) Divide both sides of the result by Δx. Now we have the ratio $\Delta y/\Delta x$, which we want. The expression then shows the rule for finding this ratio in the particular equation involved.

EXAMPLE 2. Given the function $y = x^2 + 4x + 3$, derive the equation that shows the rule for finding the ratio $\Delta y/\Delta x$. Check by actual substitution to show the ratio when x changes from 7 to 7.3.

Solution. Replacing x with $x + \Delta x$ and y with $y + \Delta y$ we get

$$y + \Delta y = (x + \Delta x)^2 + 4(x + \Delta x) + 3$$

Expanding, $\quad y + \Delta y = x^2 + 2x(\Delta x) + (\Delta x)^2 + 4x + 4(\Delta x) + 3$

We have $\quad\quad y \quad\quad = x^2 \quad\quad\quad\quad\quad\quad\quad\quad + 4x \quad\quad\quad\quad + 3$

Subtracting, $\quad\quad \Delta y = \quad\quad + 2x(\Delta x) + (\Delta x)^2 \quad\quad + 4(\Delta x)$

Dividing by Δx, $\quad \dfrac{\Delta y}{\Delta x} = \quad\quad 2x \quad + (\Delta x) \quad\quad\quad + 4$

If we now understand that x may represent any initial value of x, then the result is the rule for finding the ratio of increments, or average rate of change in y for any increment in x. As an example in applying the rule, we take $x = 7$ and $\Delta x = 0.3$. Then the ratio

$$\frac{\Delta y}{\Delta x} = 2(7) + 0.3 + 4 = 18.3$$

We check the result by direct substitution in the original equation:

when $\quad x = 7, y = 49 + 28 + 3 = 80;$

when $\quad x = 7.3, y = (7.3)^2 + 4(7.3) + 3 = 53.29 + 29.2 + 3 = 85.49$

and $\quad \Delta y = 85.49 - 80 = 5.49.$

Then the ratio $\Delta y/\Delta x = 5.49/0.3 = 18.3$.

As another application of the rule in this equation, when x changes from 9 to 9.5, the average rate of change of y with respect to x is

$$\frac{\Delta y}{\Delta x} = 2(9) + 0.5 + 4 = 22.5$$

That is, over the interval, y changes 22.5 times as much as x.

EXAMPLE 3. Derive the rule for finding the ratio $\Delta y/\Delta x$ for the function

$$y = \frac{24}{x + 3}$$

Then use this formula for finding the numerical value of this ratio when x changes from $x = 1$ to $x = 5$. Finally, check the result by computing the value of y for each value of x.

Solution. Replacing x with $(x + \Delta x)$ and y with $(y + \Delta y)$, we get

$$y + \Delta y = \frac{24}{x + \Delta x + 3}$$

Subtracting, $\quad\quad \Delta y = \dfrac{24}{x + \Delta x + 3} - \dfrac{24}{x + 3}$

10.5 A Formula for $\Delta y/\Delta x$ 143

Now we must combine the two fractions on the right into a single fraction. The common denominator is $(x + \Delta x + 3)(x + 3)$. Then we get

$$\Delta y = \frac{24(x + 3) - 24(x + \Delta x + 3)}{(x + \Delta x + 3)(x + 3)}$$

Simplifying, $\Delta y = \dfrac{-24(\Delta x)}{(x + \Delta x + 3)(x + 3)}$

Dividing both sides by Δx,

$$\frac{\Delta y}{\Delta x} = \frac{-24}{(x + \Delta x + 3)(x + 3)}$$

As x changes from 1 to 5, $\Delta x = 4$. Then

$$\frac{\Delta y}{\Delta x} = \frac{-24}{(1 + 4 + 3)(1 + 3)} = -\frac{3}{4}$$

Figure 10.4 shows these changes in x and y.

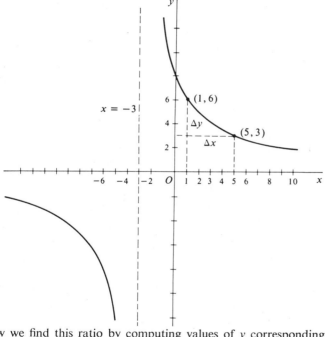

Fig. 10.4

Now we find this ratio by computing values of y corresponding to the values of x. When $x = 1$, $y = 6$; when $x = 5$, $y = 3$. For increments we have

$$\Delta x = 5 - 1 = 4; \qquad \Delta y = 3 - 6 = -3$$

For the ratio of the increments, we have

$$\frac{\Delta y}{\Delta x} = \frac{-3}{4} = -\frac{3}{4}$$

EXERCISE 10.1

1. An object fell 400 feet in 5 seconds and 1024 feet in 8 seconds. Find its average rate of fall in feet per second over the 3-sec interval.

Increments: Average Rates

2. The odometer of a car showed 3480 miles at 4 PM and 3641 miles at 7:30 PM. What was the average velocity of the car over the time period?
3. The speedometer of a car showed a velocity of 38.2 mph at exactly 2 PM. Fifteen seconds later it showed 45.7 mph. What was the average acceleration (change in velocity) in mph per second for the 15-sec period?
4. A speedometer showed 39.6 mph at 3:32 PM. At 3:33 PM it showed a velocity of 24.6 mph. What was the average acceleration in mph per second?
5. At a particular instant a car was traveling at a rate of 88 ft/sec. Twelve seconds later it was traveling at 70 ft/sec. What was its acceleration in ft/sec per second through the 12-sec interval?
6. A tank full of salt brine tested 50% salt. Fresh water was run into the tank and at the same time brine was run out at the same rate, keeping the solution continuously stirred and the tank full. At the end of 40 minutes the solution tested 34%. What was the average rate of change per minute in the per cent of salt content?
7. The temperature at a certain point on a warm day rose from 30° below zero to 18° below zero from 8 AM to 3 PM. What was the average rate of change per hour in temperature? Would it be correct to show this change by a *continuous* graph? (Is temperature change a continuum?)
8. The population of a town increased from 50,000 to 55,110 in one year. What was the average rate of increase per month? per week? per day? Would it be correct to show this increase by a continuous graph? (Is population increase a continuum?)
9. The charge on a capacitor changed from 0.00004 coulomb to 0.000005 coulomb in 0.001 second. What was the average rate of change of charge with respect to time (per second)?
10. A man ran a mile in 3 minutes 58 seconds. What was his rate per second? What was his average time per 100 yards?
11. In a particular electric circuit with a constant voltage, the power changed from 400 watts to 190 watts as resistance R changed from 20 ohms to 80 ohms. What was the average rate of change in power per ohm change in R?
12. Barometric air pressure changed from 30.04 inches to 29.88 inches in 8 hours. Find the average rate of change per hour.
13. A field measured 60 rods by 40 rods. If the length and width were changed by 5 rods each, find the average change in area per rod of change in dimensions.
14. In 5 days a certain watch lost 12 minutes. What was the average loss in time per hour?

In each of the following exercises find the rule for the ratio $\Delta y/\Delta x$ as the average rate of change and find the numerical value of this ratio over the interval indicated. Finally, check each result by computing the value of y in each case and the ratio $\Delta y/\Delta x$:

15. $y = x^2$, from $x = 6$ to $x = 7.5$.
16. $y = x^2 + 3x$, from $x = 1$ to $x = 4$.
17. $y = x^3 - 2x$, from $x = 5$ to $x = 7$.
18. $y = 8/x$, from $x = 1$ to $x = 4$.
19. $y = 16/x^2$, from $x = 2$ to $x = 4$.
20. $y = x^2$, as x changes from 4 to 7; 4 to 6; 4 to 5; 4 to 4.5; 4 to 4.1.
21. $y = 3x + 4$, as x changes from 1 to 3; 3 to 5; 5 to 7; -2 to 0.
22. $y = \dfrac{12}{(x+2)}$, as x changes from 0 to 1; 1 to 2; 2 to 4; 4 to 6.

In each of the following exercises, derive the rule for the rate of change in the function with respect to the independent variable and then compute the numerical value of this rate over the interval indicated.

23. $I = t^2 + 2t$, as t changes from 3 seconds to 3.4 seconds.
24. $i = 4 - t^2$, as t changes from 4.2 seconds to 4.5 seconds.
25. $v = 3t - 4$, as t changes from 6.5 seconds to 8 seconds.
26. $v = t^2 + 4t - 3$, as t changes from 3.6 seconds to 3.2 seconds.
27. $s = t^2 - 5t + 2$, as t changes from 4 seconds to 5.2 seconds.
28. $s = t^3 - 2t^2 - 5t$, as t changes from 5 seconds to 6 seconds.
29. $I = E/R$; $E = 120$; R changes from 10 ohms to 30 ohms.
30. $P = I^2R$; $R = 30$ ohms; I changes from 6 amperes to 7.5 amperes.

chapter

11

Limits

11.1 DEFINITION

In everyday use the word *limit* refers to a quantity that may be reached but not exceeded. When we say the speed of an automobile has a limit of 80 miles per hour, we mean that a speed of 80 miles per hour can be reached but not exceeded. This is the ordinary meaning of the word *limit*.

However, a *mathematical limit* has a different meaning. It refers to a constant that may be approached as closely as we wish but cannot actually be reached. To show the difference in the meaning of the term *limit* we use an example.

Suppose you wish to buy a particular article. The owner has set a price of, say, $64. Inwardly you decide that you will pay that much but you hope to get it for less. Your first offer is, say, $32, but the owner will not sell. You increase your offer to, say, $50, and when this is refused, you offer $60, which is also rejected. After much time taken up with offers and refusals, you finally pay $64. You originally set a limit of $64, and now you have reached it. This is the usual meaning of the word *limit*.

Now let us see the difference in the meaning of the term *mathematical limit*. Let us begin as before, but this time you set up a particular rule that you will follow in making your offers. Suppose, as before, the owner set his price at $64. Again your first offer is $32, but this is refused. Now you decide that each time you make a new offer, you will meet the owner halfway. That is, you will "split the difference." Your second offer is halfway from $32 to $64. Then your second offer is $48, but the owner sticks to his original price of $64. If you follow the rule that each new offer you make will represent an increase of one-half the difference between your previous offer and the owner's price, then your successive offers will be $32; 48; 56; 60; 62; 63; 63.50; 63.75; 63.875; 63.9375; and so on indefinitely. Your offers and his price may be shown on a horizontal scale (Fig. 11.1). In this example your offer is a variable quantity that is always approaching 64 but never reaches 64. Then we say that 64 is the *mathematical limit* of the *variable*, your offer. Note that the *difference* between

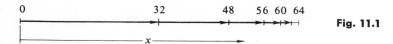

Fig. 11.1

your offer and the limit 64 becomes smaller and smaller and approaches but never reaches zero.

If we represent your offer, the variable, by the letter x, then we can say that this variable x approaches the constant in such a way that the absolute value of the difference, $|64 - x|$, becomes smaller and smaller but never reaches zero. This difference may be made as small as we wish (that is, *less than any specified value however small*), and this *difference never increases*.

We can now formulate the definition of the mathematical limit of a variable.

Definition:

If a variable x approaches a constant K in such a way that the absolute value of the difference, $|K - x|$, between the constant and the variable can be made as small as we wish but not zero, then the constant K is called the limit of the variable x. Note that $|x - K|$ is the same as $|K - x|$.

The idea of approaching is often indicated by an arrow. In the foregoing example, we can write

$$x \to 64$$

The expression is read "x approaches 64 as a limit."

A constant K is the limit of a variable x provided the following conditions hold:

1. *The variable x moves closer and closer to the constant K.*
2. *The variable may be made as close to the constant as we wish.*
3. *The variable is never equal to the constant K.*
4. *The difference between the constant and the variable becomes and remains less than any finite quantity, however small. This difference, $|K - x|$, approaches zero but is never equal to zero.*

Under these conditions, K is called the limit of the variable x. We express the idea mathematically in symbols as

$$\lim_{x \to K} x = K$$

The expression is read: "*The limit of x is K as x approaches K.*"

Warning: Do not say "the limit approaches K." *This is wrong.* The limit does *not* move. It does *not* approach K; it *is* K. *It is the variable that approaches K. The limit is K.*

Since the difference $|K - x|$ approaches zero, we can write

$$\lim_{x \to K} |K - x| = 0$$

It is possible for a variable to approach a limit by decreasing rather than increasing. In the example mentioned, if you had stuck to your original offer of $32, and if the owner had decided to meet you halfway, his second price would have been $48. If he had decided that he would come down halfway to your offer for each successive price he quoted, then his price would have been a *variable* and your offer of $32 would have been the mathematical limit. His price, the variable, would approach but never reach the constant 32. This condition is represented in Figure 11.2. Note again that the difference between the variable and the limit becomes less and less and can be made as near zero as we please, yet never equal to zero. If we write $(32 - x)$ for the difference, this difference is negative. However, since we are concerned with absolute values of the difference, we write

$$|32 - x| \to 0$$

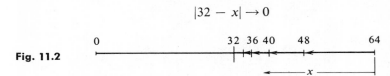

Fig. 11.2

Note: It might be well at this point to give some careful thought to the concept of *approaching*. We may think of it in one of two ways. When we say that *x approaches 64 as a limit*, do we mean that x moves along continuously and in so doing occupies every point along the way, forming a continuum? This is probably the usual understanding of the term.

However, there is another way of looking at the meaning of a limit. From the second viewpoint, we first consider the difference between the variable and its limit, such as $|64 - x|$. Now we assume some small value we might call *delta* (δ), say, 0.001. Then we say this difference, $|64 - x|$, can be made less than δ, but not zero. If the difference, $|64 - x|$, can be made less than any small δ we might wish to name, however small (but not zero), then we say the limit of x is 64. Since the difference is not zero, we can state the requirement in symbols as follows:

$$0 < |64 - x| < \delta$$

11.2 INFINITESIMAL

If the limit of a variable is zero, the variable is called an *infinitesimal*. The term *infinitesimal* in mathematics does not necessarily mean a small quantity. It is often used erroneously to refer to a very small quantity, such as 0.000001 milligram, or 0.000001 second. The term does not mean a quantity of any particular size but only a quantity that approaches zero as a limit.

If a variable x approaches some constant K as a limit, then the difference

$|K - x|$ decreases and approaches zero as a limit. The difference $|K - x|$ is therefore an infinitesimal. In symbolic notation,

$$\lim_{x \to K} |K - x| = 0$$

11.3 INFINITY

One fact should be noted here: It is possible for a variable to change in such a way that it increases or decreases without approaching a limit. In this case the variable is said to increase or decrease *without limit* or *without bound*. If the variable *increases* without bound, we say it becomes infinite *positively*. If the variable *decreases* without bound, we say it becomes infinite *negatively*. We usually indicate these conditions with arrows and the infinity sign as

$$x \to \infty \quad \text{or} \quad x \to -\infty$$

When used in this connection, the arrow should *not* be read "approaches." It is not correct to say that a variable "*approaches* infinity" because *infinity cannot be approached*. A variable cannot move nearer and nearer to infinity. If we count to a billion, we are no nearer to infinity than when we count to ten. Moreover, the *difference* between the variable and infinity does *not* approach zero as a limit, that is,

$$|\infty - x| \text{ cannot be made as small as we wish}$$

The use of the infinity sign confuses some students. Much confusion can be avoided if the expression

$$x \to \infty$$

is read "x increases without limit," or "x increases without bound," or "x becomes infinitely large."

11.4 THE LIMIT OF A FUNCTION

To understand the meaning of the limit of a function, let us take some specific function of x. Suppose we take the function, $x^2 - 1$. Now let us see what happens to this function as x approaches some constant, say, 3. We are not concerned with the value of the function when $x = 3$. We have emphasized that when we say x approaches 3, we do not mean that x is ever equal to 3. By the definition of a limit, we mean that x may get as close to 3 as we wish but never reach it.

Suppose we take some values of x that are successively closer to 3 and then compute the corresponding values of the function. To indicate that a variable approaches a limit from below (that is, through values less than the limit), we use the negative sign next to the limit. To show that x approaches 3 from below, we write

$$x \to 3^-$$

We may use these values:

if x =	0	1	1.5	2	2.5	2.8	2.9	2.99	→ 3
then $x^2 - 1$ =	−1	0	1.25	3	5.25	6.84	7.41	7.9401	→ ?

Note in the table that x approaches 3 through values *less* than 3.

To indicate that x approaches 3 from above (that is, through values greater than 3), we write

$$x \to 3^+$$

The following table shows x approaching 3 from above:

if x =	6	5	4	3.5	3.2	3.1	3.01	→ 3
then $x^2 - 1$ =	35	24	15	11.25	9.24	8.61	8.0601	→ ?

Now it appears that as x moves closer to 3, whether from below or from above, the function value moves closer to 8. In fact, in this instance, if x is exactly 3, the value of the function is exactly 8. However, we are interested in what happens to the function when x *approaches* 3, *not* when $x = 3$. In this example, it is indeed true that the function approaches 8 as a limit. That is, as x approaches 3, the function $f(x)$ approaches $f(3)$.

We might think that all we need to do to find the limit of a function is to substitute the limit of x in the function. As a matter of fact, this can be done for most functions, provided the result has meaning. For example, for the function, $f(x) = x^2 - 5x + 7$, as x approaches the limit 6, the function approaches the limit $f(6) = 36 - 30 + 7 = 13$. If x approaches the limit a, then the function approaches the limit $f(a)$. However, we shall see that simple substitution of this kind is not always possible.

The fact that the function $f(x) = x^2 - 1$ approaches the limit 8 as the variable x approaches its limit 3, is indicated as follows in symbols:

$$\lim_{x \to 3} (x^2 - 1) = 8$$

One usual way of showing the meaning of the limit of a function is by the so-called "epsilon-delta" method. We have seen that when we say x approaches 3 we mean that the quantity $(x - 3)$ approaches zero. That is to say,

$$0 < |x - 3| < \delta$$

where δ is some small quantity. When we say the function $(x^2 - 1)$ approaches 8, we mean that the *difference* between the function and 8 also approaches zero. Now we say that this difference is less than some other small quantity we call epsilon (ϵ), that is,

$$|(x^2 - 1) - 8| < \epsilon$$

Now, if we assume some small quantity for ϵ, our question is this: Can we find a δ small enough so that the function will differ from 8 by less than any ϵ we wish to name? If we can find a δ such that x differs from 3 by an absolute value less than δ, to make sure the function differs from 8 by less than any ϵ we might choose, then the limit of the function is 8. In general terms, if there is a small quantity δ such that

$$0 < |x - a| < \delta$$

for any ϵ such that $\quad |f(x) - Q| < \epsilon$

then we say that the limit of $f(x)$ is Q as x approaches a, and in symbols,

$$\lim_{x \to a} f(x) = Q$$

11.5 THE LIMIT OF A FUNCTION: A GRAPH

The meaning of the limit of a function of x may be shown graphically. Let us sketch the graph of the general function $y = f(x)$ (Fig. 11.3). Now let us suppose that x approaches some constant a from either lesser or greater values in such a way that x differs from a by some small variation less than some δ. Let us assume some quantity Q, representing the particular function value in the neighborhood of $x = a$. It may be that Q is equal to $f(a)$. However, we do not say $x = a$. We say only that x approaches a so that

$$0 < |x - a| < \delta$$

Then $f(x)$ differs from Q by some small variation less than ϵ.

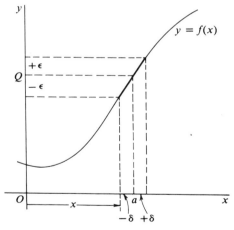

Fig. 11.3

Now suppose we wish the function $f(x)$ to have a value within some small specified variation from Q. Let us say we wish the variation to be less than any ϵ we might name. That is, we wish

$$|f(x) - Q| < \epsilon$$

If we make δ small enough, it may be that $f(x)$ will meet this requirement. If it is possible to find a δ small enough to make $f(x)$ differ from Q by less than any ε we might name, then we say the limit of $f(x)$ is Q as x approaches a. That is,

$$\lim_{x \to a} f(x) = Q$$

provided there is a δ such that

$$0 < |x - a| < \delta$$

for any ε such that

$$|f(x) - Q| < \epsilon$$

11.6 THE FUNCTION AS A MACHINE

We might think of the function of x as a machine. Let us say this machine manufactures some particular metal product, say, rivets. The machine is constructed so that the length of the rivets is controlled by a dialed mechanism (Fig. 11.4).

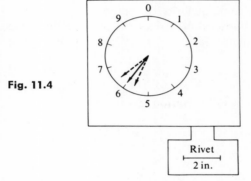

Fig. 11.4

Now suppose we are told that the rivets must be approximately 2 inches long and that any variation from this length of 2 inches must be something less than 0.03 inch. We might represent this variation in the product by *epsilon* (ε). That is, the function product may have a slight variation in length from the prescribed 2 inches, but the variation must be less than 0.03 inch, that is, less than ε.

Now the control dial is set as near as possible to produce the desired result. Let us suppose a dial setting of exactly 6 will produce a rivet exactly 2 inches long; but the dial is unsteady. Actually, the setting is a variable. Now our question is: Is it possible to set the dial so that its variation from 6 will produce a variation of less than 0.03 inch in the product? Let us call the variation in the dial setting *delta* (δ). Now we wish the dial setting x to vary from 6 by less than δ so that the product will vary from 2 inches by something less than ε, or 0.03 inch. Perhaps a dial variation of less than 0.1 will produce a rivet variation of less than 0.03 inch in length.

To summarize, the dial setting is the variable x. The function of x, $f(x)$, is the product, a rivet. The variation in x is δ. The variation in the function we call ϵ. Now, if it is possible to get a δ small enough so that the variation in the function will be less than any we wish, no matter how small an ϵ we might specify, then as the dial setting approaches 6, the rivet length approaches 2, and we have

the limit of the function is 2, as the limit of x is 6

A variation in a variable usually produces a variation in the function. However, in practice, we first select the required ϵ, the variation in the product, and then try to find a δ to meet the requirement.

11.7 THEOREMS ON LIMITS

It will be helpful at this time to know certain facts concerning limits. We have seen that as x approaches a limit, then a function of x may or may not approach a limit. Let us suppose now that we have two functions of x, say, $f(x)$ and $g(x)$, such that both these functions approach limits. Suppose further that the limit of $f(x)$ is 16 and the limit of $g(x)$ is 2. For convenience, let us call the functions simply f and g. Then we have

f approaches 16 and g approaches 2

Now our first question is: Since each of the functions f and g approaches a limit, does the sum of the functions $f + g$ approach a limit? If so, what is the limit of $(f + g)$? Our first theorem states

$$\lim_{\substack{f \to 16 \\ g \to 2}} (f + g) = \lim_{f \to 16} f + \lim_{g \to 2} g = 16 + 2 = 18$$

(1) That is, *the limit of the sum of two functions of x that approach respective limits is equal to the sum of the limits.* In general terms, if the limit of f is a and the limit of g is b, then

$$\lim_{\substack{f \to a \\ g \to b}} (f + g) = \lim_{f \to a} f + \lim_{g \to b} g = a + b$$

EXAMPLE 1. Show that the foregoing theorem is true if x approaches 5 in the two functions:

$$f(x) = x^2 - 4, \quad \text{and} \quad g(x) = x - 2$$

Solution. Each limit can be found by substitution. As x approaches 5, $f(x)$ approaches $f(5)$, or $25 - 4 = 21$, and $g(x)$ approaches $g(5) = 5 - 2 = 3$. Then the sum of the limits is $21 + 3 = 24$. Now let us find the limit of the sum of the two functions. For the sum we have

$$f(x) + g(x) = x^2 - 4 + x - 2 = x^2 + x - 6$$

Then the limit of $(f + g)$ is the limit of $(x^2 + x - 6) = 25 + 5 - 6 = 24$.

154 Limits

Note: It might be well here to mention a point that applies also to other theorems concerning limits. If one (or both) of two functions does *not* approach a limit, then the sum of the functions does *not* approach a limit. For example, if the function f increases without limit (or without bound) and if the function g approaches the limit, say, 2, then the sum of the functions, $f + g$, does *not* approach a limit but instead increases without bound. That is,

$$\lim_{\substack{f \to \infty \\ g \to 2}} (f + g) = \lim_{f \to \infty} f + \lim_{g \to 2} g = \infty + 2 = \infty$$

(2) *If two functions of x approach respective limits, then the difference between the functions approaches the difference between the limits.* That is, as x approaches a limit, if $f(x)$ approaches a and $g(x)$ approaches b, then

$$\lim (f - g) = \lim f - \lim g = a - b$$

Illustration: If $f(x) = x^2 - 4$ and $g(x) = x - 2$, then, as x approaches 5, we have

$$\lim_{x \to 5} f(x) = \lim_{x \to 5} (x^2 - 4) = f(5) = 21 \qquad \lim_{x \to 5} g(x) = \lim_{x \to 5} (x - 2) = 3$$

Then the difference between the limits is $21 - 3 = 18$. Now, if we first find the difference between the functions, we get the same result:

$$\lim_{x \to 5} (f - g) = \lim_{x \to 5} (x^2 - 4 - x + 2) = 25 - 4 - 5 + 2 = 18$$

(3) *The limit of the product of two functions of x that approach respective limits is the product of the limits.* That is, if f approaches a and g approaches b, then

$$\lim (f)(g) = (\lim f)(\lim g) = (a)(b)$$

(4) *The limit of the quotient of two functions that approach respective limits is equal to the quotient of the limits, provided the limit of the divisor is not zero.* That is, if f approaches a and g approaches b, then

$$\lim (f/g) = (\lim f) \div (\lim g) = a/b$$

(5) *The sum of a function and a constant will approach a limit if the function itself approaches a limit.* For example, if the function $f(x)$ approaches the limit 21, then $f(x) + 7$ will approach the limit $21 + 7 = 28$. In general terms, if c is any constant and f is any function of x, then

$$\lim_{f \to a} (f + c) = (\lim_{f \to a} f) + c = a + c$$

(6) *The product of a constant and a function f will approach a limit provided the function f approaches a limit.* For example, if the function f approaches the limit 21, then 3 times the function f will approach the limit $(3)(21) = 63$. In general, if c is any constant, then

$$\lim_{f \to a} cf = (c)(\lim_{f \to a} f) = (c)(a)$$

11.7 Theorems on Limits

(7) *If two functions, f and g, are always equal to each other, and if they both approach respective limits, then the limits are equal.* That is, if $f = g$, and if f approaches a, and g approaches b, then $a = b$.

(8) *If a function $f(x)$ approaches some limit L_1 and also approaches another limit L_2, then the limits are equal,* that is, $L_1 = L_2$, *provided the function is continuous.*

EXAMPLE 2. Find the limit of the following function by use of the theorems on limits:
$$\lim_{x \to 3} (x^2 + 5x + 2)$$

Solution. The function represents the *sum* of two functions and a constant. Then by Theorems 1 and 5 we have
$$\lim_{x \to 3} (x^2 + 5x + 2) = \lim_{x \to 3} x^2 + \lim_{x \to 3} 5x + 2$$
Now note that x^2 is the product of two functions and $5x$ is the product of a constant and a function. Then by Theorems 3 and 6 we have
$$\lim_{x \to 3} x^2 = (\lim_{x \to 3} x)(\lim_{x \to 3} x) \quad \text{and} \quad \lim_{x \to 3} 5x = (5) \lim_{x \to 3} x$$
Then by a combined use of limits we get
$$\lim_{x \to 3} (x^2 + 5x + 2) = (\lim_{x \to 3} x)(\lim_{x \to 3} x) + (5)(\lim_{x \to 3} x) + 2$$
$$= (3)(3) + (5)(3) + 2 = 26$$

In the foregoing example the limit of the function can be found by simple substitution of the limit of x in the function. This can be done *provided the limit of the function has meaning.*

EXAMPLE 3. Verify the first four theorems concerning limits by use of the following functions: $f(x) = x^2 + 1$; $g(x) = x - 3$; x approaches 5.

Solution. For the limits of the functions we have
$$\lim_{x \to 5} f(x) = 25 + 1 = 26 \qquad \lim_{x \to 5} g(x) = 5 - 3 = 2$$

Theorem 1 says:
$$\lim (f + g) = \lim f + \lim g$$
Now we already have the limits: $\lim f = 26$; $\lim g = 2$;

then $\lim f + \lim g = 26 + 2 = 28$, the sum of the limits

For the sum of the functions, $f + g = x^2 + 1 + (x - 3) = x^2 + x - 2$. Then
$$\lim (f + g) = \lim (x^2 + x - 2) = 25 + 5 - 2 = 28$$
the limit of the sum.

Theorem 2 says:

$$\lim (f - g) = \lim f - \lim g$$

Now $\lim f - \lim g = 26 - 2 = 24$. This should be the limit of $(f - g)$. For the difference between the functions, $f - g = (x^2 + 1) - (x - 3) = x^2 - x + 4$. Then

$$\lim_{x \to 5} (f - g) = \lim_{x \to 5} (x^2 - x + 4) = 25 - 5 + 4 = 24$$

Theorem 3 says:

$$\lim (f)(g) = (\lim f)(\lim g)$$

For the product of the limits we have $(\lim f)(\lim g) = (26)(2) = 52$. For the product of the functions, $(f)(g) = (x^2 + 1)(x - 3) = x^3 - 3x^2 + x - 3$. Then

$$\lim_{x \to 5} (f)(g) = \lim_{x \to 5} (x^3 - 3x^2 + x - 3) = 125 - 75 + x - 3 = 52$$

Theorem 4 says:

$$\lim (f \div g) = (\lim f) \div (\lim g)$$

For the quotient of the limits we have $(\lim f) \div (\lim g) = 26 \div 2 = 13$. For the quotient of the functions,

$$f \div g = (x^3 + 1) \div (x - 3) = x + 3 + \frac{10}{(x - 3)}$$

Then

$$\lim_{x \to 5} (f \div g) = \lim_{x \to 5} \left(x + 3 + \frac{10}{x - 3} \right) = 5 + 3 + \frac{10}{2} = 13$$

Note that Theorem 4 contains the provision that the limit of the divisor cannot be zero. If the limit of the divisor is zero in any problem, then the division rule fails.

EXERCISE 11.1

Verify the first four theorems concerning limits by use of the following functions and their indicated limits:

1. $f(x) = x^2 - 7$; $g(x) = x - 2$; x approaches 5 as a limit.
2. $f(x) = 3x^2 - 3x + 2$; $g(x) = x + 3$; x approaches the limit 7.
3. $f(x) = x^2 - 4x - 12$; $g(x) = x + 2$; x approaches the limit 6.
4. $f(x) = 2x^2 - 7x + 3$; $g(x) = x - 3$; x approaches the limit 3.
5. $f(x) = x - 4$; $g(x) = x^2 + 3x - 6$; x approaches the limit 4.
6. $f(x) = x^3 - 5x$; $g(x) = x - 3$; x approaches the limit 5.
7. $f(x) = 2x - 9$; $g(x) = x - 2$; x approaches the limit 0.
8. $f(x) = x^2 + 3x - 10$; $g(x) = x + 5$; x approaches the limit 0.

By taking successive values of x that approach the indicated limit from above and from below, find the limits of the following functions and show when the limit

11.8 Two Special Cases

can be found by simple substitution, that is, when the limit of $f(x)$ is equal to $f(a)$ as x approaches a:

9. $\lim\limits_{x \to 3} (2x + 5)$
10. $\lim\limits_{x \to 2} (x^2 + 3)$
11. $\lim\limits_{x \to 3} (x^2 - 2)$
12. $\lim\limits_{x \to 2} (x^3 + 1)$
13. $\lim\limits_{x \to 0} (3x + 5)$
14. $\lim\limits_{x \to -2} (x^2 + 5x)$
15. $\lim\limits_{x \to 2} \dfrac{x^2 - 4}{x - 2}$
16. $\lim\limits_{x \to 2} \dfrac{x^2 - 4}{x + 2}$
17. $\lim\limits_{x \to 2} \dfrac{x + 6}{x + 2}$
18. $\lim\limits_{x \to -3} (x^2 + 2x + 1)$
19. $\lim\limits_{x \to 2} (4 - x^2)$
20. $\lim\limits_{x \to -2} (8 - x^3)$

11.8 TWO SPECIAL CASES

There are two special cases involving limits that should be clearly understood at this time. We have seen that we can often find the limit of a function of x by simple substitution. To find the limit of the function $x^2 + 7x + 6$ when x approaches zero, we simply substitute zero for x in the function and get 6 as the limit of the function. In such cases, as x approaches a constant a, the function $f(x)$ approaches the limit $f(a)$.

However, in some cases, simple substitution has no meaning. For example, consider the following limit:

$$\lim_{x \to 0} \left(\frac{12}{x} \right)$$

In this example, if we attempt simple substitution of zero for x in the function, we get $(12/0)$, which has no meaning. We might be inclined to say that $12 \div 0$ is infinity (∞), yet this is not true because we cannot divide by zero.

To get some understanding of the meaning of the expression, let us take values of x closer and closer to zero, but not equal to zero. First, let us begin with positive values and let x approach zero from above; that is,

$$x \to 0^+$$

We may take the following corresponding sets of values:

if $x =$	12	6	4	3	2	1	0.5	0.25	0.1	0.01	0.00001	$\to 0$
then $\dfrac{12}{x} =$	1	2	3	4	6	12	24	48	1200	1200	1,200,000	$\to ?$

Note that x approaches zero from the right. The changes in x may be shown by the following line diagram:

Note that as x approaches the limit *zero*, the value of the function $12/x$ becomes larger and larger. We may take x as near to zero as we please, and the value of the function will continue to increase *positively* without limit or bound, that is,

as $x \to 0$, the function $\dfrac{12}{x}$ becomes infinitely large.

The result is sometimes shown by the following notation:

$$\lim_{x \to 0^+} \frac{12}{x} = \infty \text{ (infinity)}$$

In general terms, if c represents any constant and x represents a variable, then

$$\lim_{x \to 0^+} \frac{c}{x} = \infty \text{ (infinity)}$$

However, the expression should not be understood to mean that the function has a limit (infinity) because *infinity cannot be approached as a limit.* Instead the expression should be read: "As x approaches zero, the function $12/x$ becomes infinitely large or increases without limit."

In the function, $12/x$, it may be that x approaches zero *from the left*, that is, through negative values (or through values to the left of its limit). If we begin with some negative number, such as -12, and then let x move toward the limit *zero*, we may have values such as the following:

if $x =$	-12	-6	-4	-3	-2	-1	-0.5	-0.1	-0.01	$\to 0$
then $\dfrac{12}{x} =$	-1	-2	-3	-4	-6	-12	-24	-120	-1200	$\to\ ?$

The changes in x may be indicated graphically by the following line diagram:

$$-12 \quad -11 \quad -10 \quad -9 \quad -8 \quad -7 \quad -6 \quad -5 \quad -4 \quad -3 \quad -2 \quad -1 \quad 0$$

In this case, the successive values of x show an algebraic *increase* from -12 toward zero. The successive values of the function, $12/x$, show an algebraic *decrease*, that is, an increase in a negative direction. As the variable x approaches zero as a limit from the left, the value of the function becomes *negatively large* without limit or bound. A negative increase without limit can be indicated by the symbol, $-\infty$, a negative infinity sign.

In general, if c represents any constant and x any variable, then the foregoing limits may be indicated as follows:

$$\lim_{x \to 0^+} \frac{c}{x} = +\infty \quad \text{and} \quad \lim_{x \to 0^-} \frac{c}{x} = -\infty$$

As a *second special case* involving limits, let us consider again the function

$$\frac{12}{x}$$

11.8 Two Special Cases

Suppose now that x increases without bound. This fact can be indicated by the symbol

$$x \to \infty$$

Again, we should *not* read this "x approaches infinity," but rather as "x increases without limit or bound." Our question now is: What happens to the function $12/x$? Does the function approach a limit and, if so, what limit?

This time we let x take on larger and larger values. Suppose we begin with $x = 1$, in which case the function has a value of 12. For successive corresponding values, we may have the following:

if $x =$	1	2	3	6	12	24	120	1200	120,000	12,000,000	$\to \infty$
then $\dfrac{12}{x} =$	12	6	4	2	1	0.5	0.1	0.01	0.0001	0.000001	$\to 0$

Note that we can make the function $12/x$ as close to zero as we wish by taking values of x sufficiently large. Then, by definition, the limit of the function is *zero*. In this case the function actually has a limit *zero*, since zero is a finite number. In symbols, where c is a constant,

$$\lim_{x \to \infty} \frac{c}{x} = 0$$

If x becomes infinite *negatively*, we still have the same limit, zero, for the function, although the function approaches zero through negative values.

The following worked-out examples indicate some of the problems connected with determining the limits of functions.

EXAMPLE 4. Determine whether or not the following function has a limit as $x \to 2$ and find the limit if it exists:

$$f(x) = \frac{x^2 + 6}{x - 2}$$

Solution. We wish to find

$$\lim_{x \to 2} \frac{x^2 + 6}{x - 2}$$

If we try simple substitution of 2 for x in the function we get

$$\frac{4 + 6}{2 - 2} = \frac{10}{0}$$

The zero denominator indicates that substitution in this case has no meaning; that is, $f(2)$ does not exist. Since the numerator approaches 10 and the denominator approaches 0, it would appear that the function itself increases without bound and has a so-called limit of $+\infty$. To understand the behavior of the function as x approaches 2 from the right ($x \to 2^+$), let us set up a table of corresponding values of x and the function:

if $x =$	12	7	6	4	3	2.5	2.1	2.01	$\to 0$
then $f(x) =$	15	11	10.5	11	15	24.5	104.1	1004	$\to ?$

Note that as the value of x moves from 12 to 6 the value of the function decreases from 15 to 10.5. Between the values of $x = 6$ and $x = 4$, there is a slight further decrease until x reaches the value of approximately 5.16, at which point the function has a value of approximately 10.32. However, after passing this point, the value of the function increases and again reaches 11 when $x = 4$. From then on, the value of the function increases without bound as the value of x approaches the limit 2. Then we conclude that

$$\lim_{x \to 2^+} \frac{x^2 + 6}{x - 2} = +\infty$$

In the foregoing example, if we let x approach 2 from the left, we shall find that the function becomes infinite *negatively*. To see this happen we might use the following values of x: -1; 0; 0.5; 1; 1.5; 1.6; 1.9; and 1.99. The student should find the corresponding values of the function and then determine that the following is true:

$$\lim_{x \to 2^-} \frac{x^2 + 6}{x - 2} = -\infty$$

EXAMPLE 5. Determine whether or not the following function has a limit as x becomes infinite:

$$f(x) = (x^2 + 5x + 4)$$

Solution. We wish to find

$$\lim_{x \to \infty} (x^2 + 5x + 4)$$

In this problem substitution cannot be used because infinity cannot be substituted. We cannot operate with infinity as we operate with finite quantities. However, we can arrive at an answer by reasoning as follows: As x increases without bound, then x^2 will surely continue to increase. This can be shown by using greater and greater values of x and noting the values for x^2. We note also the increasing values of $5x$ as x continues to increase. The term x^2 may be said to represent an increase of the second order and will increase much faster than $5x$ for large values of x. The constant 4 will have little effect on the value of the function for large values of x. By similar reasoning for the sum of the terms, we can conclude that the entire function, $x^2 + 5x + 4$, will continue to increase without bound. The function therefore has no limit, and we can write

$$\lim_{x \to \infty} (x^2 + 5x + 4) = \infty$$

EXAMPLE 6. Find the following limit if it exists:

$$\lim_{x \to 3} \frac{x^2 - 9}{x - 3}$$

Solution. Here substitution of 3 in the function results in the fraction, 0/0, which is indeterminate. We might be inclined at first guess to say that the function has no limit. Yet if we substitute values of x closer and closer to 3, we shall find that the function does have a limit.

11.8 Two Special Cases

If x is equal to exactly 3, then we cannot divide the denominator into the numerator. However, *for any other value of x except 3*, the denominator is *not* equal to zero, and then we may divide and reduce the fraction. Dividing numerator and denominator by the quantity $(x - 3)$ we get the quotient $(x + 3)$. Now, for any value of x except 3, we can say the original fraction is equal to the quantity $(x + 3)$. Then we can say the limit of the original fraction is also the limit of the quotient. That is, for any value of x except 3,

$$\lim_{x \to 3} \frac{x^2 - 9}{x - 3} = \lim_{x \to 3} (x + 3) = 3 + 3 = 6$$

Moreover, if we insert values of x nearer and nearer to 3 in the function, we shall find that the original function will approach the value 6. That is, the value of the function can be made as close to 6 as we wish by taking the value of x sufficiently close to 3. Then, by definition, the limit of the function is 6. Note especially that the function is not defined for the value $x = 3$, yet it does have the limit 6. In this case we cannot say that the limit of the function is $f(3)$, as x approaches 3, since $f(3)$ is not defined.

EXAMPLE 7. Find the limit if it exists of the following function as indicated:

$$\lim_{x \to \infty} \frac{4x^2 + 5x + 6}{x^2 + 2x + 5}$$

Solution. In this function, as x becomes infinitely large, we see that the numerator $4x^2 + 5x + 6$ also increases without bound. The same is true for the denominator. In fact, we might indicate the result symbolically by the following notation:

$$\frac{\infty}{\infty} \quad \text{or} \quad \frac{\text{infinity}}{\text{infinity}}$$

Now we might be inclined to say that the limit does not exist, or we might even say: (infinity) ÷ (infinity) = 1. Yet we shall find the limit does exist and it is *not* equal to 1. Whether or not it has a limit cannot be determined by substitution.

However, let us divide the entire numerator and the entire denominator of the fraction by x^2, which is the *highest power* of x in either numerator or denominator. This operation is permissible because the divisor x is not zero but is some increasing quantity. Then the limit of the new fraction is the same as the limit of the original fraction. That is,

$$\lim_{x \to \infty} \frac{4x^2 + 5x + 6}{x^2 + 2x + 5} = \lim_{x \to \infty} \frac{\frac{4x^2}{x^2} + \frac{5x}{x^2} + \frac{6}{x^2}}{\frac{x^2}{x^2} + \frac{2x}{x^2} + \frac{5}{x^2}} = \lim_{x \to \infty} \frac{4 + \frac{5}{x} + \frac{6}{x^2}}{1 + \frac{2}{x} + \frac{5}{x^2}}$$

Now we let x increase without limit, that is, $x \to \infty$. As this happens we note that each of the four fractions, $5/x$, $6/x^2$, $2/x$, and $5/x^2$, approaches zero as a limit, as we have previously pointed out. Then the entire function approaches the fraction $4/1$, or 4, as a limit. That is,

$$\lim_{x \to \infty} \frac{4 + \frac{5}{x} + \frac{6}{x^2}}{1 + \frac{2}{x} + \frac{5}{x^2}} = \frac{4 + 0 + 0}{1 + 0 + 0} = 4$$

Limits

That a limit does exist may be seen intuitively by taking successively larger and larger values of x and then computing the value of the function

$$f(x) = \frac{4x^2 + 5x + 6}{x^2 + 2x + 5}$$

For example, we may have the following values:

if $x =$	0	1	2	3	4	5	10	100	1000	$\to \infty$
$f(x) =$	1.2	1.875	2.46	2.85	3.104	3.275	3.648	3.969	3.997	$\to 4$

EXERCISE 11.2

Determine each of the following limits if it exists:

1. $\lim\limits_{x \to 3} (x^2 - 5x - 7)$

2. $\lim\limits_{x \to 5} (3 + 4x - x^2)$

3. $\lim\limits_{x \to 2} (x^2 - 7x - 9)$

4. $\lim\limits_{x \to 3} (x^3 - 5x - 8)$

5. $\lim\limits_{x \to -3} (x^2 + 5x - 4)$

6. $\lim\limits_{x \to -4} (3x^2 - 5x - 7)$

7. $\lim\limits_{x \to 0} (3x - x^2 - 9)$

8. $\lim\limits_{x \to 0} (x^2 - 4x - 11)$

9. $\lim\limits_{x \to 3} (x^2 - 6x + 9)$

10. $\lim\limits_{x \to -3} (x^2 - 6x + 9)$

11. $\lim\limits_{x \to \infty} (x^2 + x + 4)$

12. $\lim\limits_{x \to \infty} (5 - 3x + x^2)$

13. $\lim\limits_{x \to \infty} (3 + 7x - x^2)$

14. $\lim\limits_{x \to -\infty} (x^2 + 2x + 3)$

15. $\lim\limits_{x \to 3} \dfrac{x - 3}{x^2 + 9}$

16. $\lim\limits_{x \to 3} \dfrac{x - 3}{x^2 - 9}$

17. $\lim\limits_{x \to -2} \dfrac{x + 2}{x^2 - 3x + 5}$

18. $\lim\limits_{x \to 2} \dfrac{3x^2 - 5x - 2}{x - 2}$

19. $\lim\limits_{x \to 5} \dfrac{x^2 - 25}{x + 5}$

20. $\lim\limits_{x \to 5} \dfrac{x^2 - 25}{x - 5}$

21. $\lim\limits_{x \to 0} \dfrac{x^2 - 2x - 4}{5x - 3}$

22. $\lim\limits_{x \to 0} \dfrac{x^2 + 7x + 8}{3x - 4}$

23. $\lim\limits_{x \to 2} \dfrac{3x^2 - 7x + 2}{x^2 - 4}$

24. $\lim\limits_{x \to 4} \dfrac{x^2 - 3x - 4}{2x^2 - 7x - 4}$

25. $\lim\limits_{x \to \infty} \dfrac{2x^2 + 7x - 5}{x^2 - 3x + 4}$

26. $\lim\limits_{x \to \infty} \dfrac{5x^2 + 4x - 3}{3x^2 - 5x + 7}$

27. $\lim\limits_{x \to 3^+} \dfrac{2x + 7}{x - 3}$

28. $\lim\limits_{x \to 3^-} \dfrac{2x + 7}{x - 3}$

29. $\lim\limits_{x \to \infty} \dfrac{x^2 + 5}{x^3 - 4x^2 - 1}$

30. $\lim\limits_{x \to \infty} \dfrac{x^2 - x - 5}{x + 5}$

31. $\lim\limits_{x \to \infty} \left(2 + \dfrac{1}{x}\right)$

32. $\lim\limits_{x \to \infty} \left(3x + \dfrac{1}{x} + \dfrac{1}{x^2}\right)$

33. $\lim\limits_{x \to 0} \left(2 + \dfrac{1}{x}\right)$

34. $\lim\limits_{x \to 0} \left(3x + \dfrac{1}{x}\right)$

35. $\lim\limits_{x \to 3^+} (x+1)^x$

36. $\lim\limits_{x \to 3^-} (x+1)^x$

37. $\lim\limits_{x \to 90°} \sin x$

38. $\lim\limits_{x \to 0} \tan x$

39. $\lim\limits_{x \to 0} \log x$

40. $\lim\limits_{x \to 90°} \tan x$

41. $\lim\limits_{x \to 0} (1+x)^x$

42. $\lim\limits_{x \to \infty} \left(1 + \dfrac{1}{x}\right)^x$

43. $\lim\limits_{x \to 5} \log_{10} (95 + x)$

44. $\lim\limits_{x \to 5} \log_{10} (5 - x)$

chapter

12

The Derivative: Delta Method

12.1 INSTANTANEOUS RATE OF CHANGE

We have already seen (Chap. 10) how to compute the average rate of change of a function. For example, if the function, $y = f(x)$ changes from 3 units to 9 units as x changes from 2 to 5 units, the average rate of change in y is $\Delta y/\Delta x$, or $\frac{6}{3} = 2$ units. That is, y changes *on the average* twice as much as x over the interval from $x = 2$ to $x = 5$. Now we are interested in the rate of change in y, the function, not as an average, but at the *exact instant* when x has some specific value, say, 2. We wish now to find *instantaneous rate of change*.

As another example, the distance an object falls in "free fall" is given by the formula

$$s = 16t^2 \text{ (approximately)}$$

We have seen how to compute the average rate of fall over a particular time interval. For example,

at $t = 3$ sec, $s = 144$ ft; at $t = 5$ sec, $s = 400$ ft

Then, when $\Delta t = 2$ sec, $\Delta s = 256$ feet. The average rate of fall during the given 2-seconds interval is $256/2 = 128$ ft/sec. We have emphasized that the 128 ft/sec represents the *average* rate of fall. This does not mean that the velocity was that rate every instant during the 2 seconds. In fact, we know that the velocity was changing at every instant of the time.

Now we are interested in the velocity at a particular instant, say, when t is exactly 3 seconds. This is *instantaneous rate of change*. If we remember that the velocity of a falling object continuously increases, then if we wish to find the rate of fall at a particular instant, it would almost seem that we should consider the change in distance when there is no change in time. Of course, the distance covered in zero time is also zero. To try to determine velocity using zero for distance and zero for time, of course, is useless since 0/0 is meaningless.

Consider the formula for power P in an electric circuit:

$$P = RI^2$$

We have seen how to compute average power and average rate of change of power with respect to current I. For example, if the resistance R is a steady 20 ohms, then when current I changes from 3 to 3.5 amperes, the average rate of change of power with respect to current is 130; that is, during the time when I changes from $I = 3$ to $I = 3.5$ the average change in power is 130 watts for each change of 1 ampere in current. However, what we wish now is not average rate of change but rather *instantaneous rate of change* of power at some particular instant, say, when $I = 3$ amperes.

In many problems in science it is extremely important to know the rate of change of a variable at some particular instant. If two variable voltages are impressed in a circuit, it may be necessary to know the exact and instantaneous rate of change in each so that the two together may produce the desired result. For a satellite in orbit, it is not enough that we know its average velocity; we must know the instantaneous velocity at any time. This is all the more true when two satellites are attempting to rendezvous in space. If we wish to know the force of a 5-pound weight falling to the ground from a height of 100 feet, we must know the instantaneous velocity at the exact instant of impact.

12.2 RATE OF CHANGE OF A FUNCTION

Suppose we graph the function $y = x^2$. From the graph (Fig. 12.1) it can be seen that the function, y, changes at a varying rate. For example, at $x = 0$, the increase in the function begins very slowly. At $x = 2$, the value of y increases at a much faster rate.

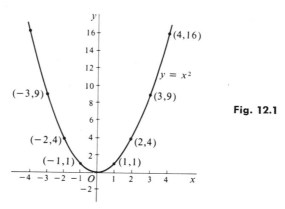

Fig. 12.1

Now, suppose we wish to know the instantaneous rate of change in y when x is equal to some particular value, say, 2. Starting with the value $x = 2$ and the corresponding value, $y = 4$, let us take some increment in x and compute the average rate of change in the function y. Usually for the increment, Δx, we take some small quantity, but it need not be small. Let us say first that we take the increment $\Delta x = 2$. Then the new value of x is 4, that is, $x + \Delta x$. The new value of y becomes 16. To get the value Δy, the increment in the function, we subtract the original y value and get

$$\Delta y = 16 - 4 = 12$$

Now we see that as x changes by 2 units, the function changes by 12. For the average rate of change we have $\Delta y / \Delta x = \frac{12}{2} = 6$. This is the average rate of change in the function as x changes from 2 to 4.

Suppose now we take smaller and smaller increments in x, such as 1.5, 1, 0.5, and so on. Then when we compute the corresponding values of the function y we shall find that the rate of change will be different for each increment in x. To get the instantaneous rate of change at the point where x equals 2, it would seem that the increments should be zero. This sounds almost like trying to find the ratio between changes when there is no change. However, we cannot take zero for Δx, for then we should get 0/0, which is impossible. Instead we let Δx *approach* zero and see what happens to the ratio $\Delta y / \Delta x$.

Let us set up a table showing smaller and smaller increments in x, with corresponding values of the function. Then we note what happens as we let Δx approach zero as a limit. We call the initial values x_1 and y_1. Starting with the function $y = x^2$, we have $x_1 = 2$ and $y_1 = 4$.

From Table 12.1 note that as Δx approaches zero as a limit, Δy also approaches zero. Now our chief concern is what happens to the ratio $\Delta y / \Delta x$. Remember, although two quantities may both approach zero as a limit, it does not follow that their ratio approaches zero. We might guess that the ratio between the increments approaches the constant 4 as a limit. In fact, this is true in this example. No matter how small we make the increment Δx, the ratio will come closer and closer to 4. It will

table 12.1

Δx	new x $x_1 + \Delta x$	new y $y_1 + \Delta y$	Δy	$\dfrac{\Delta y}{\Delta x}$
2.0	4.0	16.0	12.0	6.0
1.5	3.5	12.25	8.25	5.5
1.0	3.0	9.0	5.0	5.0
0.5	2.5	6.25	2.25	4.5
0.4	2.4	5.76	1.76	4.4
0.3	2.3	5.29	1.29	4.3
0.2	2.2	4.84	0.84	4.2
0.1	2.1	4.41	0.41	4.1
0.01	2.01	4.0401	0.0401	4.01
↓			↓	↓
0			0	?

never reach 4, but it can be made as close to 4 as we wish. Then, by the definition of a limit, 4 is the limit of the ratio $\Delta y/\Delta x$. This limit is called the *derivative* of the function.

Definition:

The derivative of a function $y = f(x)$ is the limit of the ratio $\Delta y/\Delta x$, or $\Delta f(x)/\Delta x$, as the increment Δx approaches zero as a limit. The derivative may then be called the instantaneous rate of change of a function.

Remember: The derivative is a *limit*. It is *not* any of the quotients in the fifth column of the table. It is the limit of this quotient $\Delta y/\Delta x$ as Δx approaches zero as a limit. *Do not say* the derivative approaches the limit; *the derivative is the limit itself.* In symbols, the derivative is

$$\lim_{\Delta x \to 0} \frac{\Delta y}{\Delta x}$$

12.3 DETERMINING THE DERIVATIVE: THE LIMIT OF $\Delta y/\Delta x$

The derivative is so important in mathematics that it is well to give some careful thought to this limit at this point. In fact, the entire field of differential calculus is the study of the derivative and its uses. The student would do well not to hurry through the reasoning connected with the derivative. Some students want to neglect the thinking and hurry on to the tricks in calculus. It is better to do a little extra thinking here and have a better understanding later.

To see how the limit, the derivative, may be found for different initial values of x, let us take a more general example. Instead of starting with the initial value $x = 2$, we begin with a more general initial value, $x = x_1$. The value x_1 is understood to represent some specific arbitrary constant value of x, yet it may have any value, such as 5, 8, or any other value. Then we have

$$x_1 = \text{initial value of } x \quad \text{and} \quad y_1 = \text{initial value of } y$$

Now let us begin again with the function $y = x^2$. Substituting initial values we begin with

$$y_1 = (x_1)^2$$

Again we set up a table showing a few successively smaller and smaller values of the increments, as well as the ratio between the increments (Table 12.2). Note that as the increments Δy and Δx both approach zero, the second term of the ratio also appears to approach zero. The ratio itself appears to approach the value $2x_1$ as a limit. This guess is correct. That is, the ratio approaches twice the initial value of x. This agrees with the previous example in which we began with the initial value $x = 2$ and discovered that the derivative had the value 4.

table 12.2

when $\Delta x =$	new x $x_1 + \Delta x$	new y $y_1 + \Delta y$	then Δy	$\dfrac{\Delta y}{\Delta x}$
2.0	$x_1 + 2$	$(x_1)^2 + 4x_1 + 4$	$4x_1 + 4$	$2x_1 + 2$
1.0	$x_1 + 1$	$(x_1)^2 + 2x_1 + 1$	$2x_1 + 1$	$2x_1 + 1$
0.5	$x_1 + 0.5$	$(x_1)^2 + x_1 + 0.25$	$x_1 + 0.25$	$2x_1 + 0.5$
0.1	$x_1 + 0.1$	$(x_1)^2 + 0.2x_1 + 0.01$	$0.2x_1 + 0.01$	$2x_1 + 0.1$
↓			↓	↓
0			0	?

Now we could proceed one step further. Instead of taking arithmetic increments, such as 2, 1, 0.5, and so on, we could take a general increment Δx and then try to determine what happens to the ratio when this increment approaches zero. In the foregoing example we should find that the ratio will in all cases be equal to the quantity, $2x_1 + \Delta x$, so that when Δx approaches zero, the ratio will in fact approach the value $2x_1$ as we expected. The foregoing example leads directly into the method of finding the derivative of any function, as explained in the following section.

12.4 DETERMINING THE DERIVATIVE OF ANY FUNCTION

One problem now is to determine the rule for finding the derivative of any function. Actually, the method is well indicated in the example already worked out in the preceding section. In finding the derivative we proceed exactly as we did in finding the average rate of change (Chap. 10). However, now we add one additional step. We let Δx approach zero and see what happens to the ratio $\Delta y / \Delta x$. The first three steps are exactly the same as those used in finding average rates of change (Sec. 10.5, Example 2).

It is important that we understand the meaning of the derivative. When we have found the ratio $\Delta y / \Delta x$, that is, the ratio of the increments, we note that this ratio approaches some limit. It is this *limit* that is defined as the *derivative*. *The derivative does not approach the limit; it is the limit.*

Consider again the simple quadratic function $y = x^2$. There are four steps in finding the derivative. For initial values we use x_1 and y_1. Then we begin with the particular values, $y_1 = (x_1)^2$.

Step 1. In the given function, replace x_1 with the new value, $x_1 + \Delta x$, and replace y_1 with the new value, $y_1 + \Delta y$. The result is

$$y_1 + \Delta y = (x_1 + \Delta x)^2$$

or, expanding,

$$y_1 + \Delta y = (x_1)^2 + 2x_1(\Delta x) + (\Delta x)^2$$

Step 2. Subtract the original, $\quad y_1 \quad = (x_1)^2$

The result is $\quad \Delta y = 2x_1(\Delta x) + (\Delta x)^2$

Step 3. Divide both sides by Δx: $\quad \dfrac{\Delta y}{\Delta x} = 2x_1 + \Delta x$

12.4 Determining the Derivative of Any Function

This third step shows the ratio between the increments and represents average rate of change over the interval Δx. Now we let Δx become smaller and smaller and approach zero as a limit and see what happens to the expression on the right. The increment Δy also approaches zero. Therefore, the left side of the expression approaches the limit of the ratio, and this limit we call the derivative. On the right side of the equation the term Δx approaches zero, but the term $2x_1$ is not affected. Therefore, as a final step, we have

Step 4. $$\text{The derivative} = \lim_{\Delta x \to 0} \frac{\Delta y}{\Delta x} = 2x_1$$

Note: The subscripts may be omitted if we remember that the x and y at the beginning of the process represent *any* initial values of x and y.

In Step 4 the left side of the equation is by definition the derivative. The right side is the expression for its value.

The foregoing method of determining the derivative is called by various names: the *four-step rule*, the *increment method*, and the *delta method*. The process of determining the derivative of a function is called *differentiation*, and the four-step rule is the fundamental process. Since the procedure is long and rather cumbersome, many short-cut formulas are derived, but all formulas are based on the four-step rule. Moreover, when other known formulas or rules do not apply in a particular problem, then the four-step rule is used. Since this rule is basic, it should be thoroughly understood and memorized. You will discover many "tricks" for finding the derivative, but you will do well to remember where the "tricks" come from and how they are derived.

We summarize the steps in the *delta* method:

Step 1. In the equation $y = f(x)$ replace each and every x with the quantity $(x + \Delta x)$ and replace y with the quantity $(y + \Delta y)$. Expand powers if any.

Step 2. Subtract the original function, $y = f(x)$. This isolates Δy.

Step 3. Divide both sides of the equation by Δx to get the ratio $\Delta y / \Delta x$.

Step 4. Let Δx approach zero as a limit. The left side of the equation is then the expression for the derivative. The right side shows its value. The right side can then be evaluated for any particular value of x.

Let us begin with the general function and follow the four steps. We begin with

$$y = f(x)$$

Step 1. Substituting, $\quad y + \Delta y = f(x + \Delta x)$

Step 2. Subtracting, $\quad \Delta y = f(x + \Delta x) - f(x)$

Step 3. Dividing by Δx, $\quad \dfrac{\Delta y}{\Delta x} = \dfrac{f(x + \Delta x) - f(x)}{\Delta x}$

Step 4. $\quad \lim_{\Delta x \to 0} \dfrac{\Delta y}{\Delta x} = \lim_{\Delta x \to 0} \dfrac{f(x + \Delta x) - f(x)}{\Delta x}$

Step 4 is essentially the mathematical definition of the derivative.

170   The Derivative: Delta Method

12.5   NOTATIONS FOR THE DERIVATIVE

The derivative of a function is denoted in several ways. The student should be familiar with all of them. Each notation has some particular advantage, although some forms are used more than others.

If a function is represented by the letter y, as is very common, then the derivative is usually denoted by the symbol dy/dx. However, this form is not to be considered as a fraction or as an indicated division. It should be read: "*the derivative of y with respect to x as the independent variable.*" In shortened form, it is read: "*dee y by dee x.*" Do *not* say "*dee y* divided by *dee x*" because it is *not* an indicated division. It must be understood as a *single quantity* which represents the limit of a ratio.

If y represents the function of x, then the derivative is often denoted by y', read "*y prime.*" If we use the notation $f(x)$ to represent the function, as in $f(x) = x^2 + 2x$, then we often represent the derivative by $f'(x)$, read: "*f prime of x.*" In this case, the derivative can also be denoted by the expression, $\frac{d}{dx} f(x)$, read, "*dee eff of x by dee x.*" In some cases, if y represents the function of x we use $D_x y$, or simply D to denote the derivative. To summarize,

$$\text{the derivative} = \lim_{\Delta x \to 0} \frac{\Delta y}{\Delta x} = \frac{dy}{dx} = y' = f'(x) = \frac{df(x)}{dx} = D_x y$$

EXAMPLE 1. Sketch the graph of the function $y = x^2 - 6x + 9$. Then find by the delta method the expression for the derivative dy/dx. Finally, find the instantaneous rate of change of the function at the points where x is equal to 0, 3, and 5. At what values of x is the rate of change positive and where negative?

Solution. The graph is a parabola with the vertex on the x-axis, the parabola opening upward and symmetrical with respect to the line $x = 3$ (Fig. 12.2). For the derivative we have the following steps:

Step 1.
$$y + \Delta y = (x + \Delta x)^2 - 6(x + \Delta x) + 9$$
$$y + \Delta y = x^2 + 2x(\Delta x) + (\Delta x)^2 - 6x - 6(\Delta x) + 9$$

Step 2.
$$y \quad\quad = x^2 \quad\quad\quad\quad\quad\quad - 6x \quad\quad\quad\quad + 9$$

Subtracting, $\Delta y = \quad 2x(\Delta x) + (\Delta x)^2 \quad - 6(\Delta x)$

Step 3. Dividing, $\dfrac{\Delta y}{\Delta x} = \quad 2x \quad + (\Delta x) \quad\quad - 6$

Step 4. Now as we let Δx approach zero as a limit, we get the derivative

$$\lim_{\Delta x \to 0} \frac{\Delta y}{\Delta x} = 2x - 6$$

This can be written $\dfrac{dy}{dx} = 2x - 6 \quad$ or $\quad y' = 2x - 6$

The result means that at every point on the curve the instantaneous rate of change in the function is twice the x-value minus 6. To evaluate the derivative at each point mentioned, we substitute x-values in the expression $2x - 6$. We get the

12.5 Notations for the Derivative

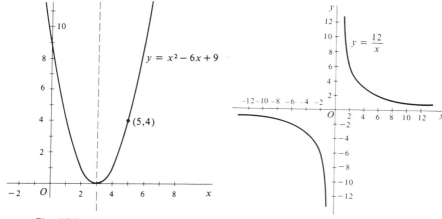

Fig. 12.2   Fig. 12.3

following values: at $x = 0, dy/dx = -6$; at $x = 3, dy/dx = 0$; at $x = 5, dy/dx = 4$. Note that the rate of change is positive (that is, the function is increasing) at all points to the right of $x = 3$. The rate of change is negative, and the function is decreasing at all points to the left of $x = 3$. The instantaneous rate of change at any point can be checked by substituting the x-value of the point in the expression for the derivative.

EXAMPLE 2. Find the instantaneous rate of change of y with respect to x when $x = 6$ in the hyperbola (Fig 12.3),

$$y = \frac{12}{x}$$

Solution. By the delta method,

$$y + \Delta y = \frac{12}{x + \Delta x}$$

Subtracting the original function,

$$\Delta y = \frac{12}{x + \Delta x} - \frac{12}{x}$$

Combining the fractions on the right [the L.C.D. is $x(x + \Delta x)$],

$$\Delta y = \frac{12x - 12(x + \Delta x)}{x(x + \Delta x)} = \frac{-12\Delta x}{x(x + \Delta x)}$$

Dividing both sides by Δx,

$$\frac{\Delta y}{\Delta x} = \frac{-12}{x(x + \Delta x)}$$

As Δx approaches zero,

$$\lim_{\Delta x \to 0} \frac{\Delta y}{\Delta x} = \frac{-12}{x^2} = \frac{dy}{dx}$$

When $x = 6$, $dy/dx = -\frac{1}{3}$. The small negative value means that at that point the function y is *decreasing slowly*. As a rough check we might note that as x changes from 6 to 7, for example, the value of the function changes by only $\frac{2}{7}$ and

172 The Derivative: Delta Method

in a *negative* direction. We might also note that y decreases for any positive change in x. The negative derivative indicates that the change in y is always in an opposite direction to a change in x.

EXAMPLE 3. One formula for power P in an electric circuit is $P = I^2R$, where P represents power in watts, I represents current in amperes, and R represents resistance in ohms. If $R = 12$ ohms, a constant, and I and P are variables, find the instantaneous rate of change of power with respect to current, that is, dP/dI, at the instant when $I = 2.5$ amperes.

Solution. Since $R = 12$ ohms and is constant, we write
$$P = 12I^2$$

Step 1.
$$P + \Delta P = 12(I + \Delta I)^2$$
or
$$P + \Delta P = 12[I^2 + 2I(\Delta I) + (\Delta I)^2]$$
$$P + \Delta P = 12I^2 + 24I(\Delta I) + 12(\Delta I)^2$$

Step 2.
$$P = 12I^2$$

Subtracting,
$$\Delta P = 24I(\Delta I) + 12(\Delta I)^2$$

Step 3. Dividing,
$$\frac{\Delta P}{\Delta I} = 24I + 12(\Delta I)$$

Step 4. As $\Delta I \to 0$,
$$\frac{dP}{dI} = 24I$$

The result means that the instantaneous rate of change of power with respect to current, that is, dP/dI, is always equal to 24 times the value of current I. At the instant, for example, when $I = 2.5$ amperes, the power is changing at the instantaneous rate $dP/dI = 24(2.5) = 60$. That is, the power is changing at a rate of 60 watts for each change of 1 ampere in current.

EXERCISE 12.1

Find the formula for the derivative of each of the following with respect to the independent variable by the *delta method*. Then evaluate the derivative for the given value of the independent variable:

1. $y = x^2 + 5, (x = -3)$
2. $y = x^2 - 3x - 4, (x = 1.5)$
3. $y = 5 - 3x^2, (x = 0.75)$
4. $y = 4 - 3x - x^2, (x = -3)$
5. $y = x^3 - 2x, (x = -2)$
6. $y = 6 - x^3, (x = 1)$
7. $y = 3x^2 - x^3, (x = -1)$
8. $y = 3 - 2x - 4x^2, (x = 0.5)$
9. $y = x^4, (x = \frac{2}{3})$
10. $y = 5x^2 + 6x - 4, (x = -2)$
11. $y = 2x^2 - 3x, (x = 0)$
12. $y = 1 - 3x - 4x^2, (x = 0)$
13. $s = t^2 - 5t + 2, (t = 2)$
14. $s = 3t - t^2, (t = 0.1)$
15. $i = t^3 - 4t^2 - 5t, (t = 3)$
16. $P = 60I^2, (I = 3)$
17. $y = \frac{2}{x}, (x = 10)$
18. $y = \frac{-12}{x+2}, (x = 3)$
19. $y = \frac{6}{x-1}, (x = 1)$
20. $y = \frac{3-2x}{x}, (x = -3)$
21. $y = \frac{4}{x^2}, (x = -2)$
22. $y = \sqrt{x+4}, (x = 5)$

23. $y = \dfrac{3}{\sqrt{x}}$, $(x = 4)$ 	 24. $y = \dfrac{4}{\sqrt{x+1}}$, $(x = 3)$

25. If $f(x) = 3x^2 - 4x - 2$, find (a) $f'(x)$; (b) $f'(1)$; (c) $f'(0)$.
26. If $f(r) = r^2 + 3r$, find (a) $f'(r)$; (b) $f'(1.5)$; (c) $f'(0)$.
27. If $Q = t^2 + 2t - 3$, find (a) $Q'(t)$; (b) $Q'(0.2)$; (c) $Q'(0)$.
28. If $y = x^2 - 4x - 5$, for what values of x is dy/dx equal to zero?
29. If $y = x^3 - 12x$, find the values of x for which y' is equal to zero.
30. If $f(x) = x^2 - 2x - 1$, for what values of x is $f'(x)$ equal to 6?
31. If $s = t^3 - 6t^2 + 9t$, for what values of t is ds/dt equal to zero?
32. When the radius of a circle changes, the diameter and the area also change. When the radius of a circle is 6 inches find (a) the rate of change of area with respect to radius; (b) the rate of change of diameter with respect to radius; (c) the rate of change of diameter with respect to area.

chapter
13

The Derivative by Formula

13.1 THE NEED FOR A FORMULA

Up to this point we have used the *four-step rule*, or *delta method*, to find the derivatives of functions. The *delta* method is fundamental to all differentiation because it is the application of the definition of the derivative. Moreover, this method is the basis for the derivation of most formulas for differentiation.

However, the delta method is often long and tedious, especially for higher powers of the variable x. There are many simple formulas that can be used to find derivatives of variables and functions. We shall now derive a few of the formulas for derivatives of single algebraic terms. The delta method in most instances is used for deriving formulas.

13.2 THE POWER RULE: DERIVATIVES OF SINGLE TERMS

One of the most powerful as well as the most useful of all formulas is the power rule. This is a rule for differentiating powers of a variable. To see how this formula comes about, consider the function

$$y = x^2$$

In this example we have already found by the delta method that the derivative is

$$\frac{dy}{dx} = 2x$$

That is, the derivative of x^2 is equal to twice the initial value of x.

Now let us consider another example with a higher power of x:

$$y = x^4$$

We shall first find the derivative by the four-step rule (delta method).

Step 1.
Substituting, $y + \Delta y = (x + \Delta x)^4$
Expanding, $y + \Delta y = x^4 + 4x^3(\Delta x) + 6x^2(\Delta x)^2 + 4x(\Delta x)^3 + (\Delta x)^4$
Original y, $\underline{y = x^4}$

13.2 The Power Rule: Derivatives of Single Terms

Step 2.
Subtracting, $\quad \Delta y = \quad 4x^3(\Delta x) + 6x^2(\Delta x)^2 + 4x(\Delta x)^3 + (\Delta x)^4$

Step 3.
Dividing by Δx, $\quad \dfrac{\Delta y}{\Delta x} = \quad 4x^3 \quad + 6x^2(\Delta x) + 4x(\Delta x)^2 + (\Delta x)^3$

Step 4.
The derivative, $\quad \dfrac{dy}{dx} = \quad 4x^3 \quad + 0 \quad\quad + 0 \quad\quad + 0$

In Step 4, note that as Δx approaches zero, the limit of the left side of the equation is, by definition, the derivative dy/dx. In the expression on the right, each term except the first contains the factor Δx, and therefore the term itself approaches zero as a limit. The only significant term remaining on the right side is the term $4x^3$, the first term.

In the result showing the derivative, the original exponent on x, which was 4, has become the coefficient of the term, and the exponent has been reduced by 1. This is exactly what happens in the second term of the expansion of any power of a binomial. A similar condition will result whatever the original power of x. If we were to begin with the power $y = x^8$ and proceed through all the four steps, we should find that

$$\dfrac{dy}{dx} = 8x^7$$

To derive a formula we begin with the general exponent n

$$y = x^n$$

Step 1. $\quad y + \Delta y = (x + \Delta x)^n$

Expanding,

$$y + \Delta y = x^n + nx^{n-1}(\Delta x) + \dfrac{n(n-1)x^{n-2}(\Delta x)^2}{2} + \cdots + (\Delta x)^n$$
$$y \quad\quad = x^n$$

Step 2. $\quad \Delta y = \quad nx^{n-1}(\Delta x) + \dfrac{n(n-1)x^{n-2}(\Delta x)^2}{2} + \cdots + (\Delta x)^n$

Step 3. $\quad \dfrac{\Delta y}{\Delta x} = \quad nx^{n-1} \quad + \dfrac{n(n-1)x^{n-2}(\Delta x)}{2} + \cdots + (\Delta x)^{n-1}$

Step 4. $\quad \dfrac{dy}{dx} = \quad nx^{n-1} \quad + \quad\quad 0 \quad\quad + \cdots + 0$

Note that in Step 3 each term on the right side, except the first, contains the factor Δx no matter how many terms there are and, therefore, each term after the first approaches zero as a limit. The result shows that for any integral positive n,

if
$$y = x^n$$
then
$$\dfrac{dy}{dx} = nx^{n-1}$$

176 The Derivative by Formula

Stated in words, the formula says, to find the derivative of any positive integral power of x,

(1) *take the exponent as the coefficient of the term;* then
(2) *decrease the exponent by* 1.

For the time being, we state without proof that the rule holds for all kinds of exponents — zero, negative, and fractional as well as positive integral.

Illustration: If $y = x^{15}$, then $dy/dx = 15x^{14}$.

13.3 THE DERIVATIVE OPERATOR

When we use some letter such as y to represent a function and then write its derivative, we need two equations. For example, if $y = x^7$, then $dy/dx = 7x^6$. If we wish to indicate the derivatives of several functions, we might place the functions in one column and their derivatives in another. Then we need not use another letter to represent the function. We might have

function	derivative
x^7	$7x^6$
x^{12}	$12x^{11}$
x^{25}	$25x^{24}$
x^{-2}	$-2x^{-3}$

One way to indicate differentiation without using a letter to represent the function is to use the notation "$\frac{d}{dx}$" before the function to be differentiated. Then the function and its derivative can be shown in the form of a single equation. For example, we can write

$$\frac{d}{dx}(x^7) = 7x^6$$

The symbol d/dx placed before a function is called the *derivative operator* or the *differentiating operator*. It signifies that the quantity following the operator is to be differentiated. Actually, when we use the letter y to represent a function and then write dy/dx to indicate the derivative, we really mean the derivative not of y but of the *function that y represents*. For example, in finding the derivative of x^7 we write:

if $\qquad\qquad\qquad y = x^7$

then $\qquad\qquad\qquad \dfrac{d(y)}{dx} = 7x^6$

This really means $\qquad\qquad \dfrac{d(x^7)}{dx} = 7x^6$

The derivative operator is sometimes denoted by the symbol D_x, or simply D. The subscript indicates the independent variable. This form is convenient and is useful in the solution of some differential equations. The subscript may be omitted if there is no mistaking the independent variable.

Illustrations: $D_x(x^8) = 8x^7$; $\quad D(x^5) = 5x^4$; $\quad D_t(t^6) = 6t^5$

13.4 SPECIAL RULES FOR DIFFERENTIATION

There are several special rules, or short-cuts, that are helpful in finding derivatives. These special rules can all be derived by the delta method and in some instances by the power rule already derived.

1. *The derivative of the product of a constant and a variable is equal to the constant times the derivative of the variable.*

Illustration: If $y = 8x^4$

then $$\frac{dy}{dx} = (8)(4x^3) = 32x^3$$

Stated in words, the rule says, to find the derivative of a constant coefficient times a power of x, we have two steps:

(1) *multiply the coefficient by the exponent*
(2) *decrease the exponent by 1.*

These two steps should be carefully identified because they are reversed in the opposite process of integration. In general terms,

if $\quad y = cx^n,\quad$ then $\quad \dfrac{dy}{dx} = cnx^{n-1}$

By use of the operator, we can write

$$\frac{d}{dx}(cx^n) = cnx^{n-1}$$

The rule can be derived by the *delta* method.

2. *The derivative of the product of a constant and a variable x is equal to the constant.* In this case the power of x is 1.

Illustration: If $\quad y = 6x$

then $\quad \dfrac{dy}{dx} = 6$

In general terms, $\quad$ if $\quad y = cx,\quad$ then $\quad dy/dx = c$.

Let us derive this rule by the delta method. We begin with $y = cx$.

The Derivative by Formula

Step 1.
$$y + \Delta y = c(x + \Delta x)$$
$$= cx + c(\Delta x)$$
$$y = cx$$

Step 2. $\qquad \Delta y = c(\Delta x)$

Step 3. $\qquad \dfrac{\Delta y}{\Delta x} = c$

Step 4. $\qquad \dfrac{dy}{dx} = c$

The result means that as Δx approaches zero as a limit, the right side of the equation is not affected, and the ratio of the increments is always equal to the constant c.

The same result can be seen by applying the power rule. For example,

if $\qquad y = 6x$

this can be written $\qquad y = 6x^1$

By the power rule, $\qquad \dfrac{dy}{dx} = (6)(1)(x^0) = 6$

3. *The derivative of a constant is zero.*

   Illustration: $\qquad$ If $y = 6$, $\qquad \dfrac{dy}{dx} = 0$

   In general terms, $\qquad$ if $y = c$, $\qquad \dfrac{dy}{dx} = 0$

By use of the derivative operator, we can write

$$\frac{d}{dx}(c) = 0$$

This rule can be derived by the delta method. It can also be shown by the power rule. If

$$y = 6$$

this can be written $\qquad y = 6(x^0)$

By the power rule, $\qquad \dfrac{dy}{dx} = (6)(0)(x^{-1}) = 0$

4. *The derivative of a variable with respect to itself is* 1.

   Illustration: If $y = x$, then $dy/dx = 1$.

By use of the differential operator we can say

$$\frac{d}{dx}(x) = 1$$

The same is true regardless of the variable. That is,

$$\frac{d}{dv}(v) = 1$$

This rule can be derived by the delta method or by the power rule.

5. *The derivative of the sum of any finite number of terms is equal to the sum of the derivatives of the separate terms.* That is, any multinomial may be differentiated term by term. For example, if we have given the function, $y = 5x^3 + 4x^2$, and work this out by the *delta* method, we shall find that the derivative is

$$\frac{dy}{dx} = 15x^2 + 8x$$

We shall work out the derivative, in general terms, of the following function of two terms (the rule also applies to three or more terms):

$$y = f(x) + g(x)$$

Step 1. $\quad y + \Delta y = f(x + \Delta x) + g(x + \Delta x)$

Step 2. $\quad \Delta y = f(x + \Delta x) - f(x) + g(x + \Delta x) - g(x)$

Step 3. $\quad \dfrac{\Delta y}{\Delta x} = \dfrac{f(x + \Delta x) - f(x)}{\Delta x} + \dfrac{g(x + \Delta x) - g(x)}{\Delta x}$

Step 4. $\quad \dfrac{dy}{dx} = \lim\limits_{\Delta x \to 0}\left[\dfrac{f(x + \Delta x) - f(x)}{\Delta x} + \dfrac{g(x + \Delta x) - g(x)}{\Delta x}\right]$

$$= \lim_{x \to 0}\left[\frac{f(x + \Delta x) - f(x)}{\Delta x}\right] + \lim_{x \to 0}\left[\frac{g(x + \Delta x) - g(x)}{\Delta x}\right]$$

Then $\quad \dfrac{dy}{dx} = \qquad f'(x) \qquad + \qquad g'(x)$

Illustration 1: If $y = 4x^5 + 7x^4 - 8x^3 + 5x - 9$,

$$\frac{dy}{dx} = 20x^4 + 28x^3 - 24x^2 + 5$$

By use of operator,

$$\frac{d}{dx}(4x^5 + 7x^4 - 8x^3 + 5x - 9) = 20x^4 + 28x^3 - 24x^2 + 5$$

Illustration 2: If $y = 6x^3 + 5x^{-3} - 12x^{1/3} - 5x^{-1/2} - x + \log_{10} 100$,

$$\frac{dy}{dx} = 18x^2 - 15x^{-4} - 4x^{-2/3} + \frac{5}{2}x^{-3/2} - 1$$

13.5. HIGHER DERIVATIVES

The derivative of a function of x is also, in general, a function of x. For example,

if

$$y = x^3 - 7x^2 + 4x - 3$$

then

$$\frac{dy}{dx} = 3x^2 - 14x + 4$$

The Derivative by Formula

Note that the derivative is also a function of x and depends for its value on the value of x.

If the derivative is a function that is differentiable, then we may also find the derivative of this derivative. The result is called the *second derivative* of the original function. The second derivative can be indicated by

$$\frac{d}{dx}\left(\frac{dy}{dx}\right)$$

If $y = f(x)$, then the second derivative is often indicated by any of the following notations:

$$\frac{d}{dx}\left(\frac{dy}{dx}\right); \quad \frac{d^2y}{dx^2}; \quad \frac{d^2}{dx^2}[f(x)]; \quad y''; \quad f''(x); \quad D_x^2 y; \quad \frac{d}{dx}[f'(x)]$$

In the notation $\frac{d^2y}{dx^2}$ it must be understood that the number 2 is *not* an exponent. It simply indicates the second derivative of the function. In a similar manner, the third derivative can be denoted by any of the following:

$$\frac{d^3y}{dx^3}; \quad \frac{d^3}{dx^3}[f(x)]; \quad y'''; \quad f'''(x); \quad D_x^3 y$$

For higher derivatives we use a similar notation. For example, the nth derivative is indicated by

$$\frac{d^n y}{dx^n}; \quad \frac{d^n}{dx^n}[f(x)]; \quad y^n; \quad f^n(x)$$

Higher derivatives, especially the second, have many important uses, as we shall see. If we say the first derivative, dy/dx, represents the rate of change in y with respect to x, then we might say the second derivative represents any change in the rate of change. Take the example

if
$$y = 3x + 2$$
then
$$\frac{dy}{dx} = 3$$

Note that here the derivative is a constant, 3. Since the rate of change of y is a constant, then there is no change in this rate of change. For,

if
$$\frac{dy}{dx} = 3$$
then
$$\frac{d^2y}{dx^2} = \frac{d}{dx}\left(\frac{dy}{dx}\right) = \frac{d}{dx}(3) = 0$$

EXAMPLE 1. Find the first four derivatives of the function
$$y = x^3 + 4x^2 - 5x - 7$$

Solution.
$$\frac{dy}{dx} = 3x^2 + 8x - 5$$

$$\frac{d^2y}{dx^2} = 6x + 8; \quad \frac{d^3y}{dx^3} = 6; \quad \frac{d^4y}{dx^4} = 0$$

13.5 Higher Derivatives

Note: If a polynomial function of x (that is, a function containing only positive integral powers of the variable x) is differentiated successively, then eventually one derivative will become zero, after which all succeeding derivatives are equal to zero. However, this is not true for algebraic functions containing negative or fractional exponents.

EXAMPLE 2. Find the first three derivatives of the function
$$y = x^{-2} + 2x^{1/2} - 6x^{1/3}$$

Solution.
$$y' = -2x^{-3} + x^{-1/2} - 2x^{-2/3}$$

$$y'' = 6x^{-4} - \frac{1}{2}x^{-3/2} + \frac{4}{3}x^{-5/3}$$

$$y''' = -24x^{-5} + \frac{3}{4}x^{-5/2} - \frac{20}{9}x^{-8/3}$$

EXAMPLE 3. Find the first two derivatives of the hyperbola
$$y = \frac{6}{x}$$

Solution. We first write the function with a negative exponent:
$$y = 6x^{-1}$$
$$\frac{dy}{dx} = -6x^{-2} = -\frac{6}{x^2}$$
$$\frac{d^2y}{dx^2} = +12x^{-3} = \frac{12}{x^3}$$

EXERCISE 13.1

Find the first derivative of each of the following functions with respect to the independent variable by the power rule:

1. $y = 3x^2 - 5x + 2$.   2. $y = 4x^5 - 7x^4 + 3x^3$.   3. $y = 30x^4 - 15x^3$.
4. $y = x^2 - 5x^3 - x^4$.   5. $y = 4 - 7t + 5t^2$.   6. $y = 3t^2 - 4t + \sqrt{2}$.
7. $s = 2t^3 - 6t^2 + 3t$.   8. $s = t^4 - t^3 - 3t^2$.   9. $x = 5t^2 + 2t - 7$.
10. $A = 8\pi r + 2\pi r^2$.   11. $C = 20v + 4v^2$.   12. $V = \frac{4}{3}\pi r^3$.
13. $y = 3x^{-2} - 6x^{1/2}$.   14. $y = 8x^{-3/4} - 5x^{1/3}$.   15. $y = 7x^{1/4} + \pi^2$.
16. $y = \frac{2}{x}$.   17. $y = \sqrt[3]{x^2}$.   18. $y = \frac{3}{x^3}$.   19. $y = \frac{-2}{x^4}$.   20. $y = \frac{-12}{x^5}$.
21. $y = 6x^{1/2} - 3x^{1/3} + 4x^{3/2} - 6x^{2/3}$.   22. $y = 3x^{-2/3} - 4x^{-1/2} + 9x^{-1/3}$.
23. $y = \frac{1}{2}x^{-4} - 4x^{-2/3} + \pi$.   24. $y = ax^3 + bx^2 - cx + d$.

Find the second and third derivatives of each of the following (No. 25–30):

25. $y = x^4 - 5x^3 + 3x^2$.   26. $y = 3x^2 + 2x - 4$.   27. $y = 3x - 4x^2 - 2x^3$.
28. $s = t^3 - 5t^2 - 2t$.   29. $y = t^{3/2} + 4t^{1/3}$.   30. $x = t^{-2} - 5t^{-1}$.
31. If $y = x^4 - 5x^3 + 4x^2 - 3x + 2$, find (a) y'. (b) y'', (c) y''', (d) y''''.

32. If $y = 4x^3 + 5x^2 - 3x + 20$, find (a) $\dfrac{dy}{dx}$, (b) $\dfrac{d^2y}{dx^2}$, (c) $\dfrac{d^3y}{dx^3}$, (d) $\dfrac{d^4y}{dx^4}$.

33. If $f(x) = 6x^4 - 2x^3$, find (a) $f'(x)$, (b) $f''(x)$, (c) $f'''(x)$.

Evaluate the derivative dy/dx for the given value of x in No. 34–37.

34. $y = x^2 - 5x + 2, (x = 1)$.
35. $y = x^3 - 2x^2 + 5x - 2, (x = -2)$.
36. $y = 3x^2 + 2x - 4, (x = -1)$.
37. $y = 2x^3 - x^2 - 3x - 1, (x = 0)$.
38. If $f(x) = 3x^2 - 4x - 2$, find (a) $f'(x)$, (b) $f'(1)$, (c) $f'(0)$.
39. If $Q(t) = t^2 + 2t - 3$, find (a) $Q'(t)$, (b) $Q'(0.2)$, (c) $Q'(0)$.
40. If $y = x^2 - 6x + 5$, for what values of x is dy/dx equal to zero?
41. If $y = 2x^3 - 9x^2 + 12x - 5$, for what values of x is y' equal to zero?
42. If $y = x^3 - 3x^2 + 15x + 2$, for what values of x is y' equal to zero?
43. If $y = x^3 - 2x^2 + 3x - 7$, for what values of x is y' equal to 5?
44. If $f(x) = x^3 - 3x^2 - 3x + 9$, for what values of x is $f'(x)$ equal to zero?
45. If $s = t^3 - 6t^2 + 12t - 5$, for what values of t is s' equal to zero?
46. If $y = x^3 - 3x^2 - 9x + 10$, find the value of y'' where y' is equal to zero.
47. Find the rate of change of the area of a circle with respect to the radius when the radius is 5 inches.
48. A certain circular cylinder has a height always equal to its diameter. Find the rate of change of the volume with respect to the radius when the radius is equal to 6 inches.
49. Find the rate of change of the total area of the cylinder in No. 48 with respect to the radius when the radius is 6 inches.
50. In a certain electric circuit the power is given by the formula $P = 25I^2$. Find the rate of change of power with respect to current I when the current is 3.5 amperes.

chapter
14

Differentiation of Algebraic Functions

14.1 THE POWER RULE FOR DIFFERENTIATION

We have seen how to find the derivatives of powers of a single variable. For example,

$$\text{if } y = x^5, \text{ then, by the power rule, } \frac{dy}{dx} = 5x^4$$

Stated in general terms, $\quad \frac{d}{dx}(x^n) = nx^{n-1}$

In this example the independent variable is x.

We have also seen that the power rule applies to any independent variable. As examples,

$$\text{if } y = t^5, \text{ then } \frac{dy}{dt} = 5t^4 \qquad \text{if } y = u^5, \text{ then } \frac{dy}{du} = 5u^4$$

In all powers of a single variable, we take the derivative of the function with respect to the independent variable, whatever that variable may be.

Now suppose we have a *power of a function* of x, such as

$$y = (x^2 + 3x)^5$$

Now we might ask: Can the power rule be applied on the function $(x^2 + 3x)$ raised to the 5th power in the same way as on a single term in x? Can we take the following two steps: (1) write the exponent 5 as the coefficient of the function; and then, (2), decrease the exponent by 1? That is,

$$\text{does } \frac{dy}{dx} = 5(x^2 + 3x)^4?$$

We shall discover that the result shown is correct as far as it goes, but there is an additional factor. First, let us see what is meant by a *function of a function*.

14.2 A FUNCTION OF A FUNCTION

Let us take again the example

$$\text{if } y = u^5, \quad \text{then } \frac{dy}{du} = 5u^4$$

Here we have the derivative of y with respect to u. Then u is the independent variable.

Now let us suppose in this example that u itself is some function of x, such as $(x^2 + 3x)$. Since y is a function of u, and u is a function of x, then y is actually a function of x through u. Then we say that y is a *function of a function*.

Our problem now is to find the derivative of y, not only with respect to u but also with respect to x as the independent variable. We wish to find dy/dx, not merely dy/du. That is, we wish to take x, not u, as the basic independent variable. Let us see how this can be done.

Consider again the increments in the variables. For any increment in x, we shall have corresponding increments in u and in y. We denote the increments by Δx, Δu, and Δy, respectively. Now as Δx approaches zero, then Δu will approach zero and Δy will also approach zero, if the functions are continuous. As long as the increments are anything other than zero, we can say

$$\left(\frac{\Delta y}{\Delta u}\right)\left(\frac{\Delta u}{\Delta x}\right) = \frac{\Delta y}{\Delta x}$$

As the increments approach zero, each fraction will approach a limit, which is a derivative, and we have

$$\left(\frac{dy}{du}\right)\left(\frac{du}{dx}\right) = \frac{dy}{dx}$$

This is the so-called *chain rule* for finding the *derivative of a function of a function*. That is, the derivative of y with respect to u times the derivative of u with respect to x is equal to the derivative of y with respect to x, provided that y and u are differentiable functions.

Now let us look again at the problem

$$y = (x^2 + 3x)^5$$

If we take the function $x^2 + 3x$ as u, then the problem is essentially

$$y = u^5 \quad \text{and} \quad \frac{dy}{du} = 5u^4$$

We can now multiply both sides of this equation by du/dx, and get

$$\left(\frac{dy}{du}\right)\left(\frac{du}{dx}\right) = 5(u)^4\left(\frac{du}{dx}\right)$$

The expression on the left side is equivalent to dy/dx, which is what we want. The result on the right means that when we have applied the power rule on the u function, we must follow this up with (du/dx), the derivative of u with respect to x. In the example we have shown, since $u = x^2 + 3x$, then $(du/dx) = 2x + 3$. This is the extra factor needed, and we get

if $$y = (x^2 + 3x)^5$$

then $$\frac{dy}{dx} = 5(x^2 + 3x)^4(2x + 3)$$

Here we identify $$\frac{dy}{dx} = 5(\quad u \quad)^4\left(\frac{du}{dx}\right)$$

We emphasize one point here that will be helpful and necessary when we come to the inverse problem of finding a function when its derivative is given. Note especially that in the example shown, the factor $(2x + 3)$, which is du/dx, appears in the derivative. One might say that this factor *emerges* from the function on which the power rule is applied. This is important to note because in the inverse process of integration we shall find that this same kind of factor must be present to begin with, but that it *reenters* or is *absorbed* by the power.

The power rule is so extremely important that it should be thoroughly understood and memorized. It is applicable to powers of all kinds of functions: algebraic, trigonometric, logarithmic, and hyperbolic. It may be stated:

Rule.

To find the derivative of a power of a function, apply the power rule on the function and then multiply the result by the derivative of the function.

The following notation may help to emphasize the rule:

if $\quad y = (\text{function of } x)^n$

then $\quad \dfrac{dy}{dx} = (n)(\text{function of } x)^{n-1} (\text{derivative of the function})$

As a formula,

$$\text{if } y = (u)^n, \quad \text{then } \frac{dy}{dx} = (n)(u)^{n-1}\left(\frac{du}{dx}\right)$$

EXAMPLE 1. Find the derivative of the function $y = (x^3 + 7x^2 - 2x + 1)^7$.

Solution. Here we identify $u = x^3 + 7x^2 - 2x + 1$.
The first part of the derivative is

$$7(x^3 + 7x^2 - 2x + 1)^6(\ldots)$$

Now we follow with the factor du/dx, the derivative of the function. The entire derivative is

$$\frac{dy}{dx} = 7(x^3 + 7x^2 - 2x + 1)^6(3x^2 + 14x - 2)$$

EXAMPLE 2. If $y = \sqrt{x^2 - 4x + 5}$, find dy/dx. Where does $dy/dx = 0$?

Solution. We rewrite the equation $y = (x^2 - 4x + 5)^{1/2}$. Applying the power rule,

$$\frac{dy}{dx} = \left(\frac{1}{2}\right)(x^2 - 4x + 5)^{-1/2}(2x - 4)$$

$$= \frac{(x-2)}{\sqrt{x^2 - 4x + 5}}$$

Now we set the derivative equal to zero and find that $x = 2$. For this value of x we find that $y = 1$. In solving for y we must be sure that this value is not imaginary.

Note: Actually it is well to think of the complete power rule as being applicable to the power on x alone. For example,

if $y = x^5$ we can say that $\dfrac{dy}{dx} = 5(x)^4 \left(\dfrac{dx}{dx}\right)$

Since $dx/dx = 1$, we usually write simply $dy/dx = 5x^4$.

In connection with powers of functions, it is important that the independent variable be correctly taken. Sometimes an entire function can be considered as the independent variable. In the following example we may consider the independent variable to be x itself, or the function $2x$.

EXAMPLE 3. Given the function $(2x)^3$, find the derivative with respect to x and also with respect to $2x$. Evaluate each derivative for $x = 1$.

Solution. Taking x as the independent variable we have

$$\frac{dy}{dx} = 3(2x)^2(2) = 6(2x)^2 = 24x^2$$

Now, if instead we take the quantity $(2x)$ as the independent variable u, we find the derivative with respect to $(2x)$, in which case we get

$$\frac{dy}{d(2x)} = 3(2x)^2, \text{ which is the same as } dy/du = 3(u)^2$$

When we evaluate the derivatives, at $x = 1$,

$$\frac{dy}{dx} = 24 \quad \text{but} \quad \frac{dy}{d(2x)} = 12$$

The difference is reasonable if we remember that we are considering the rate of change in y as compared, first, with the change in x itself, and, secondly, with the change in $2x$. That is, any change in y will bear a greater ratio to a change in x than it will to a corresponding change in $2x$. In any ratio expressed as a fraction, any increase in the denominator will result in a decrease in the value of the ratio.

EXAMPLE 4. If $y = (x^2 + 5)^3$, find dy/dx in two ways.

Solution. A polynomial raised to an integral power may be expanded and the result differentiated term by term, or it may be differentiated by the power rule.

First method. Expanding the polynomial, $y = x^6 + 15x^4 + 75x^2 + 125$.
Differentiating term by term, $dy/dx = 6x^5 + 60x^3 + 150x$.
Second method. By the power rule, $dy/dx = 3(x^2 + 5)^2(2x)$.

The second method has many advantages over the first. The most useful form of the derivative is the factored form, which is obtained as a result of the power rule. The most important advantage is, of course, the fact that many functions cannot be expanded to separate terms.

A variable denominator of a fraction may be shifted to the numerator and the sign of its exponent changed. Then the derivative can be found by applying the power rule. For example, suppose we have the function

$$y = \frac{5}{x^2 + 4}$$

This equation may be written: $y = 5(x^2 + 4)^{-1}$

Applying the power rule, we get

$$\frac{dy}{dx} = -5(x^2 + 4)^{-2}(2x); \quad \text{or} \quad \frac{-10x}{(x^2 + 4)^2}$$

The same procedure may be followed in this example:

$$y = \frac{3}{\sqrt{x^2 - 9}}$$

The equation may be written: $y = 3(x^2 - 9)^{-1/2}$

Then $\quad \dfrac{dy}{dx} = -\dfrac{3}{2}(x^2 - 9)^{-3/2}(2x); \quad \text{or} \quad \dfrac{dy}{dx} = \dfrac{-3x}{(x^2 - 9)^{3/2}}$

EXERCISE 14.1

Find the derivative of each of the following functions. Find the derivatives of the first six in two ways: by expanding and as a power.

1. $y = (4x - 3)^2$
2. $y = (2 - 5x)^2$
3. $y = (2x - 3)^3$
4. $i = (t^2 - 4)^2$
5. $y = (x^2 - 3x + 5)^2$
6. $y = (2x - 3)^4$
7. $y = (3x + 2)^5$
8. $y = (3 - 4x)^7$
9. $i = (t^2 - t + 2)^2$
10. $s = (5 - t^3)^4$
11. $s = (t^2 - 3t)^5$
12. $y = (4t - t^2)^6$
13. $u = (x^3 - 3x^2 + 4)^6$
14. $v = (6x - x^2)^{1/2}$
15. $y = (3t - 2)^{1/3}$
16. $y = \sqrt{9 - 4x^2}$
17. $u = (6x - 5)^{2/3}$
18. $v = \sqrt{3 - 4x}$
19. $y = \sqrt{8 - x^3}$
20. $i = \sqrt{3t - t^2}$
21. $\dfrac{d}{dx}(2x - 5)^4$
22. $\dfrac{d}{dv}(2 - v^2)^5$
23. $\dfrac{d}{dt}(4 - t^2)^3$
24. $\dfrac{d}{dx}(3 - 2x - x^2)^4$
25. $D_x(x^3 - 2x^2 - 5x)^4$
26. $D_t(5 + 3t - t^2)^6$

27. $y = \dfrac{6}{(x^2 + 3)^3}$

28. $y = \dfrac{4}{(5 - x^2)^7}$

29. $y = \dfrac{2}{\sqrt{x^2 + 9}}$

30. $y = \dfrac{6}{(x^2 - 8)^{2/3}}$

14.3 THE PRODUCT RULE

The product of two or more functions can sometimes be expanded before differentiating, but in most cases this is not practical. Suppose we have

$$y = (x^3 + 5x^2)(3x - 4)$$

Here we might expand the product by multiplication, but some products would be very difficult to expand, such as

$$y = (x^2 + 5x)^{1/3}\sqrt{4x - 3}$$

We therefore need a rule for differentiating a *product* of two or more factors.

It has been said that when Leibniz, one of the two inventors of calculus, first faced the problem of the derivative of a product, his first flash of thought was that the derivative was probably simply the product of the derivatives. However, he immediately saw this could not be. The derivative of a product is not quite so simple.

The product rule can be derived by the *delta* method. We use other letters, such as u and v, to represent the separate functions. For example,

$$\text{if } y = (x^3 + 5x^2)(3x - 4)$$

we shall let $\quad u = x^3 + 5x^2 \quad$ and $\quad v = 3x - 4$

Then the problem can be represented in general terms as

$$y = (u)(v)$$

Now we derive the rule by the delta method. Since u, v, and y are all functions of x, then for any Δx we shall have the corresponding increments Δu, Δv, and Δy. By the delta method:

Step 1. $\quad y + \Delta y = (u + \Delta u)(v + \Delta v)$

Expanding, $y + \Delta y = (u)(v) + (u)(\Delta v) + (v)(\Delta u) + (\Delta u)(\Delta v)$

$\phantom{\text{Expanding, }} y = (u)(v)$

Step 2. $\quad \Delta y = (u)(\Delta v) + (v)(\Delta u) + (\Delta u)(\Delta v)$

Step 3. $\quad \dfrac{\Delta y}{\Delta x} = (u)\left(\dfrac{\Delta v}{\Delta x}\right) + (v)\left(\dfrac{\Delta u}{\Delta x}\right) + \left(\dfrac{\Delta u}{\Delta x}\right)(\Delta v)$

Step 4. $\quad \dfrac{dy}{dx} = (u)\left(\dfrac{dv}{dx}\right) + (v)\left(\dfrac{du}{dx}\right) + 0$

14.3 The Product Rule

The last term approaches zero because one factor approaches zero even though the other factor may not. The formula may be stated in words:

The derivative of the product of two functions is equal to the first function times the derivative of the second, plus the second function times the derivative of the first.

The formula for the derivative of a product should be thoroughly known, not only for its important use in differentiation but also because it forms the basis for one of the most powerful methods in integral calculus. The derivative of a product should be systematically written down in exactly the order given so as to avoid errors. At first follow these steps:

1. Write the *first function* exactly as it appears, as a factor.
2. Write the complete *derivative of the second function* as a factor.
3. After the plus sign, write the *second function* exactly as given.
4. Write the *derivative of the first function* as a factor.

In some problems it may be convenient to change the order of some of these steps, but at first it is well to follow the steps as indicated.

EXAMPLE 5. If $y = (x^3 + 5x^2)(3x - 4)$, find dy/dx.

Solution. Letting $u = x^3 + 5x^2$ and $v = 3x - 4$, we have

$$du/dx = 3x^2 + 10x \quad \text{and} \quad dv/dx = 3$$

Then
$$\frac{dy}{dx} = (x^3 + 5x^2)(\ 3\ ) + (3x - 4)(3x^2 + 10x)$$

Here we identify
$$(\quad u \quad)\left(\frac{dv}{dx}\right) + (\quad v \quad)\left(\frac{du}{dx}\right)$$

Simplifying,
$$\frac{dy}{dx} = 12x^3 + 33x^2 - 40x$$
$$= x(12x^2 + 33x - 40)$$

EXAMPLE 6. If $y = (x^2 + 2)^5(4x - 7)^3$, find dy/dx.

Solution. In this example, to use the product rule for $y = (u)(v)$, we must remember that the factors, u and v, are themselves *powers*. Then we have a combination of the product rule and the power rule. Here we first identify u and v as follows:

$$\text{let} \quad u = (x^2 + 2)^5 \quad \text{and} \quad v = (4x - 7)^3$$

Then $\dfrac{dy}{dx} = (x^2 + 2)^5(3)(4x - 7)^2(4) + (4x - 7)^3(5)(x^2 + 2)^4(2x)$

We identify $(\quad u \quad)(\quad dv/dx \quad) + (\quad v \quad)(\quad du/dx \quad)$

This reduces to

$$\frac{dy}{dx} = 12(x^2 + 2)^5(4x - 7)^2 + 10x(4x - 7)^3(x^2 + 2)^4$$
$$= (x^2 + 2)^4(4x - 7)^2[12(x^2 + 2) + 10x(4x - 7)]$$
$$= (x^2 + 2)^4(4x - 7)^2(52x^2 - 70x + 24)$$

EXAMPLE 7. If $y = x^3(x^2 + 5)^{1/2}$, find dy/dx and evaluate dy/dx for $x = 2$.

Solution. Here we identify $u = x^3$ and $v = (x^2 + 5)^{1/2}$.

By the product rule
$$\frac{dy}{dx} = (x^3)\left(\frac{1}{2}\right)(x^2 + 5)^{-1/2}(2x) + (x^2 + 5)^{1/2}(3x^2)$$
$$= (x^4)(x^2 + 5)^{-1/2} + (3x^2)(x^2 + 5)^{1/2}$$

To simplify the expression, we write 1 for the denominator and then multiply numerator and denominator by the quantity $(x^2 + 5)^{1/2}$, then

$$\frac{dy}{dx} = \frac{(x^4)(x^2 + 5)^{-1/2} + (3x^2)(x^2 + 5)^{1/2}}{1}$$
$$= \frac{x^4 + 3x^2(x^2 + 5)}{(x^2 + 5)^{1/2}} = \frac{4x^4 + 15x^2}{(x^2 + 5)^{1/2}}$$

For the value $x = 2$, $y = 24$ and $dy/dx = 124/3$.

The product rule can be extended to cover the product of three or more factors. Suppose $y = (u)(v)(w)$. Then by the product rule for two factors we have

$$\frac{dy}{dx} = (uv)\left(\frac{dw}{dx}\right) + (w)\left[\frac{d(uv)}{dx}\right]$$
$$= (uv)\left(\frac{dw}{dx}\right) + (w)\left[(u)\left(\frac{dv}{dx}\right) + (v)\left(\frac{du}{dx}\right)\right]$$
$$= (uv)\left(\frac{dw}{dx}\right) + (uw)\left(\frac{dv}{dx}\right) + (vw)\left(\frac{du}{dx}\right)$$

14.4 THE QUOTIENT RULE

The rule for differentiating a quotient is also derived by the *delta* method. Suppose we have the quotient of two functions:

$$y = \frac{x^3 + 2}{x^2 + 5}$$

Representing the numerator function by u and the denominator by v we can represent the quotient of the two functions in general terms:

$$y = \frac{u}{v}$$

For any increment Δx, we have the corresponding increments Δu, Δv, and Δy.

Then
$$y + \Delta y = \frac{u + \Delta u}{v + \Delta v}$$

Subtracting,
$$\Delta y = \frac{u + \Delta u}{v + \Delta v} - \frac{u}{v}$$

14.4 The Quotient Rule

Combining fractions,

$$\Delta y = \frac{(v)(u + \Delta u) - (u)(v + \Delta v)}{(v)(v + \Delta v)}$$

$$= \frac{(u)(v) + (v)(\Delta u) - (u)(v) - (u)(\Delta v)}{(v)(v + \Delta v)}$$

$$= \frac{(v)(\Delta u) - (u)(\Delta v)}{(v)(v + \Delta v)}$$

Dividing by Δx,
$$\frac{\Delta y}{\Delta x} = \frac{(v)\left(\frac{\Delta u}{\Delta x}\right) - (u)\left(\frac{\Delta v}{\Delta x}\right)}{(v)(v + \Delta v)}$$

Then
$$\frac{dy}{dx} = \frac{(v)(du/dx) - (u)(dv/dx)}{v^2}$$

Note that the derivative of a fraction (the quotient of two functions) is another fraction. The quotient rule may be stated:

The derivative of the quotient of two functions (a fraction) is equal to the denominator times the derivative of the numerator minus the numerator times the derivative of the denominator, all divided by the square of the denominator.

Note: The important thing to remember in finding the derivative of *a fraction* (the quotient of two functions) *is to begin with the denominator.* It may be helpful to follow these steps:

1. For the numerator of the derivative, first *write the denominator of the fraction.*
2. Next, as a factor, write the *derivative of the numerator.*
3. After a minus sign, write the *numerator of the fraction.*
4. Then, as another factor, write the *derivative of the denominator.*
5. For the denominator of the derivative, write the *square* of the denominator of the fraction.

EXAMPLE 8. Given the function

$$y = \frac{x^3 + 2}{x^2 + 5}$$

find dy/dx; also find the value of dy/dx when $x = 1$.

Solution. Here we identify the denominator $v = x^2 + 5$, and the numerator $u = x^3 + 2$. Then, by the quotient rule,

$$\frac{dy}{dx} = \frac{(\ v\ )(du/dx) - (\ u\ )(dv/dx)}{v^2}$$

Therefore,
$$\frac{dy}{dx} = \frac{(x^2 + 5)(3x^2) - (x^3 + 2)(2x)}{(x^2 + 5)^2}$$

Simplifying,
$$\frac{dy}{dx} = \frac{x^4 + 15x^2 - 4x}{(x^2 + 5)^2}$$

For the value $x = 1$, $dy/dx = \frac{1}{3}$.

EXAMPLE 9. For the curve $y = \dfrac{x^3}{\sqrt{x^2 + 4}}$, find a point where $dy/dx = 0$.

Solution.
$$\frac{dy}{dx} = \frac{(x^2 + 4)^{1/2}(3x^2) - (x^3)(\tfrac{1}{2})(x^2 + 4)^{-1/2}(2x)}{x^2 + 4}$$

$$= \frac{(3x^2)(x^2 + 4)^{1/2} - (x^4)(x^2 + 4)^{-1/2}}{x^2 + 4}$$

Now we multiply numerator and denominator by the quantity $(x^2 + 4)^{1/2}$, and get

$$\frac{dy}{dx} = \frac{(3x^2)(x^2 + 4) - (x^4)(x^2 + 4)^0}{(x^2 + 4)^{3/2}}$$

Simplifying,
$$\frac{dy}{dx} = \frac{2x^4 + 12x^2}{(x^2 + 4)^{3/2}}$$

To find where $dy/dx = 0$, we set

$$\frac{2x^4 + 12x^2}{(x^2 + 4)^{3/2}} = 0$$

Solving we get $x = 0$. For this value, we have $y = 0$. Then $dy/dx = 0$ at $(0,0)$.

EXAMPLE 10. A quotient may be differentiated as a product. Let us write the problem shown in Example 9 as a product: $y = (x^3)(x^2 + 4)^{-1/2}$

Solution.
$$\frac{dy}{dx} = (x^3)\left(-\frac{1}{2}\right)(x^2 + 4)^{-3/2}(2x) + (x^2 + 4)^{-1/2}(3x^2)$$

$$= -x^4(x^2 + 4)^{-3/2} + (3x^2)(x^2 + 4)^{-1/2}$$

Now we write 1 as the denominator and multiply numerator and denominator by the quantity $(x^2 + 4)^{3/2}$. Then we get, reversing terms,

$$\frac{dy}{dx} = \frac{(3x^2)(x^2 + 4) - x^4}{(x^2 + 4)^{3/2}} = \frac{2x^4 + 12x^2}{(x^2 + 4)^{3/2}}$$

14.5 THE DERIVATIVE BY INVERSION

In some problems, especially where an equation is solved for x, it may be somewhat easier to find dx/dy rather than dy/dx. When we speak of the derivative of a function, we usually consider the variable y as representing the function and the variable x as the independent variable. Then, of course, our chief concern is the derivative dy/dx. However, if we first find the value, dx/dy, then we can find dy/dx by the relation

$$\frac{1}{dx/dy} = \frac{dy}{dx}$$

For example, if we have the relation

$$x = y^3$$

we differentiate the function of y, (y^3), with respect to y, and get

$$\frac{dx}{dy} = 3y^2; \quad \text{then} \quad \frac{dy}{dx} = \frac{1}{3y^2}$$

14.5 The Derivative by Inversion

Let us see why the foregoing relation is true. It can be shown by taking increments Δx and Δy and then using the idea of limits that

$$\frac{1}{dx/dy} = 1 \div \frac{dx}{dy} = 1 \cdot \frac{dy}{dx} = \frac{dy}{dx}$$

EXAMPLE 11. In the following relation, find dy/dx and evaluate for $x = 8$ and $y = 2$:

$$x = (y^2 - 2)^3$$

Solution. This equation would be impractical to solve for y. We therefore take y as the independent variable and differentiate the function of y with respect to y. Then

$$\frac{dx}{dy} = 3(y^2 - 2)^2(2y) = 6y(y^2 - 2)^2$$

Inverting both sides of the equation we get

$$\frac{dy}{dx} = \frac{1}{6y(y^2 - 2)^2}$$

At $y = 2$,

$$\frac{dy}{dx} = \frac{1}{6(2)(4-2)^2} = \frac{1}{48}$$

EXAMPLE 12. Find the derivative dy/dx of the following equation and evaluate it for the point (6,3):

$$x = \frac{4y}{\sqrt{y^2 - 5}}$$

Solution. This equation can be solved for y. However, let us first find dx/dy:

$$\frac{dx}{dy} = \frac{(y^2 - 5)^{1/2}(4) - 4y(\frac{1}{2})(y^2 - 5)^{-1/2}(2y)}{y^2 - 5}$$

Multiplying numerator and denominator by the quantity $(y^2 - 5)^{1/2}$, we get

$$\frac{dx}{dy} = \frac{4(y^2 - 5) - 4y^2(1)}{(y^2 - 5)^{3/2}} = \frac{4y^2 - 20 - 4y^2}{(y^2 - 5)^{3/2}} = \frac{-20}{(y^2 - 5)^{3/2}}$$

Then

$$\frac{dy}{dx} = \frac{(y^2 - 5)^{3/2}}{-20}$$

At the point (6,3),

$$\frac{dy}{dx} = \frac{(3^2 - 5)^{3/2}}{-20} = -\frac{8}{20} = -\frac{2}{5}$$

If the equation in Example 12 is solved for y we get

$$y = \frac{\sqrt{5}x}{\sqrt{x^2 - 16}}$$

Then the derivative, dy/dx, may be found directly by use of the quotient rule again. The value of the derivative at the point (6,3) is still equal to $-\frac{2}{5}$.

EXERCISE 14.2

Find the derivative of each of the following functions with respect to the independent variable:

1. $y = (4x - 3)(x^2 - 5)$
2. $y = (x^2 + 3)(4x - x^3)$
3. $y = x^2(x^3 + 5)^4$
4. $y = (x^2 + 3)^4(x^3 - 4x)^5$
5. $y = (1 - x^2)(x^3 - 2x)^4$
6. $y = x^2(3 - x^2)^{1/2}$
7. $y = (x + 4)^5(x^2 - 2x)^3$
8. $y = x^3(5 - x^2)^{1/3}$
9. $s = t^{3/2}(t^3 + 6t)^{2/3}$
10. $s = (t^2 + 4)^{3/2}(6t - 1)^{2/3}$
11. $u = \dfrac{x^2}{\sqrt{x^2 - 5}}$
12. $v = \dfrac{\sqrt{2x + 3}}{x^2}$
13. $t = \dfrac{4 - x^2}{(x^2 - 9)^2}$
14. $s = \dfrac{t^2 - 4}{\sqrt{3t^2 + 2}}$
15. $y = \dfrac{6}{(x^2 + 3)^2}$
16. $y = \dfrac{2}{\sqrt{x^2 + 9}}$
17. $y = \dfrac{\sqrt{5 - t^2}}{t^3}$
18. $x = \dfrac{(t^3 - 8)^{1/3}}{t^2}$
19. $i = \dfrac{3t^2}{(t^2 - 1)^2}$
20. $q = \dfrac{2t + 3}{t^2 - 3t}$
21. $y = \dfrac{x^2(x^2 - 4)^{1/2}}{x^2 + 4}$
22. $y = \dfrac{x^3 - 1}{x^2(8 - x^2)^{1/3}}$

Find the derivative dy/dx in each of the following:

23. $x = (2 - y - y^2)^{1/3}$
24. $x = (y^3 - 7)^{2/3}$
25. $x = (y^2 - 3)(y^3 + 1)$; evaluate dy/dx when $y = 2$.
26. $x = 3y^2(5 - y^2)^{1/2}$; evaluate dy/dx for $y = -1$.
27. $x = \dfrac{(4y - 1)}{(y - 4)^2}$; evaluate dy/dx for $y = 1$.
28. If $y = (5x)^3$, find $\dfrac{dy}{dx}$ and $\dfrac{dy}{d(5x)}$. Evaluate each for $x = -1$.
29. If $i = (at)^5$, find $\dfrac{di}{dt}$ and $\dfrac{di}{d(at)}$.
30. In the following equation, find the derivative, dy/dx, and evaluate at the points where $x = 1$ and $x = -1$: $y(x^2 + 2) = 6$.
31. Find the derivative, dy/dx, for the function $xy - y = -2$, and find the point on the curve where dy/dx is equal to $\tfrac{1}{2}$.
32. Find the derivative of the function, $y^2(x^2 - 4) = 20$, and find the value of the derivative at the point (3,2).

chapter

15

Linear Motion; Distance; Velocity; Acceleration

15.1 THE PROBLEM OF LINEAR MOTION

On an automobile trip we may be interested in such questions as the following: (a) How far have we gone? (b) How fast are we going? (c) Are we slowing down or speeding up?

When we ask how far we have gone, we are interested in *distance s*. When we ask how fast we are going, we are interested in speed or velocity *v*. When we ask whether we are slowing down or speeding up, we are interested in acceleration *a*, which represents some change in velocity. We step on the accelerator to increase velocity. We step on the brake to decrease velocity. We shall see that the accelerator and the brake may also have effects opposite to those mentioned.

Acceleration in everyday parlance usually means an *increase* in velocity. The term *deceleration* is sometimes used to indicate a *decrease* in velocity, but we shall use the term *acceleration* to refer to any change in velocity, whether positive or negative. Calculus can help to answer questions about *distance, velocity,* and *acceleration.*

15.2 AVERAGE VELOCITY

If we travel a distance of 120 miles in 4 hours, we call the 120 miles the *distance s* and the 4 hours the *time t*. To find *average velocity* we divide *distance* by *time* and get 30 miles per hour (mph). The result must be considered *average* velocity since we could not possibly drive at exactly this rate every second of the 4 hours. At some time we may have been traveling at 60 mph and at another time perhaps only 10 mph or less.

To measure average velocity, we take a carefully measured distance and divide this distance by the carefully measured time it takes to cover this distance. Suppose a car passes a particular point A at exactly 7 seconds past the hour, and it passes point B at exactly 11 seconds past the hour (Fig. 15.1). Suppose the measured distance from A to B is 240 feet. This

Fig. 15.1

$$\text{— — — — — } \overset{s}{} \overset{A}{\bullet} \overset{\Delta s}{} \overset{B}{\bullet} \text{— — —}$$

distance is covered in 4 seconds. The distance AB we call Δs, since it is an increment in the distance traveled. The time 4 seconds we call Δt, since it is an increment in the time of traveling. Then the average velocity over the distance AB is given by the formula

$$v_{(\text{average})} = \frac{\Delta s}{\Delta t} = \frac{240 \text{ ft}}{4 \text{ sec}} = 60 \text{ ft/sec}$$

15.3 INSTANTANEOUS VELOCITY

In the foregoing example the result, 60 ft/sec, shows the *average velocity*. However, the velocity at A may be different from the velocity at B. Now we are interested in the exact velocity *at the instant* when the car passes point A. This is the meaning of *instantaneous velocity*. Let us see how we can arrive at an interpretation of instantaneous velocity.

We can get a closer approximation to the instantaneous velocity at A if we make Δs smaller, say, AC (Fig. 15.2). Then Δt will also be smaller. Again, if we divide Δs by Δt, we must still say that the result represents only average velocity from A to C. We could continue to make Δs smaller and smaller, in which case Δt would also become less and less. Yet, however small we make the increment Δs, we shall get only *average velocity* between two points, not instantaneous velocity at a point.

Fig. 15.2

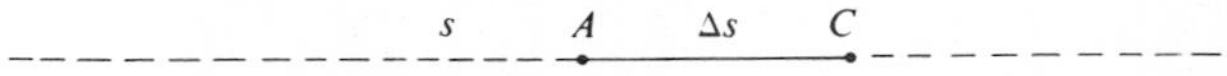

$$\text{— — — — — — } \overset{s}{} \overset{A}{\bullet} \overset{\Delta s}{} \overset{C}{\bullet} \text{— — — — —}$$

It would seem that in order to get the velocity we want, we should have $\Delta s = 0$ and $\Delta t = 0$. But then we should get $0/0$, which would be meaningless. Instead of taking $\Delta t = 0$ and $\Delta s = 0$, we let these increments *approach* zero as limits. Then we can define the instantaneous velocity v at A as the *limit of the ratio* of the increment in distance s to the increment in time t as the increments approach zero as a limit. That is,

$$v = \lim_{\Delta t \to 0} \frac{\Delta s}{\Delta t}$$

Note that this limit is the derivative of distance s with respect to time t and can then be denoted by ds/dt. It can also be denoted by s'.

EXAMPLE 1. Find the initial velocity and the velocity at the end of 1 second of a particle that moves according to the formula (distance s in feet, time t in seconds)

$$s = 3t^2 - 2t + 8$$

Solution. In the formula s represents distance (in feet) from a zero position. Differentiating,

$$v = \frac{ds}{dt} = 6t - 2 \quad \text{(velocity in feet per second)}$$

15.3 Instantaneous Velocity

At the time of initial velocity, $t = 0$. Substituting in the formula for velocity, we get

$$v = 0 - 2 = -2, \quad \text{or} \quad -2 \text{ ft/sec}$$

The result indicates that the initial velocity is in a *negative* direction. At the end of 1 second, $t = 1$. Substituting, we get

$$v = 6 - 2 = 4, \quad \text{or} \quad 4 \text{ ft/sec}$$

That is, at the end of 1 second the particle is moving in a positive direction at a rate of 4 feet per second.

EXAMPLE 2. A particle moves according to the formula $s = 3t^2 - 4t + 10$ (distance s in feet, time t in seconds). Find the position and velocity of the particle when $t = 0$ and when $t = 2$. When is the particle at rest? Show the positions on a horizontal line.

Solution. Position is given by the formula for distance in feet from a point where $s = 0$. For position we have

$$s = 3t^2 - 4t + 10$$

For velocity, we have $\quad v = \dfrac{ds}{dt} = s' = 6t - 4$

At time $t = 0$, $s = 0 - 0 + 10 = 10$ (feet from zero). That is, when $t = 0$, the particle is 10 feet in a positive direction from zero (Fig. 15.3). Using the formula, $v = 6t - 4$, for velocity, we get, at $t = 0$, $v = -4$. That is, for initial velocity, the particle is moving with a velocity of -4 ft/sec, which is, of course, in a negative direction.

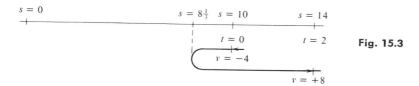

Fig. 15.3

If we take the positive direction toward the right, and the negative toward the left, then, when $t = 0$, the particle is 10 feet to the right of zero, and it is moving toward the left at a rate of 4 ft/sec. It is important that we keep clearly in mind the opposite directions, positive and negative, both for position and velocity.

At $t = 2$, we have $s = 12 - 8 + 10 = 14$ ft and $v = 12 - 4 = 8$ ft/sec. That is, at the end of 2 seconds the particle is 14 feet to the right of zero and it is moving with a velocity of $+8$ ft/sec, which is to the right.

Sometime between the times $t = 0$ and $t = 2$, the velocity of the particle changed from a negative to a positive direction. At the instant of change of direction, the particle was at rest, only instantaneously. This occurred when the velocity was equal to zero, that is,

$$\text{when} \quad 6t - 4 = 0$$

Solving for t, $\quad t = \dfrac{2}{3} \text{ second}$

Linear Motion; Distance; Velocity; Acceleration

To locate the position of the particle at the instant when velocity changed from negative to positive, we substitute $t = \frac{2}{3}$ in the formula for distance and get

$$s = 3\left(\frac{4}{9}\right) - 4\left(\frac{2}{3}\right) + 10 = \frac{26}{3} = 8\frac{2}{3} \text{ ft}$$

EXAMPLE 3. An object moves according to the formula $s = t^2 - 6t + 8$ (s in feet, v in feet per second). Find initial position and velocity, and find position and velocity when $t = 1, 5,$ and 7 seconds. When and where is the particle at rest? When is the object at the position 0? Indicate the path of motion on a horizontal line positive toward the right.

Solution. For position we have

$$s = t^2 - 6t + 8$$

Differentiating, $\qquad v = ds/dt = 2t - 6$

For initial position and velocity we let $t = 0$; then $s = 8$, $v = -6$. For the times when $t = 0, 1, 5,$ and 7 sec, we have the following table of values.

time $t =$	0	1	5	7	seconds
distance $s =$	8	3	3	15	feet from zero
velocity $v =$	-6	-4	4	8	ft/sec

To find the time when the object is at rest, we set the expression for velocity equal to zero:

$$2t - 6 = 0; \quad \text{then} \quad t = 3 \text{ seconds}$$

At that time the object is at the position $s = (3)^2 - 6(3) + 8 = -1$, which means 1 foot at the left of zero (Fig. 15.4).

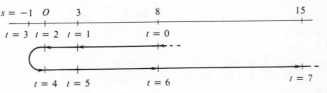

Fig. 15.4

To find the time when the object is at the position $s = 0$, we set the expression for s equal to zero: $t^2 - 6t + 8 = 0$, $t = 2$ and 4. Note that the object is at zero twice. Note also that the velocity is negative during the interval from $t = 0$ to $t = 3$. At $t = 3$ the object is at rest just instantaneously, and then changes to a positive direction of motion, which continues indefinitely.

The equation, $s = t^2 - 6t + 8$, for distance, in Example 3, may be shown more clearly on a graph by taking time t as the horizontal axis, and distance s as the vertical axis (Fig. 15.5). Taking $t = 0, 1, 2, 3, 4, 5, 6, 7,$ and so on, we find corresponding values of s. These values are then plotted on the graph. Several facts may be noted from the figure. The curve begins with a value of 8 when $t = 0$. It crosses the t-axis in two places, $t = 2$ and $t = 4$. The curve is parabolic in shape, and the lowest point of the curve is $s = -1$, when $t = 3$, called a *minimum* point. From then on, the curve continues to rise.

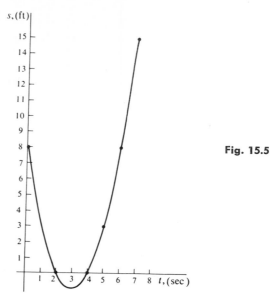

Fig. 15.5

15.4 ACCELERATION

Suppose we have two different types of cars, each traveling at, say, 10 miles per hour. Some time later each car is traveling at a rate of, say, 50 miles per hour. Each car has increased its velocity by 40 miles per hour from 10 mph to 50 mph. However, let us suppose the first car was an old one with little horsepower so that it required 20 seconds to make the change in velocity. We can say, then, that the first car changed its velocity by an average of 2 miles per hour per second. This is *average acceleration*.

If we represent first and second velocity by v_1 and v_2, respectively, then the change in velocity Δv is given by

$$\Delta v = v_2 - v_1 = 40 \text{ mph}$$

Since $\Delta t = 20$ seconds, for average acceleration we have

$$a_{\text{(average)}} = \frac{\Delta v}{\Delta t} = \frac{40 \text{ mph}}{20 \text{ sec}} = 2 \text{ mph/sec}$$

Now suppose the second car had a powerful motor and changed its velocity by 40 mph in 4 seconds. In this case,

$$\Delta v = 40 \text{ mph} \quad \text{and} \quad \Delta t = 4 \text{ sec}$$

Then

$$a_{\text{(average)}} = \frac{\Delta v}{\Delta t} = \frac{40 \text{ mph}}{4 \text{ sec}} = 10 \text{ mph/sec}$$

Each of the cars increased its velocity by the same amount, but over different periods of time. Even so, we cannot tell that the acceleration was uniform over the time interval. All we can tell so far is the average acceleration. What we wish to find next is *instantaneous acceleration a*.

Consider another example: The velocity of an automobile or other vehicle is often stated in feet per second (ft/sec). Suppose we have two cars each traveling at 15 ft/sec. Some time later each car is traveling at 75 ft/sec. Each changed its velocity by 60 ft/sec, but not in the same length of time, as shown in the following table:

	v_1	v_2	Δv	Δt	average acceleration, $\dfrac{\Delta v}{\Delta t}$
first car	15 fps	75 fps	60 fps	20 sec	3 ft/sec²
second car	15 fps	75 fps	60 fps	5 sec	12 ft/sec²

The first car required 20 seconds to make the change of 60 ft/sec in velocity, while the second required only 5 seconds. The first car increased its velocity by an average of 3 ft/sec each second. Its average acceleration was 3 ft per second per second, written 3 ft/sec/sec, or 3 ft/sec². The second car had an average acceleration of 12 ft/sec².

To find instantaneous acceleration at a point, we proceed in the same way as we did in finding instantaneous velocity. We let Δt and Δv become smaller and smaller and approach zero as a limit. We then define instantaneous acceleration a as

$$a = \lim_{\Delta t \to 0} \frac{\Delta v}{\Delta t} = \frac{dv}{dt} = v'$$

We have seen that $v = ds/dt$. Therefore,

$$a = \frac{d}{dt}(v) = \frac{d}{dt}\left(\frac{ds}{dt}\right) = \frac{d^2s}{dt^2} = s''$$

That is, acceleration is the first derivative of velocity and the second derivative of the distance. In summary,

$$\text{if } s = f(t), \quad \text{then } v = \frac{ds}{dt} = s' = f'(t)$$

and

$$a = \frac{dv}{dt} = v' = \frac{d^2s}{dt^2} = s'' = f''(t)$$

As an example, if $\quad s = t^3 - 3t^2 - 9t + 12$
for velocity we have $\quad v = 3t^2 - 6t - 9$
for acceleration $\quad a = 6t - 6$

EXAMPLE 4. A ball is thrown directly upward from the ground and its position above the ground is given by the formula $s = 80t - 16t^2$. Find initial velocity and acceleration; find s, v, and a when $t = 2$ seconds.

Solution. For distance we have $\quad s = 80t - 16t^2$
for velocity, $\quad v = 80 - 32t$
for acceleration, $\quad a = -32$

15.4 Acceleration

For initial velocity we take $t = 0$; then $v = 80 - 0 = 80$ ft/sec. In this example, note that acceleration is not affected by a change in time since it is equal to the constant, -32 ft/sec². When $t = 2$, we have, $s = 160 - 16(4) = 96$ ft, $v = 80 - 64 = 16$ ft/sec, and $a = -32$ ft/sec².

To understand the concept of acceleration it is well to think of acceleration as a force. The accelerating force can be positive or negative. Any force that tends to make velocity more positive or less negative is *positive* acceleration. Any force that tends to make velocity less positive or more negative is *negative* acceleration.

It is possible to have velocity, positive or negative, and yet have zero acceleration. It is also possible to have an accelerating force without any velocity whatever.

Suppose you are driving a car at a steady rate of 20 mph *forward*. In this case, $v = 20$ but $a = 0$. Now you step on the accelerator. Velocity increases in a positive direction, and acceleration is therefore positive. Now you step on the brake. You are still moving forward and velocity is still positive, but acceleration is negative because the accelerating force, the brake, decreases velocity.

Now let us suppose you come to a stop. Then $v = 0$ and $a = 0$.

Suppose later you are backing the car at a steady rate of 4 mph. Your velocity is now -4 mph, but acceleration is zero. While moving backward, you step on the accelerator and change velocity from -4 mph to -6 mph. In this case, the accelerator produces negative acceleration. Although your absolute motion is faster, your velocity is considered to have decreased from -4 to -6. If you now step on the brake to slow down your backward motion, the velocity becomes less negative, perhaps from -6 mph to -3 mph. The brake then acts as a positive accelerating force, producing positive acceleration.

EXERCISE 15.1

In each of the following problems in motion, distance is in feet, time is in seconds, velocity is in feet per second, and acceleration is in feet per second². In each of the first twelve problems find the velocity, distance, and acceleration when $t = 0$; also when $t = 1$:

1. $s = 3t + 7$
2. $s = 4 - 5t$
3. $s = 15 - t^3$
4. $s = 12t + 5t^4$
5. $s = 6t - 2t^2$
6. $s = t^2 - 5t - 4$
7. $s = 4t^2 - 3t + 2$
8. $s = 80 + 20t - 6t^2$
9. $s = t^3 - 6t^2 - 15t + 10$
10. $s = 8 - 3t + t^2$
11. $s = 2t^3 - 3t^2 - 12t$
12. $s = 5 - 6t + 4t^2 - 2t^3$

13. A particle moves according to the formula $s = t^3 - 6t^2 + 9t + 20$. Where is the particle with reference to the zero point when the time t is zero? How fast is it moving and in what direction (on a horizontal plane)? How is its velocity changing? Show on a horizontal scale.
14. In problem No. 13 find the position, the velocity, and the acceleration at the end of exactly 2 seconds. At what *two* times is the particle at rest? Where is it at rest? What is its acceleration when it is at rest?

15. A particle moves in a linear motion according to the formula $s = 10 - 18t^2 + t^3$. Tell exactly where the particle is and what is happening at the end of 2 seconds.
16. The motion of a particle is described by the formula $s = 15 - 27t + 9t^2 - t^3$. Describe the position, the velocity, and the acceleration at $t = 0$. When and where is the particle at rest? What is its acceleration when $v = 0$?
17. A ball is projected vertically upward from the top of a building. Its height above the ground at any time t is given by the formula

$$s = 160 + 80t - 16t^2$$

Find: (a) the initial velocity of the ball; (b) the height of the building; (c) the time required to reach its maximum height; (d) the maximum height reached by the ball; (e) the velocity with which it strikes the ground (neglect air resistance).
18. An object is projected vertically upward, and its height above the ground is given by the formula, $s = 320t - 16t^2$. Find: (a) the initial velocity of the object; (b) the maximum height attained; (c) its velocity when it strikes the ground on its return; (d) its original height above the ground when $t = 0$; (e) how long the object was in the air.
19. A 22-caliber rifle bullet is fired directly upward from the ground. Its height above the ground at any time t is given by the formula, $s = 800t - 16t^2$. Neglecting air resistance, how high will the bullet rise? How long will it be in the air before striking the ground on its return? What is the muzzle velocity of the bullet?
20. The bullet of a high-powered rifle is shot directly upward from the ground, and its height is given by the formula, $s = 2800t - 16t^2$. How high will the bullet rise (neglecting air resistance)? What is the muzzle velocity of the bullet? How long will the bullet be in the air?
21. A small boy once accidentally dropped a small marble out from near the top of the Washington Monument. It fell approximately 540 feet and struck a person in the shoulder. What was the velocity of the marble when it struck?
22. An object is projected up along a frictionless inclined plane according to the formula $s = 10t - 2t^2$, where s represents the distance (in feet) from the starting point and t represents time (in seconds). How far up the plane will the object move? What is its initial velocity? Find its velocity at the end of 2 seconds; 3 seconds; 4 seconds.
23. For a man who can high jump 6.25 ft, his distance above the ground is given approximately by the formula $s = 20t - 16t^2$. What is his initial velocity? How long does it take him to reach maximum height?
24. If the man in No. 23 were on the moon and jumped with the same initial velocity, his height above the surface would be approximately given by the formula $s = 20t - 2.6t^2$. How high could he jump on the moon?
25. Two particles start at the same time and move along the same straight line. Their distances are given, respectively, by the following formulas:

$$s_1 = 3t^2 - 2t - 3 \quad \text{and} \quad s_2 = 2t^2 + 4t - 8$$

Find: (a) the position of each particle when $t = 0$; (b) the velocity of each when $t = 0$; (c) at what two times are the particles together? (d) where?
26. Two particles start at the same time and move along the same straight line. Their distances from zero are given, respectively, by the following formulas;

$$s_1 = t^2 - 8t + 12 \quad \text{and} \quad s_2 = 3 + 4t - 2t^2$$

At what two times are the particles together? Sketch the motion of each particle along a straight line.

chapter

16

The Derivative as the Slope of a Curve

16.1 THE SLOPE OF A CURVE

We recall that in defining the slope of a straight line, we take two points, say, $P_1(x_1,y_1)$ and $P_2(x_2,y_2)$, on the line (Fig. 16.1). We then define the slope m of the straight line as

$$m = \frac{\text{rise}}{\text{run}} \quad \text{or} \quad m = \frac{y_2 - y_1}{x_2 - x_1}$$

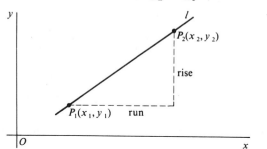

Fig. 16.1

For the slope of a curve we run into difficulties. For example, suppose we take two points P_1 and P_2 on a curve (Fig. 16.2). Then, in moving along the curve from P_1 to P_2, we get the following:

$$\text{rise} = y_2 - y_1 \qquad \text{run} = x_2 - x_1$$

However, note that in this case, if we use the formula for the slope, we get the slope of the secant line s through P_1 and P_2. That is,

$$\frac{y_2 - y_1}{x_2 - x_1} = \text{slope of secant line } s$$

Note that the slope (or steepness) of the curve is everywhere changing. For example, the slope of the curve at P_1 is different from the slope at P_2. How then shall we define the slope of the curve at one particular point? For example, what is the slope of the *curve* at P_1?

203

Fig. 16.2

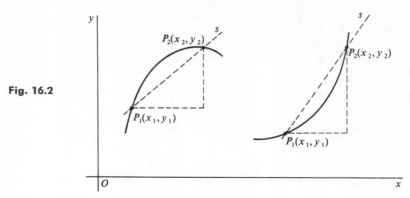

If we draw a straight line tangent to the curve at P_1 (Fig. 16.3), then we say the slope of the curve at that point is the slope of the tangent line.

Fig. 16.3

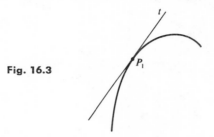

Definition:

The slope of a curve at any point is defined as the slope of the tangent line at that point.

16.2 RELATION BETWEEN DERIVATIVE AND SLOPE

If a function is graphed (in rectangular coordinates), then the derivative of the function has a simple geometric interpretation. We shall discover that the value of the *derivative* for any particular value of x is equivalent to the value of the *slope* of the curve where x has the given value. For example, let us graph the function

$$y = x^2 - 3x - 4 \quad \text{(Fig. 16.4)}$$

The derivative of the function is

$$\frac{dy}{dx} = 2x - 3$$

For any value of x, say, $x = 2$, we can now find the value of the derivative. For example, where $x = 2$,

$$\frac{dy}{dx} = 2(2) - 3 = 1$$

Now, at the point on the curve where $x = 2$, that is, at $(2, -6)$, the slope of the curve is also equal to 1, the derivative. *In general, at any point on* the curve, the slope of the curve at that point will be equal to the numerical value of the derivative at that point.

Using this rule, we can find the slope of the curve in Fig. 16.4 at any point on the curve (slope is shown by a short tangent line):

At point (4,0), $m = dy/dx = 2(4) - 3 = 5$
At point (3,−4), $m = dy/dx = 2(3) - 3 = 3$
At point (1,−6), $m = dy/dx = 2(1) - 3 = -1$
At point (−1,0), $m = dy/dx = -2 - 3 = -5$
At point (0,−4), $m = dy/dx = 2(0) - 3 = -3$
At point (1.5,−6.25), $m = 2(1.5) - 3 = 0$

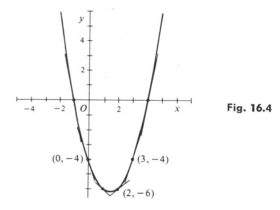

Fig. 16.4

16.3 DERIVATION OF GEOMETRIC MEANING OF DERIVATIVE

To show that the slope of a curve (in rectangular coordinates) is equivalent to the derivative of the function, as stated in Sec. 16.2, let us graph the general function

$$y = f(x) \quad \text{(Fig. 16.5)}$$

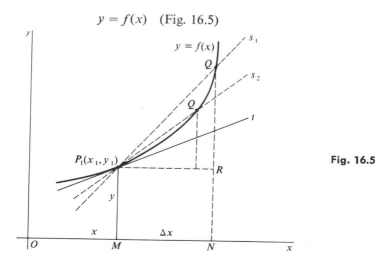

Fig. 16.5

206 The Derivative as the Slope of a Curve

Let $P_1(x_1,y_1)$ be any particular point on the curve. Now let x be increased by some increment

$$\Delta x = MN$$

Then y is increased by the increment $\Delta y = RQ$.

The straight line through P_1 and Q is the secant line s_1; and for its slope we have

$$\frac{\Delta y}{\Delta x} = \frac{RQ}{P_1 R} = \text{slope of secant line } s_1 \text{ (definition)}$$

Now let the point Q move down the curve toward P, so that Δx and Δy both become smaller. Then the slope of the new secant line is still

$$\frac{\Delta y}{\Delta x} = \text{slope of secant line } s_2$$

As Δx now approaches zero as a limit, we have the following results:

1. Δy approaches zero as a limit.
2. The secant line approaches the position of the tangent line as a limit.
3. The slope of the secant line approaches the slope of the tangent line as a limit.

Let us denote the slope of the secant line by m_s and the slope of tangent line by m_t. Then we have the following:

$$\frac{\Delta y}{\Delta x} \text{ approaches } \frac{dy}{dx} \text{ as a limit}$$

$$m_s \text{ approaches } m_t \text{ as a limit}$$

But $\Delta y/\Delta x$ is always equal to m_s. Therefore, $dy/dx = m_t$.

Theorem.

If two variables are always equal to each other and both approach limits, then the limits are equal.

We have stated that the slope of a curve at a point on the curve is defined as the slope of the tangent to the curve at that point. Therefore, the derivative dy/dx is equal to the slope of the curve at that point, that is,

$$\frac{dy}{dx} = m$$

The second derivative of the function then becomes the rate of change of the slope, that is,

$$\frac{d^2y}{dx^2} = \frac{dm}{dx}, \quad \text{rate of change of slope}$$

Note: The student is warned: Do not define the derivative of a function as the slope of the curve. When we say that the value of

16.3 Derivation of Geometric Meaning of Derivative

the derivative *happens to be* equivalent to the slope of the curve at any point, this is *not* a definition of the derivative. It just happens to be true when the function is graphed in rectangular coordinates. It is not true for polar coordinates. Moreover, we may have a derivative of a function without any curve at all. For example, when we say the rate of change of distance s with respect to time t is called velocity, we have no curve at all! All we have is

$$\text{if } s = f(t), \text{ then } v = \frac{ds}{dt}$$

Remember, the derivative is *defined* as the instantaneous rate of change of a function with respect to the independent variable, *with nothing said about a curve*. If $y = f(x)$, then dy/dx is the instantaneous rate of change of y with respect to x.

We shall now work out examples to show how the derivatives may be used to determine the slope of the curve at some particular point.

EXAMPLE 1. Find the slope of the curve $y = x^2 - 2x - 3$ at the following points: (a) where $x = 4$; (b) at point $(1, -4)$; (c) at the points where the curve crosses the x-axis, that is, where $y = 0$; (d) at the point $(2.4, -2.04)$; (e) at $(\frac{7}{3}, -\frac{20}{9})$.

Solution. First, we differentiate: $dy/dx = 2x - 2$.
(a) Where $x = 4$, $dy/dx = 2(4) - 2 = 6$. (b) At $(1, -4)$, $dy/dx = 2 - 2 = 0$.
(c) At $y = 0$, $x = -1$ and $+3$; then $dy/dx = -4$; at $x = 3$, $dy/dx = +4$.
(d) At $(2.4, -2.04)$, $dy/dx = 4.8 - 2 = 2.8$, or $\frac{14}{5}$.
(e) At $x = \frac{7}{3}$, $dy/dx = \frac{8}{3}$.

EXAMPLE 2. Find the slope of the curve $y = x^3 - x^2 - 5x + 4$ for each of the following values of x: -1; 0; 1; 1.5; 2; 3. Sketch the curve and write the equations of the tangent and the normal at the point $(2, -2)$.

Solution. Differentiating, $\dfrac{dy}{dx} = 3x^2 - 2x - 5$

Using the derivative, we find the slope for each given value of x and compute the corresponding value of y, as shown in the following table:

x	-1	0	1	1.5	2	3
y	7	4	-1	-2.375	-2	7
m	0	-5	-4	-1.25	3	13

Now we can use the slopes at the given points to aid in sketching the curve (Fig. 16.6). For the equation of the tangent at the point where $x = 2$, we have $m = 3$ and the point $(2, -2)$. The slope of the normal is $-\frac{1}{3}$. Then equation of tangent: $3x - y = 8$; normal: $x + 3y = -4$.

208 The Derivative as the Slope of a Curve

EXAMPLE 3. At what points does the following curve have horizontal tangents, if any? Also find the slope at $x = 4$ and $x = 2$ and sketch the curve:

$$y = x^3 - 3x^2 - 9x + 12$$

Solution.
$$\frac{dy}{dx} = 3x^2 - 6x - 9$$

If the tangent is to be horizontal, the slope must equal zero, that is, $m = dy/dx = 0$. We therefore set the derivative equal to zero and solve for x:

$$3x^2 - 6x - 9 = 0$$

Dividing by 3,
$$x^2 - 2x - 3 = 0$$

Factoring,
$$(x - 3)(x + 1) = 0$$
$$x = 3, \text{ and } x = -1.$$

For the coordinates of the points we have:

when $x = 3$, $y = 27 - 27 - 27 + 12 = -15$; this is the point $(3, -15)$.

when $x = -1$, $y = -1 - 3 + 9 + 12 = 17$; this is the point $(-1, 17)$.

For the value $x = 4$, $m = 3(16) - 6(4) - 9 = 15$. For $x = 2$, $m = -9$. The slopes for the various points can be used to aid in graphing (Fig. 16.7).

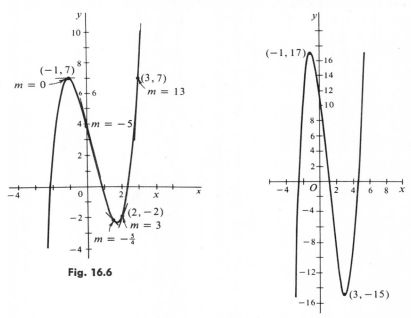

Fig. 16.6

Fig. 16.7

Note: In Example 3, in solving the equation for $m = 0$ (that is, the equation $3x^2 - 6x - 9 = 0$), we can, of course, divide both sides of the equation by 3 and obtain the equivalent equation $x^2 - 2x - 3 = 0$. However, in finding the *value* of the slope for any

16.3 Derivation of Geometric Meaning of Derivative

particular value of x, such as $x = 4$, or $x = -2$, it is important that we use the true expression for the slope; that is, $3x^2 - 6x - 9$. In fact, the expression, $x^2 - 2x - 3$, represents only $\frac{1}{3}$ of the slope value.

EXAMPLE 4. Find the slope of the curve $y = 2x - 7$ at the point $x = 3$.

Solution. Differentiating, $$dy/dx = 2$$

The result shows that the slope is the constant 2 and is the same for every value of x. This fact can, of course, be determined directly from the given equation, which is the slope-intercept form of the equation of a straight line. The curve has no horizontal tangent because $2 \neq 0$.

EXAMPLE 5. Write the equation of the tangent to the curve $y = (5 - x^2)^{1/2}$ at the point where $x = -1$.

Solution. Differentiating,

$$\frac{dy}{dx} = \left(\frac{1}{2}\right)(5 - x^2)^{-1/2}(-2x) = \frac{-x}{(5 - x^2)^{1/2}}$$

For the slope of the tangent we evaluate the derivative at $x = -1$. For this value, $dy/dx = \frac{1}{2}$. From the equation of the curve, $y = 2$ when $x = -1$. Now we have the slope $\frac{1}{2}$ and the point $(-1, 2)$. Then the equation of the tangent is

$$x - 2y = -5$$

EXAMPLE 6. Find the derivative dy/dx of the function

$$y = \frac{x^2}{\sqrt{5 - 4x}}$$

Then find the equation of the tangent where $x = -1$. Find any points where the curve has a horizontal tangent.

Solution. By the quotient rule, we get

$$\frac{dy}{dx} = \frac{10x - 6x^2}{(5 - 4x)^{3/2}}$$

Evaluating the derivative at $x = -1$, we get

$$m = -\frac{16}{27}$$

When $x = -1$, $y = \frac{1}{3}$. Then the equation of the tangent is: $16x + 27y = -7$. For a horizontal tangent the slope must be equal to zero, that is,

$$\frac{10x - 6x^2}{(5 - 4x)^{3/2}} = 0 \quad \text{or} \quad 6x^2 - 10x = 0$$

Solving the equation for x, we get

$$x = 0, \quad \text{and} \quad x = \tfrac{5}{3}.$$

However, before we can say these values represent points on the curve, we must be sure of the y-values. From the original equation we have

when $x = 0$, $y = 0$; when $x = \frac{5}{3}$, y becomes imaginary.

Then the only point where the curve has a horizontal tangent is $(0, 0)$.

EXERCISE 16.1

1. Find the slope of the graph of the following function at the given points and then use this information to aid in sketching the curve:
$$y = x^2 - 4x - 5$$
First find the slopes at the intercepts, $(-1,0)$ and $(5,0)$, and at $(0,-5)$. Then find the point (or points) where $m = 0$ and show by short horizontal tangents. Then find slope at $x = 1$; $x = 3$; $x = -1$; $x = 5$.

2. In the curve $y = x^2 - x - 2$, find the point where $m = 0$. Then find the slope at each of the following points: $x = 0$, $x = 1$, $x = 2$, $x = -2$. From this information sketch the graph.

3. In the graph of the function, $y = 4 + 3x - x^2$, find the slope at the intercepts, and find the point where $m = 0$.

4. Find the slope of the parabola, $y = x^2 - 4x - 5$ at the point $(1,-8)$, and then write the equation of the tangent to the curve at that point.

5. Write the equation of the tangent and of the normal to the curve, $y = x^2 - x - 6$ at the point where $x = 2$.

6. In the equation, $y = x^2 - 5x + 2$, find the value of x for which $dy/dx = 3$.

7. In the cubic, $y = x^3 - 3x^2 - 24x + 7$, find the slope of the curve at the following points: $x = 1$; $x = 3$; $x = -1.5$; find the points where $m = 0$.

8. In the cubic $y = 7 + 15x + 6x^2 - x^3$, find the points where the curve has horizontal tangents (two points). Find the equation of the tangent and equation of the normal at the point where $x = 2$.

9. In the cubic, $y = x^3 - 6x^2 + 12x - 9$, where is the slope equal to zero? Also find the slope at the points where $x = 1$; $x = 3$; $x = 0$; $x = 4$. Use this information to aid in graphing the curve.

10. In the parabola, $y = x^2 - 2x - 8$, find the slope of the curve at both points where $y = 0$ (x-intercepts). Graph the curve.

11. In the cubic, $y = x^3 - 3x^2 + 3x - 7$, for what values of x is the derivative equal to zero? At what points is the derivative equal to 4? Sketch the curve.

12. In the function, $s = t^3 - 6t^2 + 9t + 10$, s is in feet, t is in seconds. At what values of t is ds/dt equal to zero?

13. In the function, $i = t^2 - 3t + 2$, find the value of i when $di/dt = 5$.

14. In the graph of the relation, $y^2 = 4x$, find the slope of the curve at both points on the curve where $x = 4$. What is the slope at $x = 0$?

Evaluate the derivative of each of the following functions for the given value of x:

15. $y = (x^2 + 3)^{1/2}(5 - 6x)$; $x = 1$
16. $y = (9 - x^3)^{2/3}(x^2 - 3)^2$; $x = 1$
17. $y = x^4(2x + 3)^{2/3}$; $x = -1$
18. $y = (x^3)(13 - x^2)^{1/2}$; $x = 2$

Write the equation of the tangent to each of the following curves at the point indicated by the given value of x:

19. $y = (x^2 + 3)^{3/2}$; $x = 1$
20. $y = (3x - 2)^{3/2}$; $x = 2$
21. $y = x^2(2x + 6)^{1/2}$; $x = -1$
22. $y = (3x + 5)(3 + x^2)^{1/2}$; $x = -1$
23. $y = \dfrac{(5 - x^2)^{1/2}}{x^2}$; $x = 1$
24. $y = \dfrac{x^2}{\sqrt{x^2 - 3}}$; $x = 2$
25. $y = \dfrac{x}{(x^2 + 5)^{1/2}}$; $x = 2$
26. $y = \dfrac{(x^2 + 5)^{1/2}}{x}$; $x = 2$

Determine a point where each of the following curves has a horizontal tangent:

27. $y = (9 - x^2)^{1/2}$
28. $y = (x^2 - 4)^2(3x - 4)^3$
29. Does the following curve have a horizontal tangent: $y = \sqrt{x^2 + 6x + 4}$?

chapter

17

Maxima and Minima; Inflection Points

17.1 INCREASING AND DECREASING FUNCTIONS

We have seen that the derivative, dy/dx, of a function y is equivalent to the instantaneous rate of change of the dependent variable y with respect to the independent variable x. If the changes in the function and in x are in the *same direction*, then the rate of change, the derivative, is *positive*. That is, if the function y increases as the variable x increases, then the rate of change, dy/dx, is *positive*. If the function decreases as x increases, then the rate of change, dy/dx, is *negative*.

In considering increasing or decreasing values, we always consider x as moving from left to right and therefore increasing in value.

Increasing and decreasing values of a function can easily be seen in the graph of a function. Consider the graph of the general function, $y = f(x)$ (Fig. 17.1). In the interval from A to B, we see that x is increasing. The function y is also increasing in this interval. Therefore, dy/dx, the rate of change, is positive in this interval. This is further shown by the fact that the slope of the curve is positive.

In the interval from B to C, x is also increasing in value, but the function y is decreasing. Therefore, the rate of change, dy/dx, is negative in this interval. This is further shown by the negative slope of the curve.

At one particular point, D, the value of y, for just an instant, is neither increasing nor decreasing. The derivative at this point is therefore zero; that is, the slope m of the curve is zero.

Figure 17.1 shows that a function may be increasing over some interval and decreasing over some other interval. On the other hand, a function may be increasing for all values of x or decreasing for all values of x. A straight line with a positive slope is a continuously increasing function, whereas a straight line with a negative slope is a continuously decreasing function. There are other functions also that are continuously increasing or continuously decreasing.

212 Maxima and Minima; Inflection Points

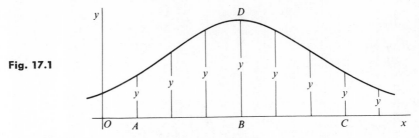

Fig. 17.1

17.2 THE SLOPE OF A CONTINUOUS CURVE, $y = f(x)$

Whereas the slope of a straight line is the same at every point, the slope of a curve, as we have seen, is continuously changing. Let us estimate the slope of a particular curve $y = f(x)$ at a number of points on the curve (Fig. 17.2). To do so, we draw a short straight line segment tangent to the curve at each point and estimate the slope of the tangent. Then the slope of the tangent line is, by definition, the slope of the curve at that point. Note that the slope m is zero at five points.

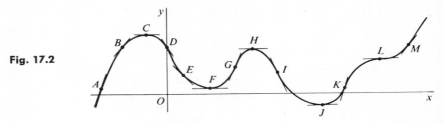

Fig. 17.2

17.3 MAXIMUM POINT AND MAXIMUM VALUE

On the curve (Fig. 17.2), a point such as C is called a *maximum* point. A maximum point is a point on a curve that is higher than a nearby point at either side, right or left. This point is not necessarily the highest point the curve ever reaches. But a maximum point, (in this case, a *relative* maximum), is higher than a neighboring point on either side. A curve may have more than one relative maximum point. The *maximum value* is the function value at the maximum point. The maximum value is actually the *ordinate* of the curve. An *absolute maximum* point is a point higher than any other point on the curve. A maximum point may be compared to the peak of a mountain.

17.4 MINIMUM POINT AND MINIMUM VALUE

On the curve (Fig. 17.2) a point such as F is called a *minimum point*. A minimum point is a point on a curve that is lower than a nearby point at either side. It is not necessarily the lowest point the curve ever reaches. But a minimum point (in this case, a *relative* minimum), is lower than a neighboring point on either side. A curve may have more than one relative minimum point. The *minimum value* of the function is the value of the ordinate at that point. An *absolute minimum* is a point lower than any other point on the curve.

17.5 MAXIMUM POINT — CHARACTERISTICS

It is often necessary to locate maximum and/or minimum points on a curve. Our problem is to see how this can be done.

Let us take a careful look at the maximum point in Figure 17.3. First, we know that the slope is zero at the maximum point; that is, $dy/dx = 0$. Suppose we have a maximum point where $x = 3$. Then, at $x = 3$, $m = 0$.

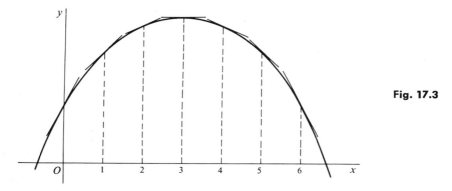

Fig. 17.3

Note that just at the left of the maximum point, say, where $x = 2$, the slope is positive. A little to the right of the maximum point, say, where $x = 4$, the slope is negative. Then, for a maximum point, the following conditions are true:

1. At the maximum point, $m = 0$.
2. A little to the left of this point, m is positive $(+)$.
3. A little to the right of this point, m is negative $(-)$.
4. As the curve passes through the maximum point, the slope changes from $(+)$ to (0) to $(-)$.

At the left of the maximum point, the function is increasing, indicated by a positive derivative. At the right of the maximum point, the function is decreasing, indicated by a negative derivative. At a maximum point the curve is said to be *concave downward*.

17.6 MINIMUM POINT — CHARACTERISTICS

Next, let us take a look at a minimum point (Fig. 17.4). Suppose we have such a point where $x = 2$. At the minimum point, the slope is zero, or $dy/dx = 0 = m$.

Note that just at the left of the minimum point, the slope is negative, say, where $x = 1$. Also, a little to the right, the slope is positive, say,

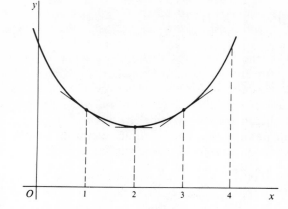

Fig. 17.4

where $x = 3$. Then, for a minimum point, the following conditions are true:

1. At the minimum point, $m = 0$.
2. A little to the left of this point, m is negative $(-)$.
3. A little to the right of this point, m is positive $(+)$.
4. As the curve passes through the minimum point, the slope changes from $(-)$ to (0) to $(+)$.

At the left of a minimum point, the function is decreasing as indicated by a negative derivative. At the right of the minimum point, the function is increasing as indicated by a positive derivative. At a minimum point the curve is said to be *concave upward*.

17.7 DETERMINING MAXIMUM AND/OR MINIMUM POINTS

The first requirement for a maximum and/or minimum point is that the slope be equal to zero, that is, $m = dy/dx = 0$. Values of x for which $dy/dx = 0$ are called *critical* values. Critical values of x *may* or *may not* represent maximum or minimum values of the function. The fact that the slope is zero at some point does not guarantee that the point is either maximum or minimum. In Figure 17.2 the slope of the curve is zero at one point (L) that is neither a maximum nor a minimum. However, the first step in determining such points is to find the values of x for which the derivative is equal to zero. We therefore first set the derivative equal to zero and solve the resulting equation for x.

As an example, consider the function

$$y = x^2 - 6x + 2$$

Differentiating, $\qquad dy/dx = 2x - 6$

Now we set the derivative equal to zero and solve:

$$2x - 6 = 0; \quad \text{or} \quad x = 3$$

17.7 Determining Maximum and/or Minimum Points

The result means that the curve has a zero slope where $x = 3$. This critical value of x represents a point that may or may not be a maximum or a minimum.

To determine whether a point is a maximum or a minimum, we now take a value of x a little less, and another value of x a little greater than the critical value of x in question. We then compute the slope for these two values of x by use of the derivative. It is not necessary to consider the exact numerical values of these slopes. By noting the change that takes place in the sign of the derivative, we can determine whether the point in question is a maximum or a minimum. We take the following steps:

(1) Find the point at which $m = 0$ (set derivative equal to zero and solve).
(2) Test: As the curve passes through this point,
    (a) If the slope changes from $(-)$ to $(+)$, the point is a minimum;
    (b) If the slope changes from $(+)$ to $(-)$, the point is a maximum;
    (c) If the slope does not change in sign, the point in question is neither a maximum nor a minimum.

The method of determining maximum and/or minimum points will be illustrated now by examples. The first example has already been mentioned in Sec. 17.7.

EXAMPLE 1. Find any maximum and/or minumum points of the following function:
$$y = x^2 - 6x + 2$$
Solution. Differentiating, $dy/dx = 2x - 6 =$ slope
For a zero slope, we must have $2x - 6 = 0$
Solving for x, $x = 3$
For the ordinate of the point, we have $y = -7$
Therefore, the point $(3, -7)$ is a possible maximum or minimum point.

To test this point, we use an x-value on either side of $x = 3$. We may use the values $x = 2$ and $x = 4$.
    At $x = 2$, $m = 2x - 6 = 4 - 6 = -2$, negative.
    At $x = 4$, $m = 2x - 6 = 8 - 6 = +2$, positive.
Since the slope (derivative) changes in sign from negative to positive, the point $(3, -7)$ is a minimum. The minimum *value* of the function is -7.

Figure 17.5 shows the graph of the function and the slopes of the curve for the values $x = 2$, $x = 3$, and $x = 4$.

EXAMPLE 2. Find any maximum and/or minimum points on the curve, $y = 3 - 2x - x^2$.
    Solution. Differentiating, $dy/dx = -2 - 2x$
For a zero slope we have $-2 - 2x = 0$
$$x = -1$$

To test this value, we use $x = -2$ and $x = 0$. At $x = -2$, $m = +2$; at $x = 0$, $m = -2$. Since the derivative changes from positive to negative, the point $(-1,4)$ is a maximum. The maximum value is 4. Figure 17.6 shows the curve and the slopes at the values $x = -2$, $x = -1$, and $x = 0$.

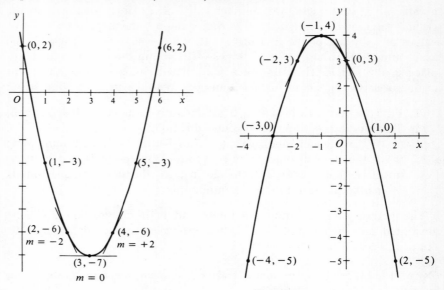

Fig. 17.5

Fig. 17.6

EXAMPLE 3. Find any maximum and/or minimum points on the graph of the function
$$y = x^3 - 6x^2 + 9x + 2$$
Solution. $\quad dy/dx = 3x^2 - 12x + 9$

Setting the derivative equal to zero and solving,
$$3x^2 - 12x + 9 = 0$$
$$x = 1, \quad \text{and} \quad x = 3$$

These two values of x represent points at which the curve may have maximum or minimum values of function. To test these values we can use the values $x = 0$, $x = 2, x = 4$.

At $x = 0$, $m = 0 - 0 + 9 = +9$, positive.

At $x = 2$, $m = 12 - 24 + 9 = -3$, negative.

At $x = 4$, $m = 48 - 48 + 9 = +9$, positive.

As the curve passes through the value $x = 1$, the derivative changes from *positive* to *negative*. The point (1,6) is therefore a *maximum* point. As the curve passes through the value $x = 3$, the derivative changes from *negative* to *positive*. The point (3,2) is therefore a *minimum* point. Figure 17.7 shows the graph of the function and the slopes at some points.

> *Note:* In choosing x-values for testing a critical point, we must be careful not to go beyond another critical value of x. For example,

if we have one critical value (slope equal to zero) at $x = 1$ and another such value at $x = 1.5$, then to test the value $x = 1$ the less value might be $x = 0$, but the greater value must be between $x = 1$ and $x = 1.5$. As long as we do not pass another critical point, we may take any x-value at the right and at the left of the critical value.

EXAMPLE 4. Find maximum and/or minimum points on the curve
$$y = 2x^3 - 15x^2 + 36x - 32$$

Solution. $dy/dx = 6x^2 - 30x + 36$

For a zero slope, $6x^2 - 30x + 36 = 0$

Solving, $x = 2$ and $x = 3$

To test, we use $x = 0$, $x = 2.5$, and $x = 4$. At $x = 0$, $m = +36$; at $x = 2.5$, $m = -\frac{3}{2}$; at $x = 4$, $m = +12$. The point $(2, -4)$ is a maximum; the point $(3, -5)$ is a minimum point. Figure 17.8 shows the graph of the function.

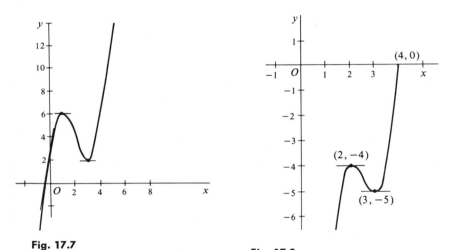

Fig. 17.7   **Fig. 17.8**

Note. In graphing a function such as in the following example (Example 5), the function values at maximum and minimum points have large absolute values. In such cases it is desirable and satisfactory to compress the vertical scale as shown in the graph (Fig. 17.9). However, the appearance of such a graph, then, will not show the correct slopes except at $m = 0$.

EXAMPLE 5. Find any maximum and/or minimum points on the curve of
$$y = 52 + 15x - 6x^2 - x^3$$

Solution. $dy/dx = 15 - 12x - 3x^2$

For $m = 0$, we have $15 - 12x - 3x^2 = 0$

Solving for x, $x = -5$ and $x = +1$

Fig. 17.9

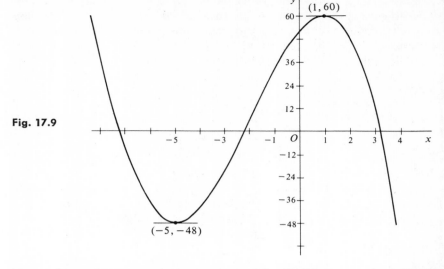

To test these values we use $x = -6$ and $x = 0$ for the value $x = -5$; we can then use $x = 0$ and $x = 2$ for the value $x = 1$. At $x = -6$, $m = -21$; at $x = 0$, $m = +15$. The point $(-5,-48)$ is therefore a minimum. At $x = 0$, $m = +15$; at $x = 2$, $m = -21$. The point $(1,60)$ is therefore a maximum. The function has a relative minimum value of -48 and a relative maximum value of 60.

17.8 MAXIMUM AND MINIMUM VALUES OF FUNCTIONS

We have seen how maximum and minimum values of a function can be represented by the height, or ordinates, of a curve. However, a maximum or a minimum value of a function need not be shown on a graph. In fact, we should not always depend on a picture for the understanding of maximum and minimum values. Any function may have a maximum or a minimum value even when no graph whatever is immediately indicated.

Let us keep in mind always that for a continuous function with a continuous derivative a maximum value is indicated where the derivative changes in sign from $(+)$ to (0) to $(-)$. To understand this meaning fully, the student should often determine maximum and minimum values without the aid of a graph of any kind. We illustrate the method by examples:

EXAMPLE 6. The position of a particle moving along a horizontal line, left and right, is given by the formula

$$s = t^3 - 9t^2 + 15t + 20$$

in which s represents the distance (in feet) from zero and t represents the time (in seconds). Find any maximum or minimum values of the function s.

Solution. Differentiating, $ds/dt = 3t^2 - 18t + 15$

We have seen that the derivative of the distance with respect to time is equivalent to the velocity, in this case, in feet per second. At any maximum or minimum value of the function s, the velocity v is zero. Setting the derivative equal to zero, we have

$$3t^2 - 18t + 15 = 0$$

Solving, $\quad t = 1 \quad \text{and} \quad t = 5$

Therefore, at the instants when $t = 1$ second and $t = 5$ seconds, the particle is not in motion. Let us assume that motion to the right is to be considered as positive and motion to the left as negative. Then we try to determine what is happening to the particle just before and just after the time of 1 second. In other words, we test the value $t = 1$ second. At $t = 0$, $ds/dt = 0 - 0 + 15$, that is, at $t = 0$, the velocity is positive. At $t = 2$, $ds/dt = 12 - 36 + 15 = -9$, that is, velocity is negative. Therefore, at $t = 1$, the function s reaches a maximum value. This maximum value is $+27$ feet to the right of zero.

As a test for the value $t = 5$, we use $t = 4$ and $t = 6$. We have at $t = 4$, $ds/dt = -9$; at $t = 6$, $ds/dt = +15$. Therefore, at $t = 5$ seconds, the function s reaches a minimum value. This value is -5 feet.

EXERCISE 17.1

Locate and test any maximum and/or minimum points on these curves:
1. $y = x^2 - 8x - 9$
2. $y = 3x^2 - 4x + 7$
3. $y = 3x - 5$
4. $y = 3 + 5x - 2x^2$
5. $y = 4x^2$
6. $y^2 = 4x$
7. $3y = x^2 - 3x - 18$
8. $4 + 4x - 3x^2 = y$
9. $y = 4x^3 - 7x^2 - 6x$
10. $y = 2x^3 - 2x^2 - 5$
11. $y = x^3 - 4x^2 - 3x + 2$
12. $y = 4 - 2x + 2x^2 - x^3$
13. $y = 6 - 48x + 8x^2 + 4x^3 - x^4$
14. $y = x^4 - 8x^3 + 18x^2 - 16x$

In each of the following equations, find the values of the independent variable for which the derivative is equal to zero; find the corresponding function value and tell whether it is a maximum or minimum:
15. $s = t^3 - 3t^2 - 24t + 12$
16. $s = t^3 - 6t^2 + 9t + 10$
17. $s = t^4 - 8t^3 + 6t^2 + 40t$
18. $i = 2t^3 - 9t^2 + 12t + 5$
19. $q = t^3 - 3t^2 - 9t + 10$
20. $q = t^3 - 9t^2 + 24t + 5$

Find any maximum and/or minimum points on each of the following curves; also find the slope of the curve for the given value of x:
21. $y = (x^2 - 2x + 5)^{1/2}$; at $x = 2$
22. $y = (x^3 - 12x + 1)^{1/3}$; at $x = 3$
23. $y = x^2(3 - x^2)^{1/2}$; at $x = 1$
24. $y = x^2(x^2 + 2)^{2/3}$; at $x = 5$
25. $y = \dfrac{3x^2}{(x^2 + 1)^2}$; at $x = 2$
26. $y = (x^2 - 2)(2x^2 - 1)^{1/2}$; at $x = \sqrt{5}$

17.9 RATE OF CHANGE OF SLOPE: dm/dx

If you drive along the curve from A to C in Figure 17.10, the steering wheel of your car is continuously turned in a clockwise direction. We call this direction of turning *negative*, since clockwise motion in mathematics is usually called *negative*. The slope of the tangent to the curve,

220 Maxima and Minima; Inflection Points

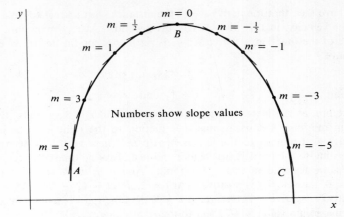

Fig. 17.10

shown at several points, is continuously changing. This means that the slope of the curve is changing as x increases from A to B to C.

We have defined the point B as a maximum point. At the left of B, the slope is positive. At the right of B, m is negative. As the value of x passes through the maximum point, the slope m changes from positive to negative.

We come now to a very important idea: As x increases from A to C, the slope m has the following values:

$$5,\ 3,\ 1,\ \tfrac{1}{2},\ 0,\ -\tfrac{1}{2},\ -1,\ -3,\ -5$$

Note especially that in this interval the slope is *continuously decreasing*. The *rate of change of slope* with respect to x is *negative*. This rate of change we can denote by

$$\frac{d(m)}{dx}$$

All the way from A to B to C, dm/dx is negative. Since $m = dy/dx$, then

$$\frac{dm}{dx} = \frac{d}{dx}\left(\frac{dy}{dx}\right) = \frac{d^2y}{dx^2}, \quad \text{the second derivative}$$

That is, the rate of change in the slope m is the second derivative of y with respect to x.

Remember: At a *maximum* point, $dy/dx = 0$, but d^2y/dx^2 is *negative*. That is to say, if we drive along a curve with the steering wheel of the car turned continuously to the right, then the second derivative of the path is negative, even if we drive all the way around a circle.

If we drive along a curve from A to C as shown in Figure 17.11, the steering wheel of the car is turned continuously in a counterclockwise, or *positive*, direction of rotation. Note that the slope of the curve throughout this interval is also continuously changing. The following slope values are shown:

$$-5,\ -3,\ -1,\ 0,\ 1,\ 3,\ 5$$

17.9 Rate of Change of Slope: dm/dx

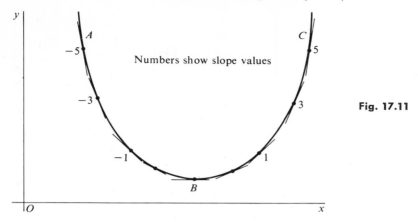

Fig. 17.11

Note that the slope is continuously *increasing* in the entire interval from A to C. That is, in this interval,

$$\frac{dm}{dx} \text{ is positive}$$

Remember: At a minimum point, $dy/dx = 0$, but d^2y/dx^2 is *positive*.
In some problems, therefore, we can use the second derivative as a test to determine whether a critical point is a maximum or minimum. At a point where $dy/dx = 0$,

(1) If d^2y/dx^2 is negative, the point is a maximum.
(2) If d^2y/dx^2 is positive, the point is a minimum.
(3) If d^2y/dx^2 is zero, this test fails. Then we resort to the slope test, taking nearby x-values at the left and right.

EXAMPLE 7. Find any possible maximum and/or minimum points on the following curve and test by the second derivative: $y = 7 + 6x - x^2$.

Solution. Differentiating, $dy/dx = 6 - 2x$ and $d^2y/dx^2 = -2$.
To find points where $m = 0$, $6 - 2x = 0$
Then $x = 3$, possible maximum or minimum.
The critical point is (3,16). Since the second derivative is always negative, -2, the point is a maximum.

EXAMPLE 8. Find any possible maximum and/or minimum points on the following curve and test by the second derivative, y''.

Solution. $y = x^3 - 3x^2 - 9x + 10$
Differentiating, $y' = 3x^2 - 6x - 9$
  $y'' = 6x - 6$
For $m = 0$, we set $y' = 0$:
  $3x^2 - 6x - 9 = 0$
Solving, $x = -1; +3$

These x-values are the abscissas of possible maximum or minimum point. The points are $(-1,15)$ and $(3,-17)$. Now we determine whether y'' is positive or negative for these values. Numerical values are not necessary.

At $x = -1$, $y'' = -6 - 6 = -$; the point $(-1,15)$ is a maximum.

At $x = 3$, $y'' = 18 - 6 = +$; the point $(3,-17)$ in a minimum.

EXAMPLE 9. Find and test possible maximum and/or minimum points of the curve $\quad y = x^3 - 3x^2 + 3x - 3$

Solution. $\quad y' = 3x^2 - 6x + 3$; and $y'' = 6x - 6$

For $m = 0$, $\quad 3x^2 - 6x + 3 = 0$

Solving for x, $\quad x = 1; 1$; [double point: $(1,-2)$]

Testing by y'', at $\quad x = 1$, $y'' = 0$; the test fails

Using the slope test, at $x = 0$, m is positive; at $x = 2$, m is positive. Therefore, since the slope is positive on both sides of $x = 1$, the point is neither maximum nor minimum.

Warning: The point in question may be a maximum or minimum even though the second derivative is zero.

EXAMPLE 10. Test for possible maximum or minimum points:
$$y = x^4 - 4x^3 + 6x^2 - 4x + 2$$

Solution.
$$y' = 4x^3 - 12x^2 + 12x - 4; \quad y'' = 12x^2 - 24x + 12$$

Setting $y' = 0$, we find a possible maximum or minimum point at $(1,1)$. At this point $y'' = 0$. The student should complete the problem and show that the point in question, $(1,1)$, is actually a minimum point.

In summary, there are three tests for a maximum or minimum point on a curve or a maximum or minimum value of a function. Once having determined a point where the curve has a horizontal tangent ($m = 0$), we can use any of the following tests:

1. *The slope test.* In most instances this is probably the best test for the slope at the left and the right of the critical point.

2. *The second derivative test.* This is good, when y'' does not equal zero. This test fails when y'' equals zero. Besides, it becomes very complicated in connection with powers, products, and quotients of functions.

3. *The ordinate test.* If the ordinate (the function value) is greater on either side of a critical point, then the point is a minimum. If the function has a smaller value on either side of the critical point, then the point in question is a maximum. If the function value is less on one side and greater on the other, the point is neither a minimum nor a maximum.

17.10 INFLECTION POINTS

If a continuous curve changes its sense of curvature from clockwise to counterclockwise, or vice versa, the point where this change takes place

17.10 Inflection Points

is called an *inflection* piont. In Figure 17.12 the curve reverses its direction of turning at point *B*. From *A* to *B* the curve turns in a clockwise, or negative, direction. From *B* to *C* it turns in a positive direction. The point *B* is a point of inflection.

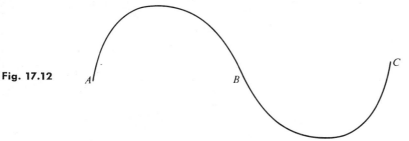

Fig. 17.12

If you drive along the curve, then from *A* to *B* the steering wheel is turned to the right. From *B* to *C* it is turned to the left. At point *B*, just for an instant, the steering wheel points the car straight ahead. We might define an inflection point as the "straight part of an *S* curve," which, of course, is a point.

At a point of inflection, just for an instant, the slope is *not* changing. Therefore, at an inflection point,

$$\frac{d(m)}{dx} = 0; \quad \text{that is,} \quad \frac{d^2y}{dx^2} = 0$$

Warning: The fact that the second derivative is zero at a point does not guarantee that the point is an inflection point. At any inflection point the second derivative must be zero, but the converse is not true. However, if d^2y/dx^2 changes sign, then we have an inflection point.

EXAMPLE 11. Find any inflection point on the curve

$$y = x^3 - 3x^2 - 9x + 10$$

Solution. Differentiating,

$$y' = 3x^2 - 6x - 9$$
$$y'' = 6x - 6$$

Setting $y'' = 0$, $\quad 6x - 6 = 0 \quad$ or $\quad x = 1$

At $x = 0$, $y'' = -$; at $x = 2$, $y'' = +$. Since the sign of y'' changes in passing through the point $(1, -1)$, this point is an inflection point. Note that the slope *m* does *not* change sign in passing through the inflection point.

EXERCISE 17.2

Find all maximum and/or minimum points on the following curves and check by the most convenient method; also find any inflection points:

1. $y = x^2 - 4x - 5$
2. $y = 2 - 6x - x^2$
3. $y = x^2 - 8x - 3$
4. $y = x^2 - 5x + 6$
5. $y = x^3 - 12x + 3$
6. $y = x^3 - 3x^2 + 12$

7. $y = x^3 - 3x^2 - 9x + 12$
8. $y = x^3 - 6x^2 - 15x + 9$
9. $y = 6 + 24x - 3x^2 - x^3$
10. $y = 2x^3 - 9x^2 - 24x$
11. $y = 9 - 12x + 6x^2 - x^3$
12. $y = 8 - 9x + 6x^2 - x^3$
13. $y = x^3$
14. $y = x^4$
15. $y = 5x + 4$
16. $y = 1 + 3x^2 - 2x^3$
17. $s = t^3 - 6t^2 + 5t + 10$
18. $3i = 4t^3 - 12t^2 + 9t$
19. $y = x^3 - 9x^2 + 24x - 12$
20. $y = x^3 - 3x^2 + 3x + 5$
21. $y = 5 - 12x + 9x^2 - 2x^3$
22. $y = 3 - 2x + 2x^2 - x^3$
23. $y = x^3 - 3x^2 + 15x + 6$
24. $y = x^4 - 8x^2 + 3$
25. $y = x^4 - 4x^3 - 8x^2 + 12$
26. $y = 4 - 12x + 2x^2 + 4x^3 - x^4$
27. $y = 3x^4 + 4x^3 - 12x^2 + 6$
28. $y = 6 - 48x + 24x^2 + 4x^3 - 3x^4$
29. $y = x^4 - 8x^3 + 24x^2 - 32x + 5$
30. $y = 3x^5 - 40x^3 + 240x$
31. $y = x^4 - 8x^3 + 24x^2 - 8x$
32. $y = 24x - 22x^2 + 8x^3 - x^4$
33. $y = (x^2 - 4x + 5)^{1/2}$
34. $y = (x^3 - 3x + 3)^{1/3}$
35. $y = x^2(6 - x^2)^{1/2}$
36. $y = \dfrac{x^2 + 6}{\sqrt{x^2 + 1}}$

chapter

18

Problems in Maxima and Minima

18.1 MEANING IN PRACTICAL PROBLEMS

We have used the terms *maximum* and *minimum* in connection with graphs to refer to a highest point or a lowest point on a curve. A maximum or a minimum value refers to the height of the curve, that is, the value of the ordinate, or the function itself.

Although a graph gives an excellent picture of the relation between a variable and its function, it should be remembered that a graph is only a pictorial representation of the relation. In many practical problems we deal with functions that have a greatest or least value. In many instances a graph will help to bring out the relation, but for many problems we can use the same reasoning with respect to greatest or least quantities without resorting to a graph. In some problems it is well to try to get the meaning of a maximum or a minimum value without the use of a graph.

If we can write the equation showing the relation between a variable and its function, then we can find the derivative without drawing a graph. We can set the derivative equal to zero. We can solve for values of the independent variable. Then we can test these values to determine whether they give maximum or minimum values of the function, and finally we can find corresponding values of the function. We can do all this without the use of a graph. At first a graph will help to show the relation between variables and their functions, but eventually we learn to get at the answers without the use of graphs.

In connection with a maximum or a minimum point on a curve, we have tested such a point to determine whether it is maximum, a minimum, or neither. In practical problems such a test is usually unnecessary since the conditions of the problem will show immediately whether the value is one or the other. Moreover, in many practical problems we may get some value of a function, such as a negative, which might not have meaning in the problem. For example, if we wish to find the dimensions of a

226   Problems in Maxima and Minima

rectangular box from certain given information, we might get a dimension that is negative, or we may get a negative volume. Such values are discarded as meaningless.

As another example, we may wish to find the number of articles to be manufactured per hour for maximum profit. If we should get two answers and one is negative, this value is discarded. Even when two positive valid answers are found, the nature of the problem will usually determine which one will produce the desired result.

However, we should not assume that simply because an answer is negative, it is meaningless. For example, we may get two values of an electric current I in an electric circuit. It is entirely possible that a negative value of I will produce a maximum value of power. After all, positive and negative currents only indicate electrons flowing in opposite directions. In the same way kinetic energy may be a maximum when velocity is negative.

18.2 GENERAL APPROACH TO PROBLEM SOLVING

In solving a practical problem it is necessary to set up an equation showing the relation between variables. This can be done only by a careful study of the given information. Then by differentiation we are able to determine any value of the independent variable that may give a possible maximum or minimum to the function. The relation is stated as an equation with the dependent variable as the quantity to be made a maximum or minimum.

For example, if we wish to find a maximum area, we write a formula for the area. If a cost is to be made a minimum in a problem, we write a formula or equation expressing the cost in terms of the variables. *The formula or relation must be a statement that is true at all times* even as the quantities vary. Do not insert in the formula or equation any particular values that happen to be true only at a particular time. Such values are substituted only *after* the differentiation has been done. The method will be shown by examples.

EXAMPLE 1. A man has 140 rods of fence with which he wishes to enclose a rectangular field of maximum area along a highway. A fence already extends along the highway, so he needs to fence only along three sides. What should be the dimensions of the field if the area is to be a maximum?

Solution. We shall make a rather careful study of this problem because it illustrates the ideas involved in practical problems in maxima and minima. Suppose we show a number of possible dimensions of the field (Fig. 18.1). We use the letter l to represent the number of rods in the length of the field along the highway, and w to represent the width or its dimension at right angles to the highway. The man wishes to get the greatest possible use of the fence along the highway. First, to make use of the highway fence as a boundary, he decides to try 130 rods for the length of the field. Then he has 10 rods left, 5 rods for each end. The area of the field is $(5)(130) = 650$ square rods, indicated in the figure by area A. He decides

18.2 General Approach to Problem Solving

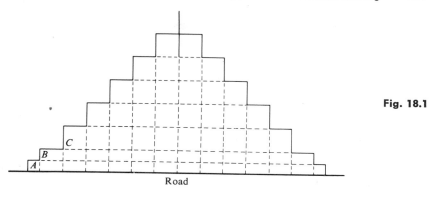

Fig. 18.1

to try a width of 10 rods, hoping to get a greater area. Now he uses 20 rods for the two ends and has 120 rods left for the length. Then area $(B) = (10)(120) = 1200$. This looks better so he decides to try a width of 20 rods, leaving 100 rods for the length. The area $(C) = (20)(100) = 2000$. This looks fine so he decides to try 30 rods for the width. The field has an area of 2400 square rods. This is getting better and better. He decides on 40 rods for the width. Then the area is 2400 square rods. The area has not increased but he decides to try a width of 60 rods. This leaves only 20 rods for the length and the area has decreased to 1200 square rods. He suddenly realizes if he uses 70 rods for the width, then the two ends require 140 rods and he has nothing left for the length.

Now he wonders what dimensions will give him a field having the greatest area. To answer this question, we first set up a table of values showing the area corresponding to each width.

if $w =$	0	5	10	20	30	40	50	60	70
$A =$	0	650	1200	2000	2400	2400	2000	1200	0

Let us graph the area A as a function of the width w (Fig. 18.2). Note that the curve representing *area* reaches a maximum at a point where the slope is equal to zero. To find this point and the corresponding value of the area we write the area as a function of the width w. Then the derivative will show the rate of change of

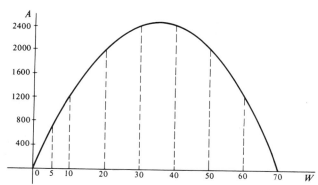

Fig. 18.2

area with respect to width. Then we set the derivative equal to zero and solve for width w. The result will show the proper width for the maximum area of the field.

Instead of taking a particular rectangular field, we sketch a general rectangle representing the field (Fig. 18.3). Since the area is the value to be maximized, we write the formula for area

$$A = lw$$

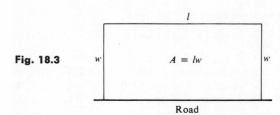

Fig. 18.3

This equation contains three variables. Therefore, we must eliminate one of them so we have only two left, one independent and one dependent variable. We choose to eliminate l. Although l and w are both variables, the following relation is always true in this problem:

$$l + 2w = 140, \quad \text{or} \quad l = 140 - 2w$$

Then, substituting in the formula we get

$$A = w(140 - 2w)$$

Now we have area A as a function of w alone. Expanding the product,

$$A = 140w - 2w^2$$

Here we have w as the independent variable and A as dependent variable. We now take the derivative of A with respect to the variable w, and get

$$\frac{dA}{da} = 140 - 4w$$

Setting the derivative equal to zero and solving for w,

$$140 - 4w = 0; \quad \text{then} \quad w = 35$$

This means a width of 35 rods will produce a maximum area. With the width equal to 35 rods, the length is 70 rods. Then area = 2450 square rods, which represents the field of greatest possible area.

In a problem of this kind it is unnecessary to test the result for a maximum or minimum, since any other dimensions for the field will produce a smaller area. Therefore, 2450 cannot be a minimum but must be a maximum area. If we wish to test the value, $w = 35$, we could use $w = 30$ and $w = 50$. When $w = 30$, the derivative is positive. For $w = 50$, the derivative is negative. Therefore, $w = 35$ gives a maximum function value.

18.2 General Approach to Problem Solving

EXAMPLE 2. A rectangular box with open top is to be formed from a rectangular sheet of metal 18 inches long and 12 inches wide by cutting a square out of each corner and turning up the four sides. Find the maximum volume of the box so formed.

Solution. We make a sketch of the rectangle indicating corners to be cut out (Fig. 18.4). Let us first consider some possible dimensions of the square. If a 1-inch square is cut out, the box would be 16 inches long, 10 inches wide, and 1 inch high. It would have a volume of 160 cu in. If a 2-inch square is cut out, the box would be 14 inches long, 8 inches wide, and 2 inches high with a volume of 224 cu in. With a 3-inch square cut out, the volume would be 216 cu in. A 4-inch square would result in a volume of 160 cu in. We can conclude that if we cut out a square of some particular size we shall have a box of maximum volume, but we had better try to determine the proper size before we start cutting.

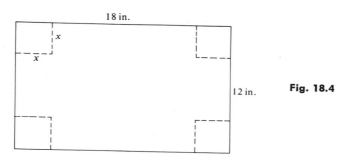

Fig. 18.4

We let x represent the number of inches in the side of the proper square to be cut out. Since we wish to maximize volume, we write the general formula for volume. In any rectangular solid, the volume is given by the formula

$$V = (l)(w)(h)$$

Here we have four variables. We eliminate two of them, leaving only the dependent variable V and one of the others. We can write the length, width, and height in terms of x, thus using x for the height, which is the side of the square. If a square with a side equal to x is cut out of each corner and the sides turned up, the box formed will have the following dimensions:

$$\text{height} = x, \quad \text{length} = 18 - 2x, \quad \text{width} = 12 - 2x$$

Using these values in the formula for volume we get

$$V = (18 - 2x)(12 - 2x)(x)$$

or

$$V = 216x - 60x^2 + 4x^3$$

Now we have V expressed in terms of the single variable x. Differentiating,

$$\frac{dV}{dx} = 216 - 120x + 12x^2$$

For a maximum or minimum volume, we set the derivative equal to zero:

$$12x^2 - 120x + 216 = 0$$

Dividing by 12,

$$x^2 - 10x + 18 = 0$$

Solving by formula,

$$x = \frac{10 \pm \sqrt{100-72}}{2} = 5 \pm \sqrt{7} = 5 \pm 2.646\ldots$$

For x, we get the two approximate values: 7.646 and 2.354. From the nature of the problem the answer could not be 7.646 inches in the side of the square because the sheet of metal is only 12 inches wide. Then the proper size of the square is 2.354 inches on a side.

To find the volume of the box it is best to use the radical form for the edge of the square. Then we have

height = $x = 5 - \sqrt{7}$
length = $18 - 2x = 18 - 2(5 - \sqrt{7}) = 18 - 10 + 2\sqrt{7} = 8 + 2\sqrt{7}$
width = $12 - 2x = 12 - 2(5 - \sqrt{7}) = 12 - 10 + 2\sqrt{7} = 2 + 2\sqrt{7}$

Then $V = (8 + 2\sqrt{7})(2 + 2\sqrt{7})(5 - \sqrt{7}) = 80 + 56\sqrt{7} = 228.2$ (approx.).

In Example 2, suppose we sketch a graph of the equation for volume

$$V = 216x - 60x^2 + 4x^3$$

with x, the edge of the square, as the independent variable, and volume V as the dependent variable (Fig. 18.5). The curve has the usual shape of a cubic equation. Note that the curve has a maximum and a minimum point. If $x = 0$, then the volume is zero. This means the curve passes through the origin.

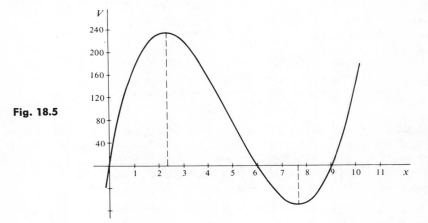

Fig. 18.5

The curve shows a maximum value for volume where $x = 5 - \sqrt{7} = 2.354$ (approx). For the value $x = 5 + \sqrt{7}$, which is approximately 7.646, the curve has a negative ordinate, which indicates a *negative volume*, a practical impossibility. From a practical standpoint the box would have a least volume, that is, a zero volume, where $x = 6$, since the sheet of metal is only 12 inches wide and a 6-inch square cut out of the corners would leave no volume.

18.3 SUMMARY OF STEPS IN PROBLEM SOLVING

The first and most important step in determining maximum or minimum values in any word problem is to express correctly the formula for the quantity that is to be maximized or minimized. This is also probably the most difficult step. After the correct equation has been stated, the matter of differentiating is almost mechanical.

Of course, it is necessary at this stage to state the equation in a form that contains only *one independent* variable. In the original statement it may be more convenient to use several variables, but all except one of the independent variables must be eliminated before differentiating. To aid in setting up the proper form of the equation and then finding the solution, the following steps may be found helpful.

(1) *If a figure is indicated by the problem, make a sketch and indicate the unknown variables by letters. Also take note of any constants.*
(2) *Write a general statement showing the relation that is always true for all values of the variables. This statement is often simply a formula for the quantity to be made a maximum or a minimum*, such as the formulas

$$A = (l)(w), \quad V = (l)(w)(h), \quad V = (\tfrac{1}{3})(\pi)(r^2)(h)$$

(3) *If there is more than one independent variable, find a relation between them and then eliminate all but one of the independent variables.*
(4) *Rewrite the formula with only one independent variable, the dependent variable being the quantity to be maximized or minimized.*
(5) *Differentiate with respect to the independent variable.*
(6) *Set the derivative equal to zero.*
(7) *Solve the resulting equation for the unknown variable.* The result will show the values of the independent variable that will produce the desired result.

EXERCISE 18.1

1. Find two numbers whose sum is 36 and whose product is a maximum.
2. Find two numbers whose sum is 30 such that the sum of their squares is a minimum. How can you tell your answer represents a minimum rather than a maximum?
3. Find two numbers whose sum is 24 such that the product of one number and the square of the other is a maximum. Which of the two numbers must be squared?
4. A farmer wishes to fence off a part of his land in the shape of a rectangle. The land lies adjacent to a river where no fence is required. If he has 360 rods of fencing for the three sides of the field, what dimensions will give the field a maximum area? What is this maximum area? How can you tell this is a minimum rather than a maximum without using the mathematical tests?
5. A rectangular field is to be fenced along two sides and two ends. It must also have two extra cross fences from side to side so as to form three plots. What dimensions of the field will give the entire field a maximum area if 180 rods of fencing are available?
6. What are the dimensions of the field in No. 5 for a maximum area if the field is to have one additional fence lengthwise through the middle of the field from end to end so as to form six plots?

7. A man wishes to fence in his yard so that it shall contain 180 square yards. If the cost is 40 cents per yard along two sides and one end and 60 cents per yard along one end, what are the dimensions for the least cost for the fence?
8. Work No. 7 if the yard is to contain 200 square yards of area.
9. A square sheet of aluminum, 24 inches on a side, is to be formed into a rectangular container without a top by cutting a square out of each corner and turning up the sides. Find the size of the square that should be cut out so that the container has a maximum volume.
10. Work No. 9 if the sheet of metal is 32 inches long and 20 inches wide.
11. A rectangular sheet of metal is 20 inches long and 16 inches wide. A rectangular container is to be formed by cutting a square out of each corner and turning up the sides. The container has no top. Find the size of the square that should be cut out so that the container has a maximum volume.
12. In No. 8, note that the dimensions of the square are irrational, as are the dimensions of the container. In any such rectangular sheet of metal, what must be the ratio of the width to the length so that the measurements of the container for maximum volume will be rational?
13. A rectangular box with a square base and no top is to be made from 64 square feet of lumber. What dimensions will give the box a maximum volume?
14. A rectangular box is twice as long as it is wide and has no top. If the total surface area is 132 square inches, what are the dimensions so that the volume of the container is a maximum?
15. A closed box (with a top) whose base is a rectangle twice as long as it is wide has a surface area of 432 square inches. What are the dimensions of the box if the volume is a maximum? What is its volume?
16. If the box in No. 15 has no top, what are its dimensions if its total surface area is 432 square inches?
17. A rectangular box without a top is two-thirds as wide as it is long. If the total surface area (outside) is 288 square inches, what are the dimensions for a maximum volume and what is the maximum volume?
18. What is the area of the largest rectangle that can be inscribed in an acute triangle having a base of 18 inches and an altitude of 12 inches with one side of the rectangle lying along the base of the triangle?
19. A right circular cone of wood has a diameter of 24 inches and an altitude of 12 inches. A right circular cylinder is to be cut from the cone. What is the volume of such a cylinder if its volume is a maximum?
20. Work No. 19 if the diameter of the cone is 20 inches and the altitude is 8 inches.
21. Same as No. 19, except that the cone has an altitude of 18 inches and a diameter of 12 inches.
22. A sphere is 24 inches in diameter. What is the volume of the cone of maximum volume that can be inscribed in the sphere? How can you tell this is a maximum rather than a minimum volume?
23. Work No. 22 in general terms, taking the radius of the sphere as r.
24. Find the volume of the cone of *least* volume that can be circumscribed about a sphere of radius r.
25. A beam having a rectangular cross section is to be sawed out of a log 12 inches in diameter. If the strength of the beam is proportional to the width and the square of the depth, what are the dimensions of the strongest beam that can be sawed from the log?
26. Work No. 25 taking the radius of the log as r.
27. Work No. 25 if the log is 10 inches in diameter, and then compare your answer with No. 28.
28. A log with a circular cross section has a diameter of 10 inches. A rectangular beam is to be sawed from the log. What are the dimensions of the strongest beam that can be cut from the log if the strength is proportional to the width and the *fourth power* of the depth?

18.3 Summary of Steps in Problem Solving

29. From your answer in No. 28, can you tell why joists in a building are set on edge rather than laid with the flat side down? Is this true also with respect to steel *I*-beams? What is the purpose of the flanges on *I-beams*?
30. A large timber has an elliptical cross section, the greatest diameter being 18 inches and the smallest 12 inches. (These are the major and the minor axes, respectively, of the ellipse.) What are the dimensions of the strongest beam of rectangular cross section that can be cut from the beam if the strength is proportional to the width and the square of the depth?
31. Two roads intersect at right angles. A spring is located 50 feet from one road and 80 feet from the other. A straight path is to be laid out to pass the spring from one road to the other. Find the least area that can be bounded by the roads and the path. Find the length of the path.
32. A tent in the shape of a pyramid with a square base is to be made from 600 square feet of canvas for the lateral area and the base. What dimensions of the tent will provide the maximum volume of the tent?
33. Find the least amount of canvas needed to make the lateral area of a conical tent if the tent has a cubic content of 288π cubic feet.
34. A cylindrical can holds 1 gallon (231 cubic inches). What dimensions of the can will require the least amount of sheet metal for the entire can, including top and bottom?
35. What is the area of the rectangle of maximum area that can be inscribed in the area bounded by the curve $y^2 = 16x$ and the line $x = 27$?
36. A rectangle is to be inscribed in the area bounded by the curves, $x^2 = 4ay$ and $y = b$. Find the area of such a rectangle if its area is a maximum.
37. Find the area of the largest isosceles right triangle that can be inscribed in the parabola $y^2 = 12x$ with the right angle at the vertex of the parabola.
38. A rectangle is to be inscribed in the area bounded by the hyperbola $x^2 - y^2 = 12$ and the line $x = 6$. Find the area of the rectangle if its area is a maximum.
39. A bridge across a road has a parabolic arch, the apex of the parabola being 12 feet above the road. The total width of the arch at road level is 16 feet. Find the cross-sectional area of the largest rectangular box that can pass under the bridge.
40. What is the area of the largest rectangle that can be inscribed in the ellipse, $9x^2 + 16y^2 = 144$?
41. What point on the parabola $y^2 = 12x$ is nearest to the point (12,0)?
42. The current i in a certain electric circuit is given by the formula $i = 4t - t^3/3$, where t denotes *time* in seconds. For what values of t is the current a maximum or a minimum? What is the current for each value of t? What is the voltage V for each value of i if the resistance in the circuit is 60 ohms?
43. The voltage V (in volts) in a certain circuit is given by $V = 80t^3 - 60t$. For what values of t is the voltage a maximum or minimum?
44. The voltage in a circuit is given by the formula $18.5 + t - t^2 = V$, where V indicates the number of volts and t indicates time in seconds. For what value of t is the voltage a maximum and what is the current then if the resistance R is equal to 25 ohms?

chapter

19

Implicit Differentiation

19.1 EXPLICIT FUNCTIONS

Let us review briefly the meaning of a function. If two variables x and y are related in some way so that for each value of x there exists a corresponding value of y, then we say y is a function of x. As we have stated, a function implies some kind of dependence between variables. Then one variable is called the *independent* and the other the *dependent* variable.

We have seen that a function is often indicated by an equation such as the following:

(a) $y = 2x + 5$ (b) $y = x^2 - 2x - 3$

In each of these examples we say y is equal to a function of x. Note that each equation is solved for y. That is, y appears alone on one side of the equation.

Whenever an equation is solved for y, we say that y is an *explicit* function of x. That is, the equation states *explicitly* that y is equal to the function. If we assign any value to x, we get the value of y *directly*. For example, in equation (a) above, if $x = 1$, then $y = 2 + 5 = 7$.

In the following equations, note that one variable is an *explicit* function of another:

(a) In the formula $P = 20i^2$, P is an explicit function of i.
(b) In the formula $A = \pi r^2$, A is an explicit function of r.
(c) In the formula $s = t^2 + 2t + 5$, s is an explicit function of t.

In any explicit function, if we assign a value to the independent variable we get the value of the dependent variable *directly*. We can represent explicit functions by general expressions such as

$$y = f(x), \quad P = f(i), \quad A = f(r), \quad s = f(t)$$

19.2 IMPLICIT FUNCTIONS

A functional relationship between x and y may not always be stated in the form

$$y = f(x)$$

19.3 Changing an Implicit Function to an Explicit Function

An equation may indicate a relation between x and y and yet not be solved for y, as in the equation

$$3x + 2y = 12$$

In this equation the value of y surely depends on what value is assigned to x. For each value of x there is a corresponding value of y. For example, if $x = 6$, then $y = -3$. Therefore, we can say that y is still a function of x.

However, the equation does not state explicitly that y is a function of x. The equation is not solved for y. Moreover, if we assign a value to x, we do not get the value of y directly. If $x = 6$, we get $18 + 2y = 12$. The equation must still be solved for the value of y.

If an equation in x and y is not solved for y, the relation between the variables is *implied*, and we say that y is then an *implicit* function of x. The following equation shows y as an *implicit function* of x:

$$x^2 + 3xy + 2y - 3x + 7 = 0$$

19.3 CHANGING AN IMPLICIT FUNCTION TO AN EXPLICIT FUNCTION

Up to this point in finding derivatives we have been dealing with explicit functions. The equations have always been solved for the dependent variable. We have found the derivative of the explicit function $y = f(x)$ by simply differentiating the expression containing x. Then we get immediately on the left side of the equation the symbol dy/dx for the derivative. For example,

$$\text{if } y = x^2 + 3x + 7, \quad \text{then} \quad \frac{dy}{dx} = 2x + 3$$

In the last statement, we can say that the equation is now solved for the symbol dy/dx since it appears alone on one side of the equation.

For the purpose of finding the derivative, it is sometimes convenient to have the equation solved for y so that we have an explicit function. An implicit function can often be changed to the explicit form. Sometimes the change is easy. However, in many instances the change is very difficult or impossible. Even when it is possible, the result is sometimes a cumbersome equation. In all cases the change from an implicit to an explicit function is accomplished by solving the equation for y. Let us consider a few examples.

EXAMPLE 1. Change the following equation to an explicit function and then find the derivative dy/dx: $3x + 4y = 17$.

Solution. We first solve for y:

$$4y = -3x + 17$$

then

$$y = -\frac{3}{4}x + \frac{17}{4}$$

Differentiating, $\quad dy/dx = -\frac{3}{4}$.

Implicit Differentiation

EXAMPLE 2. Change to explicit form and differentiate $x^2 + y^2 = 25$. Then find the slope of the curve for $x = 4$.

Solution. Solving for y, $y^2 = 25 - x^2$; then $y = \pm\sqrt{25 - x^2}$.
Note that there are two values for y:

$$y = +\sqrt{25 - x^2} \quad \text{and} \quad y = -\sqrt{25 - x^2}$$

The positive value of y represents the upper portion of the circle, $x^2 + y^2 = 25$, and the negative value represents the lower portion of the circle (Fig. 19.1). We shall then find two values of the slope where $x = 4$. Differentiating, using both signs,

$$\frac{dy}{dx} = \pm \frac{1}{2}(25 - x^2)^{-1/2}(-2x)$$

Simplifying,
$$\frac{dy}{dx} = \pm \frac{-x}{\sqrt{25 - x^2}}$$

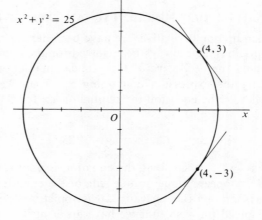

Fig. 19.1

Note that for the positive value of y, the slope is negative; and for the negative value of y, the slope is positive. Substituting the value $x = 4$, we get $dy/dx = +\frac{4}{3}$ where $y = -3$, and $dy/dx = -\frac{4}{3}$ where $y = +3$.

We shall see that even when the equation can be solved for y, differentiation is often much simpler when the equation is left in implicit form.

EXAMPLE 3. Given the function: $x^2 + 3xy - y^2 + 5x + 2y + 19 = 0$. This equation could be solved for y, but the result would be very difficult to differentiate. We shall leave this for implicit differentiation.

EXAMPLE 4. Suppose we wish to find dy/dx from the following equation:

$$x^3 + 5x^4y - 7xy^3 + xy^4 + 4x - y^2 + 7 = 0$$

This equation *cannot* be solved for y. Then we shall find that we *must* use the method of *implicit differentiation*.

19.4 FINDING THE DERIVATIVE OF AN IMPLICIT FUNCTION

Implicit differentiation is one of the most powerful methods of finding a derivative. It is accomplished simply by differentiating *each term as it stands*, without solving the equation for *y*. Each term is differentiated *with respect to x as the independent variable* (or whatever the variables may be). Two points must be observed:

1. Whenever the variable *y* appears anywhere in the equation, it is treated in the same way as any other function of *x*. For example, we recall that

$$\frac{d}{dx}(u^n) = (n)(u^{n-1})\left(\frac{du}{dx}\right)$$

In the same way,

$$\frac{d}{dx}(y^n) = (n)(y^{n-1})\left(\frac{dy}{dx}\right)$$

As an example,

$$\frac{d}{dx}(y^5) = 5y^4\left(\frac{dy}{dx}\right)$$

For a power of *x*,

$$\frac{d}{dx}(x^5) = 5x^4\left(\frac{dx}{dx}\right)$$

Since $dx/dx = 1$, this factor is omitted when we have a power of *x*.

2. A product of two factors involving both *x* and *y*, such as x^2y^3, must be differentiated by the product rule, for example,

$$\frac{d}{dx}(x^2y^3) = (x^2)(3y^2)\left(\frac{dy}{dx}\right) + (y^3)(2x)$$

Note that the factor dx/dx is omitted after the last factor $2x$.

There are two basic steps in finding the derivative by implicit differentiation:

(1) *Differentiate each term as it stands, observing products, and especially the derivative of y and powers of y,*
(2) *Solve the resulting equation for* (dy/dx).

EXAMPLE 5. Given $x^2 + 3y - 4x - 5 = 0$, differentiate implicitly.

Solution. Differentiating,

$$2x + 3\frac{dy}{dx} - 4 = 0$$

Solving for dy/dx,

$$3\frac{dy}{dx} = 4 - 2x, \qquad \frac{dy}{dx} = \frac{4}{3} - \frac{2}{3}x$$

Of course, this equation could have easily been solved for *y*.

EXAMPLE 6. Find the slope of the curve $x^2 + y^2 = 25$ at the point $(4, -3)$.

Solution. Differentiating,

$$2x + 2y\left(\frac{dy}{dx}\right) = 0$$

238 Implicit Differentiation

Solving for dy/dx,

$$2y\left(\frac{dy}{dx}\right) = -2x, \quad \text{then} \quad \frac{dy}{dx} = -\frac{x}{y}$$

At the point $(4,-3)$ the slope is equal to

$$\frac{dy}{dx} = -\frac{4}{-3} = +\frac{4}{3}$$

We have previously found the derivative in this equation by first changing to the explicit form. Note how much simpler is the method of implicit differentiation.

EXAMPLE 7. Write the equation of the tangent to the following curve at the point $(2,-1)$: $x^2 - 3xy - 4y^2 - 6 = 0$.

Solution. First we check the coordinates $(2,-1)$ to be sure the point lies on the curve: Does $2^2 - 3(2)(-1) - 4(-1)^2 - 6 = 0$? Yes. Differentiating the quantity xy as a product and taking the constant 3 as the coefficient of the derivative of this product, we get

$$2x - 3\left(x\frac{dy}{dx} + y\right) - 8y\frac{dy}{dx} = 0$$

or

$$2x - 3x\frac{dy}{dx} - 3y - 8y\frac{dy}{dx} = 0$$

Isolating dy/dx terms,

$$-3x\frac{dy}{dx} - 8y\frac{dy}{dx} = -2x + 3y$$

Solving for dy/dx,

$$\frac{dy}{dx} = \frac{2x - 3y}{3x + 8y}$$

At the point $(2,-1)$ the slope is

$$\frac{dy}{dx} = \frac{4 + 3}{6 - 8} = -\frac{7}{2}$$

For the tangent we have $m = -\frac{7}{2}$ and point $(2,-1)$; the equation is $7x + 2y = 12$.

Note: In implicit differentiation the derivative will often contain the variable y itself. If desired, the derivative can sometimes be expressed entirely in x terms by replacing y with its equivalent as obtained by solving the original equation for y. However, it is usually more convenient to leave the derivative in the form that contains y. Moreover, this form is entirely satisfactory for any way in which the derivative is to be used or evaluated.

EXAMPLE 8. Find the equation of the tangent and the equation of the normal to the following transformed hyperbola at the point $(-3,2)$:

$$x^2 + 3xy - y^2 + 5x + 2y + 24 = 0$$

Solution. Differentiating,

$$2x + 3\left(x\frac{dy}{dx} + y\right) - 2y\frac{dy}{dx} + 5 + 2\frac{dy}{dx} = 0$$

Removing parentheses,

$$2x + 3x\frac{dy}{dx} + 3y - 2y\frac{dy}{dx} + 5 + 2\frac{dy}{dx} = 0$$

19.4 Finding the Derivative of an Implicit Function

Transposing,

$$3x\frac{dy}{dx} - 2y\frac{dy}{dx} + 2\frac{dy}{dx} = -2x - 3y - 5$$

Dividing both sides of the equation by $3x - 2y + 2$, the coefficient of dy/dx, we get

$$\frac{dy}{dx} = \frac{-2x - 3y - 5}{3x - 2y + 2} = \frac{2x + 3y + 5}{2y - 3x - 2}$$

At the point $(-3, 2)$, the slope

$$m = \frac{-6 + 6 + 5}{4 + 9 - 2} = \frac{5}{11}$$

The equation of tangent: $5x - 11y = -37$; normal: $11x + 5y = -23$.

EXAMPLE 9. Find the equation of the tangent to this third degree curve at the point $(2, -1)$:

$$x^3 + x^2y - xy^2 - y^3 - 3 = 0$$

Solution. Differentiating (note especially the two products),

$$3x^2 + \left(x^2\frac{dy}{dx} + 2xy\right) - \left(2xy\frac{dy}{dx} + y^2\right) - 3y^2\frac{dy}{dx} = 0$$

or

$$3x^2 + x^2\frac{dy}{dx} + 2xy - 2xy\frac{dy}{dx} - y^2 - 3y^2\frac{dy}{dx} = 0$$

Transposing,

$$x^2\frac{dy}{dx} - 2xy\frac{dy}{dx} - 3y^2\frac{dy}{dx} = y^2 - 2xy - 3x^2$$

Dividing by coefficient of dy/dx,

$$\frac{dy}{dx} = \frac{y^2 - 2xy - 3x^2}{x^2 - 2xy - 3y^2} = \frac{y - 3x}{x - 3y}$$

At the point $(2, -1)$,

$$m = \frac{-1 - 6}{2 + 3} = \frac{-7}{5}$$

Equation of the tangent: $7x + 5y = 9$.

EXAMPLE 10. Find the slope and the equation of the tangent to the following transformed ellipse at $(1, 3)$: $3x^2 - 4xy + 2y^2 + 5x - 8y + 10 = 0$.

Solution. Differentiating,

$$6x - 4\left(x\frac{dy}{dx} + y\right) + 4y\frac{dy}{dx} + 5 - 8\frac{dy}{dx} = 0$$

or

$$6x - 4x\frac{dy}{dx} - 4y + 4y\frac{dy}{dx} + 5 - 8\frac{dy}{dx} = 0$$

Transposing,

$$-4x\frac{dy}{dx} + 4y\frac{dy}{dx} - 8\frac{dy}{dx} = -6x + 4y - 5$$

Dividing by coefficient of dy/dx,

$$\frac{dy}{dx} = \frac{-6x + 4y - 5}{-4x + 4y - 8} = \frac{6x - 4y + 5}{4x - 4y + 8}$$

At the point (1,3) we get
$$\frac{dy}{dx} = \frac{-1}{0}$$

A zero denominator and a nonzero numerator indicates a vertical tangent. The equation of the tangent is therefore $x = 1$.

EXERCISE 19.1

Differentiate implicitly; find the slope of the curve at the given point and write the equation of the tangent at the point:

1. $3x + 4y = 12$ (no tangent)
2. $x^2 + y^2 = 25$; $(4,-3)$
3. $x^2 - y^2 = 5$; $(-3,-2)$
4. $x^2 + xy - y^2 = 11$; $(5,-2)$
5. $3x^2 + 5y^2 = 68$; $(-4,2)$
6. $xy - 7x + 3y = 15$; $(-1,4)$
7. $x^2 + 5xy + 3y^2 = 1$; $(2,-3)$
8. $x^3 + 8y^2 = 16$; $(2,1)$
9. $x^2 + y^3 = 8$; $(3,-1)$
10. $3x^2 + 2xy - 4y^2 = -1$; $(-3,2)$
11. $2x^2 - 3y^2 - 4x + 5y = 6$; $(-4,-3)$
12. $x^3 + y^3 - 9xy = 0$; $(4,2)$
13. $x^2 + y^2 + 4x - 6y = 3$; $(2,3)$
14. $x^2 - 4xy + 5y^2 - 7x + 5y = 16$; $(-3,-2)$
15. $3x^2 - 5xy + 4y^2 - 6x + 7y = 9$; $(1,-2)$
16. $x^2 + 3xy + y^2 - 8x - 5y + 36 = 0$; $(5,-3)$
17. $x^{2/3} + y^{2/3} = 20$; $(64,8)$
18. $x^{2/3} - x^{1/3}y^{1/3} + y^{2/3} = 7$; $(8,-1)$
19. Find any maximum and/or minimum points on the curve

$$x^2 + 2xy + 4y^2 - 2x + 4y - 8 = 0$$

chapter
20

Related Rates

20.1 THE DERIVATIVE AS A RATE OF CHANGE

We are often concerned with the question of *how fast* a variable is changing. By the expression *how fast* we mean speed or velocity of change. In some instances, when we use this expression, we do not really refer to speed or velocity. For example, the derivative, dy/dx, by itself alone does not indicate *how fast* the variables are changing. Rather it refers to the ratio between changes in y and x. It has nothing to do with velocity.

Let us consider the equation

$$y = \frac{1}{2}x + \frac{5}{2}$$

Differentiating, we get
$$\frac{dy}{dx} = \frac{1}{2}$$

In this example the derivative, $\frac{1}{2}$, means the limit of the ratio between the increment of y and the increment of x as these increments approach zero as a limit. Note that the statement refers to the ratio between increments. It says nothing about the speed or velocity of these changes.

If we graph the equation, we see that the graph is a straight line whose slope is equal to $\frac{1}{2}$, the derivative of the function. The slope has the constant value of $\frac{1}{2}$ (Fig. 20.1).

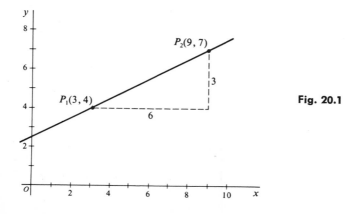

Fig. 20.1

Let us further consider the meaning of the derivative in terms of the graph. We take two points, $P_1(3,4)$ and $P_2(9,7)$, on the graph (a straight line). Note that as x changes by 6 units from 3 units to 9 units, y changes by 3 units. An increment of 6 units in x results in an increment of 3 units in y. As we move from P_1 to P_2, the increment in y is $\frac{1}{2}$ *as much as* the increment in x. In the case of this straight line, the increment of y is always $\frac{1}{2}$ *as much as* the increment in x as we move from one point to another on the line, no matter how small or large the distance between the two points. Notice the words "as much as" in the comparison of the increments. The phrase "as much as" refers to *amount* of change, not to the speed of these changes. Yet the derivative $\frac{1}{2}$ is often called the rate of change of y with respect to x. It is a comparison between amounts of change.

We sometimes say in this connection that y is changing $\frac{1}{2}$ *as fast* as x. Whenever we use the expression *fast* we are referring to velocity rather than *amount of change*. When we refer to velocity we must have reference to *time*. There is no way to define velocity except with reference to time.

In the graph of $y = \frac{1}{2}x + \frac{5}{2}$, actually the question of *time* has nothing to do with the relation between increments in going from one point to another on the line. We are not concerned with how long it takes for these changes — whether the time required is one second, one minute, one hour or one year.

20.2 TIME RATES OF CHANGE: HOW FAST?

However, it is often necessary to know *how fast* a variable is changing. We have already met the question of *how fast* in connection with velocity v as the derivative of distance s. In that connection, one of the variables was *time t*.

The question of velocity of changes is important in many practical situations. We sometimes must know how fast a current is changing in an electric circuit. We want to know how fast a capacitor is charging. We may want to know how fast the surface area of a balloon is changing. We may find it necessary to know, for example, how fast any of the following are changing: voltage or power in an electric circuit, circumference or area of a circle, depth of liquid in a conical or other container, total surface area of a cylinder, moment of inertia, distance between two moving objects, length of a shadow, slope of a curve, and many others. In many situations like these, we are interested in not only the varying amounts of change or the ratio of changes, but very often *how fast* these changes are taking place.

If we wish to know how fast such changes take place we refer to time. We are then concerned with the rate of change with respect to *time*, or the *time-rate* of change.

In the example we have mentioned (Fig. 20.1) we find the derivative, dy/dx, is equal to the constant $\frac{1}{2}$. This means that the change in y is $\frac{1}{2}$

20.2 Time Rates of Change: How Fast?

as much as the change in x at any instant. Now if we are told that the variable x is changing at a rate of 8 units per second, for example, then we can state logically that the variable y is changing $\frac{1}{2}$ as fast as x, or 4 units per second.

The *time-rates* of change of x and y are denoted by the derivatives with respect to time:

(a) the rate of change of x with respect to time is denoted by $\frac{dx}{dt}$,

(b) the rate of change of y with respect to time is denoted by $\frac{dy}{dt}$.

In any problem in which y is a function of x, the relationship between the variables, x and y, also extends to their respective *time-rates* of change. Since there is a relationship between x and y, there is also a relationship between the *time-rates*, dy/dt and dx/dt. The *time-rates* are therefore called *related time-rates*, or simply *related rates*. In the example mentioned, the time-rate of y is $\frac{1}{2}$ the time-rate of x.

Note especially that the ratio of the time-rates of change of y and x is equal to the derivative of y with respect to x. That is,

$$\frac{dy/dt}{dx/dt} = \frac{dy}{dx}$$

Remember: Whenever we wish to find the time-rate of change of any variable, we differentiate with respect to time t.

Let us consider another example, the equation $y = x^2 - 4x + 5$. Suppose we wish to find the value of the derivative at the point (3,2). Differentiating, we get

$$\frac{dy}{dx} = 2x - 4$$

At the point (3,2), $dy/dx = 2$. If we graph the equation (Fig. 20.2), we find that the derivative 2 means that the slope of the curve at (3,2) is 2. At that particular point we can say the rate of change of y with respect to x is 2. That is, the instantaneous change in y is *twice as much as* the change in x.

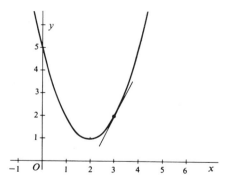

Fig. 20.2

Here again we sometimes say that at the point (3,2), the value of the ordinate y is changing twice *as fast* as the abscissa x. Yet we do not mean velocity of these changes. In fact, all we see is a graph, a picture, with no motion whatever. We know nothing about the velocity of these changes. However, if we are told that x is changing at a rate of 5 inches per second, then we can properly say that y is changing twice *as fast* as x, or 10 inches per second. Then we have

$$\frac{dx}{dt} = 5 \text{ in./sec} \quad \text{and} \quad \frac{dy}{dt} = 10 \text{ in./sec}$$

Here again the time-rates of change are stated as a certain number of units of change *per unit of time*. Note also $(dy/dt)/(dx/dt) = dy/dx$.

20.3 FINDING A PARTICULAR TIME-RATE

In a problem involving time-rates of change we must consider not only the variables themselves, such as y and x, but their related time-rates. The four variables to be considered are

$$x, y, \frac{dy}{dt}, \frac{dx}{dt}$$

When one of the time-rates is given, together with a particular set of values of x and y, our problem becomes that of finding the unknown time-rate. If other variables appear in a problem, some are usually eliminated through some relation existing between them.

EXAMPLE 1. Given the equation $y = 2x^2 - 7x + 8$. If x is changing at a rate of 1.5 units per minute, find the time-rate of change in y when $x = 3$; that is, we are asked to find dy/dt.

Solution. Differentiating with respect to time, we get

$$\frac{dy}{dt} = (4x)\left(\frac{dx}{dt}\right) - (7)\left(\frac{dx}{dt}\right)$$

Now we substitute known values: $x = 3$; $dx/dt = 1.5$ units/min. Then

$$\frac{dy}{dt} = (4)(3)(1.5) - (7)(1.5) = 7.5 \text{ units/min}$$

Note that the ratio between the time rates is equal to the derivative value when $x = 3$: $\dfrac{7.5 \text{ units/min}}{1.5 \text{ units/min}} = 5$, the value of the derivative.

Warning. Do not substitute specific values for x or y before differentiating. The given equation $y = 2x^2 - 7x + 8$, states a relation that is true at all times. *This must be differentiated before inserting specific values.*

EXAMPLE 2. In the relation $x^2 - y^3 = 17$, if x is changing at a constant rate of 3 units per second, find the time-rate of change in y when y is equal to 2.

20.3 Finding a Particular Time-Rate

Solution. Differentiating with respect to time,

$$(2x)\left(\frac{dx}{dt}\right) - (3y^2)\left(\frac{dy}{dt}\right) = 0$$

Solving for dy/dt,

$$\frac{dy}{dt} = \left(\frac{2x}{3y^2}\right)\left(\frac{dx}{dt}\right)$$

Note that the expression for dy/dt includes a value of x. Therefore, we find the value of x that corresponds to the value $y = 2$. Solving the original equation for x we find that there are two values of x: $+5$ and -5. If we sketch the graph of the equation, we see at once there are two values of x that correspond to $y = 2$ (Fig. 20.3). To find dy/dt we first use the value $x = +5$. Then we get (using $y = 2$)

$$\frac{dy}{dt} = \frac{(2)(5)}{(3)(4)}(3) = \frac{5}{2} = 2.5 \text{ units per second}$$

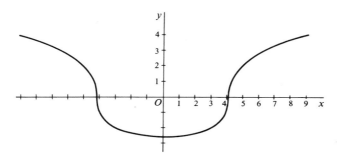

Fig. 20.3

For $x = -5$ we have

$$\frac{dy}{dt} = \frac{(2)(-5)}{(3)(4)}(3) = -\frac{5}{2} = -2.5 \text{ units per second}$$

The result means that at the point (5,2), y is *increasing* at a rate of 2.5 units per second, while at the point $(-5,2)$, y is *decreasing* at a rate of 2.5 units per second. Again we shall find that the derivative dy/dx is equal to the ratio of the time-rates, or the slopes of the curve at the two points:

$$\frac{dy}{dx} = \frac{dy/dt}{dx/dt} = \frac{\pm 2.5}{3} = \pm \frac{5}{6}$$

EXAMPLE 3. A ladder 26 feet long leans up against a vertical wall. The foot of the ladder is being pulled away from the wall at a constant rate of $\frac{1}{2}$ foot per second. How fast is the top of the ladder descending when the foot of the ladder is 10 feet from the wall (Fig. 20.4)?

Solution. In a problem of this kind we make a sketch or figure showing the variables in any general position. In this problem we let x represent the number of feet from the wall to the foot of the ladder at any time t. We let y represent the number of feet from the bottom of the wall to the top of the ladder at any time t. Then x and y are variables. The rate of change in x, or dx/dt, is $\frac{1}{2}$ ft/sec. We are to find dy/dt.

As our next step we write a general equation that expresses the correct relation between the variables x and y *at any time t*. In this example we have

$$x^2 + y^2 = 26^2$$

Fig. 20.4

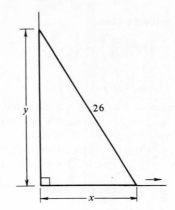

This is a relation that is always true for all positions of the ladder. Now we differentiate with respect to time:

$$(2x)\left(\frac{dx}{dt}\right) + (2y)\left(\frac{dy}{dt}\right) = 0$$

At this point we at once substitute known values or we can first solve for dy/dt:

$$(2y)\left(\frac{dy}{dt}\right) = -(2x)\left(\frac{dx}{dt}\right); \quad \text{then} \quad \frac{dy}{dt} = -\left(\frac{x}{y}\right)\left(\frac{dx}{dt}\right)$$

When $x = 10$, by the Pythagorean rule, $y = 24$. Then

$$\frac{dy}{dt} = -\left(\frac{10}{24}\right)(0.5) = -\frac{5}{24}$$

The negative sign means that the top of the ladder is *descending* at a rate of $\frac{5}{24}$ ft/sec; that is, the value of y is decreasing. The specific values must *not* be substituted before differentiating.

EXAMPLE 4. In an electric circuit the current I is given by the formula $I = E/R$, where E is voltage (in volts) and R is resistance (in ohms). If R is a variable and is changing at 0.5 ohm per second, and $E = 120$ volts, how fast is the current I changing when $R = 20$ ohms?

Solution. Our general equation is

$$I = \frac{120}{R}, \quad \text{or} \quad I = 120\,R^{-1}$$

Differentiating,
$$\frac{dI}{dt} = -120\,R^{-2}\left(\frac{dR}{dt}\right)$$

Substituting known values,

$$\frac{dI}{dt} = \frac{-120}{400}(0.5) = -\frac{60}{400} = -0.15$$

The result means that the current I is changing at a rate of -0.15 ampere per second, which represents a decreasing current.

EXAMPLE 5. One formula for power in an electric circuit is $P = i^2 R$. If resistance R is equal to the constant $6(10^5)$ ohms and current i is changing at a rate of 3(10⁻⁶) ampere per second, find the time-rate of change in power P when the current i is equal to $4(10^{-5})$ ampere.

Solution. Differentiating,

$$\frac{dP}{dt} = (2)(R)(i)\left(\frac{di}{dt}\right)$$

Supplying values,

$$\frac{dP}{dt} = (2)(6)(10^5)(4)(10^{-5})(3)(10^{-6})$$

$$= 144(10^{-6}) \text{ watt per second}$$

EXAMPLE 6. In the parabola $x^2 - 40x = -40y$, if the abscissa is changing at a rate of 3 ft/sec, how fast is the ordinate changing when $x = 30$?

Solution. A sketch of the parabola shows (Fig. 20.5) that it passes through the origin, and the vertex is at (20,10). Differentiating,

$$2x\left(\frac{dx}{dt}\right) - 40\left(\frac{dx}{dt}\right) = -40\left(\frac{dy}{dt}\right)$$

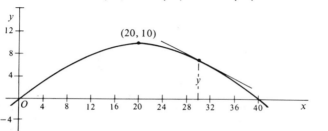

Fig. 20.5

Supplying values,

$(60)(3) - (40)(3) = -(40)(dy/dt)$. Solving, $dy/dt = -\frac{3}{2}$ ft/sec

EXAMPLE 7. A vat in the shape of an inverted circular cone whose diameter is 4 feet and whose altitude is 3 feet is full of water. The water is running out through a hole in the apex at a rate of 8 cu in./sec. How fast is the surface of the water lowering when the depth of the water is 15 inches? How fast is the area of the surface changing?

Solution. We make a sketch of the inverted cone, showing dimensions (Fig. 20.6). Now at any time t we have the formulas for volume and surface area:

$$V = \frac{\pi r^2 h}{3} \quad \text{and} \quad A = \pi r^2$$

In the formulas r refers to the radius of the surface of the water and h refers to the depth of the water at any time t. Then r and h are variables.

We are given the fact that the volume of water in the cone is changing at the rate $dV/dt = -8$ cu in./sec, since the amount of water is decreasing. Since we wish to find dh/dt and dA/dt, we try to eliminate r through some relation between r and h. By similar triangles we have, at any time t,

$$\frac{r}{h} = \frac{2}{3}$$

Fig. 20.6

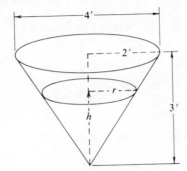

The fraction $\frac{2}{3}$ comes from the fact that the altitude of the cone is 3 ft and the radius of the top is 2 ft. Then, solving for r,

$$r = \frac{2h}{3} \quad \text{and} \quad r^2 = \frac{4h^2}{9}$$

Substituting, $\quad V = \frac{1}{3}(\pi)\left(\frac{4h^3}{9}\right) = \frac{4}{27}\pi h^3$

Differentiating, $\quad \dfrac{dV}{dt} = \left(\dfrac{4}{9}\right)(\pi)(h^2)\left(\dfrac{dh}{dt}\right)$

Substituting values $dV/dt = -8$, and $h = 15$ we get

$$-8 = \frac{4\pi(15)(15)}{9}\left(\frac{dh}{dt}\right); \quad \text{then} \quad \frac{dh}{dt} = \frac{-2}{25\pi}$$

This means that $dh/dt = -2/(25\,\pi)$ in./sec; that is, the surface is descending at a rate of about $\frac{1}{40}$ inch per second.

For the rate of change in surface area we begin with $A = \pi r^2$. To simplify the solution we substitute the equivalent of r^2 and get

$$A = \left(\frac{4}{9}\right)(\pi)(h^2)$$

Differentiating with respect to time,

$$\frac{dA}{dt} = \frac{8}{9}\pi h\left(\frac{dh}{dt}\right)$$

Substituting known values,

$$\frac{dA}{dt} = \frac{8}{9}\pi(15)\left(\frac{-2}{25\pi}\right) = -\frac{16}{15}$$

That is, the surface area is *decreasing* by $\frac{16}{15}$ sq. in./sec.

EXAMPLE 8. Two cars leave a point O at the same time, one going east at 50 mph, the other north at 30 mph. How fast are they separating at the end of 30 minutes?

Solution. In this problem we have three variables: x and y for the varying distances traveled by each car, respectively, and we let s represent the distance between them at any time t, another variable. We first set up a relation that is always true. Since the cars are traveling at right angles, we have

$$s^2 = x^2 + y^2$$

In this example it is not necessary to eliminate one of the variables. We differentiate with respect to time:

$$2s\frac{ds}{dt} = 2x\frac{dx}{dt} + 2y\frac{dy}{dt}$$

When $t = \frac{1}{2}$ hr, $x = 25$, $y = 15$, $s = \sqrt{850}$. Substituting known values, we have

$$(2\sqrt{850})\frac{ds}{dt} = 2(25)(50) + 2(15)(30)$$

Solving for ds/dt, $ds/dt = 10\sqrt{34}$, or 58.3 mph (approx.)

EXERCISE 20.1

1. A ladder 30 feet long leans up against a vertical wall. The foot of the ladder is being pulled away from the wall at a constant rate of 1.5 feet per second. How fast is the top of the ladder descending when the foot of the ladder is 18 feet from the wall?
2. A ladder 32 feet long leans up against a vertical wall. The foot of the ladder is being pushed toward the wall at a constant rate of 0.5 foot per second. How fast is the top of the ladder rising when the ladder reaches a point on the wall 16 feet from the ground?
3. A ladder 24 feet long leans up against a vertical wall. The foot of the ladder is being pulled out from the wall at a constant rate of 1 foot per second. How fast is the top of the ladder descending when the top is 2 feet from the ground?
4. Sand is falling through a hole onto a level surface forming a conical pile in which the radius of the base is always $\frac{2}{3}$ of the altitude of the cone. If the sand is falling at the rate of 1.5 cu ft per minute, how fast is the altitude changing when the radius of the pile is 3 feet?
5. A vat in the shape of an inverted circular cone is full of water. The vat has an altitude of 10 feet, and the radius of the cone is 6 feet. If the water is running out of a hole in the apex at the bottom at the rate of 1.5 cu ft per minute, how fast is the top of the water descending when the radius of the surface is 4 ft?
6. Same as No. 5, except that the conical container has a diameter of 5 feet and an altitude of 6 feet and the water is running out at 24 cu in. per sec.
7. A conical vat similar to that in No. 5 is 10 feet in diameter and has an altitude of 12 feet. If water is running into the tank at the rate of 2 gallons per second, how fast is the surface increasing in area when the water is 8 feet deep?
8. A tank in the shape of a paraboloid with the vertex at the bottom has a circular top 6 feet in diameter. The tank has a total depth of 8 feet. If water is running into the tank at a constant rate of 4.8 cu ft per minute, how fast is the surface of the water rising when the water is 3 feet deep?
9. An hourglass has the shape of two circular cones with the two joined at the apex. Each half of the hourglass is a cone whose altitude is 12 centimeters, and the diameter of the circular base is 10 centimeters. The sand falls through the apex of one cone into the other cone at a steady rate of 3 cu cm per minute. When the sand in the bottom cone is 4 centimeters deep, how fast is the surface of the sand rising?
10. One car leaves a particular point and travels directly east at an average rate of 36 miles per hour. At the same time a second car leaves the same point and travels directly north at an average rate of 27 miles per hour. How fast are the cars separating at the end of (a) 1 hour; (b) 2 hours?
11. In No. 10, if the second car starts a half hour after the first car has left, how fast are the two separating when the second car has traveled 2 hours?

12. A man 6 feet tall starts walking from a point directly under a street light that hangs 20 feet above the street. He walks directly away from the light at a steady rate of 4 miles per hour. How fast is his shadow lengthening when the man has walked 28 feet?
13. The side of a square is increasing at a steady rate of 2 inches per minute. How fast is the area increasing when a side is 3 feet?
14. The area of a square is decreasing at 4 square inches per second. How fast is a side decreasing when the side is 20 feet?
15. How fast is the volume of a cube changing when the edge is 24 inches and the edge is increasing at a rate of 0.5 inch per minute.
16. How fast is the diameter of a sphere increasing when the diameter is 18 inches and the radius is changing at a rate of 3 inches per minute?
17. How fast is the radius of a sphere increasing when the diameter is 18 inches and the volume is increasing at 1.2 cu in. per second?
18. In No. 17, how fast is the surface of the sphere increasing?
19. The power P in a particular electric circuit is given by the formula $P = 20I^2$, where I represents current in amperes. If the current is changing at a rate of 0.002 ampere per second, how fast is the power changing when the current is 3 amperes?
20. Boyle's law states that with a constant temperature the volume of a gas varies inversely as the pressure p. If the volume is 120 cu in. when the pressure is 3 pounds per square inch, and the pressure is decreasing at a steady rate of 2 pounds per minute, how fast is the volume changing when the pressure is 24 pounds per square inch?
21. Kinetic energy K is given by the formula $K = (\frac{1}{2})mv^2$, where mass m is in grams, velocity v is in centimeters per second, and energy K is in ergs. If $m = 20$ grams, and $v = 10$ cm/sec, how fast is K changing if v is changing by 4 cm/sec^2.
22. In the formula for kinetic energy, $K = (\frac{1}{2})mv^2$, if velocity is measured in ft/sec, mass in slugs, and energy in foot-pounds, and if v is changing at -1.5 ft/sec^2? how fast is K changing when $v = 15$ ft/sec?
23. Electrical resistance in a certain conductor is given by the formula $R = 30 + 0.03T^2$, where T represents temperature in degrees centigrade (°C). If T is increasing at a rate of 0.15 C° per second, find the rate of change in the resistance when $T = 60°$ C. (Note the difference in the meaning between *degrees centigrade* and *centigrade degrees*. The notation 15° C refers to a particular *point* on the scale, that is, a particular temperature reading. On the other hand, 15 C° means 15 centigrade degrees, which is a span of 15 degrees anywhere on the centigrade scale.)

chapter

21

The Differential

21.1 GEOMETRIC DEFINITION

The *differential of an independent variable* x and the *differential of a function* are important concepts in calculus. The *differential of* x, the independent variable, is usually denoted by dx, and the *differential of the function* by the notation $df(x)$, or dy.

The differential of a function dy involves the derivative, but the two should not be confused. For the meaning of the differential of the function, we have two definitions, one algebraic the other geometric. The most useful is probably the algebraic, but the most easily understood is the geometric. We begin therefore with the geometric definition.

Let us graph the function of x (Fig. 21.1), represented by the equation $y = f(x)$. If we take a particular value of x, as at A, then point P, on the curve has the coordinates (x,y), and the value of the function, y, is the ordinate AP.

Now let x take on an increment, $\Delta x = AB$. For the new value of x

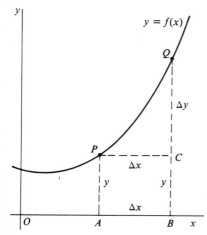

Fig. 21.1

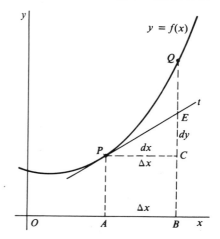

Fig. 21.2

252 The Differential

(that is, $x + \Delta x$), the function is represented by the ordinate BQ; that is, $BQ = y + \Delta y$. The increment Δy is equivalent to CQ.

The graph of the function $y = f(x)$ is repeated in Fig. 21.2, with the additional line t, the tangent to the curve at P, and intersecting CQ at E. The line segment PC (which is equal to AB) represents Δx, the increment in x. This is equivalent to dx, the *differential of x*. That is, $dx = \Delta x$. The line segment CQ represents Δy, as we have seen. However, the line segment CE represents the *differential of y*, denoted by dy.

From Fig. 21.2 we have

$PC = \Delta x = dx$, the differential of x.
$CQ = \Delta y$, the actual increment in y.
$CE = dy$, the differential of y, or differential of the function.

Note especially the following:

$$dx = \Delta x, \quad \text{but } dy \text{ does not equal } \Delta y \text{ (usually).}$$

Note also, that for a very small value of dx, dy is very nearly equal to Δy.

21.2 ALGEBRAIC DEFINITION

We are now in a position to derive the algebraic definition of the *differential of the function*, or dy. We recall that the slope of any line is defined by the equation

$$m = \frac{\text{rise}}{\text{run}}$$

From point P to point E, rise $= CE = dy$, the differential of y;

$$\text{run} = PC = dx, \text{ the differential of } x$$

Then the slope of the tangent line t is given by the ratio

$$m = \frac{\text{differential of } y}{\text{differential of } x}$$

If we write this expression by use of the notations, dy and dx, for the differentials, it is important at this point that we consider each differential as a separate and distinct quantity. We circle them to indicate this fact. Then

$$m = \frac{\boxed{dy}}{\boxed{dx}}$$

Although the expression now resembles the derivative in appearance, the two should not be confused. The expression here means *division* involving two differentials.

Now we recall that the slope of the tangent line t at point P is equivalent to the value of the derivative at that point. That is,

$$m = \text{the derivative} = \frac{dy}{dx}$$

21.2 Algebraic Definition

Since the slope m is equal to the derivative and also to the quotient of the differentials, we can state that the *quotient of the differentials is equal to the derivative*. Then we make this amazing discovery:

$$\frac{\widehat{dy}}{\widehat{dx}} = \frac{dy}{dx}$$

The left side of the foregoing equation is a fraction showing the ratio between the differentials, dy and dx, considered as *two separate quantities*. The right side is *not a fraction*, but instead is simply the expression for the derivative, which is *the limit of a ratio*.

We can now state the algebraic meaning of dy, the differential of the function. We begin with the statement

$$\frac{\widehat{dy}}{\widehat{dx}} = \text{the derivative}$$

Since the left side is a fraction or ratio between two separate quantities, we can multiply both sides of the equation by dx and get

$$dy = (\text{the derivative})\,(dx)$$

This is the algebraic definition of the differential of y, or differential of the function. *The differential of y is defined as the derivative of the function multiplied by the differential of x.* In simpler notation,

$$dy = \left(\frac{dy}{dx}\right)(dx)$$

Other notations for the differential of a function are

$$df(x) = f'(x)\,dx \quad \text{or} \quad dy = y'\,dx$$

The differential of a function represents the change that would take place in the function over an interval from $x = A$ to $x = B$ if the instantaneous rate of change at A continued constant over the interval.

We have seen that the quotient of the differentials dy and dx is equivalent to the derivative. Therefore, hereafter, whenever and wherever the derivative appears, *we may immediately consider it as the quotient of differentials*, if we wish to do so. Moreover, *the quotient of any two differentials may be considered as a derivative.* This change in meaning occurs constantly in integral calculus and in differential equations.

EXAMPLE 1. Find the differential of the function $y = x^3 + 5x^2 - 7x + 6$.

Solution. We first find the derivative:

$$\frac{dy}{dx} = 3x^2 + 10x - 7$$

By definition, $\qquad dy = (\text{derivative})\,(dx)$

Therefore, $\qquad dy = (3x^2 + 10x - 7)\,dx$

In an example of this kind we might be inclined at first sight to say that all we need to do to get dy is to multiply both sides of the equation by the quantity dx. This is not permissible as long as we think of the derivative, dy/dx, as being a single quantity. We cannot split up a derivative in this manner. However, as we have stated, we can think of the derivative as being the quotient of the differentials. Then we have a right to multiply both sides of the equation by dx and get immediately the result

$$dy = (3x^2 + 10x - 7)\,dx$$

21.3 USE OF THE DIFFERENTIAL OF A FUNCTION

The differential is used to find the approximate change in a function when it might be very difficult to find the actual change. The value of the differential lies in the fact that it is much easier to compute than Δy, the actual change in the function.

We have seen that dy is not equal exactly to Δy. However, from the graph (Fig. 21.2) we observe that if the change in x is small, then dy is approximately equal to Δy. For very small changes in x, the differential of y is often satisfactory as an approximation to Δy, the actual change in y.

EXAMPLE 2. Find the actual change, Δy, and the approximate change, dy, in the following function as x changes from 5 to 5.02:

$$y = x^2 + 3x + 7$$

Solution. To find the actual change in y, we compute the values of the function for $x = 5$ and $x = 5.02$ and then subtract:

$$f(5) = 25 + 15 + 7 = 47$$

$$f(5.02) = (5.02)^2 + 3(5.02) + 7 = 47.2604$$

Actual change in the function, $\Delta y = 47.2604 - 47 = 0.2604$. For the approximate change we have

$$dy = (dy/dx)(dx)$$

or

$$dy = (2x + 3)(dx)$$

Taking $x = 5$, and $dx = 0.02$,

$$dy = (10 + 3)(0.02) = 0.26$$

Note the difference in the amount of computation involved in finding the differential dy as compared with finding the actual change in y. The error in using the differential is only 0.0004.

As another use of differentials, let us consider the area of a ring. The area of a circular ring may be considered as the increase or decrease in the area of a circle. To find the exact area we should need to find the area of the large outer circle, then the area of the smaller inner circle, and then subtract the smaller area from the larger. However, we can get the approximate area of the ring by considering its area as the *change* in the

21.3 Use of the Differential of a Function

area of a circle. This approximate area can be found by differentials, provided the ring represents a small change in the area of the circle, that is, the width of the ring represents a small change in the radius. In a similar way the differential can be used to approximate changes in volume and other functions.

EXAMPLE 3. Find the approximate area of a circular ring having an inside diameter of 16 inches and an outside diameter of 16.6 inches.

Solution. We consider the area of the ring as the increase in the area of a circle whose diameter increases from 16 inches to 16.6 inches. For the area of any circle we have the formula

$$A = \pi r^2$$

Differentiating

$$\frac{dA}{dr} = 2\pi r$$

or, as differentials,

$$dA = (2\pi r)(dr)$$

In this equation, the quantity dA represents the approximate change in area. Taking $r = 8$ and $dr = 0.3$ we have

$$dA = (2\pi)(8)(0.3) = 4.8\pi$$

If we compute the actual change ΔA we shall find it is equal to 4.809π. The error by using differentials is only $0.009\,\pi$, which is less than 0.2% of the area of the ring.

The differential can also be used to determine the approximate error in a function, as shown in the following example.

EXAMPLE 4. A cube is measured as having each of its dimensions equal to 12 centimeters. If the error is stated as ± 0.02 cm, find the maximum possible error in the volume.

Solution. The possible error, ± 0.02 cm, in the edge may be considered as representing a possible change in the measurement. If we let x represent the number of centimeters in the edge, then $dx = 0.02$. (The error may be considered positive.) For the function we have

$$V = x^3$$

Differentiating,

$$dV = (3x^2)(dx)$$

Taking $x = 12$ and $dx = 0.02$, we get

$$dV = 3(144)(0.02) = 8.64 \text{ cm}^3$$

By using differentials we can approximate roots of numbers if the number whose root is to be found is near a perfect power corresponding to the index of the root, as shown in the following example.

EXAMPLE 5. Find the approximate square root of 4.03 by differentials.

Solution. As a general statement, we take the following equation to represent the square root of a perfect square:

$$y = \sqrt{x}$$

The Differential

We note that the answer is easy when $x = 4$. Then $y = 2$. Now let us say that x changes from 4 to 4.03, that is, $dx = 0.03$. We wish to find dy, the differential of y. Differentiating,

$$\text{since } y = x^{1/2}, \text{ then } dy = \frac{1}{2\sqrt{x}}(dx)$$

Substituting numerical values, $dy = \dfrac{1}{2(2)}(0.03) = 0.0075$

Since dy represents a change in y, we add this change to the original y, that is, the value of y when $x = 4$; we get

$$\sqrt{4.03} = 2.0075$$

A table shows the square root of 4.03 as 2.007486, approximately.

EXERCISE 21.1

1. In the equation $y = x^4$, find dy and Δy if $x = 4$, and $dx = 0.02$.
2. If $y = 3x^2 - 5x + 4$, find dy and Δy if $x = 5$ and $dx = -0.002$.
3. If $y = x^3 + 5x^2 + 7x + 8$, find dy and Δy if $x = 3$ and $dx = -0.0004$.
4. A particle moves according to the formula $s = t^3 - 3t^2 + 12t + 4$, in which distance s is in feet, time t is in seconds. Find ds, the approximate change in the distance traveled when t changes from 6 to 6.03 seconds.
5. A freely falling body falls according to the formula $s = 16t^2$. How far does it fall in 5 seconds? How far approximately does it fall in the next 0.2 second? (use differentials).
6. One formula for *power* in an electric circuit is the following: $P = I^2 R$. If the resistance R in a circuit is 20 ohms, find the approximate change in power (watts) when the current I changes from 3.5 amperes to 3.54 amperes.
7. One formula for *power* in an electric circuit is $P = E^2/R$. If the resistance R is always equal to 40 ohms, find the approximate change in power (watts) when the voltage E changes from 80 volts to 79.4 volts.
8. Using the formula in No. 7, if the voltage E is always equal to 40 volts, find the approximate change in power when the resistance R changes from 80 ohms to 81.2 ohms.
9. Using the formula in No. 6, if the current I remains constant at 5 amperes, find the approximate change in power when the resistance R changes from 60 ohms to 58.4 ohms.
10. Find the approximate change in the area of a circle when the diameter changes from 10 inches to 9.6 inches
11. Find the approximate area of a ring having an outside diameter of 14 inches and an inside diameter of 13.5 inches.
12. Find the approximate change in the volume of a sphere when its diameter increases from 24 inches to 25 inches. What is the actual change (ΔV)?
13. Find the approximate number of cubic inches of iron in a hollow iron sphere if the outside diameter is 10 inches and the metal shell is $\frac{1}{8}$ inch thick.
14. What is the approximate change in the volume of a cube if the edges are changed from 8 inches to 7.8 inches?
15. The measured diameter of a circle is stated as being 20 inches. If the maximum error in this measurement is ± 0.05 inch, what is the approximate maximum error in the area of the circle? This error is what percent of the total area? (This is percentage error.)
16. If a steel ball is measured as having a diameter of 3.2 inches with a maximum possible error of ± 0.0024 inch in the diameter, what is the approximate maximum error in the volume of the ball?

21.3 Use of the Differential of a Function 257

17. In an electric circuit containing a resistance of 20 ohms, the current I is measured as being 2.745 amperes with a maximum possible error of ± 0.0004 ampere, What is the approximate maximum error in power?
18. In a circuit in which power P is given by the formula $P = E^2/80$, find the approximate change in power when voltage E changes from 60 to 58.5 volts.
19. In a circuit in which power is given by the formula $P = 20I^2$, find the approximate *average change* in power *per second* when the current I changes from 8 amperes to 7.4 amperes in 1.5 seconds.
20. In a certain electric circuit the voltage E is given by the formula $E = 60I$. Find the approximate change in voltage when I changes from 0.02 ampere to 0.0196 ampere.
21. If $y = (x^2 - 3x)^4$, find dy when x changes from 4 to 4.02.
22. If $y = (x^2 + 5)^{1/2}$, find dy when x changes from 2 to 2.06.
23. If $y = (5 - x^3)(x^2 + 5)^{1/2}$, find the approximate change in y when x changes from 2 to 2.03.
24. If $y = (2x - 3)^2(x + 1)^{1/2}$, find the approximate change in y when x changes from 3 to 3.04.
25. If $y = (10 - x^2)^{1/2}(x^3 - 3)$, find the approximate change in y when x changes from 1 to 0.94.
26. If $y = \dfrac{x}{(4 + 3x)^{1/2}}$, find the approximate change in y when x changes from 4 to 3.84.
27. If $y = \dfrac{\sqrt{2x + 3}}{x^2}$, find dy when x changes from 3 to 3.06.
28. If $4x^2 - 5xy + 5y^2 = 80$, find dy when x changes from 5 to 4.94.
29. If $2xy^2 - 5x^2 + 21 = 0$, find dy when x changes from 3 to 2.88.
30. If $x^2 - 4xy + 3y^2 + 8 = 0$, find dy when x changes from 5 to 4.995.
31. If $x^2 - 3xy - 2y^2 + 4x - 5y + 10 = 0$, find dy when $x = 4$ and $dx = -0.003$.
32. Find by differentials the approximate roots indicated:
    (a) $\sqrt{16.05}$ (b) $\sqrt{8.98}$ (c) $\sqrt{24.96}$ (d) $\sqrt[3]{8.08}$ (e) $\sqrt[4]{80.6}$

chapter
22

Integral as Antiderivative

22.1 THE ANTIDERIVATIVE

We have seen how to find the derivative and the differential of a function. For example,

$$\text{if } y = x^4 + 7, \quad \text{then } dy/dx = 4x^3, \quad \text{and } dy = 4x^3\, dx$$

We have also noted that the derivative is the rate of change of a function with respect to the independent variable. We have discovered that the derivative is equivalent to the slope of a curve (in rectangular coordinates).

We now come to the inverse process, that of finding the function itself when its derivative (or differential) is given. This type of problem is as important as that of differentiation. If we know the rate of change of a function our problem often is to find the function itself. If we know the expression representing the slope of a curve, we wish to find the equation of the curve itself.

The problem of finding a function whose derivative (or differential) is given is analogous to that of finding the antilogarithm when a logarithm is given. For this reason, if the derivative of a function is given, then the function itself may be called the *antiderivative*.

We shall also use the customary name, the *integral*, for the function to be found, that is, the function whose derivative is given. The process of finding the antiderivative or integral is called *integration*. Just as multiplication and division are two inverse processes, so are the two processes of *differentiation* and *integration*.

We shall find that integration is one of the most powerful tools in the mathematics of science, as well as in other fields. As an example involving motion, we have seen that we differentiate an expression representing distance s to find the expression for velocity v. Now, if we have an expression representing *velocity*, we reverse the process of differentiation and get the expression for *distance* by integration. In like manner, the expression for velocity is differentiated to get *acceleration*. Then, if we know the expression for acceleration a, we integrate to find the expression for velocity;

then we can integrate once more and get the expression for distance. We shall also see that we can go one step further and find, through integration, the *total* distance traveled over a period of time. If we know the rate of change of current, di/dt, in an electric circuit, we can integrate to find the expression for current itself. Then by another integration we can find the total charge on a capacitor or the total transport of charge. Problems of these kinds and many others can be solved by integration.

22.2 THE PROCESS OF INTEGRATION

The process of integrating a given function is, in some respects, a matter of memory. To some extent it depends on a thorough knowledge of differentiation. This is like saying division depends on a good knowledge of multiplication. For example, if we are asked to find 63 divided by 9, we should know immediately that the answer is 7 because we recall instantly that 7 times 9 is 63.

As an example, suppose we have the function

$$y = x^4$$

Then the derivative is

$$\frac{dy}{dx} = 4x^3$$

and the differential is

$$dy = 4x^3 \, dx$$

Now, suppose we know the derivative of a function is $4x^3$, or, in differential form

$$dy = 4x^3 \, dx$$

If we wish to find the function itself, in this simple example, we should know immediately that

$$y = x^4$$

To indicate integration, we use the integral sign, $\int$, an elongated *S*. The sign really asks a question. For example, when we write

$$\int 4x^3 \quad \text{or} \quad \int 4x^3 \, dx$$

we are really asking: What function is it whose derivative is $4x^3$ or whose differential is $4x^3 \, dx$?

Note: In connection with integration, the quantity to be integrated is usually the *differential* rather than simply the *derivative*. The answer is the same. We can write

$$\int 4x^3 = x^4, \quad \text{but more often} \quad \int 4x^3 \, dx = x^4$$

The integral sign asks us to find the integral, which is the antiderivative.

The quantity *under* (to the right of) the integral sign is called the *integrand*. It consists of the derivative and the factor dx. We shall see later that the differential dx is included because it has important uses. In

260 Integral as Antiderivative

the foregoing example the integrand is $4x^3\,dx$. We can check the work of integration, of course, by differentiating the result, the integral.

22.3 STEPS IN FINDING THE INTEGRAL, OR ANTIDERIVATIVE

Finding a function whose derivative is given is sometimes called "educated guessing." We look at a function to be integrated and then recall that this is the differential of some other function we happen to know. For example, we may have the problem

$$\int 5x^4\,dx$$

Now we recognize that the integrand is the differential of x^5. Yet this is probably not really guessing any more than when we are asked to divide 63 by 9 and we guess the answer is 7. It is just a matter of knowing. Some functions can be integrated by just knowing the integral through a knowledge of derivatives.

However, for integrating most functions we need some specific steps to follow, some procedures that are superior to guessing. Let us therefore derive one of the most basic and powerful formulas for integrating many functions.

Consider carefully the steps in this example in differentiation:

$$\frac{d}{dx}(7x^4)$$

We shall first perform the *differentiation* and note carefully the steps. Then we shall simply *reverse* the steps for the process of *integration*. In differentiation, we recall the following steps:

1. First, *multiply* the function, $7x^4$, by the exponent 4 getting the result:

$$28x^4$$

2. Next, *decrease* the exponent by 1. We get $28x^3$, or the differential

$$28x^3\,dx$$

Now suppose we reverse the problem to one of integrating the differential:

$$\int 28x^3\,dx$$

For integration we *reverse* the differentiating procedure. We reverse not only the *process* in each step but also the *order* of the steps.

1. First, reversing step 2, we *increase* the exponent by 1 and get $28x^4$.
2. Next, reversing step 1, we *divide* the entire quantity by the new exponent 4 and get

$$\frac{28x^4}{4} = 7x^4$$

Note that the factor dx disappears in the process of integration.

22.4 THE INDEFINITE INTEGRAL

We have seen that many different functions, such as the following, have the same derivative because they differ by a constant only:

$$\frac{d}{dx}(x^4) \quad \frac{d}{dx}(x^4 + 5) \quad \frac{d}{dx}(x^4 - 3) \quad \frac{d}{dx}(x^4 + C)$$

For all four functions, the derivative is $4x^3$, and the differential is $4x^3\,dx$. (We recall that the derivative of a constant is zero.)

Now, if we have given the differential, $4x^3\,dx$, as the integrand, we cannot determine definitely the integral function. The integral will contain the term x^4, but it may also contain a constant term. Any of the following would be correct for the integral:

$$x^4 + 6;\ x^4 - 2;\ x^4 + \frac{1}{2};\ x^4 + 3^2;\ x^4 - \sqrt{2};\ x^4 + \pi.$$

In order to show that the integral may also contain some constant term, we say that the integral is equal to x^4 *plus some constant*. This constant is denoted by the letter C (or any other letter understood to be a constant). The constant C may be positive, negative, or zero. Then we write

$$\int 4x^3\,dx = x^4 + C$$

The result is called the *indefinite integral* because the constant C is as yet unknown. The C is called the *constant of integration.*

As another illustration,

$$\int 30x^4\,dx = 6x^5 + C$$

The result can be checked by differentiation:

$$d(6x^5 + C) = 30x^4\,dx$$

In general terms, if K represents a constant coefficient in the integrand, then we have the following rule for integrating a single term containing a variable, such as x, raised to any power:

The Power Rule:

$$\int K x^n\,dx = \frac{Kx^{n+1}}{n+1} + C \quad (n \neq -1)$$

In words, the rule says:

(1) *Increase the exponent by* 1.
(2) *Divide the term by the new exponent,* $n + 1$.
(3) *Add the constant of integration,* C.

Note that in the formula we have mentioned the fact that n, the exponent on the variable, cannot be equal to -1. In this case the power rule fails, for then we should have division by zero, which is impossible.

The following examples illustrate the use of the power rule in the integration of single terms of the variable. (Later we shall see how the same power rule is applied to the powers of functions.) Notice that the variable itself, such as x, y, t, and r, has associated with it the proper differential, dx, dy, dt, and dr, respectively. The variable indicated by the differential is called the *variable of integration:*

(a) $\int 200\, i^4\, di = \dfrac{200\, i^5}{5} + C = 40\, i^5 + C$ (b) $\int 5\, t\, dt = \dfrac{5t^2}{2} + C$

(c) $\int 12\, y^2\, dy = 4y^3 + C$ (d) $\int 2\pi r\, dr = \pi r^2 + C$

Although the differential factor, such as dx in the integrand, plays an important role in a problem, as we shall see, yet in integrating the function, we consider chiefly the *derivative* portion of the integrand. Note, however, that the factor dx indicates the variable of integration, that is, it refers to the variable on which the power is placed. We cannot integrate directly, for instance, the following:

$$\int 3x^2\, dt$$

The differential dt indicates that the variable of integration is t, not x, and, therefore, in this example we cannot apply the power rule on x. To integrate an expression of this kind we must replace x with some equivalent expression in t.

22.5 CONSTANT FACTOR IN THE INTEGRAND

There is one point that should be especially noted in connection with a constant factor, or coefficient, in the integrand. Consider the following example. By the power rule, we have

$$\int x^2\, dx = \dfrac{x^3}{3} + C$$

Now suppose that the integrand contains a constant factor, say, 12, as in $12x^2\, dx$. Then by the power rule we get

$$\int 12\, x^2\, dx = \dfrac{12x^3}{3} + C = 4x^3 + C$$

Now, before integrating, let us transfer the constant factor, 12, to the outside (the left) of the integral sign, so that we have

$$12 \int x^2\, dx$$

Now the expression means "12 times the integral of $x^2\, dx$." Then we get

$$12 \int x^2\, dx = 12\left(\dfrac{x^3}{4}\right) + C = 4x^3 + C$$

The answer is the same as before when the factor *12* was a part of the integrand. It makes no difference whether the constant factor is outside or under the integral sign. Then we can state this important fact:

Rule. *A constant factor in the integrand may be transferred to the left of the integral sign.*

Warning: A *variable* factor *cannot* be transferred across the integral sign in this manner.

The foregoing rule will be found extremely useful when we get to the integration of more complicated functions.

22.6 SPECIAL RULES FOR INTEGRATION

A few special rules at this point will be helpful in the integration of some simple forms. Each result can be checked by differentiation.

1. *The integral of a constant is equal to the constant times the variable to the first power.* This refers to the variable of integration, which is indicated by the differential, such as dx. As an example,

$$\int 5 \, dx = 5x + C$$

This rule is easily seen as a consequence of the power rule. For example,

$$\int 5 \, dx \quad \text{can be written} \quad \int 5x^0 \, dx$$

Then, by the power rule, we get

$$\frac{5x^1}{1} + C = 5x + C$$

2. *The integral of a differential*, such as dx, *is the first power of the variable of integration.* Note the following examples:

(a) $\int dx = x + C$ (b) $\int dy = y + C$ (c) $\int dA = A + C$

(d) $\int dv = v + C$ (e) $\int ds = s + C$ (f) $\int dQ = Q + C$

Actually, in the foregoing examples we should consider that dx is really $1 \, dx$. Therefore, we integrate the constant 1, not the dx. Then the rule is really covered in Rule 1.

3. *The integral of zero is a constant.* For example,

$$\int 0 \, dx = 0(x) + C = C$$

This rule agrees with an earlier one in differentiation that the derivative of a constant is zero.

4. *We state without proof at this time that the power rule is valid for negative and fractional powers*, as in the following examples:

(a) $\int 8x^{-3} \, dx = \frac{8x^{-2}}{-2} + C = -\frac{4}{x^2} + C$ (b) $\int 8x^{1/3} \, dx = 6x^{4/3} + C$

The results can be checked by differentiation.

5. *A polynomial (or a multinomial) can be integrated term by term.* In the following example we apply the power rule on each term individually:

$$\int (6x^5 - 8x^3 + 3x - 4x^{-2} + 6x^{-1/3}) \, dx =$$
$$x^6 - 2x^4 + \frac{3}{2}x^2 + \frac{4}{x} + 9x^{2/3} + C$$

The single constant C represents the combined value of the several C's that would arise in the integration of the separate terms.

EXAMPLE 1. Find

$$\int (4x^2 - 5x)(3x + 2) \, dx$$

Solution. A product of two or more terms can often be expanded before integrating. Expanding the multiplication we get

$$\int (12x^3 - 7x^2 - 10x) \, dx = 3x^4 - \frac{7}{3}x^3 - 5x^2 + C$$

EXAMPLE 2. Find

$$\int (x^2 - 2)^3 \, dx$$

Solution. Expanding, we get

$$\int (x^6 - 6x^4 + 12x^2 - 8) \, dx = \frac{x^7}{7} - \frac{6x^5}{5} + 4x^3 - 8x + C$$

EXAMPLE 3. The slope of a curve is given by the equation $m = 3 - 2x$. What is the general equation of the curve? Describe the curve.

Solution. We know that the slope of a curve is equivalent to the derivative. Therefore,

$$\frac{dy}{dx} = 3 - 2x$$

To find the equation of the curve itself, we integrate the expression for the slope, $3 - 2x$. The integrand is written in the form of a differential:

$$\int (3 - 2x) \, dx = 3x - x^2 + C$$

Then we can say the curve is represented by the function $3x - x^2 + C$. Using y to represent the function we get a quadratic equation, $y = C + 3x - x^2$. Now we recognize the equation as that of a parabola that opens downward. In fact, the general equation, the indefinite integral, represents an entire family of parabolas whose axis of symmetry is the line $x = 1.5$.

EXERCISE 22.1

Perform the following indicated integrations:

1. $\int 4 \, dx$
2. $\int 4x \, dx$
3. $\int 4x^2 \, dx$
4. $\int 3y \, dy$
5. $\int dq$
6. $\int dm$
7. $\int dw$
8. $\int 8t \, dt$
9. $8 \int t \, dt$

10. $\int 12 x^2 \, dx$
11. $\int (x - 2) \, dx$
12. $\int (5 - y) \, dy$
13. $\int (t^2 - 2t) \, dt$
14. $\int (x^{1/2} + x^{-3}) \, dx$
15. $\int (t^{-2} - t^{1/3}) \, dt$
16. $\int (3t^2 + 4t) \, dt$
17. $\int (i^2 - i) \, di$
18. $\int (ax + b) \, dx$
19. $\int (a_1 + a_2 y) \, dy$
20. $\int (2x - 3)^2 \, dx$
21. $\int (y - 3)(y + 5) \, dy$
22. $\int (t^2 - t^{-2}) \, dt$
23. $\int (x^{-4} - x^{-1}) x \, dx$
24. $\int (3x - 5) x^2 \, dx$

25. $\int \dfrac{1}{x^2} \, dx$
26. $\int \dfrac{6 \, dx}{x^3}$
27. $\int \dfrac{dx}{(3x^2 - x)^{-2}}$

28. $\int (3t^3 - 1)^2 \, t \, dt$
29. $\int x^4 (1 - 3x^3) \, dx$
30. Try $\int \dfrac{dx}{x}$

31. $\int (x^{-3} + 5x^{-2} + 3x^2 - x + 3) \, dx$
32. $\int (8x^7 - 4x^5 + 2x^3 - x^2 + \pi) \, dx$
33. $\int (x^{-1/2} - x^{-1/3} - 5x^{2/3} - x) \, dx$
34. $\int (2y^{-3/2} - 8y^{5/3} - 6y^{1/4}) \, dy$

35. If $y = \dfrac{x^2}{3}$, find $\int y \, dx$.
36. If $y = x^2 + 3x + 2$, find $\int y \, dx$.
37. If $y = 2x - 2x^{1/2}$, find $\int y \, dx$.
38. If $i = t^2 - 2t + 3$, find $\int i \, dt$.
39. If the slope of a curve is given by the equation, $m = 2x + 3$, what is the general equation of the curve? Describe the curve.
40. If the acceleration of an object is shown by the equation, $a = 6t + 2$, what is the general equation for velocity?
41. If the velocity of an object is given by $v = 3t^2 - 2t + 5$, what is the general equation for the distance s?

22.7 THE POWER RULE FOR SINGLE TERMS

We have seen how to apply the power rule in the integration of single terms containing a variable, as in these examples:

(a) $\int x^3 \, dx = \dfrac{x^4}{4} + C$ (b) $\int 15 t^4 \, dt = 3t^5 + C$ (c) $\int u \, du = \dfrac{u^2}{2} + C$

Let us restate the steps in finding the integral of a single term by the power rule:

Step 1. We increase the exponent by 1.

Step 2. We then divide the entire quantity by the new exponent.

Of course, for the indefinite integral, we add the constant C.

In general terms, if K is a constant coefficient in the integrand, then

$$\int K x^n \, dx = \dfrac{K x^{n+1}}{n+1} + C$$

Note especially *four* facts:

(1) The differential factor, dx, indicates the variable of integration.
(2) This factor, dx, is the differential of the variable on which the new power is to be placed.
(3) This factor, dx, must be present in the integrand before we can perform the integration.
(4) The factor dx disappears in the process of integration; that is, it is *absorbed* in the integral.

266 Integral as Antiderivative

Note how the factor dx appears when we differentiate:

$$\text{if} \quad y = 7x^4, \quad \text{then} \quad dy = 28x^3\, dx$$

Then, in the process of integration, the differential dx disappears.

The four facts stated concerning the differential dx applies equally to the differential of any variable on which the power rule is to be applied. The rule for integrating single terms by the power rule applies to *any* variable. That is,

$$\int K\, t^n\, dt = \frac{K\, t^{n+1}}{n+1} + C \quad \text{and} \quad \int 7\, u^5\, du = \frac{7u^6}{6} + C$$

Note that the differentials, dt and du, refer to the variables t and u, respectively, on which the power is increased. If the power rule is to be applied on any variable, say, v, in the integrand, then the integrand must also contain the differential, dv, of that variable.

22.8 INTEGRATION OF FUNCTIONS BY THE POWER RULE

We come now to the integration of entire functions by the power rule. In such problems we must be sure that the integrand contains not only dx, but the entire differential of the function. For example, suppose we have the problem

$$\int (x^2 + 5x + 4)^3\, dx$$

Now our question is: Can we simply use the power rule on the entire function $(x^2 + 5x + 4)$? Can we: (1) increase the exponent by 1 and then (2), divide by the new exponent 4, and get

$$\frac{(x^2 + 5x + 4)^4}{4}?$$

That is our question. If we let u represent the function $(x^2 + 5x + 4)$, then the problem is really

$$\int (u)^3\, dx$$

Note that the factor dx is *not* the differential of the function. The complete differential du must be present in the integrand. If we let u represent the function, $x^2 + 5x + 4$, then $du = (2x + 5)\, dx$. The du, the differential of u, is equal to the *derivative times dx*; that is,

$$du = \left(\frac{du}{dx}\right) dx$$

To apply the power rule on the function we must have the *complete du* present in the integrand. That is, we must have the factor $(2x + 5)$ as well as dx. The factor $(2x + 5)$ may be called an *integrating factor*. It must be present before we can apply the power rule. In our problem, this factor is missing. Therefore, we cannot use the power rule.

22.8 Integration of Functions by the Power Rule

The only way to work the foregoing problem as it stands is to expand the cube $(x^2 + 5x + 4)^3$, and then integrate term by term. The expansion will contain seven terms and would entail considerable labor. If we had the complete *du*, everything would be fine, and we could immediately apply the power rule.

On the other hand, let us look at another problem:

$$\int (x^2 - 7x + 4)^5 (2x - 7) \, dx$$

Our first thought is to apply the power rule on the function $(x^2 - 7x + 4)$ and get

$$\frac{(x^2 - 7x + 4)^6}{6}$$

However, before we do so, we must be sure we have the necessary integrating factor *du*. If we identify

$$u = x^2 - 7x + 4, \quad \text{then} \quad du = (2x - 7) \, dx$$

Note that the entire *du* is present in the integrand. Then the problem can be called

$$\int (\ u\ )^5 \, du$$

In this case we can apply the power rule on the function and get

$$\int (x^2 - 7x + 4)^5 (2x - 7) \, dx = \frac{(x^2 - 7x + 4)^6}{6} + C$$

We identify $\int (\quad u\quad)^5 (\quad du\quad) = \dfrac{(\ u\ )^6}{6} + C$

Many students have trouble identifying the factor *du*. We face the same problem when we come to the integration of powers of other functions, such as the trigonometric and others. A correct identification of the factor *du* is the secret to success in much work in integration. Remember, *u* represents the function on which the power is to be increased by 1. Remember also that *du* consists of the *derivative of this function* times the *factor dx;* that is, $du = (du/dx) \, dx$. The complete differential *du* must be present in the integrand. Upon integration the factor *du* disappears, or, as one student remarked, it is *absorbed* into the integral.

The process is reasonable if we recall what happens in differentiating. For example,

if

$$y = \frac{(x^2 - 7x + 4)^6}{6} + C$$

then

$$dy = (x^2 - 7x + 4)^5 (2x - 7) dx$$

In the process of differentiation the factor $(2x - 7)$ *emerges* from the function. Then in the inverse process of integration, this factor $(2x - 7)$ *reenters the function and disappears.*

268 Integral as Antiderivative

The power rule applies to powers of all kinds of functions. Whenever we wish to use the rule to integrate a function, u^n, then *the integrand must contain the perfect du as a factor.*

EXAMPLE 4. Integrate
$$\int (x^3 + 5x^2 + 3x - 4)^7 \, (3x^2 + 10x + 3) \, dx$$

Solution. We immediately think of applying the power rule on the function $(x^3 + 5x^2 + 3x - 4)$. Then the first thing to do is to look for the derivative of this function. It is $(3x^2 + 10x + 3)$, and it is present in the integrand. That is, we identify

$$u = (x^3 + 5x^2 + 3x - 4); \quad \text{then} \quad du = (3x^2 + 10x + 3) \, dx$$

Since the du factor is present in the integrand, we immediately apply the power rule on the function and get the indefinite integral,

$$\int (x^3 + 5x^2 + 3x - 4)^7 \, (3x^2 + 10x + 3) \, dx = \frac{(x^3 + 5x^2 + 3x - 4)^8}{8} + C$$

The answer can be checked by differentiation.

22.9 SUPPLYING A CONSTANT FACTOR

In some problems the perfect differential du may not be present, but we may be able to make it perfect by a legal operation. If the only thing missing is a constant factor, this factor can be supplied as in the following example.

EXAMPLE 5. Find
$$\int (x^2 - 6x + 7)^4 \, (x - 3) \, dx$$

Solution. In this example our first thought is to apply the power rule on the function $(x^2 - 6x + 7)$. However, we must first be sure the derivative of this function is present. If we

$$\text{let} \quad u = x^2 - 6x + 7, \quad \text{then} \quad du = (2x - 6) \, dx$$

Before the power rule can be applied, the integrand must contain the factor $(2x - 6) \, dx$. We do not have the factor $(2x - 6)$. However, we do have the factor $(x - 3)$, which can be multiplied by 2 to provide the perfect derivative $(2x - 6)$. Then we supply the factor 2 and compensate for this by multiplying at the same time by $\frac{1}{2}$. We get

$$\int \left(\frac{1}{2}\right)(2)(x^2 - 6x + 7)^4 \, (x - 3) \, dx$$

If we now transfer the constant factor $\frac{1}{2}$ to the outside of the integral sign, we shall have the perfect differential du, or $(2)(x - 3) \, dx$, under the integral sign. Then we get

$$\frac{1}{2} \int 2(x^2 - 6x + 7)^4 (x - 3) \, dx$$

22.9 Supplying a Constant Factor

Now, in the process of applying the power rule on the function, the differential, $du = 2(x - 3) \, dx$, disappears. However, the constant coefficient $\frac{1}{2}$ remains as a coefficient of the integral. We get

$$\left(\frac{1}{2}\right)\left(\frac{(x^2 - 6x + 7)^5}{5}\right) + C = \frac{(x^2 - 6x + 7)^5}{10} + C$$

Rule.
If the only part lacking in the integrand for a perfect differential is a constant factor, this factor may be supplied and then compensated for by writing its reciprocal outside the integral sign.

Warning. *A variable factor cannot be supplied in this manner.*

EXAMPLE 6. Find the indefinite integral

$$\int x^2 (8 - x^3)^4 \, dx$$

Solution. Here we identify

$$u = 8 - x^3$$

and look for

$$du = -3x^2 \, dx$$

We need $(-3x^2)$ as the integrating factor. We have only (x^2). Therefore, we supply the constant factor (-3) and write its reciprocal $(-\frac{1}{3})$ outside the integral sign and get

$$-\frac{1}{3}\int -3x^2(8 - x^3)^4 \, dx = \left(-\frac{1}{3}\right)\left(\frac{(8 - x^3)^5}{5}\right) = -\frac{(8 - x^3)^5}{15} + C$$

EXAMPLE 7. Find the indefinite integral

$$\int 5x\sqrt{25 - x^2} \, dx$$

Solution. First we write the radical as a power. Then we have

$$\int 5x(25 - x^2)^{1/2} \, dx$$

To apply the power rule, we identify $u = 25 - x^2$, and look for $du = -2x \, dx$. We note the integrand contains the factor $5x$. We have $5x$; we need $-2x$. The best way to get the correct constant factor is to transfer the 5 at once to the left of the integral sign, and get

$$5 \int x(25 - x^2)^{1/2} \, dx$$

Now we supply the constant factor (-2) in the integrand and write its reciprocal at the left of the integral sign. We then have

$$-\frac{5}{2} \int -2x(25 - x^2)^{1/2} \, dx$$

Applying the power rule, we get

$$\left(-\frac{5}{2}\right)\left(\frac{(25 - x^2)^{3/2}}{3/2}\right) = -\frac{5}{3}(25 - x^2)^{3/2} + C$$

EXAMPLE 8. Find the indefinite integral

$$\int \frac{(3x - 6) \, dx}{\sqrt{3 + 4x - x^2}}$$

Solution. First we write the radical as a power in the numerator and at the same time transfer the constant factor 3 to the left of the integral sign:

$$3 \int (x-2)(3+4x-x^2)^{-1/3}\,dx$$

Now we identify $u = 3 + 4x - x^2$, and look for $du = (4-2x)dx$. We have the factor $(x-2)$; we need the factor $(4-2x)$. We supply the constant factor (-2), write its reciprocal outside the integral sign and get

$$-\frac{3}{2}\int (-2)(x-2)(3+4x-x^2)^{-1/3}\,dx$$

Applying the power rule, we get

$$\left(-\frac{3}{2}\right)\left(\frac{(3+4x-x^2)^{2/3}}{2/3}\right) + C = -\frac{9}{4}(3+4x-x^2)^{2/3} + C$$

EXERCISE 22.2

Find the indefinite integral for each of the following:
1. $\int x(x^2+5)^3\,dx$
2. $\int 3x(2-x^2)^4\,dx$
3. $\int 5y\sqrt{8-y^2}\,dy$
4. $\int 6t(t^2-5)^{1/3}\,dt$
5. $\int (x-2)(x^2-4x+3)\,dx$
6. $\int (x-3)(5+6x-x^2)^3\,dx$
7. $\int t^2(8-t^3)^2\,dt$
8. $\int (3x-6)(5+4x-x^2)^{1/2}\,dx$
9. $\int (2-4x)(x^2-x-2)^3\,dx$
10. $\int (t-3)(6t-t^2)^{1/4}\,dt$
11. $\int (3x-2)^5\,dx$
12. $\int 5(y-3)^4\,dy$
13. $\int 3(2-x)^6\,dx$
14. $\int (4x+1)^{1/2}\,dx$
15. $\int (3-2x)^{2/3}\,dx$
16. $\int (4-y)^{3/2}\,dy$
17. $\int 4t(5-2t)\,dt$
18. $\int 3x^2(x+1)^2\,dx$
19. $\int y^2(y-2)^3\,dy$
20. $\int x^3(3-2x)^2\,dx$
21. $\int \dfrac{7x\,dx}{\sqrt{x^2-7}}$
22. $\int \dfrac{3t\,dt}{\sqrt{5-t^2}}$
23. $\int \dfrac{5x\,dx}{(2-5x^2)^{3/2}}$
24. $\int \dfrac{x(3x^2-5)^4\,dx}{5}$
25. $\int \dfrac{x(3-x^2)^3\,dx}{2}$
26. $\int \dfrac{(v-2)\,dv}{\sqrt{3+4v-v^2}}$

22.10 A SIMPLE DIFFERENTIAL EQUATION

A problem calling for the integration of a function can often be stated in the convenient form of an equation. For example, suppose we have the problem

$$\int (2x+3)\,dx$$

This is not an equation but only a direction to integrate. The integral is, of course, $x^2 + 3x + C$. To represent the integral function we can write

$$y = x^2 + 3x + C$$

Now let us see how the integrand, $2x + 3$, came about in the first place. Suppose we begin with the equation

$$y = x^2 + 3x + C$$

22.10 A Simple Differential Equation

This equation represents a parabola opening upward. Now if we differentiate the function we get

$$\frac{dy}{dx} = 2x + 3$$

Note that we now have an *equation containing a derivative*. If we write this in differential form we have

$$dy = (2x + 3)\, dx$$

Again we have an *equation*, this time *containing differentials*. In either case we have a *differential equation*.

Definition:

A differential equation is an equation containing a derivative or differentials. The following are differential equations:

$$\frac{dy}{dx} = 2x + 3 \quad \text{or} \quad dy = (2x + 3)\, dx$$

Let us begin our original problem in the form of a differential equation:

$$dy = (2x + 3)\, dx$$

Now, if we integrate *both sides* of this equation, we eliminate the differentials and get

$$\int dy = \int (2x + 3)\, dx$$

On the left side we get $y + C_1$; on the right side we get $x^2 + 3x + C_2$ or

$$y + C_1 = x^2 + 3x + C_2$$

In the result we have shown two constants C_1 and C_2. However, we can transpose one constant, say, C_1, and get

$$y = x^2 + 3x + C_2 - C_1$$

Whatever the values of C_1 and C_2, the two can be combined into a single constant that we may call C. Then the result can be written

$$y = x^2 + 3x + C$$

When we have eliminated the differentials, the result is called a *solution* of the original differential equation.

Definition:

A solution of a differential equation is an equation devoid of derivatives or differentials and one that satisfies the original differential equation. If we begin with the differential equation

$$dy = (2x + 3)\, dx,$$

the *general solution* is

$$y = x^2 + 3x + C$$

It is called a *general solution* because it contains at least one arbitrary constant C whose value is undetermined. However, we see that the solution represents a parabola, in fact, an entire family of parabolas, opening upward. To check the solution we differentiate to see whether the original equation is satisfied. Differentiating we get

$$dy = (2x + 3)\,dx$$

This is exactly the original differential equation, and we have the correct solution.

Note that solving a differential equation does not mean the same as solving an ordinary algebraic equation. For example, suppose we have the equation

$$3x + 7 = 25$$

Solving, $\qquad\qquad\qquad\qquad x = 6$

In this case we say the solution is 6. However, the solution of a differential equation is another *equation* obtained by eliminating the differentials.

If a differential equation contains a derivative, we usually change the form so as to show differentials instead. Many differential equations show the *differential form* rather than the *derivative*. For example, if we have the form

$$\frac{dy}{dx} = 3x^2 - 4x + 5$$

we take the fraction dy/dx to be the quotient of differentials and then multiply both sides of the equation by dx. Then we get the form

$$dy = (3x^2 - 4x + 5)\,dx$$

Now we integrate both sides of the equation and get

$$y = x^3 - 2x^2 + 5x + C$$

If a differential equation contains some power of y as well as x and dy/dx, then we attempt to arrange the equation so that the differential dy will be associated with the terms containing y itself, and the differential dx will be associated with the terms containing x. This procedure is called *separating the variables*.

EXAMPLE 9. Solve the differential equation $\dfrac{dy}{dx} = \dfrac{x}{y}$ by first separating the variables. What kind of curve does the solution represent?

Solution. Multiplying both sides of the equation by the quantity $y\,dx$ we get

$$y\,dy = x\,dx$$

Now that the variables are separated, we simply integrate both sides of the equation and get

$$\frac{y^2}{2} = \frac{x^2}{2} + C_1$$

If we eliminate the denominators by multiplying both sides of the equation by 2, we get
$$y^2 = x^2 + 2C_1$$
Since the constant $2C_1$ is equivalent to a single constant, we can write
$$y^2 = x^2 + C$$
or
$$y^2 - x^2 = C$$
Now we recognize the curve as a hyperbola. If C is positive, the hyperbola opens upward and downward; if C is negative, the hyperbola opens to the right and to the left.

EXAMPLE 10. The slope of a certain curve is given by the formula $m = x^2y^2$. What is the general equation of the curve?

Solution. We begin with the statement
$$\frac{dy}{dx} = x^2y^2$$
Now we multiply both sides of the equation by dx and divide by y^2 to separate the variables. Then we have
$$\frac{dy}{y^2} = x^2\, dx$$
This can be written
$$y^{-2}\, dy = x^2\, dx$$
Integrating both sides of this differential equation we get the general solution
$$-\frac{1}{y} = \frac{x^3}{3} + C_1$$
Clearing of fractions and transposing, we get $-3C_1 y - 3 = x^3 y$. Since $(-3C_1)$ is a constant, we can write the solution: $Cy - 3 = x^3 y$.

EXERCISE 22.3

Find the general solution for each of the following differential equations. Write the constant as a single C without a denominator:

1. $dy = (6x - 5)\, dx$
2. $dy = (x - 2)\, dx$
3. $di = (2t - 1)\, dt$
4. $dy = (3x - 4)\, dx$
5. $ds = (5t - 3)\, dt$
6. $di = (7 - 3t)\, dt$
7. $dy = (3x^2 - 2x)\, dx$
8. $dy = (x^2 + x)\, dx$
9. $dq = (2t^2 - 3t)\, dt$
10. $y\, dy = x^2\, dx$
11. $y^2\, dy = x^3\, dx$
12. $y^4\, dy = 5x^2\, dx$
13. $\dfrac{dy}{dx} = -\dfrac{x}{y}$
14. $\dfrac{dy}{dx} = \dfrac{x^3}{y}$
15. $\dfrac{dy}{dx} + \dfrac{x^2}{y^4} = 0$
16. $\dfrac{dx}{dy} = \dfrac{2y - 1}{3x + 4}$
17. $\dfrac{dy}{dx} = \dfrac{x^2 - x}{3y - 5}$
18. $\dfrac{dx}{dt} = \dfrac{2t + 3}{5x - 2}$

Write the equation of the family of curves whose slope is equal to each of the following expressions:

19. $2x + 5$
20. $4x - 7$
21. $3x^2 - 2x + 1$
22. $3 - 4x$
23. $2 - 6x - x^2$
24. $x^2 + 5x - 3$
25. $8x - x^2 - x^3$
26. $x^3 - 2x^2 - 5x$
27. $2 - 7x - 8x^2$
28. $x^4 - 3x^2 - 2$
29. $5 - x^2 - x^3$
30. 3

31. Find the general equation of the curve whose slope is always equal to twice the abscissa.
32. If the acceleration of an object is given by the formula, $a = 6t + 5$, find the general equation for the velocity. Then find the general equation for the distance s.

chapter

23

Integrals: Particular and Definite

23.1 DEFINITION: THE PARTICULAR INTEGRAL

We have seen that when we integrate a certain function, the integrand, we get another function, which we may call the *integral function*. To indicate that this integral function may have many particular values which differ by only a constant, we add the arbitrary constant C. The integral therefore represents an entire family of functions. For example,

$$\int (2x - 2)\, dx = x^2 - 2x + C$$

We have called the result, $x^2 - 2x + C$, the *indefinite* integral because of the undetermined value of the constant C. The result represents any number of functions, such as the following, and many others:

$$x^2 - 2x + 2;\quad x^2 - 2x - 3;\quad x^2 - 2x + 0;\quad \text{and others}$$

Whatever the value of the constant the result represents a parabola whose equation may be like one of the following:

$$y = x^2 - 2x + 2;\quad y = x^2 - 2x - 3;\quad y = x^2 - 2x$$

The general form $\qquad y = x^2 - 2x + C$

represents an entire family of parabolas, some of which are shown in Figure 23.1.

Now if there is any way we can determine the specific value of C, we shall have the equation of a *particular* parabola. For example, if $C = -3$, then we have the *particular integral*, $x^2 - 2x - 3$. This particular integral represents the particular parabola whose equation is $y = x^2 - 2x - 3$.

Actually, in practical work, it is the particular integral that is of value. For example, there is little value in knowing only that an electric current in a circuit is represented by the indefinite integral

$$i = t^2 + t + C$$

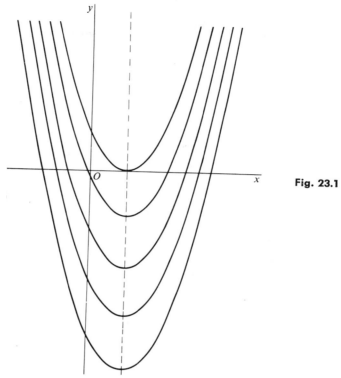

Fig. 23.1

When $t = 2$ seconds, we still do not know the current. However, if we know that $C = 3$, then we can tell the current at any time t. When $t = 2$ seconds, we know the current is exactly 9 amperes. Now let us see how the value of C can be determined in a particular problem.

23.2 THE VALUE OF THE CONSTANT C

Let us look again at the integral function, the indefinite integral,

$$x^2 - 2x + C$$

Setting y equal to the function, we get the general equation of a family of parabolas. The equation represents parabolas opening upward with the principal axis of the parabolas along the line $x = 1$. Some of the parabolas are shown in Figure 23.2.

If we now wish the equation for a particular parabola, we must have more information. If we specify that our particular parabola must pass through the point $(-2,5)$, this point will determine one particular parabola. Since the point must satisfy the equation, we substitute the coordinates for x and y in the general equation and get

$$5 = 4 + 4 + C \quad \text{or} \quad C = -3$$

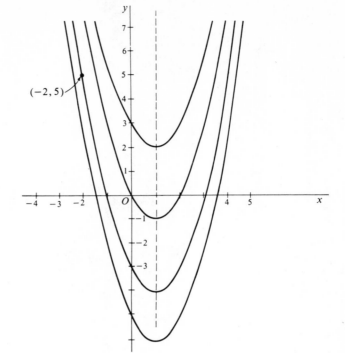

Fig. 23.2

Then we know that this value of C will satisfy the equation of the particular parabola,

$$y = x^2 - 2x - 3$$

The function itself, $x^2 - 2x - 3$, is called the *particular integral*.

Note especially the name *particular integral*. The term *indefinite* integral is used to refer to the integral which still contains the arbitrary constant C. One might suppose that when the constant is once determined the result would be called the definite integral. Instead it is called the *particular* integral. Note that the particular integral is still a function of x. For convenience, let us call the indefinite integral $I(x)$ and the particular integral $P(x)$. Note that both of these are functions of x:

$$I(x) = x^2 - 2x + C, \quad \text{the } \textit{indefinite integral}$$
$$P(x) = x^2 - 2x - 3, \quad \text{the } \textit{particular integral}$$

The term *definite* integral is reserved for another meaning, as we shall see in Section 23.5. The definite integral is not a function of x, but instead refers to a specific value for the entire integral.

23.3 BOUNDARY OR INITIAL CONDITIONS

The given conditions that determine the value of a constant in the particular integral are called *boundary conditions* or *boundary values*. They

23.3 Boundary or Initial Conditions

are sometimes called *initial conditions*. The term, *initial* conditions, is often used when the independent variable is time t. If a particular curve passes through a given point, then the coordinates of the point constitute the *boundary* conditions.

EXAMPLE 1. Find the indefinite integral as indicated:
$$\int (2x - 7)\, dx$$
Then find the particular integral if the following boundary conditions are given: When $x = 2$, the value of the integral is -4.

Solution. For the indefinite integral we have
$$\int (2x - 7)\, dx = x^2 - 7x + C$$
The expression $x^2 - 7x + C$ is the indefinite integral. Now we have given the information that the integral is equal to -4 when $x = 2$. Substituting $x = 2$ in the integral we get
$$4 - 14 + C = -4$$
Solving for C we get $\qquad C = 6$
Then the particular integral becomes $\qquad x^2 - 7x + 6$

EXAMPLE 2. Find the following indicated integral. Then find the particular integral from the following boundary values: when $x = 3$, the integral is equal to 7:
$$\int (3x^2 - 8x + 8)\, dx$$

Solution. Integrating,
$$\int (3x^2 - 8x + 8)\, dx = x^3 - 4x^2 + 8x + C$$
Now we substitute the boundary values:
$$27 - 36 + 24 + C = 7; \quad \text{then} \quad C = -8$$
The particular integral is $\qquad x^3 - 4x^2 + 8x - 8$

If we set the function equal to y we have the equation of a third-degree curve. The boundary values, $x = 3$ and the integral $= 7$, are equivalent to saying that the curve passes through the point (3,7).

EXAMPLE 3. Find the indefinite integral as indicated. Then find the particular integral by determining the value of the constant C from the boundary conditions: when $x = -2$, the integral value I equals 5:
$$\int (3 - 2x - 3x^2)\, dx$$

Solution. Integrating we get the indefinite integral
$$\int (3 - 2x - 3x^2)\, dx = 3x - x^2 - x^3 + C$$
If we call the integral function $I(x)$, then $I(x) = 5$ when $x = -2$. Substituting these values we get
$$-6 - 4 + 8 + C = 5, \quad \text{from which} \quad C = 7$$
Then the particular integral is
$$P(x) = 3x - x^2 - x^3 + 7$$

Integrals: Particular and Definite

EXAMPLE 4. The slope of a certain curve is given by the formula $m = 2x - 4$. Find the general equation of the family of curves having this slope. Then find the equation of the particular curve passing through the point $(1,-6)$.

Solution. Since the slope m is equivalent to the derivative dy/dx, we get the equation of the curve by integration. Since $dy/dx = 2x - 4$, we have

$$\int (2x - 4)\, dx = x^2 - 4x + C$$

If we represent the integral by y, we have the equation $y = x^2 - 4x + C$. This equation represents a family of parabolas opening upward. The point $(1,-6)$ on one parabola determines the value of C and the equation of the particular parabola. Substituting the coordinates $(1,-6)$ which represent boundary conditions, we get

$$-6 = 1 - 4 + C; \quad \text{then} \quad C = -3$$

The equation of the particular parabola passing through the point $(1,-6)$ is

$$y = x^2 - 4x - 3$$

EXAMPLE 5. Find the particular equation of the curve passing through the point $(2,-3)$ and having a slope given by $m = 6x^2 - 2x - 5$.

Solution. For the indefinite integral we have

$$\int (6x^2 - 2x - 5)\, dx = 2x^3 - x^2 - 5x + C$$

The indefinite integral represents a family of third-degree curves whose equation is

$$y = 2x^3 - x^2 - 5x + C$$

For the boundary conditions we have $y = -3$ when $x = 2$. Substituting values,

$$-3 = 16 - 4 - 10 + C; \quad \text{then} \quad C = -5$$

Then the particular integral is

$$2x^3 - x^2 - 5x - 5$$

The equation of the particular third-degree curve through the point $(2,-3)$ is

$$y = 2x^3 - x^2 - 5x - 5$$

EXAMPLE 6. In a certain electric circuit the rate of change of current is given by the formula

$$\frac{di}{dt} = 2t - 3$$

Find the indefinite integral and then find the particular integral from the initial values: when $t = 1.5$ sec, the integral $I(t) = 6$. Finally, find the current at the instant when $t = 2.5$ sec.

Solution. Integrating,

$$\int (2t - 3)\, dt = t^2 - 3t + C$$

For the initial conditions we have this: when $t = 1.5$, $I(t) = 6$. Substituting values,

$$2.25 - 4.5 + C = 6; \quad \text{then} \quad C = 8.25$$

The particular integral is $\quad t^2 - 3t + 8.25$

23.4 A Differential Equation: Particular Solution

The particular equation for the current becomes
$$i = t^2 - 3t + 8.25$$
This equation can now be used to find the instantaneous current at any time t. When $t = 2.5$ seconds,
$$i = 6.25 - 7.5 + 8.25 = 7 \text{ amperes}$$

EXERCISE 23.1

Find the indefinite integral of each of the following; then find the value of the constant C from the given boundary conditions; finally, write the particular integral (we represent the integral function by I):

1. $\int (2x - 5)\, dx; x = 1; I(x) = 8$
2. $\int (4x + 7)\, dx; x = 0, I(x) = 3$
3. $\int (5 - 6x)\, dx; x = -2; I(x) = -8$
4. $\int (3t - 4)\, dt; t = -1; I(t) = 10$
5. $\int (x - 3)\, dx; x = 1.5; I(x) = 9$
6. $\int (2s - 7)\, ds; s = 1; I(s) = -4$
7. $\int (4 - 5y)\, dy; y = 3; I(y) = -6$
8. $\int (x + 2)\, dx; x = -3; I(x) = 8$
9. $\int 7x\, dx; x = 1; I(x) = 2$
10. $\int 3\, dt; t = 1.2; I(t) = 8$
11. $\int 5\, ds; s = -2.5; I(s) = 0$
12. $\int -4\, dy; y = 0; I(y) = 0$

In the following examples the value of the independent variable and the corresponding value of the integral function are shown as a pair of numbers in parentheses, respectively. Find the particular integral.

13. $\int (3x^2 - 4x + 5)\, dx;\ (0,4)$
14. $\int (6t^2 - 2t - 7)\, dt;\ (-1,9)$
15. $\int (4 + 5s - 3s^2)\, ds;\ (-2,8)$
16. $\int (y^2 - y - 2)\, dy;\ (-3,7)$
17. $\int (x^2 - 3)\, dx;\ (1.25, 3.5)$
18. $\int (t^2 - 2t + 3)\, dt;\ (-1,-5)$
19. $\int (s^2 - 3s + 5)\, ds;\ (0,0)$
20. $\int (x^3 - 3x^2 + 4x - 5)\, dx;\ (1,-2)$

Find the equation of the particular curve whose slope is shown by the given function of x, if the curve passes through the indicated point:

21. Slope $= 2x - 3$; point $(1,-7)$
22. Slope $= 4x + 7$; point $(2,10)$
23. $m = x - 3$; point $(3,-2)$
24. $m = 5 - 2x$; point $(-2,-4)$
25. $m = -1 - 6x$; point $(-2,7)$
26. $m = 2x + 7$; point $(2,9)$
27. $m = x^2 - 4x + 1$; point $(3,-1)$
28. $m = 3x^2 - 6x - 9$; point $(-2,0)$
29. $m = x^2 - x - 3$; point $(0,0)$
30. $m = t^3 - 6t^2 - t + 3$; point $(-2,0)$
31. $m = x^{1/2} - x^{-1/2}$; point $(4,10)$
32. $m = x^{-3} - 2x^{-2}$; point $(3,7)$

23.4 A DIFFERENTIAL EQUATION: PARTICULAR SOLUTION

We have seen how a function to be integrated can be expressed in the form of a *differential equation*. For example, suppose we have the problem
$$\int (2x + 5)\, dx$$
This is not an equation. The integral sign simply tells us to do something. However, we can say that the expression can be written in the form of an equation containing differentials:
$$dy = (2x + 5)\, dx$$
Now we integrate *both sides* of the equation and get
$$y = x^2 + 5x + C$$

280 Integrals: Particular and Definite

We call the resulting equation a *general solution* of the differential equation. Note that the general solution corresponds to the indefinite integral, each one containing an unknown constant C.

Now suppose we also have the added information that $y = -12$ when $x = -3$. These are the boundary conditions that determine the value of the constant C. Substituting these values we get

$$-12 = 9 - 15 + C; \quad \text{then} \quad C = -6$$

When we have determined the value of the constant C in the general solution, we get what is called the *particular solution* of the differential equation. In this example, the particular solution is

$$y = x^2 + 5x - 6$$

Note that the *particular solution* of a differential equation corresponds to the *particular integral*. In each case the value of the constant C has been determined by boundary or initial conditions.

EXAMPLE 7. Find the general solution of the following differential equation; then find the particular solution from the initial conditions: when $t = 1.5$ seconds, $i = 4$ amperes:

$$di = (t^2 - 3t + 2) \, dt$$

Solution. Integrating both sides of the equation,

$$\int di = \int (t^2 - 3t + 2) \, dt$$

then

$$i = \frac{t^3}{3} - \frac{3t^2}{2} + 2t + C$$

Substituting boundary values,

$$4 = 1.125 - 3.375 + 3 + C$$
$$C = 3.25$$

Now we have the particular solution:

$$i = \frac{t^3}{3} - \frac{3t^2}{2} + 2t + 3.25$$

EXAMPLE 8. Find the particular solution of the following differential equation and describe the curve passing through the point (0,2):

$$\frac{dy}{dx} = -\frac{4x}{9y}$$

Solution. Separating the variables we get

$$9y \, dy = -4x \, dx$$

Integrating,

$$\int 9y \, dy = -\int 4x \, dx$$

then

$$\frac{9y^2}{2} = \frac{-4x^2}{2} + \frac{C}{2}$$

Multiplying both sides of the equation by 2,
$$9y^2 = -4x^2 + C$$
Transposing,
$$4x^2 + 9y^2 = C$$
Now we substitute the coordinates of the point on the curve and get
$$0 + 36 = C; \text{ then } C = 36$$
The particular curve has the equation $4x^2 + 9y^2 = 36$.

Now we recognize the curve as an ellipse with major axis of 6 along the x-axis, and minor axis of 4 along the y-axis. The center is at the origin.

EXERCISE 23.2

Find the general solution of each of the following differential equations by first separating the variables; then, from the given boundary or initial conditions, find the particular solution of each:

1. $\dfrac{dy}{dx} = 2x - 6; y = -2,$ when $x = 3$
2. $\dfrac{di}{dt} = 4t - 3; t = 0; i = 4$
3. $\dfrac{ds}{dt} = 5 - 6t; t = 2; s = 20$
4. $\dfrac{dy}{dt} = t + 2; t = 0; y = -4$
5. $\dfrac{dr}{dt} = 3 - 2t; t = 0; r = 20$
6. $\dfrac{dq}{dt} = 2t - 3; t = 2.5; q = 0.04$
7. $\dfrac{ds}{dt} = 3t^2 - 2t + 2; t = 1.2; s = 5$
8. $\dfrac{dy}{dx} = 4x = 9; x = -3; y = 20$
9. $\dfrac{dy}{dx} = 3x^2 - \dfrac{x}{2} - 4; x = 2; y = 60$
10. $\dfrac{dV}{dt} = \dfrac{t^2}{2} + t - 3; t = 3; V = 30$

Solve the following differential equations; describe the family of curves and find the equation of the particular curve through the given point:

11. $\dfrac{dy}{dx} = \dfrac{x}{y}$; point $(4, -2)$
12. $\dfrac{dy}{dx} = -\dfrac{x}{y}$; point $(-6, 3)$
13. $\dfrac{dy}{dx} = -\dfrac{3x}{5y}$; point $(5, -3)$
14. $\dfrac{dy}{dx} = \dfrac{5x}{2y}$; point $(-3, 4)$
15. $\dfrac{dy}{dx} = \dfrac{x^2}{y}$; point $(3, -4)$
16. $\dfrac{dy}{dx} = \dfrac{y^3}{x^2}$; point $(8, -2)$
17. $\dfrac{dx}{dy} = \dfrac{3y^2}{x}$; point $(-4, 2)$
18. $\dfrac{dx}{dy} = 8y$; point $(2.25, 3.5)$
19. $\dfrac{dy}{(3x - 2)^2} = \dfrac{dx}{5 - 4y}; (-1, -3)$
20. $\dfrac{dx}{3 - y} = \dfrac{dy}{2x - 5}$; point $(0.5, -1)$

23.5 THE DEFINITE INTEGRAL

To see what is meant by the definite integral, let us first restate the steps in finding the indefinite and particular integrals. We begin with some function of x, say, $f(x)$, which together with dx forms the integrand, the function to be integrated, such as
$$\int f(x)\, dx$$

When we integrate the function $f(x)$, we get another function, which we may call $F(x)$. To make the integration complete we add the constant C, which represents any constant, and we get the indefinite integral:

$$\int f(x)\,dx = F(x) + C$$

As a specific example,

$$\int (2x + 3)\,dx = x^2 + 3x + C$$

Now it may be that we happen to know the value of the integral corresponding to some value of x, that is, we may know the integral value is, say, 33, when x is, say, 4. From this additional information we can compute the value of the constant C:

$$33 = 4^2 + (3)(4) + C; \quad \text{then} \quad C = 5$$

Now we can write the particular integral:

$$x^2 + 3x + 5$$

The advantage of knowing the numerical value of C is simply this: Now we can compute the value of the integral for any value of x. For example,

if $x = 7$, the value of the integral is 75

if $x = 2$, the value of the integral is 15

Now it often happens that we wish to find the *difference between two integral-values*. In fact, one of the most important uses of integral calculus involves finding the *difference between the integral values corresponding to two different values of x*. In the example we have used, we found that

if $x = 7$, the integral value is 75

if $x = 2$, the integral value is 15

The difference between the two integral-values is $75 - 15 = 60$. This difference is called the *definite integral*.

The integral values are found, of course, by inserting the values of x in the particular integral $x^2 + 3x + 5$. The difference between the integral values might have been shown without combining the terms:

$$(7^2 + 21 + 5) - (2^2 + 6 + 5)$$
$$= (49 + 21 + 5) - (4 + 6 + 5)$$
$$= 49 + 21 + 5 - 4 - 6 - 5$$

At this point let us make one important observation: In combining the terms, the $+5$ and the -5 cancel each other. In fact, we could have omitted the constant 5 entirely in computing the *difference* between the two integral values. If we use only the integral, $x^2 + 3x$, without the constant and then compute the difference between the integral values, for $x = 7$ and $x = 2$, we get

$$(7^2 + 21) - (2^2 + 6) = 49 + 21 - 4 - 6 = 60$$

The answer is the same as before.

Our question now might well be: can we always omit the value of the constant C in computing the *difference between two integral values?*

Let us try this in more general terms. We begin with the integrand, $f(x)\,dx$, and get

$$\int f(x)\,dx = F(x) + C$$

Now we evaluate the integral for any two values of x, say $x = 7$ and $x = 2$:

If $x = 7$, the value of the integral is $F(7) + C$

if $x = 2$, the value of the integral is $F(2) + C$

For the difference we have

$$[F(7) + C] - [F(2) + C] = F(7) + C - F(2) - C = F(7) - F(2)$$

Note that the constant C drops out. Moreover, since 7 and 2 are definite numbers, the net result is a *definite* value for the expression. *This result is called the Definite Integral.*

The *definite integral* then represents the *numerical difference between values of the integral function for two different values of the independent variable.*

23.6 LIMITS OF INTEGRATION

In evaluating the definite integral, we find the difference by using two different values of x. These two values of x are called the *limits of integration*. The word *limit* as here used is not the same as when used in connection with the study of a variable approaching a constant. The present use refers simply to the two values of x used in evaluating the definite integral.

The limit having the greater algebraic value is called the *upper* limit; the smaller x-value is called the *lower* limit. In the preceding example the upper limit is 7; the lower limit is 2.

The definite integral is indicated by writing the two limits near the integral sign, the upper limit near the top and the lower limit near the bottom, as shown here:

$$\int_2^7 f(x)\,dx$$

This notation means that the integral is to be evaluated over the interval from $x = 2$ to $x = 7$.

Note that in evaluating the definite integral, we first find the integral value at the *upper* limit. Then from this we subtract the *lower*-limit value.

The first step in any problem is to perform the integration, omitting the constant C of integration. The integral as found is usually enclosed in brackets (sometimes only one bracket is used), and the limits are also usually shown near the integral. The final step is to evaluate the integral, first at the *upper* limit, then at the *lower* limit, showing the subtraction. The procedure may be shown in general terms as

$$\int_a^b f(x)\,dx = F(x)\Big]_a^b = F(b) - F(a)$$

The following examples illustrate the fact that the definite integral may have a value that is positive, negative, fractional, or zero.

EXAMPLE 9. Evaluate the integral

$$\int_2^5 x\,dx$$

Solution. The first step is to perform the integration. The integral is then evaluated as follows:

$$\int_2^5 x\,dx = \frac{x^2}{2}\Big]_2^5 = \frac{25}{2} - \frac{4}{2} = \frac{21}{2} = 10.5$$

EXAMPLE 10. Evaluate the integral

$$\int_{-3}^1 (3t^2 + 6t - 5)\,dt$$

Solution. Note that the greater limit value is placed near the top, that is, $1 > -3$. After the integration has been done, the integral is evaluated:

$$\int_{-3}^1 (3t^2 + 6t - 5)\,dt = t^3 + 3t^2 - 5t\Big]_{-3}^1$$
$$= (1 + 3 - 5) - (-27 + 27 + 15) = -16$$

EXAMPLE 11. Evaluate the integral

$$\int_{-2}^0 (5 - 2y - 3y^2)\,dy$$

Solution.

$$\int_{-2}^0 (5 - 2y - 3y^2)\,dy = 5y - y^2 - y^3\Big]_{-2}^0 = 0 - 0 - 0 - (-10 - 4 + 8) = +6$$

EXAMPLE 12. Evaluate the integral

$$\int_1^3 (3v^2 - 4v - 5)\,dv$$

Solution. Note that the value of the definite integral is not affected by the particular variable of the integrand. The solution follows:

$$\int_1^3 (3v^2 - 4v - 5)\,dv = v^3 - 2v^2 - 5v\Big]_1^3 = 27 - 18 - 15 - 1 + 2 + 5 = 0$$

EXAMPLE 13. Taking acceleration for a free fall to be 32 ft/sec², how far will an object fall during the fifth second of its fall?

23.6 Limits of Integration

Solution. The velocity at any time t (sec) is given by the formula $v = 32t$. The velocity is equivalent to ds/dt. To find s we integrate *velocity* v. Then we evaluate the result to find s (distance) between the limits $t = 4$ and $t = 5$:

$$\int_4^5 32\, t\, dt = 16t^2 \Big]_4^5 = 400 - 256 = 144$$

Therefore, the object will fall 144 feet during the fifth second.

We have seen that we cannot integrate directly an expression such as the following

$$\int_1^5 x\, dt$$

The differential dt indicates that the variable of integration is t, not x. In order to perform the indicated integration we must have given the relation between x and t, that is, x must be a known function of t. Moreover, the limits are t-limits, not x-limits. It may be that $x = t^2 - 3$. Then we get

$$\int_1^5 x\, dt = \int_1^5 (t^2 - 3)\, dt = \frac{t^3}{3} - 3t \Big]_1^5 = \frac{125}{3} - 15 - \frac{1}{3} + 3 = \frac{88}{3}$$

In some problems we may have a combination of two functions, as in the following example, a type that occurs often:

EXAMPLE 14. Find

$$\int_2^6 (x_2 - x_1)\, dy;\ x_2 = y - 3;\ \text{and}\ x_1 = \frac{y^2}{4} - y.$$

Solution. We first find the value of $(x_2 - x_1)$ in terms of y:

$$x_2 - x_2 = (y - 3) - \left(\frac{y^2}{4} - y\right) = y - 3 - \frac{y^2}{4} + y = 2y - 3 - \frac{y^2}{4}$$

Replacing the quantity $(x_2 - x_1)$ with its equivalent, $2y - 3 - \frac{y^2}{4}$, we get

$$\int_2^6 (x_2 - x_1)\, dy = \int_2^6 \left(2y - 3 - \frac{y^2}{4}\right) dy = y^2 - 3y - \frac{y^3}{12} \Big]_2^6$$

$$= 36 - 18 - 18 - \left(4 - 6 - \frac{8}{12}\right) = 36 - 18 - 18 - 4 + 6 + \frac{2}{3} = \frac{8}{3}$$

EXERCISE 23.3

Evaluate each of the following definite integrals:

1. $\int_3^5 dx$
2. $\int_3^5 dy$
3. $\int_3^5 dt$
4. $\int_3^5 dv$
5. $\int_1^4 (4x - 3)\, dx$
6. $\int_1^4 (4y - 3)\, dy$
7. $\int_1^4 (4t - 3)\, dt$
8. $\int_{-2}^3 (2x + 5)\, dx$
9. $\int_{-2}^3 (2t + 5)\, dt$
10. $\int_{-2}^3 (2y + 5)\, dy$
11. $\int_2^3 (2x - 5)\, dx$
12. $\int_1^3 (3x^2 - 4x + 7)\, dx$
13. $\int_{-1}^2 (6x^2 + 8x - 9)\, dx$

14. $\int_0^2 (3x^2 - 6x)\,dx$ 15. $\int_0^2 (3t - t^2)\,dt$ 16. $\int_1^4 (3 - x - x^2)\,dx$

17. $\int_{-2}^1 (5 - 4t - t^2)\,dt$ 18. $\int_{-3}^{-1} (8t - 6t^2)\,dt$ 19. $\int_0^3 (x^2 - 2x)x^2\,dx$

20. $\int_{-3}^1 (2v - v^2)\,dv$ 21. $\int_5^{10} (4q^3 - 6q^2)\,dq$ 22. $\int_{-1}^1 (2x + 3)(x - 5)\,dx$

23. $\int_1^3 (2 - 3x)^2 x\,dx$ 24. $\int_1^{64} (x^{1/3} - x^{1/2})\,dx$ 25. $\int_1^4 (x^{1/2} - x^{-3/2})\,dx$

26. $\int_1^8 (t^{2/3} + t^{1/3})\,dt$ 27. $\int_1^3 (x^{-2} - x^{-3})\,dx$ 28. $\int_{50}^{100} (x^2 + 3x)\,dx$

29. $\int_{-2}^2 (t^3 - 3t^2 - 9t + 10)\,dt$ 30. $\int_{-2}^0 (x^2 - x^{-2} - x + 1)\,dx$

31. $\int_1^3 y\,dx;\ \text{if}\ y = \dfrac{x^2}{4}$ 32. $\int_0^2 3x\,dy;\ \text{if}\ y = x^2$

33. $\int_1^2 x^2\,dy;\ \text{if}\ 3x = 2y$ 34. $\int_{-1}^2 y^2\,dx;\ \text{if}\ y = 2x + 3$

35. $\int_1^4 y\,dt;\ \text{if}\ y = 3t^2 - 2t$ 36. $\int_0^4 y\,dx;\ \text{if}\ y^2 = 4x$

37. $\int_{-2}^4 (x_2 - x_1)\,dy;\ \text{if}\ x_2 = \dfrac{y}{2} + 2,\ \text{and}\ x_1 = \dfrac{y^2}{4}$

38. $\int_{-2}^4 (y_2 - y_1)\,dx;\ \text{if}\ x = 2y_2,\ \text{and}\ x^2 = 4y_1 + 8$

chapter
24

The Definite Integral as an Area

24.1 THE AREA ROUNDED BY CURVES

We shall see that the definite integral is a powerful means of computing the exact area bounded by one or more curves. Of course, in analytic geometry, the term *curve* is usually taken to include straight lines. The area to be found may be bounded in part by one or both of the coordinate axes. We shall consider several cases.

We know that a rectangular area is easily computed by multiplying the length by the width. For example, a rectangle 7 inches long and 4 inches wide has an area of 28 square inches. Even in the case of a triangle, a parallelogram, or a trapezoid, we can use the geometric formulas for the areas of such figures bounded by straight lines.

However, the exact area bounded by a curve cannot be computed in this manner. For example, in Figure 24.1 we have an area bounded by the lines $x = 2$, $x = 7$, $y = 0$, and the curve $y = f(x)$. This area is called the area *under* the curve from $x = A$ to $x = B$.

To approximate the area we might block it off into squares, count the number of complete squares, and estimate the combined area of the partial squares. But even so, the area obtained in this way will not be exact.

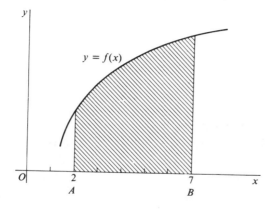

Fig. 24.1

288 The Definite Integral as an Area

24.2 EXACT AREA UNDER A CURVE

As a first step toward determining the exact area under the curve, we divide the line segment AB into a number of segments, say, 5, of uniform length Δx. Call the points of division on the x-axis x_1, x_2, x_3, x_4, with the values x_0 and x_5 corresponding to points A and B, respectively. Now we divide the desired area into five strips by drawing vertical lines through the points on AB, intersecting the curve at the points E, F, G, H (Fig. 24.2). The total area under the curve is, of course, exactly equal to the sum of the areas of these 5 strips.

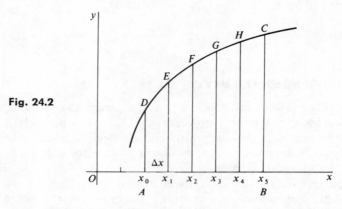

Fig. 24.2

To estimate the areas of the strips, we form perfect rectangles by drawing horizontal lines through the points D, E, F, G, H, as shown in Figure 24.3. Call the rectangles I, II, III, IV, V. The combined area of these 5 rectangles is slightly less than the total area under the curve.

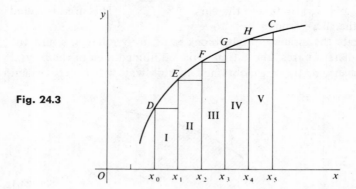

Fig. 24.3

Now we write the area of each rectangle. The width of each rectangle is Δx; the length is the ordinate, or $f(x)$. For rectangle I, the area is *length* times *width*, or $f(x_0)\,\Delta x$. The sum of the areas of the 5 rectangles can be expressed as

$$f(x_0)\Delta x + f(x_1)\Delta x + f(x_2)\Delta x + f(x_3)\Delta x + f(x_4)\Delta x$$

This sum may be expressed in a short form (using the summation sign):

$$\text{Area of 5 rectangles} = \sum_{i=0}^{i=4} f(x_i)\Delta x$$

This symbol is read: "The sum of all the terms in the series $f(x_i)\Delta x$, as i varies by integral values from $i = 0$, to $i = 4$."

Since the combined area of the 5 rectangles is slightly less than the exact area under the curve, we can write

$$\sum_{i=0}^{i=4} f(x_i)\Delta x < \text{total area under the curve}$$

Suppose now we divide the area into 10 strips instead of 5, and form rectangles as before (Fig. 24.4). We find that the area of the inscribed rectangles is still slightly less than the exact area under the curve. However, the *difference* would be less than before. The combined area of the 10 rectangles will be more nearly equal to the exact area under the curve.

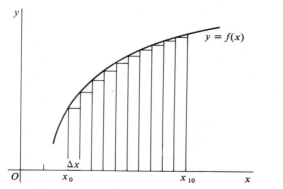

Fig. 24.4

If we increase n the number of strips of area, to, say, 20, or 40, or 80, or any number, the combined area of the rectangles is more nearly equal to the exact area under the curve (Fig. 24.5). But there would always be a slight difference represented in the figure by the jagged, rough edge along the curve.

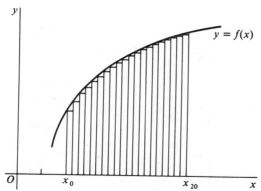

Fig. 24.5

Now, suppose we let the number of strips increase without limit (letting n tend to infinity). Then the width Δx of the strips approaches zero. We shall discover this very important fact:

The combined area of the inscribed rectangles approaches nearer and nearer the exact area under the curve. For any finite number of strips we should always have a rough, jagged edge along the curve. But *as the number of strips increases without bound, and the width of the strips approaches zero as a limit, the combined area of all the rectangles approaches the exact area under the curve as a limit.*

Symbolically,

$$\lim_{n \to \infty} \sum_{i=1}^{i=n} f(x_i)\Delta x = \text{exact area under the curve}$$

Our problem involves one more step: *The fundamental theorem of integral calculus states that the limit of a sum of this kind is equal to the definite integral between any two given values of the independent variable;* that is,

$$\lim_{n \to \infty} \sum_{i=1}^{i=n} f(x_i)\Delta x = \int_{x=a}^{x=b} f(x)\,dx$$

How then can we add up a hundred strips of area or a thousand or an infinite number? Answer: *by the definite integral.* The beauty of it is that with a single stroke, $\int$, we can add up an *infinite number* of strips, we get the *exact area*, and *all rough edges are made smooth*.

24.3 THE ELEMENT OF AREA

We have seen that the problem of computing the exact area under a curve is a matter of summation, that is, the summing up of an infinite number of strips. A single strip is called an *element of area*. This element, sometimes called a *differential of area*, is denoted by dA.

To compute a particular area by integration, we follow a few definite steps:

(1) Our first step is to show the area to be found (Fig. 24.6).

(2) Next, we suppose this area divided into many strips, vertical or horizontal. *We show one of these strips.*

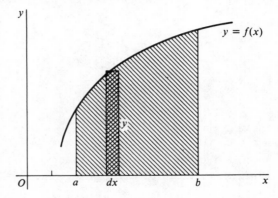

Fig. 24.6

24.3 The Element of Area

(3) *We state the length and the width of this strip.* In this case the length is the height, or y, of the curve. The width is a differential; in this case dx. Remember, however, that the width dx is an infinitesimal approaching zero as a limit as the number of strips increases without limit (becomes infinite).

Note: Although we show only one strip, we must remember that this element represents all the infinite strips into which the area is imagined to be divided.

(4) *We express the area of the element.* Since the element is a rectangle, the area is (length) × (width). For the element shown in this example,

$$\text{length} = y \qquad \text{width} = dx$$

Then the area of the element itself is given by: $dA = (y)(dx)$.

(5) Finally, *we sum up all the infinite elements by integration,* computing the definite integral between limits:

$$\text{total area} = \int_{x=a}^{x=b} dA = \int_{x=a}^{x=b} y\, dx$$

Note that the *integrand* is the *single element*.

The key to success in finding the exact area by integration is

(1) *Determine correctly the element of area.*
(2) *State correctly the area of this element.*
(3) *Integrate the area of the element.*
(4) *Evaluate the definite integral between given limits.*

Note: In all cases of integration as a process of summation, the first step is to determine correctly the *element of the quantity to be found.* Whatever total quantity we wish to find, whether it be total area, total volume, total mass, total work done, total force, total first moment, total moment of inertia, total electrical charge on a capacitor, or any other quantity, *the element is always a small quantity or small bit of the quantity to be found.* It may, in general, be called dQ. *The total quantity* is then found by *integrating the element* and *evaluating the integral between the given limits.*

EXAMPLE 1. Find the total area under the curve $x = 2y$ from $x = 4$ to $x = 10$.

Solution. **Step 1.** Sketch the curve and show the area to be found (Fig. 24.7).

Step 2. Imagine the area divided into strips, vertical or horizontal. In this case, let us take vertical strips. Show one of these strips as a rectangle. Call its area dA.

Step 3. The length of the strip is y, the height of the curve. The width is dx.

Step 4. The area of the element is given by

$$dA = (y)(dx)$$

Step 5. Now we sum up the infinite number of elements by integrating the element and evaluating between $x = 4$ and $x = 10$. Since the factor dx indicates the variable of integration, we must change the y in the integrand to its equivalent, $y = x/2$, from the *equation of the curve.* Then

$$\text{Area} = \int_{x=4}^{x=10} y\, dx = \int_{x=4}^{x=10} \frac{x}{2}\, dx = \frac{x^2}{4}\Big]_4^{10} = 21$$

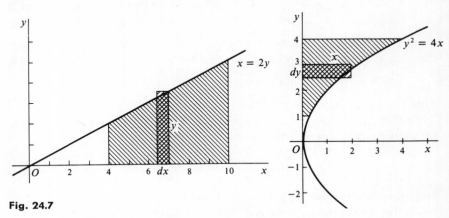

Fig. 24.7

Fig. 24.8

EXAMPLE 2. Find the area between the y-axis and the curve $y^2 = 4x$ from $y = 1$ to $y = 4$.

Solution. We sketch the curve and show the area to be found. In this example, let us assume the area divided into *horizontal* strips. We show the element of area as one of these strips, and state its area (Fig. 24.8):

length of strip = the x of the curve;
width of strip = dy, differential of y.

Then $\quad\quad\quad\quad dA = x\, dy; \quad \text{or} \quad dA = \frac{y^2}{4}\, dy$

For total area we integrate the element and evaluate from $y = 1$ to $y = 4$:

$$\text{Area} = \int_{y=1}^{y=4} x\, dy = \int_{y=1}^{y=4} \frac{y^2}{4}\, dy = \frac{y^3}{12}\Big]_1^4 = \frac{1}{12}(64 - 1) = \frac{21}{4}$$

24.4 TOTAL AREA AS THE SUM OF SEPARATE PORTIONS

Sometimes it is necessary to compute separate portions of a given area and then add these portions. We then make use of the following idea.

In Example 1, in the previous section (24.3), we found the area to be 21 square units from $x = 4$ to $x = 10$. The area could have been considered in two or more portions (Fig. 24.9).

24.4 Total Area as the Sum of Separate Portions

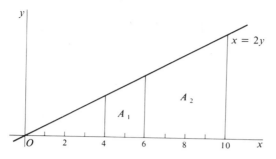

Fig. 24.9

Let A_1 be the area from $x = 4$ to $x = 6$
Let A_2 be the area from $x = 6$ to $x = 10$.

For A_1 we get
$$\int_4^6 \frac{x}{2} dx = \frac{x^2}{4}\Big]_4^6 = 5 \text{ units}$$

For A_2 we get
$$\int_6^{10} \frac{x}{2} dx = \frac{x^2}{4}\Big]_6^{10} = 16 \text{ units}$$

The total area, A_t, from $x = 4$ to $x = 10$, is $A_t = A_1 + A_2 = 5 + 16 = 21$.
For this problem, this principle is illustrated as follows:

$$\int_4^{10} \frac{x}{2} dx = \int_4^6 \frac{x}{2} dx + \int_6^{10} \frac{x}{2} dx = \frac{x^2}{4}\Big]_4^6 + \frac{x^2}{4}\Big]_6^{10} = 5 + 16 = 21$$

Stated in general terms,

$$\int_a^b f(x)\, dx + \int_b^c f(x)\, dx = \int_a^c f(x)\, dx$$

The foregoing principle *must* be used in some problems, of which the following is an example:

EXAMPLE 3. Find the total area bounded by the x-axis and the straight line $x + 2y = 6$, from $x = 0$ to $x = 10$.

Solution. The required area and the element are shown (Fig. 24.10). The element of area is

$$dA = (\text{length})(\text{width})$$

or
$$dA = y\, dx$$

Solving the equation for y, $y = 3 - \dfrac{x}{2}$

If we integrate directly from $x = 0$ to $x = 10$ we get

$$A = \int_0^{10} y\, dx = \int_0^{10} \left(3 - \frac{x}{2}\right) dx = 3x - \frac{x^2}{4}\Big]_0^{10} = 5 \text{ sq. units}$$

Yet if we integrate the two portions of the area separately, A_1 from $x = 0$ to $x = 6$, and A_2 from $x = 6$ to $x = 9$, we get

$$A_1 = \int_0^6 \left(3 - \frac{x}{2}\right) dx = 3x - \frac{x^2}{4}\bigg]_0^6 = 18 - 9 = 9, \text{ sq. units}$$

$$A_2 = \int_6^{10} \left(3 - \frac{x}{2}\right) dx = 3x - \frac{x^2}{4}\bigg]_6^{10} = 30 - 25 - 18 + 9 = -4, \text{ sq. units}$$

The definite integral from $x = 6$ to $x = 10$ turns out to be *negative*. Since no area can be negative, we must take the *absolute value* of the second portion. The total area is therefore 13 square units whereas the definite integral from 0 to 10 is only 5. Using signed values we have $A_1 + A_2 = 9 + (-4) = 5$, which agrees with the definite integral but does not show total actual area.

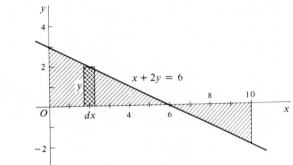

Fig. 24.10

Warning. The fact that a portion of desired area is on one side of an axis and another portion on the other side will not always give the two areas opposite signs. It is only when the element *reverses* itself in moving from one portion of the area to the other. Note in Figure 24.10 if we show the element in the first portion of the area, the *top* of the element lies on the curve. In the second portion of area, the *bottom* of the element would lie on the curve. *Whenever an element reverses itself from one portion of an area to the other, then one portion will subtract from the other.* For this reason it is extremely important that the element be correctly determined. *The key to success is the element.*

EXERCISE 24.1

In each problem, sketch the area and show the element.
1. Find the area in the first quadrant bounded by the curves $y^2 = 4x$, $y = 0$, and $x = 4$.
2. Find the area in the first quadrant bounded by the curves $y^2 = 4x$, $x = 0$, and $y = 4$. Use horizontal element.
3. Find the area bounded by the curves $y = x^3$, $y = 0$, and $x = 2$.
4. Find the area bounded by the curves $y = x^4$, $y = 0$, and $x = 1$.
5. Find the area under the curve $y^2 = 2x$, from $x = 0$ to $x = 8$.
6. Find the area bounded by the curves $y^2 = 2x$, $x = 0$, and $y = 4$.
7. Find the area bounded by the curves $x^2 = 8y$, $x = 0$, and $y = 2$ in the first quadrant.

8. Find the area bounded by the curves $y^2 = 4x$, $x = 0$, $y = 4$, $y = -2$.
9. Find the area under the curve $y^2 + 2y - 4x - 7 = 0$, from $x = -1$ to $x = 2$.
10. Find the total area enclosed by the curves $x = 0$ and $y^2 + 4x - 16 = 0$.
11. The lines $x = 4$ and $y = 4$ form a square with the coordinate axes. What portion of this square is inside the parabola $y^2 = 4x$, and what portion of the square is outside the parabola?
12. The lines $y = 4$ and $x = 8$ form a rectangle with the coordinate axes. What portion of this rectangle is inside and what portion outside the parabola $y^2 = 2x$?
13. What tentative hypothesis might be assumed from Problems 11 and 12?
14. Find the total area bounded by the x-axis, the y-axis, and the lines $x - 2y = 8$.
15. Find the total area between the line $x - 2y = 5$ and the x-axis, from $x = 0$ to $x = 11$.
16. Find the total area bounded by the lines $2x + y = 8$, $x = 0$, $y = 0$, and $y = 12$.

24.5 AREA BETWEEN CURVES

The problem of computing the area *between* two curves is somewhat more complicated than that of simply finding the area under a curve. There are two ways this might be done. The simplest way is to consider only the given area itself as divided into vertical or horizontal strips. The method is illustrated by an example:

EXAMPLE 4. Find the area between the two curves $x - 2y + 8 = 0$, and $x^2 - 4y + 8 = 0$ (Fig. 24.11).

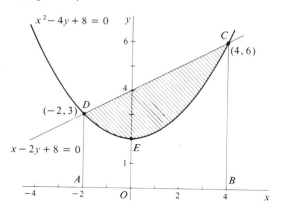

Fig. 24.11

Solution. Solving the two equations as a system, we find their points of intersection are $(-2,3)$ and $(4,6)$.

The area could be found in two different ways. We could first disregard the parabola and find the area between the x-axis and the straight line from $x = -2$ to $x = 4$. This is the trapezoid $ABCD$, whose area is 27 square units. We could compute the area between the parabola and the x-axis between the same limits. This area, $ABCED$, is 18 square units. Then the desired area between the two curves is found by subtraction:

$$27 - 18 = 9$$

A second method is somewhat simpler. We imagine the given area itself divided into vertical strips, and take one of these strips as the element, dA (Fig. 24.12).

The Definite Integral as an Area

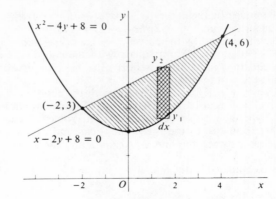

Fig. 24.12

Here we must be careful to state correctly the length of the element. Note that the length is given by the formula

$$\text{top } y - \text{bottom } y, \quad \text{or} \quad y_2 - y_1$$

The width is dx. Then, for the element we have

$$dA = (y_2 - y_1)\, dx$$

Now we sum up the infinite number of elements by the definite integral, evaluating between the limits $x = -2$ and $x = 4$:

$$A = \int_{x=-2}^{x=4} (y_2 - y_1)\, dx$$

Since the factor dx indicates the variable of integration, we replace the y_2 and y_1 with their equivalent x's by solving the equations for y. Since y_2 is the y of the straight line,

$$y_2 = \frac{x}{2} + 4$$

Since y_1 is the y of the parabola, $\quad y_1 = \dfrac{x^2}{4} + 2$

Then the *length* of the element is

$$y_2 - y_1 = \left(\frac{x}{2} + 4\right) - \left(\frac{x^2}{4} + 2\right) = \frac{x}{2} + 4 - \frac{x^2}{4} - 2 = \frac{x}{2} + 2 - \frac{x^2}{4}$$

Then

$$A = \int_{-2}^{4} \left(\frac{x}{2} + 2 - \frac{x^2}{4}\right) dx = \frac{x^2}{4} + 2x - \frac{x^3}{12} \Big]_{-2}^{4} = 9$$

Note: In all cases involving a vertical element, the length is given by the formula

$$\text{top } y - \text{bottom } y$$

This is true whether the element cuts across the x-axis or is entirely above or below it. In all cases involving a horizontal element, the length is correctly expressed by

$$\text{right } x - \text{left } x, \quad \text{indicated by} \quad x_2 - x_1$$

This is true whether the element cuts across the y-axis or is entirely on the right side or on the left side of the axis.

24.6 CHOOSING A VERTICAL OR HORIZONTAL ELEMENT

A critical problem is that of choosing the best type of element, vertical or horizontal. Remember, the given area extending from one limit to the other is assumed to be divided into an infinite number of strips. As we sketch one of these strips, we must remember that this one particular strip shown *must represent all possible positions for the element*. It is a good idea to draw the element in several different positions.

In Figure 24.13, if we take a horizontal element, note that the element has its right end on the straight line and the left end on the parabola. This is true no matter where the element is drawn. Each end of the element remains on the same curve.

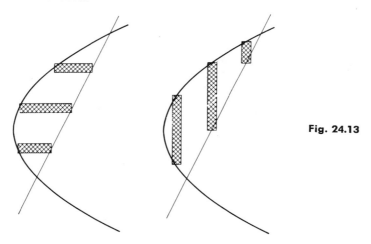

Fig. 24.13

Now, instead, suppose we assume the given area divided into vertical strips. Note that the bottom of the element may be on the straight line *or* on the parabola. In this instance we should use a horizontal element. That is, we should imagine the area divided into horizontal strips.

It is possible, of course, to compute an area of this kind by using vertical strips. But then the area must be computed in two portions. In one portion the element has both ends on the parabola; in the other portion the element extends from the straight line to the parabola. Two different elements must be used. In most instances it is best to assume the area divided in such a way that only one element is used and that *each* end of the element is always on the same curve, regardless of where it is shown.

EXERCISE 24.2

Find the area bounded by the curves indicated:

1. The curves $y^2 = 4x$ and $2x - y = 4$.
2. The parabola $x^2 = y$ and the straight line $x + y = 2$.
3. The curves $x^2 = 4y$; $y = 0$; $x = 4$ (use horizontal element).
4. The curves $y^2 = 2x$; $x = 2$; $y = 4$ (use vertical element).

5. The curves $y^2 = 2x$; $y = 0$; $x = 8$ (first quadrant; horizontal element).
6. The curves $y^2 - 4y + 4x - 12 = 0$; $x = 0$.
7. The curves $y^2 - 4y - 2x = 0$; $x = 0$.
8. The curves $y = x^2$; $y = x^3$; (a) use vertical element; (b) use horizontal element.
9. The curves $y = x^2$; $y = x^4$ (first quadrant); (a) use vertical element; (b) use horizontal element.
10. The curves $y = x^3$; $y = x^4$.
11. The curves $y^2 + 2y - 4x - 7 = 0$; $2x + y + 1 = 0$.
12. The curves $x^2 - 4x - 4y - 8 = 0$; $x - 2y = 4$.
13. The curves $x^2 = 6y$; $2x + 3y - 6 = 0$.
14. The curves $x^2 - 6x - 3y = 0$ and the x-axis.
15. The curves $x^2 - 4x + y = 0$; $x - y = 0$.
16. The curves $x^2 - 4x + 4y = 0$; $x - y = 0$; $x = 4$.
17. The curves $y^2 - 4x - 12 = 0$; $y^2 + 8x - 24 = 0$.
18. The curves $x^2 - 8y + 8 = 0$; $x^2 - 4y - 4 = 0$.
19. The curves $y = x^3$; $4x - y = 0$. (Find total area.)
20. The curves $y = x^3 - 2x^2 - 5x + 6$ and x-axis, from $x = -2$ to $x = 3$. (This total area must be found in two portions.)
21. The curves $x^2 = 4y$; $x + 2y = 4$; $y = 0$.
22. The curves $x^2 = 4y$; $y^2 = 4x$.
23. The curves $y^2 = x$; $x^2 = 8y$.

chapter

25

Volume of a Solid

25.1 A SOLID OF REVOLUTION

If a given area is rotated about a straight line as an axis, it generates a geometric solid called a *solid of revolution*. For example, if the area bounded by the curves $y = 0$, $x = 2y$, and $x = 8$, is rotated about the x-axis, it generates a cone (Fig. 25.1a). If the same area is rotated about the y-axis, it generates a sort of bowl-shaped solid with vertical sides (Fig. 25.1b).

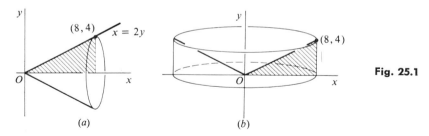

Fig. 25.1

One of our problems in connection with solids is to determine the volume. This is the subject of this chapter. Although the problem here is chiefly one dealing with solids of revolution, it should be mentioned that there are other types of solids whose volume we sometimes wish to find.

25.2 THE ELEMENT OF VOLUME

We recall that in computing a given area, we imagine the area divided systematically into strips, vertical or horizontal. One such strip is called an element of area, dA. In a similar manner we imagine a given volume divided systematically into *elements of volume*. The *element* of volume is denoted by dV, and is a portion of the volume to be found.

Just as we found the total area by the summation of an infinite number of elements or strips, in the same way we find the total volume of a solid by the summation of an infinite number of the elements of volume. This *summation* is accomplished by evaluating the *definite integral* between limits.

There are two standard ways of dividing a given volume into elements:
(1) The *disc*, or slice, method
(2) The *shell*, or hollow cylinder, method.

These methods will now be explained and illustrated by examples.

25.3 THE DISC METHOD

Consider the area bounded by the lines $x = a$, $x = b$, $y = 0$ and the curve $y = f(x)$. If this area is rotated about the x-axis, a solid is generated (Fig. 25.2). We shall use the disc method to find the volume of the solid.

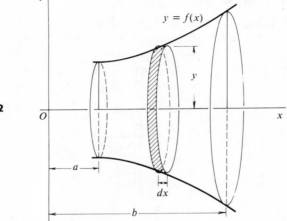

Fig. 25.2

In the *disc*, or slice, method, we imagine the solid divided into slices perpendicular to the axis of rotation. The slices are all similar though not of the same size. In any solid of revolution such a slice is a disc in the form of a thin cylinder, such as a coin. We take one typical slice, or disc, as an element of volume and call it dV. The total volume of the solid is, of course, the sum of all the infinite number of slices.

We recall that in considering a strip of area, we assume it to be a rectangle even though one end is on a curve, because the width, dx, approaches zero as the number of strips increases without limit. In the same way, we assume an element of volume, the disc, has the form of a right circular cylinder. To express the volume of the element, we use the general formula for the volume of any right circular cylinder: $V = \pi r^2 h$. For the element shown in this example, the radius $= y = f(x)$, and the thickness, or height, is $dx = h$. Then the volume of the element is

$$dV = \pi y^2\, dx$$

25.3 The Disc Method

Now we sum up the infinite number of slices or discs by integration and evaluate the definite integral from $x = a$ to $x = b$. Then

$$V = \pi \int_a^b y^2\, dx$$

Of course, the particular form of the volume of the element will depend upon which way the slices are taken, vertical or horizontal.

EXAMPLE 1. Find the volume of the solid generated by the rotation about the x-axis of the area bounded by the lines $y = 0$, $x = 2y$, and $x = 6$ (Fig. 25.3).

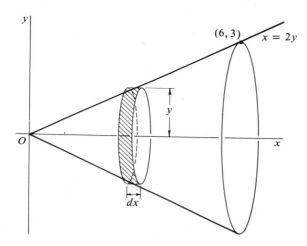

Fig. 25.3

Solution. Imagine the solid to be divided into slices perpendicular to the axis of rotation. We show one of these slices, in this case a disc, and call it the element of volume dV. The radius of the disc is the y of the line. The height (thickness) of the disc is dx.

For the volume of the element we have

$$dV = \pi y^2\, dx$$

Since the y of the line is $\dfrac{x}{2}$, we have

$$dV = \pi (x/2)^2\, dx$$

Then

$$V = \pi \int_0^6 \frac{x^2}{4}\, dx = \pi \left[\frac{x^3}{12}\right]_0^6 = 18\pi \text{ cu units}$$

The volume can be checked by the familiar geometric formula for the volume of a cone:

$$V = \frac{1}{3} \pi r^2 h$$

This formula is itself derived by integration.

EXAMPLE 2. Find the volume of the solid formed by rotating about the y-axis the area between the curves $x^2 = 2y$ and $y = 2x$.

Volume of a Solid

Solution. The solid is shown in Fig. 25.4. Note that it resembles a sort of bowl whose outside is a paraboloid and whose interior is cone-shaped. The two curves intersect at the points (0,0) and (4,8).

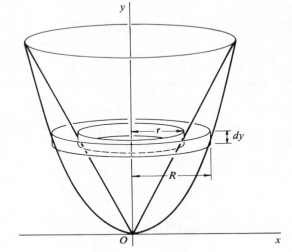

Fig. 25.4

In this case we divide the solid into slices horizontally perpendicular to the y-axis. However, in this instance the element resembles a washer, a disc with a hole in the center. To express the volume of this *washer element*, we must consider both the outside radius and the inside radius. Calling the outer radius R and the inner radius r, we have for the volume of the element

$$dV = \pi(R^2 - r^2)\,dy$$

The outer radius R is the x of the parabola, which we shall call x_2. The inner radius is the x of the line, which we call x_1. Then $x_2 = \sqrt{2y}$, and $x_1 = y/2$. Restating the volume of the element we have

$$dV = \pi(x_2^2 - x_1^2)\,dy = \pi\left(2y - \frac{y^2}{4}\right)dy$$

For the total volume we sum up the elements from $y = 0$ to $y = 8$:

$$V = \pi \int_0^8 \left(2y - \frac{y^2}{4}\right)dy = \pi\left[y^2 - \frac{y^3}{12}\right]_0^8 = \pi\left(64 - \frac{128}{3}\right) = \frac{64}{3}\pi$$

Note that the first part of the integral, 64π, represents the volume of the paraboloid as a solid, while the second part, $\frac{128}{3}\pi$, represents the volume of the cone-shaped interior of the bowl.

EXERCISE 25.1

In each of the following problems find the volume of the solid generated by rotating the indicated area about the axis named:

1. Area bounded by $y^2 = 4x$; $x = 4$; in first quadrant; (a) x-axis; (b) y-axis.
2. Area bounded by $x^2 = 4y$; $x = 4$; $y = 0$; (a) x-axis; (b) y-axis.

3. Area bounded by $x^2 = 4y$; $x = y$; (a) x-axis; (b) y-axis.
4. Area bounded by $y^2 = 4x$; $x^2 = 4y$; (a) x-axis; (b) y-axis.
5. Area bounded by $y^2 = x^3$; $x = 2$; first quadrant; (a) x-axis; (b) y-axis.
6. Area bounded by $2y - 3x = 6$; $y = 6$; $x = 0$; (a) x-axis; (b) y-axis.
7. Area bounded by $y = 8x - x^2$; $y = 0$; (a) x-axis; (b) y-axis.
8. Area bounded by $x^2 - 4x + y = 0$; $x = y$; about x-axis.
9. Area bounded by $y = 2x$; $x = 2y$; $x = 4$; about x-axis.
10. Area bounded by $3y - x = 3$; $x = 6$; $x = 0$; $y = 0$; about x-axis.
11. Area bounded by $y^2 = 4x - 4$; $y = 4$; $x = 0$; $y = 0$; about y-axis.
12. Area bounded by $y^2 + 2x = 4$; $x = 0$; about y-axis.
13. Area bounded by $y = x^2$; $y = x^3$; (a) x-axis; (b) y-axis.
14. Area bounded by $y = x^3$; $y = x^4$; (a) x-axis; (b) y-axis.
15. Area in Problem No. 1, about the line $x = 4$.
16. Area in Problem No. 2, about the line $x = 4$.
17. Area in Problem No. 6, about the line $y = 6$.
18. Area in Problem No. 3, about the line $y = 4$.
19. Show in general terms that the volume of a cone inscribed in a circular cylinder is equal to $\frac{1}{3}$ the volume of the cylinder.
20. Show that the volume of a paraboloid inscribed in a right circular cylinder is equal to one-half the volume of the cylinder.

25.4 THE SHELL METHOD

A solid of revolution can be divided systematically into portions called *shells*. Suppose we have a cylinder of radius equal to b and length equal to x, formed from the trunk of a tree with the center of the tree as the axis of the cylinder. At one end of the cylinder we note the annual rings, each ring the end of a shell, and each shell representing a yearly growth (Fig. 25.5). Then the volume of the entire cylinder is made up of the summation of the volume of all the shells.

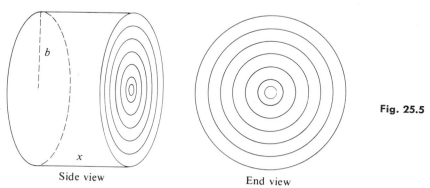

Fig. 25.5

Side view   End view

Now we consider one of the shells as the *element of volume, dV*. Note especially that the shell extends all around the center of the cylinder. The shell is really a hollow cylinder. The amount of material in the shell is given approximately by the formula for the approximate volume of a hollow cylinder:

$$\text{Volume} = (\text{circumference})(\text{length})(\text{thickness})$$

or,
$$V = 2\pi\, r(h)\,(\text{thickness})$$

304 Volume of a Solid

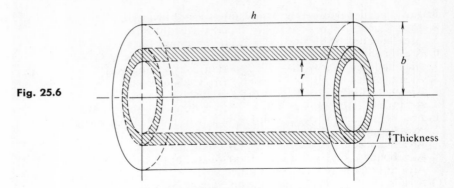

Fig. 25.6

The element of volume, a single shell, is shown in Fig. 25.6.

If we sum up the volume of all the shells from the center to the outer rim, we get the total volume of the cylinder. We do not use the limits from $-b$ to $+b$ but only from $y = 0$ to $y = b$, because each shell extends all around the center. The method is shown by examples.

EXAMPLE 3. Let us consider again the volume of the solid of *Example 1*, formed by rotating about the x-axis the area bounded by the lines $y = 0$, $x = 2y$, and $x = 6$ (Fig. 25.7). Find the volume by the shell method.

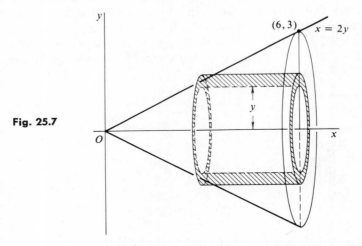

Fig. 25.7

Solution. We imagine this solid, the cone, formed from a tree trunk with the center of the tree as the axis of the cone. At the base of the cone we see the annual rings, each ring now representing a shell of the total volume. The shells of course have different lengths and different radii. We take one shell as representative of all the shells and call it an *element of volume*, dV. The sum of the infinite number of shells of infinitesimal thickness is the volume of the solid.

The *volume of this element*, dV, (a shell) must first be stated correctly. We again use the formula

$$dV = 2\pi(r)(h)(\text{thickness})$$

25.4 The Shell Method

From the position of the shell, its radius = y, its thickness = dy, and its height (length) = $x_2 - x_1$. Then

$$dV = 2\pi y (x_2 - x_1) \, dy$$

Substituting equivalent values, $x_2 = 6$ and $x_1 = 2y$, we have

$$dV = 2\pi y (6 - 2y) \, dy$$

Once we have stated correctly the volume of the element, we sum up the volume of all the shells by integration and evaluate between the proper limits:

$$V = 2\pi \int_0^3 (6y - 2y^2) \, dy = 2\pi \left[3y^2 - \frac{2y^3}{3} \right]_0^3 = 2\pi (27 - 18) = 18\pi$$

Note especially that evaluation is done from $y = 0$ to $y = 3$, that is, from *center* to outer rim. The answer checks with the answer in Example 1.

EXAMPLE 4. Find by the shell method the volume generated by rotating about the y-axis the area bounded by the lines $y = 0$, $x = 6$, $x = 2y$.

Solution. Imagine the solid divided into shells, one of which is shown (Fig. 25.8). For this shell (dV) the volume is

$$dV = 2\pi \, rh(\text{thickness})$$

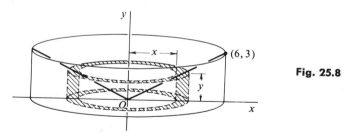

Fig. 25.8

For the element, thickness = dx, height = the y of the line, and radius = x. Substituting these values,

$$dV = 2\pi xy \, dx; \quad \text{or} \quad dV = 2\pi x \left(\frac{x}{2} \right) dx$$

Then

$$V = 2\pi \int_0^6 \frac{x^2}{2} \, dx = \pi \int_0^6 x^2 \, dx = \pi \left[\frac{x^3}{3} \right]_0^6 = 72\pi$$

EXAMPLE 5. Find the volume by the shell method of the solid formed by rotating about the y-axis the area bounded by the curves $2y = x^2$ and $y = 2x$. This is the solid of Example 2.

Solution. Imagine the solid divided into shells, one of which is shown as an element of volume, with its center on the y-axis (Fig. 25.9). The radius of this shell is x, the height (length) is $y_2 - y_1$, and the thickness is dx. Then

$$dV = 2\pi x(y_2 - y_1) \, dx$$

or

$$dV = 2\pi x \left(2x - \frac{x^2}{2} \right) dx$$

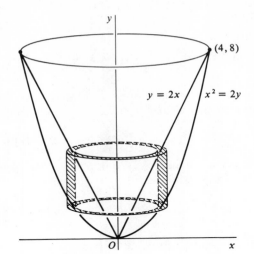

Fig. 25.9

Then

$$V = 2\pi \int_0^4 \left(2x^2 - \frac{x^3}{2}\right) dx = 2\pi \left[\frac{2x^3}{3} - \frac{x^4}{8}\right]_0^4 = 2\pi\left(\frac{128}{3} - 32\right) = \frac{64}{3}\pi$$

It is interesting to note that the portion of the integral $\frac{256}{3}\pi$ represents the volume between the x-axis and the cone (that is, outside the cone), and the remainder of the integral, 64π, represents the volume outside the paraboloid and above the x-axis.

EXERCISE 25.2

In each of the following problems find by the shell method the volume of the solid generated by rotating the indicated area about the axis named:

1. Area bounded by $x^2 = 4y$; $y = 4$; $x = 0$; first quadrant; about both axes.
2. Area bounded by $x^2 = 4y$; $x = 4$; $y = 0$; (a) x-axis; (b) y-axis.
3. Area bounded by $y^2 = 4x$; $y = x$; (a) x-axis; (b) y-axis.
4. Area bounded by $y^2 = 4x$; $x^2 = 4y$; (a) x-axis; (b) y-axis.
5. Area bounded by $y = x^3$; $x = 2$; $y = 0$; (a) x-axis; (b) y-axis.
6. Area bounded by $x^2 = y^3$; $x = 8$; $y = 0$; (a) x-axis; (b) y-axis.
7. Area bounded by $2x - 3y = 6$; $x = 6$; $y = 0$; (a) x-axis; (b) y-axis.
8. Area bounded by $x^2 - 2x - y = 0$; $x = 0$; $y = 3$; about y-axis.
9. Area bounded by $x^2 - 4x + y = 0$; $x = y$; about y-axis.
10. Area bounded by $y = 0$; $3y - x = 3$; $x = 6$; $x = 0$; y-axis (a dish).
11. Area bounded by $x^2 = 4y - 4$; $x = 4$; $y = 0$; $x = 0$; y-axis (a very stable soup bowl).
12. A vase: $y^2 - 8x + 8$; $y = -2$; $y = 4$; y-axis. (Use disc method.)
13. A punch bowl: $y = x^2$; $y = x^3$; y-axis.
14. A fragile punch bowl: $8y = x^3$; $16y = x^4$; y-axis.
15. A lens: $y^2 = 8x + 16$; $y^2 + 16x = 16$; in first and second quadrant; about x-axis.

chapter

26

Linear Motion: Integration

26.1 THE INTEGRAL IN LINEAR MOTION

We have seen that if the distance s traveled by a particle is stated as a function of time t, then the derivative ds/dt is the expression for velocity. That is,

$$\text{if} \quad s = f(t)$$
$$\text{then} \quad v = \frac{ds}{dt} = f'(t)$$

If *distance* s is understood to be measured in feet and *time* in seconds, then *velocity* is measured in *feet per second*, written (ft/sec).

As an example,

$$\text{if} \quad s = t^2 + 3t + 2 \quad \text{(ft)}$$
$$\text{then} \quad v = s' = 2t + 3 \quad \text{(ft/sec)}$$

From these equations we can find the values of s and v for any t. For instance, when $t = 2$ sec, then $s = 12$ ft and $v = 7$ ft/sec. The conditions may be represented on a horizontal scale (Fig. 26.1). At the end of 2 seconds the particle is 12 feet at the right of zero, and it is moving in a positive direction at a rate of 7 feet per second.

Fig. 26.1

Now, suppose we know the velocity of a particle is given by the function

$$v = 2t + 3$$

Then we reverse the procedure and *integrate velocity* to get *distance* s. Symbolically,

$$\int \text{velocity} = \text{distance}, \quad \text{or} \quad \int v = s$$

If
$$v = f(t), \quad \text{then} \quad s = \int f(t)dt.$$

26.2 THE INDEFINITE INTEGRAL IN LINEAR MOTION

In finding the integral in connection with motion, as usual we first arrive at the indefinite integral, which includes the general constant C. For example, if the velocity of a particle is given by the formula

$$v = 2t - 1 \quad \text{(ft/sec)}$$

then the distance s from zero is

$$s = \int (2t - 1)\, dt = t^2 - t + C$$

If we wish to know the velocity at any particular time, say, when $t = 3$ seconds, we have

$$v = 2(3) - 1 = 5 \quad \text{(ft/sec)}$$

The result means that when $t = 3$ seconds, the particle is moving in a positive direction at 5 ft/sec. However, this does not tell us the position of the particle. We still do not know its distance from zero. The particular distance is determined by the constant C in the integral.

Again, if we represent motion along a horizontal scale (Fig. 26.2), the point O denotes the position $s = 0$. We might indicate that when $t = 3$ sec, $v = 5$ ft/sec. But, s is still unknown. We know how the particle is moving but we do not know where it is.

Fig. 26.2

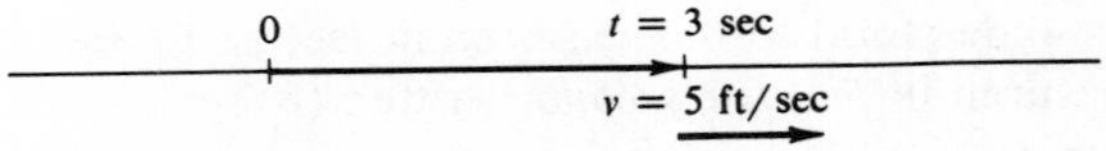

26.3 THE PARTICULAR INTEGRAL IN MOTION

Suppose now we have found the indefinite integral representing distance

$$s = t^2 - t + C$$

If we know the position of the particle at some instant, say, when $t = 3$, we can find the value of C and get the particular integral. Then this particular integral will show the position as well as velocity for this particular particle.

Let us assume that at the instant when $t = 3$, the particle is 8 feet at the left of zero. This information enables us to find C. When $t = 3$, $s = -8$. Substituting in the indefinite integral we get

$$-8 = (3)^2 - (3) + C$$

Solving for C,
$$C = -14$$

Having found the value of C, we use this value in the formula for distance and get the particular integral

$$s = t^2 - t - 14$$

Now that we have the equations for the distance and the velocity of the particle we can tell its exact position as well as its velocity at any instant. On a horizontal scale we can now represent position as well as velocity at the end of 3 seconds, or at any other time (Fig. 26.3). Of course, to find any particular integral, we must always have enough information to determine C.

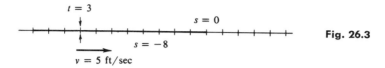

Fig. 26.3

26.4 A DIFFERENTIAL EQUATION

It is often convenient to state a function, such as velocity, in the form of a *differential equation*. A *differential equation* is an equation containing a derivative or differentials, as already defined in section 22.10.

Let us look again at the following function expressing velocity:

$$v = 2t - 1$$

As we have seen, when we integrate the expression for velocity, we get the expression for distance, s. Then

$$s = t^2 - t + C$$

Now, let us first state the velocity equation in the form of a differential equation, showing a derivative or differentials; that is, since $v = \dfrac{ds}{dt}$, we can write

$$v = 2t - 1$$

$$\frac{ds}{dt} = 2t - 1$$

Now we have a differential equation. To solve this equation, we attempt to eliminate the derivative. We could integrate both sides of the equation as it stands. However, let us make one minor change and express the same meaning in terms of differentials. We multiply both sides of the equation by dt and get

$$ds = (2t - 1)\, dt$$

This is a differential equation containing differentials. Now we integrate *both* sides of the equation and get

$$\int ds = \int (2t - 1)\, dt$$

or

$$s = t^2 - t + C$$

The answer is, of course, the same as we obtained before, but it is often convenient and desirable to state a function first as a differential equation. When we integrate both sides of the differential equation, we get a new equation called the *solution* of the differential equation. The equation $s = t^2 - t + C$ is a solution of the equation $ds = (2t - 1) \, dt$.

As long as the solution of a differential equation contains the arbitrary constant C it is called the *general solution* since it represents many particular cases. When we have determined the value of C we have the *particular solution*. We have already seen how to find the value of C. For example, in our general solution as shown, if $s = -8$ when $t = 3$, then we find that $C = -14$. Then the particular solution is

$$s = t^2 - t - 14$$

There is another way that is sometimes convenient for finding the particular solution. That is to evaluate both integrals (on both sides of the equation) between *corresponding limits*. For evaluating the two integrals of the equation

$$\int ds = \int (2t - 1) \, dt$$

we can use the following corresponding limits: when $t = 3$, $s = -8$, and when $t = t$, $s = s$. Then we can write

$$\int_{-8}^{s} ds = \int_{3}^{t} (2t - 1) \, dt$$

and get

$$s \Big]_{-8}^{s} = \left(t^2 - t \right) \Big]_{3}^{t}$$

Then, evaluating, $s - (-8) = t^2 - t - (9 - 3)$

or $s = t^2 - t - 14$

Note especially that lower and upper limits are corresponding values for s and t.

26.5 FROM ACCELERATION TO VELOCITY

We have seen that if distance is given as a function of time, then velocity is the first derivative, and acceleration the second derivative of the distance. Acceleration is the first derivative of velocity.
That is,

if $s = f(t)$

then $v = \dfrac{ds}{dt} = s' = f'(t)$

and $a = \dfrac{dv}{dt} = v' = \dfrac{d^2 s}{dt^2} = s'' = f''(t)$

For example, let us assume the velocity of a particle is given by

$$v = 6t + 5 \text{ (ft/sec)}$$

then $a = \dfrac{dv}{dt} = 6 \text{ (ft/sec}^2)$

26.5 From Acceleration to Velocity

The result means that the velocity is changing at a rate of 6 feet per second every second. In this example, acceleration is a constant, while velocity is a variable function of time t.

Now, if we have given the acceleration, 6 ft/sec², then we reverse the process of differentiation and integrate acceleration to get the expression for velocity. In general we might say

$$\int \text{acceleration} = \text{velocity}$$

That is,
$$\text{if } a = f(t)$$
$$\text{then } v = \int f(t)\, dt$$

When we integrate the expression for acceleration, we again first arrive at the indefinite integral, which includes the constant C. For example, suppose we have the acceleration given by the formula

$$a = 6$$

Then the velocity at any instant is given by

$$v = \int 6\, dt = 6t + C$$

Up to this point we cannot tell from the formula the exact velocity at any instant. We know only that it is a function of time and that it is changing at a rate of 6 ft/sec/sec. The exact velocity at any instant can be determined only when we know the value of C.

Now let us suppose that at the end of 2 seconds the velocity is 8 feet per second. This information enables us to evaluate the constant C. We have

$$v = 6t + C$$

When $t = 2$, $v = 8$. Then $\quad 8 = (6)(2) + C$

Solving for C, $\quad C = -4$

Then the particular formula for the velocity is

$$v = 6t - 4$$

Before integrating to find the expression for velocity, let us first write the equation for acceleration in the form of a *differential equation:* since

$$a = 6$$

we can write
$$\frac{dv}{dt} = 6$$

Now we have a differential equation. Multiplying both sides of the equation by dt we get

$$dv = 6\, dt$$

We *integrate both sides* of this equation, using only one constant C:

$$\int dv = \int 6\, dt$$

or,
$$v = 6t + C$$

This is the *general solution* of the differential equation, $dv = 6\,dt$. Evaluating for $v = 8$ when $t = 2$, we again find $C = -4$. Then the particular solution is

$$v = 6t - 4$$

Instead, we might find the particular solution by integrating both sides and then evaluating between limits. When $t = 2$, $v = 8$. When $t = t$, $v = v$. Then

$$\int_8^v dv = \int_2^t 6\,dt$$

Integrating, $\qquad\qquad\qquad v\Big]_8^v = 6t\Big]_2^t$

Evaluating, $\qquad\qquad\qquad v - 8 = 6t - 12$

The particular solution is $\qquad v = 6t - 4$

In the above example, when we have the expression for velocity, $v = 6t - 4$, we can integrate once more and get the formula for distance s. We write the formula for velocity in the form of a differential equation. That is,

for $\qquad\qquad\qquad\qquad v = 6t - 4$

we write $\qquad\qquad\qquad\quad \dfrac{ds}{dt} = 6t - 4$

Multiplying by dt, $\qquad\quad ds = (6t - 4)\,dt$

Then $\qquad\qquad\qquad\quad \int ds = \int (6t - 4)\,dt$

Once again we encounter an arbitrary constant, which we shall call C_2. Integrating we get

$$s = 3t^2 - 4t + C_2$$

Here again we have the general solution of the differential equation.

Now we need additional information to find the value of the second constant C_2. Let us suppose that at the time when $t = 2$ seconds, the particle was 20 feet to the positive side of zero. Then we have

$$20 = 3(2^2) - 4(2) + C_2$$

Solving for C_2 we get $\qquad C_2 = 16$

Then the formula for the distance s at any time t is the particular solution

$$s = 3t^2 - 4t + 16$$

In general terms, if a denotes acceleration, we have

$$\dfrac{dv}{dt} = a, \quad \text{or} \quad dv = a\,dt; \quad \text{then} \quad v = at + C_1$$

26.5 From Acceleration to Velocity

Since $v = \dfrac{ds}{dt}$, we write $\dfrac{ds}{dt} = at + C_1$

or, $ds = (at + C_1)\, dt$

Integrating both sides, $s = \dfrac{at^2}{2} + C_1 t + C_2$

These two C's *cannot* be combined into a single constant. The reason that we get two independent constants, C_1 and C_2, is that we started with the *second* derivative of the distance: $d^2s/dt^2 = a$.

EXAMPLE 1. A ball is projected upward from the top of a building 80 feet high (above the ground level) with an initial velocity of 96 ft/sec. The acceleration due to the force of gravity of the earth is approximately 32 ft/sec². Find the particular formulas for velocity v and distance a of the ball above the ground at any time t (s is measured in feet and t in seconds).

Solution. If we take an upward direction to be positive, then the acceleration of 32 ft/sec² is downward, or negative. Then we begin with

$$a = -32 \text{ ft/sec}^2$$

As a differential equation, $\dfrac{dv}{dt} = -32$

or $dv = -32\, dt$

Integrating both sides, $v = -32t + C_1$

Since the initial velocity was 96 ft/sec, when $t = 0$, $v = 96$. With this information, we have

$$96 = -32(0) + C_1$$

Solving for C_1, $C_1 = 96$

The particular solution for velocity is

$$v = -32t + 96$$

Now we write the formula for velocity in the form of a differential equation:

$$\dfrac{ds}{dt} = 96 - 32t$$

or, $ds = (96 - 32t)\, dt$

Integrating both sides, $s = 96t - 16t^2 + C_2$

Since the ball started at a point 80 feet above the ground, when $t = 0$, $s = 80$. With this information substituted in the general solution we find that $C_2 = 80$. Then we have the particular solution:

$$s = 80 = 96t - 16t^2$$

In the solution for Example 1, note that we have two variables, s and t, and that one variable s is of the first degree while the second variable t is of the second degree. This is the form of the equation of a parabolic curve. Therefore, the motion of the ball is parabolic. Although the motion is along a straight line, the motion itself has the characteristic of a parabola. If the initial thrust had been at a slight angle to the vertical, the curve of the path of the ball would have been seen as a parabola (Fig. 26.4).

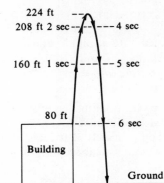

Fig. 26.4

EXAMPLE 2. An object is thrust upward along a frictionless inclined plane with an initial velocity of 18 feet per second. The plane is inclined at an angle of 22° from the horizontal, and as a result the downward acceleration because of gravity is approximately 12 feet per second per second. Write the equations for velocity and acceleration. How far up along the plane will the object travel? How long will it be from the time of the initial thrust until it returns to the starting point (Fig. 26.5)?

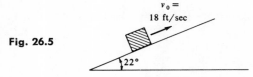

Fig. 26.5

Solution. Calling the upward direction positive, we begin with the downward acceleration:
$$a = -12$$
Writing this as a differential equation we have
$$dv = -12\, dt$$
Integrating both sides of the equation,
$$v = -12t + C_1$$
Since the initial velocity is 18 ft/sec, we get
$$18 = 0 + C_1; \quad \text{then} \quad C_1 = 18$$
For the particular solution for the velocity, we get
$$v = 18 - 12t$$
Since $v = ds/dt$, we have the differential equation
$$ds = (18 - 12t)\, dt$$
Solving the differential equation,
$$s = 18t - 6t^2 + C_2$$

26.5 From Acceleration to Velocity

Since $s = 0$ when $t = 0$, then we find that $C_2 = 0$, and the particular solution for s becomes

$$s = 18t - 6t^2$$

To find the distance the object will travel upward along the plane, we set the expression for velocity equal to zero and solve for t:

$$18 - 12t = 0; \quad \text{then} \quad t = 1.5 \text{ seconds}$$

Substituting this value of t in the distance formula, we get

$$s = 18(1.5) - 6(1.5)^2 = 27 - 13.5 = 13.5 \text{ feet}$$

To find the time until the object returns to its initial position, we note that at the instant it arrives at its initial position, $s = 0$. Then we set the formula for distance s equal to zero and solve for t:

$$18t - 6t^2 = 0; \quad \text{solving for } t \text{ we get} \quad t = 0 \quad \text{and} \quad t = 3$$

The total time the object was moving was 3 seconds, since the value $t = 0$ represents the time of starting.

EXAMPLE 3. If a ball were projected directly upward from the surface of the moon with an initial velocity of 80 ft/sec, how high would it rise? The gravity of the moon causes an acceleration of approximately 5 ft/sec².

Solution. We begin with

$$a = -5$$

As a differential equation

$$dv/dt = -5; \quad \text{or,} \quad dv = -5 \, dt$$

We integrate both sides and evaluate between limits: when $t = 0$, $v = 80$. Then

$$\int_{80}^{v} dv = \int_{0}^{t} -5 \, dt$$

Integrating both sides,

$$v \Big]_{80}^{v} = -5t \Big]_{0}^{t}$$

Evaluating,

$$v - 80 = -5t - 0, \quad \text{or,} \quad v = 80 - 5t$$

This is the particular solution for velocity. To find the equation for distance, we write, since $v = ds/dt$,

$$\frac{ds}{dt} = 80 - 5t; \quad \text{or} \quad ds = (80 - 5t) \, dt$$

Let us again integrate between limits: when $t = 0$, $s = 0$. Then

$$\int_{0}^{s} ds = \int_{0}^{t} (80 - 5t) \, dt$$

Integrating,

$$s \Big]_{0}^{s} = 80t - \frac{5t^2}{2} \Big]_{0}^{t}$$

Evaluating,

$$s - 0 = 80t - \frac{5t^2}{2}$$

or

$$s = 80t - \frac{5t^2}{2}$$

This particular solution gives the formula for the distance at any time.

To find the maximum height, as usual we set the velocity expression equal to zero:
$$80 - 5t = 0$$
Solving for t,
$$t = 16$$
Substituting 16 (seconds) in the formula for s, we get, for maximum height,
$$s = 80(16) - 2.5(16)^2 = 640 \text{ (feet)}$$
Therefore, on the moon, the ball would reach a height of 640 feet.

EXAMPLE 4. If a man can high jump 6 feet on earth, how high could he jump on the moon?

Solution. First we must find his initial velocity. Again, we begin with the acceleration on earth:
$$a = -32$$
As a differential equation,
$$\frac{dy}{dt} = -32; \quad \text{or} \quad dv = -32 \, dt$$
Solving the equation, $\quad v = -32t + C_1$

Since we do not know his initial velocity, let us call it v_0. Then, when $t = 0$, $v = v_0$; substituting, we get
$$C_1 = v_0$$
The particular solution becomes: $\quad v = v_0 - 32t$

This is the specific formula for velocity on earth for the high jumper.

We must have the formula for distance, so we write the expression for velocity as a differential equation:
$$\frac{ds}{dt} = v_0 - 32t; \quad \text{or} \quad ds = (v_0 - 32t) \, dt$$
The general solution is $\quad s = v_0 t - 16t^2 + C_2$

From the initial conditions, $s = 0$ when $t = 0$, we get $C_2 = 0$. Then the particular solution is
$$s = v_0 t - 16t^2$$
For maximum height we go back to the formula for velocity and set it equal to zero:
$$v_0 - 32t = 0$$
Solving for t,
$$t = \frac{v_0}{32}$$
This means the maximum height is reached when t is equal to $v_0/32$ seconds. Substituting this value in the formula for s we get
$$s = v_0 \left(\frac{v_0}{32}\right) - 16\left(\frac{v_0}{32}\right)^2$$
This simplifies to $\quad s = \dfrac{(v_0)^2}{64}$

Now we substitute the known maximum height, 6 feet, and get
$$6 = \frac{(v_0)^2}{64}, \quad \text{or} \quad 384 = v_0^2$$

Solving for v_0 we get
$$v_0 = \sqrt{384} = 8\sqrt{6} = 19.6 \text{ ft/sec (approx.)}$$
Therefore the man's initial velocity on earth is about 19.6 ft/sec. We shall use this also for the initial velocity on the moon.

For maximum height on the moon, we begin with
$$a = -5$$
As a differential equation,
$$\frac{dv}{dt} = -5; \quad \text{or} \quad dv = -5 \, dt$$
Solving,
$$v = -5t + C_1$$

Now we take the same initial velocity as for the earth, 19.6 ft/sec. Then for initial conditions we have $v = 19.6$ when $t = 0$. The particular formula for velocity is
$$v = 19.6 - 5t$$
To find the formula for s, we have
$$\frac{ds}{dt} = 19.6 - 5t$$
or
$$ds = (19.6 - 5t) \, dt$$
Solving the equation,
$$s = 19.6t - \frac{5t^2}{2} + C_2$$
Since $s = 0$ when $t = 0$, then C_2 becomes 0, and we get the particular solution
$$s = 19.6t - \frac{5t^2}{2}$$
Now we have both formulas for v and s on the moon. To find maximum height, we set
$$19.6 - 5t = 0$$
then
$$t = 3.92 \text{ (approx.)}$$
Substituting 3.92 sec in the formula for s we get
$$s = 19.6(3.92) - \frac{5}{2}(3.92)^2 = 38.4 \text{ ft (approx.)}$$

This means that a man who can high jump 6 feet on earth could high jump a distance of 38.4 feet on the moon.

After some practice we discover that we can omit some of the steps in finding formulas for velocity and distance. For example, if we know the initial velocity v_0 and the acceleration a, we can write immediately
$$v = v_0 + at$$
To find s we can write
$$ds = (v_0 + at) \, dt$$
Solving,
$$s = v_0 t + \frac{at^2}{2} + C_2$$
Moreover, C_2 will represent s_0 when $t = 0$. Then the formula for s becomes
$$s = s_0 + v_0 t + \frac{at^2}{2}$$

EXERCISE 26.1

Find the general equation for the distance s for each of the following by the indefinite integral; find the particular equation where the initial conditions are given:

1. $v = 2t - 1$
2. $v = 3 - 4t$
3. $v = 3t^2 - 2t$
4. $v = 4t - 6t^2$
5. $v = 60 - 32t$
6. $v = 3t^2 + 6t - 2$
7. $v = 20 - 8t; s = -6$, when $t = 0$
8. $v = 10t + t^2; s = 5; t = 1$
9. $v = t^2 - 2t - 3; s = 0; t = 0$
10. $v = t - t^2; s = 20; t = 1$
11. $v = t^2 + t - 2; t = 6; s = 100$
12. $v = 3t - t^2; s = 30; t = 3$

By use of the differential equation form, find the particular solution and the particular expression for velocity and distance for each of the following with the given initial conditions:

13. $a = 20$ ft/sec^2; $t = 0$; $v = 16$; $s = 20$
14. $a = 30$ ft/sec^2; when $t = 0$, $v = -12$; $s = 18$
15. $a = -10$ ft/sec^2; when $t = 0$, $v = -24$; $s = 60$
16. $a = -6$ ft/sec^2; when $t = 0$; $v = 36$; $s = 80$
17. $a = 12$ ft/sec^2; when $t = 0$, $v = -6$; $s = 10$
18. $a = -8$ ft/sec^2; when $t = 0$, $v = 0$; $s = 0$
19. $a = -4$ ft/sec^2; when $t = 0$, $v = 10$; $s = 0$
20. $a = 0$; when $t = 0$, $v = 20$; $s = -30$
21. An object is projected upward from ground level with an initial velocity of 160 ft/sec. Find the formula for the position at any time t.
22. An object is projected directly downward from a balloon with an initial velocity of 20 ft/sec. What is the formula for distance from the balloon?
23. An object is projected directly upward from a tower 640 feet high with an initial velocity of 160 ft/sec. How high will it rise above the ground level? How high above the top of the tower? With what velocity will it strike the ground?
24. An object is projected upward along a frictionless inclined plane with an initial velocity of 15 ft/sec. If the force of gravity produces an acceleration downward along the plane of 6 ft/sec^2, what is the formula for the distance of the object from the starting point? How far up along the plane will the object move?
25. An object is projected upward along a frictionless inclined plane with an initial velocity of 20 ft/sec. The force of gravity produces an acceleration downward along the plane of 8 ft/sec^2. If the object is started at a distance of 11 feet upward from the lower edge of the plane, how far up along the plane will the object move?
26. A rifle bullet with a muzzle velocity of 2400 ft/sec is fired directly upward from ground level. How high will the bullet go (neglecting air resistance)? How long will it be in the air until it returns to the ground?
27. At a carnival an iron disc is thrust upward along a pole by the blow of a heavy mallet. The iron disc is supposed to strike a gong at the top 40 feet from the bottom where the blow is struck. What must be the initial velocity of the disc in order to ring the gong?
28. The accelerating force of gravity on each of the following planets is approximately as shown: Venus, 28.3 ft/sec^2; Jupiter, 85.2 ft/sec^2; Mars, 12.5 ft/sec^2. A man who can high jump 6.25 feet on earth has an initial upward velocity of approximately 20 ft/sec. How high could he jump on each of the three planets mentioned?

26.6 THE DEFINITE INTEGRAL IN RECTILINEAR MOTION

We have seen how the definite integral is used to find the exact area under a curve. In a similar way it can be used to find the total distance traveled by an object if the motion is along a straight line.

26.6 The Definite Integral in Rectilinear Motion

As an example, let us consider the motion of an object in a free fall. At the earth's surface the velocity of the falling object is approximately given by the formula

$$v = 32t$$

To get the formula for distance, we integrate the expression for velocity:

$$s = \int v \, dt = \int 32 \, t \, dt = 16t^2 + C$$

If $s = 0$ when $t = 0$, we get the particular integral $s = 16t^2$.

Now, if we wish to find the distance covered in a particular interval of time, we can find the answer by means of the definite integral. For example, suppose we wish to find the distance covered in the interval from $t = 4$ seconds to $t = 6$ seconds. This means we can evaluate the definite integral between these two limits. That is, the distance covered is

$$s = \int_4^6 32 \, t \, dt = 16t^2 \Big]_4^6 = 16(36 - 16) = 320 \text{ feet}$$

To find the distance covered during the eighth second, we take the interval of time from $t = 7$ to $t = 8$ and get

$$16 \left[t^2 \right]_7^8 = 16(64 - 49) = 240 \text{ feet}$$

EXAMPLE 5. The velocity of a certain moving object is given by the formula, where t is in seconds and v is in ft/sec:

$$v = 8t + 5$$

How far does the object travel in the interval from $t = 3$ to $t = 5$ sec?

Solution. We first write $ds = (8t + 5) \, dt$

Then we have $s = \int_3^5 (8t + 5) \, dt = 4t^2 + 5t \Big]_3^5 = 100 + 25 - 36 - 15 = 74$ (ft)

In a problem involving the use of the definite integral to determine the distance traveled by an object in linear motion, we must be careful about a *reversal of motion*. If the object *reverses its direction* of motion, the definite integral will show the *net difference* in distance between the beginning and the end of the time interval. Yet it may not show the actual total distance traveled. If an object moves toward the left a distance of, say, 20 feet and then to the right a distance of 15 feet, the definite integral will show a net distance of -5 feet, that is, from the starting point to the finishing point. This is shown in a study of the following example.

EXAMPLE 6. A ball is projected directly upward from ground level with an initial velocity of 128 feet per second. Because of the effect of gravity, its vertical velocity at any time t (in seconds) is given by the formula $v = 128 - 32t$. Find the distance covered in the following intervals of time: from $t = 1$ to $t = 3$; from $t = 3$ to $t = 5$; from $t = 2$ to $t = 6$; from $t = 1$ to $t = 7$; $t = 0$ to $t = 8$.

Linear Motion: Integration

Solution. The motion of the ball is shown in Fig. 26.6. Integrating for distance s, first using the limits $t = 1$ and $t = 3$,

$$s = \int_1^3 (128 - 32t)\, dt = 128t - 16t^2 \Big]_1^3 = 128$$

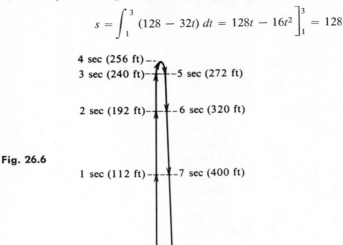

Fig. 26.6

In the interval from $t = 1$ to $t = 3$, the ball traveled a distance of 128 feet. For the limits $t = 3$ and $t = 5$, we get $s = 0$. It would appear that in the interval from $t = 3$ to $t = 5$ the object did not move at all. However, this was not the case. The zero value of the definite integral in this interval means that at the end of the 5 seconds, the object was at the same distance from zero, or at the same position that it was at the end of 3 seconds.

As we evaluate the definite integral between the limits 2 and 6, 1 and 7, and 0 and 8, we get a value of zero for the definite integral in each case. In these intervals the ball reversed its direction of motion. At the end of each interval of time it was at the same position as at the beginning of the interval.

In order to find the actual total distance traveled in this example, we must first find the point where the object reversed its direction of motion. Setting the velocity equal to zero to find the maximum point, we have

$$128 - 32t = 0; \quad \text{then} \quad t = 4$$

This means that when $t = 4$ seconds the ball reached its greatest height and any distance from then on must be considered negative. To find the total actual distance traveled in an interval, such as $t = 2$ to $t = 6$, we find first the distance covered from $t = 2$ to $t = 4$ seconds, and then from $t = 4$ to $t = 6$ seconds. From $t = 2$ to $t = 4$ we have

$$128t - 16t^2 \Big]_2^4 = 64 \text{ feet; this is upward distance}$$

From $t = 4$ to $t = 6$ seconds, we have

$$128t - 16t^2 \Big]_4^6 = 768 - 576 - 512 + 256 = -64 \text{ feet}$$

That is, during the interval from $t = 4$ to $t = 6$, the ball traveled downward 64 feet, so that its total actual distance traveled during the 4-second interval was 128 feet.

EXERCISE 26.2
Use the definite integral in the following problems:
1. Find the distance traveled by an object in free fall in the first three seconds. Find the distance traveled in the fourth second. Find the distance traveled in the tenth second.
2. Compare the distances traveled by an object in free fall during the two time-intervals (in seconds): $t = 0$ to $t = 5$, and $t = 5$ to $t = 10$.
3. An object is projected directly upward from a tower 256 feet high with an initial velocity of 96 ft/sec. How far does it travel in the time interval from $t = 0$ to $t = 3$ seconds? from $t = 2$ to $t = 9$ seconds? from $t = 1$ to $t = 5$ seconds?
4. A ball is projected directly upward from ground level with an initial velocity of 160 ft/sec. Find the actual distance traveled during the time-intervals: $t = 2$ to $t = 4$ sec; $t = 4$ to $t = 6$ sec; $t = 2$ to $t = 8$ sec.
5. An object is projected upward along a frictionless inclined plane with an initial velocity of 20 ft/sec. If the force of gravity produces an acceleration downward along the plane of 5 ft/sec^2, how far does the object travel in the first 4 seconds? from $t = 4$ to $t = 10$ seconds?
6. In No. 5, if the initial velocity is 15 ft/sec and the acceleration is 4 ft/sec^2, find the distance traveled in the following time intervals: $t = 3$ to $t = 6$ seconds; $t = 4$ to $t = 10$ seconds.
7. An object is projected upward from ground level with an initial velocity of 128 ft/sec. Compare the distance traveled in the third and fourth seconds.
8. Work No. 7 if the object is projected upward from the top of an 80-foot building.
9. A rifle bullet with a muzzle velocity of 2800 ft/sec is fired directly upward from ground level. Compare the distances traveled in the fifth and sixth seconds.

26.7 COMPONENTS OF VELOCITY AND DISTANCE

According to Newton's first law, if an object is in motion or put into motion it will continue with the same velocity and in a straight line unless acted on by some other force. If an object is projected directly upward at 80 ft/sec, it would continue to rise at that velocity if it were not for some other force such as gravity or air resistance. The same is true if the initial motion is along a horizontal direction or at any angle with the horizontal.

However, if a ball is projected upward, horizontally, downward, or at an angle, we must insert the effect of gravity, which is always downward. This force is sufficient, at the earth's surface, to cause a downward velocity of $32t$ feet per second. Note especially that this downward velocity due to gravity is a linear function of time. If an object is projected directly upward from ground level at an initial velocity of 80 ft/sec, then, as we have seen,

$$v = 80 - 32t \quad \text{and} \quad s = 80t - 16t^2$$

In the formula for velocity, $v = 80 - 32t$, the constant 80 represents the initial upward velocity. The negative $32t$ represents downward velocity caused by the force of gravity.

If an object is projected horizontally at 80 ft/sec, it will continue horizontally at that same velocity (except for slight air resistance). However, gravity will pull it downward still with a velocity of 32t ft/sec. The result downward will be exactly the same as in a free fall.

Now let us take the case of an object projected at an angle with the horizontal. Let us suppose an object is projected with an initial velocity of 80 ft/sec at an angle of 60° with the horizontal. We let v represent

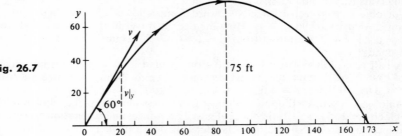

Fig. 26.7

the initial velocity as shown by the arrow in the figure (Fig. 26.7). However, since the initial velocity v of 80 ft/sec is at an angle of 60° with the horizontal and not directly upward, then the upward velocity is less than 80 ft/sec. If we let v_y denote the upward velocity, then

$$\frac{v_y}{v} = \sin 60°$$

or $\qquad v_y = 80 \sin 60° = 80(0.866) = 69.28$ (approx.)

We call v_y the vertical component of the initial velocity v. However, for the vertical velocity of the object at any time t we must still consider the force of gravity and the resulting downward velocity. Then the actual vertical velocity at any time becomes

$$v_y = 69.28 - 32t$$

In more general terms, if we let v represent the initial velocity and let θ represent the angle of elevation, then the velocity in the vertical direction is given by the formula

$$v_y = v \sin \theta - 32t$$

For the horizontal component of velocity we have a similar formula. If we let v_x represent the horizontal component, we have

$$v_x = v \cos \theta$$

The horizontal component is not affected by any other force (except a slight air resistance). In the example shown, where $v = 80$ and $\theta = 60°$,

$$v_x = 80 \cos 60° = 80\left(\frac{1}{2}\right) = 40$$

To find the vertical and horizontal components of velocity at any instant, therefore, we use the formulas

$$v_x = v \cos \theta \quad \text{and} \quad v_y = v \sin \theta - 32t$$

To find the horizontal and vertical components of the distance traveled,

26.7 Components of Velocity and Distance

we can integrate each of the expressions for the velocity components. Letting x represent the horizontal component of distance and y the vertical component of distance, we have

$$x = \int v_x \, dt = \int v \cos \theta \, dt = (v)(t)(\cos \theta) + C_1$$

$$y = \int v_y \, dt = \int (v \sin \theta - 32t) \, dt = (v)(t)(\sin \theta) - 16t^2 + C_2$$

If x and y both equal zero when $t = 0$, then C_1 and C_2 both become zero and we have

$$x = vt \cos \theta; \quad y = vt \sin \theta - 16t^2$$

Remember, that in these two equations, the angle θ is not a variable but is a constant representing the angle of projection.

The two equations for the components of the distance are actually *parametric equations* of the curve representing the path of the object. The variable t is the *parameter*. With the two parametric equations we can pinpoint the exact location of the object at any instant. If we wish, we can eliminate the parameter t and get the cartesian equation of the curve. The result would be an equation clearly indicating a parabola.

EXAMPLE 7. A rifle bullet having a muzzle velocity of 2000 ft/sec is fired from ground level at an upward angle of 20° from the vertical — (70° from the horizontal) (Fig. 26.8). What is the formula for its elevation at any time t and the formula for its vertical velocity? How high will the bullet rise? How long will it be in the air? How far will it travel horizontally? (Disregard air resistance.)

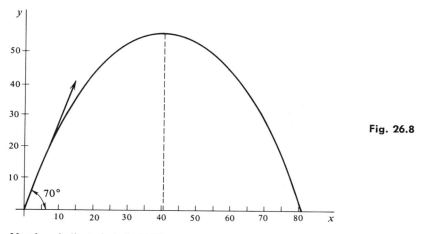

Fig. 26.8

Numbers indicate feet, in 1000's

Solution. Let y represent the number of feet in the height at any time, a variable. Let v_y represent the vertical velocity at any time. After the bullet is fired, the only accelerating force is the force of gravity. We begin with

$$a = -32$$

Writing this as a differential equation,

$$\frac{d(v_y)}{dt} = -32; \text{ or } (d\,v_y) = -32\,dt$$

Integrating
$$v_y = -32t + C_1$$

To find the constant C_1 we must know the initial upward velocity. This is not 2000, but only 2000 sin 70°, or about 1880 ft/sec. Then we have, when $t = 0$, $v_y = 1880$. This gives us the value $C_1 = 1880$. The particular solution of the differential equation is then

$$v_y = 1880 - 32t$$

To find the formula for y, we write the equation for v_y as the differential equation

$$v_y = \frac{dy}{dt} = 1880 - 32t$$

or
$$dy = (1880 - 32t)\,dt$$

Solving the equation we get
$$y = 1880t - 16t^2 + C_2$$

Using the initial values, when $t = 0$, $y = 0$, we get $C_2 = 0$. Then the particular solution is

$$y = 1880t - 16t^2$$

This result is now the specific formula for the vertical distance y in feet at any time t in seconds.

When the maximum height is reached, the bullet still has a horizontal velocity but no vertical velocity. That is,

$$v_y = 0; \text{ or } 1880 - 32t = 0$$

Solving for t we get
$$t = 58.75 \text{ seconds}$$

To find the maximum height, we use the value $t = 58.75$ in the formula for y and get

$$y = 1880\,(58.75) - 16\,(58.75)^2 = 55{,}225 \text{ ft (about 10.46 miles)}$$

Now we let x represent the horizontal distance in feet at any time. We note there is no horizontal acceleration. Then

$$v_x = 2000 \cos 70° = 2000\,(0.3420) = 684 \text{ ft/sec}$$

Since $v_x = \dfrac{dx}{dt}$, we write the formula for v_x as a differential equation:

$$\frac{dx}{dt} = 684; \text{ or, } dx = 684\,dt$$

Solving the differential equation,

$$x = 684t + C_2$$

For the initial conditions, we have, when $t = 0$, $x = 0$. Then $C_2 = 0$, and we get the particular solution,

$$x = 684t$$

At the end of 58.75 seconds the bullet has reached its peak and has traveled a horizontal distance of

$$x = (684)(58.75) = 40{,}185 \text{ feet}$$

26.7 Components of Velocity and Distance

To find the horizontal distance the bullet has traveled when it hits the ground, we note that at that time, $y = 0$. Therefore, we set the expression for y equal to zero, and solve for t:

$$1880t - 16t^2 = 0; \quad \text{then} \quad t = 0, \quad \text{and} \quad t = 117.5$$

The result means that the bullet is at ground level twice, when $t = 0$ and when $t = 117.5$ seconds. The first value represents the time at the instant of firing. To find the horizontal distance traveled, we use 117.5 seconds in the formula for x and get

$$x = 684(117.5) = 80{,}370 \text{ ft, or about } 15.2 \text{ miles}$$

EXERCISE 26.3

1. A rifle bullet has an initial velocity of 2000 ft/sec. Find its maximum height and maximum distance if it is fired from ground level at an angle of 40° with the horizontal.
2. The projectile of a gun has a muzzle velocity of 3000 ft/sec. If it is fired from ground level, find its maximum height and maximum distance when it is fired at each of the following angles with the horizontal: (a) 15°, (b) 30°, (c) 45°, (d) 60°, (e) 75°, (f) 90°.
3. Work No. 2 if the initial velocity is 2400 ft/sec.
4. In No. 2, describe exactly the position of the projectile at the end of 4 seconds (height and horizontal distance).
5. A projectile is fired from ground level at an angle of elevation of 30°. Compare the range and the maximum height of the projectile for the following initial velocities: (a) 1000 ft/sec; (b) 2000 ft/sec; (c) 3000 ft/sec.
6. If the muzzle velocity of a rifle bullet is 2000 ft/sec, what is the average velocity throughout the length of the gun barrel? If the length of the rifle barrel is 30 inches (2.5 feet), how long does it take the bullet to travel the length of the barrel from the time it leaves the cartridge casing until it reaches the muzzle of the rifle? What is the average acceleration of the bullet over this distance in ft/sec²?
7. At the instant when the thrust in a satellite is changing at a rate of 1 g per 16 seconds, what is the formula for the distance s? (One $g = 32$ ft/sec².) How would you express the rate of change of acceleration?
8. A boy throws a ball at a vertical wall 100 feet from him with an initial velocity of 80 ft/sec. The ball leaves his hand 5 feet above the ground. If the angle of elevation of the throw is 30°, how high on the wall will the ball strike? What should be the angle of elevation of the throw for the ball to strike the wall 55 feet above the ground?
9. A gun has a muzzle velocity of 2000 ft/sec. At what angle of elevation should it be fired to strike a target on the same level as the gun at the following distances: (a) 10 miles; (b) 15 miles; (c) 20 miles?

chapter
27

Approximate Integration

27.1 LIMITATION OF THE DEFINITE INTEGRAL METHOD

We have seen how to find the value of a definite integral by integrating a function, $f(x)$, and then evaluating between limits. That is,

$$\int_a^b f(x)\, dx = F(x) \Big]_a^b = F(b) - F(a)$$

The result is the exact value of the definite integral.

However, this method is not always practical simply because the integration of some functions is very difficult or impossible by ordinary methods. In such cases, certain methods of approximation are useful.

Of course, if a function can be integrated by the usual methods, then the integral can be evaluated between limits. If this can be done, then this is the simplest thing to do. For example, the exact value of the following definite integral can be found because the function can easily be integrated:

$$\int_2^5 x^2\, dx = \frac{x^3}{3}\Big]_2^5 = \frac{125}{3} - \frac{8}{3} = \frac{117}{3} = 39$$

If we could find all definite integrals in this manner, the problem would be simple. However, for many functions the problem is not so easy. For example, suppose we wish to evaluate the following definite integral:

$$\int_1^4 e^{-x^2}\, dx$$

The definite integral does have an exact value, as we can see if we sketch the curve $y = e^{-x^2}$. However, the function cannot be integrated by any of the usual methods. In such problems we can get an approximation to the definite integral by other methods. One such method is the *trapezoidal rule*.

27.2 THE TRAPEZOIDAL RULE

The trapezoidal rule is based on the formula for the area of a trapezoid. Although this formula refers to an area, the principle is also used to evaluate other quantities.

A trapezoid, you recall, is a quadrilateral having two parallel sides, called *bases* b_1 and b_2 (Fig. 27.1). The perpendicular distance between the bases is called the *altitude h*. Then the area of the trapezoid is given by the formula

$$A = \frac{1}{2} h(b_1 + b_2)$$

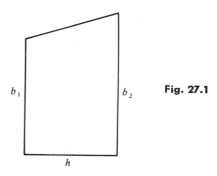

Fig. 27.1

That is, to find the area we add the two bases b_1 and b_2, multiply this sum by the altitude, and then take one half of the result. For example, if a trapezoid has bases of 6 and 9 and an altitude of 5, the area is $(\frac{1}{2})(5)(6 + 9) = 75/2$.

Now suppose we have several adjacent trapezoids (Fig. 27.2) each one with an altitude of 3, and bases as shown. Then the total area is the sum of the areas of the trapezoids. The areas are as follows:

$$\text{first trapezoid,} \quad A_1 = \frac{1}{2}(3)(5 + 7)$$

$$\text{second trapezoid,} \quad A_2 = \frac{1}{2}(3)(7 + 8)$$

$$\text{third trapezoid,} \quad A_3 = \frac{1}{2}(3)(8 + 10)$$

$$\text{fourth trapezoid,} \quad A_4 = \frac{1}{2}(3)(10 + 11)$$

Then the sum of the areas of the trapezoids can be written

$$A = \left(\frac{1}{2}\right)(3)(5 + 7) + \left(\frac{1}{2}\right)(3)(7 + 8) + \left(\frac{1}{2}\right)(3)(8 + 10) + \left(\frac{1}{2}\right)(3)(10 + 11)$$

Factoring out the common factor, $(\frac{1}{2})(3)$, we can write

$$A = \left(\frac{1}{2}\right)(3)(5 + 7 + 7 + 8 + 8 + 10 + 10 + 11) = 99$$

328 Approximate Integration

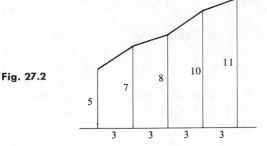

Fig. 27.2

Notice that in the polynomial *all the bases appear twice* except the first and the last.

Now suppose we have the general function, $y = f(x)$, and wish to find the area under the curve from $x = a$ to $x = b$ (Fig. 27.3). That is,

$$A = \int_a^b f(x)\,dx$$

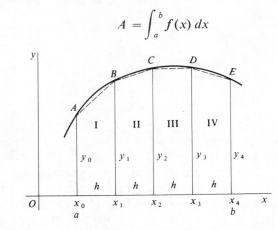

Fig. 27.3

To derive the formula for the trapezoidal rule, we divide the interval $[a,b]$, into any number of subintervals of uniform width. Let us call the width h because this width becomes the altitude of the trapezoids to be formed. Suppose we have four segments. Let us call the points of division $x_0(=a)$, x_1, x_2, x_3, and x_4 ($=b$). We erect perpendiculars at the points on $[a,b]$, cutting the curve at A, B, C, D, and E. Then we have four strips of area. The exact area under the curve is, of course, the sum of the areas of the four strips.

Now we connect the successive points, A, B, C, D, and E with straight lines. Then we have four trapezoids, I, II, III, and IV. The sum of the areas of the four trapezoids is approximately but not exactly equal to the area under the curve. Now we find the total area of the trapezoids. Note that the bases of the trapezoids are the vertical line segments, which are the function-values, or the y-values, corresponding to each point on OX. These ordinates we may call y_0, y_1, y_2, y_3, and y_4. The altitude h of each trapezoid is the length of each subinterval on OX.

Now let us express the area of each trapezoid, representing the separate

27.2 The Trapezoidal Rule

areas by A_1, A_2, A_3, and A_4. For example, the bases of trapezoid I are y_0 and y_1 and the altitude is h. Then

$$A_1 = \left(\frac{1}{2}\right)(h)(y_0 + y_1), \quad A_2 = \left(\frac{1}{2}\right)(h)(y_1 + y_2), \text{ and so on.}$$

The total area is

$$A = \left(\frac{1}{2}\right)(h)(y_0 + y_1) + \left(\frac{1}{2}\right)(h)(y_1 + y_2) + \left(\frac{1}{2}\right)(h)(y_2 + y_3)$$
$$+ \left(\frac{1}{2}\right)(h)(y_3 + y_4)$$

or $$A = \left(\frac{1}{2}\right)(h)(y_0 + y_1 + y_1 + y_2 + y_2 + y_3 + y_3 + y_4)$$

Now we combine and get *twice* each ordinate except the first and the last, and we have the trapezoidal rule:

$$A = \left(\frac{1}{2}\right)(h)(y_0 + 2y_1 + 2y_2 + 2y_3 + y_4)$$

Of course, in using the trapezoidal rule we must compute the ordinate value or function value for each value of x along the interval $[a,b]$. This is the tedious part of the process, but it must be done. To show how the rule is used, let us take an example that can also be evaluated by the usual method of finding the definite integral. Then we shall compare the results.

EXAMPLE 1. Find the area under the curve $x^2 = 4y$, from $x = 1$ to $x = 4$. Find the approximate area by the trapezoidal rule and then the exact area by the usual method of integration (Fig. 27.4).

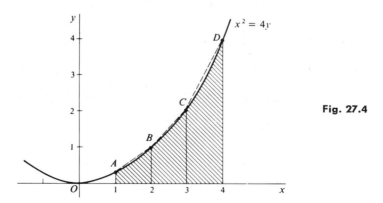

Fig. 27.4

Solution. If we divide the segment from $x = 1$ to $x = 4$ into three segments, each segment has a width of 1 unit, taking the points of division at $x = 2$ and $x = 3$. (Of course, if we use more subintervals, we would get a closer approximation.) Erecting perpendiculars at $x = 1, 2, 3,$ and 4, we get three strips of area. The perpendiculars intersect the curve at points A, B, C, and D. Now we connect these points in succession with straight-line segments and get three trapezoids. The area of the trapezoids is approximately equal to the area under the curve. The ordinates of the curve are the bases of the trapezoids. Now we must compute the lengths of these ordinates for each value of x, from the function $y = x^2/4$. We get at $x = 1$, $y = \frac{1}{4}$; at $x = 2$, $y = 1$; at $x = 3$, $y = \frac{9}{4}$; at $x = 4$, $y = 4$.

By the trapezoidal rule we have

$$A = \left(\frac{1}{2}\right)(1)(y_1 + 2y_2 + 2y_3 + y_4)$$

Then

$$A = \left(\frac{1}{2}\right)(1)\left(\frac{1}{4} + 2 + \frac{9}{2} + 4\right) = \frac{43}{8} = 5.375$$

If we use more intervals, we get a closer approximation to the true area. For example, let us take the width of each strip as $\frac{1}{2}$ unit. Then the points on the interval [1,4] are 1, 1.5, 2, 2.5, 3, 3.5, and 4. Now we must compute the ordinate y, or function value, for each value of x. We get

$f(1) = \frac{1}{4}$; $f(1.5) = 2.25/4$; $f(2) = 1$; $f(2.5) = 6.25/4$; $f(3) = \frac{9}{4}$; $f(3.5) = 12.25/4$;

$f(4) = 4$. Now we have seven values of the ordinate y.

Then

$$A = \left(\frac{1}{2}\right)\left(\frac{1}{2}\right)(y_1 + 2y_2 + 2y_3 + 2y_4 + 2y_5 + 2y_6 + y_7)$$

or

$$A = \left(\frac{1}{4}\right)(0.25 + 1.125 + 2 + 3.125 + 4.5 + 6.125 + 4) = 5.281$$

By using the definite integral, we get the exact area:

$$\int_1^4 x^2 \, dx = \frac{x^3}{12}\bigg]_1^4 = \frac{64}{12} - \frac{1}{12} = \frac{63}{12} = 5.25, \text{ exact area}$$

Now let us compare the areas obtained by each method:
  Using 3 strips, we get   5.375, an error of approximately 2.4%.
  Using 6 strips, we get   5.281, an error of approximately 0.57%.
  Using the definite integral, we get the exact area, 5.25.

Note that in this example the area of each trapezoid is slightly more than the exact area of the corresponding strip. Whether it is less or more depends on the nature of the curve. If the curve has an inflection point in the interval, then the trapezoidal rule gives a still closer approximation.

In the following example the function cannot be integrated by the usual method.

EXAMPLE 2. Find the area under the curve $y = e^{-x^2}$ from $x = 0$ to $x = 3$ (Fig. 27.5).

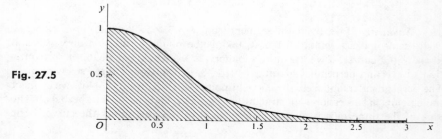

Fig. 27.5

27.2 The Trapezoidal Rule

Solution. Again we point out that although the problem refers to area, the approximation to the value of the definite integral could be computed without any regard to area. That is, we wish to find the approximate value of the integral

$$\int_0^3 e^{-x^2}\, dx$$

If we use six subintervals along the interval $x = 0$ to $x = 3$, then the length of each interval is $\frac{1}{2}$ unit. Then we must compute the ordinate values for the following x-values: $x = 0$; 0.5; 1.0; 1.5; 2.0; 2.5; and 3. Then the total area (approximate) is given by the trapezoidal rule.

$$\begin{aligned}
A &= \left(\frac{1}{2}\right)\left(\frac{1}{2}\right)[f(0) + 2f(0.5) + 2f(1) + 2f(1.5) + 2f(2) + 2f(2.5) + f(3)] \\
&= \left(\frac{1}{2}\right)\left(\frac{1}{2}\right)[1 + 2e^{-0.25} + 2e^{-1} + 2e^{-2.25} + 2e^{-4} + 2e^{-6.25} + e^{-9}] \\
&= \left(\frac{1}{4}\right)[1 + 1.55760 + 0.73576 + 0.21080 + 0.03664 + 0.00386 + 0.00012] \\
&= 0.886195
\end{aligned}$$

In this problem, if we take only three subintervals, we get 0.88626 as the approximate value of the definite integral.

Although the trapezoidal rule is directly related to the area of trapezoids, the principle is applicable to any problem involving a summation between limits. It can just as well be applied to a problem that has no relation to area.

In using the trapezoidal rule we must first decide how many subintervals we wish to use. The number of subintervals is usually denoted by n. The number we take will depend to some extent on the total interval between the given limits of summation. It will also depend on the degree of accuracy desired. The use of more subintervals will usually give a higher degree of accuracy.

To find the length of each subinterval, we divide the total interval by n, the number of subintervals we wish to use. For example, if we wish to find the approximate integral of $f(x)$ from $x = 3$ to $x = 6$, the length of the total interval is 3 units. Now, if we wish to have six subintervals (that is, $n = 6$), we divide and get the length of each subinterval: $\frac{1}{2}$. Since each subinterval is $\frac{1}{2}$ unit, we shall have to compute the following:

$$f(3);\quad 2f(3.5);\quad 2f(4);\quad 2f(4.5);\quad 2f(5);\quad 2f(5.5);\quad f(6)$$

As another example, if we wish to evaluate the integral of $f(t)$ from $t = 1$ to $t = 3$ by this rule, we note that the interval is 2 units in length. If we wish to use 4 subintervals, each one will be $\frac{1}{2}$ unit in length. If we use 6 subintervals, each one will be $\frac{1}{3}$ unit in length.

For example, suppose we wish to find the total transport of charge in an electric circuit in a particular time interval. The charge is measured in *coulombs* and is usually denoted by Q. The rate of change of the charge, or rate of transport of charge, is called *current*, usually denoted by i. That is,

$$\frac{dQ}{dt} = i$$

Approximate Integration

In the following problem we use the trapezoidal rule to find the total transport of charge in a time interval.

EXAMPLE 3. In a particular electric circuit, for a short interval of time, the instantaneous current i is given by the formula $i = t^3$, where t is in seconds and i is measured in amperes. Find the approximate total transport of charge during the time interval from $t = 0$ to $t = 2$, by use of the trapezoidal rule. Use $n = 6$.

Solution. In this example the time interval is 2 seconds. Dividing by 6, the number of subintervals, we get $\frac{1}{3}$ as the length of each subinterval. Then, by the trapezoidal rule, we must compute the following:

$$Q = \left(\frac{1}{2}\right)\left(\frac{1}{3}\right)[f(0) + 2f\left(\frac{1}{3}\right) + 2f\left(\frac{2}{3}\right) + 2f(1) + 2f\left(\frac{4}{3}\right) + 2f\left(\frac{5}{3}\right) + f(2)]$$

Computing,

$$Q = \left(\frac{1}{6}\right)\left[0 + \frac{2}{27} + \frac{16}{27} + 2 + \frac{128}{27} + \frac{250}{27} + 8\right] = \frac{37}{9} = 4\frac{1}{9}$$

Of course, in this example we could find the exact value by use of the definite integral. For the exact Q we have

$$Q = \int_0^2 t^3 \, dt = \left.\frac{t^4}{4}\right]_0^2 = 4$$

EXAMPLE 4. The velocity of a certain object in motion is given by the formula

$$v = \sqrt{t + 5} \quad (t \text{ in seconds}; v \text{ in ft/sec})$$

Find by the trapezoidal rule the approximate total distance covered in the time interval from $t = 3$ to $t = 5$. Use $n = 4$.

Solution. The total interval of time is 2 seconds. The number of subintervals is 4. Then the length of each subinterval is $2 \div 4 = \frac{1}{2}$. Then we must evaluate the function of t for the following values of t:

$$t = 3; \ 3.5; \ 4; \ 4.5; \ 5$$

Then by the trapezoidal rule, the total approximate distance s becomes

$$s = \left(\frac{1}{2}\right)\left(\frac{1}{2}\right)[f(3) + 2f(3.5) + 2f(4) + 2f(4.5) + f(5)]$$

Computing values

$$s = \left(\frac{1}{4}\right)(\sqrt{8} + 2\sqrt{8.5} + 6 + 2\sqrt{9.5} + \sqrt{10})$$

$$s = \left(\frac{1}{4}\right)(2.8284 + 5.8309 + 6 + 6.1644 + 3.1622) = 5.9965 \text{ (approx.)}$$

The definite integral in this case can be evaluated and will be found to be approximately 5.997.

EXAMPLE 5. In a certain electric circuit the instantaneous current is given by the formula $i = \sin \theta$. Find the total transport of charge Q in the interval as θ varies from $\theta = 0$ to $\theta = \pi(180°)$. Use $n = 4$.

Solution. If we use 4 subintervals from 0 to π, each subinterval will have a length of $\frac{\pi}{4}$, or 45°. Then we evaluate the function for the following values of θ: 0°, 45°, 90°, 135°, 180°. By the trapezoidal rule,

$$Q = \left(\frac{1}{2}\right)\left(\frac{\pi}{4}\right)[f(0) + 2f(45°) + 2f(90°) + 2f(135°) + f(180°)]$$

Evaluating,

$$Q = \frac{\pi}{8}[0 + \sqrt{2} + 2 + \sqrt{2} + 0] = 1.89611 \text{ (approx.)}$$

If we had used 6 subintervals, the answer would have been 1.95408 (approx.). By the definite integral, as we shall see later, the exact answer is 2.

Note: Another method sometimes used in finding an approximation to a definite integral is the so-called *Simpson's Rule*. This rule gives a slightly closer approximation to the true answer, but it is more involved and has certain disadvantages. Briefly, Simpson's rule depends on taking a parabolic arc rather than a straight line from one point to another on the curve. It can be used only by taking an even number of intervals. The trapezoidal rule is usually sufficiently accurate for most work.

EXERCISE 27.1

Find by the trapezoidal rule the approximate total value of the integral of the given function between the given limits:

1. $y = x^2$; $x = 0$ to $x = 2$; $n = 4$
2. $y = 2\sqrt{x}$; $x = 0$ to $x = 4$; $n = 4$
3. $y = x^3$; $x = 0$ to $x = 3$; $n = 6$
4. $Q = t^2$; $t = 0$ to $x = 3$; $n = 6$
5. $y = \sqrt{9 - x^2}$; $x = 0$ to $x = 3$; $n = 3$
6. $y = 2^{-x}$; $x = 0$ to 2; $n = 4$
7. $y = \dfrac{2}{x^2 + 1}$; $x = 0$ to $x = 4$; $n = 4$
8. $y = 2^x$; $x = 0$ to $x = 2$; $n = 4$
9. $y = \sqrt{x^2 - 4}$; $x = 2$ to $x = 5$; $n = 3$
10. $y = x^{3/2}$; $x = 0$ to 3; $n = 3$
11. $i = 4^x - 4^{-x}$; $x = 0$ to 2; $n = 4$
12. $s = 6/t$; $t = 1$ to 3; $n = 4$
13. $\displaystyle\int_0^6 t^2 \, dt$; $n = 6$
14. $\displaystyle\int_1^4 \frac{dt}{t}$; $n = 3$; and $n = 6$
15. $\displaystyle\int_0^4 \sqrt{9 + t^2} \, dt$; $n = 4$
16. $\displaystyle\int_0^2 \sqrt{x^2 + 2x} \, dx$; $n = 4$
17. $\displaystyle\int_0^3 t^2\sqrt{t^2 + 1} \, dt$; $n = 6$
18. $\displaystyle\int_{-1}^3 t\sqrt{1 + t} \, dt$; $n = 4$
19. $\displaystyle\int_1^8 \frac{dx}{x}$; $n = 7$
20. $\displaystyle\int_{-1}^4 \sqrt{16 - x^2} \, dx$; $n = 5$
21. $Q = \displaystyle\int_0^3 (t^2 + 2t + 1) \, dt$; $n = 6$
22. $s = \displaystyle\int_{-1}^3 (2 + 2t - t^2) \, dt$; $n = 4$
23. $\displaystyle\int_1^4 (t^{2/3} - t^{1/3}) \, dt$; $n = 3$
24. $Q = \displaystyle\int_0^\pi \sin\theta \, d\theta$; $n = 6$
25. $\displaystyle\int_1^6 \log_{10} x \, dx$; $n = 5$
26. $\displaystyle\int_1^5 \ln x \, dx$; $n = 4$
27. $\displaystyle\int_0^{\pi/2} \cos x \, dx$; $n = 6$
28. $\displaystyle\int_0^4 \left(\frac{x^2}{4} - 2\right) dx$; $n = 4$

chapter

28

Trigonometric Functions: Differentiation

28.1 TRANSCENDENTAL FUNCTIONS

In our study of calculus up to this point we have dealt with only the so-called *algebraic* functions. These are functions that involve only the basic operations of addition, subtraction, multiplication, division, and finding powers and roots of the independent variable x, such as x^2, $(x^2 + 5x)^3$, $\sqrt{3x - 5}$, and so on. We have already seen how to find their derivatives and how to integrate some of these forms.

In addition to the algebraic functions, calculus must also take into account the so-called *transcendental* functions. These include the *trigonometric*, the *inverse trigonometric*, the *exponential*, and the *logarithmic* functions. We often need to determine the derivatives of such functions as $\sin x$, $\log 3x$, 10^{2x}, e^x, $\arctan 5x$, and so on. These functions are called *transcendental* because they cannot be expressed in purely algebraic terms.

28.2 RADIAN MEASURE

Before determining the derivatives of the trigonometric functions, let us review the meaning of the term *radian*. A radian is defined as an angle of such size that if the vertex of the angle is placed at the center of a circle, the two sides of the angle will intercept on the circumference an *arc* equal in length to the radius of the circle, whatever the size of the circle (Fig. 28.1). Stated in another way, if the radius is laid off on the circumference and the ends connected to the center of the circle, the angle formed at the center is one radian. Actually, a radian is an angle of approximately 57.3°.

To determine the size of a radian, we recall the formula for the circumference of a circle:

$$C = 2\pi r$$

That is,   circumference = 2π (radius)

28.3 Graphs of the Trigonometric Functions

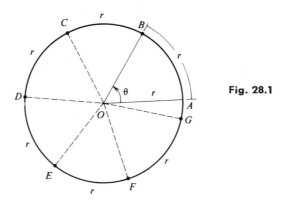

Fig. 28.1

Since each radius length laid off on the circumference subtends one radian at the center of the circle, it follows that

$$2\pi \text{ radians} = 1 \text{ revolution, or } 360°$$

Therefore, $\pi \text{ radians} = 180°$

or, $1 \text{ radian} = \dfrac{180°}{\pi} = 57.3° \text{ (approx.)}$

In calculus, angles are usually understood to be measured in radians rather than in degrees because the formulas for the derivatives of the trigonometric functions are much simpler when radian measure is used. When we write *sin x*, the number represented by *x* is the number of radians in the angle. For example,

$$\sin 1 = \sin 1 \text{ radian} = \sin 57.3° = 0.8415 \text{ (approx.)}$$

Also, $\tan \dfrac{\pi}{3} = \tan 1.0472 \text{ radians} = \tan 60° = \sqrt{3}$

and $\cos 1.7 = \cos 1.7 \text{ radians} = \cos 97.4° = -0.1288 \text{ (approx.)}$

28.3 GRAPHS OF THE TRIGONOMETRIC FUNCTIONS

In graphing the trigonometric functions, such as $y = \sin x$, we proceed as usual to find corresponding values of x and y, and then plot these values. However, if the curves are to have the correct form so that the slope of the curve will everywhere be equivalent to the derivative, then the following points must be observed in graphing:

(1) The units on the *x*- and *y*-axes must be equal in length.
(2) The units on the *x*-axis must represent the angle in *radian* measure.

Let us see what this second point means. As we mark off and number the units 1, 2, 3, and so on on the *x*-axis, these numbers refer to radians. Then, in the equation $y = \sin x$, when we let x equal 1, we must take

$$y = \sin 1 = \sin 1 \text{ radian} = \sin 57.3° = 0.8415$$

Trigonometric Functions: Differentiation

It will be convenient to state the angle in terms of π radians even though we may show the angle in degrees. For example,

$$\sin \frac{\pi}{6} = \sin 30° = 0.5000$$

For this reason we first show the location on the x-axis of points represented by values of π. After locating the points 1, 2, 3, and so on, we should next show the location of π, which is approximately $3\frac{1}{7}$ units from zero. Then we can show the divisions of π, such as $\frac{\pi}{6}, \frac{\pi}{4}, \frac{\pi}{3}, \frac{\pi}{2}, \frac{2\pi}{3}$, and so on, up to 2π.

To graph the function $y = \sin x$ we may use the following values from zero to π (the angle is shown in degrees as well as in radians).

x, radians	0	$\frac{\pi}{6}$	$\frac{\pi}{4}$	$\frac{\pi}{3}$	$\frac{\pi}{2}$	$\frac{2\pi}{3}$	$\frac{3\pi}{4}$	$\frac{5\pi}{6}$	π
x, degrees	0	30°	45°	60°	90°	120°	135°	150°	180°
y	0	0.5	0.707	0.866	1	0.866	0.707	0.5	0

The values of x may be continued on to 2π. When we have once established the correct length on the x-axis for $\pi/6$, or a 30° interval, we can use this length as a measuring unit for the distance to 2π and farther.

Now we plot the corresponding values of x and y and connect them with a smooth curve (Fig. 28.2).

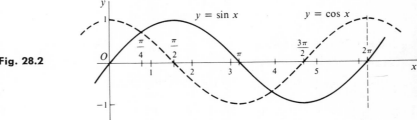

Fig. 28.2

We could continue to take values of x beyond 2π, but we should find that the curve would simply repeat itself. In fact, from $x = 0$ to $x = 2\pi$ we get a complete pattern of the curve. Since the trigonometric curves repeat their patterns, they are called *periodic* curves. One complete pattern of any curve is called one *cycle*. The curve, $y = \sin x$, has one cycle from $x = 0$ to $x = 2\pi$. Then we say the *period* of this curve is 2π.

The graph of the cosine curve, $y = \cos x$, is similar in shape to the sine curve, but it is displaced by a distance of 90° (Fig. 28.2, broken line). This displacement is called a *phase difference* or *phase angle*. The phase angle between the sine and cosine curves is 90°.

28.3 Graphs of the Trigonometric Functions

It often happens that we get more than one cycle of a curve in the interval from zero to 2π, as in the function

$$y = \sin 2x$$

In this example we must double the angle before we find the sine values. Therefore, we shall find that the curve completes one cycle in the interval from zero to π. Its period is π, and there are two complete cycles from zero to 2π (Fig. 28.3). In general, in the equation, $y = \sin nx$, there will be n cycles in the interval from zero to 2π.

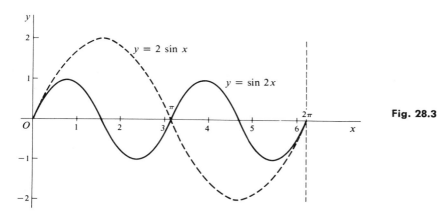

Fig. 28.3

Note that in the sine curve, $y = \sin x$, the function values (y) were limited to the range from -1 to $+1$, since $\sin x$ can never be greater than 1. However, if we have some coefficient other than 1, the range will depend on this coefficient, as in the equation

$$y = 2 \sin x$$

In this equation the value of the sine is multiplied by 2 and the value of y can then range from -2 to $+2$. This coefficient indicates the *amplitude* of the curve (Fig. 28.3, broken line). In general, in the function, $y = m \sin x$, we call m the *amplitude* factor. For example, in the function

$$y = 20 \sin 30 \, x$$

there are 30 cycles in the interval from zero to 2π, and the curve extends 20 units above and below the x-axis.

The graph of the equation, $y = \tan x$ (Fig. 28.4) intersects the x-axis at the points where $x = 0$, $x = \pi$, and $x = 2\pi$. Note that this curve completes one cycle in the interval from $x = 0$ to $x = \pi$. The period is π. The graph shows a rapid increase in the function as x approaches 90°. However, for 90° the tangent is not defined. The lines, $x = \dfrac{\pi}{2}$, and $x = \dfrac{3\pi}{2}$ are asymptotes. Just at the left of 90°, the tangent has a large

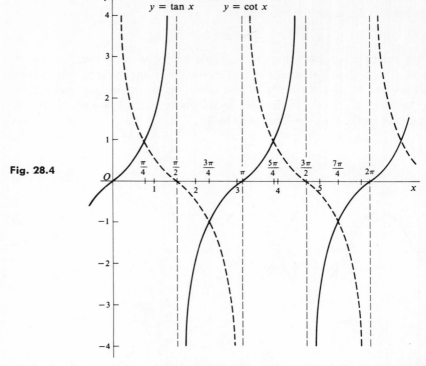

Fig. 28.4

positive value. Just past 90° it has a large negative value, and then rises to cross the x-axis again at $x = \pi$. Note that the function, for all values for which it is defined, is an increasing function. The slope is everywhere *positive*.

The curve $y = \cot x$ (Fig. 28.4, broken line) is similar in shape to the tangent curve except that it is a decreasing function for all values for which it is defined. The slope is everywhere *negative*. For the cotangent curve, the asymptotes are the lines $x = 0$ and $x = \pi$.

The secant and cosecant curves do not intersect the x-axis. Whereas the *sine* and *cosine* curves have absolute values *never greater than* 1, the *secant* and *cosecant* curves have absolute values *never less than* 1. This stems from the fact that the secant and cosecant functions are the reciprocals of the cosine and sine functions, respectively. Note that the period of the secant and cosecant functions is 2π. Also note that each curve is continuous at points where the slope changes sign (Fig. 28.5).

EXERCISE 28.1

Sketch the following curves using values of x from zero to 2π. State the period for each and the number of cycles in the interval 2π.

1. $y = \sin x$
2. $y = \cos x$
3. $y = \tan x$
4. $y = \cot x$
5. $y = \sec x$
6. $y = \csc x$
7. $y = \sin 3x$
8. $y = \cos 2x$
9. $y = 2 \sin 2x$

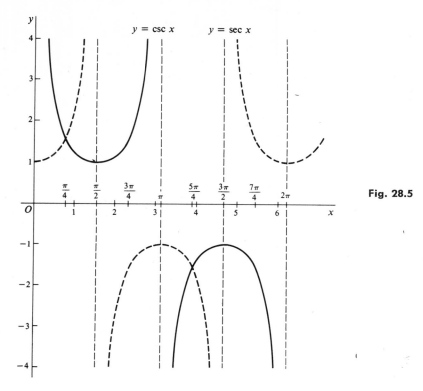

Fig. 28.5

10. $y = 5 \sin 3x$
11. $y = -\cos x$
12. $y = \sin \frac{x}{2}$
13. $y = \sin x + 1$
14. $y = 2 - \cos x$
15. $y = \sin^2 x$
16. $y = \cos^3 x$
17. $y = \frac{1}{2} \sin 2x$
18. $y = \sin x + \cos x$
19. $y = x + \sin x$
20. $y = \tan 2x$
21. $y = \sin x + \sin 2x$
22. $y = \tan^2 x + 1$
23. $y = 0.1 \tan x$
24. $y = \sin^2 x + \cos^2 x$
25. $y = (\sin x)^{1/2}$
26. $y = x \sin x$
27. $y = (\sin x)(\cos x)$
28. $y = \sin^3 x \cos^2 x$
29. $y = (\sin^4 x)(\cos^2 x)$
30. $y = \sin x + \sin 3x + \sin 5x$

28.4 LENGTH OF ARC OF A CIRCLE; AREA OF A SECTOR

Let us derive two simple formulas that we shall need in finding the derivative of sin u. One formula is for the *length of an arc s* of a circle. The other is for the *area of a sector* of a circle.

We have seen that the number of radians in a complete revolution is 2π. In the circle (Fig. 28.6), note that the arc s has the same ratio to the entire circumference as the angle θ has to 2π, that is,

$$\frac{s}{2\pi r} = \frac{\theta}{2\pi}$$

Multiplying both sides by $2\pi r$, we get the formula $\quad s = r\theta$

340 Trigonometric Functions: Differentiation

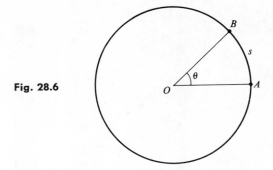

Fig. 28.6

To get a formula for the area of the sector AOB, we note that this area has the same ratio to the area of the entire circle as the arc s has to the entire circumference, that is, letting A represent the area of the sector, we have

$$\frac{A \text{ (of sector)}}{\pi r^2} = \frac{r\theta}{2\pi r}$$

Multiplying both sides by πr^2, we get the formula $\quad A = \tfrac{1}{2} r^2 \theta$

28.5 THE LIMIT OF (SIN θ)/θ

In deriving the formula for the derivative of the sine function, we run into the problem of evaluating the following limit:

$$\lim_{\theta \to 0} \frac{\sin \theta}{\theta}$$

We cannot evaluate this limit by substituting zero for θ, because then we should get the meaningless form, 0/0. To evaluate this limit, let us take the sector AOB of a circle with center at O (Fig. 28.7). Then we have

$$\angle AOB = \theta; \quad AB = s, \quad r = OA = OB$$

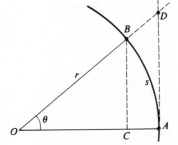

Fig. 28.7

At A we erect a perpendicular to OA, meeting OB extended at D. Also we draw $BC \perp OA$. Let

$\quad A_1 =$ area of small triangle OCB
$\quad A_2 =$ area of sector AOB
$\quad A_3 =$ area of large triangle OAD

Then we have the following inequalities: $A_1 < A_2 < A_3$

Now we express each of these three areas using trigonometry:
For the area A_1 (of the small triangle) we have $A_1 = \frac{1}{2}(OB)(OC) \sin \theta$
Since $OB = r$ and $OC = r \cos \theta$, we have $\quad A_1 = \frac{1}{2}r^2 \cos \theta \sin \theta$
For the area A_2 (of the sector), we have $\quad A_2 = \frac{1}{2}r^2 \theta$
For the area A_3 (of the large triangle), we have $\quad A_3 = \frac{1}{2}(OA)(AD)$
Since $OA = r$ and $AD = r \tan \theta$, we have $\quad A_3 = \frac{1}{2}r^2 \tan \theta$
For the inequalities we get

$$\tfrac{1}{2}r^2 \cos \theta \sin \theta < \tfrac{1}{2}r^2 \theta < \tfrac{1}{2}r^2 \tan \theta$$

Dividing through the inequality by $(\frac{1}{2}r^2 \sin \theta)$ we get

$$\cos \theta < \frac{\theta}{\sin \theta} < \frac{1}{\cos \theta}$$

Inverting the terms and reversing the inequality signs we get

$$\frac{1}{\cos \theta} > \frac{\sin \theta}{\theta} > \cos \theta$$

Now we are ready to let θ approach zero as a limit. As this happens, the limit of the first term is 1, the limit of the third term is 1, and the limit of the middle term must then be squeezed in between the two limits, 1 and 1. Therefore,

$$\lim_{\theta \to 0} \frac{\sin \theta}{\theta} = 1$$

The result is true for any form of the angle θ. For example,

(a) $\lim\limits_{t \to 0} \dfrac{\sin 3t}{3t} = 1$ $\qquad$ (b) $\lim\limits_{x \to 0} \dfrac{\sin \frac{x}{2}}{\frac{x}{2}} = 1$

28.6 DERIVATIVE OF THE SINE FUNCTION

In the function, $\sin u$, the u represents an angle stated in radians. We may further think of the angle as being represented by x alone, or by some function of x, represented by u.

Our question now is: what is the derivative of the function $\sin u$? If we let the sine function be represented by y, we have

$$y = \sin u$$

We shall show that the following statement is true: If u is some function of x, and

$$\text{if } y = \sin u, \text{ then } \frac{dy}{dx} = (\cos u)\left(\frac{du}{dx}\right)$$

Trigonometric Functions: Differentiation

Stated in another way,

$$\frac{d}{dx}(\sin u) = (\cos u)\left(\frac{du}{dx}\right)$$

Stated in words: The derivative of the sine of an angle is equal to the cosine of that angle times the derivative of the angle.

For example, if $y = \sin 4x$, then $dy/dx = (\cos 4x)(4) = 4 \cos 4x$.

To derive the formula for the derivative of sin u, we use the delta method. We begin with

$$y = \sin u$$

Step 1. $\qquad y + \Delta y = \sin(u + \Delta u)$

Step 2. (subtraction) $\qquad \Delta y = \sin(u + \Delta u) - \sin u$

At this point we must simplify the right side of the equation. We transform the right side by use of a trigonometric identity involving the difference between the sines of two angles:

$$\sin M - \sin N = 2\left(\cos \frac{M+N}{2}\right)\left(\sin \frac{M-N}{2}\right)$$

In this formula, we let $M = u + \Delta u$, and $N = u$. Thus we are led to

$$\Delta y = 2\left[\cos\left(u + \frac{\Delta u}{2}\right)\right]\left[\sin \frac{\Delta u}{2}\right]$$

Step 3.

Dividing by Δu, $\qquad \dfrac{\Delta y}{u\Delta} = \dfrac{2\left[\cos\left(u + \frac{\Delta u}{2}\right)\right]\left[\sin \frac{\Delta u}{2}\right]}{\Delta u}$

Now we divide numerator and denominator of the right side by 2, and get

$$\frac{\Delta y}{\Delta u} = \frac{\left[\cos\left(u + \frac{\Delta u}{2}\right)\right]\left[\sin \frac{\Delta u}{2}\right]}{\frac{\Delta u}{2}}$$

The denominator, $\Delta u/2$, can be considered the denominator of the second factor of the numerator.

Step 4. Now when Δx approaches zero, the increments Δu and Δy also approach zero, and the left side of the equation approaches the limit which is the derivative, dy/du. On the right side, the factor $\cos\left(u + \dfrac{\Delta u}{2}\right)$ approaches cos u as a limit. Moreover, we have shown that

$$\frac{\sin \frac{\Delta u}{2}}{\frac{\Delta u}{2}} \text{ approaches the limit 1}$$

Since u represents some function of x, we multiply both sides by du/dx and get

$$\frac{dy}{dx} = (\cos u)\left(\frac{du}{dx}\right)$$

Stated in another way,
$$\frac{d}{dx}(\sin u) = \cos u \frac{du}{dx} \qquad (1)$$

EXAMPLE 1.   In the following function, find dy/dx:
$$y = \sin(5 - 3x)$$

Solution.   In this example, the angle is $(5 - 3x)$, which is equivalent to the u in the formula. Therefore,

$$\frac{dy}{dx} = [\cos(5 - 3x)][-3]$$
$$= -3\cos(5 - 3x)$$

EXAMPLE 2.   If $i = \sin \omega t$, find $\dfrac{di}{dt}$, and also $\dfrac{di}{d(\omega t)}$.

Solution.   If the entire angle ωt is taken as the independent variable, then the derivative of ωt with respect to ωt is 1. However, the derivative of ωt with respect to t is ω. Therefore,

$$\frac{di}{dt} = (\cos \omega t)(\omega) \quad \text{and} \quad \frac{di}{d(\omega t)} = \cos \omega t$$

Note: In finding any derivative it is important that we understand clearly just what we are to consider as the independent variable.

28.7   THE DERIVATIVE OF THE COSINE FUNCTION

We shall show that

if $y = \cos u$ (u is a function of x)

then $\dfrac{dy}{dx} = (-\sin u)\dfrac{du}{dx}$

Stated in another way,

$$\frac{d}{dx}(\cos u) = (-\sin u)\frac{du}{dx} \qquad (2)$$

That is, the derivative of the cosine of an angle is equal to the negative of the sine of the angle times the derivative of the angle.

To show that this is the formula for the derivative of the cosine function, we need not use the *delta* method. Instead, we refer to the formula already derived for the sine function. We begin with

$$y = \cos u$$

Whatever the angle u, we recall from trigonometry that the cosine of any angle is equal to the sine of the complementary angle. The equation

$$y = \cos u$$

is changed to the form

$$y = \sin (90° - u)$$

Now we differentiate by the formula for the derivative of the sine function:

$$\frac{dy}{dx} = [\cos (90° - u)](-1)\frac{du}{dx}$$

or

$$\frac{dy}{dx} = -[\cos (90° - u)]\frac{du}{dx}$$

At this point we change the expression $\cos (90° - u)$ to its equivalent $\sin u$ and obtain

$$\frac{dy}{dx} = -(\sin u)\frac{du}{dx}$$

EXAMPLE 3. In the function $y = \cos (x^2 - 3)$, find dy/dx.

Solution. In this example we identify the expression $(x^2 - 3)$ as the u of the formula. Then

$$\frac{dy}{dx} = -[\sin (x^2 - 3)](2x)$$

$$= -2x \sin (x^2 - 3)$$

28.8 THE DERIVATIVES AS THE SLOPES OF THE TRIGONOMETRIC CURVES

We have found the following derivatives:

$$\frac{d}{dx}(\sin x) = \cos x, \quad \text{and} \quad \frac{d}{dx}(\cos x) = -\sin x$$

The first of these formulas means that for any value of x, the slope of the sine curve, $y = \sin x$, is always equal to the algebraic value of the cosine curve at that point. For example, at the value $x = 0$ the slope of the sine curve is 1, which is the value of the $\cos 0$ (Fig. 28.8).

Note especially the slope of the sine curve at the following values of x. At the value, $x = \pi/2$, the slope of the curve is zero; this is the numerical

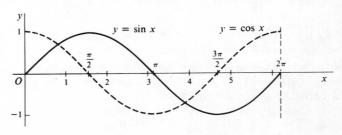

Fig. 28.8

value of the cosine curve for that value of x. At the value, $x = \pi$, the slope of the sine curve is -1, which is the value of $\cos \pi$. In fact, if we plot the *slope* values of the sine curve, we get the cosine curve.

Consider now the slope of the cosine curve. We have seen that the derivative of *cosine* x is $-\text{sine } x$. This means that for any value of x, the slope of the cosine curve is equal to the *negative* of the algebraic value of the sine curve at that point. For example, at the value, $x = \pi/2$, the slope of the cosine curve is -1, which is the negative of the sine value. For the value, $x = (3\pi)/2$, or 270°, the slope of the cosine curve is $+1$, which is the *negative* of the sine of 270°.

EXAMPLE 4. For the sine curve, $y = \sin x$, find the following: (a) the angles at which the curve crosses the x axis from 0 to 2π, (b) the slope of the curve where $x = 1$, (c) values of x where the curve has horizontal tangents.

Solution. We first find the derivative (which is equivalent to the slope):

$$\frac{dy}{dx} = \cos x; \quad \text{then} \quad m = \cos x$$

(a) The curve crosses the x-axis at the values where $y = 0$, that is,

$$\sin x = 0$$

Solving for x, $\qquad x = 0; \pi; 2\pi$

For $x = 0$, we have $m = \cos 0 = 1$. This is also true for $x = 2\pi$. Therefore, the curve passes through the origin at the angle, arctan 1, or 45°. At the point where $x = \pi$, or 180°, we have

$$m = \cos \pi = -1$$

Therefore, the curve crosses the x-axis at the angle, arctan $(-1) = 135°$.

(b) For the value $x = 1$, we have

$$m = \cos 1 = 0.540 \text{ (approx.)}$$

(c) To determine the points where the curve has a horizontal tangent, we recall that the derivative must equal zero, and we set

$$\cos x = 0$$

Solving for x values from 0° up and including 360°, we find

$$x = 90° \quad \text{and} \quad x = 270°$$

To test the value $x = 90°$ for maximum or minimum value, we take values of x slightly less than and slightly more than 90°. For $x < 90°$, the slope, $\cos x$, is positive. For $x > 90°$, m is negative. Since the slope changes from positive to negative, the curve reaches a maximum at $x = 90°$. A similar test shows the sine curve reaches a minimum at $x = 270°$.

28.9 DERIVATIVES OF OTHER TRIGONOMETRIC FUNCTIONS

So far we have derived the formulas for the derivatives of (1) the sine function and (2) the cosine function. The formulas for the derivatives of the four remaining trigonometric functions are found by making use

346 Trigonometric Functions: Differentiation

of the known sine and cosine derivatives, without recourse to the *delta* method.

The derivatives of these four functions are as follows:

(3) $\dfrac{d}{dx}(\tan u) = (\sec^2 u)\dfrac{du}{dx}$

(4) $\dfrac{d}{dx}(\cot u) = (-\csc^2 u)\dfrac{du}{dx}$

(5) $\dfrac{d}{dx}(\sec u) = (\sec u)(\tan u)\dfrac{du}{dx}$

(6) $\dfrac{d}{dx}(\csc u) = -(\csc u)(\cot u)\dfrac{du}{dx}$

Notice the similarity between the derivatives of a function and its co-function. Also note that the derivatives of the cofunctions are all negative.

It might be pointed out that the derivative of any trigonometric function is also a trigonometric function. This is not true with regard to the inverse functions, which we shall study later.

We shall now derive the formulas for the derivatives of the tangent and secant functions. The other two are left as exercises for the student.

28.10 DERIVATIVE OF THE TANGENT FUNCTION

First we let

$y = \tan u$ (where u is some function of x)

Now we recall from trigonometry that

$$\tan u = \frac{\sin u}{\cos u}$$

The equation $y = \tan u$ then becomes

$$y = \frac{\sin u}{\cos u}$$

Now we apply the quotient rule and get

$$\frac{dy}{dx} = \frac{(\cos u)(\cos u)\dfrac{du}{dx} - (\sin u)(-\sin u)\dfrac{du}{dx}}{\cos^2 u}$$

Combining and factoring, we get

$$\frac{dy}{dx} = \frac{\cos^2 u + \sin^2 u}{\cos^2 u}\left(\frac{du}{dx}\right)$$

Since $(\cos^2 u + \sin^2 u) = 1$ we get

$$\frac{dy}{dx} = \frac{1}{\cos^2 u}\left(\frac{du}{dx}\right); \quad \text{or} \quad \frac{dy}{dx} = (\sec^2 u)\left(\frac{du}{dx}\right)$$

Stated in words, the derivative of the tangent of an angle is equal to the secant squared times the derivative of the angle.

28.11 DERIVATIVE OF THE SECANT FUNCTIONS

To find the derivative of the secant function we first let

$$y = \sec u$$

This is changed to its equivalent,

$$y = \frac{1}{\cos u}$$

By the quotient rule,

$$\frac{dy}{dx} = \frac{(\cos u)(0) - (1)(-\sin u)}{\cos^2 u} \left(\frac{du}{dx}\right)$$

or,

$$\frac{dy}{dx} = \frac{\sin u}{\cos^2 u} \left(\frac{du}{dx}\right)$$

This can be written

$$\frac{dy}{dx} = \left(\frac{\sin u}{\cos u}\right)\left(\frac{1}{\cos u}\right)\left(\frac{du}{dx}\right)$$

By trigonometric identities,

$$\frac{dy}{dx} = (\tan u)(\sec u)\frac{du}{dx}$$

Stated in words, the rule says: The derivative of the secant of an angle is equal to the secant of the angle times the tangent of the angle times the derivative of the angle.

EXERCISE 28.2

In the first 12 exercises find dy/dx:

1. $y = \sin 3x$
2. $y = \cos \frac{x}{2}$
3. $y = \tan 5x$
4. $y = \sec 2x$
5. $y = \csc 4x$
6. $y = \cot \frac{2x}{3}$
7. $y = \sin x^2$
8. $y = \tan \sqrt{x}$
9. $y = \sin(6 - x)$
10. $y = \cos x^{1/3}$
11. $y = \csc x^3$
12. $y = \cot(3 - x^2)$

Find the following derivatives:

13. $\frac{d}{dt}(50 \sin 8t)$
14. $\frac{d}{dt}(20 \cos 30t)$
15. $\frac{d}{dt}(6 \tan 50t)$
16. $\frac{d}{dt}(30 \sin 377t)$
17. $\frac{d}{dt}(60 \sec 20t)$
18. $\frac{d}{dt}(100 \sin 120t)$

19. If $y = \sin at$ (a is a constant), find dy/dt; also $dy/d(at)$.
20. If $i = 10 \sin \omega t$, find di/dt; also $di/d(\omega t)$.
21. If $e = 100 \sin (2\pi f)t$, find de/dt; also $de/d(2\pi f)t$.
22. If $e = 20 \sin 5t$, find de/dt; also $de/d(5t)$.

Find the derivative of each of these functions (No. 23–34) with respect to the independent variable:

23. $i = 20 \sin 30t + 15 \cos 30t$
24. $e = 50 \sin 377t + 20 \cos 377t$
25. $e = 30 \sin 10t - 15 \cos 10t$
26. $q = 5 \sin 100t - 4 \cos 100t$
27. $y = 3 \tan 5t + 4 \cot 5t$
28. $r = 10 \sec \phi - 20 \csc \phi$
29. $x = 20 \sin 15t + \cos 15t$
30. $i = 0.05 \sin 20t - 0.1 \cos 20t$
31. $y = \sin 5x \cos 5x$
32. $y = \tan 3x \sec 3x$
33. $i = t^2 \cos 3t$
34. $e = (t^2 + 1) \sin (3 - 2t)$

Find the second derivative of each of the following (No. 35–38):

35. $y = x^2 \sin 5x$
36. $y = (x^2 + 3) \cos 2x$
37. $y = (4 - x) \sin (3 - x)$
38. $i = t^3 \sin 10t$
39. What is the slope of the curve, $y = \sin 2x$, at $x = 0$? at $x = \pi/2$? At what points between 0 and 360° does the curve have $m = 0$?
40. Where does the curve, $y = \sin 3x$, have horizontal tangents?
41. Where does the curve, $y = \cos 4x$, have horizontal tangents?
42. What is the slope of the curve, $i = \cos 2t$, at $t = 0$? at $t = 30°$? at $t = 90°$? at $t = 180°$?
43. What is the slope of the curve, $y = \tan x$, at $x = 0$? at $x = 60°$?
44. Derive the formula for the derivative of the cotangent function.
45. Derive the formula for the derivative of the cosecant function.
46. In a certain electric circuit, the instantaneous current i is given by the formula $i = I_{max} \sin (t + \phi)$. If $I_{max} = 20$ amperes and $\phi = 15°$, find the instantaneous current and the instantaneous rate of change of current when $t = \pi/4$.

28.12 POWERS OF TRIGONOMETRIC FUNCTIONS

In finding the derivatives of powers of trigonometric functions, we use the same power rule that is used for powers of algebraic functions. Let us look again at the power rule for the function

$$y = u^n$$

in which the u represents some function of x.

Up to this point the u has represented an algebraic function, such as

$$x^2 + 5x + 2$$

However, u may well represent a trigonometric function. For example, we may have a power such as

$$\sin^4 x$$

The power rule is valid for powers of *all* functions.

We recall the steps in applying the power rule on an algebraic function. Suppose we have the function

$$y = (\ u\ )^n$$

28.12 Powers of Trigonometric Functions

The first step is to take the exponent n as the coefficient. Then we reduce the exponent by 1 and get

$$\frac{dy}{dx} = n(u)^{n-1}\left(\frac{du}{dx}\right)$$

Note again that the complete derivative must also include the derivative of the u function, or du/dx.

Let us take the following example of a power on an algebraic function:

$$y = (x^2 + 5x + 2)^4$$

Here the u function is $(x^2 + 5x + 2)$. As the first step in applying the power rule we get

$$4(x^2 + 5x + 2)^3$$

Now we must follow this up with the factor, du/dx, or the derivative of the function on which the power appears. In this example, $du/dx = 2x + 5$. Therefore, the complete derivative is

$$\frac{dy}{dx} = 4(x^2 + 5x + 2)^3 (2x + 5)$$

Here we identify $\quad 4(\quad u \quad)^3 \left(\dfrac{du}{dx}\right)$

Now suppose u represents some trigonometric function on which a power appears, such as

$$y = (\sin 5x)^4 \quad \text{(usually written } \sin^4 5x\text{)}$$

As a first step in finding the derivative, we apply the power rule in the same way as with algebraic functions, and get

$$4(\sin 5x)^3$$

Now, we must follow this up with the derivative of the u function, $\sin 5x$, which is $(\cos 5x)(5)$. For the complete derivative we get

$$\frac{dy}{dx} = 4(\sin 5x)^3(\cos 5x)(5)$$

This simplifies to $\quad \dfrac{dy}{dx} = 20(\sin 5x)^3(\cos 5x)$

or, as usually written, $\quad \dfrac{dy}{dx} = 20 \sin^3 5x \cos 5x$

Note that there are *three* separate steps in finding the complete derivative of a power of a trigonometric function. This procedure may be called the

Three-Step Rule for derivatives of powers of trigonometric functions:

Step 1.   *Apply the power rule with regard to the exponent.*
Step 2.   *Find the derivative of the trigonometric function itself.*
Step 3.   *Finally, find the derivative of the angle.*

EXAMPLE 5. Given $y = \tan^6 2x$, find dy/dx.

Solution. First step: $6 (\tan^5 2x)$
Second step: $6 (\tan^5 2x)(\sec^2 2x)$
Third step: $6 (\tan^5 2x)(\sec^2 2x)(2)$

Simplifying,
$$\frac{dy}{dx} = 12 \tan^5 2x \sec^2 2x$$

EXAMPLE 6. Given $y = \csc^4 \sqrt{x^2 + 5}$, find dy/dx.

Solution. First step: $4 (\csc^3 \sqrt{x^2 + 5})$
Second step: $4 (\csc^3 \sqrt{x^2 + 5})(-\csc \sqrt{x^2 + 5})(\cot \sqrt{x^2 + 5})$

As the third step, the derivative of the angle is $\frac{1}{2}(x^2 + 5)^{-1/2} (2x)$. Therefore, the complete derivative is, after simplifying,

$$\frac{dy}{dx} = \frac{-4x}{\sqrt{x^2 + 5}} (\csc^4 \sqrt{x^2 + 5})(\cot \sqrt{x^2 + 5})$$

EXERCISE 28.3

Find the derivative of each of the following functions with respect to the independent variable, such as x, t, or θ:

1. $\sin^4 7x$
2. $\cos^5 3t$
3. $\tan^3 x^2$
4. $\cot^4 (x^2 + 2)$
5. $\sec^3 (5 - 2t)$
6. $\csc^6 (x + 4)^2$
7. $\sin^4 \sqrt{t^2 + 3}$
8. $\cos^2 \sqrt{x}$
9. $\tan^5 (\theta - 2)$
10. $2x + \cot^2 2x$
11. $\frac{x}{2} + \frac{1}{4} \sin 2x$
12. $\tan x - x$
13. $x^2 + x^2 \sin^2 2x$
14. $\sin^3 2t + \cos^3 2t$
15. $(\sin^3 10x)(\cos^3 10x)$
16. $(\tan^2 3t)(\sec^2 3t)$
17. $(\sin 3x)(\sin 6x)$
18. $(\csc^3 5t)(\cot^2 5t)$
19. $\frac{1}{6} \sec^2 3t - \frac{1}{3} \sec 3t$
20. $\theta + \sin^2 \theta + \cos^2 \theta$
21. $3x^2 - 3x \sin 3x - \cos^2 3x$
22. $(\sec^3 \sqrt{x})(\tan^3 \sqrt{x})$
23. $x \cos^2 x - x \sin^2 x$
24. $(x^2 + 5) \sin^3 (x^2 + 5)$
25. $(3 - x^2)^3 \sin^3 x^2$
26. $(4 - 3x^2)^3 \cos^2 (4 - 3x^2)$

Simplify each of the following expressions and then find the derivative of each:

27. $\dfrac{\sin^2 x}{\cos^3 x}$
28. $\dfrac{\sec^3 t}{\tan^4 t}$
29. $\dfrac{\cot^4 x}{\csc^2 x}$
30. $\dfrac{1 - \sin^2 x}{\cot^3 x}$
31. $\dfrac{\csc^2 t - \cot^2 t}{\sec^3 t}$
32. $\dfrac{\sec^2 \theta - \tan^2 \theta}{\sin^3 \theta}$
33. $\dfrac{\sin^4 3x}{3x}$
34. $\dfrac{\tan \sqrt{x}}{\sqrt{x}}$
35. $\dfrac{\cos^3 (x^2 + 4)}{(x^2 + 4)}$

Find the second derivative of each of the following:

36. $y = x^2 \sin 3x$
37. $y = \sin 5x \cos 5x$
38. $y = \tan x \sec x$
39. $y = \cos^2 x - \sin^2 x$
40. $y = x^2 \tan^2 3x + \tan^3 3x$
41. $y = \cos x + \sin x \cos x$
42. $y = \theta^2 \sin \theta + \cos^3 \theta$

Find any maximum or minimum points and values of the following curves; also locate any inflection points; sketch the curves:
43. $y = \sin^2 x$
44. $y = (\sin x)(\cos x)$
45. $y = \sin^3 x \cos^4 x$
46. Show that the curve, $y = \tan x$, has no horizontal tangent. Does it have a point of inflection? if so, where?
47. Show that the curve, $y = \cot x$, has no horizontal tangent. Does it have a point of inflection? if so, where?
48. Find any possible maximum or minimum point of the curve, $y = \sec x$. Does it have a point of inflection? if so, where?
49. Find any possible maximum or minimum point of the curve, $y = \tan^2 x$.
50. Find any maximum or minimum points of the curve, $y = x \sin^2 x$.

chapter
29

Inverse Trigonometric Functions: Differentiation

29.1 DEFINITION OF INVERSE FUNCTIONS

When we say
$$\sin x = 0.5$$
we understand that x is some angle (or number if we prefer) whose sine value is 0.5. Here we shall consider x as an angle although the same reasoning applies to x as a number. For example, an angle of 30° can be considered as the number $\pi/6$ radian; that is, $\sin 30° = \sin \pi/6$.

To solve the foregoing equation for x means to find the angle x whose sine is 0.5. Now, we immediately recall some angles (or numbers) that have a sine value of 0.5. Two such angles are 30° and 150°. However, there are other angles that have a sine value of 0.5, some of which are 390°, 510°, −210°, and −330°. Actually, there is an infinite number of angles having this sine value. In most instances, however, we confine our discussion to angles within certain limits.

If we have the equation
$$\sin x = 0.5,$$
we can say $\qquad x =$ the angle whose sine is 0.5

This statement says nothing new. The two equations are equivalent. The second statement can be written
$$x = \arcsin 0.5$$
The word *arcsin* (pronounced *arc sine*) means *angle* and is equivalent to the phrase "an angle whose sine is." Another notation for arcsin is
$$x = \sin^{-1} 0.5$$
in which the "−1" is *not* to be taken as a negative exponent. The reciprocal of sin x with a negative exponent must be written
$$\frac{1}{\sin x} = (\sin x)^{-1}$$

Since the "−1" used to indicate the inverse functions is often confused with a negative exponent, we use the form *arcsin* for the inverse sine function. A similar notation is used for all inverse trigonometric functions.

In general terms, we can say

$$\text{if } y = \sin x$$
$$\text{then } x = \arcsin y$$

These two statements say the same thing. The second is *not* the inverse of the first.

29.2 INVERSE FUNCTIONS

It is important to remember that in any function we take x as the independent and y as the dependent variable. If we wish to have the inverse function of $\sin x$, we must still have x as the independent variable. This cannot be done simply by writing the equivalent form, $x = \arcsin y$. The two statements $y = \sin x$ and $x = \arcsin y$ mean exactly the same. Instead, to get the inverse of $y = \sin x$ we must find the form with the inverse meaning and with y as the dependent variable.

To find the inverse of an algebraic function, $y = f(x)$, we interchange the variables, x and y, and then solve for y. We do the same for trigonometric functions and their inverses. We begin with

$$y = \sin x \quad \text{(a)}$$

Interchanging variables,

$$x = \sin y$$

Solving for y, $\quad y = \arcsin x \quad \text{(b)}$

The two equations (a) and (b) represent inverse trigonometric functions. Note that, in both equations, the independent variable is x and the dependent variable is y. In the same way, the following are inverses:

$$y = \tan x \quad \text{and} \quad y = \arctan x$$

Note especially that inverse functions are *not* equivalent.

29.3 GRAPHS OF INVERSE TRIGONOMETRIC FUNCTIONS

As we have pointed out, the equations $y = \sin x$ and $y = \arcsin x$ are not equivalent. However, they do say the same thing with the two variables interchanged. For this reason we should expect their graphs to be similar but interchanged with reference to the x-axis and the y-axis. Figure 29.1 shows the two curves on the same set of axes. Whereas the graph of $y = \sin x$ lies along the x-axis, the graph of $y = \arcsin x$ lies along the y-axis. The two graphs are symmetrical with respect to the 45° line $y = x$. If we draw the sine curve, $y = \sin x$, then turn the paper through a 90° angle, and hold the sketch up to a mirror, we shall see the arcsine curve.

Fig. 29.1

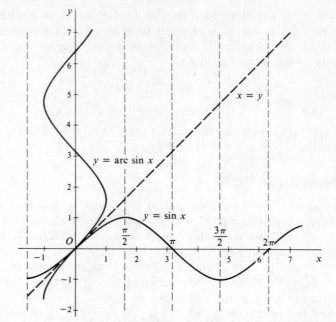

The graphs of the six inverse trigonometric functions are all similar in shape, respectively, to the corresponding trigonometric function curves, but they are reversed with respect to the axes. The 45° line, $y = x$, is the axis of symmetry for each set of curves for a trigonometric function and its inverse. This is, of course, true for algebraic and all other functions. Figure 29.2 shows the graphs of the six inverse trigonometric functions.

29.4 SINGLE-VALUED FUNCTIONS

In the function, $y = \arcsin x$, we have seen that the angle y may have an infinite number of values. This may be seen from the graph. The domain (x-values) is limited from -1 to $+1$, since the sine value of any angle is confined between these limits. However, the range of values for y is from negative infinity to positive infinity. The graph is infinite vertically and it will be found to have any number of values for any particular value of x (Fig. 29.3). For example, corresponding to the value $x = 0.5$, the graph has many values of y, some of which are 30° ($\pi/6$); 150° ($5\pi/6$); 390° ($13\pi/6$); $-210°$ ($-7\pi/6$); and so on. In fact, the angle y can be stated as

$$y = \frac{\pi}{6} + 2\pi n \quad \text{and} \quad y = \frac{5\pi}{6} + 2\pi n$$

where n is any integer.

However, we usually wish to confine our attention to a particular portion of the curve so that we may have a single-valued function. We call this

29.4 Single-Valued Functions

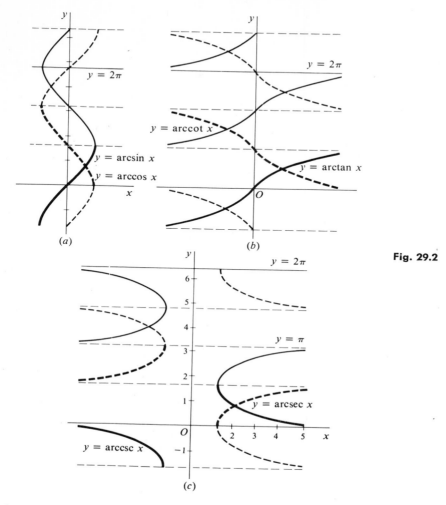

Fig. 29.2

the *principal value* of the function. The principal values of the inverse trigonometric functions are usually indicated by capitalizing the name. For example, the principal value of the arcsine is written

$$y = \text{Arcsin } x$$

In the case of the arcsine, we usually take the principal value of y to be from $-\pi/2$ to $+\pi/2$; that is from $-90°$ to $+90°$. This, of course, places the angle, the arcsin x, in the fourth and the first quadrants. Note that in this section of the curve, the slope is always *positive;* thus, when we come to the derivative of the arcsin x, the derivative will always be positive if we take the angle in these quadrants.

For each of the inverse trigonometric functions, we select a particular range for the principal value. This corresponds to a certain portion of the curve. Now, we recall that the derivative of a function (in rectangular

Fig. 29.3

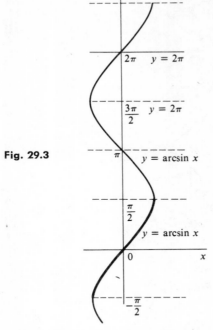

coordinates) is equivalent to the slope of its curve. Then, in selecting the particular range for a principal value, we take the range in which the derivative *does not change sign*.

For the principal value of $y = $ Arcsin x, we select the range of values from $-90°$ to $+90°$ (from $-\pi/2$ to $+\pi/2$) since the slope of this portion of the curve is everywhere *positive*. We select Arccos x values from 0 to π, since the slope of the curve there is *everywhere negative*. The slope of the Arctangent curve is *everywhere positive*, so we choose the *continuous* portion from $-\pi/2$ to $+\pi/2$. The Arccotangent curve is *everywhere negative*, so we take the continuous section from 0 to π.

For the principal values of the Arcsecant and Arccosecant, there is no general agreement, and we may choose one way or another. If we select a continuous portion, the signs will change. It is probably better to take a portion in which the slope does not change sign, even though the graph is not continuous. We take the Arcsecant from 0 to π because in this portion of the curve the slope is everywhere *positive*. We take the Arccosecant in the range from $-\pi/2$ to $+\pi/2$ in which the slope of the curve is negative. For the last two, the curves are not continuous in the range of the principal values. Figure 29.2 shows the principal values in heavy lines.

For all positive values of x, the principal values of the inverse functions lie in the first quadrant. For example, Arccot $1 = 45°$. However, for negative values of x, the quadrants vary for the principal values of different functions. The following examples may help to make clear the selection of principal values for negative values of x.

$$\text{Arcsin}\,(-0.5) = -30° \qquad \text{Arccot}\,(-1) = 135°$$
$$\text{Arccos}\,(-0.5) = 120° \qquad \text{Arcsec}\,(-2) = 120°$$
$$\text{Arctan}\,(-1) = -45° \qquad \text{Arccsc}\,(-2) = -30°$$

29.5 RATE OF CHANGE OF AN ANGLE

It is sometimes necessary to find the derivative of the inverse trigonometric functions. Let us see what this implies.

Since we take an inverse function to mean an angle, its derivative means the instantaneous rate of change of an angle. We have already found the derivatives of the direct trigonometric functions. For example, if $y = \sin x$, then dy/dx means the rate of change of the sine value. Now we come to the *rate of change in the angle itself.* This involves the derivative of an inverse trigonometric function.

The rate of change of an angle is not unusual in many everyday experiences. For example, suppose you stand 30 ft from a street, and a car passes you going at a steady rate of 20 mph. As you watch the car approaching, note how slowly your head turns until the car is nearest you. Then, just as it passes you, note how fast your head turns. Now we might ask: How fast is your head turning at the instant when the car has passed your nearest point by, say, 40 ft?

Or another example: You see the words of a famous patriot chiseled in a wall on one side of his memorial. The lower edge of the engraving is 12 ft above the level of the floor and the top edge is 26 ft above floor level. How far from the wall should you stand so as to get the maximum angle between your lines of sight to the bottom and the top of the engraving? This angle will give you the best position for viewing.

The problems we have mentioned represent a type in which it is necessary to find the derivative of an angle.

In finding the rate of change of an angle, we run into trouble in the matter of units of measurement. When we say

$$y = \arctan x$$

then $\Delta y/\Delta x$ represents the ratio of the number of units of change in y and the number of units of change in x. To emphasize that y is an angle, let us use θ for the angle. Then we have

$$\theta = \arctan x$$

Now our question is: In what kind of units shall we measure the angle θ?

For the simplest results in calculus, we measure an angle in radians. This will avoid the use of a conversion factor. Of course, an angle may

358   Inverse Trigonometric Functions: Differentiation

be first measured in degrees, but for use in formulas in calculus the degree measurement is converted to radian measure.

As an example, suppose we have t (time) measured in seconds, and the angle θ measured in radians. Then if we get

$$\frac{d\theta}{dt} = 0.23$$

this means that the angle θ is changing at the constant rate of 0.23 rad/sec. Of course, if we wish, we can convert this to degree measurement and say that angle θ is changing at about 13.2°/sec.

29.6 DERIVATIVE OF ARCSIN u

The derivatives of the inverse trigonometric functions can all be determined without recourse to the *delta* method. In each case the equation in terms of the inverse function is first stated as a direct function of the angle itself. We illustrate the method in finding the formula for the derivative of the arcsin function. Let

$$y = \arcsin u$$

where u is some function of x. Reversing the form of the equation, we get the equivalent statement

$$u = \sin y$$

Now we have y as the *independent* variable and u as the *dependent* variable. We therefore differentiate $\sin y$ with respect to y, to find du/dy:

$$\frac{du}{dy} = \cos y$$

Since du/dy may be considered as the quotient of differentials, we invert both sides of the equation and get

$$\frac{dy}{du} = \frac{1}{\cos y}$$

From trigonometry we have the identity: $\cos^2 y + \sin^2 y = 1$, or

$$\cos y = \pm \sqrt{1 - \sin^2 y}$$

Now if we choose to take the angle y in the fourth or first quadrants, in which the cosine function is positive, the radical will have the positive sign, and the derivative will always be positive. We have seen that the slope of the arcsin curve is positive for these quadrants. Then, using the positive form of the radical and substituting, we get

$$\frac{dy}{du} = \frac{1}{\sqrt{1 - \sin^2 y}}$$

Now we note that the second step in our derivation says: $u = \sin y$. Substituting u for $\sin y$, we have

$$\frac{dy}{du} = \frac{1}{\sqrt{1 - u^2}}$$

Multiplying both sides of the equation by du/dx, we get the formula:

if $y = \arcsin u$, then $\dfrac{dy}{dx} = \dfrac{du/dx}{\sqrt{1 - u^2}}$

For the case in which $u = x$, we have simply, in derivative operator form,

$$\frac{d}{dx}(\arcsin x) = \frac{1}{\sqrt{1 - x^2}}$$

Note again that the slope of the arcsine curve is positive for all values of the angle y in the range from $-\pi/2$ to $+\pi/2$. This agrees with our taking the positive radical. Note also that when $x = \pm 1$, the derivative has a denominator of zero. This is also shown by the curve, in which the slope is infinite for these values of x.

We may note further that the derivative cannot be set equal to zero, for then we should have $1 = 0$, which is impossible. We can then conclude that the curve has no maximum or minimum point and no horizontal tangent, as the curve itself clearly shows.

EXAMPLE 1. Find dy/dx for the function

$$y = \arcsin 3x$$

Solution. In this equation we identify the following:

$$u = 3x \quad \text{and} \quad du/dx = 3$$

Then,
$$\frac{dy}{dx} = \frac{3}{\sqrt{1 - 9x^2}}$$

Note. The derivative as usual indicates the slope of the curve of the function, $y = \arcsin 3x$. Whatever the value of x, the slope is positive. Moreover, the value of x must be less than $\frac{1}{3}$ for any value of the slope. If $x = \frac{1}{3}$, the denominator of the derivative becomes zero, which means that the curve has an infinite slope at that point. If the absolute value of x is greater than $\frac{1}{3}$, then the curve does not exist since the sine of the angle would then be greater than 1, which is impossible.

29.7 DERIVATIVE OF ARCCOS u

The formula for the derivative of the arccos u can be found in the same way as for arcsin u. First we write

$$y = \arccos u$$

That is,
$$u = \cos y$$

Differentiating,
$$\frac{du}{dy} = -\sin y$$

Inverting,
$$\frac{dy}{du} = \frac{1}{-\sin y}$$

Substituting $\sqrt{1 - \cos^2 y}$ for sin y, we get

$$\frac{dy}{du} = -\frac{1}{\sqrt{1 - \cos^2 y}}$$

Here we take the negative value in the denominator so the derivative will have a negative value. The slope of the *arccosine* curve will not change sign but will be negative throughout the range if we take the angle y in the range from 0 to π. This agrees with a negative derivative.

Again we note that we have the equation, $u = \cos y$. Substituting and then multiplying both sides of the equation by du/dx, we get the formula:

$$\text{if} \quad y = \arccos u, \quad \text{then} \quad \frac{dy}{dx} = -\frac{du/dx}{\sqrt{1 - u^2}}$$

29.8 DERIVATIVE OF ARCTAN u

The derivatives of the remaining inverse trigonometric functions are found in the same way as the derivatives of the arcsine and arccosine functions. The inverse function is first restated in the equivalent direct trigonometric form and then differentiated by known methods. For the function

$$y = \arctan u$$

we write

$$u = \tan y$$

Differentiating,

$$\frac{du}{dy} = \sec^2 y$$

Inverting both sides,

$$\frac{dy}{du} = \frac{1}{\sec^2 y}$$

Now we make use of the identity, $\sec^2 y = \tan^2 y + 1$.

Substituting,

$$\frac{dy}{du} = \frac{1}{\tan^2 y + 1}$$

Now we note in the first step we have,

$$u = \tan y$$

Substituting, we get

$$\frac{dy}{du} = \frac{1}{u^2 + 1}$$

Multiplying both sides by du/dx,

$$\frac{dy}{dx} = \frac{du/dx}{u^2 + 1}$$

EXAMPLE 2. What is the slope of the following curve at the point where $x = 2$? Also find the value of y at that point.

$$y = \arctan(x^2 - 1)$$

Solution. Here we identify the following:
$$u = x^2 - 1; \text{ then } du/dx = 2x; \text{ and } u^2 = (x^2 - 1)^2$$
Using the formula, we get
$$\frac{dy}{dx} = \frac{2x}{(x^2 - 1)^2 + 1} = \frac{2x}{x^4 - 2x^2 + 2}$$
Substituting the value $x = 2$,
$$\frac{dy}{dx} = \frac{4}{16 - 8 + 2} = \frac{2}{5}$$
To find the value of y that corresponds to the value $x = 2$, we substitute this value in the original equation and get
$$y = \arctan(4 - 1) = \arctan 3 = 1.25 \text{ (approx.)}$$
If we use a table showing degree measure, we shall find that arctan 3 is an angle of approximately 71° 34'. In radians this becomes 1.25. Note by the formula the slope of the arctan curve is always positive. Whatever the value of x, the value of u^2 will be positive.

29.9 DERIVATIVE OF ARCSEC u

For the derivative of the inverse secant function we begin with
$$y = \text{arcsec } u$$
Rewriting the equation, $\quad u = \sec y$

Differentiating, $\quad \dfrac{du}{dy} = \sec y \tan y$

Inverting both sides, $\quad \dfrac{dy}{du} = \dfrac{1}{\sec y \tan y}$

Now we make use of the identity, $\tan^2 y = \sec^2 y - 1$.

Then $\quad \tan y = \sqrt{\sec^2 y - 1}$

Substituting, we get
$$\frac{dy}{du} = \frac{1}{\sec y \sqrt{\sec^2 y - 1}}$$
Substituting $u = \sec y$,
$$\frac{dy}{du} = \frac{1}{u \sqrt{u^2 - 1}}$$
Multiplying both sides of the equation by du/dx, we get the formula
$$\frac{dy}{dx} = \frac{du/dx}{u \sqrt{u^2 - 1}}$$
If we always take the angle y in the first and second quadrants, the sign of the derivative (the slope of the curve) will be positive.

362 Inverse Trigonometric Functions: Differentiation

The derivation of the formulas for the derivatives of the functions arccot u and arccsc u will be left as exercises for the student.

Note: Recall that the derivatives of the trigonometric functions are also trigonometric functions. However, this is not the case with the inverse trigonometric functions. The derivatives of the inverse functions are *algebraic*. Therefore, we should expect that the integrals of some algebraic functions will turn out to be inverse trigonometric functions.

EXAMPLE 3. A picture 10 ft tall hangs on a vertical wall with the lower edge of the picture 11 ft above the floor level (Fig. 29.4). If a viewer's eyes are 5 ft above the floor level, how far should the viewer stand from the wall so as to get the best view of the picture?

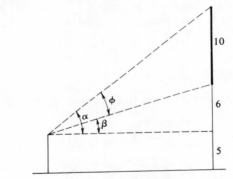

Fig. 29.4

Solution. The angle formed between the viewer's lines of sight to the bottom and the top of the picture should be a maximum. We let x represent the number of feet from the viewer to the wall. We call his angle of elevation to the bottom of the picture beta (β) and the angle to the top of the picture alpha (α). Then the angle we wish to maximize is the angle represented by the difference between these two angles. Let us call this angle ϕ. Then $\phi = \alpha - \beta$. Also $\tan \phi = \tan(\alpha - \beta)$. From the given information, we have

$$\tan \alpha = 16/x \quad \text{and} \quad \tan \beta = 6/x$$

Then

$$\tan \phi = \tan(\alpha - \beta) = \frac{\tan \alpha - \tan \beta}{1 + (\tan \alpha)(\tan \beta)}$$

$$= \frac{\dfrac{16}{x} - \dfrac{6}{x}}{1 + \dfrac{96}{x^2}} = \frac{10x}{x^2 + 96}$$

or

$$\phi = \arctan \frac{10x}{x^2 + 96}$$

Differentiating and simplifying,

$$\frac{d\phi}{dx} = \frac{-10x^2 + 960}{100x^2 + 96(x^2 + 96)^2}$$

Setting the derivative equal to zero and simplifying, we get

$$10x^2 = 960 \quad \text{or} \quad x = \sqrt{96} = 9.8 \text{ ft (approx.)}$$

EXAMPLE 4. A man stands 30 ft from a straight road. A car passes along the road at a rate of 15 mph. As the man watches the car, how fast is his eyesight turning when the car is nearest to him? (Fig. 29.5.)

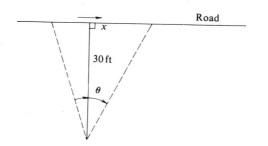

Fig. 29.5

Solution. Since the distance from the road is expressed in feet, we state the velocity of the car as 22 feet per second. We let θ be the positive angle between the perpendicular distance to the road and his line of sight to the car at any time t. Then what we wish to know is the rate of change of the angle with reference to time t; or $d\theta/dt$. We let x represent the positive distance the car has traveled from the point nearest the man. Then we have

$$\frac{x}{30} = \tan \theta \quad \text{or} \quad \theta = \arctan \frac{x}{30}$$

Differentiating with respect to time, we get, after simplifying,

$$\frac{d\theta}{dt} = \frac{30}{x^2 + 900} \frac{dx}{dt}$$

Now we have given the fact that $dx/dt = 22$ feet per second. Then, when $x = 0$, we have

$$\frac{d\theta}{dt} = \frac{30}{900}(22) = \frac{11}{15}$$

This fraction, 11/15, represents radians per second. If we wish, we may convert the rate of turning to degrees and call it approximately 42°/sec.

EXERCISE 29.1

Find the derivative of each of the following functions with respect to the independent variable.

1. $y = \arcsin 5x$
2. $y = \arccos 4x$
3. $y = \arctan 3x$
4. $y = \text{arccot } x^2$
5. $y = \text{arcsec } 2x$
6. $y = \text{arccsc } 6x$
7. $y = \arcsin \dfrac{t}{3}$
8. $y = \arccos \dfrac{2t}{3}$
9. $y = \arctan (t + 3)$
10. $y = \arcsin (x - 1)$
11. $y = \arccos (2x + 1)$
12. $y = \arctan (x^2 + 5)$
13. $y = \arctan \dfrac{3x}{5}$
14. $y = \arcsin \dfrac{3x}{4}$
15. $y = \text{arcsec } \dfrac{4x}{3}$

16. $y = \operatorname{arcsec} x^3$
17. $y = \arctan \dfrac{x+3}{5}$
18. $y = \arcsin(1 - x^2)$
19. $y = \arcsin \sqrt{x}$
20. $y = \arctan \sqrt{x+1}$
21. $y = \arcsin(x+1)^{3/2}$
22. $y = x \arctan x$
23. $y = x^2 \arcsin x$
24. $y = \sqrt{x} \arccos x$
25. $y = \dfrac{\arctan 3x}{x}$
26. $y = \dfrac{x}{\arcsin x}$
27. $y = \dfrac{\arctan x}{(x^2+1)^{-1}}$
28. $y = \arcsin^2 3x$
29. $y = \arctan \dfrac{2}{x}$
30. $y = \sin(\arctan x)$
31. $y = \arcsin 2x - \sqrt{1 - 4x^2}$
32. $y = \dfrac{1}{1 + 9x^2} - \arctan 3x$

33. Show that the curve, $y = \arcsin x$, can have no maximum or minimum point. Find the point of inflection of the principal value. Find m at $x = 0.5$.
34. Show that the curve, $y = \arccos x$, can have no maximum or minimum point. Find the point of inflection of the principal value. Find m at $x = 0.5$.
35. A picture 15 ft tall hangs on a vertical wall so that the bottom edge is 12 ft above the level of the viewer's eyes. How far from the wall should the viewer stand to get the best view of the picture?
36. A searchlight 200 ft from a straight road follows with its light a car traveling along the road at a steady rate. At the instant when the car is nearest the light, the light is turning at a rate of 0.45 rad/sec. How fast is the car traveling?
37. A searchlight 600 ft from a railroad track follows with its light a train traveling along the track at 72 miles per hour. How fast is the light turning when the train has passed the nearest point by 900 ft?
38. A trough having a cross section the shape of an isosceles trapezoid is to be made from a strip of metal that is 26 in. wide by bending up the sides. If the base of the trough is to be 14 in. wide, what should be the width across the top for the greatest volume of the trough?

chapter

30

Exponential and Logarithmic Functions

30.1 DEFINITIONS

An *exponential function* is a function in which the independent variable x appears in the exponent, such as in the following:

$$y = b^x \quad y = 10^{3x} \quad y = 2^x \quad y = 10^{3x+2}$$

The number on which the exponent is placed is called the *base*. In the function $y = b^x$, the base is b, where $b > 0$.

The base may be any number except 1 or zero. Although the base may contain a variable, we shall for the present consider only functions in which the base is a constant, such as 10, 2, e (the natural base), or any other positive constant except 1.

An exponential function may be changed to the form of a logarithm. A logarithm is best defined by the statement: If y is equal to the base b raised to the power x, then we define x as the logarithm of y to the base b. That is,

$$\text{if} \quad y = b^x, \quad \text{then} \quad x = \log_b y$$

These two statements have the same meaning.

However, to define a *logarithmic function* correctly, we want x as the independent variable. Then a *logarithmic function* is one in which we have the logarithm of some function of x, such as the following:

$$y = \log_{10}(3x + 2) \quad y = \log_e \sin 3x; \quad y = \log_e(x^2 - 4x + 5)$$

Note that in a logarithmic function, the dependent variable is y.

As we have said, $b > 0$ in the exponential, $y = b^x$. Then $y > 0$ for any value of x. The same condition exists in the logarithmic form:

$$x = \log_b y$$

Then $y > 0$ for all values of x. That is, zero and negative numbers have no real logarithms.

366   Exponential and Logarithmic Functions

In connection with logarithms, therefore, we must make certain assumptions. Suppose we have the logarithmic function

$$y = \log_{10}(x - 3)$$

First, we can assume that we shall consider only absolute values of the expression $(x - 3)$, in which case we write

$$y = \log_{10}|x - 3|$$

On the other hand, we can indicate what values are to be excluded in the domain of x. In this example, we must exclude all values except those for which $x - 3 > 0$; that is, $x > 3$.

30.2 INVERSES

A logarithmic function is the *inverse* of an exponential function. Let us see what this means. Consider the exponential function

$$y = b^x$$

Here we have y stated as a function of x. The independent variable is x; the dependent variable is y. As we have seen, this can be written

$$x = \log_b y$$

Now we have x stated as a function of y. The independent variable is y, not x. The two equations,

$$y = b^x \quad \text{and} \quad x = \log_b y$$

are *not* inverses. They say the same thing; they are identical in meaning.

As we have previously stated in connection with inverse trigonometric functions, we do not get the inverse of a function simply by writing the same meaning in another form. We must have x as the independent variable in both. To get the inverse of any function, $y = f(x)$, we *interchange the variables*, x and y, and then *solve the resulting equation* for y.

Suppose we have the equation, representing a logarithmic function,

$$y = \log_b x$$

Now let us interchange the variables and get

$$x = \log_b y$$

Solving this equation for y,

$$y = b^x$$

Therefore the two inverse forms are

$$y = \log_b x \quad \text{and} \quad y = b^x$$

In both equations, x is the independent variable. Other examples:

1. The equation $y = 2^x$ is the inverse of the equation $y = \log_2 x$.
2. The equation $y = \log_3 5x$ is the inverse of the equation $y = 3^x/5$.

30.3 GRAPHS

The graphs of the logarithmic and exponential functions show the inverse relationship. To graph the function

$$y = 2^x$$

we might use the following values:

$x =$	-4	-3	-2	-1	0	1	2	3
$y =$	$\frac{1}{16}$	$\frac{1}{8}$	$\frac{1}{4}$	$\frac{1}{2}$	1	2	4	8

The graph is shown in Figure 30.1. Note that y cannot be negative or zero. The equation $2^x = 0$ therefore has no root. The function y will be positive for all values of x. We can make y as small as we wish by taking x very large *negatively*. The value of y will then approach zero, and the curve will approach the x-axis toward the left. The x-axis is therefore an asymptote. The curve has no x-intercept, but it has a y-intercept at (0,1).

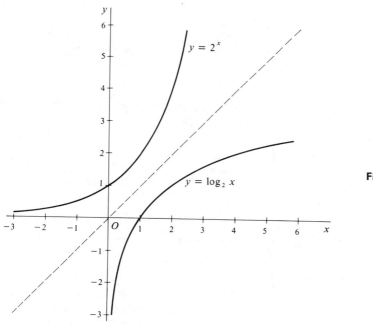

Fig. 30.1

Now let us graph the equation $y = \log_2 x$ on the same set of axes. We take some values of x and find corresponding values of y. For x we must omit zero and negative values since they have no real logarithms. The domain is $0 < x < +\infty$. We may use the following pairs of values:

$x =$	$\frac{1}{4}$	$\frac{1}{2}$	1	2	3	4	6	8	16
$y =$	-2	-1	0	1	1.58	2	2.58	3	4

Exponential and Logarithmic Functions

The graph is shown in Figure 30.1. Note that this curve has an x-intercept at (1,0), but it has no y-intercept. The y-axis is an asymptote.

Note the similarity in the shape of the graphs of the two equations:

$$y = 2^x \quad \text{and} \quad y = \log_2 x$$

The two equations are equivalent *except that the variables are reversed*. Therefore, we should expect that one curve has the same position relative to the y-axis as the other curve has to the x-axis. This is precisely what happens. The two curves are symmetrical with respect to the line $y = x$. One curve is a reflection of the other in the line. In fact, the line $y = x$ is the axis of symmetry of any curve and its inverse, provided the function has an inverse.

30.4 THE NATURAL BASE e

In deriving the formulas for the derivatives of logarithmic and exponential functions, we run into the problem of evaluating the following type of limit:

$$\lim_{v \to 0} (1 + v)^{1/v}$$

This limit cannot be evaluated by substituting zero for v in the function. The result would be

$$(1 + 0)^{1/0} \quad \text{or} \quad 1^\infty$$

The result is meaningless. We might erroneously suppose that the number 1 raised to an infinite power would still be 1. However, the function does *not* approach 1 as a limit.

We can get an idea of the limit by taking values of v successively closer and closer to zero. Let us take v equal to 4, 3, 2, 1, 0.5, and so on, and see what happens to the value of the function. The following table shows v approaching zero from above:

$v =$	4	3	2	1	0.5	0.1	0.01	$\to 0$
$(1+v)^{1/v} =$	1.495	1.587	1.732	2	2.25	2.594	2.705	$\to e$

Note that as v decreases and approaches zero, the value of the function increases. We might guess that it increases without bound and has no limit. Yet it can be proved that the limit does exist and that it is an irrational number somewhere between 2.71828 and 2.71829. This limit we call e.

If we let v approach zero from the left (through negative values), we shall find that the function approaches the value 2.71828... from above. For example, we might take the following values:

$v =$	-0.9	-0.8	-0.5	-0.2	-0.1	$\to 0$
$(1+v)^{1/v} =$	13	7.5	4	3.1	2.8	$\to e$

The limit e appears in deriving the formulas for the derivatives of logarithmic and exponential functions.

EXERCISE 30.1

Graph the following equations:
1. $y = 2^x$
2. $y = 3^x$
3. $y = 10^x$
4. $y = e^x$
5. $y = \log_2 x$
6. $y = \log_3 x$
7. $y = \log_{10} x$
8. $y = \log_e x$
9. $y = e^{-x}$
10. $y = e^{x+1}$
11. $y = (\tfrac{1}{2})^x$
12. $y = -(2^x)$
13. $y = 8^{-x}$
14. $y = 4^{x/2}$
15. $y = 4^{2-x}$
16. $y = 2^{x^2}$
17. $y = 10^{x/4}$
18. $y = 2^{\sqrt{x}}$
19. $y = x + 2^x$
20. $y = (x)(2^x)$
21. $y = \log_2 3x$
22. $y = x + \log_2 x$
23. $y = 5 \log_{10} x$
24. $y = 2 \log_{10} 10x$
25. $y = x \log_2 x$
26. $y = \log_2 (x + 2)$
27. $y = 2 + \log_2 x$
28. $y = \log_3 (x + 3)$
29. $y = \log_e (x + 3)$
30. $y = 4 \log_e (x/4)$
31. $y = \log_e (2x - 3)$
32. $y = 2 - \log_e x$

30.5 THE DERIVATIVE OF A LOGARITHMIC FUNCTION

We shall first show the use of the rule for finding the derivative of a logarithmic function and then later we shall derive the formula. If the base of the logarithm is the natural base e (2.71828...), then the derivative is very simple. As an example, suppose we have

$$y = \log_e (3x + 2)$$

The derivative of a logarithmic function is a fraction. The denominator of the fraction is the given function of x — in this case, $3x + 2$. The numerator of the fraction is the derivative of this function — in this case, 3. Therefore,

if $\quad y = \log_e (3x + 2)$

then $\quad \dfrac{dy}{dx} = \dfrac{3}{3x + 2}$

We simply write the *function* of x itself as the *denominator* of the fraction, and *its derivative* as the *numerator*. The rule may be stated in general terms:

if $\quad y = \log_e u \quad$ (u being some function of x)

then $\quad \dfrac{dy}{dx} = \dfrac{du/dx}{u}$

Illustration:

if $\quad y = \log_e (2x^3 - 5x^2 + 3x - 4)$

then $\quad \dfrac{dy}{dx} = \dfrac{6x^2 - 10x + 3}{2x^3 - 5x^2 + 3x - 4}$

If the base of the logarithm is some number other than e, the derivative is found in exactly the same way as for base e, but there is one additional factor. After finding the derivative in the same way as shown for base

370 Exponential and Logarithmic Functions

e, we follow it up with the logarithm of e to the *given* base. As an example, suppose we have

$$y = \log_{10}(3x + 2)$$

The first part of the derivative is $3/(3x + 2)$, as before. However, now we must follow this up with the factor, $\log_{10} e$, which has a decimal value of approximately 0.4343. The complete derivative is

$$\frac{dy}{dx} = \frac{3}{3x + 2}(\log_{10} e)$$

The foregoing rule holds true for any base used for the logarithm. In general, let us take the logarithmic function

$$y = \log_b u$$

Here b represents any base, and u represents a function of x. Then

$$\frac{dy}{dx} = \frac{du/dx}{u}(\log_b e)$$

Now, if we use e itself as the base, then the last factor reduces to 1. For this reason, the base e is used almost exclusively in calculus. In this way we avoid the need for the extra multiplying factor. If no base is shown, then the base e is always understood. Logarithms to the base e are called natural logarithms. A natural logarithm is often denoted by *ln*; that is, $\log_e(3x + 2)$ is usually written $\ln(3x + 2)$.

EXAMPLE 1. If $y = \ln(x^2 + 3x + 5)$,

$$\text{then } \frac{dy}{dx} = \frac{2x + 3}{x^2 + 3x + 5}$$

EXAMPLE 2. If $y = \log_{10}(x^2 + 3x + 5)$,

$$\text{then } \frac{dy}{dx} = \frac{2x + 3}{x^2 + 3x + 5}(\log_{10} e)$$

The foregoing examples show *how* to find the derivative of a logarithmic function. Now let us see *why* by deriving the formula.

30.6 DERIVATION OF THE FORMULA

The formula is derived by the *delta* method, as usual. We begin with the function

$$y = \log_b u$$

Here, b represents the base of logarithms used, and u represents some function of x. For any increment Δx, we have corresponding increments Δu and Δy. Then

$$y + \Delta y = \log_b(u + \Delta u)$$

Subtracting $y = \log_b u$, we get

$$\Delta y = \log_b (u + \Delta u) - \log_b u$$

By use of the principle of logarithms, $\log M - \log N = \log (M \div N)$, the right side is simplified and we get

$$\Delta y = \log_b \left(1 + \frac{\Delta u}{u}\right)$$

Now we divide both sides of the equation by Δu. The right side is written as shown here:

$$\frac{\Delta y}{\Delta u} = \frac{1}{\Delta u} \cdot \log_b \left(1 + \frac{\Delta u}{u}\right)$$

Now we multiply the right side by 1, written u/u, and get

$$\frac{\Delta y}{\Delta u} = \frac{u}{u} \cdot \frac{1}{\Delta u} \cdot \log_b \left(1 + \frac{\Delta u}{u}\right)$$

The right side of the equation is rewritten to show the factor $u/\Delta u$:

$$\frac{\Delta y}{\Delta u} = \frac{1}{u} \cdot \frac{u}{\Delta u} \cdot \log_b \left(1 + \frac{\Delta u}{u}\right)$$

At this point we recall a principle of logarithms that states that the coefficient of the logarithm of a number may be shifted from its position as a coefficient to the position of an exponent on the number. For example, $3 \log x = \log x^3$. In the last of the foregoing equations, the coefficient $u/\Delta u$ is shifted to the position of an exponent as shown here:

$$\frac{\Delta y}{\Delta u} = \frac{1}{u} \cdot \log_b \left(1 + \frac{\Delta u}{u}\right)^{u/\Delta u}$$

Now we let Δu approach zero and see what happens to the right side of the equation. The fraction $1/u$ is not changed. Our problem is to determine what happens to the expression

$$\left(1 + \frac{\Delta u}{u}\right)^{u/\Delta u}$$

As Δu approaches zero, the fraction $\Delta u/u$ approaches zero, and the exponent $u/\Delta u$ increases without limit. The entire expression is similar in nature to the expression we have previously considered; that is,

$$(1 + v)^{1/v}$$

We have seen that this expression approaches the limit we denote by e. Therefore,

$$\frac{dy}{du} = \frac{1}{u} \cdot \log_b e$$

Since u is a function of x, we multiply both sides by du/dx, and get

$$\frac{dy}{dx} = \frac{1}{u} \cdot \frac{du}{dx} \cdot \log_b e$$

If the base used is e, we have $\log_e e$, which is 1, and the formula becomes

$$\frac{dy}{dx} = \frac{1}{u} \cdot \frac{du}{dx} \quad \left(\text{often written } \frac{du/dx}{u}\right)$$

Stated in words, *the derivative of the logarithm (base e) of a function u is equal to a fraction having the function u as the denominator and the derivative of u as the numerator. If the base for the logarithm is any number other than e, then the derivative must further include the extra factor, the logarithm of e to the given base, or $\log_b e$.*

EXAMPLE 3. If $y = \ln(3x^2 + 5)$, find dy/dx; also evaluate dy/dx at $x = 1$.

Solution.
$$\frac{dy}{dx} = \frac{6x}{3x^2 + 5}$$

At $x = 1$,
$$\frac{dy}{dx} = \frac{6}{3 + 5} = \frac{3}{4}$$

EXAMPLE 4. If $y = \log_{10}(3x^2 + 5)$, find dy/dx, and evaluate dy/dx at $x = 1$.

Solution.
$$\frac{dy}{dx} = \frac{6x}{3x^2 + 5}(\log_{10} e)$$

At $x = 1$,
$$\frac{dy}{dx} = \frac{3}{4}(0.4343) = 0.3257$$

EXAMPLE 5. If $y = \ln(3 - 7x - 5x^2)$, find dy/dx.

Solution.
$$\frac{dy}{dx} = \frac{-7 - 10x}{3 - 7x - 5x^2} = \frac{10x + 7}{5x^2 + 7x - 3}$$

EXAMPLE 6. If $y = \ln(x^5 - 4x^3 + 3x)$, find dy/dx.

Solution.
$$\frac{dy}{dx} = \frac{5x^4 - 12x^2 + 3}{x^5 - 4x^3 + 3x}$$

Note. Notice that in Examples 3–6, we have the logarithm of an algebraic polynomial. In these examples the derivative is a fraction in which the degree of the numerator is *one less* than the degree of the denominator. For instance, in Example 6, the denominator is of degree *five*, whereas the numerator is of degree *four*. This is an important fact to recall when we come to the integration of algebraic fractions leading to logarithms.

EXAMPLE 7. If $y = \ln \sin 5x$, find dy/dx; evaluate dy/dx at $x = \pi/6$.

Solution. $\quad \dfrac{dy}{dx} = \dfrac{5 \cos 5x}{\sin 5x} = 5 \cot 5x$

At $x = \pi/6$, $\quad \dfrac{dy}{dx} = 5 \cot 5(30°) = 5 \cot 150° = -8.66$

EXAMPLE 8. If $y = x^4 \ln(3x + 2)$, find dy/dx; evaluate dy/dx at $x = 1$.

Solution. By the product rule, we get

$$\dfrac{dy}{dx} = (x^4)\left(\dfrac{3}{3x+2}\right) + 4x^3 \ln(3x+2) = (x^3)\left[\dfrac{3x}{3x+2} + 4 \ln(3x+2)\right]$$

At $x = 1$,

$$\dfrac{dy}{dx} = (1)\left[\dfrac{3}{5} + 4 \ln 5\right] = 0.6 + 4(1.60944) = 7.03776 \text{ (approx.)}$$

30.7 LOGARITHM OF A POWER

The logarithm of a power can be differentiated in two different ways, as shown by the following example:

EXAMPLE 9. If $y = \ln(x^2 + 5)^3$, find dy/dx.

Solution.

First method: If we take $u = (x^2 + 5)^3$, then by the formula

$$\dfrac{dy}{dx} = \dfrac{du/dx}{u}$$

we get

$$\dfrac{dy}{dx} = \dfrac{3(x^2+5)^2(2x)}{(x^2+5)^3}$$

However, this fraction can be reduced to

$$\dfrac{dy}{dx} = \dfrac{6x}{x^2+5}$$

Second method: Since $y = \ln(x^2 + 5)^3$, we can write the exponent 3 as the coefficient of the logarithm and get

$$y = 3 \ln(x^2 + 5)$$

Then

$$\dfrac{dy}{dx} = (3)\dfrac{2x}{x^2+5} = \dfrac{6x}{x^2+5}$$

30.8 THE SLOPE OF THE LOGARITHMIC CURVE

Figure 30.2 shows the graphs of the curves $y = \ln x$ and $y = \log_{10} x$. Differentiating these two functions, we have

$$\dfrac{d}{dx}(\ln x) = \dfrac{1}{x}$$

$$\dfrac{d}{dx}(\log_{10} x) = \left(\dfrac{1}{x}\right)(\log_{10} e) = \left(\dfrac{1}{x}\right)(0.4343)$$

374 Exponential and Logarithmic Functions

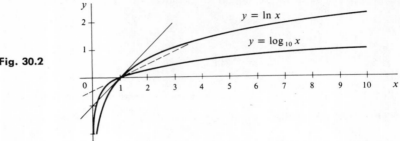

Fig. 30.2

Note especially that the derivatives of these two functions can never equal zero. Therefore, there are no maximum or minimum points of these curves.

The slope of either curve can be found from the value of the derivative for any value of x except zero and negative values. At the value $x = 0$, the derivative does not exist for then we should have $1/0$, which is not possible.

Note especially the slope of each curve where it crosses the x-axis, that is, at $x = 1$. Consider the curve $y = \ln x$. For this curve $dy/dx = 1/x$. For the value, $x = 1$, $dy/dx = 1/1 = 1$. Therefore, this curve crosses the x-axis at a 45° angle.

For the curve $y = \log_{10} x$, we have $dy/dx = 0.4343/x$. For the value $x = 1$, the derivative is equal to $0.4343/1 = 0.4343$. Therefore, this curve crosses the x-axis at an angle, arctan $0.4343 = 23.5°$ (approx.).

30.9 EXCLUDED VALUES

In connection with logarithmic functions, we must be careful to exclude values of x for which the function or its derivative does not exist. The following example shows a problem of this kind.

EXAMPLE 10. If $y = \ln (x^2 - 6x + 5)$, find dy/dx. Evaluate dy/dx at $x = 2$. Also find the point on the curve where $m = 0$.

Solution.
$$\frac{dy}{dx} = \frac{2x - 6}{x^2 - 6x + 5}$$

At $x = 2$,
$$m = \frac{dy}{dx} = \frac{-2}{-3} = \frac{2}{3}$$

From the result, we might at first glance say that at $x = 2$, $m = \frac{2}{3}$. However, if we try to locate the point, we get $y = \ln (-3)$. Since a negative number has no real logarithm, the point does not exist.

Now let us try to determine any x for which $m = 0$. As usual we set the derivative equal to zero and solve for x.

$$\frac{2x - 6}{x^2 - 6x + 5} = 0$$

30.9 Excluded Values

Multiplying both sides by $x^2 - 6x + 5$, we get

$$2x - 6 = 0 \quad \text{or} \quad x = 3$$

The value $x = 3$ would seem to indicate a value for which the slope is zero. However, if we insert this value in the original function, we get

$$y = \ln(9 - 18 + 5) = \ln(-4)$$

Therefore, this curve has no point where $m = 0$.

In this example, the function $(x^2 - 6x + 5)$ cannot be less than or equal to zero for any real points. For any real value of y, the following condition must hold:

$$x^2 - 6x + 5 > 0$$

To solve this inequality, we factor the left side and get $(x - 1)(x - 5) > 0$. If the product of the two factors is to be positive, then both factors must be positive or both negative. If both factors are positive, we have $x - 1 > 0$ and $x - 5 > 0$; then $x > 1$ *and* $x > 5$. The two conditions are true if and only if $x > 5$.

If both factors are negative, then $x - 1 < 0$ *and* $x - 5 < 0$. Solving for x, we get $x < 1$ *and* $x < 5$. These conditions are satisfied if $x < 1$.

Therefore, for this function, x must be less than 1 or more than 5. Of course, if we consider the absolute value of the expression in x, then we need not exclude values except those for which the expression is zero.

EXAMPLE 11. If

$$y = \ln \frac{(2x - 1)^3 (x^2 - 2)^4}{x^3 \sqrt{6x - 2}}$$

find dy/dx. Find the slope of the curve where $x = 1$; also state values of x that must be excluded.

Solution. The function can be written in expanded log form as follows:

$$y = 3\ln(2x - 1) + 4\ln(x^2 - 2) - 3\ln x - \frac{1}{2}\ln(6x - 2)$$

$$\frac{dy}{dx} = (3)\frac{2}{2x - 1} + (4)\frac{2x}{x^2 - 2} - \frac{3}{x} - \left(\frac{1}{2}\right)\frac{6}{6x - 2}$$

$$= \frac{6}{2x - 1} + \frac{8x}{x^2 - 2} - \frac{3}{x} - \frac{3}{6x - 2}$$

At $x = 1$,

$$\frac{dy}{dx} = \frac{6}{1} + \frac{8}{-1} - \frac{3}{1} - \frac{3}{4} = -\frac{23}{4}$$

To determine values of x that must be excluded, we first exclude $x = 0$, since the denominator of the function cannot equal zero. The factor $(6x - 2)$ must be greater than zero; if negative, we should have the square root of a negative quantity, which would be imaginary. Since $6x - 2 > 0$, then $x > \frac{1}{3}$.

Now when we consider the numerator of the function, we see that the factor $(x^2 - 2)^4$ will be positive for all values of x. Since we cannot have the logarithm of a negative number, the factor $(2x - 1)^3$ also must be positive. Therefore,

$$2x - 1 > 0 \quad \text{or} \quad x > \frac{1}{2}$$

We can now conclude that the function and its derivative are possible for all values of x greater than $\frac{1}{3}$. For the value $x = 1$, $m = -23/4$, and $y = \ln 0.5 = -0.69315$ (approx.).

EXERCISE 30.2

Find dy/dx in equations involving x and y. In other equations find the derivative with respect to the independent variable.

1. $y = \ln(x^2 - 7x)$
2. $y = \ln(3x - 2)^3$
3. $y = \ln \cos 2t$
4. $y = \ln(3x - x^3)$
5. $y = \ln \sin^3 \theta$
6. $y = \ln \sin x^3$
7. $y = \ln \tan t$
8. $x = \ln(3y - 1)$
9. $x = \ln(y^2 - 4)$
10. $x = \ln(3y + 4)^5$
11. $y = x^3 \ln(x^2 + 3)$
12. $y = x \ln x - x$
13. $y = \ln \sqrt{4 - 3x}$
14. $y = \ln(3x - 3)$
15. $y = \ln(7x - 7)$
16. $y = x^4 \ln(x^2 - 1)$
17. $y = \ln(5 - x)^{1/3}$
18. $y = \ln(2x + 5)^{3/4}$
19. $i = \ln(5 - 3t)^4$
20. $E = \ln \sin t^2$
21. $r = \ln \cot 3\theta$
22. $y = \ln(x + \sqrt{x^2 + 9})$
23. $y = \ln(\sec t + \tan t)$
24. $y = \ln(\csc \theta + \cot \theta)$
25. $y = \ln(x \tan x)$
26. $y = \ln(3x - 2)^3(4x - 3)^2$
27. $y = \ln(x^2 - 1)^3 \sqrt{4x + 2}$
28. $y = \log_{10}(5x + 2)^2$
29. $y = \log_7(x^3 - 4)$
30. $y = \ln \dfrac{5}{4x + 7}$
31. $y = \ln \dfrac{x^2 \sqrt{2x + 4}}{(3x - 6)^{1/3}}$

32. Compare dy/dx for the following: $y = \ln x$; $y = \ln 2x$; $y = \ln 5x$; $y = \ln cx$.
33. Compare dy/dx for $y = \ln x$; $y = \ln x^2$; $y = \ln x^3$; $y = \ln x^n$.
34. If $y = \ln \dfrac{3x - 1}{x^2 + 1}$, find dy/dx and evaluate at $x = 2$ and $x = 3$.
35. If $y = \ln \dfrac{x^2 \sqrt{x + 2}}{(x - 1)^3 \sqrt{3x + 3}}$, find dy/dx and evaluate at $x = 2$.
36. If $y = \ln \dfrac{(5 - 2x)^3 \sqrt{x^2 + 4x}}{(3x + 4)^2 (x^2 - 3)}$, find the value of dy/dx at $x = 2$.
37. Find any maximum and/or minimum points of the curve $y = x^2 \ln x$.
38. Find any maximum and/or minimum points of the curve $y = x + \ln \sin x$.
39. Find any maximum and/or minimum points of the curve $y = x^2 - \ln x^4$.

Find dy/dx for the following relationships:

40. $y = \ln \dfrac{1}{x}$
41. $x = \ln \dfrac{1}{y}$
42. $y = (\ln x)^2$
43. $y = \dfrac{\ln x}{x}$

30.10 LOGARITHMIC DIFFERENTIATION

We have seen how to differentiate a logarithmic function. However, *logarithmic differentiation* can often be used to advantage even though no logarithm appears in the given function, as in the example

$$y = (2 - x^2)^4 (5x - 3)^2$$

Of course, this function can be differentiated by the product rule. Yet the procedure is simpler if we use logarithmic differentiation. The method consists essentially of injecting a logarithm into a function where none is given. There are three basic steps to the process:

30.10 Logarithmic Differentiation

1. *Take the logarithm of both sides of the equation, and write the result in expanded* log *form by use of the principles of logarithms.*
2. *Differentiate both sides of the equation with respect to x.*
3. *Solve the resulting equation for dy/dx.*

One point must be carefully observed. The logarithm of y must be differentiated in the same way as the logarithm of any other function of x. We have seen that

$$\frac{d}{dx}(\ln u) = \frac{du/dx}{u}$$

In the same way,

$$\frac{d}{dx}(\ln y) = \frac{dy/dx}{y}$$

EXAMPLE 12. If $y = (2 - x^2)^4(5x - 3)^2$, find dy/dx; evaluate at $x = 1$.

Solution. Taking the logarithm of both sides of the equation, we get

$$\ln y = \ln (2 - x^2)^4(5x - 3)^2$$

Expanding,

$$\ln y = 4 \ln (2 - x^2) + 2 \ln (5x - 3)$$

Differentiating both sides of the equation with respect to x, we get

$$\frac{dy/dx}{y} = (4)\left(\frac{-2x}{2-x^2}\right) + (2)\left(\frac{5}{5x-3}\right)$$

or

$$\frac{dy/dx}{y} = \frac{-8x}{2-x^2} + \frac{10}{5x-3}$$

To solve for dy/dx, we multiply both sides of the equation by y and get

$$\frac{dy}{dx} = \left(\frac{-8x}{2-x^2} + \frac{10}{5x-3}\right)(y)$$

Note that the derivative in this form includes the factor y. In most instances this form of the derivative is entirely satisfactory. It can be evaluated for any value of x by including the corresponding value of y; for example, when $x = 1$, $y = 4$, in this problem. To evaluate the derivative, we use the values $x = 1$ and $y = 4$, and get

$$\frac{dy}{dx} = \left(\frac{-8}{2-1} + \frac{10}{5-3}\right)(4) = -12$$

If we wish, we can first express the derivative entirely in terms of x by replacing y with the original function:

The derivative

$$\frac{dy}{dx} = \left(\frac{-8x}{2-x^2} + \frac{10}{5x-3}\right)(y)$$

becomes

$$\frac{dy}{dx} = \left(\frac{-8x}{2-x^2} + \frac{10}{5x-3}\right)(2-x^2)^4(5x-3)^2$$

Simplifying,

$$\frac{dy}{dx} = -2(2-x^2)^3(5x-3)(25x^2 - 12x - 10)$$

This is the same expression we should have obtained by using the product rule and simplifying the result. If we evaluate this form of the derivative for $x = 1$, we get -12, as before.

378 Exponential and Logarithmic Functions

Logarithmic differentiation is especially useful and convenient when the function is very complicated, as in the following example:

EXAMPLE 13. If
$$y = \frac{x^3 \sqrt{x^2 - 3}}{(2x - 5)^{1/3}}$$
evaluate dy/dx at $x = 2$.

Solution. Taking the log of both sides and writing in expanded form,
$$\log y = 3 \log x + \frac{1}{2} \log (x^2 - 3) - \frac{1}{3} \log (2x - 5)$$

Differentiating both sides of the equation, we get
$$\frac{dy/dx}{y} = \frac{3}{x} + \left(\frac{1}{2}\right)\left(\frac{2x}{x^2 - 3}\right) - \left(\frac{1}{3}\right)\left(\frac{2}{2x - 5}\right)$$
$$\frac{dy}{dx} = \left[\frac{3}{x} + \left(\frac{1}{2}\right)\left(\frac{2x}{x^2 - 3}\right) - \left(\frac{1}{3}\right)\left(\frac{2}{2x - 5}\right)\right][y]$$

Simplifying,
$$\frac{dy}{dx} = \left[\frac{3}{x} + \frac{x}{x^2 - 3} - \frac{2}{3(2x - 5)}\right][y]$$

To evaluate the derivative at $x = 2$, we note that when $x = 2$, $y = -8$, and get
$$\frac{dy}{dx} = \left[\frac{3}{2} + \frac{2}{4 - 3} - \frac{2}{3(4 - 5)}\right][-8] = -\frac{100}{3}$$

EXERCISE 30.3

Find dy/dx by logarithmic differentiation. Evaluate for given x-values.

1. $y = x^2 \sqrt{x^2 - 3}$ ($x = 2$)
2. $y = x^3 \sqrt{5 + x}$ ($x = -1$)
3. $y = x^2(3x + 2)^3$ ($x = -1$)
4. $y = x^4 \sqrt{5 - x^2}$ ($x = -1$)
5. $y = x^2(1 - x^2)^{1/3}$ ($x = 3$)
6. $y = x^3(7 - x^3)^{2/3}$ ($x = 2$)
7. $y = \dfrac{x^3}{(x^2 - 4)^2}$ $\left(x = \dfrac{3}{2}\right)$
8. $y = \dfrac{x(3x + 5)^2}{\sqrt{5 - 2x}}$ ($x = -2$)
9. $y = \dfrac{\sqrt{x^2 - 3}}{\sqrt{x(2x - 7)^2}}$ ($x = 4$)
10. $y = \dfrac{x^2 \sqrt{8 - x^3}}{(x^2 + 4)^{1/3}}$ ($x = -2$)
11. $y = \dfrac{x^2(3 - 2x)^{2/3}}{(x^2 - 3)^{3/2}}$ ($x = 2$)
12. $y = \dfrac{x^3(3 - x)^{1/2}}{(2x + 1)^2}$ ($x = -1$)
13. $y = \dfrac{(8 - x^2)^{3/2}}{x^3(2x - 5)^{2/3}}$ ($x = 2$)
14. $y = \dfrac{x^2 \sqrt{x^3 + 3}}{(x^3 + 7)^{1/3}}$ ($x = 1$)
15. Write the equations of the tangent and the normal to the following curve at the point where $x = -1$:
$$y = \frac{(8 - x^3)^{1/2}}{x^2(9 - x^2)^{2/3}}$$

30.11 THE DERIVATIVE OF THE EXPONENTIAL FUNCTION

The rule for the derivative of an exponential function is very simple. We shall first show the rule applied to an example and then later derive the formula. Suppose we have the function
$$y = 10^{3x+2}$$

The first part of the derivative is always exactly *the same as the function itself*. In this example, the first factor of the derivative is

$$10^{3x+2}$$

Next, this quantity is multiplied by the derivative of the exponent. We get

$$(10^{3x+2})(3)$$

There is still another multiplying factor. That is the natural logarithm of the base; in this example, ln 10. For the complete derivative we have

$$\frac{dy}{dx} = (10^{3x+2})(3)(\ln 10) \qquad (\ln 10 = 2.3026...)$$

The rule for the derivative of an exponential function may be stated as a formula. If b represents the base, and u represents the exponent as a function of x, then

$$\text{if } y = b^u, \quad \frac{dy}{dx} = (b^u)\left(\frac{du}{dx}\right) \ln b$$

If the natural base, e, is used in the exponential function, the final factor becomes (ln e), which is 1. Then the formula simplifies as follows:

$$\text{if } y = e^u, \text{ then } \quad \frac{dy}{dx} = (e^u)\left(\frac{du}{dx}\right)(\ln e)$$

or

$$\frac{dy}{dx} = (e^u)\left(\frac{du}{dx}\right)$$

30.12 DERIVATION OF THE FORMULA

We need not use the *delta* method to derive the formula for the derivative of the exponential function. Instead, the exponential function is first stated as a logarithm. Then it is differentiated by the known method used for the logarithmic function. Let us begin with the general exponential function, in which b represents the base and u represents the exponent:

$$y = b^u \quad (u \text{ is a function of } x)$$

We first rewrite the equation in logarithmic form:

$$u = \log_b y$$

Differentiating, $\qquad \dfrac{du}{dy} = \dfrac{1}{y}(\log_b e) = \dfrac{\log_b e}{y}$

Inverting both sides of the equation,

$$\frac{dy}{du} = (y)\left(\frac{1}{\log_b e}\right)$$

Now, in the study of logarithms, we discovered that

$$\frac{1}{\log_b e} = \log_e b$$

Exponential and Logarithmic Functions

Moreover, $y = b^u$. Therefore, we can write

$$\frac{dy}{du} = (b^u)(\log_e b)$$

Multiplying both sides of the equation by du/dx, we get

$$\frac{dy}{dx} = (b^u)\left(\frac{du}{dx}\right)(\ln b)$$

If the natural base is used, we have $b = e$, and the formula reduces to

$$\frac{dy}{dx} = (e^u)\left(\frac{du}{dx}\right)$$

Stated in words, *the derivative of an exponential function consists of three factors: (1) the function itself; (2) the derivative of the exponent; and (3) the natural logarithm of the base. If the base is e, the third factor is 1.*

EXAMPLE 14. If $y = e^{x^2-2x-1}$, find dy/dx. Evaluate at $x = 0$ and at $x = 2$. For what value of x does the curve have a horizontal tangent?

Solution. Differentiating,

$$\frac{dy}{dx} = (e^{x^2-2x-1})(2x - 2)$$

At $x = 0$, $\frac{dy}{dx} = (e^{-1})(-2) = -\frac{2}{e} = -0.74$ (approx.). This is at $(0, e^{-1})$.

At $x = 2$, $\frac{dy}{dx} = (e^{-1})(+2) = +\frac{2}{e} = +0.74$ (approx.). This is at $(2, e^{-1})$.

For a horizontal tangent, the slope must equal zero. Setting the derivative equal to zero, we have

$$(e^{x^2-2x-1})(2x - 2) = 0$$

The first factor can never be equal to zero. Therefore, we must have

$$2x - 2 = 0$$

Solving for x, $\qquad x = 1$

Therefore, at $x = 1$, $m = 0$. This is at the point $(1, e^{-2})$. Since the slope changes from negative to positive, this point is a minimum.

EXAMPLE 15. If $y = e^{\tan x}$, find dy/dx; evaluate dy/dx at $x = \pi/4$.

Solution. $\qquad \frac{dy}{dx} = e^{\tan x}(\sec^2 x)$

At $x = \pi/4$,

$$\frac{dy}{dx} = (e^{\tan 45°})(\sec^2 45°) = (e^1)(2) = 2e = 5.43656 \text{ (approx.)}$$

EXAMPLE 16. Find the first three derivatives of $y = e^x$ and $y = e^{-x}$. Show the two curves on the same graph. Find the intercepts if any. At what angles do the two curves cross the y-axis?

30.12 Derivaton of the Formula

Solution. Figure 30.3 shows the curves. For the curve $y = e^x$, we have $y' = e^x$, $y'' = e^x$, $y''' = e^x$. In terms of slope, the graph of the function $y = e^x$ has a slope that is always equal to the function itself. At $x = 0$, $y = 1$, and $m = 1$. Therefore, the curve crosses the y-axis at an angle of 45°. There are no x-intercepts because e^x is positive for all values of x, positive or negative. Then y can never be negative or zero. Note that the slope is positive for all values of x.

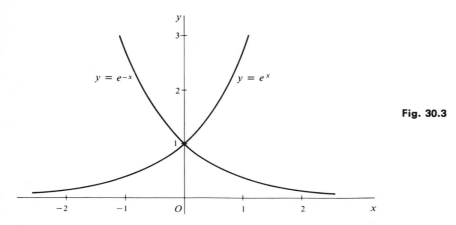

Fig. 30.3

For the curve $y = e^{-x}$, we have the derivatives

$$y' = -e^{-x} \quad y'' = +e^{-x} \quad y''' = -e^{-x}$$

Note that the curve has a negative slope at all points, since y' is negative. At $x = 0$, $y = 1$ and $m = -1$. Therefore, the curve crosses the y-axis at $y = 1$, and at an angle of 135° with respect to the positive portion of the x-axis. For this curve, also, there are no x-intercepts since e^{-x} is never negative or zero. Note that successive derivatives alternate in sign.

The function, $y = e^x$, is the only function that is equal to its own derivative and to all successive derivatives. For example, at $x = 0$, the value of the function is 1, and the slope is also 1. For the value $x = 1$, the value of the function is e (about 2.71828), and the slope of the curve is also equal to e.

EXAMPLE 17. If $y = e^{(3x+2)^4}$, find dy/dx.

Solution. We must be especially careful with regard to powers of the exponent in exponential functions. In this example, we identify the exponent u as

$$u = (3x + 2)^4; \quad \text{then} \quad du/dx = 4(3x + 2)^3(3) = 12(3x + 2)^3$$

Therefore,

$$\frac{dy}{dx} = (e^{(3x+2)^4})(12)(3x + 2)^3$$

which is usually written

$$\frac{dy}{dx} = 12(3x + 2)^3 e^{(3x+2)^4}$$

EXAMPLE 18. Find any possible maximum and/or minimum points on the following curve; also sketch the curve:

$$y = x^3 e^{-2x}$$

Solution. Applying the product rule, we get

$$\frac{dy}{dx} = (x^3)(e^{-2x})(-2) + (e^{-2x})(3x^2)$$

Rearranging and factoring,

$$\frac{dy}{dx} = (x^2)(e^{-2x})(3 - 2x)$$

For $dy/dx = 0$, we have $\quad (x^2)(e^{-2x})(3 - 2x) = 0$

The factor e^{-2x} is always positive for all values of x and can never be equal to zero. Therefore, the only factors that can equal zero are $x^2 = 0$, and $(3 - 2x) = 0$. Therefore the slope is equal to zero where $x = 0$ and $x = \frac{3}{2}$. To test these values, we use a value of x less than 0, one value between 0 and $\frac{3}{2}$, and a value greater than $\frac{3}{2}$. Moreover, all we need to do is to determine whether the factors of dy/dx are positive or negative, and whether their product is positive or negative. Writing the factors again,

$$(x^2)(e^{-2x})(3 - 2x)$$

At the value $x = -1$, we have $\quad (+)(+)(\ + \ ) = +$
At the value $x = 1$, we have $\quad (+)(+)(\ + \ ) = +$
At the value $x = 2$, we have $\quad (+)(+)(\ - \ ) = -$

Therefore, the point (0,0) is an inflection point, since dy/dx does not change sign. At $x = \frac{3}{2}$, $y = 27/8e^3$. This is a maximum point (Fig. 30.4). Note that in the figure, the units of length are not the same on the x-axis and the y-axis. One unit on the x-axis is equal in length to only 0.2 of a unit on the y-axis. This is done to show more clearly the changes in the curve, especially the maximum point and the points of inflection.

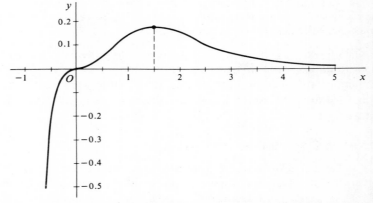

Fig. 30.4

To locate the points of inflection we find the second derivative and set it equal to zero. For the first derivative, we have

$$\frac{dy}{dx} = 3(x^2 e^{-2x}) - 2x^3 e^{-2x}$$

Then

$$\frac{d^2y}{dx^2} = 3(2xe^{-2x} - 2x^2 e^{-2x}) - 2(3x^2 e^{-2x} - 2x^3 e^{-2x})$$

$$= 6x\,e^{-2x} - 12x^2 e^{-2x} + 4x^3 e^{-2x} = 2xe^{-2x}(3 - 6x + 2x^2)$$

We set $\quad 2xe^{-2x}(3 - 6x + 2x^2) = 0$

30.12 Derivation of the Formula

The only possibility for this expression to equal zero is that
$$x = 0 \quad \text{or} \quad 2x^2 - 6x + 3 = 0$$
that is, $\quad x = 0, \quad x = 0.634, \quad x = 2.366$

Note that each of these values does correspond to an inflection point. The inflection points are at (0,0) and at the points whose approximate coordinates are (0.634, 0.072) and (2.366, 0.116).

For plotting the curve we may use the following values:

$x =$	-0.5	-0.25	0	0.25	0.5	0.634	0.75	1	1.5	2	2.37
$y =$	-0.348	-0.026	0	0.01	0.046	0.072	0.094	0.135	0.168	0.147	0.116

$x =$	3	5
$y =$	0.067	0.006

EXERCISE 30.4

In the following functions, find the derivative of the function with respect to the independent variable:

1. $y = e^{5x}$
2. $y = e^{-7x}$
3. $y = e^{x^2-4x+2}$
4. $y = 10^{6x}$
5. $y = 2^{-4x}$
6. $i = e^{3+5t-2t^2}$
7. $i = e^{3t^4}$
8. $q = e^{\sqrt{t}}$
9. $y = e^{\sin 3t}$
10. $i = te^{3t}$
11. $x = t^2 e^{-2t}$
12. $y = (x^2 + 1)e^{-5t}$
13. $y = e^{\tan t}$
14. $x = e^{\sin t}$
15. $q = e^{\cos(t^2-3)}$
16. $y = \dfrac{e^{2x}}{3x}$
17. $i = \dfrac{\cos 3t}{e^t}$
18. $y = \dfrac{\sin 5t}{e^{-4t}}$
19. $x = e^{2t} - e^{-2t}$
20. $y = x^2 e^{\cos x}$
21. $y = \arctan e^{4x}$
22. $y = e^x \ln 2x$
23. $y = e^x \arctan x$
24. $y = e^{x \sin 3x}$
25. $y = e^{\sqrt{3x-2}}$
26. $y = e^x \ln(e^x + 1)$
27. $y = \sqrt{x}\, e^{(x+2)^2}$
28. $y = e^{2x} \cos e^{2x}$
29. $y = e^x \arctan e^x$
30. $y = (e^x + 1)^3$
31. $i = e^{5t}(\sin 30t + \cos 30t)$
32. $q = e^{-20t}(\sin 15t - \cos 15t)$
33. $i = e^{-10t}(20 \sin 5t - 30 \cos 5t)$
34. $i = e^{-30t}(10 \sin 20t + 5 \cos 20t)$

Find the second derivative, d^2y/dx^2, of each of the following functions:

35. $y = x e^x$
36. $y = x^2 e^{-3x}$
37. $y = e^x \sin 5x$
38. $y = e^x \cos 3x$
39. $y = \arctan e^x$
40. $y = e^x \ln x$
41. Find d^2q/dt^2 if $q = e^{-10t}(\sin 30t + \cos 30t)$.
42. Find d^2i/dt^2 if $i = e^{-5t}(4 \sin 10t - 3 \cos 10t)$.
43. Find the slope of the curve $x = e^{y^2}$ at the point where $y = 1$.
44. Find the slope of the curve $x = e^y + e^{-y}$ at the point where $y = -1$.
45. Evaluate the derivative, dy/dx, for the curve $x = y^{-1}e^{2y}$, for the following values of y: $y = 0$; $y = \frac{1}{2}$; $y = -1$; $y = 1$.
46. Find dy/dx for the relation $x^2 y^3 + e^{-y} + x = 0$.
47. If $y = e^{-x}$, what is the value of $dy/dx + y$?
48. If $y = 4e^{-5x}$, show that $y'' + 5y' = 0$.

Find any maximum, minimum, and/or inflection points in the following curves:

49. $y = x^2 e^{-x}$
50. $y = x^3 e^{-x}$

chapter

31

Integration: Power, Logarithmic, and Exponential Forms

31.1 THE POWER RULE: ITS LIMITATIONS

We have already seen (Chapter 22) how the power rule is applied in the integration of some algebraic forms. As a formula, the rule is

$$\int u^n \, du = \frac{u^{n+1}}{n+1} + C \quad (n \neq -1)$$

The rule can be applied in the case of some fractional integrands as in the following examples:

(a) $\int \frac{2x \, dx}{(x^2 - 1)^2} = \int (2x)(x^2 - 1)^{-2} \, dx = -\frac{1}{x^2 - 1} + C$

(b) $\int \frac{x \, dx}{\sqrt{9 - x^2}} = -\frac{1}{2}\int (-2)(x)(9 - x^2)^{-1/2} \, dx = -(9 - x^2)^{1/2} + C$

However, the power rule cannot be applied in the case of some algebraic fractions. It cannot be used, for example, to integrate the following:

(c) $\int \frac{x \, dx}{x^2 + 4}$ (d) $\int \frac{dx}{\sqrt{4 - x^2}}$ (e) $\int \frac{dx}{x^2 + 4}$ (f) $\int \frac{dx}{x^2 - 4}$

In (c), if we attempt to apply the power rule, we transfer the denominator to the numerator with the exponent, -1, and get

$$\int x(x^2 + 4)^{-1} \, dx = \frac{(x^2 + 4)^0}{0}$$

This leads to division by zero, which of course is impossible. Thus we see that the power rule cannot be applied where the power is -1. If we attempt to use the power rule on the very simple fraction,

$$\int \frac{1}{x} \, dx$$

we get the same result.

In example (d), if we attempt to apply the power rule, we get

$$\int (4 - x^2)^{-1/2} \, dx$$

Here we run into another kind of difficulty. The integrand does not contain the necessary integrating factor, $-2x$, and this factor cannot be supplied since it is a variable. In example (e) we have both of the difficulties involved in (c) and (d). The same is true for example (f).

31.2 THE ORIGIN OF ALGEBRAIC FRACTIONS AS DERIVATIVES

To integrate algebraic fractions, let us first see how they originate. Of course, one way in which derivatives come about is through the power rule in differentiation. For example,

$$\text{if } y = \sqrt{x^2 - 4}, \text{ then } \frac{dy}{dx} = \frac{1}{2}(x^2 - 4)^{-1/2}(2x) = \frac{x}{\sqrt{x^2 - 4}}$$

In examples of this kind we can apply the power rule for integrating fractions; that is, since the differential of the power of a function may be a fraction, then the integral of such a differential can be found by the power rule for integration. This is the rule used for examples (a) and (b) above.

However, another function whose derivative is an algebraic fraction is the logarithm of an algebraic function. Take the following example:

$$\text{if } y = \ln(x^2 + 3), \text{ then } \frac{dy}{dx} = \frac{2x}{x^2 + 3}$$

We have seen that the derivative of the logarithm of a function is a fraction (Chapter 30). The denominator of the fraction is the function itself, and the numerator of the fraction is the derivative of the denominator. As another example, if

$$y = \ln(x^3 + 5x^2 + 2x - 4)$$

then the derivative, dy/dx, is the fraction whose denominator is the function $(x^3 + 5x^2 + 2x - 4)$ and whose numerator is the derivative of the denominator; that is,

$$\frac{dy}{dx} = \frac{3x^2 + 10x + 2}{x^3 + 5x^2 + 2x - 4}$$

Therefore, we should expect that the integration of some algebraic fractions will lead to logarithms. Of course, the fraction must be of the correct form.

A third kind of function whose derivative is a fraction is an *inverse trigonometric* function, such as *arcsine, arctangent*, and so on. For example,

$$\text{if } y = \arctan 3x, \text{ then } \frac{dy}{dx} = \frac{3}{9x^2 + 1}$$

or, as another example,

$$\text{if } y = \arcsin 4x, \text{ then } \frac{dy}{dx} = \frac{4}{\sqrt{1 - 16x^2}}$$

Note that the derivatives of these inverse trigonometric functions are algebraic fractions. Then we should expect that the integration of some algebraic fractions leads to inverse trigonometric functions.

In short, the integration of fractions leads to *powers, logarithms,* and *inverse trigonometric functions.* We have already considered the integration of powers (Chapter 22). We shall now consider the type leading to logarithmic functions. Fractions leading to inverse trigonometric functions will be considered in the next chapter in connection with the integration of trigonometric functions.

31.3 INTEGRATION LEADING TO LOGARITHMS

Let us look again at the derivative (or differential) of a simple logarithm:

$$\text{if} \quad y = \ln x, \quad \text{then} \quad \frac{dy}{dx} = \frac{1}{x} \quad \text{or} \quad dy = \frac{dx}{x}$$

Then it is reasonable that we can reverse the procedure and get

$$\int \frac{dx}{x} = \ln x + C$$

In general terms, if u is a function of x, and

$$\text{if} \quad y = \ln u, \quad \text{then} \quad \frac{dy}{dx} = \frac{1}{u}\left(\frac{du}{dx}\right) \quad \text{or} \quad dy = \frac{1}{u}(du)$$

Reversing the procedure, we have the formula

$$\int \frac{du}{u} = \ln u + C$$

Stated in words, the formula says: *If an integrand is a fraction whose numerator is the perfect derivative (or differential) of the denominator, then the integral is the natural logarithm of the denominator.*

Let us consider again an example in which we have the logarithm of an algebraic polynomial function:

$$\text{if} \quad y = \ln(x^3 + 5x^2 + 2x - 4), \quad \text{then} \quad \frac{dy}{dx} = \frac{3x^2 + 10x + 2}{x^3 + 5x^2 + 2x - 4}$$

In the derivative, note especially that the *degree of the numerator* is one *less* than *the degree of the denominator.* The denominator is a *third-degree* polynomial; the numerator is of the *second degree.* As a result, we can say: In any integrand consisting of a rational algebraic fraction, if the degree of the numerator is one *less* than the degree of the denominator, then we can expect the integral to be a *natural logarithm.* If the degree of the numerator is *equal to* or *greater than* the degree of the denominator, then we first use long division to reduce the degree of the numerator.

If the integrand is such a fraction as we have described, then the integral is the *natural logarithm of the denominator.* Note especially that the integral is the *natural* and not the common logarithm.

31.3 Integration Leading to Logarithms

EXAMPLE 1. Integrate as indicated:

$$\int \frac{2x+5}{x^2+5x-3}\,dx$$

Solution. Here we note that in the integrand, which is an algebraic fraction, the numerator is the perfect derivative of the denominator. Then the integral is the natural logarithm of the denominator. That is,

$$\int \frac{2x+5}{x^2+5x-3}\,dx = \ln(x^2+5x-3) + C$$

Note concerning the constant C: If we wish, we can call the constant $\ln C$. This is sometimes convenient. Then the integral can be written as a single logarithm as

$$\ln(x^2+5x-3) + \ln C = \ln C(x^2+5x-3)$$

Of course, the C in this case is not equivalent to the C shown in the first form of the indefinite integral.

In the integration leading to logarithms, it is also sometimes necessary to supply a constant factor, as in the next example.

EXAMPLE 2. Find the indefinite integral

$$\int \frac{(3x-2)\,dx}{5+5x-3x^2}$$

Solution. In the integrand, the denominator of the fraction is of the second degree and the numerator is of the first degree. Therefore, we expect the integral to be a logarithm. The numerator of the fraction is not the perfect derivative of the denominator, but it can be made so by a constant factor. For the numerator we should have $(4-6x)$. Instead we have $(3x-2)$. Then we supply the constant factor (-2), and write its reciprocal outside the integral sign. Then we have

$$-\frac{1}{2}\int \frac{-2(3x-2)\,dx}{5+4x-3x^2} = -\frac{1}{2}\ln(5+4x-x^2) - \ln C$$

$$= -[\ln(5+4x-3x^2)^{1/2} + \ln C] = -\ln C\sqrt{5+4x-3x^2}$$

Recall that a coefficient of a logarithm of a function can be written as the power of the function. For example, $3\ln x = \ln x^3$.

EXAMPLE 3. Evaluate

$$\int_1^2 \frac{5x\,dx}{10-2x^2}$$

Solution. Recalling the formula

$$\int \frac{du}{u} = \ln u + C$$

we identify the following:

if $u = 10 - 2x^2$, then $du = -4x\,dx$

In this example the numerator should contain the quantity $du = -4x\,dx$. Instead, we have $5x\,dx$. Note that the numerator is almost the perfect differential of the denominator. Instead of the constant factor 5, we should have the factor -4. We first move the constant factor 5 to the left of the integral sign, then supply the constant factor -4, and write its reciprocal outside the integral sign, as shown here:

$$5\int_1^2 \frac{x\,dx}{10-2x^2} = -\frac{5}{4}\int_1^2 \frac{-4x\,dx}{10-2x^2} = -\frac{5}{4}[\ln(10-2x^2)]_1^2$$

$$= -\frac{5}{4}[\ln 2 - \ln 8] = +\frac{5}{4}[\ln 8 - \ln 2] = \frac{5}{4}\ln 4 = \frac{5}{4}(1.3863) = 1.733\ldots$$

EXAMPLE 4. Find

$$\int \frac{(2x^3 - 9x^2 + 6x + 13)\,dx}{x^2 - 2x - 3}$$

Solution. Dividing,

$$\frac{2x^3 - 9x^2 + 6x + 13}{x^2 - 2x - 3} = 2x - 5 + \frac{2x-2}{x^2 - 2x - 3}$$

Then the problem becomes

$$\int \left(2x - 5 + \frac{2x-2}{x^2 - 2x - 3}\right) dx = x^2 - 5x + \ln(x^2 - 2x - 3) + C$$

Warning. Note especially that in the integration of algebraic fractions leading to logarithms, the denominator taken *as a whole* has the exponent 1. Separate terms of the denominator may have other exponents, but the entire denominator has the exponent 1. If the denominator, *taken as a whole*, has any exponent other than 1, then the integral is not a logarithm unless the numerator is the derivative of the entire denominator. This is a point that causes students trouble in many instances. For example, the following integrations do not lead to logarithms:

(a) $\quad\displaystyle\int \frac{(x+3)\,dx}{(x^2+6x+1)^3}\quad$ (This leads to a power)

(b) $\quad\displaystyle\int \frac{(2x-10)\,dx}{\sqrt{x^2-10x+3}}\quad$ (This leads to a power)

Moreover, we have seen that separate *factors* can be transferred from denominator to numerator provided the sign of the exponent is changed. However, separate *terms* of a polynomial denominator *cannot* be transferred in this manner. For example, suppose we have the fraction

$$\frac{3x-5}{x^2-x^3}$$

The separate *terms*, x^2 and x^3, *cannot* be moved to the numerator and the signs of their exponents changed. If the denominator is to be transferred to the numerator, then the entire denominator must be transferred, and we get

$$(3x-5)(x^2-x^3)^{-1}$$

31.3 Integration Leading to Logarithms

The rule for integration leading to logarithms can be extended to functions other than algebraic. For example, in the formula

$$\int \frac{du}{u} = \ln u + C$$

the "u" function may be exponential, trigonometric, or some other function. In any fractional integrand, if the numerator is the perfect derivative of the denominator, then the integral is the natural logarithm of the denominator, whatever the kind of function. This is shown in the following examples.

EXAMPLE 5. Find

$$\int \frac{\cos x}{\sin x} dx$$

Solution. Note that the numerator is the perfect derivative of the denominator. Therefore,

$$\int \frac{\cos x}{\sin x} dx = \ln \sin x + C$$

EXAMPLE 6. Find

$$\int \frac{e^{2x} dx}{e^{2x} + 1}$$

Solution. In this example, the numerator is the differential of the denominator with the exception of the factor 2, which can be supplied. Then

$$\int \frac{e^{2x} dx}{e^{2x} + 1} = \frac{1}{2} \int \frac{2 e^{2x} dx}{e^{2x} + 1} = \frac{1}{2} \ln (e^{2x} + 1) + C$$

EXERCISE 31.1

Perform the following integrations. Evaluate as indicated.

1. $\int \dfrac{dx}{3 - 2x}$
2. $\int \dfrac{2\, dx}{x}$
3. $\int \dfrac{-5\, dx}{x}$
4. $\int \dfrac{(2 - x)\, dx}{x^2 - 4x - 4}$
5. $\int \dfrac{5x\, dx}{(x^2 + 4)^3}$
6. $\int \dfrac{5x\, dx}{x^2 + 4}$
7. $\int \dfrac{3x\, dx}{4 - x^2}$
8. $\int \dfrac{3x\, dx}{(4 - x^2)^2}$
9. $\int \dfrac{x\, dx}{\sqrt{9 - x^2}}$
10. $\int \dfrac{x\, dx}{9 - x^2}$
11. $\int \dfrac{4\, dx}{2 - x}$
12. $\int \dfrac{7x\, dx}{(4 - x^2)^{1/3}}$
13. $\int 3x(x^2 + 3)^{-1}\, dx$
14. $\int \dfrac{(x - 4)\, dx}{(8x - x^2)^3}$
15. $\int \dfrac{\sin 2x\, dx}{\cos 2x}$
16. $\int \dfrac{\sec^2 x\, dx}{\tan x}$
17. $\int \dfrac{(1 + \sin x)\, dx}{x - \cos x}$
18. $\int \dfrac{\csc^2 3x\, dx}{4 + \cot 3x}$

19. $\displaystyle\int \frac{e^{3x}\,dx}{e^{3x}-2}$
20. $\displaystyle\int \frac{e^{3x}\,dx}{\sqrt{e^{3x}-2}}$
21. $\displaystyle\int \frac{e^{-2x}\,dx}{e^{-2x}+5}$

22. $\displaystyle\int \frac{4\,e^{-4x}\,dx}{\sqrt{5+e^{-4x}}}$
23. $\displaystyle\int \frac{\sec x \tan x\,dx}{\sec x}$
24. $\displaystyle\int \frac{\cos 2x - \sin 2x}{\sin 2x + \cos 2x}\,dx$

25. $\displaystyle\int \frac{(1+2x-x^2)\,dx}{x^3 - 3x^2 - 3x + 9}$
26. $\displaystyle\int \frac{(3+4x-3x^2)\,dx}{2x^3 - 4x^2 - 6x - 3}$

27. $\displaystyle\int \frac{(\sec^2 x + \sec x \tan x)\,dx}{\sec x + \tan x}$
28. $\displaystyle\int \frac{(1 + 2\sin x \cos x)\,dx}{x + \sin^2 x}$

29. $\displaystyle\int \frac{x\,dx}{x+2}$
30. $\displaystyle\int \frac{(2x-3)\,dx}{x-2}$
31. $\displaystyle\int \frac{(6x^2 + 3x - 2)\,dx}{2x - 3}$

32. $\displaystyle\int \frac{(8x^2 + 4x - 7)\,dx}{2x+3}$
33. $\displaystyle\int \frac{(3x^3 - 17x^2 + 21x - 1)\,dx}{x^2 - 4x + 1}$

34. $\displaystyle\int_0^2 \frac{3x\,dx}{x^2+4}$
35. $\displaystyle\int_3^5 \frac{4x\,dx}{\sqrt{25-x^2}}$
36. $\displaystyle\int_0^2 \frac{5x^2\,dx}{9-x^3}$

37. $\displaystyle\int_0^1 \frac{e^x\,dx}{e^x+4}$
38. $\displaystyle\int_0^{\pi/6} \frac{\sin 2x\,dx}{4 + \cos 2x}$

31.4 INTEGRATION OF EXPONENTIAL FUNCTIONS

Let us recall just what happens in differentiating an exponential function. Suppose we have the function

$$y = e^{3x}$$

We have seen that in an exponential function, if the base is e, then the derivative consists of only two factors:

(1) The function itself.
(2) The derivative of the exponent.

For example, if $y = e^{3x}$, then $dy/dx = (e^{3x})(3)$; or $dy = (e^{3x})(3)\,dx$.

For the reverse process of integration, it is reasonable then that

$$\int 3\,e^{3x}\,dx = e^{3x} + C$$

Note that the factor 3 appears as the derivative of the exponent when we differentiate. Then in integration, this factor disappears.

In general terms, if u is any function of x, and

$$\text{if } y = e^u, \text{ then } \frac{dy}{dx} = (e^u)\left(\frac{du}{dx}\right)$$

Stated as differentials,

$$dy = (e^u)\left(\frac{du}{dx}\right)(dx) = (e^u)(du)$$

Then, in integration, the factor du must be present in the integrand, but it disappears in the integral; that is, in general,

$$\int e^u\,du = e^u + C$$

31.4 Integration of Exponential Functions

In words, this formula says: *The integral of an exponential function is the exponential function itself, provided the integrand contains the complete factor du, and provided the base is e.*

Again, let us remember that the differential du is equal to the derivative of u times the differential of x: that is, $du = (du/dx)(dx)$.

As another example, take the exponential function

$$y = e^{x^2+3x}$$

Here we identify the quantity $(x^2 + 3x)$ as the exponent u in e^u. Then $du/dx = 2x + 3$ and $du = (2x + 3)\,dx$. Differentiating the exponential, we get

$$\frac{dy}{dx} = (e^{x^2+3x})(2x + 3)$$

or, as differentials, $\quad dy = (e^{x^2+3x})(2x + 3)\,dx$

For integrating this exponential function, the entire factor, $(2x + 3)\,dx$, which is equivalent to du, must be present in the integrand. It disappears in the integral. That is,

$$\int (2x + 3)\, e^{x^2+3x}\, dx = e^{x^2+3x} + C$$

Let us take an exponential function with a trigonometric exponent:

$$y = e^{\cos 3x}$$

Differentiating, $\quad \dfrac{dy}{dx} = (e^{\cos 3x})(-3 \sin 3x)$

or as differentials, $\quad dy = -3 \sin 3x\, e^{\cos 3x}\, dx$

The entire factor $(-3 \sin 3x\, dx)$ is equivalent to du. This factor disappears in integrating; that is,

$$\int -3 \sin 3x\, e^{\cos 3x}\, dx = e^{\cos 3x} + C$$

If the base of an exponential function is any number other than e, the derivative contains a third factor, the *natural logarithm of the base*. If the base is represented by b, if u is some function of x, and

$$\text{if } y = b^u, \quad \text{then } \frac{dy}{dx} = (b^x)\left(\frac{du}{dx}\right)(\ln b)$$

As an example,

$$\text{if } y = 10^{x^2-5x}, \quad \text{then } \frac{dy}{dx} = (10^{x^2-5x})(2x - 5)(\ln 10)$$

Note. The natural logarithm of 10 is a constant equal to 2.30259 (approx.).

Some students have trouble identifying the differential du. Remember:

(1) In the case of exponential functions, the complete du is the differential of the exponent; that is, $du = (du/dx)\,dx$.

(2) It must be present in the integrand.
(3) It disappears in the process of integration.

As we have emphasized before, the differential du always emerges from the function in differentiation. Then, in integration, it reenters the function.

EXAMPLE 7. Find
$$\int (4x + 5)e^{2x^2+5x+3}\, dx$$

Solution. In the integrand, we identify u as the exponent on e; that is, $u = 2x^2 + 5x + 3$. Therefore, $du = (4x + 5)\, dx$. Now we look for the factor $(4x + 5)$ in the integrand. Since the integrand contains the complete du as a factor, we integrate at once and get
$$\int (4x + 5)e^{2x^2+5x+3}\, dx = e^{2x^2+5x+3} + C$$

The result can be checked by differentiation.

EXAMPLE 8. Find
$$\int \cos x \, e^{\sin x}\, dx$$

Solution. We first look for the derivative of the exponent $\sin x$. It is $\cos x$, which is contained in the integrand as a factor. We identify
$$u = \sin x, \quad du/dx = \cos x, \quad \text{and} \quad du = \cos x\, dx$$

Since we have the complete du in the integrand, we apply the exponential rule and get at once
$$\int \cos x \, e^{\sin x}\, dx = e^{\sin x} + C$$

and the du factor disappears.

31.5 SUPPLYING A CONSTANT FACTOR

As we have pointed out in connection with the integration of algebraic functions, if the factor du is not present in the integrand, then the integration cannot be performed directly. The same is true in connection with exponential functions. However, again, if only a constant factor is lacking for a perfect du, then this factor can be supplied as usual and its reciprocal written outside the integral sign as a coefficient.

EXAMPLE 9. Find
$$\int e^{-5x}\, dx$$

Solution. Before we apply the exponential rule, we look for the differential of the exponent, $-5x$. If we identify: $u = -5x$, then $du = -5\, dx$. We lack the constant factor, -5. This factor is supplied as usual:
$$\int e^{-5x}\, dx = -\frac{1}{5}\int (-5)\, e^{-5x}\, dx = -\frac{1}{5}e^{-5x} + C$$

EXAMPLE 10. Find
$$\int e^{-20t}\, dt$$

Solution. Supplying the constant factor, -20, we shall have the coefficient, $-\frac{1}{20}$, at the left of the integral sign. Then the integral becomes

$$\int e^{-20t} \, dt = \frac{e^{-20t}}{-20} + C$$

Note. It might be helpful at this point to contrast the effects of differentiating and integrating an exponential function, such as e^{-5x}. In differentiation, the function is *multiplied* by the constant, -5:

$$\frac{d}{dx}(e^{-5x}) = (-5)(e^{-5x})$$

In integration, the function is *divided* by the constant, -5:

$$\int e^{-5x} \, dx = \frac{e^{-5x}}{-5} + C$$

In general, $\quad \frac{d}{dx}(e^{ax}) = (a)(e^{ax}), \quad$ and $\quad \int e^{ax} \, dx = \frac{e^{ax}}{a}$

where a is a constant. It is well to remember this difference in working with exponential functions, especially in connection with integration by *parts* and in differential equations.

31.6 BASE OTHER THAN e

We have seen that if the base of an exponential function is any number other than e, then the derivative contains the extra factor, the *natural logarithm of the base*. For example, if the base is 10, the extra factor in the derivative is the *natural logarithm* of 10. If the base is any constant b, then the natural logarithm of this constant is also a constant. Then this constant factor, $\ln b$, can also be supplied the same as any other constant factor.

EXAMPLE 11. Find the indefinite integral

$$\int 10^{-4x} \, dx$$

Solution. In this example, the base of the exponential function is 10. When we differentiate the function, 10^{-4x}, we get the factor, -4, the derivative of the exponent, and we also get the factor $\ln 10$. This factor, $\ln 10$, is also a constant (approximately, 2.30259). The two constant factors, -4, and $\ln 10$, are not present in the integrand but can be supplied in the usual way. Then we get

$$\left(-\frac{1}{4}\right)\left(\frac{1}{\ln 10}\right) \int (-4)(\ln 10) \, 10^{-4x} \, dx = -\frac{10^{-4x}}{4 \ln 10} + C$$

EXAMPLE 12. Integrate and evaluate

$$\int_0^2 e^{-3x} \, dx$$

394 Integration: Power, Logarithmic, and Exponential Forms

$$-\frac{1}{3}\int_0^2 (-3)\, e^{-3x}\, dx = -\frac{1}{3}[e^{-3x}]_0^2 = -\frac{1}{3}[e^{-6} - e^0]$$

$$= \frac{1}{3}\left[e^0 - \frac{1}{e^6}\right] = \frac{1}{3}[1 - 0.00248] = 0.3326 \text{ (approx.)}$$

Solution. We note that the factor, -3, must be supplied. Then we get

EXAMPLE 13. Integrate and evaluate

$$\int_0^{\pi/6} \sin 3t \; e^{\cos 3t}\, dt$$

Solution. To apply the exponential rule, we look for the differential of the exponent. If we identify

$$u = \cos 3t$$

then
$$du = -3 \sin 3t \, dt$$

The constant factor, -3, can be supplied, and we get

$$-\frac{1}{3}\int_0^{\pi/6} -3 \sin 3t \; e^{\cos 3t}\, dt = -\frac{1}{3}[e^{\cos 3t}]_0^{\pi/6} = -\frac{1}{3}[e^{\cos \pi/2} - e^{\cos 0}]$$

$$= -\frac{1}{3}[e^0 - e^1] = \frac{1}{3}[e - 1] = \frac{1}{3}[2.71828 - 1] = 0.57276$$

EXAMPLE 14. Integrate and evaluate

$$\int_{-1}^{1} 10^{-2x}\, dx$$

Solution. Here we must supply the constant factors, -2 and $\ln 10$. Then we get

$$\left(-\frac{1}{2}\right)\left(\frac{1}{\ln 10}\right) \int (-2)(\ln 10)(10^{-2x})\, dx = -\frac{1}{2 \ln 10}[(10^{-2x})]_{-1}^{1}$$

$$= -\frac{1}{4.6052}[10^{-2} - 10^2] = \frac{1}{4.6052}[10^2 - 10^{-2}] = 0.215 \text{ (approx.)}$$

In this problem we could use the fact that

$$\frac{1}{\ln 10} = \log_{10} e = 0.43429 \text{ (approx.)}$$

EXAMPLE 15. Find the area under the curve $y = e^{-x}$ from $x = -1$ to $x = 2$. If this area is rotated about the x-axis, find the volume of the solid generated. Sketch the figure, showing area and volume.

Solution. In this example, we take a vertical element of area (Fig. 31.1). The length of this element is y and the width is dx. Then we have

$$dA = y\, dx \quad \text{or} \quad dA = e^{-x}\, dx$$

To sum up the elements of area from $x = -1$ to $x = 2$, we integrate the element and evaluate the integral between limits:

$$A = \int_{-1}^{2} e^{-x}\, dx = -[e^{-x}]_{-1}^{2} = -[e^{-2} - e^1] = e - e^{-2} = 1.84969 \text{ (approx.)}$$

31.6 Base Other Than e

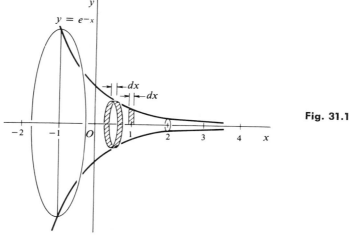

Fig. 31.1

To find the volume we take the element of volume as a vertical slice, which is a disk in the form of a cylinder. For the element of volume, we have

$$dV = \pi y^2 \, dx \text{ (since the radius is } y)$$

Since $y = e^{-x}$, then
$$y^2 = (e^{-x})^2 = e^{-2x}$$

Therefore,

$$V = \pi \int_{-1}^{2} e^{-2x} \, dx = -\frac{\pi}{2}[e^{-2x}]_{-1}^{2} = -\frac{\pi}{2}[e^{-4} - e^{2}]$$

$$= \frac{\pi}{2}[e^{2} - e^{-4}] = \frac{\pi}{2}[7.3891 - 0.0183] = 3.6854\pi$$

EXAMPLE 16. The current i in an electric circuit is changing at the rate of $0.08e^{-20t}$ ampere/sec, where t represents *time* in seconds. If the current is equal to zero when $t = 0$, find the particular formula for this current at any time t. What is the current at the end of 0.5 sec?

Solution. Since the rate of change of current is $0.08e^{-20t}$, we write

$$\frac{di}{dt} = 0.08e^{-20t} \quad \text{or} \quad di = 0.08e^{-20t} \, dt$$

Now we have a differential equation. Integrating both sides of the equation, we get

$$\int di = \int 0.08e^{-20t} \, dt$$

or

$$i = \frac{0.08}{-20} e^{-20t} + C$$

Simplifying,
$$i = -0.004e^{-20t} + C$$

Substituting values corresponding to initial conditions, $t = 0$, and $i = 0$, we find the value of C:

$$0 = -0.004e^{0} + C \quad \text{or} \quad C = 0.004$$

The particular solution is

$$i = -0.004e^{-20t} + 0.004$$

Factoring and reversing terms,

$$i = 0.004(1 - e^{-20t})$$

This equation now represents the current in this particular circuit at any time t (in seconds).

To find the current at the instant when $t = 0.5$ sec, we insert this value in the formula and get

$$i = 0.004\,(1 - e^{-10}) = 0.004\left(1 - \frac{1}{e^{10}}\right)$$

The quantity, $1/e^{10}$, is extremely small (approximately 0.00005). For all practical purposes, at the end of $\frac{1}{2}$ sec, the current is the steady-state constant, 0.004 ampere, or 4 milliamperes.

EXERCISE 31.2

Integrate the following functions as indicated. Evaluate the definite integral where the limits of integration are shown.

1. $\int e^{5x}\, dx$
2. $\int e^{-30t}\, dt$
3. $\int (x-2)e^{x^2-4x}\, dx$
4. $\int 10^{3x}\, dx$
5. $\int t\, e^{4t^2}\, dt$
6. $\int (e^{ax} + e^{-ax})\, dx$
7. $\int 10^{-5x}\, dx$
8. $\int x^2\, e^{x^3}\, dx$
9. $\int (x-3)e^{6x-x^2}\, dx$
10. $\int \dfrac{dx}{e^{4x}}$
11. $\int \dfrac{t^3\, dt}{e^{t^4}}$
12. $\int \dfrac{e^{\cos 2x}}{\csc 2x}$
13. $\int \dfrac{dt}{e^{3-t}}$
14. $\int \dfrac{e^{\tan x}\, dx}{\cos^2 x}$
15. $\int \dfrac{e^{2/x}\, dx}{x^2}$
16. $\int \dfrac{\cos x\, dx}{e^{\sin x}}$
17. $\int (\sin 2x)(e^{\sin^2 x})\, dx$
18. $\int \sec^2 4x\, e^{\tan 4x}\, dx$
19. $\int_0^2 e^{5x}\, dx$
20. $\int_0^{\sqrt{3}} x\, e^{-x^2}\, dx$
21. $\int_0^{\pi/4} \sin 2x\, e^{\cos 2x}\, dx$
22. $\int_0^2 \dfrac{(e^x - e^{-x})\, dx}{2}$
23. $\int_{-1}^1 \dfrac{(e^x + e^{-x})\, dx}{2}$
24. $\int_0^{\pi/4} \sec^2 x\, e^{\tan x}\, dx$

25. Find the area bounded by the curves, $y = e^{2x}$, $y = 0$, $x = 0$, $x = 2$.
26. Find the area bounded by the curves, $y = e^{-2x}$, $y = 0$, $x = -3$, $x = +3$.
27. Find the area under the curve $y = e^{3x}$ from $x = -3$ to $x = +3$.
28. Find the area under the curve $y = e^{3x}$ from $x = -4$ to $x = -3$.
29. Find the area bounded by the curves, $y = e^x$, $y = e^{-x}$, $y = 0$, $x = -2$, $x = +2$.
30. The current i in a certain electric circuit is changing according to the formula $di/dt = 0.04e^{-10t}$, where t represents *time* in seconds. What is the instantaneous current when $t = 2$ sec?

31.7 IMPROPER INTEGRALS

Up to this point, in evaluating a definite integral between two limits, say, a and b, we have assumed that the limits are constants. We have further assumed that the function to be integrated is continuous on the closed interval $[a,b]$. If the definite integral involves *infinite limits*, or if

the function to be integrated is *not continuous*, then the integral is called an *improper integral*. There are essentially two types of improper integrals:
(1) Type 1 is an integral involving *infinite limits*.
(2) Type 2 is an integral in which the function to be integrated is *discontinuous* at some point on the interval of evaluation.

In problems involving improper integrals, the definite integral may or may not have a meaning. We shall consider first an example of Type 1.

EXAMPLE 17. Find the total area under the curve, $y = 6/x$, from $x = 1$ to $x = \infty$.

Solution. Let us first sketch the curve, indicating the limits (Fig. 31.2). Now we might ask: What is meant by the entire area to an infinite limit? If the area approaches a limit as x becomes infinitely large, then we say this limit is the total area under the curve. On the other hand, the area may continue to increase without limit.

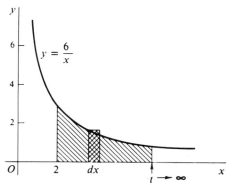

Fig. 31.2

From the graph we see that the curve will never touch the x-axis, for then y would be equal to zero. From the equation we see this is impossible, for then we should have: $0 = 6/x$. Let us try to compute the area in the usual way. Taking a vertical element of area, we have

$$dA = y\, dx, \quad \text{or} \quad dA = \frac{6}{x} dx$$

Then our problem becomes

$$A = \int_{2}^{\infty} \frac{6}{x} dx$$

Now, we know we cannot substitute infinity as a limit in the integral. We cannot operate with infinity as we do with finite quantities. Instead, we make use of the notion of a limit.

For the upper limit we arbitrarily take any letter, say t, and then we let t increase without limit and see what happens. We first write the expression to indicate a limit. We say that the total area from $x = 2$ to $x = \infty$ is the *limit* of the area from 2 to t as t becomes infinite. That is,

$$A = \int_{2}^{\infty} \frac{6}{x} dx = \lim_{t \to \infty} \int_{2}^{t} \frac{6}{x} dx$$

Following through with the integration, we carry the idea of a limit all the way to the end. We get

$$A = \lim_{t \to \infty} \int_2^t \frac{6}{x} dx = \lim_{t \to \infty} [6 \ln x]_2^t = \lim_{t \to \infty} [6 \ln t - 6 \ln 2]$$

At this point we see that as t continues to increase without bound, the logarithm of t will also increase without bound. Therefore, the value of the integral continues to increase and has no limit. The fact that the lower limit results in subtracting the logarithm of 64 will not alter the fact that the integral is infinite. Therefore, the area is infinite and does not approach a finite limit. In this case, the integral is said to *diverge*.

If the limit of the integral is some finite quantity, then this quantity is said to be the value of the definite integral. In the case of the area under a curve, we say then that the area approaches a limit, and this limit is called the total area under the curve. Then the integral is said to *converge* to this limit, as shown in the following example:

EXAMPLE 18. Find the area under the curve, $y = 6/x^2$, from $x = 2$ to $x = \infty$.

Solution. We first sketch the curve in the first quadrant (Fig. 31.3). From the graph we see that the curve does not touch the x-axis. Now we ask: Does the area converge to a limit as x increases, or does it diverge? If it converges to a limit, this limit is called the total area under the curve. Again we take a vertical element of area. Then

$$dA = y \, dx = \frac{6}{x^2} dx$$

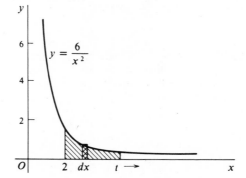

Fig. 31.3

Again we express the total area as the limit of the definite integral, using t as the upper limit and letting t increase without limit. Then

$$A = \int_2^\infty \frac{6}{x^2} dx = \lim_{t \to \infty} \int_2^t \frac{6}{x^2} dx = \lim_{t \to \infty} \left[-\frac{6}{x} \right]_2^t = \lim_{t \to \infty} \left[-\frac{6}{t} + \frac{6}{2} \right]$$

At this point we see that as t continues to increase without bound, the value of the first fraction in the integral approaches zero as a limit. However, the second fraction remains, and the value of the integral is therefore 3. Thus the definite integral converges to the value 3, which can be said to be the total area under the curve from $x = 2$ to $x = \infty$.

31.7 Improper Integrals

EXAMPLE 19. Find the area, if it exists, under the curve, $y = e^x$, from $x = -\infty$ to $x = 0$. The graph is shown in Figure 31.4.

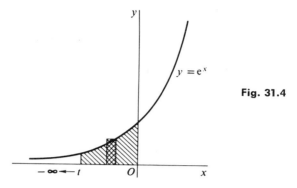

Fig. 31.4

Solution. Again taking a vertical element,

$$dA = y\, dx \quad \text{or} \quad dA = e^x\, dx$$

$$A = \int_{-\infty}^{0} e^x\, dx = \lim_{t \to -\infty} \int_{t}^{0} e^x dx$$

$$= \lim_{t \to -\infty} [e^x]_{t}^{0} = \lim_{t \to -\infty} [e^0 - e^t] = 1$$

As t becomes infinite negatively, e^t becomes smaller and smaller and approaches zero as a limit. Then the definite integral converges to the value indicated by e^0, which is 1. From the graph we see that the curve, $y = e^x$, approaches the x-axis in a negative direction. Although it never touches the x-axis, the area under this portion of the curve does approach a limit, 1, which is called the total area under the curve.

EXAMPLE 20. Find the total area under the following curve from $x = 0$ to $x = 4$ if it exists (Fig. 31.5):

$$y = \frac{1}{\sqrt{x}}$$

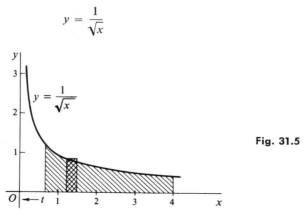

Fig. 31.5

Solution. The graph shows that the curve approaches both the x-axis and the y-axis as asymptotes. Note that at one limit, $x = 0$, the curve is discontinuous.

Integration: Power, Logarithmic, and Exponential Forms

Now we wish to see whether the area approaches a limit as the curve approaches the y-axis. For a vertical element of area, we have
$$dA = y\,dx \quad \text{or} \quad dA = x^{-1/2}$$
Since the curve is discontinuous at $x = 0$, we cannot use zero as a lower limit. However, again we use a letter, say t, to represent the lower limit, and then let t approach zero as a limit from the right. Then we have
$$A = \int_0^4 x^{-1/2}\,dx = \lim_{t \to 0^+} \int_t^4 x^{-1/2}\,dx = \lim_{t \to 0^+} [2x^{1/2}]_t^4 = \lim_{t \to 0^+} \left[4 - 2\sqrt{t}\right]$$
Now, as t moves toward zero, the second term, $2\sqrt{t}$, also approaches zero, and the integral approaches the value 4. Since the integral converges to 4, this value is called the total area under the curve from $x = 0$ to $x = 4$.

In evaluating an integral over an interval, it is important to note whether the curve is continuous throughout the interval. Otherwise the evaluation can lead to wrong results. A function may exist at both limits, yet may be discontinuous in the interval, as shown in the following example.

EXAMPLE 21. Find the total area under the following curve from $x = 0$ to $x = 4$, if it exists (Fig. 31.6):
$$y = \frac{1}{(x - 2)^2}$$

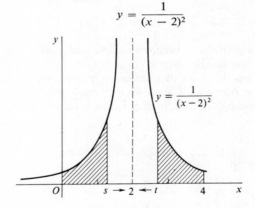

Fig. 31.6

Solution. Note that the function exists at both limits: when $x = 0$, $y = \frac{1}{4}$; and when $x = 4$, $y = \frac{1}{4}$. However, we must take into account the discontinuity at $x = 2$. Again we use the notion of a limit for each portion of the area. For the left portion we take some letter, say s, as the upper limit, and then let s approach 2 from the left. For the right portion, we may use another letter, say t, as the lower limit, and then let t approach 2 from the right. For the total area we take the sum of these two integrals, if they exist. For the element of area for each portion we have
$$dA = (x - 2)^{-2}\,dx$$
$$A = \int_0^4 (x - 2)^{-2}\,dx = \lim_{s \to 2^-} \int_0^s (x - 2)^{-2}\,dx + \lim_{t \to 2^+} \int_t^4 (x - 2)^{-2}\,dx$$
$$= \lim_{s \to 2^-} \left[-\frac{1}{x - 2}\right]_0^s + \lim_{t \to 2^+} \left[-\frac{1}{x - 2}\right]_t^4$$
$$= \lim_{s \to 2^-} \left[\frac{1}{2 - s} - \frac{1}{2}\right] + \lim_{t \to 2^+} \left[-\frac{1}{2} + \frac{1}{t - 2}\right] = ?$$

Since the first integral contains a fraction whose denominator approaches zero as s approaches 2, the fraction increases without limit, and the integral diverges. The second integral diverges, and therefore the area does not exist. What would be the answer obtained if the integral were evaluated directly between the two given limits without observing the discontinuity?

EXERCISE 31.3

Evaluate the following integrals or show that the integral has no meaning. Compare results in Problems 1, 2, and 3; also in Problems 4, 5, and 6.

1. $\int_1^\infty \dfrac{dx}{x^2}$
2. $\int_2^\infty \dfrac{dx}{x^2}$
3. $\int_0^1 \dfrac{dx}{x^2}$

4. $\int_1^\infty \dfrac{dx}{x^3}$
5. $\int_2^\infty \dfrac{dx}{x^3}$
6. $\int_0^1 \dfrac{dx}{x^3}$

Compare Problems 7 and 8: 7. $\int_0^1 \dfrac{dx}{\sqrt{x}}$ 8. $\int_1^\infty \dfrac{dx}{\sqrt{x}}$

Compare Problems 9 and 10: 9. $\int_1^\infty \dfrac{dx}{x^{1/3}}$ 10. $\int_0^1 \dfrac{dx}{x^{1/3}}$

11. $\int_0^\infty xe^{-2x}\, dx$
12. $\int_0^\infty xe^{-4x}\, dx$
13. $\int_0^\infty 2x\, e^{-x^2}\, dx$

14. $\int_0^\infty \dfrac{dx}{x+1}$
15. $\int_2^\infty \dfrac{dx}{(x-1)^2}$
16. $\int_0^\infty \dfrac{dx}{x^2+4}$

17. $\int_0^3 \dfrac{dx}{\sqrt{3-x}}$
18. $\int_1^4 \dfrac{x\, dx}{\sqrt{4-x}}$
19. $\int_0^2 \dfrac{dx}{(2x-1)^{1/3}}$

20. $\int_{-2}^2 \dfrac{x\, dx}{\sqrt{4-x^2}}$
21. $\int_{-1}^1 \dfrac{dx}{\sqrt{x^2-1}}$
22. $\int_0^1 \dfrac{dx}{\sqrt{x-1}}$

23. Find the total area under the curve, $y = e^x$, from $x = -\infty$ to $x = 1$.
24. If the area in the first quadrant under the curve, $y = e^{-x}$, is rotated around the x-axis, find the volume of the solid of revolution generated.
25. The morning-glory problem: If the area under the curve $xy = 6$ from $x = 1$ to $x = +\infty$, is rotated about the x-axis, find (a) the volume of the solid generated and (b) the surface area of this solid. Does this mean that if the solid were hollow in the shape of a morning glory flower and filled with paint, you would still not have enough paint to cover the inside?

chapter
32

Integration of Trigonometric and Inverse Trigonometric Functions

32.1 INTEGRATION AS THE REVERSE OF DIFFERENTIATION

The integration of many trigonometric functions can be performed through a knowledge of the differentiation formulas. Since integration is the reverse of differentiation, some elementary forms can be integrated by recalling certain derivatives. Integration then becomes simply a matter of remembering derivatives.

32.2 INTEGRATION OF COS u

In differentiating trigonometric functions we have seen that

$$\text{if} \quad y = \sin x, \quad \text{then} \quad \frac{dy}{dx} = \cos x \quad \text{and} \quad dy = \cos x \, dx$$

Therefore, we can easily reverse the process and write the integral

$$\int \cos x \, dx = \sin x + C$$

Of course, the angle may be a more complicated function of x, which we may call u. Then the integrand must contain the entire differential du, not merely dx. The derivative of u is du/dx, and the differential of u is

$$du = \left(\frac{du}{dx}\right) dx$$

In this case the formula for the indefinite integral becomes

$$\int \cos u \, du = \sin u + C$$

EXAMPLE 1. Find $\quad \int \cos 5x \, dx$

32.5 Integration of tan u and cot u

Solution. In this example, the angle u is represented by $5x$. We have seen that in differentiating $\sin 5x$, we get the extra factor 5. That is,

if $y = \sin 5x$, then $dy = (\cos 5x)(5)\, dx$

The entire differential du must be present in the integrand before we can perform the integration. In this example we lack the extra factor 5. This factor, being a constant, can be supplied under the integral sign, and its reciprocal, $\frac{1}{5}$, written before the integral sign; thus,

$$\int \cos 5x\, dx = \frac{1}{5} \int 5 \cos 5x\, dx = \frac{1}{5} \sin 5x + C$$

Here we identify: $\cos u = \cos 5x$; $u = 5x$; $du = 5\, dx$. Note that du disappears. However, the coefficient $\frac{1}{5}$ remains in the integral.

32.3 INTEGRATION OF SIN u

From differentiation, we recall that

if $y = \cos u$, then $dy = -\sin u\, du$

Therefore, the reverse process becomes

$$\int -\sin u\, du = \cos u + C$$

However, if the negative sign is lacking in the integrand, we supply the factor, -1, and write its reciprocal outside the integral sign. Therefore,

$$\int \sin u\, du = -\int -\sin u\, du = -\cos u + C$$

32.4 SUMMARY OF FORMULAS FOR INTEGRATION

The first six of the following formulas follow directly by memory from the differentiation formulas; the remaining four will be derived.

1. Since $d(\sin u) = \cos u\, du$, therefore $\int \cos u\, du = \sin u + C$
2. Since $d(\cos u) = -\sin u\, du$, therefore $\int \sin u\, du = -\cos u + C$
3. Since $d(\tan u) = \sec^2 u\, du$, therefore $\int \sec^2 u\, du = \tan u + C$
4. Since $d(\cot u) = -\csc^2 u\, du$, therefore $\int \csc^2 u\, du = -\cot u + C$
5. Since $d(\sec u) = \sec u \tan u\, du$,

   therefore $\int \sec u \tan u\, du = \sec u + C$
6. Since $d(\csc u) = -\csc u \cot u\, du$,

   therefore $\int \csc u \cot u\, du = -\csc u + C$
7. $\int \tan u\, du = -\ln \cos u + C = \ln \sec u + C$
8. $\int \cot u\, du = \ln \sin u + C$
9. $\int \sec u\, du = \ln (\sec u + \tan u) + C$
10. $\int \csc u\, du = -\ln (\csc u + \cot u) + C$

32.5 INTEGRATION OF tan u AND cot u

To find the integral of tan u, we first rewrite tan u by use of a trigonometric identity and get

$$\int \tan u\, du = \int \frac{\sin u}{\cos u}\, du$$

Now we note that the numerator is almost the exact derivative of the denominator. This fact suggests a logarithm. It lacks the factor (-1). The factor (-1) is supplied, with its reciprocal written outside the integral sign. We get

$$-\int \frac{-\sin u}{\cos u}\, du = -\ln \cos u + C$$

This integral may be written

$$\ln (\cos u)^{-1} = \ln \sec u + C$$

The integral of $\cot u\, du$ is derived in a similar manner. The derivation is left to the student.

32.6 INTEGRATION OF sec u AND csc u

In one example in differentiation, we came across the following:

$$y = \ln (\sec u + \tan u)$$

Differentiating, $\quad \dfrac{dy}{du} = \dfrac{\sec u \tan u + \sec^2 u}{\sec u + \tan u}$

Factoring, $\quad \dfrac{dy}{du} = \dfrac{\sec u (\tan u + \sec u)}{\tan u + \sec u}$

Reducing, $\quad \dfrac{dy}{du} = \sec u \quad \text{or} \quad dy = \sec u\, du$

Now let us reverse the process to find

$$\int \sec u\, du$$

If we remember how the differential $\sec u\, du$ came about, we may think of the trick of multiplying the integrand by this equivalent of 1:

$$\frac{\sec u + \tan u}{\sec u + \tan u} = 1$$

We get $\quad \displaystyle\int \sec u\, du = \int \frac{\sec u (\sec u + \tan u)}{\sec u + \tan u}\, du$

$$= \int \frac{\sec^2 u + \tan u \sec u}{\sec u + \tan u}\, du$$

Now the numerator is the exact derivative of the denominator. Therefore,

$$\int \sec u\, du = \ln (\sec u + \tan u) + C$$

The integral of $\csc u\, du$ is derived in a similar manner. The derivation is left to the student.

The ten foregoing formulas for the integration of simple trigonometric functions may be called the *ten elementary forms*. They should be memorized because they are used in further integration of more difficult trigonometric functions. As a student you should know how formulas 1–6 originate and how formulas 7–10 are derived.

EXAMPLE 2. Find $\int x \cos x^2 \, dx$

Solution. We first identify the angle $u = x^2$; then $du = 2x \, dx$. Note that the integrand *does* include the factor $x \, dx$. The extra constant factor, 2, is supplied and its reciprocal written as a coefficient of the integral. Then the differential, $du = 2x \, dx$, disappears in integration, but the coefficient $\frac{1}{2}$ remains. We get

$$\int x \cos x^2 \, dx = \frac{1}{2} \int 2x \cos x^2 \, dx = \frac{1}{2} \sin x^2 + C$$

EXAMPLE 3. Evaluate the integral

$$\int 5 \csc 2x \, dx$$

from $x = \pi/6$ to $x = \pi/4$.

Solution. Our first problem is to get a constant factor 2 instead of 5. The best way to do this is to move the 5 outside the integral first, and then supply the 2, as shown here; then the integration can be completed.

$$5 \int \csc 2x \, dx = \frac{5}{2} \int 2 \csc 2x \, dx = -\frac{5}{2} \ln (\csc 2x + \cot 2x)$$

To evaluate the integral between limits, we first note that the function is continuous on the interval. Then the value of the integral is

$$-\frac{5}{2} [\ln (\csc 90° + \cot 90°) - \ln (\csc 60° + \cot 60°)]$$

$$= -\frac{5}{2} [\ln 1 - \ln \sqrt{3}] = \frac{5}{2} \ln 1.732 = 1.373 \text{ (approx.)}$$

EXERCISE 32.1

Integrate and evaluate as indicated.

1. $\int \sin 7x \, dx$
2. $\int 3x \cos 5x^2 \, dx$
3. $\int \tan (3x + 2) \, dx$
4. $\int \csc 3x \, dx$
5. $\int \sec 4x \, dx$
6. $\int \cot (2 - 5x) \, dx$
7. $\int_0^{\pi/2} \sin \theta \, d\theta$
8. $\int_0^{\pi} \cos \phi \, d\phi$
9. $\int_0^{\pi/4} \tan t \, dt$
10. $\int_0^{\pi/6} \sin 2t \, dt$
11. $\int_{\pi/4}^{\pi/2} \cot \theta \, d\theta$
12. $\int_{-\pi/4}^{\pi/4} \sec t \, dt$
13. $\int \sin \omega t \, dt$
14. $\int \cos 2\pi f t \, dt$
15. $\int \csc u \, du$ (derive)
16. Find the area under the sine curve, $y = \sin x$, from $x = 0$ to $x = \pi$. What is the *average height* of the curve?
17. Find the area under the tangent curve, $y = \tan x$, from $x = 0$ to $x = \pi/6$.
18. Find the area between the tangent curve and the secant curve, $y = \sec x$, between the limits $x = 0$ and $x = \pi/4$.
19. Find the area between the sine curve and the cosine curve between $x = 0$ and their first point of intersection.

20. The instantaneous current in an electric circuit is given by the formula $i = 20 \sin 5t$, where i is current in amperes, and $t =$ time in seconds. Find the maximum current. Find the total transport of charge (coulombs) from $t = 0$ to $t = \pi/10$. Find the average current over the interval.
21. In a certain electric circuit, the current at any instant is given by the formula $i = 10 \sin (2\pi f) t$. Find the total transport of charge during the interval from $t = 0$ to $t = 0.003$ sec. Find the average current over the interval.

32.7 THE POWER RULE IN THE INTEGRATION OF TRIGONOMETRIC FUNCTIONS

Recall the application of the power rule in the integration of algebraic functions:

$$\int (u)^n \, du = \frac{u^{n+1}}{n+1} + C$$

The integrand must contain, not only dx, but the entire factor du. Consider the example

$$\int (x^2 + 3x + 4)^5 \, dx$$

Here, u represents $(x^2 + 3x + 4)$; then $du = (2x + 3) \, dx$. In this example, we do have dx but we do *not* have the factor $(2x + 3)$, which may be called the *integrating factor*. To apply the power rule we must have the complete factor, du. Therefore, we cannot apply the power rule on the foregoing example.

Now, consider the following example:

$$\int (x^2 + 3x + 4)^5 (2x + 3) \, dx$$

In this example, we identify the following: $u^5 = (x^2 + 3x + 4)^5$; then,

$$u = x^2 + 3x + 4 \quad \text{and} \quad du = (2x + 3) \, dx$$

Note that the integrand contains the complete integrating factor, $du = (2x + 3) \, dx$. Integrating, we get

$$\frac{(x^2 + 3x + 4)^6}{6} + C$$

Note that the entire du disappears (or is absorbed into the integral).

Now suppose the u function is some trigonometric function on which we have a power, as in

$$\int \sin^5 x \, dx$$

If u represents $\sin x$, then $du = \cos x \, dx$. Since the integrand does not contain the integrating factor, $\cos x$, we cannot apply the power rule.

However, suppose we have the following problem:

EXAMPLE 4. Find $\int \sin^5 x \cos x \, dx$

Solution. Here the power rule can be applied directly. The entire factor, $du = \cos x \, dx$, disappears in the process and we get

$$\int \sin^5 x \cos x \, dx = \frac{\sin^6 x}{6} + C$$

EXAMPLE 5. Find $\quad \int \tan^3 5x \sec^2 5x \, dx$

Solution. If we take $u = \tan 5x$, then $du = (\sec^2 5x)(5) \, dx$. Hence the integrand contains the complete du except for the factor 5, which can be supplied. We get

$$\frac{1}{5}\int 5 \tan^3 5x \sec^2 5x \, dx = \frac{1}{5}\left(\frac{\tan^4 5x}{4}\right) + C = \frac{1}{20} \tan^4 5x + C$$

32.8 ALTERING THE FORM OF THE INTEGRAND

In some problems involving trigonometric functions, the power rule cannot be applied directly but can be applied if certain changes are made in the form of the integrand. In some cases the change is very simple; in others it is more complicated. Consider the problem

$$\int \tan x \sec^5 x \, dx$$

In this example the power rule cannot be applied on either $\tan x$ or $\sec x$, since we do not have the necessary integrating factor. However, we make a change in the integrand by separating $\sec^5 x$ into $\sec x \sec^4 x$. Then we get

$$\int \tan x \sec^5 x \, dx = \int \tan x \sec x \sec^4 x \, dx$$

We can now apply the power rule on $\sec x$, since we have the integrating factor $(\sec x)(\tan x)$. The result is

$$\int \tan x \sec x \sec^4 x \, dx = \frac{\sec^5 x}{5} + C$$

In general, in a similar manner,

$$\int \tan u \sec^n u \, du = \frac{\sec^n u}{n} + C$$

32.9 ODD POWERS OF SINE AND COSINE

We have pointed out that the power rule cannot be applied on any function u if the integrand lacks the necessary integrating factor du. Such is the case in the problem

$$\int \sin^3 x \, dx$$

In order to apply the power rule on sin x we must have the derivative of sin x present in the integrand. Here we identify

$$u = \sin x$$

and look for
$$du = \cos x \, dx$$

In many instances the integrand can be separated in such a way that the necessary integrating factor can be made to appear. This can always be done in the case of *odd powers of sines or cosines*, whether each of these appears alone or the two as factors in the same integrand. The method is shown by examples.

EXAMPLE 6.   Find $\int \sin^3 x \, dx$

Solution.   We separate $\sin^3 x$ into the factors $(\sin^2 x)(\sin x)$ and get

$$\int \sin^2 x \sin x \, dx$$

Now we make use of the trigonometric identity: $\sin^2 x = 1 - \cos^2 x$, and get

$$\int (1 - \cos^2 x) \sin x \, dx = \int (\sin x - \cos^2 x \sin x) \, dx$$

or, as two integrations,

$$= \int \sin x \, dx - \int \cos^2 x \sin x \, dx$$

The two integrations can now be performed. For the second, we note that the integrating factor is $(-\sin x)$. We transfer the negative sign to the right of the integral sign and then apply the power rule on cos x. We get

$$\int \sin x \, dx + \int (-\sin x) \cos^2 x \, dx = -\cos x + \frac{\cos^3 x}{3} + C$$

The procedure in the foregoing example can be employed in all cases where an odd power of sine or cosine appears in the integrand. For any odd power, such as $\cos^7 x$, we separate the function into factors showing squares. Then it is possible to *replace each second power with its equivalent* by using the trigonometric identity, $\sin^2 x + \cos^2 x = 1$. A *first power* of sine or cosine will always remain as an *integrating factor*.

EXAMPLE 7.   Find $\int \cos^7 x \, dx$

Solution.   Factoring the integrand we get

$$\int \cos^2 x \cos^2 x \cos^2 x \cos x \, dx \quad \text{or} \quad \int (\cos^2 x)^3 \cos x \, dx$$

Now we replace $\cos^2 x$ with $1 - \sin^2 x$, and get

$$\int (1 - \sin^2 x)^3 \cos x \, dx$$

Expanding, then writing as separate integrations, and finally applying the power rule, we get

$$\int (1 - 3 \sin^2 x + 3 \sin^4 x - \sin^6 x) \cos x \, dx$$
$$= \int \cos x \, dx - 3 \int \sin^2 x \cos x \, dx + 3 \int \sin^4 x \cos x \, dx - \int \sin^6 x \cos x \, dx$$
$$= \sin x - \sin^3 x + \frac{3 \sin^5 x}{5} - \frac{\sin^7 x}{7} + C$$

EXAMPLE 8. Find $\int \sin^4 x \cos^3 x \, dx$

Solution. Remember, if the integrand contains an *odd power and an even power*, it is the *odd* power that should be split into factors. If both powers are odd, then either one may be split into factors. In this example, we get

$$\int \sin^4 x \cos^2 x \cos x \, dx = \int \sin^4 x \, (1 - \sin^2 x) \cos x \, dx$$
$$= \int \sin^4 x \cos x \, dx - \int \sin^6 x \cos x \, dx = \frac{\sin^5 x}{5} - \frac{\sin^7 x}{7} + C$$

32.10 EVEN POWERS OF SINE OR COSINE

The integration of an even power of sine or cosine requires a substitution different from the one we have used for odd powers. It will not help to use the identity, $\sin^2 x + \cos^2 x = 1$. If we replace $\sin^2 x$ with the binomial $(1 - \cos^2 x)$, we only get another square, which does not help. Instead, we use another identity which comes from the half-angle formulas. We make use of the following substitutions:

$$\sin^2 x = \frac{1}{2}(1 - \cos 2x) \quad \text{and} \quad \cos^2 x = \frac{1}{2}(1 + \cos 2x)$$

EXAMPLE 9. Find the indefinite integral

$$\int \sin^2 x \, dx$$

Solution.

$$\int \sin^2 x \, dx = \frac{1}{2}\int (1 - \cos 2x) \, dx = \frac{1}{2}\int dx - \frac{1}{2}\int \cos 2x \, dx$$

Now we supply the constant factor 2 for the second integral and get an extra $\frac{1}{2}$ outside the integral sign. Then we have

$$\frac{1}{2}\int dx - \frac{1}{4}\int 2 \cos 2x \, dx = \frac{x}{2} - \frac{1}{4}\sin 2x + C$$

This type of problem occurs in electric circuits in connection with finding the effective value of an alternating current.

32.11 EVEN POWERS OF SECANTS AND COSECANTS

For even powers of secants and cosecants the method is similar to that used for odd powers of sines and cosines. Recall that $\sec^2 x$ is a necessary integrating factor for applying the power rule on $\tan x$. In the same way, $-\csc^2 x$ is a necessary integrating factor for applying the power rule on $\cot x$. Therefore, we split up even powers of secants (or cosecants) into factors containing second powers. Then, *leaving one second power as an integrating factor*, we substitute for the others by use of the identity, $\sec^2 \theta = \tan^2 \theta + 1$.

EXAMPLE 10. Find $\int \sec^6 7t \, dt$

410 Integration of Trigonometric and Inverse Trigonometric Functions

Solution. We separate the integrand into factors containing second powers of sec $7t$, and then *leave one second power as an integrating factor.* Then, for the other second powers we use the identity

$$\sec^2 7t = \tan^2 7t + 1$$

We get

$$\int \sec^6 7t \, dt = \int (\sec^2 7t)^2 \sec^2 7t \, dt = \int (\tan^2 7t + 1)^2 \sec^2 7t \, dt$$
$$= \int (\tan^4 7t + 2 \tan^2 7t + 1) \sec^2 7t \, dt$$
$$= \int \tan^4 7t \sec^2 7t \, dt + 2 \int \tan^2 7t \sec^2 7t \, dt + \int \sec^2 7t \, dt$$

Now we have the necessary integrating factor except for the constant 7, which is supplied, and then the power rule is applied to all but the last integral. The indefinite integral is

$$\frac{1}{7}\left[\frac{\tan^5 7t}{5} + \frac{2 \tan^3 7t}{3} + \frac{\tan 7t}{1}\right] + C$$

or

$$\frac{\tan^5 7t}{35} + \frac{2 \tan^3 7t}{21} + \frac{\tan 7t}{7} + C$$

Note: For the integration of *odd* powers of secant and cosecant, we must wait for the method called *integration by parts*, which will be explained in Chapter 34.

32.12 ANY POWER OF TANGENT OR COTANGENT

In the integration of powers of tangent (or cotangent), the power may be odd or even, and as a result, the integral takes a slightly different form. Whether the power is odd or even, however, we first split the power into factors showing as many *second powers* as possible. If the power is odd there will be one factor of the first power. Then for the other factors (second powers) we make use of the identity, $\tan^2 \theta = \sec^2 \theta - 1$.

EXAMPLE 11. Find $\int \tan^5 x \, dx$

Solution.

$$\int \tan^5 x \, dx = \int \tan^2 x \tan^2 x \tan x \, dx$$
$$= \int (\tan^2 x)^2 \tan x \, dx = \int (\sec^2 x - 1)^2 \tan x \, dx$$
$$= \int (\sec^4 x - 2 \sec^2 x + 1) \tan x \, dx$$
$$= \int \sec^4 x \tan x \, dx - 2 \int \sec^2 x \tan x \, dx + \int \tan x \, dx$$

Now the first two integrations can be performed by the method of Section 32.8. The last integral is one of the elementary forms. Then we get the integral,

$$\frac{\sec^4 x}{4} - \sec^2 x + \ln \sec x + C$$

If the power on tangent (or cotangent) is even, we split the power into factors, showing all second powers. From here on, we have a choice. Sometimes it is simpler to leave one second power factor and substitute $(\sec^2 x - 1)$ for the others.

However, in some cases it may be simpler to substitute for all the factors, $\tan^2 x$. In any event, we usually must substitute *twice*.

EXAMPLE 12. Find $\quad \int \tan^4 t \, dt$

Solution.

$$\int \tan^4 t \, dt = \int \tan^2 t \tan^2 t \, dt = \int (\sec^2 t - 1) \tan^2 t \, dt$$
$$= \int \sec^2 t \tan^2 t \, dt - \int \tan^2 t \, dt$$

Now again we substitute $(\sec^2 t - 1)$ for the $\tan^2 t$ in the second integrand. We get

$$\int \sec^2 t \tan^2 t \, dt - \int (\sec^2 t - 1) \, dt$$
$$= \int \sec^2 t \tan^2 t \, dt - \int \sec^2 t \, dt + \int dt$$
$$= \frac{\tan^3 t}{3} - \tan t + t + C$$

32.13 A PRODUCT OF POWERS OF TANGENT AND SECANT (OR COTANGENT AND COSECANT)

If the integrand consists of a product of powers of tangent and secant, there are four possibilities. Suppose we have the general product

$$\tan^m x \sec^n x \quad (\text{or} \quad \cot^m x \csc^n x)$$

The four possibilities are:

(1) m and n, both odd 　　(2) m and n, both even
(3) m odd, n even 　　　　(4) m even, n odd

In the first three cases the integration can be performed by methods we have already shown. The fourth case must wait until we have taken up integration by the method of *parts* (Chapter 34).

(1) If the power on tangent and the power on secant are both odd, then we split the powers of tangent into as many second-power factors as possible and replace each such factor with $(\sec^2 x - 1)$.

EXAMPLE 13. Find $\quad \int \tan^3 x \sec^3 x \, dx$

Solution.

$$\int \tan^3 x \sec^3 x \, dx = \int \tan x \tan^2 x \sec^3 x \, dx$$
$$= \int \tan x (\sec^2 x - 1) \sec^3 x \, dx = \int (\tan x \sec^5 x - \tan x \sec^3 x) \, dx$$
$$= \int \sec^5 x \tan x \, dx - \int \sec^3 x \tan x \, dx = \frac{\sec^5 x}{5} - \frac{\sec^3 x}{3} + C$$

(2) If m and n are both even, we split up $\sec^n x$ into factors of second powers of secant and replace each, except one factor, with $(\tan^2 x + 1)$.

EXAMPLE 14. Find $\quad \int \tan^6 u \sec^4 u \, du$

Solution.

$$\int \tan^6 u \sec^4 u \, du = \int \tan^6 u \sec^2 u \sec^2 u \, du$$
$$= \int \tan^6 u (\tan^2 u + 1) \sec^2 u \, du = \int (\tan^8 u \sec^2 u + \tan^6 u \sec^2 u) \, du$$
$$= \int \tan^8 u \sec^2 u \, du + \int \tan^6 u \sec^2 u \, du = \frac{\tan^9 u}{9} + \frac{\tan^7 u}{7} + C$$

(3) Suppose m is odd and n is even, as in

$$\int \tan^3 x \sec^4 x \, dx$$

Then either power may be split up into factors of second powers and the same type of substitutions made. The results will turn out to appear different but they are essentially the same.

(4) As we have said, if m is even and n is odd, then we must wait for Chapter 34, integration by the method of parts.

EXERCISE 32.2

Perform the following indicated integrations. (Evaluate as indicated.)

1. $\int \cos^3 5t \, dt$
2. $\int \sin^5 3x \, dx$
3. $\int \sin^3 4t \cos^6 4t \, dt$
4. $\int \cos^2 3x \, dx$
5. $\int \sin^4 x \, dx$
6. $\int \sin^5 \frac{x}{2} \cos^4 \frac{x}{2} \, dx$
7. $\int \cos^4 \omega t \, dt$
8. $\int \sin^2 2\pi f t \, dt$
9. $\int \sin^2 x \cos^2 x \, dx$
10. $\int \sec^4 3x \, dx$
11. $\int \csc^6 2t \, dt$
12. $\int \tan^3 \theta \sec^4 \theta \, d\theta$
13. $\int \tan^2 \frac{t}{3} \, dt$
14. $\int \cot^3 2\phi \, d\phi$
15. $\int \csc^4 \omega t \cot^2 \omega t \, dt$
16. $\int \tan^5 10t \, dt$
17. $\int \cot^4 3t \, dt$
18. $\int \tan^3 2\phi \sec^3 2\phi \, d\phi$
19. $\int \tan^6 5t \, dt$
20. $\int \csc^4 4x \, dx$
21. $\int \cot^5 3x \csc^6 3x \, dx$
22. $\int \sin^3 2x \, d(2x)$
23. $\int \cos^2 \omega t \, d(\omega t)$
24. $\int \sin^5 3t \cos 3t \, d(3t)$
25. $\int_0^\pi \sin^3 x \, dx$
26. $\int_0^\pi \tan^2 \frac{x}{3} \, dx$
27. $\int_0^{\pi/2} \sin^3 x \cos^2 x \, dx$
28. $\int_{\pi/4}^{\pi/2} \csc^4 x \, dx$
29. $\int_{\pi/2}^\pi \cos^2 t \, dt$
30. $\int_0^{\pi/4} \tan^3 x \sec x \, dx$

31. Find the total area under the curve, $y = \sin^2 x \cos^3 x \, dx$ from 0 to π. Sketch the graph of the curve. Find the area under the curve $y = \sin^3 x \cos^2 x$, from 0 to π. Sketch the curve. Explain the results of the integration for the area under each curve and explain the difference.

32. What is the average value of the ordinate of the curve, $y = \sin^2 x$ from $x = 0$ to $x = \pi$?

32.14 INTEGRATION LEADING TO INVERSE FUNCTIONS

Although the expression *inverse function* refers also to functions other than the trigonometric functions, in the present connection we mean the term *inverse* to refer specifically to the inverse trigonometric functions. We have seen that the derivative of an inverse function, such as arcsine, arctangent, and so on, is algebraic. Note especially the derivatives of the following inverse functions:

$$\frac{d}{dx}(\arcsin x) = \frac{1}{\sqrt{1-x^2}}$$

$$\frac{d}{dx}(\arctan x) = \frac{1}{x^2+1} \quad \left(\text{or } \frac{1}{1+x^2}\right)$$

$$\frac{d}{dx}(\text{arcsec } x) = \frac{1}{x\sqrt{x^2-1}}$$

All of the foregoing derivatives are algebraic fractions. Therefore, we can expect that the integration of some algebraic fractions will lead to inverse functions.

In order to be able to integrate fractions of this kind it is necessary to recognize the form of the fraction for each inverse function as distinguished from other types of algebraic fractions. We shall concentrate our attention on the arcsine and the arctangent.

32.15 INTEGRATION LEADING TO ARCSINE

Since
$$\frac{d}{dx}(\arcsin x) = \frac{1}{\sqrt{1-x^2}}$$

then we can reverse the procedure and get at once

$$\int \frac{1}{\sqrt{1-x^2}} dx = \arcsin x + C$$

In the same way, for any other variable, u,

if $y = \arcsin u$, then $\dfrac{dy}{du} = \dfrac{1}{\sqrt{1-u^2}}$ and $dy = \dfrac{du}{\sqrt{1-u^2}}$

Now suppose u represents some function of x, say $4x$; then u^2 is the square of the function, or $16x^2$, and the numerator will contain du, the differential of u. This is seen in differentiating the inverse function:

if $y = \arcsin 4x$, then $\dfrac{dy}{dx} = \dfrac{4}{\sqrt{1-16x^2}}$ and $dy = \dfrac{4\, dx}{\sqrt{1-16x^2}}$

Note especially the differential, $4\, dx$, in the numerator. Remember, the numerator is not the differential of $16x^2$, but of $4x$, the square root of $16x^2$. This point must be carefully noted in all integration leading to inverse functions.

If u represents some function of x, and if we have the complete du in the numerator of such a fraction, then we can write the integral at once by the following formula:

$$\int \frac{du}{\sqrt{1-u^2}} = \arcsin u + C$$

Of course, if the numerator lacks only a constant factor, this factor can be supplied in the usual manner.

EXAMPLE 15. Find $\int \dfrac{3\, dx}{\sqrt{1 - 9x^2}}$

Solution. To see whether this is the proper form for an arcsine, we identify:
$$u^2 = 9x^2, \quad u = 3x, \quad \text{and} \quad du = 3\, dx$$

Note that the numerator is the differential of $3x$, not of $9x^2$. Since we have the complete du in the numerator, we can say at once that

$$\int \frac{3\, dx}{\sqrt{1 - 9x^2}} = \arcsin 3x + C$$

We pause here to point out the difference between two examples in which the integrands look very much alike:

$$\text{(a)} \int \frac{3\, dx}{\sqrt{1 - 9x^2}} \qquad \text{(b)} \int \frac{3x\, dx}{\sqrt{1 - 9x^2}}$$

We have seen that the first example (a) leads to an arcsine. In the second example (b), the numerator is essentially the differential of $9x^2$ except for a constant factor. It is *not* the differential of $3x$. In (b) we transfer the denominator to the numerator and use the power rule:

$$\int 3x(1 - 9x^2)^{-1/2}\, dx$$

Now all that is needed in order to apply the power rule is the constant factor, -6, which can be supplied. Then we get the necessary integrating factor, $-18x$. By the power rule we get

$$-\frac{1}{6}\int -18x(1 - 9x^2)^{-1/2}\, dx = -\frac{\frac{1}{6}(1 - 9x^2)^{1/2}}{\frac{1}{2}} = -\frac{1}{3}(1 - 9x^2)^{1/2} + C$$

EXAMPLE 16. Find $\int \dfrac{dx}{\sqrt{1 - 16x^2}}$

Solution. Here we identify:
$$u^2 = 16x^2 \ (\text{not } -16x^2), \quad u = 4x, \quad \text{and} \quad du = 4\, dx$$

The numerator should contain $4\, dx$, the complete du. The constant factor 4 can be supplied as usual and the result is

$$\frac{1}{4}\int \frac{4\, dx}{\sqrt{1 - 16x^2}} = \frac{1}{4} \arcsin 4x + C$$

Note that the coefficient, $\frac{1}{4}$, of the arcsine $4x$ came about through supplying the necessary constant factor for the complete du.

In the two examples shown so far, the constant term in the radicand is 1. Now let us see what happens in the case of a constant term other than 1 in the radicand. For example, take the problem

$$\int \frac{3\, dx}{\sqrt{4 - 9x^2}}$$

32.15 Integration Leading to Arcsine

Let us represent this type of integrand in general terms by

$$\int \frac{du}{\sqrt{a^2 - u^2}}$$

In this form,

a^2 represents the square of a constant
a then represents the constant itself
u^2 represents the square of a function of x
u then represents the function itself
du represents the differential of u, not of u^2

Then, in whatever way we work out the integration, we shall get

$$\int \frac{du}{\sqrt{a^2 - u^2}} = \arcsin \frac{u}{a} + C$$

The foregoing five elements should always be identified in any integrand that has the appearance of leading to an arcsine. Remember that the numerator must contain the complete du. Now we apply this formula to the problem we have already mentioned.

EXAMPLE 17. Find $\displaystyle\int \frac{3\,dx}{\sqrt{4 - 9x^2}}$

Solution. Here we identify the following

$$a^2 = 4 \qquad u^2 = 9x^2$$
$$a = 2 \qquad u = 3x \qquad du = 3\,dx$$

Since we have the complete du in the numerator, we apply the formula directly and get

$$\int \frac{3\,dx}{\sqrt{4 - 9x^2}} = \arcsin \frac{3x}{2} + C$$

Let us work out the same example by first putting it into the form in which the constant term in the radicand is 1, to see whether we get the same answer. We begin with the given problem

$$\int \frac{3\,dx}{\sqrt{4 - 9x^2}}$$

If we factor the 4 out of the radicand, we get 2 outside the radical and we have

$$\frac{3\,dx}{2\sqrt{1 - \frac{9x^2}{4}}} \quad \text{or} \quad \frac{(3/2)\,dx}{\sqrt{1 - \frac{9x^2}{4}}}$$

Now we identify $\quad u^2 = \dfrac{9x^2}{4} \quad u = \dfrac{3x}{2} \quad du = \dfrac{3}{2}dx$

The numerator is now the perfect differential du, and the integration is completed:

$$\int \frac{(3/2)\,dx}{\sqrt{1 - (9x^2)/4}} = \arcsin \frac{3x}{2} + C$$

EXAMPLE 18. Find $\int \dfrac{dx}{\sqrt{16 - 25x^2}}$

Solution. We identify

$$a^2 = 16 \quad u^2 = 25x^2$$
$$a = 4 \quad u = 5x \quad du = 5\, dx$$

The numerator lacks the constant factor 5, which can be supplied. Then

$$\frac{1}{5}\int \frac{5\, dx}{\sqrt{16 - 25x^2}} = \frac{1}{5} \arcsin \frac{5x}{4} + C$$

Note that the coefficient $\tfrac{1}{5}$ is *not part of the formula* but came about only through supplying the constant factor 5.

EXAMPLE 19. Find $\int \dfrac{2\, dx}{\sqrt{5 - 3x^2}}$

Solution. In this problem we run into radicals. We identify

$$a^2 = 5 \quad u^2 = 3x^2$$
$$a = \sqrt{5} \quad u = \sqrt{3}\, x \quad du = \sqrt{3}\, dx$$

The constant factor $\sqrt{3}$ can be supplied in the numerator. The most convenient way is first to transfer the factor 2 to the left of the integral sign:

$$\frac{2}{\sqrt{3}} \int \frac{\sqrt{3}\, dx}{\sqrt{5 - 3x^2}} = \frac{2}{\sqrt{3}} \arcsin \frac{\sqrt{3}\, x}{\sqrt{5}} + C$$

If we would rather rationalize the denominators, we can do so; in most instances, the most useful form is a decimal form:

$$1.155 \arcsin 0.775x + C$$

EXAMPLE 20. Evaluate $\int_1^2 \dfrac{3\, dx}{\sqrt{25 - 4x^2}}$

Solution. We identify

$$a^2 = 25 \quad u^2 = 4x^2$$
$$a = 5 \quad u = 2x \quad du = 2\, dx$$

To complete the differential du in the numerator, we move the factor 3 outside the integral sign and supply the constant factor 2. Then we get

$$\frac{3}{2} \int_1^2 \frac{2\, dx}{\sqrt{25 - 4x^2}} = \frac{3}{2} \arcsin \frac{2x}{5} \bigg]_1^2$$
$$= \frac{3}{2}\left[\arcsin \frac{4}{5} - \arcsin \frac{2}{5}\right] = \frac{3}{2}[0.9273 - 0.4116] = 0.774 \text{ (approx.)}$$

Up to this point we have seen how some algebraic functions lead to arcsines. However, the same principle we have used here can be applied to some fractions that already involve trigonometric functions.

EXAMPLE 21. Evaluate from $x = 30°$ to $x = 60°$:

$$\int \frac{\sin x \, dx}{\sqrt{9 - 4\cos^2 x}}$$

Solution. We identify

$$a^2 = 9 \qquad u^2 = 4\cos^2 x$$
$$a = 3 \qquad u = 2\cos x \qquad du = -2\sin x \, dx$$

Supplying the constant factor -2, we get

$$-\frac{1}{2}\int \frac{-2\sin x \, dx}{\sqrt{9 - 4\cos^2 x}} = -\frac{1}{2}\left[\arcsin\left(\frac{2\cos x}{3}\right)\right]$$

Evaluating, we have

$$-\frac{1}{2}\left[\arcsin\left(\frac{2}{3}\cdot\frac{1}{2}\right) - \arcsin\left(\frac{2}{3}\cdot\frac{\sqrt{3}}{2}\right)\right] = +\frac{1}{2}\left[\arcsin\frac{\sqrt{3}}{3} - \arcsin\frac{1}{3}\right]$$
$$= \frac{1}{2}[0.61552 - 0.33976] = \frac{1}{2}(0.27576) = 0.1379 \text{ (approx.)}$$

32.16 INTEGRATION LEADING TO ARCTANGENT

We have seen that $\quad \dfrac{d}{dx}(\arctan x) = \dfrac{1}{x^2 + 1}$

Then we can reverse the procedure and get immediately

$$\int \frac{1}{x^2 + 1} \, dx = \arctan x + C$$

In the same way, for any other variable u,

if $y = \arctan u$, then $\dfrac{dy}{du} = \dfrac{1}{u^2 + 1}$ and $dy = \dfrac{du}{u^2 + 1}$

Now suppose u represents some function of x, say $5x$. Then u^2 is the square of this function, or $25x^2$. Then the numerator will contain du, the perfect differential of u, for whatever function is represented by u. This is seen in differentiating the arctangent function:

if $y = \arctan 5x$, then $\dfrac{dy}{dx} = \dfrac{5}{25x^2 + 1}$ and $dy = \dfrac{5 \, dx}{25x^2 + 1}$

Note especially that the differential, $5 \, dx$, in the numerator is the differential not of $25x^2$, but of $5x$, the square root of $25x^2$.

Now suppose we have an integrand consisting of a fraction similar to the one shown. The denominator contains the square of some function of x and a second term, the constant 1. Then if we let u represent the function of x, and if the numerator contains du, the perfect differential of u, then we can write the integral immediately by the following formula:

$$\int \frac{du}{u^2 + 1} = \arctan u + C$$

If the numerator lacks only a constant factor to make the perfect differential, this factor can be supplied in the usual manner.

EXAMPLE 22. Find $$\int \frac{4\,dx}{16x^2 + 1}$$

Solution. To see whether the integrand is the proper form for an arctangent, we identify the following:

$$u^2 = 16x^2, \quad u = 4x, \quad \text{and} \quad du = 4\,dx$$

Note that the numerator is the differential of $4x$, *not* of $16x^2$. Now that we have the complete du in the numerator, we integrate directly by the formula and get

$$\int \frac{4\,dx}{16x^2 + 1} = \arctan 4x + C$$

It might be well at this point to call attention to the difference between two examples in which the integrands look very much alike:

$$\text{(a)} \int \frac{4\,dx}{16x^2 + 1} \qquad \text{(b)} \int \frac{4x\,dx}{16x^2 + 1}$$

We have seen that the first example (a) leads to an arctangent. In the second example (b), the numerator is essentially the differential of the denominator except for a constant factor. It is not the differential of $4x$. We have seen that when the numerator is the differential of the denominator, then the integral is a logarithm. All we need to do in the second example is to supply the constant factor, 8, making the numerator $32x\,dx$, which is the perfect differential of the denominator. Then we have

$$\frac{1}{8}\int \frac{32x\,dx}{16x^2 + 1} = \frac{1}{8}\ln(16x^2 + 1) + C$$

EXAMPLE 23. Find $$\int \frac{dx}{9x^2 + 1}$$

Solution. Here we identify:

$$u^2 = 9x^2, \quad u = 3x, \quad \text{and} \quad du = 3\,dx$$

For the proper form for an arctangent, the numerator should contain the factor $3\,dx$. The constant factor, 3, can be supplied in the usual way. Then we have

$$\frac{1}{3}\int \frac{3\,dx}{9x^2 + 1} = \frac{1}{3}\arctan 3x + C$$

Note especially that the coefficient, $\frac{1}{3}$, of $\arctan 3x$ came about through supplying the necessary constant factor for the complete du.

In the two examples we have just worked out, in which the integrals are arctangents, note especially the constant term, 1, in the denominator of the integrand. Now let us see what happens if this constant is something other than 1. For example, suppose we have the problem

$$\int \frac{4\,dx}{16x^2 + 9}$$

32.16 Integration Leading to Arctangent

Let us represent this type of integrand in general terms by

$$\int \frac{du}{u^2 + a^2}$$

In this form,

a^2 represents the square of some constant
a represents the constant itself
u^2 represents the square of some function of x
u then represents the function itself
du represents the differential of u, not of u^2

When, however, we work out the integration, we obtain the formula

$$\int \frac{du}{u^2 + a^2} = \frac{1}{a} \arctan \frac{u}{a} + C$$

The coefficient, $1/a$, of the arctangent, is a *part of the formula*. It does not come about through supplying a constant factor for the numerator. In the formula for the arcsine, this coefficient does not appear. Now let us apply the formula to the problem we have already mentioned.

EXAMPLE 24. Find

$$\int \frac{4 \, dx}{16x^2 + 9}$$

Solution. Here we identify

$$a^2 = 9 \qquad u^2 = 16x^2$$
$$a = 3 \qquad u = 4x \qquad du = 4 \, dx$$

Since we have the complete du in the numerator, we apply the formula and get at once

$$\int \frac{4 \, dx}{16x^2 + 9} = \frac{1}{3} \arctan \frac{4x}{3} + C$$

The answer has the coefficient $\frac{1}{3}$, which is equivalent to $1/a$ in the formula. The fraction, $4x/3$, is equivalent to u/a.

EXAMPLE 25. Find the integral and evaluate

$$\int_0^5 \frac{dx}{9x^2 + 25}$$

Solution. We identify

$$a^2 = 25 \qquad u^2 = 9x^2$$
$$a = 5 \qquad u = 3x \qquad du = 3 \, dx$$

If we had the perfect differential, $3 \, dx$, in the numerator, we should get at once the integral: $\frac{1}{5} \arctan (3x/5)$. However, the constant factor 3 must be supplied, and the coefficient $\frac{1}{3}$ outside the integral sign:

$$\frac{1}{3} \int_0^5 \frac{3 \, dx}{9x^2 + 25} = \frac{1}{3} \left(\frac{1}{5} \arctan \frac{3x}{5} \right) \Big]_0^5 = \frac{1}{15} \arctan \frac{3x}{5} \Big]_0^5$$

Note that the coefficient $\frac{1}{3}$, which came about through supplying the constant factor 3 for the perfect du, remains as a coefficient of the result. In addition, there is the extra coefficient $\frac{1}{5}$, which is *part of the formula*, and is equivalent to the coefficient $1/a$. Evaluating, we get

$$\frac{1}{15}[\arctan 3 - \arctan 0] = \frac{1}{15}[1.25 - 0] = 0.083$$

EXERCISES 32.3

Find the indefinite integral of each of the following. Evaluate as indicated by limits.

1. $\int \dfrac{x\, dx}{x^2 + 25}$
2. $\int \dfrac{dx}{x^2 + 25}$
3. $\int \dfrac{x\, dx}{\sqrt{16 - x^2}}$

4. $\int \dfrac{dx}{\sqrt{4 - 25x^2}}$
5. $\int \dfrac{5x\, dx}{4x^2 + 9}$
6. $\int \dfrac{5\, dx}{4x^2 + 9}$

7. $\int \dfrac{e^{5x}\, dx}{e^{5x} + 4}$
8. $\int \dfrac{e^{5x}\, dx}{e^{10x} + 4}$
9. $\int \dfrac{e^{3x}\, dx}{\sqrt{1 - e^{3x}}}$

10. $\int \dfrac{e^{4x}\, dx}{\sqrt{1 - e^{8x}}}$
11. $\int \dfrac{\cos x\, dx}{\sqrt{4 - 9\sin^2 x}}$
12. $\int \dfrac{\csc^2 x\, dx}{\cot^2 x + 9}$

13. $\int_0^1 \dfrac{dx}{9x^2 + 4}$
14. $\int_0^2 \dfrac{dx}{\sqrt{25 - 4x^2}}$
15. $\int_0^2 \dfrac{dx}{\sqrt{9 - 2x^2}}$

16. $\int_0^3 \dfrac{dx}{x^2 + 9}$
17. $\int_0^4 \dfrac{dx}{3x^2 + 16}$
18. $\int_{-1}^1 \dfrac{dx}{\sqrt{4 - x^2}}$

19. $\int \dfrac{5\, dx}{\sqrt{4 - 9x^2}}$ First put this into the form: $\int \dfrac{du}{\sqrt{1 - u^2}}$

20. $\int \dfrac{5\, dx}{9x^2 + 16}$ First put this into the form: $\int \dfrac{du}{u^2 + 1}$

21. Find the area in the first quadrant under the curve, $y^2(9 - x^2) = 16$.

32.17 MORE COMPLICATED FRACTIONS

In some instances the fraction in the integrand is such that the integration leads to a combination of integrals, as in the following example:

EXAMPLE 26. Find $\displaystyle\int \dfrac{5x - 3}{x^2 + 4}\, dx$

Solution. Since the power of x in the numerator is 1 less than the power in the denominator, then at least a portion of the integral is a logarithm. For a logarithm as the integral, the numerator of the integrand should contain the term $2x$, the derivative of the denominator. Then we split up the numerator and write each portion over the denominator. Then the integrand is written as two fractions instead of one:

$$\int \frac{5x\, dx}{x^2 + 4} - \int \frac{3\, dx}{x^2 + 4}$$

32.17 More Complicated Fractions

Now we integrate each fraction and get the answer

$$\frac{5}{2} \ln (x^2 + 4) - \frac{3}{2} \arctan \frac{x}{2} + C$$

Note that a *logarithm* appears along with an *arctangent*. These two functions often appear together in the integral of fractions.

The same procedure is used in the following example:

EXAMPLE 27. Find $\displaystyle\int \frac{3x - 7}{\sqrt{9 - 16x^2}} \, dx$

Solution. We shall see that the integration of this fraction leads to a combination of a power and an arcsine. If the numerator contained only a constant, then the integration would lead to an arcsine. We have seen, however, that with an *x*-term in the numerator of such a fraction, the integral is a power. We therefore separate the integrand into two fractions:

$$\int \frac{3x \, dx}{\sqrt{9 - 16x^2}} - \int \frac{7 \, dx}{\sqrt{9 - 16x^2}}$$

$$= \int 3x \, (9 - 16x^2)^{-1/2} \, dx - \frac{7}{4} \int \frac{4 \, dx}{\sqrt{9 - 16x^2}}$$

$$= -\frac{3}{16} (9 - 16x^2)^{1/2} - \frac{7}{4} \arcsin \frac{4x}{3} + C$$

Note in this case that the integral consists of a *power* and an *arcsine*. These two functions often appear together in the same integral.

In some problems in integration leading to inverse functions, the elements u^2 and u are somewhat more complicated functions of x, as in the next example.

EXAMPLE 28. Find $\displaystyle\int \frac{5 \, dx}{(x + 3)^2 + 16}$

Solution. Here we have the square of a function of x in the denominator together with a constant term, and no *x*-term in the numerator. The fraction has all the earmarks of leading to an arctangent. Then we identify

$$a^2 = 16 \qquad u^2 = (x + 3)^2$$
$$a = 4 \qquad u = x + 3 \qquad du = dx$$

Then we get $\displaystyle\int \frac{5 \, dx}{(x + 3)^2 + 16} = \frac{5}{4} \arctan \frac{x + 3}{4} + C$

This problem might have seemed more complicated if the denominator had originally been written in the expanded form. In that case, it is first necessary to complete the square in x, using as much of the constant term as necessary for the square. This procedure is shown next.

EXAMPLE 29. Find $\displaystyle\int \frac{dx}{x^2 - 10x + 29}$

Solution. The denominator is first rearranged to show a perfect square which shall be equivalent to u^2 in the formula. We write

$$\int \frac{dx}{x^2 - 10x + 25 + 4} = \int \frac{dx}{(x-5)^2 + 4} = \frac{1}{2} \arctan \frac{x-5}{2} + C$$

Completing a square is sometimes a little more difficult, as shown in the next example.

EXAMPLE 30. Find $\displaystyle\int \frac{dx}{3x^2 - 7x + 6}$

Solution. First we factor a 3 out of the denominator so that the coefficient of x^2 is 1. This factor is placed outside the integral sign:

$$\frac{1}{3}\int \frac{dx}{x^2 - (7/3)x + 2}$$

To complete the square in x, we need the square of one half of the coefficient of x; that is, 49/36. We take the 49/36 out of the constant 2, which leaves 23/36 for the constant term. Then we write

$$\frac{1}{3}\int \frac{dx}{x^2 - \frac{7}{3}x + \frac{49}{36} + \frac{23}{36}} = \frac{1}{3}\int \frac{dx}{\left(x - \frac{7}{6}\right)^2 + \frac{23}{36}}$$

Now we identify $\quad a^2 = \dfrac{23}{36} \quad u^2 = \left(x - \dfrac{7}{6}\right)^2$

$$a = \frac{\sqrt{23}}{6} \quad u = x - \frac{7}{6} \quad du = dx$$

The integral is $\quad \dfrac{1}{3}\left(\dfrac{6}{\sqrt{23}}\right) \arctan \dfrac{x - (7/6)}{\sqrt{23}/6} + C$

or $\quad \dfrac{2}{\sqrt{23}} \arctan \dfrac{6x - 7}{\sqrt{23}} + C$

EXAMPLE 31. Solve $\displaystyle\int \frac{7x + 3}{3x^2 + 5x + 8}\,dx$

Solution. The x-term in the numerator and the x^2 term in the denominator indicate a logarithm for at least a portion of the integral. In that case the numerator should be $(6x + 5)$, the derivative of the denominator. We have the numerator $(7x + 3)$ and we need $(6x + 5)$. Our problem now is how to get the quantity $(6x + 5)$ into the numerator.

First, let us factor a 7 out of the numerator and get

$$7\int \frac{(x + 3/7)}{3x^2 + 5x + 8}\,dx$$

(We do not factor a 3 out of the denominator as long as we are trying to establish the form for a logarithm.)

Since we should have $(6x + 5)$ in the numerator, we call it $6(x + \frac{5}{6})$. The constant factor 6 can be supplied later provided we have $(x + \frac{5}{6})$. To change

32.17 More Complicated Fractions

the given numerator, $x + \frac{3}{7}$, into $x + \frac{5}{6}$, we simply add and then subtract the fraction $\frac{5}{6}$, writing the numerator, $x + \frac{3}{7}$, as

$$x + \frac{5}{6} - \frac{5}{6} + \frac{3}{7}$$

Now we take the portion, $x + \frac{5}{6}$ by itself and say this quantity constitutes the numerator of one fraction of the integrand. Then we combine the remaining portion of the numerator for the other fraction: we get

$$7\int \frac{x + \frac{5}{6} - \frac{5}{6} + \frac{3}{7}}{3x^2 + 5x + 8} dx = 7\int \frac{x + \frac{5}{6} - \frac{17}{42}}{3x^2 + 5x + 8} dx$$

$$= 7\int \frac{(x + 5/6)\, dx}{3x^2 + 5x + 8} + 7\int \frac{(-17/42)\, dx}{3x^2 + 5x + 8}$$

$$= \frac{7}{6}\int \frac{(6x + 5)\, dx}{3x^2 + 5x + 8} - \frac{17}{6}\int \frac{dx}{3x^2 + 5x + 8}$$

Now let us work with the two fractions separately. The first fraction integrates into a logarithm:

$$\frac{7}{6} \ln(3x^2 + 5x + 8)$$

The second fraction leads to an arctangent. Now we are ready to factor a 3 out of the denominator and complete the square in x. For the second fraction, we get

$$-\frac{17}{6}\int \frac{dx}{3x^2 + 5x + 8} = -\frac{17}{18}\int \frac{dx}{x^2 + \frac{5}{3}x + \frac{8}{3}}$$

$$= -\frac{17}{18}\int \frac{dx}{x^2 + \frac{5}{3}x + \frac{25}{36} + \frac{71}{36}} = -\frac{17}{18}\int \frac{dx}{\left(x + \frac{5}{6}\right)^2 + \frac{71}{36}}$$

Now we identify $\quad a^2 = \frac{71}{36} \quad u^2 = \left(x + \frac{5}{6}\right)^2$

$$a = \frac{\sqrt{71}}{6} \quad u = x + \frac{5}{6} \quad du = dx$$

Then we get the integral,

$$\left(-\frac{17}{18}\right)\left(\frac{6}{\sqrt{71}}\right) \arctan \frac{x + 5/6}{\sqrt{71}/6} = -\frac{17}{3\sqrt{71}} \arctan \frac{6x + 5}{\sqrt{71}}$$

The complete integral is

$$\frac{7}{6}\ln(3x^2 + 5x + 8) - \frac{17}{3\sqrt{71}} \arctan \frac{6x + 5}{\sqrt{71}} + C$$

A problem in integration leading to an arcsine can be even more difficult if the quantity identified as u^2 under the radical is the square of a binomial or other polynomial. Let us consider two examples of this kind. If the u^2 quantity is already expressed as the square of a binomial, then the problem is less difficult, as in the following example:

EXAMPLE 32. Find the indefinite integral

$$\int \frac{dx}{\sqrt{25 - (3x - 2)^2}}$$

Solution. First we recall the formula for the integration leading to an arcsine:

$$\int \frac{du}{\sqrt{a^2 - u^2}} = \arcsin \frac{u}{a} + C$$

In this example we identify the following:

$$a^2 = 25 \quad u^2 = (3x - 2)^2$$
$$a = 5 \quad u = 3x - 2 \quad du = 3\,dx$$

Then, supplying the constant 3, we have at once

$$\frac{1}{3}\int \frac{3\,dx}{\sqrt{25 - (3x - 2)^2}} = \frac{1}{3} \arcsin \frac{3x - 2}{5} + C$$

Note that in the foregoing example the quantity u^2 is already expressed in the form of the square of a binomial, $(3x - 2)^2$. Then the problem is comparatively easy. However, let us suppose the original integrand contains the expanded form in which the radical is written

$$\sqrt{21 + 12x - 9x^2}$$

This is the result of simplifying the radicand. Now, if we have the problem in this expanded form, we must try to get it back into the standard form for the arcsine.

Our first clue is the fact that we have the negative term, $-9x^2$, under the radical sign. The following example shows the difficulty in this kind of problem.

EXAMPLE 33. Find the definite integral:

$$\int_1^4 \frac{3\,dx}{\sqrt{11 + 20x - 4x^2}}$$

Solution. We could possibly get the radicand into the proper form for an arcsine by inspection or by guessing. But there is a more direct way. We use a trick that is a favorite with mathematicians. If they do not know the answer to a problem, they assume they already have it.

Let us assume we know the radicand. It is a quantity that contains the square of a constant, a^2, then *minus* the square of a binomial containing an x term, which we call $(bx + c)^2$. Then we know the radicand is

$$a^2 - (bx + c)^2$$

Now all we need to do is to find a, b, and c. So we say this quantity must be equal to the given radicand in the problem. That is,

$$a^2 - (bx + c)^2 = 11 + 20x - 4x^2$$

or
$$a^2 - b^2x^2 - 2bcx - c^2 = 11 + 20x - 4x^2$$

If one side of the equation is to be exactly equal to the other side, then the coefficients of like powers of x must be equal to each other. The constant term on the left side consists of the quantity $a^2 - c^2$. This must equal the constant 11 on the right side; then

$$a^2 - c^2 = 11 \quad \text{or} \quad a^2 = 11 + c^2$$

The coefficient of x on the left is $(-2bc)$. The coefficient of x on the right is 20. Therefore,
$$-2bc = 20 \quad \text{or} \quad bc = -10$$
The coefficients of x^2 are also equated and we get
$$-b^2 = -4 \quad \text{or} \quad b = \pm 2$$
If we take the b as $+2$, we have
$$c = \frac{-10}{2} = -5$$

Since $a^2 = 11 + c^2$, we get $a^2 = 36$, or $a = 6$.

Now that we know the values of a, b, and c, we can write the radicand in the proper form for an arcsine:

$$\int \frac{3 \, dx}{\sqrt{36 - (2x - 5)^2}}$$

Now we identify the following:

$$a^2 = 36 \quad u^2 = (2x - 5)^2$$
$$a = 6 \quad u = 2x - 5 \quad du = 2 \, dx$$

Then we have

$$\frac{3}{2} \int_1^4 \frac{2 \, dx}{\sqrt{36 - (2x - 5)^2}} = \frac{3}{2} \arcsin \frac{2x - 5}{6} \Big]_1^4 = \frac{3}{2} \left[\arcsin \frac{1}{2} - \arcsin \left(-\frac{1}{2} \right) \right]$$
$$= \frac{3}{2} \left[\frac{\pi}{6} - \left(-\frac{\pi}{6} \right) \right] = \frac{\pi}{2} = 1.5708 \text{ (approx.)}$$

Note that for $\arcsin(-\frac{1}{2})$ we take the principal value, $-\pi/2$.

EXERCISE 32.4

Integrate the following and evaluate as indicated by limits.

1. $\int \frac{(3x - 7) \, dx}{x^2 + 16}$

2. $\int \frac{(5x - 3) \, dx}{\sqrt{4 - x^2}}$

3. $\int \frac{(3x + 5) \, dx}{9x^2 + 16}$

4. $\int \frac{(3x - 2) \, dx}{\sqrt{5 - 2x^2}}$

5. $\int \frac{(5x + 3) \, dx}{3x^2 + 5}$

6. $\int \frac{(5x - 2) \, dx}{9x^2 + 25}$

7. $\int \frac{3 \, dx}{x^2 - 10x + 29}$

8. $\int \frac{2x \, dx}{x^2 + 4x + 40}$

9. $\int \frac{(x - 5) \, dx}{x^2 - 4x + 16}$

10. $\int \frac{(x + 5) \, dx}{x^2 - 3x + 3}$

11. $\int \frac{(x + 3) \, dx}{\sqrt{25 - 16x^2}}$

12. $\int \frac{(3x - 7) \, dx}{\sqrt{4 - 25x^2}}$

13. $\int \frac{dx}{x^2 + 6x + 9}$

14. $\int \frac{dx}{4x^2 + 12x + 25}$

15. $\int \frac{(x - 3) \, dx}{\sqrt{7 - 6x + x^2}}$

16. $\int_0^1 \frac{(5x - 3) \, dx}{9x^2 + 16}$

17. $\int_0^2 \frac{(4x - 3) \, dx}{\sqrt{16 - 3x^2}}$

18. $\int_0^1 \frac{(3x + 2) \, dx}{5x^2 + 4}$

19. $\int_0^1 \dfrac{(4x-3)\,dx}{\sqrt{9-5x^2}}$
20. $\int_1^5 \dfrac{dx}{x^2-6x+13}$
21. $\int_3^5 \dfrac{(x-3)\,dx}{x^2-4x+7}$
22. $\int_0^2 \dfrac{(5x-3)\,dx}{x^2+2x+10}$
23. $\int_0^4 \dfrac{(x-3)\,dx}{x^2-10x+25}$
24. $\int_{-1}^0 \dfrac{(5x-3)\,dx}{4x^2+12x+13}$
25. $\int \dfrac{3\,dx}{2x^2-6x+11}$
26. $\int \dfrac{(2-5x)\,dx}{3x^2+4x+4}$
27. $\int \dfrac{2x\,dx}{3x^2-5x+7}$
28. $\int \dfrac{(3x+5)\,dx}{\sqrt{5+4x-x^2}}$
29. $\int \dfrac{(4-3x)\,dx}{\sqrt{16-6x-x^2}}$
30. $\int \dfrac{x\,dx}{\sqrt{20x-4x^2-21}}$

chapter

33

Partial Fractions

33.1 RATIONAL FRACTIONS

A rational fraction is a fraction in which the numerator and the denominator are both rational functions of x; that is, neither numerator nor denominator may contain a quantity irrational in x. An irrational quantity in x is one in which x appears in a radicand. The following are illustrations of rational fractions:

$$\frac{x}{x^2 + 9} \qquad \frac{3x - 5}{x^2 + 16} \qquad \frac{5x + 14}{x^2 - 4x - 12} \qquad \frac{3x(2x - 5)^2}{x^2(x - 3)(x^2 + 1)}$$

The following is not a rational fraction since the denominator contains a quantity with x in a radicand. The radical is irrational in x.

$$\frac{x}{\sqrt{9 - x^2}}$$

If $F(x)$ and $G(x)$ represent two rational functions of x, then a rational fraction can be represented by the fraction $F(x)/G(x)$.

33.2 INTEGRATION OF FRACTIONS

Some rational fractions can be integrated by methods already mentioned. For example, the following fractions are types of those we have already integrated by the application of elementary formulas. The first four are rational fractions. The fifth is irrational but can still be integrated as shown.

(a) $\displaystyle\int \frac{x\,dx}{x^2 + 9} = \frac{1}{2} \ln(x^2 + 3) + C$

(b) $\displaystyle\int \frac{dx}{x^2 + 9} = \frac{1}{3} \arctan \frac{x}{3} + C$

(c) $\displaystyle\int \frac{(x^2 + 4x + 1)\,dx}{x^3 + 6x^2 + 3x + 4} = \frac{1}{3} \ln(x^3 + 6x^2 + 3x + 4) + C$

(d) $\int \frac{(3x - 5)\,dx}{x^2 + 16} = \frac{3}{2}\ln(x^2 + 16) - \frac{5}{4}\arctan\frac{x}{4} + C$

(e) $\int \frac{x\,dx}{\sqrt{9 - x^2}} = -(9 - x^2)^{1/2} + C$

Now, it often happens that a rational fraction cannot be integrated directly in its given form. For example, the following problem cannot be worked out by any of the methods used so far:

$$\int \frac{(3x + 14)\,dx}{x^2 + 7x + 12}$$

Then other methods can be attempted. One technique is the use of *partial fractions*.

33.3 PARTIAL FRACTIONS: DEFINITION

In some problems involving fractions, a rational fraction can be separated into two or more *partial fractions*. Partial fractions are the separate fractions that have been combined by addition to form the given fraction. For example, consider again the following problem:

$$\int \frac{(3x + 14)\,dx}{x^2 + 7x + 12}$$

We cannot integrate this fraction in its present form. However, the fraction can be separated into two distinct fractions. In fact, this fraction came about through the addition of the following fractions:

$$\frac{5}{x + 3} + \frac{-2}{x + 4}$$

The given fraction cannot be integrated as it stands, but the two separate fractions can easily be integrated by known methods. Then the problem

$$\int \frac{(3x + 14)\,dx}{x^2 + 7x + 12}$$

is the same as

$$\int \frac{5\,dx}{x + 3} - \int \frac{2\,dx}{x + 4} = 5\ln(x + 3) - 2\ln(x + 4) + \ln C$$
$$= \ln \frac{C(x + 3)^5}{(x + 4)^2}$$

Our first problem, then, is simply this: Given a rational fraction that cannot be integrated in its present form, can it be separated into two or more partial fractions that can be integrated as separate terms? If so, how can we find the partial fractions that have been combined to form the given fraction? This is not so difficult as it might seem.

33.4 FINDING THE PARTIAL FRACTIONS

In order better to understand the problem of determining the partial fractions, let us first consider a problem in arithmetic. Suppose we add the following fractions:

$$\frac{3}{5} + \frac{2}{7}$$

Since neither denominator can be factored, then the lowest common denominator is their product, (5)(7), or 35. Now each fraction is changed to 35ths and then added:

$$\frac{3}{5} + \frac{2}{7} = \frac{21}{35} + \frac{10}{35} = \frac{31}{35}$$

Now, suppose we know that the sum of two fractions is $\frac{31}{35}$ but do not know the original fractions. Then our problem is to find two fractions whose sum is $\frac{31}{35}$. In other words, we try to work back to the original fractions.

In arithmetic this is not always simple. However, in algebra, it is not so difficult. Even in arithmetic, if the sum of two fractions is given as $\frac{31}{35}$, we can logically assume that the original fractions probably had denominators of 5 and 7. As another example, if the sum of two fractions is $\frac{17}{21}$, then it can be assumed that the original fractions had the denominators, 3 and 7, respectively. This is the theory that forms the basis of partial fractions. From the given denominator, we determine all the possible denominators of the two or more original fractions.

In order to see how this theory works out, let us return to the problem

$$\int \frac{(3x + 14)\,dx}{x^2 + 7x + 12}$$

Since we cannot perform the integration of the fraction, we attempt to find two or more fractions that may have been combined to form the given fraction.

We begin by factoring the given denominator and get $(x + 3)(x + 4)$. (While determining the partial fractions, we omit the integral sign.) Now we assume that the partial fractions must have had the two denominators, respectively, $(x + 3)$ and $(x + 4)$. That is the only way in which the lowest common denominator could be $(x + 3)(x + 4)$.

In this example, then, we can say that the two partial fractions have the following forms with denominators as shown and with the numerators still unknown:

$$\frac{?}{x + 3} + \frac{?}{x + 4}$$

Now if we knew the numerators, we could integrate the separate fractions.

At this point we do something that is often done in mathematics. *We assume the problem is already done.* Let us assume that we have the two partial fractions. We represent the numerators by some letters, such as A and B, respectively. (Any letters can be used, since these are only "dummy" letters whose numerical values will be determined.) Then the sum of these two partial fractions can be said to be equal to the given fraction; that is,

$$\frac{A}{x+3} + \frac{B}{x+4} = \frac{3x+14}{x^2+7x+12}$$

The numerators are as yet undetermined. Now all we need to do is to find their numerical values. Then we can integrate the two partial fractions.

There are several methods of finding the values of the numerators. One method that can always be used is the so-called *classical* method, which we shall call Method 1. We show this method first, and then later briefly show the other methods that can often be used to advantage.

Method 1. Equating coefficients of like powers. In this method, we add the two fractions, using the undetermined numerators, A and B. We get the sum of the fractions:

$$\frac{A}{x+3} + \frac{B}{x+4} = \frac{A(x+4) + B(x+3)}{(x+3)(x+4)} = \frac{Ax + 4A + Bx + 3B}{(x+3)(x+4)}$$

We have said that the two fractions we have set up represent the two partial fractions. Therefore their sum is equal to the original fraction with which we began; that is,

$$\frac{Ax + 4A + Bx + 3B}{(x+3)(x+4)} = \frac{3x+14}{(x+3)(x+4)}$$

Since these two fractions are identically equal for all values of x, and since the denominators are equal, the numerators also must be equal to each other. This means that the coefficient of x on one side of the equation must be equal to the coefficient of x on the other side. Also, the constant term on one side must equal the constant on the other side.

On the left side, the coefficient of x is the quantity, $A + B$. On the right side the coefficient of x is 3. Therefore, $A + B$ must equal 3. On the left side the constant term is the quantity, $4A + 3B$. On the right side the constant term is 14. Therefore, $4A + 3B$ must equal 14. We now have the two equations to solve as a system:

coefficient of x, $\quad A + B = 3$
constant term, $\quad 4A + 3B = 14$

Solving this system, we get $A = 5$ and $B = -2$.

We have now determined the numerical values of the numerators, A and B, respectively, in the partial fractions. We insert these values for A and B in the fractions we set up, and get

$$\frac{5}{x+3} + \frac{-2}{x+4}$$

We have already integrated these fractions.

33.4 Finding the Partial Fractions

EXAMPLE 1. Perform the following integration by first finding partial fractions:

$$\int \frac{(29 - 8x)\, dx}{x^2 - 5x - 14}$$

Solution. We omit the integral sign while finding the partial fractions. We first write the fraction with factored denominator:

$$\frac{29 - 8x}{(x + 2)(x - 7)}$$

Now we set up the form of the partial fractions, each with one of the factors as its denominator; we use A and B for the numerators.

$$\frac{A}{x + 2} + \frac{B}{x - 7}$$

Combining the fractions,

$$\frac{A}{x + 2} + \frac{B}{x - 7} = \frac{Ax - 7A + Bx + 2B}{(x + 2)(x - 7)}$$

Since this fraction is identically equal to the original fraction, the numerators must be equal to each other. Equating coefficients, we have

$$\text{coefficients of } x, \qquad A + B = -8$$
$$\text{constant terms,} \qquad -7A + 2B = 29$$

Solving this system of two equations, we get $A = -5$, and $B = -3$. Therefore, the original fraction can be written

$$\frac{29 - 8x}{x^2 - 5x - 14} = \frac{-5}{x + 2} - \frac{3}{x - 7}$$

Now our problem in integration becomes

$$\int \left(\frac{-5}{x + 2} + \frac{-3}{x - 7} \right) dx = -5 \ln (x + 2) - 3 \ln (x - 7) - \ln C$$
$$= -[\ln (x + 2)^5 + \ln (x - 7)^3 + \ln C] = -\ln C(x + 2)^5 (x - 7)^3$$

Note. The procedure of equating numerators of two equal fractions, and then equating coefficients of *like powers* of x in the two fractions depends on an important principle. *If two polynomials are identically equal for all values of the variable x, then the coefficients of corresponding powers of x must be respectively equal.* For example,

$$\text{if} \quad Ax^3 + Bx^2 + Cx + D \equiv 5x^3 - 3x^2 + 7x - 8,$$
$$\text{then} \quad A = 5, B = -3, C = 7, \text{ and } D = -8$$

This is the principle we use in finding the values of undetermined numerators.

EXAMPLE 2. Integrate by first finding the partial fractions:

$$\int \frac{(-2x^2 + 11x + 31)\, dx}{x^3 + 2x^2 - 5x - 6}$$

432 Partial Fractions

Solution. By synthetic division we discover the denominator can be factored and the fraction written in the form

$$\frac{-2x^2 + 11x + 31}{(x + 1)(x - 2)(x + 3)}$$

Now we set up the form of the partial fractions, each with one of the factors as its denominator; we use A, B, and C for the numerators. The C used here is not the constant of integration, but is simply one of the numerator constants whose value is yet undetermined. After all, the A, B, and C are only symbols to represent the numerators temporarily and will be replaced with their numerical values.

As to the order of the denominators, it makes no difference which denominator is written first or which numerator is called A, B, or C. Whatever the order, or whatever the numerators are called, any fraction with a particular denominator will eventually get its proper numerator with the correct numerical value.

We set up the fractions

$$\frac{A}{x + 1} + \frac{B}{x - 2} + \frac{C}{x + 3}$$

Adding the fractions with the numerators as shown, we get

$$\frac{A(x - 2)(x + 3) + B(x + 1)(x + 3) + C(x + 1)(x - 2)}{(x + 1)(x - 2)(x + 3)}$$

Since we are now interested in only the numerator, we omit the denominator and write the numerator

$$A(x^2 + x - 6) + B(x^2 + 4x + 3) + C(x^2 - x - 2)$$
$$= Ax^2 + Ax - 6A + Bx^2 + 4Bx + 3B + Cx^2 - Cx - 2C = \text{numerator}$$

This numerator must now be equal to the numerator of the original fraction since we have equated the fractions. This means that the corresponding coefficients of like powers of x must be equal; that is,

$$\text{coefficients of } x^2, \quad A + B + C = -2$$
$$\text{coefficients of } x, \quad A + 4B - C = 11$$
$$\text{constant terms,} \quad -6A + 3B - 2C = 31$$

Solving this system, we get $A = -3$; $B = 3$; $C = -2$.

Now that we have discovered the values of the numerators of the fractions, our problem becomes

$$\int \left(\frac{-3}{x + 1} + \frac{3}{x - 2} + \frac{-2}{x + 3} \right) dx$$
$$= -3 \ln (x + 1) + 3 \ln (x - 2) - 2 \ln (x + 3) + \ln C$$
$$= \ln \frac{C(x - 2)^3}{(x + 1)^3 (x + 3)^2}$$

EXERCISES 33.1

Integrate each of the following by first separating into partial fractions. In some cases long division is first necessary.

1. $\int \frac{(3x + 19) \, dx}{x^2 + x - 6}$

2. $\int \frac{(2x + 17) \, dx}{x^2 - 3x - 4}$

3. $\int \dfrac{(2x + 3)\,dx}{x^2 + 6x + 8}$

4. $\int \dfrac{(2x + 4)\,dx}{x^2 - 6x + 5}$

5. $\int \dfrac{(-2x + 9)\,dx}{x^2 + 3x}$

6. $\int \dfrac{(3x - 20)\,dx}{x^2 - 5x}$

7. $\int \dfrac{(3x + 7)\,dx}{x^2 - 1}$

8. $\int \dfrac{(9x + 7)\,dx}{x^2 + x - 6}$

9. $\int \dfrac{(x - 5)\,dx}{x^2 - 9}$

10. $\int \dfrac{(x + 14)\,dx}{x^2 - 36}$

11. $\int \dfrac{1}{x^2 - 64}\,dx$

12. $\int \dfrac{1}{x^2 - 7x}\,dx$

13. $\int \dfrac{dx}{x^2 - 3}$

14. $\int \dfrac{dx}{x^2 + 5x}$

15. $\int \dfrac{(2x^2 + 9x + 30)\,dx}{x^3 + 3x^2 - 10x}$

16. $\int \dfrac{(-x^2 - 8x - 24)\,dx}{x^3 - 2x^2 - 8x}$

17. $\int \dfrac{(x^2 - 11x - 18)\,dx}{x^3 + x^2 - 12x}$

18. $\int \dfrac{(3x^2 - 12x - 18)\,dx}{x^3 - 9x}$

19. $\int \dfrac{(x^2 + 25x - 36)\,dx}{x^3 + 2x^2 - 24x}$

20. $\int \dfrac{(3x^2 + 2x - 6)\,dx}{x^3 + 6x^2 + 8x}$

21. $\int \dfrac{(3x^2 + 23x + 12)\,dx}{x^3 + 6x^2 - x - 30}$

22. $\int \dfrac{(2x^2 + x + 5)\,dx}{x^3 - 7x + 6}$

23. $\int \dfrac{(11x + 13)\,dx}{x^3 + 2x^2 - x - 2}$

24. $\int \dfrac{6x\,dx}{x^3 - x^2 - 4x + 4}$

25. $\int \dfrac{(3x - 9)\,dx}{2x^2 - x - 10}$

26. $\int \dfrac{(6x + 11)\,dx}{3x^2 - 10x - 8}$

27. $\int \dfrac{(x + 7)\,dx}{2x^2 - 7x + 3}$

28. $\int \dfrac{x^2\,dx}{x^2 - 2x - 8}$

29. $\int \dfrac{(x^2 - 3x + 5)\,dx}{x^2 - 4}$

30. $\int \dfrac{x^3\,dx}{x^2 - x - 12}$

33.5 ALTERNATE METHODS

The method we have shown for finding the numerical values of the undetermined numerators can always be used. However, sometimes other methods are simpler. We shall briefly describe two other methods.

Method 2. Substituting values. This method consists essentially of clearing the equation of fractions and then substituting various values for x on both sides of the resulting identity. For example, take the fraction:

$$\dfrac{3x - 26}{(x - 4)(x + 3)}$$

Again we set up the partial fractions and equate to the given fraction:

$$\dfrac{A}{x - 4} + \dfrac{B}{x + 3} = \dfrac{3x - 26}{(x - 4)(x + 3)}$$

Clearing of fractions,

$$A(x + 3) + B(x - 4) = 3x - 26$$

Now, since this is an identity, it is true for all values of x. We could substitute any value for x, and the equation would be true. However, we choose a value that will make one factor equal to zero. Setting $x = -3$, we get

$$0 + B(-7) = -35 \quad \text{or} \quad B = 5$$

Setting $x = 4$, we get

$$7A + 0 = -14; \quad \text{or} \quad A = -2$$

Method 2 works equally well in the case of three or more linear factors. The following example has a denominator of three linear factors:

EXAMPLE 3. Integrate by first finding partial fractions:

$$\int \frac{(6x^2 - 25x - 24) \, dx}{x(x - 3)(x + 2)}$$

Solution. We set up the form of the partial fractions and equate to the given fraction:

$$\frac{A}{x} + \frac{B}{x - 3} + \frac{C}{x + 2} = \frac{6x^2 - 25x - 24}{x(x - 3)(x + 2)}$$

Clearing the equation of fractions,

$$A(x - 3)(x + 2) + Bx(x + 2) + Cx(x - 3) = 6x^2 - 25x - 24$$

Now we substitute values of x that will make some terms equal to zero. It happens that the values we choose are those obtained by setting each denominator equal to zero. Then if $x = 0$,

$$-6A + 0 + 0 = -24 \quad \text{or} \quad A = 4$$

if $x = 3$, $\quad 0 + 15B + 0 = -45 \quad \text{or} \quad B = -3$

if $x = -2$, $\quad 0 + 0 + 10C = 50 \quad \text{or} \quad C = 5$

Then the problem in integration becomes

$$\int \left(\frac{4}{x} + \frac{-3}{x - 3} + \frac{5}{x + 2} \right) dx = 4 \ln x - 3 \ln (x - 3) + 5 \ln (x + 2) + C$$

Method 3. A variation of Method 2. To show this method, we use the same examples used for Method 2. For the example,

$$\frac{3x - 26}{(x - 4)(x + 3)}$$

we begin in the same way as usual by setting up the equation

$$\frac{A}{x - 4} + \frac{B}{x + 3} = \frac{3x - 26}{(x - 4)(x + 3)}$$

Now, instead of multiplying both sides of the equation by the entire denominator, we multiply by *only the denominator under A*, and get

$$A + \frac{(x-4)B}{x+3} = \frac{3x-26}{x+3}$$

This is an identity and is therefore true for all values of x. Let us first substitute the value $x = 4$. This value came about through letting the first denominator, $x - 4$, approach zero as a limit. When $x = 4$, the second fraction becomes zero. We indicate the substituted values of x as we would a limit. Then

$$A = \frac{3x-26}{x+3}\bigg]_{x=4} = \frac{-14}{7} = -2$$

Now, going back to the first equation, we multiply both sides by the *second denominator*, and get

$$\frac{A(x+3)}{x-4} + B = \frac{3x-26}{x-4}$$

If we now substitute the value $x = -3$ (which came from $x + 3 = 0$), the first fraction becomes zero, and we have

$$B = \frac{3x-26}{x-4}\bigg]_{-3} = \frac{-35}{-7} = +5$$

Then we have the same values for the numerators as before.

As another example, take the following fraction which we have already used in Example 3:

$$\frac{A}{x} + \frac{B}{x-3} + \frac{C}{x+2} = \frac{6x^2-25x-24}{x(x-3)(x+2)}$$

Multiplying by each denominator separately in turn, and substituting the corresponding values of x found by letting each denominator approach zero, we get

$$A = \frac{6x^2-25x-24}{(x-3)(x+2)}\bigg]_0 = \frac{-24}{-6} = 4$$

$$B = \frac{6x^2-25x-24}{x(x+2)}\bigg]_3 = \frac{54-75-24}{(3)(5)} = \frac{-45}{15} = -3$$

$$C = \frac{6x^2-25x-24}{x(x-3)}\bigg]_{-2} = \frac{24+50-24}{(-2)(-5)} = \frac{50}{10} = 5$$

EXERCISE 33.2

Integrate as indicated by first finding the partial fractions. Use Method 2 or Method 3 as directed by the instructor.

1. $\int \frac{(x-8)\,dx}{x^2-6x+8}$

2. $\int \frac{(x+9)\,dx}{x(x-3)}$

Partial Fractions

3. $\int \dfrac{(5 - 3x)\,dx}{x(x + 2)}$

4. $\int \dfrac{(11 - x)\,dx}{x^2 + 3x - 4}$

5. $\int \dfrac{(3x + 4)\,dx}{x^2 + x - 12}$

6. $\int \dfrac{(2x + 2)\,dx}{x^2 + 7x + 10}$

7. $\int \dfrac{(11 - 7x)\,dx}{x^2 - 4x - 5}$

8. $\int \dfrac{(5 - 3x)\,dx}{2x^2 + 5x - 3}$

9. $\int \dfrac{(4x - 3)\,dx}{3x^2 - 7x + 2}$

10. $\int \dfrac{(5x - 2)\,dx}{2x^2 + x - 6}$

11. $\int \dfrac{(2x - 7)\,dx}{3x^2 - 7x - 6}$

12. $\int \dfrac{(3x - 4)\,dx}{6x^2 + 5x - 6}$

13. $\int \dfrac{(3x^2 - 25x - 12)\,dx}{x^3 + x^2 - 6x}$

14. $\int \dfrac{(6 - 11x)\,dx}{x^3 + 3x^2 - 4x}$

15. $\int \dfrac{(2x^2 - 23x + 6)\,dx}{x^3 - x^2 - 6x}$

16. $\int \dfrac{6\,dx}{4x^3 + 4x^2 - 3x}$

17. $\int \dfrac{(x^2 - 10x + 13)\,dx}{x^3 - 2x^2 - x + 2}$

18. $\int \dfrac{(2x^2 + 7)\,dx}{x^3 - 2x^2 - 11x + 6}$

19. $\int \dfrac{(x^3 - 6x + 12)\,dx}{x^4 - 3x^3 - 4x^2 + 12x}$

20. $\int \dfrac{(3x^3 - 9x^2 - 9x - 2)\,dx}{x^4 - 5x^2 + 4}$

33.6 REPEATED FACTORS IN DENOMINATOR

If a denominator as factored contains two or more equal factors, such a factor is said to be *repeated*. In this case, the various fractions that are set up must contain denominators that involve *all* powers of the repeated factor. For example, if the denominator contains the factor $(x + 3)^2$, in which the factor $(x + 3)$ is repeated, then the fractions set up must show one denominator of $(x + 3)$ and another denominator of $(x + 3)^2$. If there are three such factors, the denominators of the fractions set up must show $(x + 3)$, $(x + 3)^2$, and $(x + 3)^3$.

In general, if the denominator of the given fraction contains a factor of the form $(x + a)^n$, then the separate fractions set up must show the several denominators $(x + a)$, $(x + a)^2$, $(x + a)^3$, and so on to the final form $(x + a)^n$. The numerators can then be represented by constants such as A, B, C, and so on.

Note. We pause here to note an important fact. If a denominator contains any number of repeated factors as indicated by powers such as x^2, x^3, x^4, and so on, these factors are all *linear factors*. For example, although the term x^2 is itself a quadratic term, nevertheless the separate factors of x^2 are *linear*. Whenever the fractions as set up have denominators that contain only linear factors, then only constants will appear in the numerators. Of course, it may happen that some constant numerators turn out to be zero when their numerical values have been determined. Yet the fractions set up must

33.6 Repeated Factors in Denominator

provide for *all powers* of repeated factors in the original denominator. The following example must make use of this principle.

EXAMPLE 4. Integrate

$$\int \frac{(7x^2 - 18x + 9)\,dx}{x^3 - 3x^2}$$

Solution. The integral sign is omitted until we have found the partial fractions. The fraction is first written with a factored denominator. Then the forms of all possible fractions are set up.

$$\frac{7x^2 - 18x + 9}{x^3 - 3x^2} = \frac{7x^2 - 18x + 9}{x^2(x-3)} = \frac{A}{x} + \frac{B}{x^2} + \frac{C}{x-3}$$

It is necessary to include the two fractions having the denominators of x and x^2, respectively, because it is entirely possible that there may be two such partial fractions.

Now we add the fractions whose numerators are A, B, and C. The lowest common denominator is still the quantity $x^2(x-3)$, the same as in the given fraction. Adding the fractions we get

$$\frac{A}{x} + \frac{B}{x^2} + \frac{C}{x-3} = \frac{Ax(x-3) + B(x-3) + Cx^2}{x^2(x-3)}$$

$$= \frac{Ax^2 - 3Ax + Bx - 3B + Cx^2}{x^2(x-3)}$$

The final fraction is the sum of the three fractions and is therefore equal to the given fraction; that is,

$$\frac{Ax^2 - 3Ax + Bx - 3B + Cx^2}{x^2(x-3)} = \frac{7x^2 - 18x + 9}{x^2(x-3)}$$

Since the two denominators are equal to each other, and the two fractions are identically equal for all values of x, then *the numerators must also be equal to each other.* Therefore, the coefficients of like powers of x must be equal.

The coefficients of the various powers of x and the constant terms of the two numerators can now be equated as shown here:

coefficients of x^2, $A + C = 7$
coefficients of x, $-3A + B = -18$
constant terms, $-3B = 9$

Solving this system in three unknowns, we get: $A = 5$; $B = -3$; $C = 2$. These numerical values are now inserted in place of A, B, and C, respectively, in the numerators of the fractions that were set up. The problem in integration then becomes

$$\int \frac{7x^2 - 18x + 9}{x^3 - 3x^2}\,dx = \int \left(\frac{5}{x} - \frac{3}{x^2} + \frac{2}{x-3}\right) dx$$

$$= 5 \ln x + \frac{3}{x} + 2 \ln(x-3) + \ln C$$

$$= \frac{3}{x} + \ln Cx^5(x-3)^2$$

Note. The partial fractions can always be checked by addition. Their sum should be equal to the given fraction.

Method 3 may be used only partially in the case of repeated factors. Again, let us take the example (4) that we have used:

$$\frac{A}{x} + \frac{B}{x^2} + \frac{C}{x-3} = \frac{7x^2 - 18x + 9}{x^2(x-3)}$$

In this example we use Method 3 to find B and C, but not A. This method can be used for the highest power of a repeated factor, but not for lower powers. To find B, we multiply both sides by x^2 and get

$$B = \frac{7x^2 - 18x + 9}{x-3}\bigg]_0 = \frac{9}{-3} = -3$$

To find C we multiply the original equation by $(x - 3)$, and get

$$C = \frac{7x^2 - 18x + 9}{x^2}\bigg]_3 = \frac{63 - 54 + 9}{9} = 2$$

We cannot find A in this manner. Instead, we clear of fractions and get

$$Ax(x-3) + B(x-3) + Cx^2 = 7x^2 - 18x + 9$$

Now we let x equal any convenient value except zero or 3 (which we have already used). Suppose we let $x = 1$. Then

$$A(-2) + B(-2) + C = 7 - 18 + 9$$

We already know that $B = -3$, and $C = 2$. Substituting these values, we get

$$-2A + 6 + 2 = 7 - 18 + 9 \quad \text{or} \quad A = 5$$

EXERCISE 33.3

Integrate by first separating into partial fractions:

1. $\int \frac{(2x^2 + x + 6)\,dx}{x^3 - 2x^2}$

2. $\int \frac{(x^2 + 11x - 4)\,dx}{x^3 + 4x^2}$

3. $\int \frac{(3x^2 - 10x - 12)\,dx}{x^3 + 3x^2}$

4. $\int \frac{(x^2 + 11x - 20)\,dx}{x^3 + 5x^2}$

5. $\int \frac{(x^2 + 2x - 18)\,dx}{x^3 - 6x^2 + 9x}$

6. $\int \frac{(x^2 + 3x + 6)\,dx}{x^3 + 4x^2 + 4x}$

7. $\int \frac{(x^2 + 8x - 15)\,dx}{x^3 - 3x + 2}$

8. $\int \frac{(x^2 - 22x - 43)\,dx}{x^3 + x^2 - 8x - 12}$

9. $\int \frac{(2x^2 - 5x - 9)\,dx}{x^3 - 6x^2 + 32}$

10. $\int \frac{(2x^3 - 2x^2 + 4x - 8)\,dx}{x^4 - 4x^3 + 4x^2}$

11. $\int \frac{(3x^3 - 7x^2 + 5x + 2)\,dx}{x^4 - 2x^3}$

12. $\int \frac{(x^3 + 8x^2 - x + 6)\,dx}{x^4 + 3x^3}$

13. $\displaystyle\int \frac{(4x^2 - 5x^3 + 5x - 10)\,dx}{x^4 - 2x^3}$

14. $\displaystyle\int \frac{(x^3 - 5x^2 - 12x - 8)\,dx}{x^4 + 4x^3 + 4x^2}$

15. $\displaystyle\int \frac{(3x^3 + 11x^2 - 27)\,dx}{x^4 + 6x^3 + 9x^2}$

16. $\displaystyle\int \frac{(x^3 - 6x^2 + 11x - 12)\,dx}{x^4 - 6x^3 + 9x^2}$

17. $\displaystyle\int \frac{(3x^3 + 8x^2 - 3x + 12)\,dx}{x^4 - 4x^3}$

18. $\displaystyle\int \frac{(2 + 3x^2 - 2x^3 - 7x)\,dx}{x^4 - 2x^3}$

19. $\displaystyle\int \frac{(48 - 24x - 3x^2 - x^3)\,dx}{x^4 - 8x^3 + 16x^2}$

20. $\displaystyle\int \frac{(x^3 - 5x^2 + 7x - 3)\,dx}{x^4 - 4x^3 + 4x^2}$

21. $\displaystyle\int \frac{(2x^3 + 3x^2 + 12x - 81)\,dx}{(x^2 - 9)^2}$

22. $\displaystyle\int \frac{(10x^4 - 32x^2 + 23x + 32)\,dx}{x(x^2 - 4)^2}$

23. $\displaystyle\int \frac{(x^4 - 3x^2 + 2)\,dx}{x^3 + 2x^2}$

24. $\displaystyle\int \frac{(3x^5 - 2x^4 + 2)\,dx}{x^4 - x^2}$

25. Compare the following three integrations:

(a) $\displaystyle\int \frac{12\,dx}{x^2 - 6x + 8}$ (b) $\displaystyle\int \frac{12\,dx}{x^2 - 6x + 9}$ (c) $\displaystyle\int \frac{12\,dx}{x^2 - 6x + 10}$

33.7 DENOMINATORS WITH QUADRATIC FACTORS

If the denominator of a fraction contains a quadratic factor that cannot be split up into real linear factors, then *this quadratic factor must have appeared as the denominator of one of the partial fractions.* For example, if the denominator of a given fraction contains the factor $(x^2 + 4)$, then this factor must have been the denominator of one of the partial fractions. Otherwise it would not have appeared in the LCD.

Consider the fraction

$$\frac{7x^2 - 3x + 8}{x^3 + 4x}$$

Here the denominator can be factored into $x(x^2 + 4)$. Then we can assume that one of the partial fractions must have had the denominator $(x^2 + 4)$.

We must be especially careful about numerators of such fractions. If the denominator of a fraction is a nonfactorable quadratic factor, then it is possible that the numerator may contain an x-term as well as a purely constant term. In fact, the numerator may contain a single term in x, or a single constant term, or *both*. For example, in the above fraction, if the denominator of a partial fraction is the binomial $x^2 + 4$, then the numerator could be some single term as $3x$, or a single constant such as 5, or a binomial such as $3x + 5$. It is possible to have any one of these three conditions. In setting up the form of the numerator, we must make allowance for all possibilities. The procedure is shown by an example.

EXAMPLE 5. Integrate by first changing to partial fractions:

$$\int \frac{7x^2 - 3x + 8}{x^3 + 4x}\,dx$$

440 Partial Fractions

Solution. We first write the fraction showing the factors of the denominator and then set up the form of the partial fractions:

$$\frac{7x^2 - 3x + 8}{x(x^2 + 4)} = \frac{?}{x} + \frac{?}{x^2 + 4}$$

The first denominator is a *linear* factor, x. Then the numerator will be simply a *constant*, which we can represent by A. The second fraction has the denominator $x^2 + 4$, a nonfactorable quadratic factor. Therefore, we must make allowance for a binomial numerator consisting of a possible x-term with some numerical coefficient, and a purely constant term. The numerator of the second fraction we represent by $Bx + C$. Then A, B, and C are the undetermined constants.

Now we write the forms of the partial fractions using the constants, and have

$$\frac{7x^2 - 3x + 8}{x(x^2 + 4)} = \frac{A}{x} + \frac{Bx + C}{x^2 + 4}$$

Adding the two partial fractions we have set up, we get

$$\frac{A}{x} + \frac{Bx + C}{x^2 + 4} = \frac{A(x^2 + 4) + x(Bx + C)}{x(x^2 + 4)} = \frac{Ax^2 + 4A + Bx^2 + Cx}{x(x^2 + 4)}$$

Since this final fraction is equal to the original fraction, then the numerators must be equal. Comparing the two fractions, we find the coefficients of like powers of x, as well as constant terms, in the numerators, and equate coefficients of like powers:

$$\text{coefficients of } x^2, \quad A + B = 7$$
$$\text{coefficients of } x, \quad C = -3$$
$$\text{constant terms}, \quad 4A = 8$$

Solving, $A = 2$, $B = 5$, $C = -3$. Using these values in the numerators, the problem in integration becomes

$$\int \left(\frac{2}{x} + \frac{5x - 3}{x^2 + 4} \right) dx = 2 \ln x + \frac{5}{2} \ln (x^2 + 4) - \frac{3}{2} \arctan \frac{x}{2} + C$$

33.8 REPEATED QUADRATIC FACTORS

If a quadratic factor is repeated in the denominator of a fraction, we use the same idea as for repeated linear factors. For example, if a factor such as $(x^2 + 4)$ is repeated so that we have the form $(x^2 + 4)^n$, then we must set up fractions having denominators that correspond to $(x^2 + 4)$, $(x^2 + 4)^2, \ldots, (x^2 + 4)^n$. The same rule holds if the quadratic factors are trinomials of the form $(x^2 + px + q)^n$. In each case the numerator must provide for the possibility of two constants, one as the coefficient of x and the other a purely constant term.

EXAMPLE 6. Find

$$\int \frac{-x^5 + 12x^4 - x^3 + 19x^2 + 3x + 7}{x^2(x^2 + 1)^2} dx$$

Solution. The following forms of fractions are set up:

$$\frac{A}{x} + \frac{B}{x^2} + \frac{Cx + D}{x^2 + 1} + \frac{Ex + F}{(x^2 + 1)^2}$$

Adding the fractions and expanding and combining the numerator we get

$$\frac{Ax^5 + 2Ax^3 + Ax + Bx^4 + 2Bx^2 + B + Cx^5 + Dx^4 + Cx^3 + Dx^2 + Ex^3 + Fx^2}{x^2(x^2 + 1)^2}$$

Now we look for coefficients of like powers of x and equate. The coefficients are (K denotes constant)

for x^5: $A + C = -1$ x^3: $2A + C + E = -1$ x: $A = 3$
x^4: $B + D = 12$ x^2: $2B + D + F = 19$ K: $B = 7$

Solving the system, $A = 3$; $B = 7$; $C = -4$; $D = 5$; $E = -3$; $F = 0$. With these numerical values inserted for A, B, C, D, E, and F, in the partial fractions, the problem in integration becomes

$$\int \left(\frac{3}{x} + \frac{7}{x^2} + \frac{-4x + 5}{x^2 + 1} + \frac{-3x}{(x^2 + 1)^2} \right) dx$$

$$= 3 \ln x - \frac{7}{x} - 2 \ln (x^2 + 1) + 5 \arctan x + \frac{3}{2(x^2 + 1)} + C$$

33.9 IMPROPER FRACTIONS

If the numerator of the given fraction in the integrand contains a power of x equal to or greater than the highest power of the denominator, then the first step to be done is long division so as to reduce the fraction to the form of a whole number and a proper fraction.

EXAMPLE 7.   Find $\int \dfrac{x^3 - 5x^2 + 4x + 1}{x^2 - 3x - 10} dx$

Solution. By long division the fraction is changed to

$$x - 2 + \frac{8x - 19}{x^2 - 3x - 10}$$

Now the rational proper fraction is changed by the method already shown into the two partial fractions

$$\frac{3}{x - 5} + \frac{5}{x + 2}$$

The integration problem then becomes

$$\int \left(x - 2 + \frac{3}{x - 5} + \frac{5}{x + 2} \right) dx$$

$$= \frac{x^2}{2} - 2x + 3 \ln (x - 5) + 5 \ln (x + 2) + \ln C$$

The result may be written

$$\frac{x^2 - 4x}{2} + \ln C(x - 5)^3 (x + 2)^5$$

EXAMPLE 8.   Find $\int \dfrac{x^2 \, dx}{x^2 - 9}$

442 Partial Fractions

Solution. Here we must first use long division. As a result the fraction becomes

$$1 + \frac{9}{x^2 - 9}$$

Now the proper fraction is reduced to partial fractions:

$$\frac{9}{(x-3)(x+3)} = \frac{A}{x-3} + \frac{B}{x+3} = \frac{Ax + 3A + Bx - 3B}{(x-3)(x+3)}$$

For coefficients of x, $A + B = 0$; for the constant term, $3A - 3B = 9$. Solving for A and B, we get $A = 3/2$ and $B = -3/2$. Inserting these numerical values in the partial fractions, we get

$$\int \left(1 + \frac{3/2}{x-3} + \frac{-3/2}{x+3}\right) dx = x + \frac{3}{2} \ln \frac{x-3}{x+3} + C$$

The question is sometimes asked: What is to be done if the denominator contains a cubic (third degree) factor that cannot be reduced to factors of lower degree? It can be proved that every polynomial can be reduced to the product of linear and/or quadratic factors. It is sometimes difficult to find the factors, but all the fractions with which we shall here be concerned can be factored rather easily into factors of the first or second degree. For polynomials of the third degree and higher, synthetic division is very helpful.

EXERCISES 33.4

Integrate by first separating into partial fractions:

1. $\displaystyle\int \frac{(2x^2 + 2x + 45)\, dx}{x^3 + 9x}$

2. $\displaystyle\int \frac{(x^2 + 4x - 6)\, dx}{x^4 + 4x^2}$

3. $\displaystyle\int \frac{(2x^3 - 5x^2 + 12)\, dx}{x^4 + 4x^2}$

4. $\displaystyle\int \frac{(5x^4 - 6x^3 - 2x^2 - 16x - 8)\, dx}{x^5 + 4x^3}$

5. $\displaystyle\int \frac{(8x^2 + 28)\, dx}{x^3 - 8}$

6. $\displaystyle\int \frac{(3x^3 - 20x + 65)\, dx}{x^4 - 4x^3 + 13x^2}$

7. $\displaystyle\int \frac{(6x - 4x^3 - 6x^2 - 4)\, dx}{x^5 - x}$

8. $\displaystyle\int \frac{(5x^3 - 18x^2 + 18x - 45)\, dx}{4x^4 + 9x^2}$

9. $\displaystyle\int \frac{(x^4 - x^3 - 9x - 18)\, dx}{x^4 + 9x^2}$

10. $\displaystyle\int \frac{(3x^4 - 3x^3 - 15x^2 - 12)\, dx}{x^5 + 9x^3}$

11. $\displaystyle\int \frac{(3x^4 - 2x^3 + 5x^2 + 8x - 20)\, dx}{x^4 + 4x^2}$

12. $\displaystyle\int \frac{(5x^4 + 24x^3 - 9x^2 + 12x - 4)\, dx}{9x^5 + 4x^3}$

13. $\displaystyle\int \frac{(5x^3 + 5x^2 + 20x + 32)\, dx}{x^5 + 8x^3 + 16x}$

14. $\displaystyle\int \frac{(x^4 - 5x^3 - x^2 - 2)\, dx}{x^6 + 2x^4 + x^2}$

15. $\displaystyle\int \frac{(2x^4 - 9x^2 + 4x - 95)\, dx}{x^3 + 3x^2 + 4x + 12}$

16. $\displaystyle\int \frac{(2x^3 - 3x^2 - 7x - 7)\, dx}{x^4 + 13x^2 + 36}$

17. $\displaystyle\int \frac{(2x^3 - 7x^2 + 32x - 28)\, dx}{x^4 - 16}$

18. $\displaystyle\int \frac{(3x^5 - 2x^4 - 3x^3 - 5x^2 - 12)\, dx}{x^6 + 5x^4 + 4x^2}$

19. $\displaystyle\int \frac{(2x^3 + 7x^2 - 7x)\, dx}{x^4 - 3x^3 + 3x^2 - 3x + 2}$

20. $\displaystyle\int \frac{(2x^3 - 2x + 16)\, dx}{x^4 + 2x^3 + 4x^2 - 2x - 5}$

21. $\int \dfrac{7e^{4x}\, dx}{12 - e^{4x} - e^8}$

22. $\int \dfrac{5e^{-3x}\, dx}{e^{-6x} - e^{-3x} - 6}$

23. $\int \dfrac{\cos x\, dx}{\sin^2 x + 6 \sin x + 8}$

24. $\int \dfrac{4 \sec^2 x\, dx}{1 - 2 \tan x - 3 \tan^2 x}$

25. $\int \dfrac{2x^5 + 4x^4 - 12x^3 + 64x^2 - 76x + 160}{x^6 - 64}\, dx$

33.10 THE DISCRIMINANT OF A QUADRATIC DENOMINATOR

In Chapter 32 we saw how a fraction with a quadratic denominator leads to an arctangent or to a power. Take the two examples:

$$\int \frac{dx}{x^2 - 4x + 4} \quad \text{and} \quad \int \frac{dx}{x^2 - 4x + 5}$$

In the first example, we rewrite the problem and apply the power rule:

$$\int \frac{dx}{(x - 2)^2} = \int (x - 2)^{-2}\, dx = \frac{(x - 2)^{-1}}{-1} = -\frac{1}{x - 2} + C$$

In the second example we get an arctangent:

$$\int \frac{dx}{x^2 - 4x + 5} = \int \frac{dx}{x^2 - 4x + 4 + 1}$$

$$= \int \frac{dx}{(x - 2)^2 + 1} = \arctan(x - 2) + C$$

Now, in connection with partial fractions, we have a third situation. Take the example

$$\int \frac{dx}{x^2 - 4x + 3}$$

Here the denominator can be factored into two unlike factors. Then we have the partial fractions:

$$\frac{1}{x^2 - 4x + 3} = \frac{A}{x - 3} + \frac{B}{x - 1} = \frac{Ax - A + Bx - 3B}{(x - 3)(x - 1)}$$

Solving for A and B, we get $A = 1$ and $B = -1$. Then the problem becomes

$$\int \frac{dx}{x - 3} - \int \frac{dx}{x - 1} = \ln(x - 3) - \ln(x - 1) = \ln \frac{x - 3}{x - 1} + C$$

Now let us compare the three fractions and the results:

(a) $\int \dfrac{dx}{x^2 - 4x + 3}$ leads to a logarithm.

(b) $\int \dfrac{dx}{x^2 - 4x + 4}$ leads to a power.

(c) $\int \dfrac{dx}{x^2 - 4x + 5}$ leads to an arctangent.

Note the similarity between the integrands.

Now we might ask: Under what conditions will such a fraction lead to each of these results? Moreover, how can the types of integral be determined by inspection of the denominator?

In example (b), in which the denominator is a quadratic perfect square, the discriminant of this function is zero. This means that if we try to complete a square for the form of an arctangent, we take the entire constant 4 for the square and we have nothing left for the constant term required for the form leading to an arctangent.

In example (a), with the denominator $x^2 - 4x + 3$, if we attempt to complete the square in x, we not only use up the entire constant, 3, but we need more. The result is that any remaining constant is negative, and the result cannot be an arctangent. In the quadratic denominator, the discriminant is positive: $16 - 12 = 4$. In the case of such a quadratic denominator, the fraction is first separated into partial fractions, and the integral is a logarithm.

In example (c) the quadratic function, $x^2 - 4x + 5$, in the denominator contains the constant 5, which is more than enough to complete the square. We have a positive quantity left after completing the square. If we represent any remaining positive quantity by q, we get a denominator of the form

$$(ax + b)^2 + q$$

The integral is then an arctangent. With a constant greater than enough to complete the square, the discriminant of the quadratic denominator will be negative.

In brief, if we compute the discriminant of the quadratic denominator of a fraction in which the numerator is some constant, we can always tell from the discriminant the nature of the integral.

(a) If the discriminant of the quadratic denominator is *positive*, the zeros of this function are real and unequal, and the integral is a *logarithm*.

(b) If the discriminant is *zero*, the zeros of the denominator are equal, and the integral is a *power*.

(c) If the discriminant is *negative*, the zeros of the denominator are imaginary, and the integral is an *arctangent*.

The foregoing principles are exemplified in the following illustrations:

(a) $\int \dfrac{dx}{3x^2 - 2x - 4}$ (b) $\int \dfrac{dx}{4x^2 + 12x + 9}$ (c) $\int \dfrac{dx}{5x^2 - 4x + 2}$

In (a) the discriminant of the denominator quadratic function is 52. Therefore, the integral is a logarithm. In (b) the discriminant is zero and the integral is therefore obtained by the power rule. In (c) the value of the discriminant is -24. Therefore, the integral is an arctangent.

chapter

34

Special Techniques of Integration

34.1 LIMITATION OF ELEMENTARY FORMS

The methods of integration we have used up to this point may be called the *elementary forms*. They include the power rule, the integration of trigonometric functions, integration leading to logarithms and inverse trigonometric functions, and the integration of exponential functions.

The elementary forms should be used whenever possible. However, there are many problems in integration in which these forms cannot be applied directly. Then certain other techniques may be attempted. It might be mentioned that in integration there is no direct method that corresponds to the "delta" method in differentiation. However, probably the most direct method is that of *integration by parts*. This technique does not always work, but it is one of the most powerful of all methods.

34.2 INTEGRATION BY PARTS

The formula for integration by parts comes about through the rule for the differentiation of a product. You will recall that if

$$y = (u)(v)$$

then $$dy = u\,dv + v\,du$$

which is the same as $$d(uv) = u\,dv + v\,du$$

Transposing, we get $$u\,dv = d(uv) - v\,du$$

Now, if we integrate both sides of this equation, we get

$$\int u\,dv = uv - \int v\,du$$

This is the formula for integration by parts. We assume the *left* side of the equation is the given function. Then it may be possible to find the desired integral by means of the *right* side of the equation.

In applying this method, we first separate mentally the given integrand into two *factors*, or *parts*. We identify one of the factors as *u* and the other factor as *dv* of the formula. However, the formula also calls for *du* and *v*. The method is shown by examples.

EXAMPLE 1. Find $\int xe^x \, dx$

Solution. Note that this integration cannot be done by any of the elementary forms. We therefore try integration by parts. First, we consider the integrand as consisting of two factors: one we call *u*; the other, *dv*. The factor *x* may be taken as the *u* part of the formula; then the factor $e^x \, dx$ is taken as the part *dv*. (Of course, the factors may be taken differently, but let us take them as stated.) Now we must also have *du* and *v*. We arrange the information as follows:

Let	$u = x$	$dv = e^x \, dx$
Then	$du = dx$	$v = e^x$

Note that the *given* information is arranged across the top. This is done for convenience. Then we must *differentiate* the factor *u* and *integrate* the factor *dv*, since the formula also calls for *du* and *v*. Now we have all four elements required by the formula. (It is not necessary to add the constant *C* in the integral of *dv*.)

The proper values are now inserted into the formula

$$\int u \, dv = uv - \int v \, du$$

We get $\quad\quad\quad\quad \int x \, e^x \, dx = x \, e^x - \int e^x \, dx$

The left side of this equation represents the given problem. On the right side we have the answer to the problem. Of course, this still contains an indicated integration, but this integration is easily performed. The entire integral becomes

$$\int x \, e^x \, dx = x \, e^x - e^x + C$$

Note. In the foregoing example, we might have taken the factor e^x as the *u* part of the formula, and the factor $x \, dx$ as the *dv* part. However, this selection would not have been successful. The way in which we choose the parts *u* and *dv* will often mean success or failure. In many problems, the correct choice can be discovered only by trial and error. In some problems the choice must be made in a certain way; in others, it does not matter. However, the correct choice is usually not difficult.

In choosing the correct factors as *u* and *dv* for integration by parts the following two rules will help:

(1) *The factor dv must always include the factor dx in the integrand.*

(2) *The part taken as dv must always be something that can be integrated by some elementary method.*

EXAMPLE 2. Find $\int x^2 \, e^{-3x} \, dx$

Solution.

Let	$u = x^2$	$dv = e^{-3x} \, dx$
Then	$du = 2x \, dx$	$v = -\tfrac{1}{3}e^{-3x}$

34.2 Integration by Parts

Substituting in the formula,

$$\int x^2 e^{-3x} \, dx = -\frac{1}{3} x^2 e^{-3x} + \frac{2}{3} \int x e^{-3x} \, dx$$

Now the "parts" rule is applied a second time, this time on the last term.

Let	$u = x$	$dv = e^{-3x} \, dx$
Then	$du = dx$	$v = -\frac{1}{3} e^{-3x}$

The final term then becomes

$$+ \frac{2}{3} \left[-\frac{1}{3} x e^{-3x} + \frac{1}{3} \int e^{-3x} \, dx \right]$$

Supplying another "-3" to the final integral, we perform the integration and get the entire result

$$\int x^2 e^{-3x} \, dx = -\frac{1}{3} x^2 e^{-3x} - \frac{2}{9} x e^{-3x} - \frac{2}{27} e^{-3x} + C$$

The result can be reduced to

$$-\frac{1}{27} e^{-3x} (9x^2 + 6x - 2) + C$$

Note. In Example 2, the original integrand contained the factor x^2. Note that the *power* of x is reduced by 1 each time the parts rule is applied. In general, given any integral of the form

$$\int x^n e^{ax} \, dx$$

the integration can then be performed by applying the parts rule n times.

EXAMPLE 3. Find $\int x^4 \arctan x \, dx$

Solution. In general, if an integrand contains the factor x^n, we would seek to reduce the power of x rather than increase it. However, this does not always work, as in this example, for then we should take $\arctan x \, dx$ as dv. We have seen that the factor dv must be a factor that can easily be integrated into v. Therefore, we take the following factors, or parts:

Let	$u = \arctan x$	$dv = x^4 \, dx$
Then	$du = \dfrac{1}{x^2 + 1} \, dx$	$v = \dfrac{x^5}{5}$

Substituting in the formula,

$$\int x^4 \arctan x \, dx = \frac{x^5}{5} \arctan x - \frac{1}{5} \int \frac{x^5}{x^2 + 1} \, dx$$

By long division the last term becomes

$$-\frac{1}{5} \int \left(x^3 - x + \frac{x}{x^2 + 1} \right) dx$$

448 Special Techniques of Integration

The entire integral becomes

$$\frac{x^5}{5} \arctan x - \frac{1}{20} x^4 + \frac{x^2}{10} - \frac{1}{10} \ln (x^2 + 1) + C$$

Note. In Example 3, we take $dv = x^4 \, dx$ even though the integration of dv increases the power of x. In general, if the integrand is a product of any power of x and an *arctangent* or a *logarithm*, the integration can be performed by only one application of the *parts* rule.

There is a somewhat different type of problem in which the parts rule can be used. In general, in applying the parts rule, we hope to get a simpler integral, which can be evaluated. However, in some problems we do not get a simpler integral but instead we get an integral exactly like the original, but with a different coefficient. The next two examples illustrate this type of problem and the method of integration.

EXAMPLE 4. Find $\quad \int \csc^3 x \, dx$

Solution.

Let $\quad u = \csc x$	$dv = \csc^2 x \, dx$
Then $\quad du = -\csc x \cot x \, dx$	$v = -\cot x$

Therefore,

$$\int \csc^3 x \, dx = -\csc x \cot x - \int \csc x \cot^2 x \, dx$$
$$= -\csc x \cot x - \int \csc x \, (\csc^2 x - 1) \, dx$$
$$= -\csc x \cot x - \int \csc^3 x \, dx + \int \csc x \, dx$$

Transposing the second term on the right side and integrating the third term,

$$2 \int \csc^3 x \, dx = -\csc x \cot x - \ln (\csc x + \cot x)$$

Dividing both sides by 2, we get the answer we seek:

$$\int \csc^3 x \, dx = \frac{1}{2} [-\csc x \cot x - \ln (\csc x + \cot x)] + C$$

EXAMPLE 5. Find $\quad \int e^{3x} \cos 5x \, dx$

Solution.

Let $\quad u = \cos 5x$	$dv = e^{3x} \, dx$
Then $\quad du = -5 \sin 5x \, dx$	$v = \tfrac{1}{3} e^{3x}$

Therefore,

$$\int e^{3x} \cos 5x \, dx = \frac{1}{3} e^{3x} \cos 5x + \frac{5}{3} \int e^{3x} \sin 5x \, dx$$

Now we again apply the parts rule on the last term.

Let $\quad u = \sin 5x$	$dv = e^{3x} \, dx$
Then $\quad du = 5 \cos 5x \, dx$	$v = \tfrac{1}{3} e^{3x}$

Then

$$\int e^{3x} \cos 5x\, dx = \frac{1}{3} e^{3x} \cos 5x + \frac{5}{3}\left[\frac{1}{3} e^{3x} \sin 5x - \frac{5}{3}\int e^{3x} \cos 5x\, dx\right]$$

$$= \frac{1}{3} e^{3x} \cos 5x + \frac{5}{9} e^{3x} \sin 5x - \frac{25}{9}\int e^{3x} \cos 5x\, dx$$

Transposing the last term, and combining on the left side,

$$\frac{34}{9}\int e^{3x} \cos 5x\, dx = \frac{1}{3} e^{3x} \cos 5x + \frac{5}{9} e^{3x} \sin 5x$$

Multiplying both sides of the equation by 9/34 and simplifying, we get

$$\int e^{3x} \cos 5x\, dx = \frac{e^{3x}}{34}(3 \cos 5x + 5 \sin 5x) + C$$

EXERCISE 34.1

Integrate by parts:

1. $\int x e^{-3x}\, dx$
2. $\int t e^{5t}\, dt$
3. $\int x^3 \arctan x\, dx$
4. $\int x^3 e^{2x}\, dx$
5. $\int x^5 \ln x\, dx$
6. $\int x \arctan x\, dx$
7. $\int x^{-3} \ln x\, dx$
8. $\int \ln x\, dx$
9. $\int \arctan x\, dx$
10. $\int x \sin x\, dx$
11. $\int x^2 \cos 3x\, dx$
12. $\int t^2 \sin 5t\, dt$
13. $\int \sec^3 \theta\, d\theta$
14. $\int \sec^5 x\, dx$
15. $\int \sec^2 t \csc t\, dt$
16. $\int \sin x \cos x\, dx$
17. $\int t^3 \sin 6t\, dt$
18. $\int e^{4x} \sin 2x\, dx$
19. $\int e^{-5x} \cos 3x\, dx$
20. $\int e^{-10t} \sin 40\, t\, dt$
21. $\int e^{\pi\theta} \sin \theta\, d\theta$
22. $\int \sin^2 x \cos^3 x\, dx$
23. $\int \sec^3 x \csc^2 x\, dx$
24. $\int e^{-10x} (\sin 15x - \cos 15x)\, dx$
25. $\int e^{-5t}(\sin \omega t + \cos \omega t)\, dt$
26. $\int e^{ax} \sin bx\, dx$
27. $\int e^{mx} \cos nx\, dx$
28. State the results of Problems 26 and 27 as formulas.
29. If the curve $y = e^{-x}$ from $x = -1$ to $x = 2$ is rotated about the x-axis, what is the volume of the solid formed?
30. If the area bounded by the curve $y = e^{2x}$, the coordinate axes, and the line $x = 2$, is rotated about the x-axis, what is the volume of the solid formed?
31. What is the area under the curve $y = \ln x$ from $x = 1$ to $x = 4$?
32. If the area under the curve $y = \cos x$ from $x = 0$ to $x = 90°$ is rotated about the y-axis, what is the volume of the solid formed?

34.3 ALGEBRAIC SUBSTITUTION

Another technique that might be used in the integration of certain functions is that of *algebraic substitution*. This method is often useful where the integrand contains an irrational factor — that is, a radical involving x.

Algebraic substitution consists of the use of a single letter to represent a complicated function. An entire radical may be represented by a single letter, say v. Then each x, including dx, must be replaced by its equivalent v form. We shall see that the factor dx is usually *not* simply dv.

The advantage of algebraic substitution lies in the fact that after the integrand has been altered by a change of variable, the integration can often be carried out in terms of the new variable. Of course, the resulting integral must be converted back to a form involving the original variable.

Special Techniques of Integration

If the integrand contains a radical we can follow these steps:

(1) Let some letter, such as v, represent the root indicated by the radical. For example, if we have the radical $(x^2 + 4)^{3/2}$, it is best to let the letter v represent $(x^2 + 4)^{1/2}$; that is, the root.

(2) Find the expression in v-terms representing every form of x, including dx, in the radicand, and make the proper substitutions.

(3) Integrate the result in terms of the new variable v.

(4) Convert the result back to x-terms, or whatever the original variable.

(5) For a definite integral, evaluate between the original x-limits. The fourth step may be omitted if the limits are also changed to v-limits. Then evaluation can be done immediately after step 3.

The technique of algebraic substitution will be shown by examples.

EXAMPLE 6. Find $\displaystyle\int_{-1}^{4} \frac{x\,dx}{\sqrt{x+5}}$

Solution. We shall omit the limits until we have found the integral. Then we shall evaluate the result in two ways. Note that we cannot apply any of the elementary formulas. Then we try algebraic substitution. We first let

$$v = (x + 5)^{1/2}$$

Now we must replace each x including dx with its equivalent v-form. To find dx, we must first solve for x.

$$v^2 = x + 5, \qquad x = v^2 - 5, \qquad dx = 2v\,dv$$

Note that dx does *not* become simply dv. In fact, getting the correct expression for dx is usually the most troublesome. Now we make the substitutions in the original integrand. We replace x with its equivalent $v^2 - 5$; dx with its equivalent $(2v\,dv)$; and the radical with v. The problem becomes

$$\int \frac{(v^2 - 5)2v\,dv}{v}$$

This reduces to $\int 2(v^2 - 5)\,dv$

The integration now is simple and leads to

$$2\left(\frac{v^3}{3} - 5v\right) = \frac{2}{3}(v)(v^2 - 15)$$

Now we shall evaluate the integral in two ways. First, we make the reverse substitution to x-terms and get

$$\frac{2(x+5)^{1/2}}{3}(x + 5 - 15) = \frac{2}{3}(x+5)^{1/2}(x - 10)\Big]_{-1}^{4}$$

$$= \frac{2}{3}[(4+5)^{1/2}(4-10) - (-1+5)^{1/2}(-1-10)]$$

$$= \frac{2}{3}[(3)(-6) - (2)(-11)] = \frac{2}{3}[-18 + 22] = \frac{8}{3}$$

34.3 Algebraic Substitution

We have noted that if the definite integral is desired, then the evaluation can be done without reconverting the integral to x-terms, *provided the limits are changed from x-limits to v-limits.* For example, we have already let

$$v = (x+5)^{1/2}$$

Then, for the limits,

when $x = 4$, $v = 3$

when $x = -1$, $v = 2$

Now we can use the new limits and the problem becomes

$$\int_{x=-1}^{x=4} \frac{x\,dx}{\sqrt{x+5}} = \int_{v=2}^{v=3} 2(v^2 - 5)\,dv = \frac{2}{3}\left[v(v^2 - 15)\right]_2^3$$

$$= \frac{2}{3}[3(-6) - 2(-11)] = \frac{2}{3}[-18 + 22] = \frac{8}{3}$$

EXAMPLE 7. Evaluate $\int_0^4 x^3(25 - x^2)^{3/2}\,dx$

Solution. In this example, we let v represent the root, $(25 - x^2)^{1/2}$. It would be possible to let v represent the entire radical, but it is usually much better to start with the root alone. This will avoid many fractional exponents. Then

$$v^2 = 25 - x^2, \qquad x^2 = 25 - v^2, \qquad x = (25 - v^2)^{1/2}$$

then

$$dx = \frac{1}{2}(25 - v^2)^{-1/2}(-2v)\,dv$$

For the new limits:

when $x = 4$, $v = (25 - 16)^{1/2} = 3$

when $x = 0$, $v = (25 - 0)^{1/2} = 5$

Making the substitutions,

$$\int_5^3 (25 - v^2)^{3/2}(v^3)\left(\frac{1}{2}\right)(25 - v^2)^{-1/2}(-2v)\,dv$$

$$= \int_5^3 (25 - v^2)(-v^4)\,dv = \int_5^3 (v^6 - 25v^4)\,dv = \left[\frac{v^7}{7} - 5v^5\right]_5^3$$

or

$$\left[\frac{v^5}{7}(v^2 - 35)\right]_5^3 = 3561\frac{5}{7}$$

If we wish to reconvert the integral back to x-terms, we get

$$\left[\frac{(25 - x^2)^{5/2}}{7}(25 - x^2 - 35)\right]_0^4 = -\left[\frac{(25 - x^2)^{5/2}}{7}(x^2 + 10)\right]_0^4$$

The result can be checked by differentiation by the product rule. The differential of the integral should be equal to the original integrand.

EXAMPLE 8. Integrate $\int \frac{dx}{x + 2\sqrt{x} + 10}$

Solution. We make the substitutions: Let $v = \sqrt{x}$; then $v^2 = x$; $dx = 2v\,dv$. The problem becomes

$$\int \frac{2v\,dv}{v^2 + 2v + 10} = \int \frac{(2v + 2 - 2)\,dv}{v^2 + 2v + 10}$$

$$= \int \frac{(2v + 2)\,dv}{v^2 + 2v + 10} - \int \frac{2\,dv}{v^2 + 2v + 10}$$

$$= \ln(v^2 + 2v + 10) - \frac{2}{3}\arctan\frac{v + 1}{3} + C$$

$$= \ln(x + 2\sqrt{x} + 10) - \frac{2}{3}\arctan\frac{\sqrt{x} + 1}{3} + C$$

EXAMPLE 9. Integrate and evaluate

$$\int_0^4 \frac{x\,dx}{x + 2\sqrt{x} + 2}$$

Solution. Let $v = \sqrt{x}$; then $v^2 = x$; $dx = 2v\,dv$; to evaluate, we note the limits: when $x = 4$, $v = 2$; when $x = 0$, $v = 0$. Substituting, we get

$$\int_0^2 \frac{v^2(2v)\,dv}{v^2 + 2v + 2} = \int_0^2 \frac{2v^3\,dv}{v^2 + 2v + 2}$$

By long division, the problem becomes

$$2\int_0^2 \left(v - 2 + \frac{2v + 4}{v^2 + 2v + 2}\right)dv$$

$$= 2\int_0^2 \left(v - 2 + \frac{2v + 2}{v^2 + 2v + 2} + \frac{2}{v^2 + 2v + 2}\right)dv$$

$$= 2\left[\frac{v^2}{2} - 2v + \ln(v^2 + 2v + 2) + 2\arctan(v + 1)\right]_0^2$$

$$= 2[2 - 4 + \ln 10 + 2\arctan 3 - 0 + 0 - \ln 2 - 2\arctan 1]$$

$$= 2[-2 + \ln 5 + 2(\arctan 3 - \arctan 1)]$$

$$= -4 + 2\ln 5 + 2\arctan\frac{1}{2} = -4 + 2(1.6094) + 2(0.4637) = 0.1462 \text{ (approx.)}$$

Note. To find the value of arctan 3 − arctan 1, we use the formula for the tangent of the difference between two angles. Calling the angle for the difference, ϕ, we have

$$\tan\phi = \frac{3 - 1}{1 + 3} = \frac{1}{2}$$

Then $\phi = \arctan\frac{1}{2} = 0.4637$ radian. We must also note that the function is continuous between limits.

34.4 NECESSARY CONDITIONS FOR ALGEBRAIC SUBSTITUTION

In the foregoing examples we have used a single letter to represent an irrational expression. Then the new expression has become rational in terms of the new variable. We have seen that the integration is then fairly simple, depending only on the use of the elementary formulas.

34.4 Necessary Conditions for Algebraic Substitution

Now, we might well ask: Why not use this method of substitution for all radicals in an integrand? However, then we must also ask: Will this kind of substitution always result in a rational integrand? If not, under what conditions will it not? To answer these questions, let us begin with an integrand containing very general terms with irrational expressions. Remember, we hope to get a rational integrand in terms of the new variable.

Let us assume the integrand contains a radical, which we shall represent in general terms. We assume the radical contains a power of x, such as x^n, with some constant coefficient a. The radicand may also contain a constant term b, and the radical may be indicated by a general exponent, p/q. Let us also assume that the integrand contains another factor involving another general power of x, such as x^m. The integrand then has the form

$$x^m(ax^n + b)^{p/q}\, dx$$

We can omit any possible constant factor in the integrand itself since any such factor could be placed outside the integral sign.

Now our question is: under what conditions of a, b, m, n, p, and q will a substitution of the kind we have used result in a rational integrand? Let us make the substitution in the general integrand and see what happens.

Starting with the problem

$$\int x^m(ax^n + b)^{p/q}\, dx$$

as a first step, we let v represent the root; that is,

$$v = (ax^n + b)^{1/q}$$

Now we must replace every form of x, including dx, with its equivalent form of the variable v. We must solve for x so that we can find the proper expression for x^n, x^m, and dx. Raising both sides to the qth power, we get

$$v^q = ax^n + b, \qquad ax^n = v^q - b, \qquad x^n = \frac{v^q - b}{a}$$

Then

$$x = \left(\frac{v^q - b}{a}\right)^{1/n} \qquad x^m = \left(\frac{v^q - b}{a}\right)^{m/n}$$

and

$$dx = \frac{1}{n}\left(\frac{v^q - b}{a}\right)^{(1/n)-1}\left(\frac{1}{a}\right)(q)(v^{q-1})\, dv$$

Now we have everything we need to make the replacement in the integrand. Substituting, and noting that $(ax^n + b)^{p/q} = v^p$, we get

$$\int \underbrace{\left(\frac{v^q - b}{a}\right)^{m/n}}_{R}\underbrace{(v^p)}_{K}\underbrace{\left(\frac{1}{n}\right)}_{}\underbrace{\left(\frac{v^q - b}{a}\right)^{(1/n)-1}}_{K}\underbrace{\left(\frac{1}{a}\right)}_{K}\underbrace{(q)(v^{q-1})}_{R}\, dv$$

If we use R to indicate a rational quantity, and K to indicate a constant, the only factor that is neither constant nor rational is, after combining,

$$\left(\frac{v^q - b}{a}\right)^{[(m+1)/n]-1}$$

Special Techniques of Integration

This factor also will be rational *if the exponent is zero or an integer.* That is,

$$\frac{m+1}{n} - 1$$

must be an integer. If we omit the -1, we still can say that $(m+1)/n$ *must be zero or an integer.* Now we take careful note of the exponents, m and n, in the original integrand.

We can now test any integrand containing an irrational expression to determine whether or not the expression can be made rational by the kind of substitution we have used. We note the power on x *outside* the radical and call it m. Then we note the power on x inside the radical and call it n. Then, if the quantity $(m+1)/n$ is zero or an integer, positive or negative, the integrand can be made rational by substituting a single letter for the radical.

Note. If x^m appears in the denominator of a fractional integrand, then the sign of m must be taken as it would appear if the factor is transferred to the numerator. However, the sign of n is not affected by transferring the radical from denominator to numerator.

Tests are here shown for a few integrands of this kind.

(a) $\int x^5(9-x^2)^{5/2}\, dx \quad m=5, n=2; \quad \dfrac{m+1}{n} = 3$, integer

(b) $\int \dfrac{(x^3-8)^{2/3}\, dx}{x^5} \quad m=-5, n=3; \quad \dfrac{m+1}{n} = \dfrac{-4}{3}$, fraction

(c) $\int x^5(x^3-8)^{2/3}\, dx \quad m=5, n=3; \quad \dfrac{m+1}{n} = 2$, integer

(d) $\int \dfrac{(4x^2-9)^{3/2}\, dx}{x} \quad m=-1, n=2; \quad \dfrac{m+1}{n} = 0.$

(e) $\int \dfrac{dx}{x^3(9+4x^2)^{2/3}} \quad m=-3, n=2; \quad \dfrac{m+1}{n} = -1$, integer

(f) $\int \dfrac{(x^3-8)^{3/4}\, dx}{x^7} \quad m=-7, n=3; \quad \dfrac{m+1}{n} = -2$, integer

EXERCISE 34.2

Perform the following integrations by means of a suitable substitution unless one of the elementary formulas applies. Evaluate as indicated.

1. $\int x^3\sqrt{x^2-9}\, dx$
2. $\int x^5\sqrt{4-9x^2}\, dx$
3. $\int x\sqrt{16-9x^2}\, dx$
4. $\int \dfrac{x^3\, dx}{\sqrt{9-x^2}}$
5. $\int \dfrac{(x^2-9)^{3/2}\, dx}{x}$
6. $\int \dfrac{x^3\, dx}{\sqrt{x^2-16}}$
7. $\int x\sqrt{x+9}\, dx$
8. $\int x^3(4+x^2)^{1/3}\, dx$
9. $\int x^5(x^2+4)^{1/2}\, dx$

34.5 Trigonometric Substitution

10. $\int x^3(x^2-4)^{1/3}\,dx$
11. $\int x^5\sqrt{16-9x^2}\,dx$
12. $\int x^3(x^2+1)^{5/2}\,dx$
13. $\int \dfrac{x\,dx}{\sqrt{9-x^2}}$
14. $\int \dfrac{\sqrt{(x^2-4)}\,dx}{x}$
15. $\int \dfrac{dx}{\sqrt{9-25x^2}}$
16. $\int \dfrac{(x^2-1)^{3/2}\,dx}{x^5}$
17. $\int \dfrac{dx}{x+2\sqrt{x}+1}$
18. $\int \dfrac{dx}{2+\sqrt{x}}$
19. $\int \dfrac{(4+x^2)^{3/2}\,dx}{x}$
20. $\int \dfrac{(1-x^2)^{3/2}\,dx}{x^5}$
21. $\int \dfrac{(x^2-1)^{2/3}\,dx}{x^3}$
22. $\int \dfrac{(x^2+4)^{5/2}\,dx}{x}$
23. $\int \dfrac{(x^4+4)^{3/2}\,dx}{x^5}$
24. $\int \dfrac{x^3\,dx}{(1-x^2)^{5/2}}$
25. $\int_0^4 x^3(x^2+9)^{1/2}\,dx$
26. $\int_0^3 x^3(x^2+16)^{3/2}\,dx$
27. $\int_0^2 x^5(x^3-8)^{1/3}\,dx$
28. $\int_0^4 x\sqrt{x+9}\,dx$
29. $\int_4^5 \dfrac{\sqrt{x^2-16}\,dx}{x}$
30. $\int_0^1 \dfrac{dx}{x-3\sqrt{x}-10}$
31. $\int_0^9 \dfrac{dx}{x-4\sqrt{x}+4}$
32. $\int_0^2 \dfrac{dx}{x+4\sqrt{x}+13}$
33. $\int_1^4 \dfrac{x\,dx}{x+2\sqrt{x}+5}$
34. $\int_3^4 \dfrac{(25-x^2)^{3/2}\,dx}{x}$

34.5 TRIGONOMETRIC SUBSTITUTION

In some problems in integration it may be possible and convenient to make use of a device called *trigonometric substitution*. In this method we inject a trigonometric function into a problem that does not originally contain such a function. The technique is especially useful where an irrational function of x appears in the integrand. It can often be used when other methods fail. By a proper substitution, an algebraic function that is irrational may become rational as a trigonometric function. It should be pointed out, however, that this method also has limitations.

To understand the purpose of this device, suppose an integrand contains the expression, $\sqrt{1-x^2}$, which is irrational in x. Now, if we let x be represented by a proper trigonometric function, the expression becomes rational. In this example, if we let x equal the sine of some angle θ (where θ is any angle whatever), we have

$$x^2 = \sin^2\theta$$

Then the radical becomes $\sqrt{1-\sin^2\theta}$

The expression can now be rationalized by use of the trigonometric identity, $1-\sin^2\theta = \cos^2\theta$. The result is

$$\sqrt{1-\sin^2\theta} = \sqrt{\cos^2\theta} = \cos\theta$$

and the radical has conveniently disappeared!

The advantage of trigonometric substitution lies in the fact that it is often possible to integrate the resulting rational trigonometric function whereas there might be no way to integrate the original function. The integration is performed in trigonometric terms. Then, of course, after the integration has been accomplished, the result must be reconverted back to the equivalent algebraic form.

34.6 CHOOSING THE PROPER COEFFICIENT

It is often necessary to take some number other than 1 for the coefficient of the trigonometric function to be used. The proper coefficient is determined by the constant term in the radical. For example, suppose we wish to rationalize the expression

$$\sqrt{9 - x^2}$$

In this example, we must let $x^2 = 9 \sin^2 \theta$. Substituting, we get

$$\sqrt{9 - 9\sin^2 \theta} = \sqrt{9(1 - \sin^2 \theta)} = \sqrt{9 \cos^2 \theta} = 3 \cos \theta$$

Note that the coefficient of $\sin^2 \theta$ must be 9, the same as the constant term. This is necessary in order that the factor 9 may be factored out of the entire radicand. This leaves the factor $(1 - \sin^2 \theta)$ to be replaced by $\cos^2 \theta$.

Even though the constant term is not what we usually call a perfect arithmetic square, yet we take this number as the coefficient of the trigonometric function. For example, consider the expression

$$\sqrt{2 - x^2}$$

Here we let $x^2 = 2 \sin^2 \theta$. Then we get

$$\sqrt{2 - 2\sin^2 \theta} = \sqrt{2(1 - \sin^2 \theta)} = \sqrt{2 \cos^2 \theta} = \sqrt{2} \cos \theta$$

The expression is still rational in the variable.

Choosing the proper coefficient is not so difficult as it might seem. Suppose we have the radical

$$\sqrt{25 - 4x^2}$$

Here the constant term is 25. In order that we may be able to factor 25 out of the entire radicand, we must have

$$\sqrt{25 - 25 \sin^2 \theta}$$

Therefore, we must let $\quad 4x^2 = 25 \sin^2 \theta$

Then

$$\sqrt{25 - 4x^2} = \sqrt{25 - 25 \sin^2 \theta} = \sqrt{25(1 - \sin^2 \theta)} = \sqrt{25 \cos^2 0} = 5 \cos \theta$$

It might be wondered whether we could not also use $\cos \theta$ instead of $\sin \theta$ in our substitution of a trigonometric function. In examples such as

34.6 Choosing the Proper Coefficient

those mentioned, either $\sin\theta$ or $\cos\theta$ might be used. However, it is common practice to use $\sin\theta$. Two other substitutions that we shall use for certain types of integrands are $\tan\theta$ and $\sec\theta$. These will be explained and shown in examples.

In using trigonometric substitution, it is well to observe the following steps carefully. When x (or any variable) in the integrand is replaced by some function of an angle θ, it must be remembered that:

(1) Each and every form of x, including dx, must be replaced with its equivalent in terms of the angle θ. It will be seen that dx does not ordinarily become simply $d\theta$.

(2) The integration itself is performed in terms of the trigonometric function.

(3) Finally, the resulting integral is reconverted to terms involving the original variable.

If a definite integral is indicated, evaluation can be done without changing the integral back to original terms, provided the limits are changed to the corresponding θ-limits. However, in the case of trigonometric substitution, it is usually best to change the integral back to terms involving the original variable.

The method of trigonometric substitution will be shown by examples.

EXAMPLE 10. Find by the trigonometric substitution

$$\int x\sqrt{9 - 4x^2}\,dx$$

Solution. Although this problem can be worked out by the power rule, we use it to illustrate trigonometric substitution. We make the following substitutions: Let

$$4x^2 = 9\sin^2\theta$$

To find dx, we first solve for x:

$$2x = 3\sin\theta$$

$$x = \frac{3}{2}\sin\theta$$

$$dx = \frac{3}{2}\cos\theta\,d\theta$$

Then $\int x\sqrt{9 - 4x^2}\,dx$

becomes
$$\int \frac{3}{2}\sin\theta\sqrt{9 - 9\sin^2\theta}\,\frac{3}{2}\cos\theta\,d\theta$$

$$= \int \frac{3}{2}\sin\theta\,\sqrt{9(1 - \sin^2\theta)}\,\frac{3}{2}\cos\theta\,d\theta$$

$$= \int \frac{9}{4}\sin\theta\sqrt{9\cos^2\theta}\,\cos\theta\,d\theta = \frac{27}{4}\int \sin\theta\cos^2\theta\,d\theta$$

We can now apply the power rule on $\cos\theta$. The integrating factor is the negative ($-\sin\theta$). The factor, -1, can be supplied, and the integral becomes

$$-\frac{27}{4}\left(\frac{\cos^3\theta}{3}\right) = -\frac{9}{4}\cos^3\theta$$

We omit the constant of integration, C, until the result has been changed back to x-terms.

The reverse substitution back to x-terms is most easily done by means of a *transformation triangle*. We sketch a convenient right triangle and show θ as one of the acute angles (Fig. 34.1). The values for the sides of the triangle are taken from one of the first substitution equations. Note that we have stated

$$2x = 3 \sin \theta$$

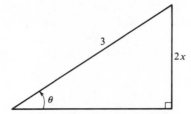

Fig. 34.1

Dividing both sides of the equation by 3, we get

$$\frac{2x}{3} = \sin \theta$$

Now the two values, $2x$ and 3, of the fraction are placed *on the proper sides of the triangle* so that their ratio will correctly represent the sine of the angle θ. Then the third side of the triangle is computed by the Pythagorean rule. It is

$$\sqrt{9 - 4x^2}$$

Note that for the integral in trigonometric form we obtained

$$-\frac{9}{4} \cos^3 \theta$$

From the transformation triangle, we see that

$$\cos \theta = \frac{\sqrt{9 - 4x^2}}{3}$$

then

$$\cos^3 \theta = \frac{1}{27} (9 - 4x^2)^{3/2}$$

The integral becomes

$$\left(-\frac{9}{4}\right)\left(\frac{1}{27}\right)(9 - 4x^2)^{3/2} + C = -\frac{1}{12}(9 - 4x^2)^{3/2} + C$$

The result can be checked by differentiation. If

$$y = -\frac{1}{12}(9 - 4x^2)^{3/2}$$

then

$$\frac{dy}{dx} = -\frac{1}{12}\left(\frac{3}{2}\right)(9 - 4x^2)^{1/2}(-8x) = x(9 - 4x^2)^{1/2}$$

34.7 SUBSTITUTION OF TANGENT AND SECANT FUNCTIONS

In some radicals we must make use of other substitutions besides the sine of an angle. For instance, the substitution of $\sin \theta$ in

$$\sqrt{x^2 + 1}$$

34.7 Substitution of Tangent and Secant Functions

would not help. The result would be

$$\sqrt{\sin^2 \theta + 1}$$

The radicand is not equal to any convenient trigonometric form. Instead, we look for a function whose square plus 1 is equal to the square of another trigonometric function. We recall the identity

$$\tan^2 \theta + 1 = \sec^2 \theta$$

Then we make the substitution, $x^2 = \tan^2 \theta$, and the radical becomes

$$\sqrt{\tan^2 \theta + 1} = \sqrt{\sec^2 \theta} = \sec \theta$$

The irrational algebraic quantity becomes rational when expressed as the *secant of an angle*.

The following example requires still a different substitution:

$$\sqrt{x^2 - 1}$$

Here the substitution of $\sin \theta$ or $\tan \theta$ would not help. Instead, we look for a function whose square minus 1 is equal to the square of another function. We recall the identity:

$$\sec^2 \theta - 1 = \tan^2 \theta$$

Then we make the substitution, $x^2 = \sec^2 \theta$, and the radical becomes

$$\sqrt{\sec^2 \theta - 1} = \sqrt{\tan^2 \theta} = \tan \theta$$

Again, the irrational quantity has become rational in terms of an angle.

EXAMPLE 11. Integrate by trigonometric substitution:

$$\int \frac{x^3 \, dx}{\sqrt{x^2 - 4}}$$

Solution. For our substitution, we let

$$x^2 = 4 \sec^2 \theta$$

then

$$x = 2 \sec \theta$$

and

$$dx = 2 \sec \theta \tan \theta \, d\theta$$

Note that in all cases the coefficient of the trigonometric function must correspond to the constant term in the radicand, which in this case is 4. Making the substitutions, we get

$$\int \frac{(8 \sec^3 \theta)(2 \sec \theta \tan \theta \, d\theta)}{\sqrt{4 \sec^2 \theta - 4}} = \int \frac{16 \sec^4 \theta \tan \theta \, d\theta}{\sqrt{4(\sec^2 \theta - 1)}}$$

$$= \int \frac{16 \sec^4 \theta \tan \theta \, d\theta}{\sqrt{4 \tan^2 \theta}} = \int \frac{16 \sec^4 \theta \tan \theta \, d\theta}{2 \tan \theta} = \int 8 \sec^4 \theta \, d\theta$$

$$= 8 \int \sec^2 \theta \sec^2 \theta \, d\theta = 8 \int \sec^2 \theta (\tan^2 \theta + 1) \, d\theta$$

$$= 8 \int \sec^2 \theta \tan^2 \theta \, d\theta + 8 \int \sec^2 \theta \, d\theta = 8 \left[\frac{\tan^3 \theta}{3} + \tan \theta \right]$$

To change the integral back to x-terms we make use of the transformation triangle, in which we show one acute angle as angle θ (Fig. 34.2). The values to be placed on the sides are taken from one of the first statements in our substitution:

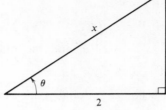

Fig. 34.2

Dividing both sides by 2, we get

$$\frac{x}{2} = \sec \theta$$

Now the two values, x and 2, of the fraction are placed on the sides of the triangle in such a position that their ratio will correctly represent the secant of θ. Then the third side is found to be

$$\sqrt{x^2 - 4}$$

Now we note that the integral calls for the tangent of the angle as well as $\tan^3 \theta$. From the triangle,

$$\tan \theta = \frac{(x^2 - 4)^{1/2}}{2}$$

then

$$\tan^3 \theta = \frac{(x^2 - 4)^{3/2}}{8}$$

Then the integral, in x-terms, becomes

$$(8)\left[\frac{(x^2 - 4)^{3/2}}{(8)(3)} + \frac{(x^2 - 4)^{1/2}}{2}\right] = \left(\frac{8}{24}\right)(x^2 - 4)^{1/2}(x^2 - 4 + 12)$$

$$= \frac{1}{3}(x^2 - 4)^{1/2}(x^2 + 8) + C$$

The result can be checked by differentiation:

$$\frac{d}{dx}\left[\left(\frac{1}{3}\right)(x^2 - 4)^{1/2}(x^2 + 8) + C\right]$$

$$= \frac{1}{3}(x^2 - 4)^{1/2}(2x) + (x^2 + 8)\left(\frac{1}{2}\right)(x^2 - 4)^{-1/2}(2x)$$

$$= \frac{1}{3}\left(\frac{2x(x^2 - 4) + x(x^2 + 8)}{(x^2 - 4)^{1/2}}\right) = \frac{1}{3}\left(\frac{2x^3 - 8x + x^3 + 8x}{(x^2 - 4)^{1/2}}\right)$$

$$= \frac{1}{3}\left(\frac{3x^3}{(x^2 - 4)^{1/2}}\right) = \frac{x^3}{(x^2 - 4)^{1/2}}$$

EXAMPLE 12. Integrate by trigonometric substitution:

$$\int \frac{\sqrt{x^2 + 9}\, dx}{x^2}$$

34.8 Three Trigonometric Substitutions

Solution. (Incidentally, a quick check of the powers of x will show that algebraic substitution could not be used in this problem.) For trigonometric substitution, we note the plus (+) in the radicand. Then we must replace the x^2 with a function whose square plus 1 is equal to the square of another function. We therefore, let

$$x^2 = 9 \tan^2 \theta$$

then
$$x = 3 \tan \theta$$
and
$$dx = 3 \sec^2 \theta \, d\theta$$

Then the problem becomes

$$\int \frac{\sqrt{9 \tan^2 \theta + 9} \, (3)(\sec^2 \theta) \, d\theta}{9 \tan^2 \theta} = \int \frac{\sqrt{9(\tan^2 \theta + 1)} \sec^2 \theta \, d\theta}{3 \tan^2 \theta}$$

$$= \int \frac{3 \sec \theta \sec^2 \theta \, d\theta}{3 \tan^2 \theta} = \int \frac{\sec^3 \theta \, d\theta}{\tan^2 \theta} = \int \sec \theta \csc^2 \theta \, d\theta$$

Integrating by parts, we let

$$u = \sec \theta \qquad dv = \csc^2 \theta \, d\theta$$
$$du = \sec \theta \tan \theta \, d\theta \qquad v = -\cot \theta$$

Then the integral becomes

$$\int \sec \theta \csc^2 \theta \, d\theta = -\sec \theta \cot \theta + \int \sec \theta \tan \theta \cot \theta \, d\theta$$
$$= -\csc \theta + \int \sec \theta \, d\theta = -\csc \theta + \ln (\sec \theta + \tan \theta)$$

Now we sketch the transformation triangle showing the acute angle θ (Fig. 34.3). From our statement, $x = 3 \tan \theta$, we get $x/3 = \tan \theta$. Then the numbers, x and 3, are placed on the triangle so that their ratio represents the tangent of the angle θ. The third side is then found to be

$$\sqrt{x^2 + 9}$$

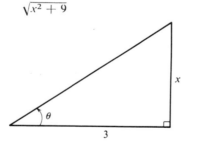

Fig. 34.3

Choosing the correct values from the triangle, we get the integral in x-terms:

$$-\frac{(x^2 + 9)^{1/2}}{x} + \ln \left(\sqrt{x^2 + 9} + x\right) + C$$

The constant, $-\ln 3$, is included in the general constant C.

34.8 THREE TRIGONOMETRIC SUBSTITUTIONS

In rationalizing an irrational algebraic expression through the substitution of a trigonometric function, we need make use of only one of the following identities:

$$1 - \sin^2 \theta = \cos^2 \theta$$
$$\tan^2 \theta + 1 = \sec^2 \theta$$
$$\sec^2 \theta - 1 = \tan^2 \theta$$

The result comes about because of an important fact concerning squares. For example, the following expression is not rational in r or s:

$$\sqrt{r^2 + s^2}$$

However, if we know the sum, $r^2 + s^2$, is equal to a third square, say t^2, then we can say $\sqrt{r^2 + s^2} = \sqrt{t^2} = t$, a rational quantity. In like manner, the expression $\sqrt{u^2 - v^2}$ is not rational. However, if we know that the difference, $u^2 - v^2$, is equal to a third square, say w^2, then we can say $\sqrt{u^2 - v^2} = \sqrt{w^2} = w$, a rational quantity.

To be able to rationalize an irrational expression in this way, it is lucky for us that the following relations are true:

$$1 - \sin^2 \theta = \cos^2 \theta \qquad \tan^2 \theta + 1 = \sec^2 \theta \qquad \sec^2 \theta - 1 = \tan^2 \theta$$

The three types of substitution may be stated in general terms. If u^2 represents an algebraic variable, and a^2 represents a constant, then

(1) if we have the form $\sqrt{a^2 - u^2}$, we substitute $u^2 = a^2 \sin^2 \theta$;
(2) if we have the form $\sqrt{u^2 + a^2}$, we substitute $u^2 = a^2 \tan^2 \theta$;
(3) if we have the form $\sqrt{u^2 - a^2}$, we substitute $u^2 = a^2 \sec^2 \theta$.

In some types of problem, of course, trigonometric substitution cannot be used. It is usually used where the form x^2 (or u^2) appears under the radical sign and then only if the radical has the index 2; that is, the indicated square root. It cannot be used where we have an indicated cube root. It cannot be used for the cube root of a cube form, for the cube root of a square form, or for the square root of a cube form. To rationalize a radical, it is used only for the *square root* of a *square form*. The reason is that its use depends on the trigonometric identities involving squares as stated above. However, it should be mentioned that there are other situations in which certain other trigonometric substitutions are used, but those we have described are some of the most important ones.

EXERCISE 34.3

Integrate by trigonometric substitution:

A. *Very easy.*

1. $\int \dfrac{dx}{x^2(x^2 - 4)^{1/2}}$
2. $\int \dfrac{dx}{x^2(x^2 + 4)^{1/2}}$
3. $\int \dfrac{dx}{x^2(4 - x^2)^{1/2}}$
4. $\int \dfrac{(9 - x^2)^{1/2} \, dx}{x}$
5. $\int \dfrac{(x^2 - 16)^{1/2} \, dx}{x^4}$
6. $\int \dfrac{(4 - x^2)^{1/2} \, dx}{x^4}$

B. *Easy.*

7. $\int \dfrac{dx}{x(x^2 - 25)^{1/2}}$
8. $\int \dfrac{x^3 \, dx}{(9 - 4x^2)^{1/2}}$
9. $\int \dfrac{x^2 \, dx}{(4 - 9x^2)^{3/2}}$
10. $\int x^3(4 - x^2)^{1/2} \, dx$
11. $\int x^3(x^2 - 9)^{1/2} \, dx$
12. $\int x^3(4 - x^2)^{3/2} \, dx$

34.8 Three Trigonometric Substitutions

13. $\int \dfrac{(x^2 - 4)^{1/2} \, dx}{x}$

14. $\int \dfrac{(9 - 4x^2)^{1/2} \, dx}{x^3}$

15. $\int \dfrac{(x^2 - 4)^{3/2} \, dx}{x}$

16. $\int \dfrac{(1 - x^2)^{3/2} \, dx}{x^4}$

17. $\int \dfrac{x^3 \, dx}{(x^2 - 9)^{1/2}}$

18. $\int \dfrac{dx}{x^4(1 - x^2)^{1/2}}$

C. Fairly easy.

19. $\int \dfrac{(x^2 - 4)^{1/2} \, dx}{x^3}$

20. $\int \dfrac{dx}{x^3(x^2 - 9)^{1/2}}$

21. $\int \dfrac{x^2 \, dx}{(4 - 9x^2)^{1/2}}$

22. $\int x^3(x^2 - 4)^{3/2} \, dx$

23. $\int x^3(4 + x^2)^{1/2} \, dx$

24. $\int x^3(1 + x^2)^{3/2} \, dx$

25. $\int \dfrac{(1 - x^2)^{3/2} \, dx}{x^4}$

26. $\int \dfrac{(x^2 + 1)^{1/2} \, dx}{x^4}$

27. $\int \dfrac{x^3 \, dx}{(4 + x^2)^{5/2}}$

D. Difficult.

28. $\int \dfrac{x^2 \, dx}{(x^2 - 4)^{1/2}}$

29. $\int \dfrac{(x^2 + 4)^{1/2} \, dx}{x^3}$

30. $\int \dfrac{(x^2 + 4)^{1/2} \, dx}{x}$

31. $\int \dfrac{(x^2 + 1)^{1/2} \, dx}{x^2}$

32. $\int \dfrac{x^2 \, dx}{(x^2 - 4)^{3/2}}$

33. $\int \dfrac{x^2 \, dx}{(x^2 + 1)^{1/2}}$

Integrate and evaluate:

34. $\int_0^2 \dfrac{(x^2 - 4)^{3/2} \, dx}{x}$

35. $\int_0^4 \dfrac{dx}{x^2(x^2 + 9)^{1/2}}$

36. $\int_3^4 \dfrac{x^2 \, dx}{(25 - x^2)^{1/2}}$

chapter
35

Some Applications: Liquid Pressure; Work

35.1 PROBLEMS IN PHYSICS

In this chapter we consider two types of problems in physics whose solutions are simplified by the use of integral calculus. One type of problem is that of finding the total force on a surface due to the pressure of a liquid. The other type has to do with computing the total work done by a variable force. We shall see how such problems are often conveniently solved by calculus.

35.2 LIQUID PRESSURE

In designing tanks, ships, dams, and other structures, engineers must take into account the amount of force on a surface due to liquid pressure. This is true whether the pressure is from the inside as in tanks, the outside as in the case of ships, or on only one side as in dams. They must consider such problems as the following:

1. What is the total force on one side or on one end of a tank full of water?
2. What is the force on one face of a dam? How does this force vary at different water levels?
3. A patching steel plate has been welded over a hole in a ship's hull. What is the total force on this patch when the ship is afloat at sea?
4. A cylindrical tank 6 ft in diameter lies on its side. What is the force on one end when the tank is half full of oil?

35.3 FORCE ON A HORIZONTAL SURFACE

The force of liquid pressure is due to the weight of the liquid. One cubic foot of water weighs approximately 62.4 lb. Therefore, the downward force of one cubic foot of water is 62.4 lb. Imagine that we have this amount of water in the shape of a cube (Fig. 35.1). Then 62.4 lb rests on an area of 1 sq ft.

35.3 Force on a Horizontal Surface

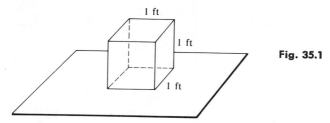

Fig. 35.1

Now let us suppose a tank has a base of 12 sq ft and contains water to a depth of 1 ft. (Fig. 35.2) The downward force is then

$$(12)(62.4) = 748.8 \text{ pounds}$$

Note that the force is equal to the area of the base times the weight per unit of volume. The weight per unit of volume is called *density*.

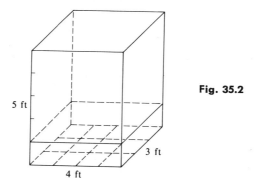

Fig. 35.2

Finally, consider the same tank filled with water to a depth of 5 ft. Then we have a 5-ft column of water resting on 12 sq ft of area (Fig. 35.3). The downward force on the bottom of the tank is the total weight of the water, or

$$\text{Force} = (12)(5)(62.4) = 3744 \text{ pounds}$$

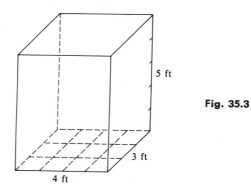

Fig. 35.3

We can now state the rule as a formula:

$$\text{Force} = (\text{area})(\text{height})(\text{weight})$$

or
$$F = (A)(h)(wt)$$

Note that the force *per square foot* on the bottom is

$$(5)(62.4) \text{ pounds} = 312 \text{ pounds}$$

Since the height is everywhere 5 ft, the force *per unit of area* on the bottom of the tank is equal to the *height times* the *density*.

35.4 VARIABLE HEIGHT

One important principle in physics states that at a particular point in a liquid, the pressure on a particle is the same in all directions, up, down, and from side to side (Fig. 35.4). This pressure is determined by the height of the column of liquid above the particle. If the pressure is 3 lb at the top of the particle, it is also 3 lb at the bottom, assuming the particle to be an infinitesimal. The pressure on any side of the particle is also 3 lb.

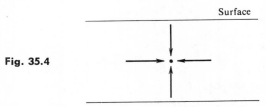

Fig. 35.4

The force of liquid pressure on the side wall of a tank is not everywhere uniform per unit of area since the height of the water column above is a variable. In the tank mentioned and shown in Figure 35.3, containing water to a depth of 5 ft, the force on the bottom is everywhere 312 lb/sq ft. However, the force on the side wall is not uniformly 312 lb/sq ft because the water column above is not everywhere 5 ft high.

It is true the pressure at the bottom is the same as the pressure on the side wall right next to the bottom. But the height of the water column is a variable. If we consider the entire side wall of the tank, the height of the water column varies from 5 ft to 0 ft. In Figure 35.5, this variable

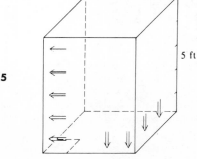

Fig. 35.5

35.4 Variable Height

force is indicated by arrows where the force corresponds to the width of the arrows. Even for one square foot of the side wall near the bottom of the tank, the height varies from 5 ft to 4 ft.

Consider a strip of the side wall 1 ft wide extending horizontally near the bottom of the tank (Fig. 35.6). If this strip is next to the bottom, then the height of the water column pressing on the strip varies from 4 ft to 5 ft. If this strip, 1 ft wide and 3 ft long, were at the bottom of the tank, the water column above it would everywhere be 5 ft high, and the total force on the strip would be $(3) \times (5) \times (62.4)$ lb = 936 lb. However, since the height varies from 4 ft to 5 ft, we take its *average* height and get

$$f = (3)(4.5)(62.4) = 842.4 \text{ pounds}$$

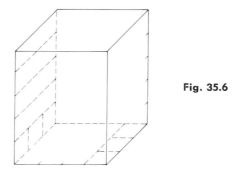

Fig. 35.6

To get the total force on the end, we can take the average height of each of the five horizontal strips (4.5, 3.5, 2.5, 1.5, and 0.5 ft), compute the force on each strip, and find the total force is 2340 lb. This is essentially what we do in computing the total force by calculus. We shall now work the same problem by calculus.

EXAMPLE 1. Find the total force on one end of a tank 3 feet wide and 5 feet high when the tank is full of water (Fig. 35.7).

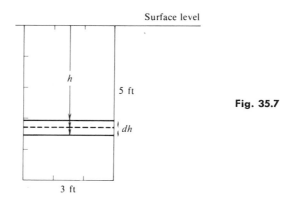

Fig. 35.7

Solution. We imagine the end wall divided into horizontal strips of area, just as we do in finding total areas by calculus. One strip, the element of area, dA, is shown. The length of this element is the constant 3 ft. If we let h represent the height of the water column above the element, then the width of the element we denote by *dh*, or differential of *h*. The area of the element is therefore 3 *dh*.

For the height of the water column above the strip, we can take *h* to either the bottom of the strip, the top, or the middle, since we consider the quantity *dh* an infinitesimal approaching zero as the number of strips becomes infinite.

Now we express the force on the element of area and call this the element of force, or *dF*. Then

$$dF = (dA)(h)(wt)$$

or
$$dF = (3\,dh)(h)(62.4)$$

To find the total force on the end of the tank, we sum up the elements of force by integration from $h = 0$ to $h = 5$, and get

$$F = (3)(62.4)\int_0^5 h\,dh = (187.2)\left[\frac{h^2}{2}\right]_0^5 = 93.6(25) = 2340 \text{ pounds}$$

EXAMPLE 2. A rectangular steel plate, 8 ft long and 4 ft wide, is submerged vertically in a liquid weighing 55 lb/cu ft so that the 8-ft edges are parallel to the surface of the liquid, one edge 3 ft below the surface and the other edge 7 ft below the surface. Find the total force on one side of the plate (Fig. 35.8).

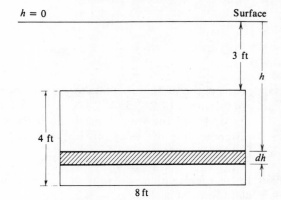

Fig. 35.8

Solution. We imagine the area of the plate is divided into horizontal strips, one of which is shown. Taking *h* as the height of the liquid column above the strip, we have the line $h = 0$ at the surface of the liquid. Then the width of the strip is *dh*. Since the length of the strip is the constant 8 ft, we have for the element of area,

$$dA = 8\,dh$$

Then the element of force becomes

$$dF = (55)(8)(h)(dh)$$

Integrating, from $h = 3$ to $h = 7$

$$F = 440\int_3^7 h\,dh = 440\left[\frac{h^2}{2}\right]_3^7 = 220(49 - 9) = 8800 \text{ lb}$$

Note. Since the force will depend on the weight of the liquid, we include the weight per unit volume in the formula for force and usually denote this weight by δ (small delta). The formula for the element of force then becomes

$$dF = (\delta)(h)(dA)$$

35.4 Variable Height 469

EXAMPLE 3. A triangular sheet of metal has a base of 10 ft and an altitude of 4 ft. It is submerged in water vertically, with the base of the triangle parallel to and 6 ft below the surface, and the vertex of the triangle 2 ft below the surface. Find the force on one side of the plate (Fig. 35.9).

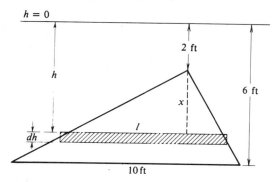

Fig. 35.9

Solution. Take a horizontal element of area as shown. The area of this element is, of course, (length) times (width) or

$$dA = (l)(dh)$$

For the element of force we have

$$dF = (62.4)(h)(l)(dh)$$

Now we must eliminate l by expressing it as a function of h.

To find the relation between h and l, we introduce x as the distance from the element of area to the vertex of the triangle. Then by similar triangles we have

$$\frac{x}{l} = \frac{4}{10} \quad \text{or} \quad l = \frac{5}{2}x$$

Now we note that $x = h - 2$. Substituting, we get

$$l = \frac{5}{2}(h - 2)$$

Substituting in the formula for the element of force, we get

$$dF = (62.4)(h)\frac{5}{2}(h - 2)(dh)$$

or

$$dF = 156h(h - 2)\, dh$$

Integrating,

$$F = 156 \int_{h=2}^{h=6} (h^2 - 2h)\, dh = 156 \left[\frac{h^3}{3} - h^2 \right]_2^6 = 5824 \text{ lb}$$

EXAMPLE 4. A tank has a parabolic end 6 ft across the top. The tank is 8 ft deep. (Fig. 35.10.) It is filled with oil weighing 58 lb/cu ft. Find the force on one end of the tank.

Solution. We could take the horizontal axis at the surface of the liquid, but it is sometimes more convenient to take it at some other place. In this example, writing the equation of the parabola is somewhat more simple if the axis is taken through the vertex of the parabola. We shall therefore use the x-axis and the y-axis for writing the equation of the parabola with the vertex at the origin.

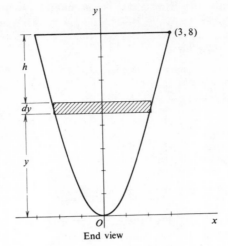

Fig. 35.10

For the horizontal element of area we have

$$\text{length} = 2x \quad \text{width} = dy$$

Then
$$dA = 2x\, dy \quad \text{and} \quad dF = (58)(h)(2x)\, dy$$

We must now express the variables, x and h, in terms of y, since dy denotes y as the variable of integration. First, note that $h = 8 - y$.

For the equation of the parabola, we have $x^2 = Ky$. To find the value of K, we note that the point $(3,8)$ lies on the parabola. Substituting these values, $9 = 8K$, or $K = 9/8$. Then the equation of the parabola becomes $8x^2 = 9y$. Solving for x,

$$x = \frac{3}{2\sqrt{2}} y^{1/2}$$

Substituting values,

$$dF = (58)(8 - y)\left(\frac{3}{\sqrt{2}}\right)(y^{1/2})\, dy$$

or
$$dF = 87\sqrt{2}(8 - y)(y^{1/2})\, dy$$

Integrating,
$$F = 87\sqrt{2} \int_{y=0}^{y=8} (8y^{1/2} - y^{3/2})\, dy$$

$$F = 87\sqrt{2} \left[\frac{16y^{3/2}}{3} - \frac{2y^{5/2}}{5}\right]_0^8 = 5939\tfrac{1}{5} \text{ lb}$$

EXERCISE 35.1

1. A rectangular plate, 11 ft long and 7 ft wide, is submerged with its plane vertical, in a liquid weighing 60 lb/cu ft, so that one 11-ft edge is parallel to and 5 ft below the surface of the liquid, and the other 11-ft edge 12 ft below the surface. Find the total force on one side of the plate.
2. Find the force on the top half of the plate in Problem 1; also find the force on the bottom half.
3. A triangular plate has a base of 10 ft and an altitude of 6 ft. It is submerged

vertically in water so that the base is 8 ft below and parallel to the surface, and the vertex is 2 ft below the surface. Find the total force on one side of the plate.

4. Find the force on one side of the plate in Problem 3, if the water level is lowered to the vertex of the triangle.
5. Find the total force on the plate in Problem 3 if the water level is lowered to a point 1 ft below the vertex.
6. To what point should the water level in Problem 3 be lowered so that the force on the side of the plate be half as much as in Problem 3?
7. If the plate in Problem 3 is inverted, so that the 10-ft base is parallel to and 2 ft below the surface, and the vertex is 8 ft below the surface, what is the total force on the plate?
8. A ship has been damaged and has a hole in one side in the shape of a triangle with a horizontal base of 6 ft and an altitude of 4 ft. A steel patching plate has been welded over the hole. Find the total force on the patch when the vertex of the triangle is 10 ft below the water level. (Take the weight of salt water to be 64.5 lb/cu ft.)
9. If the ship's damage in Problem 8 is a square hole, 4 ft by 4 ft, with a diagonal of the square horizontal and the top vertex of the square 6 ft below the water line, find the total force on the patch. What is the difference in force on the top half and the bottom half of the square?
10. A rectangular gate in a dam is 20 ft long and 16 ft high. When the water level is at the top of the gate, what is the total force on the gate?
11. How much must the water level in Problem 10 be lowered in that the force on the gate be halved?
12. A horizontal tank has parabolic ends, 6 ft across the top and 6 ft deep. What is the force on one end of the tank when full of water? What is the force when the water is 3 ft deep in the tank?
13. What is the force on the dam gate in Problem 10 if the water level is 6 ft above the top of the gate?
14. An irrigation ditch, 6 ft deep, has a cross section in the shape of a trapezoid. The ditch is 6 ft wide at the bottom and 12 ft at the top. A gate is placed vertically at one end of the ditch. Find the force on the gate when the water level is at the top of the gate.
15. Find the force on the gate in Problem 14, when the water is only 4 ft deep.
16. A swimming pool is 60 ft long with a sloping bottom. It is 4 ft deep at one end and 10 ft deep at the other. Find the total force on one side when the water level is 1 ft below the top of the pool.
17. A rectangular tank has a square opening, 3 ft by 3 ft, on one side. The top edge of the square is parallel to and 6 ft below the top of the tank. This square hole is closed by two triangular gates, closing along a diagonal of the square. Find the total force on each gate when the tank is full of water.
18. Find the total force of wet concrete (120 lb/cu ft) on one side of a wall 10 ft long if the concrete extends 8 ft up the wall.
19. The ends of a trough are in the shape of equilateral triangles 1.5 ft on a side. Find the force on one end of the trough when the trough is full of water.
20. A dam has the shape of an isosceles trapezoid with parallel sides horizontal. The top is 1000 ft long and the bottom 600 ft long. If the dam is 400 ft high, find the total force on the side of the dam when the water level is at the top of the dam.

35.5 WORK

Work is produced as the result of some force operating through some distance. Work implies motion. If a force of 20 lb is applied but produces no motion, then no work has been done. Work, then, is the result of force and motion.

If a force of one pound operates through a distance of one foot, the work done is defined as one *foot-pound*. If a constant force of 5 pounds operates through a distance of 6 feet, the amount of work done is

$$(6 \text{ ft})(5 \text{ lb}) = 30 \text{ ft-lb}$$

Thus the definition of work is given by the formula

$$(\text{work}) = (\text{force})(\text{distance})$$

or

$$w = Fs$$

Lifting a weight of 50 lb through a vertical distance of 20 ft against gravity represents 1000 ft-lb of work done. Note that if both force and distance are constants, then the amount of work done can be found simply by arithmetic.

When either force or distance is a variable, the problem becomes more complicated. Then its solution is simplified by using calculus. A variable force or distance appears in many types of problems, such as, for example, (1) in the compression or elongation of a spring, (2) in pumping water into or out of a tank; and (3) in winding up a chain onto a drum or windlass from which the chain is suspended.

35.6 A SPRING PROBLEM

In the compression or elongation of a spring, the force applied is a variable since the force at any particular point depends on the amount of distortion. In fact, the distortion is proportional to the distorting force, according to *Hooke's law*. This means that if one end of a spring is fixed and the free end displaced a distance of 2 in. by a 5-lb force, then it will be displaced 8 in. by a 20-lb force, provided that the spring is not stretched beyond its elastic limit.

The amount of force necessary to stretch or compress a spring one *unit of length* is called the *spring modulus* (sometimes called the *spring constant*). This modulus, often denoted by K, is a constant that is a characteristic of any particular spring. For example, if a certain spring is stretched 1 in. by a force of 2 lb, we say the modulus K for that particular spring is 2 lb/in.; that is, $K = 2$. If another spring is stretched 2 in. by a force of 6 lb, the modulus of that spring can be found by the ratio

$$K = \frac{6 \text{ pounds}}{2 \text{ inches}} = 3 \text{ lb/in.}$$

Note especially that the spring modulus is the number of *pounds per inch*, not the number of inches per pound. The single unit is *length*. In general, K can be found by the formula

$$K = \frac{\text{force}}{\text{length}}$$

35.6 A Spring Problem

Since *work* is often computed in foot-pounds rather than in inch-pounds, we may express the modulus in pounds per foot. If a force of 3 lb is required to stretch a spring 1 in., this force is equivalent to 36 lb/ft. The modulus can then be stated

$$3 \text{ lb/in.} = 36 \text{ lb/ft}$$

In finding the modulus in pounds per foot directly, it is often best to express values in fractional form. For example, suppose a spring is stretched 2.5 in. by a force of 3.5 lb. If we wish the modulus in pounds per foot, we can first change 2.5 inches to feet:

$$2.5 \text{ in.} = \frac{2.5}{12} \text{ ft}$$

Then
$$4.5 \div \frac{2.5}{12} = (4.5)\left(\frac{12}{1.5}\right) = 16 \text{ lb/ft}$$

It is essential in any problem that the distances taken correspond to the unit of distance in the spring modulus.

We have said that the force exerted in stretching a spring varies with the amount of distortion. For example, if a spring has a modulus K of 3 lb/in., the force required to maintain the distortion at any length x is equal to $3x$ pounds. In general, the force required is given by

$$F = Kx$$

The force is therefore a variable.

To find the amount of work done in stretching a spring over a certain distance, we let x represent the distance of elongation, a variable. In Figure 35.11, AB represents the normal length of the spring. Then the force required to hold the spring stretched to any point is

$$Kx \text{ pounds}$$

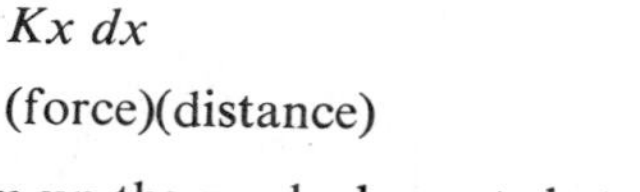

Fig. 35.11

Now we take this force, Kx, as operating through an increment of distance, dx. Then the work done through the distance dx is called the *element of work*, denoted by dw. For the element of work we have

$$dw = Kx\, dx$$

Again, note that work = (force)(distance)

Now, all we need to do is to sum up the work elements between the given limits of distortion. In general, for the total work done, we have

$$w = \int_a^b Kx\, dx$$

where a and b are the limits of distortion.

EXAMPLE 5. A certain spring whose normal length is 20 in. is stretched 1.5 in. by a force of 4.5 lb. Find the modulus in lb/in. and in lb/ft. How much work is done in stretching the spring from its normal length to a length of 24 in.? From 24 to 28 in.?

Solution. For the modulus in lb/in., we have

$$4.5 \div 1.5 = 3 \text{ lb/in.}$$

For the modulus in lb/ft, we have

$$4.5 \div \frac{1.5}{12} = (4.5)\left(\frac{12}{1.5}\right) = 36 \text{ lb/ft}$$

Let us first find the work in inch-pounds and then in foot-pounds. In Figure 35.12, AB represents the normal length of the spring, 20 in. Then let

$x = $ the number of inches elongated

then $3x = $ force (in pounds) exerted at any point

and $dx = $ the increment of elongation

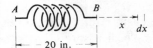

Fig. 35.12

Then the element of work, dw, is given by

$$dw = 3x \, dx$$

For stretching the spring from its normal length, 20 in., to a length of 24 in., the limits of elongation are zero and 4. Integrating,

$$\text{work} = \int_0^4 3x \, dx = \frac{3x^2}{2}\bigg]_0^4 = 24 \text{ in.-lb}$$

In stretching the spring from 24 to 28 in., the limits are 4 and 8. Then

$$w = \frac{3x^2}{2}\bigg]_4^8 = 96 - 24 = 72 \text{ in.-lb.}$$

If we take the modulus in pounds per *foot*, we must take the limits also in *feet*. For the element of work, we have

$$dw = 36x \, dx$$

For the first part of the problem, we have the limits of zero and $\frac{1}{3}$. Then

$$w = \int_0^{1/3} 36x \, dx = 18x^2 \bigg]_0^{1/3} = 2 \text{ ft-lb}$$

For the second part,

$$w = 18x^2 \bigg]_{1/3}^{2/3} = 8 - 2 = 6 \text{ ft-lb}$$

The answers in foot-pounds of work correspond to the answers in inch-pounds.

EXERCISE 35.2

1. A spring whose normal length is 12 in. has a modulus of 2.8 lb/in. What is the modulus in lb/ft? Find the amount of work done in stretching the spring from its normal length to a length of 15 in.

2. Find the amount of work done in stretching the spring in Problem 1 from a length of 15 in. to a length of 18 in.
3. A spring whose normal length is 15 in. is elongated to a length of 18 in. by a 5-lb force. Find the modulus in lb/ft and the work done in stretching the spring to a length of 21 in.
4. A spring is suspended at one end. (Take the weight of the spring itself as negligible.) Now a 5.25-lb weight is attached to the bottom end of the spring, which stretches the spring a distance of 2 and $\frac{1}{3}$ in. Find the modulus. Then find the amount of work done by gravity in stretching the spring.
5. A spring 24 in. long is stretched 1.4 in. by a force of 5.6 lb. Find the work done in stretching the spring from its normal length to a length of 28 in.
6. Find the work done by stretching the spring in Problem 5 from its normal length to a length of 27 in. How much work is done in stretching it from 27 in. to 30 in.?
7. A force of 3.6 lb stretches a spring 15 in. long to a length of 17.7 in. How much work is done in stretching the spring to 18.5 in.?
8. A force of 3.6 lb stretches a 20-in. spring 2.4 in. How much work is done in stretching it to a length of 23 in.? How much work is done in stretching it an additional 3 in.?
9. A force of 3.5 lb stretches a spring 2.1 in. How much work is done in stretching it 3 in.? How much work is done in stretching it an additional 3 in.?
10. Compare the amount of work done in elongating a spring a units with the amount of work done in elongating it $2a$ units. Take K as the modulus.
11. A 24-in. spring requires 60 lb to stretch it 2.4 in. Find the modulus. Then find the work done in stretching the spring to 30 in.
12. A railroad car bumper has a spring modulus of 48,000 lb/in. Find the work done when the spring is compressed $\frac{3}{16}$ in.
13. If 24 in.-lb of work is done in stretching a spring 4 in., find the spring modulus.
14. If 4.5 ft-lb of work is done in stretching a 20-in. spring to a length of 26 in., find the spring modulus.
15. A certain spring is elongated 3 in. from its normal length. Now if 3 ft-lb of work is done in stretching it another 3 in., find the spring modulus.
16. A spring has a modulus of 28.8 lb/ft. If 2.5 ft-lb of work is done in stretching the spring to a total length of 20 in., find the normal length of the spring.
17. A spring has a modulus of 38.4 lb/ft. How much will 4.8 ft-lb of work stretch the spring?
18. A certain spring whose modulus is 2.4 lb/in., is first stretched 3 in. Then how much farther will the spring be stretched by 4 ft-lb of work?

35.7 A TANK PROBLEM

When water, or any liquid, is pumped into or out of a tank, work is done. In this type of problem, the variable is the distance each particle of the liquid must be raised. We therefore take an element of volume, denoted by dV, and express the distance this element must be raised. The element of volume is usually taken as a horizontal slice of the entire volume so that each particle in the slice will be raised the same distance. The weight per unit of volume of water (or other liquid) must also be taken into account. The method of finding the total work done is shown by examples.

EXAMPLE 6. A tank in the shape of a circular cylinder is 8 ft in diameter and 20 ft high (standing on end). Find the work done in filling the tank with water if the source of supply is at a level with the bottom of the tank (Fig. 35.13).

476   Some Applications: Liquid Pressure; Work

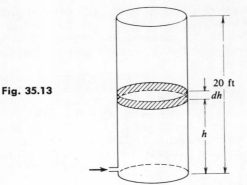

Fig. 35.13

Solution. Imagine the water in the tank divided into thin horizontal slices, or disks. First note that the disk at the bottom of the tank can be pumped into the tank with very little effort; therefore, little work is done. However, the disk at the very top must be raised approximately 20 ft.

We take a typical disk anywhere and call this the element of volume, dV. If this volume is expressed in cubic feet, then the weight of this element is (62.4)(dV) pounds. The work done in raising this disk of water through the distance h is called the element of work, dw, and we have

$$dw = (h)(62.4)(dV)$$

To find the equivalent of dV, we note that this disk is really a cylinder with radius = 4, and thickness equal to dh. For the element of work we have

$$dw = (62.4)(h)(\pi)(4^2)(dh)$$

or

$$dw = (998.4)(\pi)h\, dh$$

Integrating, between the limits, $h = 0$ to $h = 20$, we get

$$w = 998.4\pi \int_{h=0}^{h=20} h\, dh = 199{,}680\pi \text{ ft-lb}$$

Note. In a problem such as Example 1, if the pump is at a source of water supply some distance, say 5 feet, below the bottom of the tank, we usually place the axis, $h = 0$, at the water supply level and then integrate between the proper limits — say, from $h = 5$ to $h = 25$.

If the radius is a variable, as in a tank having the shape of a cone or a paraboloid, then the radius must be expressed in terms of the other variable. In such instances it may be more convenient to place the axis, $h = 0$ (or $y = 0$), through the vertex.

EXAMPLE 7. A tank in the shape of a paraboloid has a circular top 8 ft in diameter and a depth of 6 ft. If it is full of oil weighing 60 lb/cu ft, find the work done in pumping the oil to a distance of 3 ft above the top of the tank (Fig. 35.14).

Solution. For the simplest equation of the parabola, we make use of the x and y-axes, taking the x-axis through the vertex of the parabola and the y-axis along the axis of the paraboloid. Then we take a horizontal slice, or disk, as an element of volume. For this volume, we have

$$dV = \pi x^2\, dy$$

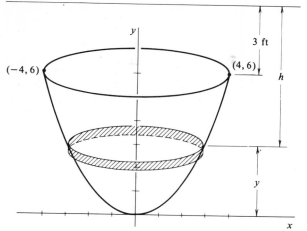

Fig. 35.14

Each element of volume must be raised through a distance h, 3 feet above the top of the tank. Then

$$dw = (60)(h)(\pi)(x^2)\, dy$$

Now we must express h and x^2 in terms of y. First note that $h = 9 - y$. To find the relation between x and y, we write the general equation of the parabola,

$$x^2 = Ky$$

Since the parabola passes through the point (4,6) we can find the value of K.

$$16 = 6K \quad \text{or} \quad K = \frac{8}{3}$$

Then the parabola is

$$x^2 = \left(\frac{8}{3}\right)y$$

Now we substitute for h and x^2 in the work element and get

$$dw = (60)(9 - y)(\pi)\left(\frac{8}{3}\right)(y)\, dy$$

or

$$dw = 160\pi(9y - y^2)\, dy$$

Integrating between the limits, $y = 0$ to $y = 6$, we get

$$w = 160\pi \int_0^6 (9y - y^2)\, dy = 14{,}400\pi \text{ ft-lb}$$

35.8 A CHAIN PROBLEM

Suppose a 100-ft chain weighing 3 lb/ft is suspended from a drum and is to be wound up on the drum. At the first instant the entire chain of 300 lb must be lifted against gravity. As the chain continues to be wound on the drum, the amount to be lifted decreases. For example, when 90 ft of the chain have been wound up, only 10 ft or 30 lb remain to be lifted. In this type of problem the length of the chain, and hence the weight, to be lifted is a variable. Finding the total work done in a problem of this type is illustrated in the following example:

EXAMPLE 8. A 60-ft chain weighing 1.6 lb/ft hangs from a drum on which it is to be wound. Two men have agreed to take turns winding it up. One says, "You wind up the first 30 ft, and I'll wind up the remaining 30 ft." How much work does each man do?

Solution. In Figure 35.15, we let x represent the part of the chain to be lifted at any time. Note that at the start the entire chain must be lifted. As the chain is wound on the drum, the weight to be lifted decreases. Since x represents the number of feet remaining to be lifted, the weight of this part is always $1.6x$ pounds.

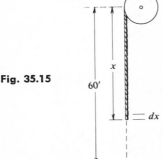

Fig. 35.15

Now we let this weight be lifted through the distance dx. The element of work is then

$$dw = 1.6 \, x \, dx$$

Summing up the elements by integration between proper limits, we have, for the first man,

$$w = 1.6 \int_{x=30}^{x=60} x \, dx = (1.6) \frac{x^2}{2} \Big]_{30}^{60} = 2160 \text{ ft-lb}$$

For the second man we have the limits, 0 to 30, and get 720 ft-lb. The results may be checked by computing the total work done between the limits, $x = 0$ to $x = 60$, which becomes 2880 ft-lb.

Note. If an extra constant weight is attached to the bottom end of the chain, this weight must be included in the weight lifted. In the foregoing example, if a 50-lb weight is attached to the bottom end of the chain, we should have

$$dw = (1.6x + 50) \, dx$$

EXERCISE 35.3

1. A cylindrical tank standing on end has a diameter of 6 ft and a height of 12 ft. If it is full of water, find the work done in pumping all the water to the top of the tank.
2. Find the work done in pumping all the water in the tank in Problem 1 to a height of 8 ft above the top of the tank.

3. A rectangular tank 8 ft long, 6 ft wide, and 10 ft deep is half full of water. Find the work done in pumping the water in the tank to a distance of 20 ft above the top of the tank.
4. A cylindrical tank 16 ft in diameter and 30 ft high stands on end. Find the work done in filling half the tank from a source 10 ft below the bottom of the tank.
5. A conical tank is 20 ft deep and has a circular top 12 ft in diameter. Find the work done in pumping this water into the tank to a depth of 10 ft through a vent in the bottom of the tank if the water supply is at the level of the bottom of the tank.
6. How much work is done in pumping the remaining top 10 ft of water into the tank in Problem 5.
7. When the tank in Problem 5 is full of water, how much work is done pumping the top 10 ft of water to the top of the tank?
8. How much work is done in pumping the bottom 10 ft of water in the tank in Problem 5 to the top of the tank?
9. A trough 12 ft long has triangular ends, 5 ft across the top and is 3 ft deep from the top to the vertex at the bottom. Find the amount of work done in filling the trough with water from a source of supply 8 ft below the bottom of the trough.
10. An oil tank in the shape of a paraboloid with vertex at the bottom is 10 ft deep and has a circular top 8 ft in diameter. It is full of oil weighing 60 lb/cu ft. The top 4 ft of oil is to be pumped to a height of 20 ft above the top of the tank. The remaining 6 ft of oil is to be pumped to a height of 12 ft above the top of the tank. Find the total work done in pumping the oil out of the tank.
11. A chain 80 ft long and weighing 2.2 lb/ft hangs from a drum on which it is to be wound. Find the work done in winding up 20 ft of the chain. Find the work done in winding up the entire chain.
12. Work Problem 11 if a 200-lb weight is attached to the bottom of the chain.
13. From a crane hangs a 100-ft cable weighing 1.2 lb/ft. Find the work done in lifting the cable alone a distance of 90 ft.
14. Work Problem 13 if a 500-lb weight is attached to the bottom of the cable and this weight is lifted 90 ft.
15. A chain 60 ft long and weighing 1.8 lb/ft hangs from a drum on which it is to be wound. A 200-lb weight is attached to the bottom end and this weight raised 20 ft. Then an additional 100-lb weight is attached to the bottom and all raised to a distance of 10 ft from the drum. Find the total amount of work done.
16. An icy stairway is to be sprinkled with salt. The stairway has a total rise of 40 ft and requires approximately 50 lb of salt to cover the treads sufficiently. A man weighing 170 lb takes a 60-lb bag of salt, opens the top slightly and lets the salt run out gradually as he walks from the bottom to the top. When he reaches the top of the stairway, he has 10 lb of salt left. If the salt ran out uniformly, how much work did he do in walking up the stairway?

chapter
36

Centroids

36.1 THE MEANING OF CENTROID

The term *centroid* is used in connection with areas, solids, arcs, and curved surfaces. It may have either two-dimensional or three-dimensional meaning. Although the problem of finding the centroid is much simpler in connection with a plane area, the meaning of the term is probably more easily understood in connection with a solid. We shall therefore begin the explanation with reference to a solid.

Suppose we have a solid of some particular mass (or weight). It may have a somewhat regular or an irregular shape (Fig. 36.1). Somewhere, usually within the solid, is a point called the *centroid* or *center of mass*. The centroid is the point around which the entire mass of the substance might be said to balance.

Fig. 36.1

If an object is suspended by a single, thin cord, the straight vertical line through the point of suspension will pass through the centroid of the solid. This fact provides a mechanical method of determining the location of the centroid. The object may be suspended in two or more positions.

The centroid is sometimes called the *center of gravity*, although the two terms are not entirely synonymous. If an object is taken out into space into the area of weightlessness, there would of course be no effect of gravity, and hence the term *center of gravity* would have no meaning. Yet the object would still have a center of mass, or *centroid*.

The centroid is often within the material substance of an object itself but it need not be so. For example, the centroid of a metal ring is not in a portion of the metal. In a circular ring the centroid is at the center of the circle provided the ring is of uniform size and density. The centroid is not in any portion of the material of a horseshoe, a coffee cup, or probably a kitchen stool. However, for each of these, there is a point in space that represents the centroid or center of gravity of the object such that all the mass of the object is completely balanced about this point.

36.2 FIRST MOMENT

To understand how to find the centroid, we must first understand what is meant by *moment*. The concept of *moment* can be conveniently shown by the common seesaw (Fig. 36.2). This is, of course, simply the law of the lever, the seesaw being a lever of the first class with the fulcrum at the balancing or pivoting point.

Fig. 36.2

Suppose John, weighing 160 lb, sits at one end of the seesaw 6 ft from the fulcrum, and James, weighing 120 lb, sits at the other end 8 ft from the fulcrum. Now, we ask, will the seesaw balance if no other force is applied? (Let us here disregard the weight of the lever.)

To find the answer we multiply: (weight) times (arm), on each side. On the left side we have

$$(160 \text{ lb})(6 \text{ ft}) = 960 \text{ lb-ft}$$

On the right side we have

$$(120 \text{ lb})(8 \text{ ft}) = 960 \text{ lb-ft}$$

The product on each side, 960 lb-ft, is called the *first moment*, or *moment of force*. The first moment may be defined as *a measure of the force tending to produce rotation about an axis*. It is often called *torque*. In the foregoing example, since the two first moments are equal but opposite in direction of rotation, the seesaw will balance. (Of course, the plank is assumed to be weightless.)

In mathematics it is usual to consider *counterclockwise* rotation *positive* and *clockwise* rotation *negative*.[1] Then the downward force on the left is a positive moment, or $+960$ lb-ft; while the force downward on the right is a negative moment, or -960 lb-ft. If the two moments are added, the sum is zero, a necessary condition for equilibrium:

$$-960 \text{ lb-ft} + 960 \text{ lb-ft} = 0$$

In all cases, if the positive and negative moments add to zero, the result is equilibrium, and no motion takes place.

[1] This is the usual notation for positive and negative rotation in mathematics. However, in mechanics the reverse order is sometimes used.

The moment on one side of the fulcrum (or axis of rotation) can be computed without regard to the opposite side. For example (Fig. 36.3), if an object weighing 8 lb is at a distance of 20 in. from the axis of rotation, the moment is

$$(8 \text{ lb})(20 \text{ in.}) = 160 \text{ lb-in.}$$

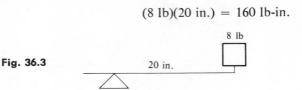

Fig. 36.3

Note. When we say that an 8-lb object is placed 20 in. from the fulcrum, we must understand that the distance, 20 in., must be measured from the fulcrum to the *center of gravity* of the object. When we say a 160-lb man sits 6 ft from the fulcrum of a seesaw, we must understand that the distance of 6 ft must be measured from the fulcrum to a point within the man which represents his center of gravity or centroid.

In all cases for any particle of mass, the moment M_s, with respect to a given axis s, is equal to the product

$$M_s(\text{moment}) = (\text{mass}) \times (\text{arm length})$$

36.3 TOTAL MOMENT

If two or more separate weights are placed at various positions on one side of the fulcrum, the moment must be computed separately for each weight, and then the total moment is found by taking the sum of the individual moments. As an example, consider a seesaw with two weights on one side and three on the other, with the distance of each as shown in Figure 36.4. We have

$$w_1 = 150 \text{ lb}; \, d_1 = 7 \text{ ft} \quad M_1 = 1050$$
$$w_2 = 130 \text{ lb}; \, d_2 = 4 \text{ ft} \quad M_2 = 520$$
$$w_3 = 70 \text{ lb}; \, d_3 = 3 \text{ ft} \quad M_3 = 210$$
$$w_4 = 80 \text{ lb}; \, d_4 = 5 \text{ ft} \quad M_4 = 400$$
$$w_5 = 120 \text{ lb}; \, d_5 = 8 \text{ ft} \quad M_5 = 960$$

For the total moment on the left,

$$w_1 d_1 + w_2 d_2 = 1570$$

On the right,

$$w_3 d_3 + w_4 d_4 + w_5 d_5 = 1570$$

Fig. 36.4

36.3 Total Moment

Since the two moments are equal but opposite in direction of force, the seesaw will balance.

Consider now the situation in which the weights on a side of the seesaw are not separate but rather form a continuous substance. Suppose we have a block of wood 6 ft long, with a cross section of 1 sq ft, placed lengthwise along one side of the seesaw (Fig. 36.5). Suppose the block of wood weighs 300 lb and is of uniform thickness, size, and density. Let us imagine the block divided into cubes one foot on the edge. Now we compute the moment of the separate cubes and then find the total moment of the six cubes. Each cube weighs 50 lb. Note that the distance d for each block is taken to the center of the block. Here the blocks all have the same weight, but this would not be necessary provided we know the weight of each separate block.

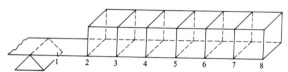

Fig. 36.5

For the total moment, we have the following:

$w_1 = 50$ lb; $d_1 = 2.5$ ft; $w_1 d_1 = 125$

$w_2 = 50$ lb; $d_2 = 3.5$ ft; $w_2 d_2 = 175$

$w_3 = 50$ lb; $d_3 = 4.5$ ft; $w_3 d_3 = 225$

$w_4 = 50$ lb; $d_4 = 5.5$ ft; $w_4 d_4 = 275$

$w_5 = 50$ lb; $d_5 = 6.5$ ft; $w_5 d_5 = 325$

$w_6 = 50$ lb; $d_6 = 7.5$ ft; $w_6 d_6 = 375$

Total moment = 1500

Now, to find the distance from the fulcrum to the centroid of the entire block, we divide the total moment M_t by the total weight w_t, and get

$$\frac{M_t}{w_t} = \frac{1500 \text{ lb-ft}}{300 \text{ lb}} = 5 \text{ ft}$$

The centroid of the entire block is at a point 5 ft from the fulcrum.

Note that the centroid with reference to an axis s is found by use of the following steps:

1. Find the sum of the individual moments of the individual particles; this is the total moment.
2. Divide this total moment by the sum of the individual weights.

The rule may be stated as a formula. If we let r represent the distance from the axis s to the centroid, we have

$$r = \frac{w_1 d_1 + w_2 d_2 + \cdots + w_n d_n}{w_1 + w_2 + \cdots + w_n}$$

This formula is essentially that of the so-called "weighted average." For example, suppose we have the following four weights placed at the indicated distances from an axis, all on the same side:

10 lb at 5 ft   15 lb at 7 ft

40 lb at 9 ft   60 lb at 11 ft

Now suppose we wish to find the average distance of the weights. Adding the distances *alone* we get 32. Dividing by 4 we get the average distance, 8 ft. However, in this simple average distance, we have neglected the different weights. For the *weighted average* we first multiply each weight by its corresponding distance and get the moment of each:

(10)(5) = 50   (15)(7) = 105   (40)(9) = 360   (60)(11) = 660

Adding these products we get a total moment of 1175. The total weight is 125 pounds. For the weighted average distance, we have

1175 ÷ 125 = 9.4 ft

36.4 CENTROID OF AREA

It may seem a little confusing to speak of the center of gravity of an area, since an area, being of two dimensions only, has no weight. Yet any area has associated with it a point called the *centroid* around which the entire area can be said to balance.

For example, the centroid of a square is its geometric center. The centroid of a rectangle is a point midway between the ends and midway between its sides. The centroid of any triangle is the point of intersection of its medians (Fig. 36.6). A thin sheet of material of uniform thickness and density can be balanced at its centroid.

Fig. 36.6

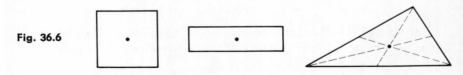

36.5 MOMENT OF AREA

For any given area, one problem is to locate its centroid. To find the centroid, we must first see what is meant by the *moment of area* with respect to an axis. Suppose we have any given area A some distance from an axis s. We indicate the centroid by the point C, and let l represent the distance or arm length from the axis to the centroid C (Fig. 36.7).

Fig. 36.7

36.5 Moment of Area

Then the moment of the area with respect to the axis s is denoted by M_s, and is defined as

$$M_s = (\text{arm})(\text{area})$$

When an area is placed on the xy coordinate plane, the moment of area is considered with respect to each of these axes. The centroid is then indicated by the two coordinates of the point, usually denoted by $(\bar{x},\bar{y})$ (Fig. 36.8). The point is read "x-bar, y-bar." Then the moments of the area with respect to the two axes are defined as follows:

$$M_y = \bar{x}A \quad M_x = \bar{y}A$$

The subscripts denote the axes of reference in each case.

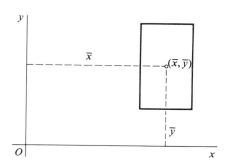

Fig. 36.8

Let us take a specific problem: the area bounded by the lines, $x = 8$, $x = 14$, $y = 3$, $y = 7$ (Fig. 36.9). Note that the area is 24 square units. To find $\bar{x}$, we take

$$\tfrac{1}{2}(x_2 + x_1) = \tfrac{1}{2}(14 + 8) = 11 \text{ units}$$

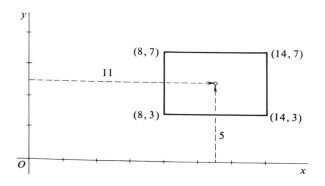

Fig. 36.9

To find $\bar{y}$, we take

$$\tfrac{1}{2}(y_2 + y_1) = \tfrac{1}{2}(7 + 3) = 5 \text{ units}$$

The centroid is the point (11,5). Then

$$M_y = (\bar{x})(\text{area}) = (11)(24) = 264 \quad \text{and} \quad M_x = (\bar{y})(\text{area}) = (5)(24) = 120$$

486 Centroids

The reason M_y is greater than M_x is that the centroid is farther from the y-axis. Whenever a given area is moved to a greater distance from an axis, its moment about that axis is increased. As a boy moves farther from the fulcrum of a seesaw, the moment increases.

EXERCISE 36.1

In each of the first three problems below, find the distance from the axis to the centroid. Each of the weights is at the indicated distance from the axis, and all on the same side.

1. 20 lb at 9 ft; 30 lb at 8 ft; 40 lb at 6 ft; 70 lb at 2 ft.
2. 3 lb at 19 in.; 4 lb at 15 in.; 6 lb at 17 in.; 7 lb at 13 in.
3. 90 lb at 10 ft; 120 lb at 6 ft; 80 lb at 5 ft; 60 lb at 4 ft.
4. On one side of a seesaw a weight of 90 lb is at 10 ft from the fulcrum, and another weight of 120 lb is at 6 ft from the fulcrum. On the other side a weight of 160 lb is at 8 ft from the fulcrum. Where should a weight of 80 lb be placed to make the seesaw balance?
5. On one side of a seesaw a weight of 120 lb is placed 6 ft from the fulcrum and another weight of 80 lb at 10 ft from the fulcrum. On the other side a weight of 140 lb is at 7 ft from the fulcrum. Where should a weight of 60 lb be placed to make the seesaw balance?
6. On the left side of a seesaw, weights of 80 and 180 lb are placed at distances of 8 and 4 ft, respectively, from the fulcrum. On the right side a weight of 160 lb is 10 ft from the fulcrum. Where on the right side should a weight of 120 lb be placed to make the seesaw balance?

Find the centroid of the area bounded by the lines in each of the following:

7. $y = 0; x = 0; y = 6; x = 4$
8. $y = 0; x = 0; x = -7; y = 3$
9. $y = 0; x = 4; x = 11; y = 3$
10. $x = 0; x = 5; y = 4; y = 7$
11. $y = 6; y = -2; x = 5; x = -1$
12. $y = 0; y = 3; x = 0; x = 8$

36.6 FINDING THE CENTROID

In problems involving moments and centroids, our main objective is to find the centroid — that is, the arm lengths, $\bar{x}$ and $\bar{y}$. In working a problem, we first find the *area* by integration; next, we find the *moments* by integration; finally we find each *arm length* by *division*. The formulas are used in reverse, as follows:

$$\bar{x} = \frac{M_y}{A} \qquad \bar{y} = \frac{M_x}{A}$$

The method is shown by examples.

EXAMPLE 1. Find the centroid of the area bounded by $x = 2y; y = 0; x = 8$.

Solution. First we sketch the area (Fig. 36.10). The element of area is shown, with length y and width dx. We first find the *area* by integration:

$$A = \int_0^8 y\, dx = \int_0^8 \frac{x}{2}\, dx = 16$$

To find the moments of the area, let us start with the y-axis, or M_y. Using the element of area shown, we first express the moment of this element. For every

36.6 Finding the Centroid   487

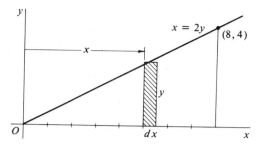

Fig. 36.10

element of area, dA, we can say the arm length is x, since dx is an infinitesimal. Then the moment of the element is

(arm)(dA)   or   (x)(dA)

The moment of the element of area is called the *element of moment*, denoted by dM_y, and represents a small portion of the total moment. Then we have

$$dM_y = (x)(dA)$$

To find the total moment of the entire area, *we sum up all the individual moments of the infinite number of elements of area by integration.* Then

$$M_y = \int_0^8 dM_y = \int_0^8 x\,dA = \int_0^8 x(y\,dx)$$
$$= \int_0^8 x\left(\frac{x}{2}\right)dx = \int_0^8 \frac{x^2}{2}\,dx = \frac{x^3}{6}\bigg]_0^8 = \frac{256}{3}$$

The number, 256/3, represents the total moment of the entire area with respect to the y-axis. To find the arm length, $\bar{x}$, we divide the total moment by the area, 16:

$$\bar{x} = \frac{M_y}{A} = \left(\frac{256}{3}\right) \div (16) = \frac{16}{3}$$

To state definitely the centroid, we must now find its ordinate as well as its abscissa. This can be done in either of two ways:

1. We can use the same vertical element; or
2. We can use a horizontal element.

We shall work this problem both ways. Either method can be used. However, in some problems, one *or* the other may be more convenient.

We have said the centroid of a rectangle is midway between the ends. The centroid of the element shown (Fig. 36.10) must be taken midway between the ends of the element. Since one end of the element is on the x-axis, the *arm length* to the centroid is $y/2$.

To find $\bar{y}$ we first express the element of moment, dM_x. The arm length is $y/2$; the element of area is dA. Then

$$dM_x = \left(\frac{y}{2}\right)(dA) = \left(\frac{y}{2}\right)(y\,dx) = \frac{y^2}{2}\,dx$$

Substituting $y = x/2$, we have

$$dM_x = \frac{x^2}{8}\,dx$$

Then
$$M_x = \int_0^8 \frac{x^2}{8}\,dx = \frac{x^3}{24}\bigg]_0^8 = \frac{64}{3}$$

Then
$$\bar{y} = \frac{M_x}{A} = \left(\frac{64}{3}\right) \div (16) = \frac{4}{3}$$

The centroid is the point (16/3, 4/3).

Second method for finding M_x: horizontal element (Fig. 36.11). Sketch the same area and show a horizontal element. In this case the length of the element is $(8 - x)$. (Note: In all cases involving a horizontal element, the length is given by $x_2 - x_1$. Here, $x_2 = 8$ and x_1 is the x of the line, which is $2y$.) Since the width of the element is now dy, we have

$$dA = (8 - x)\, dy = (8 - 2y)\, dy$$

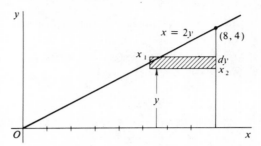

Fig. 36.11

We now write the expression for the *element of moment* with respect to the x-axis, dM_x. In this case the arm length is simply y, since the entire element is located at the end of the arm, y. Therefore, we have

$$dM_x = (y)(8 - 2y)\, dy = (8y - 2y^2)\, dy$$

Then

$$M_x = \int_0^4 (8y - 2y^2)\, dy = 4y^2 - \frac{2y^3}{3}\Big]_0^4 = 64 - \frac{128}{3} = \frac{64}{3}$$

Note. If a vertical element has neither end on the x-axis, the distance to the centroid of the element is the *average* of the y-values of the two ends (Fig. 36.12); that is,

$$\text{arm length} = \tfrac{1}{2}(y_2 + y_1)$$

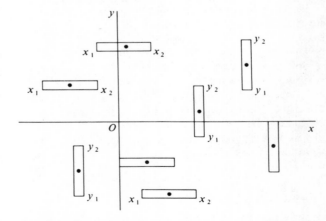

Fig. 36.12

If a horizontal element has neither end on the y-axis, the distance to the centroid of the element is the *average* of the x-values of the ends (Fig. 36.12); that is,

$$\text{arm length} = \tfrac{1}{2}(x_2 + x_1)$$

The formulas hold even when one end of the element is on an axis.

36.6 Finding the Centroid

EXAMPLE 2. Find the centroid of the area (Fig. 36.13) bounded by the curves
$y^2 = 4x$ and $2x - y = 4$

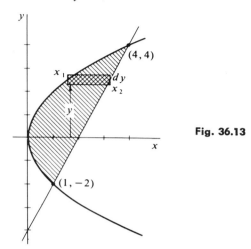

Fig. 36.13

Solution. We sketch the curves showing the indicated area and an element. In this example we use a horizontal element for area and also for both moments. To find the limits of integration we solve the equations as a system. The points of intersection are $(1, -2)$ and $(4, 4)$.

We first find the area. For the element we have

$$dA = (x_2 - x_1) \, dy$$

Solving each equation for x, we get

$$x_2 = \frac{y}{2} + 2 \qquad x_1 = \frac{y^2}{4}$$

Then

$$dA = \left(\frac{y}{2} + 2 - \frac{y^2}{4}\right) dy$$

Integrating,

$$A = \int_{-2}^{4} \left(\frac{y}{2} + 2 - \frac{y^2}{4}\right) dy = \frac{y^2}{4} + 2y - \frac{y^3}{12} \Big]_{-2}^{4} = 9 \text{ square units}$$

Now to find the moment we go back to the element of area, dA. For the moment with reference to the y-axis, the arm length is

$$\frac{1}{2}(x_2 + x_1)$$

Since

$$dA = (x_2 - x_1) \, dy$$

then

$$dM_y = (\text{arm})(dA) = (\text{arm})(x_2 - x_1) \, dy = \frac{1}{2}(x_2 + x_1)(x_2 - x_1) \, dy$$

or

$$dM_y = \frac{1}{2}(x_2^2 - x_1^2) \, dy$$

Note that if the element is expressed in this form before substituting the y-equivalents, the result always shows the *difference between two squares*.

Now we substitute y-equivalents and integrate. Then,

$$M_y = \frac{1}{2}\int_{-2}^{4}(x_2^2 - x_1^2)\,dy = \frac{1}{2}\int_{-2}^{4}\left(\frac{y^2}{4} + 2y + 4 - \frac{y^4}{16}\right)dy$$

$$= \frac{1}{2}\left[\frac{y^3}{12} + y^2 + 4y - \frac{y^5}{80}\right]_{-2}^{4} = \frac{72}{5}$$

Then $\bar{x} = (M_y) \div (A) = \frac{72}{5} \div 9 = \frac{8}{5}$

Next, to find M_x we use the same element. For the arm length we can use y since the entire element is assumed to be at the end of the arm. Then, since

$$dA = (x_2 - x_1)\,dy \quad \text{and} \quad \text{arm length} = y$$

we have

$$dM_x = y(x_2 - x_1)\,dy = y\left(\frac{y}{2} + 2 - \frac{y^2}{4}\right)dy = \left(\frac{y^2}{2} + 2y - \frac{y^3}{4}\right)dy$$

Integrating,

$$M_x = \int_{-2}^{4}\left(\frac{y^2}{2} + 2y - \frac{y^3}{4}\right)dy = \frac{y^3}{6} + y^2 - \frac{y^4}{16}\bigg]_{-2}^{4} = 9$$

Then $\bar{y} = M_x \div A = 9 \div 9 = 1$, the ordinate of the centroid. The centroid of the area is the point $(8/5, 1)$.

EXERCISE 36.2

Find the centroid of the area determined by each of the following sets of curves:

1. The curves $y^2 = 4x$; $x = y$
2. The curves $x^2 = 4y$; $y = 4$
3. $y^2 = 4x$; $x = 4$; first quadrant
4. $x^2 = 4y$; $x = 4$; $y = 0$
5. $y^2 = 2x$; $x = 2y$
6. $y^2 = 4x$; $x = 0$; $y = 4$; $y = -4$
7. $2x - y - 4 = 0$; $y^2 = 4x$; first quadrant
8. $x - 2y + 4 = 0$; $x = 0$; $y = 0$; $x = 6$
9. $x^2 = 4y$; $y^2 = 4x$
10. $y^2 = 4x + 16$; $y^2 + 8x = 16$
11. $y = x^2$; $y = x^3$
12. $x + 2y = 0$; $y = 0$; $x = -6$
13. $x + 4 = 2y$; $x^2 = 4y$; $y = 0$
14. $x = 2y - 6$; $x = 3$; $y = 0$
15. $y + x^3 = 0$; $x = -1$; $y = -1$
16. $x^2 = 4y$; $y^3 = 8x$
17. $x + 2y - 4 = 0$; $x^2 - 4y = 0$
18. $x + 2y = 4$; $x^2 = 4y$; $y = 0$
19. $x^2 = 4y$; $y = 0$; $x = -2$; $x = 4$
20. $x^2 - 4x - y = 0$; $y = -x$
21. $y = e^x$; $y = 0$; $x = 0$; $x = 2$
22. $y = e^{-x}$; $y = 0$; $x = 0$; $x = 2$
23. $y = e^{-x}$; $y = 0$; $x = -1$; $x = 1$
24. $y = \sin x$; $y = 0$; $x = 0$; $x = \pi$
25. $y = \cos x$; $y = 0$; $x = 0$; $x = \pi/2$
26. $x^2 + y^2 = 25$, first quadrant
27. $4x^2 + 9y^2 = 36$, first quadrant
28. $x^2 - y^2 = 9$; $x = 5$, first quadrant

36.7 CENTROID OF A SOLID

We have said (Section 36.1) that for every solid there is a point, usually within the solid itself, called the *center of mass* or *centroid*, around which the entire mass of the solid may be said to balance. Our problem now is to find the centroid of a solid.

When we show the position of a solid on the x- and y-axis system, then we attempt to find the centroid and express it as a point with the coordinates $(\bar{x}, \bar{y})$. This is simply a convenient way of locating the centroid.

36.8 FIRST MOMENT OF A SOLID

To find the centroid or center of mass of a geometric solid, we proceed in much the same way as for the centroid of an area. We first find the weight (or mass), then the moment with respect to an axis, and finally the arm length or distance from the axis to the centroid.

To review the meaning of first moment, let us take again the example of the lever, or seesaw. If we have a weight of 20 lb at one end of the lever arm, 8 ft from the fulcrum, then the corresponding moment is

$$(\text{arm})(\text{weight}) = (8 \text{ ft})(20 \text{ lb}) = 160 \text{ lb-ft}$$

On the other hand, if we know that the moment is 360 lb-ft for a 40-lb weight, we find the arm length by division:

$$\text{arm} = \frac{360 \text{ lb-ft}}{40 \text{ lb}} = 9 \text{ ft}$$

Note: A quantity such as the moment of 160 lb-ft should be expressed to read "pounds-feet" to distinguish it from *work* done, such as 160 *foot-pounds*. If this moment is written *160 ft-lb*, it can easily be confused with 160 foot-pounds, a form used to indicate amount of work done.

36.9 DENSITY OR WEIGHT IMPORTANT IN MOMENTS

In connection with moments of solids, there is one important point of difference from moments of areas; that is, an area has no weight, whereas a solid has weight or mass. The moment itself must take into account the weight or mass as well as the volume of the solid.

The usual method of indicating the mass of a solid is first to state its volume, V. We then multiply the number of units of volume by the density to indicate the total mass (or weight). By *density* we mean the mass (or weight) per unit of volume. We shall use k to denote density, a constant for a particular substance.

36.10 WEIGHT VERSUS MASS

In working problems involving moments and centroids, we may use either weight or mass. Since force is often measured in pounds, we can use weight in pounds. However, from a strictly physics standpoint, it is sometimes preferable to use mass.

Yet there need be no confusion. For example, suppose we find by integration that the volume of a solid is 5 cu ft. If the density of the solid is, say, 48 lb/cu ft, then the weight of the solid is

$$(48)(5) = 240 \text{ lb}$$

Note that $\qquad$ weight = (density)(volume)

492   Centroids

Of course, we may consider mass instead of weight. Then the number indicating density may be called (48) ÷ (32) = 1.5 units of mass instead of 48. In this case the mass of the solid is

$$(1.5)(5) = 7.5 \text{ units of mass}$$

The following formula is still true:

$$\text{mass} = (\text{density})(\text{volume})$$

Another point should be mentioned here. The centroid, or center of mass, being a point, may be computed without regard to density and by considering only volume. However, strictly speaking, the moment itself should also take into account the density of the solid. The difference will be pointed out in examples.

36.11 FORMULAS FOR MOMENT AND CENTROID

For any weight (or mass) at a particular distance (arm) from an axis, s, the first moment is given by the formula

$$M_s = (\text{arm})(\text{weight})$$

In Figure 36.14, we express the moments with respect to the coordinate axes as follows:

$$M_y = (\bar{x})(\text{weight}) \qquad M_x = (\bar{y})(\text{weight})$$

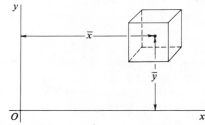

Fig. 36.14

Since weight is equal to (density)(volume), we have

$$M_y = (\bar{x})(k)(V) \qquad M_x = (\bar{y})(k)(V)$$

To find the distance from an axis to the centroid, we first find mass by integration, then moments by integration, and finally we use the formulas in reverse to find the arm lengths, $\bar{x}$ and $\bar{y}$. That is,

$$\bar{x} = \frac{M_y}{kV} \qquad \bar{y} = \frac{M_x}{kV}$$

Note: If the solid has an axis of symmetry, the centroid lies on this axis.

To find the moments with respect to the x and y axes, we may use either the *disk* or the *shell* method, just as in finding volume. The two methods will be shown by examples.

36.12 THE DISK METHOD

EXAMPLE 3. Find the centroid $(\bar{x},\bar{y})$ of the solid generated by rotating about the x-axis the area in the first quadrant bounded by the curves, $4x = y^2$, $x = 4$, and $y = 0$. (Fig. 36.15.) Let k represent the density of the solid generated.

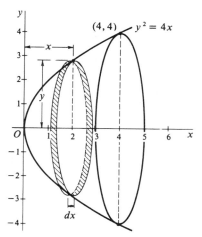

Fig. 36.15

Solution. Since the x-axis is the axis of symmetry, we have $\bar{y} = 0$. To find $\bar{x}$, we first find volume and mass. We have already used the disk method to find the volume of a solid. We imagine the solid divided into slices or disks, one of which is shown as an element of volume:

$$dV = \pi y^2\, dx = \pi(4x)\, dx$$

At this point, if we wish we can find the volume itself, which turns out to be 32π. However, instead, we can multiply the element of volume, dV, by the density and state immediately the *element of mass, dm.* Then

$$dm = (k)\, dV \quad \text{or} \quad dm = (k)\pi(4x)\, dx$$

That is, the *element of mass* is the mass of the *element of volume.* For the mass of the entire solid, we integrate the element of mass and get

$$m \text{ (mass)} = k\pi \int_0^4 4x\, dx = k\pi\, [2x^2]_0^4 = 32\pi k$$

Note that the total mass of the solid is simply (density) $\times$ (volume).

To find M_y we first express the *moment of the element* and call it dM_y, the element of moment. The moment of this element of volume is its mass multiplied by the arm length, which in this case is x. Then

$$dM_y = (x)(k)\pi y^2\, dx = k\pi x y^2\, dx = k\pi(x)(4x)\, dx$$

$$M_y = \pi k \int_0^4 4x^2\, dx = \pi k \left[\frac{4x^3}{3}\right]_0^4 = \frac{256}{3}(\pi)(k)$$

For $\bar{x}$, we have
$$\bar{x} = \frac{M_y}{\text{total mass}} = \frac{\pi k(256/3)}{32\pi k} = \frac{8}{3}$$

If the distance is stated in inches, and mass in pounds, we have

$$\frac{\text{moment (lb)(in.)}}{\text{mass (lb)}} = (\bar{x})(\text{in.})$$

494 Centroids

Note that the density disappears in finding the *distance* to the centroid. However, the moment itself should retain the density for correct meaning.

36.13 THE SHELL METHOD

Recall that we have used the shell method for finding the volume of solids of revolution. The same method may be extended to find the first moment of a solid. We simply express the mass of the shell, then the moment of the shell, and finally integrate to find total moment. The method is shown by examples.

Let us first recall that the centroid of a rectangular strip of area is midway between the ends of the rectangle. In Figure 36.16 the centroid of the strip of area, $x \, dy$, is at a point whose distance from the y-axis is $x/2$. Now if this strip is rotated about the x-axis, it generates a hollow cylinder, or shell. Then the centroid of the shell is at a distance $x/2$ from the y-axis. As we express the moment of a shell, we take the arm length as the distance from the axis to the midpoint of the shell.

Fig. 36.16

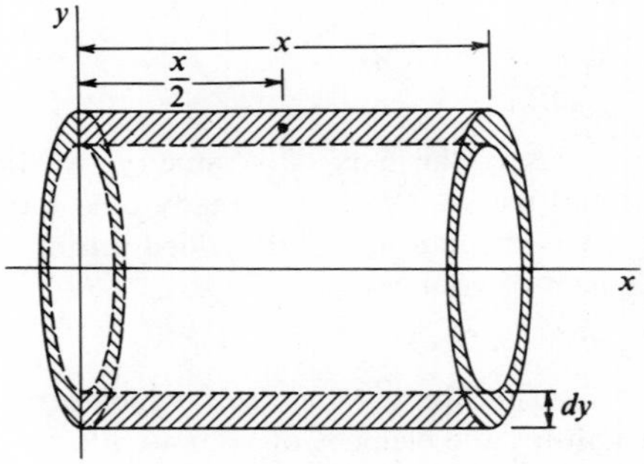

EXAMPLE 4. Using shells, find the centroid of the cone generated by rotating about the x-axis the area bounded by the curves $2x + 3y = 12$; $y = 0$; $x = 0$.

Solution. Let us disregard the density of the solid formed. Imagine the solid divided into shells one of which is shown as an element of volume with center on the x-axis (Fig. 36.17). The volume of the shell is

$$dV = 2\pi y x \, dy = 2\pi y \left(6 - \frac{3y}{2}\right) dy$$

Integrating and evaluating from $y = 0$ to $y = 4$, we get $V = 32\pi$. Now if we wish to state the mass of the solid, we can multiply volume by density and get

$$\text{mass } (m) = 32\pi k$$

Next, to find the total moment of the solid with respect to the y-axis, we first express the element of mass, dm. This element of mass is simply the mass of the element of volume; that is,

$$dm = k(2\pi y x \, dy)$$

However, here again we may disregard the density and express simply the moment of the element of volume as

$$dM_y = (\text{arm})(dV) = \left(\frac{x}{2}\right)(2\pi y x \, dy) = \pi y x^2 \, dy = \pi y \left(6 - \frac{3y}{2}\right)^2 dy$$

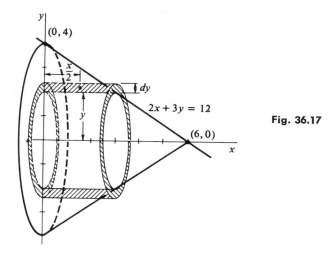

Fig. 36.17

That is,

$$dM_y = \pi y\left(36 - 18y + \frac{9y^2}{4}\right)dy = \pi\left(36y - 18y^2 + \frac{9y^3}{4}\right)dy$$

Then

$$M_y = \pi \int_0^4 \left(36y - 18y^2 + \frac{9y^3}{4}\right)dy = \pi\left[18y^2 - 6y^3 + \frac{9y^4}{16}\right]_0^4 = 48\pi$$

The quantity, 48π, expresses the moment of volume. If we wish to express the moment of mass, we multiply by density and get $48\pi(k)$. To find $\bar{x}$,

$$\bar{x} = \frac{M_y}{\text{mass}} = \frac{48\pi(k)}{32\pi(k)} = \frac{3}{2}$$

Since $\bar{y} = 0$, the centroid is at the point (3/2, 0).

EXAMPLE 5. Using shells, find the centroid of the solid generated by rotating around the x-axis the area in the first quadrant between the curves $y^2 = 6x$ and $x = 6$ (Fig. 36.18).

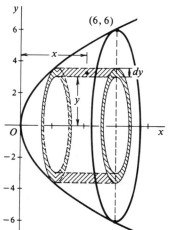

Fig. 36.18

Centroids

Solution. We see at once that $\bar{y} = 0$, since the x-axis is the axis of symmetry of the solid. Taking a shell as element of volume, we have

$$dV = 2\pi y (x_2 - x_1) dy$$

or

$$dV = 2\pi y \left(6 - \frac{y^2}{6}\right) dy = \pi\left(12y - \frac{y^3}{3}\right) dy$$

Integrating and evaluating from 0 to 6,

$$\pi \int_0^6 \left(12y - \frac{y^3}{3}\right) dy = \pi\left[6y^2 - \frac{y^4}{12}\right]_0^6 = 108\pi$$

Now using the shell as the element of volume, we take the arm length from the y-axis as the distance to the midpoint of the shell, which is $\frac{1}{2}(x_2 + x_1)$. Then for the element of moment we have

$$dM_y = \frac{1}{2}(x_2 - x_1) dV = \frac{1}{2}(x_2 + x_1)(2\pi y)(x_2 - x_1) dy$$

This reduces to

$$dM_y = \pi y(x_2^2 - x_1^2) dy = \pi y\left(36 - \frac{y^4}{36}\right) dy$$

Integrating and evaluating,

$$M_y = \pi \int_0^6 \left(36y - \frac{y^5}{36}\right) dy = \pi\left[18y^2 - \frac{y^6}{216}\right]_0^6 = 432\pi$$

Then

$$\bar{x} = \frac{\text{moment of volume}}{\text{total volume}} = \frac{M_y}{V} = \frac{432\pi}{108\pi} = 4$$

Since $\bar{y} = 0$, the centroid is at the point (4,0).

Note: It can be shown in general terms that the centroid of a paraboloid of revolution, as in Example 3, lies *two thirds of the distance from the vertex to the flat face.*

EXERCISE 36.3

Find the centroid of the solid of revolution determined by each of the following areas rotated about the indicated axes:
1. The first quadrant area bounded by the curves $y^2 = 4x$; $x = 4$; $y = 0$; axis of rotation: $y = 0$; $x = 0$.
2. Area bounded by the curves $y^2 = 4x$; $y = x$; axis: $y = 0$.
3. Area bounded by the curves $x^2 = 4y$; $x = 4$; $y = 0$; axis: $y = 0$.
4. Area bounded by the curves $y^2 = 2x$; $x = 2y$; axis: $y = 0$; $x = 0$.
5. Area bounded by the curves $x - 2y + 4 = 0$; $y = 0$; $x = 0$; $x = 6$; axis: $y = 0$.
6. Area in the first quadrant bounded by the curves $y^2 = 4x$; $2x - y = 4$; axis: $y = 0$; also axis: $x = 0$.
7. Area bounded by the curves $x^2 = 4y$; $x = -4$; $x = 4$; $y = -2$; axis: $y = -2$.
8. Area bounded by the curves $y = x^2$; $y = x^3$; axis: $y = 0$.
9. Second quadrant portion of area bounded by the curves $y^2 = 4x + 16$ and the coordinate axes; axis: $y = 0$; also axis: $x = 0$.
10. First and second quadrant portion of the area bounded by the curves, $y^2 = 4x + 16$; $y^2 + 8x + 16 = 0$; axis: $y = 0$.

11. Area bounded by the curves $x = 2y - 6$; $x = 3$; $y = 0$; axis: $y = 0$.
12. Area bounded by the curves $y + x^3 = 0$; $x = -1$; $y = -1$; axis: $x = -1$.
13. Area bounded by the curves $x^2 = y$; $y^2 = 8x$; axes: $y = 0$; $x = 0$.
14. Area bounded by the curves $x^2 = 4y$; $x + 2y = 4$; first quadrant portion; axis: $x = 0$.
15. Area bounded by the curves $y = 0$; $x^2 = 4y$; $x + 2y = 4$; axis: $y = 0$.
16. Area bounded by the curves $x^2 - 4x = y$; $y = 0$; axis: $y = 0$.
17. Find the centroid of the cone generated by rotating the following area about the x-axis: area bounded by $y = 0$; $y = mx$; $x = a$; axis: $y = 0$.
18. Find the centroid of the cone whose base radius is r and whose altitude is h. (Assume a right circular cone.)
19. Find the centroid of the frustum of a cone having bases whose diameters are 10 in. and 6 in., respectively, and whose altitude is 8 in.
20. Find the centroid of the frustum of a right circular cone, if the radii of the bases of the frustum are R and r, respectively, and the altitude of the frustum is h.
21. Find the centroid of the volume generated by rotating about the x-axis the first quadrant portion of the ellipse $4x^2 + 9y^2 = 36$.
22. Find the centroid of the volume generated by rotating the area of Problem 21 about the y-axis.
23. Find the centroid of the volume generated by rotating about the x-axis the area bounded by the curves $x^2 - y^2 = 9$; $y = 0$; $x = 0$; $y = 4$.
24. Find the centroid of the volume generated by rotating the area of Problem 23 about the y-axis.
25. Find the centroid of the volume generated by rotating about the x-axis the area bounded by the curves $xy = 12$; $x = 2$; $x = 6$; $y = 0$.
26. Find the centroid of the volume generated by rotating about the x-axis the area bounded by the curves $y = e^x$; $y = 0$; $x = 0$; $x = 2$.
27. Find the centroid of the volume generated by rotating the same area indicated in Problem 26 about the y-axis.
28. Find the centroid of the volume generated by rotating about the x-axis the area bounded by the curves $y = \sin x$; $y = 0$; $x = 0$; $x = \pi$.

chapter
37

Moment of Inertia

37.1 DEFINITION: MOMENT OF INERTIA, OR SECOND MOMENT

We have mentioned that the *first moment* of an area or a solid is sometimes called the *moment of force*. Also associated with an area, as well as with a solid, is another quantity, called the *second moment* or the *moment of inertia*. This quantity has many important applications in science, especially in connection with rotational motion. The moment of inertia is usually denoted by I.

We have said that the first moment is a measure of the force tending to produce rotation about an axis. Then the second moment, or moment of inertia, may be defined as *a measure of the resistance to motion or to a change in motion*. It refers to the force required to start a motion, or to stop a motion once it has started.

Suppose we have a 3-lb weight at the end of an 8-in. arm on one side of an axis (Fig. 37.1). Then for the first moment M, we have

$$M = (\text{arm})(\text{weight}) = (8 \text{ in.})(3 \text{ lb}) = 24 \text{ lb-in.}$$

Fig. 37.1

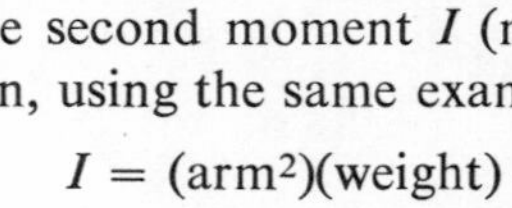

Note that the first moment is directly *proportional to the arm length*. If the arm length is doubled, the moment is doubled. If the arm length is 8 in., the moment is 24. If the arm length is 16 in., the moment is 48.

To find the second moment I (moment of inertia), we *square* the arm length. Then, using the same example, we have

$$I = (\text{arm}^2)(\text{weight}) = (8^2)(3) = 192 \text{ lb-in.}^2$$

Note that if the arm length is doubled, then I is multiplied by 4. That is, the *second moment I is proportional to the square of the arm length*.

The moment of inertia is best defined by a formula. For a particle of area about an axis, we have

$$I_s = (\text{arm}^2)(\text{area})$$

37.1 Definition: Moment of Inertia, or Second Moment

For a particle of mass,

$$I_s = (\text{arm}^2)(\text{mass})$$

In words, *the moment of inertia of a particle with respect to an axis s is equal to the mass of the particle multiplied by the square of its distance from the axis.*

To find the second moment we first state the element of area or mass, just as we do for the first moment. However, for the second moment, we *square the arm length.*

In setting up the expression for the element of the second moment, we must be sure the entire element of area or mass is at the end of the arm. The method is shown by the following example:

EXAMPLE 1. Find I_y and I_x, the moments of inertia with reference to the y-axis and the x-axis, respectively, of the area bounded by the curves $x = 2y$, $y = 0$, and $x = 8$.

Solution. To find I_y, the moment with reference to the y-axis, we take a vertical element of area (Fig. 37.2). Note that the entire element then lies at the end of the arm. Note especially the similarity and the difference between the elements of first and second moments. For the elements, we have

element of area	$dA = y\,dx$
element of first moment	$dM_y = (\text{arm})(dA) = (x)(y\,dx)$
element of second moment	$dI_y = (\text{arm}^2)(dA) = (x^2)(y\,dx)$

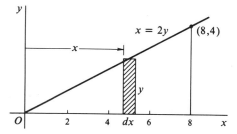

Fig. 37.2

To find the total moment of inertia of the entire area, we sum up the elements by integration, and get

$$I_y = \int_0^8 x^2 y\,dx = \int_0^8 x^2\left(\frac{x}{2}\right)dx = \int_0^8 \frac{x^3}{2}dx = \frac{x^4}{8}\bigg]_0^8 = 512$$

The result represents units of area multiplied by the square of the arm length.

To find I_x, the moment of inertia with reference to the x-axis, we run into a little difficulty if we use the same vertical element. Note especially that the entire element is not at the end of the arm. You will recall that for the first moment, we can take the arm length as the distance to the midpoint of the element. We cannot use the same method for the second moment. There is a way to use the same vertical element, but the method is a little more complicated, as will be explained in the next section.

Moment of Inertia

Instead, for this problem, let us take a horizontal element (Fig. 37.3). For the element of area we have

$$dA = (x_2 - x_1)\, dy = (8 - 2y)\, dy$$

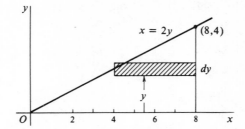

Fig. 37.3

Since this entire element is at a distance y from the x-axis, we have, for the element of the second moment,

$$dI_x = (\text{arm}^2)(dA) = (y^2)(8 - 2y)\, dy$$

Then

$$I_x = \int_0^4 y^2(8 - 2y)\, dy = \int_0^4 (8y^2 - 2y^3)\, dy = \frac{8y^3}{3} - \frac{2y^4}{4}\Big]_0^4 = \frac{128}{3}$$

37.2 MOMENT OF INERTIA OF A RECTANGLE

We have seen that if the element of area is perpendicular to an axis, then the entire element is not at the end of the arm. For the first moment we can take the distance to the midpoint of the element as the arm length. However, we cannot proceed in this way for the second moment. There is a formula, however, that can be used in such a situation.

To derive the formula, let us begin with a rectangle of width w and length l. We place the rectangle with one end along the x-axis and with a side parallel to the y-axis (Fig. 37.4). Now we wish to find the moment of inertia of this rectangle with reference to the x-axis.

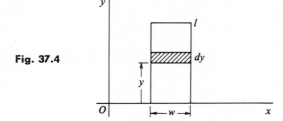

Fig. 37.4

We take a horizontal element of area. This element has a length w and a width dy. Then we have

$$dA = w\, dy$$

For this element of area, the arm length is y. Then, for the element of moment of inertia we have, since w is a constant,

$$dI_x = (\text{arm}^2)\, dA = (y^2)(w\, dy)$$

37.2 Moment of Inertia of a Rectangle

Integrating between the limits $y = 0$ and $y = l$, we get

$$I_x = \int_0^l wy^2 \, dy = (w)\frac{y^3}{3}\Big]_0^l = (w)\frac{l^3}{3}$$

The result has the following important application to any rectangle: *The moment of inertia of any rectangular area with reference to an axis along one end is equal to one third the width multiplied by the cube of the length.* The rule is equally applicable if the axis lies along one side of the rectangle.

Let us now see how the foregoing rule applies to the vertical element of area in Example 1 (shown again in Fig. 37.5). The element is a rectangle with one end along the x-axis. Its length is equal to y, and its width is dx. Then, for the moment of inertia of this element with reference to the x-axis, we have

$$dI_x = \frac{1}{3}(dx)y^3 \quad \text{or} \quad dI_x = \frac{1}{3}y^3 \, dx$$

Then

$$I_x = \frac{1}{3}\int_0^8 y^3 \, dx = \frac{1}{3}\int_0^8 \frac{x^3}{8} \, dx = \frac{1}{3}\left[\frac{x^4}{32}\right]_0^8 = \frac{128}{3}$$

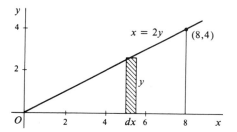

Fig. 37.5

The same result might have been obtained by integrating for the element itself. We take a small horizontal element of the vertical element (Fig. 37.6). This small element has a length dx, which we assume is constant. The width of the element is dy. The small element has an area of $(dx)(dy)$. Now, taking y as the arm length, we express the moment of inertia of the small element with reference to the x-axis, as

$$d(dI_x) = y^2 \, dx \, dy$$

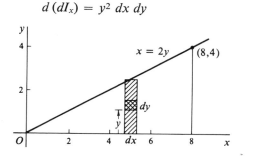

Fig. 37.6

502 Moment of Inertia

Now, taking dx as a constant, we integrate with respect to y, and then evaluate between the y-limits, 0 and y. Then

$$dI_x = \int_0^y (dx)y^2\,dy = \frac{1}{3}y^3\,dx$$

Now we have the element dI_x expressed in the same form as before.

37.3 RADIUS OF GYRATION

We recall that in connection with the first moment, there is a point called the *centroid*. The centroid is a point such that if the entire area or mass were concentrated at that point, the first moment with reference to some axis would remain unchanged. We have also seen that if we know the area and the first moment, we can compute the arm length.

Now, in connection with the moment of inertia, we also have a distance called the *radius of gyration*. *The radius of gyration is a positive distance from an axis such that if an entire area or mass were concentrated at this distance, the moment of inertia would remain unchanged.* We shall also find that, if we know a particular area or mass together with its moment of inertia with reference to an axis, we can compute the radius of gyration.

Suppose, in some particular problem we find that an area of, say, 20 sq in. has a moment of inertia of, say, 1280, with respect to some axis s (Fig. 37.7). Now our question is: At what distance from s could the entire area be concentrated so that the total moment of inertia would still be 1280? Assume this distance is R. Then we wish to find the distance R (the radius of gyration) such that

$$(R^2)(\text{area}) = 1280 \quad \text{or} \quad (R^2)(20) = 1280$$

Fig. 37.7

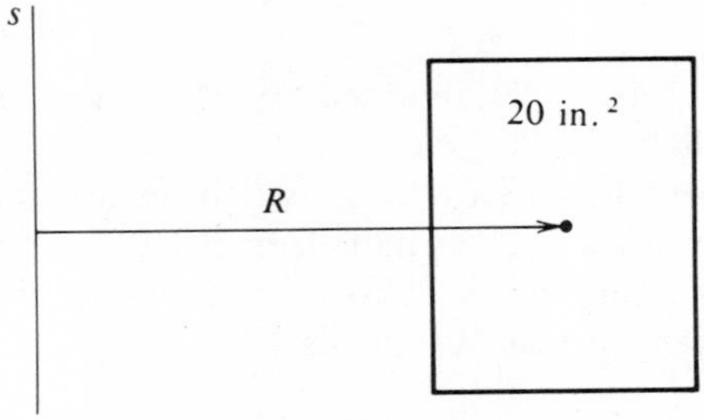

Solving, $\qquad R^2 = 1280/20 = 64 \qquad R = 8$

This means that if the entire area were concentrated at 8 units from the axis, the moment of inertia of the area would be 1280.

The radius of gyration is always positive. Moreover, the moment of inertia is always a positive number. Note especially the formulas

$$R^2 A = I \qquad R^2 = \frac{I}{A} \qquad R = \sqrt{\frac{I}{A}}$$

37.3 Radius of Gyration

EXAMPLE 2. A horizontal rectangle 4 in. wide has one end on the line $x = 2$ and the other end on the line $x = 14$ (Fig. 37.8). Find the moment of inertia with respect to the y-axis and the radius of gyration with respect to the same axis.

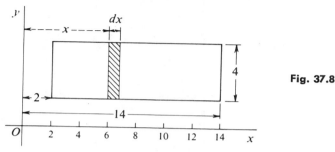

Fig. 37.8

Solution. First we take a vertical element of area dA, as shown, with length of 4, and width dx. Then $dA = 4\,dx$. To find the area we can integrate between the limits, $x = 2$ and $x = 14$. However, since the figure itself is a rectangle, we can find its area by multiplication: $(4)(12) = 48$. To express the element I_y, note that the arm length to the element of area is x. Then, the element of second moment is

$$dI_y = (\text{arm}^2)(dA) = (x^2)(4\,dx) = 4x^2\,dx$$

Now we sum up by integration the infinite number of elements of second moment from $x = 2$ to $x = 14$, and get

$$I_y = \int_2^{14} 4x^2\,dx = \frac{4x^3}{3}\Big]_2^{14} = \frac{4}{3}(2744 - 8) = 3648$$

Then $\quad R^2 = \dfrac{I}{A} = \dfrac{3648}{48} = 76 \quad$ or $\quad R = \sqrt{76} = 8.72$ (approx.)

Note that the radius of gyration is greater than the distance to the centroid.

EXAMPLE 3. Find the moment of inertia and the radius of gyration with respect to the x-axis and the y-axis of the area bounded by $x = 2y$; $x = 6$; $y = 0$.

Solution. Taking a vertical element of area (Fig. 37.9), we have

$$dA = y\,dx = \frac{x}{2}\,dx$$

Then $\quad A = \displaystyle\int_0^6 \frac{x}{2}\,dx = \frac{x^2}{4}\Big]_0^6 = 9$

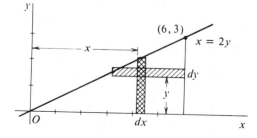

Fig. 37.9

504 Moment of Inertia

For the moment of inertia with respect to the y-axis, we have the element

$$dI_y = (\text{arm})^2(dA) = x^2\left(\frac{x}{2}\right)dx$$

Then
$$I_y = \int_0^6 \frac{x^3}{2}dx = \frac{x^4}{8}\bigg]_0^6 = 162$$

For the radius of gyration with respect to the y-axis, we have

$$R_y^2 = I_y \div A = 162 \div 9 = 18$$

then
$$R_y = \sqrt{18} = 4.243 \text{ (approx.)}$$

For the moment of inertia with respect to the x-axis, we can use a horizontal element of area. Then we have

$$dA = (6 - 2y)\,dy$$

For the element of moment of inertia,

$$dI_x = (\text{arm})^2(dA) = y^2(6 - 2y)\,dy = (6y^2 - 2y^3)\,dy$$

$$I_x = \int_0^3 (6y^2 - 2y^3)\,dy = 2y^3 - \frac{y^4}{2}\bigg]_0^3 = \frac{27}{2}$$

Then
$$R_x^2 = I_x \div A = \frac{27}{2} \div 9 = 1.5$$

$$R_x = \sqrt{1.5} = 1.225 \text{ (approx.)}$$

EXAMPLE 4. Find the centroid and the radius of gyration with respect to each axis of the area bounded by the curves: $x^2 = 4y$; $y = 0$; $x = 4$.

Solution. We take a vertical element of area (Fig. 37.10). For the element of area, we have

$$dA = y\,dx = \frac{x^2}{4}\,dx$$

Then
$$A = \int_0^4 \frac{x^2}{4}\,dx = \frac{x^3}{12}\bigg]_0^4 = \frac{16}{3}$$

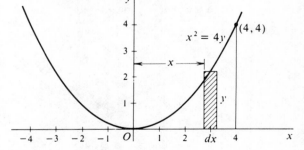

Fig. 37.10

For the first moment with respect to the y-axis,

$$dM_y = x\,(dA) = (x)\left(\frac{x^2}{4}\right)dx = \frac{x^3}{4}\,dx$$

$$M_y = \int_0^4 \frac{x^3}{4}\,dx = \frac{x^4}{16}\bigg]_0^4 = 16$$

then
$$\bar{x} = M_y \div A = 16 \div \frac{16}{3} = 3$$

For the first moment with respect to the x-axis, using the same element, we have

$$dM_x = \frac{1}{2}(y)(dA) = \frac{1}{2}\left(\frac{x^2}{4}\right)\left(\frac{x^2}{4}dx\right) = \frac{1}{32}x^4\,dx$$

$$M_x = \frac{1}{32}\int_0^4 x^4\,dx = \frac{1}{160}x^5\Big]_0^4 = \frac{32}{5}$$

$$\bar{y} = M_x \div A = \frac{32}{5} \div \frac{16}{3} = \frac{6}{5}$$

The centroid is therefore at the point (3,1.2).

For the moment of inertia with respect to the y-axis (I_y), we use the same vertical element of area. Then the element of I_y is

$$dI_y = (\text{arm})^2(dA) = (x^2)(y\,dx) = (x^2)\left(\frac{x^2}{4}\right)dx = \frac{x^4}{4}dx$$

Thus

$$I_y = \int_0^4 \frac{x^4}{4}dx = \frac{x^5}{20}\Big]_0^4 = \frac{256}{5}$$

$$R_y^2 = I_y \div A = \frac{256}{5} \div \frac{16}{3} = \frac{48}{5}$$

$$R_y = \sqrt{\frac{48}{5}} = 3.0984\ (\text{approx.})$$

For the moment of inertia of the vertical element with respect to the x-axis, we cannot take one-half of the arm length. Instead we must use another method. We might take a horizontal element in this case, but this sometimes runs into more difficulties. Instead we shall make use of a fact we have previously shown: *The moment of inertia of a rectangle with respect to an axis along one end (or side) is equal to one third of the cube of the length, multiplied by the width.* Now, the length of this vertical element is y and the width is dx. Then the moment of this element is

$$dI_x = \frac{1}{3}y^3\,dx$$

$$I_x = \int_0^4 \frac{1}{3}y^3\,dx = \frac{1}{3}\int_0^4 \frac{x^6}{64}dx = \frac{1}{3}\frac{x^7}{448}\Big]_0^4 = \frac{256}{21}$$

$$R_x^2 = I_x \div A = \frac{256}{21} \div \frac{16}{3} = \frac{16}{7}$$

$$R_x = \frac{4\sqrt{7}}{7} = 1.512\ (\text{approx.})$$

EXAMPLE 5. Compare the first moment, M_x, of the area in the first quadrant bounded by the curves, $y^2 = 4x$ and $x = 4$, with M_x of the entire area between the two curves (first and fourth quadrants). Do the same for I_x; that is, compare I_x for the same two areas (Fig. 37.11).

Solution. A horizontal element of area is probably most convenient in this case. Then

$$dA = (x_2 - x_1)\,dy = \left(4 - \frac{y^2}{4}\right)dy$$

$$dM_x = (y)\left(4 - \frac{y^2}{4}\right)dy,\ \text{element of first moment}$$

$$dI_x = (y^2)\left(4 - \frac{y^2}{4}\right)dy,\ \text{element of second moment}$$

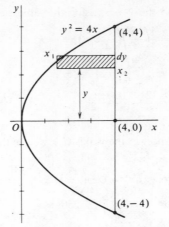

Fig. 37.11

Integrating first from $y = 0$ to $y = 4$, we have

$$A = \int_0^4 \left(4 - \frac{y^2}{4}\right) dy = 4y - \frac{y^3}{12}\Big]_0^4 = \frac{32}{3}$$

$$M_x = \int_0^4 \left(4y - \frac{y^3}{4}\right) dy = 2y^2 - \frac{y^4}{16}\Big]_0^4 = 32 - 16 = 16$$

Then
$$\bar{y} = \frac{3}{2}$$

Now if we consider the entire area between the two curves, we use the same element of area and the same element of first moment. The only difference is that we shall then integrate from $y = -4$ to $y = +4$. For the entire area we get

$$A = 4y - \frac{y^3}{12}\Big]_{-4}^4 = 16 - \frac{64}{12} + 16 - \frac{64}{12} = \frac{64}{3}$$

Note that the area has doubled, just as we should have expected. For the first moment, we have

$$M_x = 2y^2 - \frac{y^4}{16}\Big]_{-4}^{+4} = 32 - 16 - 32 + 16 = 0$$

For $\bar{y}$ we now have

$$\bar{y} = M_x \div A = 0 \div A = 0$$

This means that the centroid lies on the x-axis. Note that the area in the fourth quadrant has offset the moment in the first quadrant, so that the net first moment with respect to the x-axis is *zero*.

For the second moment of the area in the first quadrant only, we have

$$I_x = \int_0^4 \left(4y^2 - \frac{y^4}{4}\right) dy = \frac{4y^3}{3} - \frac{y^5}{20}\Big]_0^4 = \frac{256}{3} - \frac{256}{5} = \frac{512}{15}$$

If we take the entire area, we shall find that

$$I_x = \int_{-4}^4 \left(4y^2 - \frac{y^4}{4}\right) dy = \frac{4y^3}{3} - \frac{y^5}{20}\Big]_{-4}^4 = \frac{256}{3} - \frac{256}{5} + \frac{256}{3} - \frac{256}{5} = \frac{1024}{15}$$

Note that I_x for the entire area on both sides of the x-axis is *twice* I_x for the portion in the first quadrant. That is, for equal areas on opposite sides of an axis, the total moment of inertia is twice as much as the moment of the area on just one side. However, the radius of gyration remains the same, since the area is also doubled. On the other hand, *with equal areas on both sides of an axis of symmetry, the first moment becomes zero.* In either case, $R_x = 1.789$ (approx).

EXERCISE 37.1

Find the moments of inertia with respect to the coordinate axes and the radius of gyration in each case for the areas bounded by the following sets of curves:

1. $x^2 = 6y$; $y = 0$; $x = 6$
2. $y^2 = 2x$; $x = 8$; first quadrant
3. $x + 2y = 6$; $y = 0$; $x = 0$
4. $x - 2y = 2$; $y = 0$; $x = 0$
5. $x - 2y = 2$; $y = 0$; $x = 8$
6. $x^2 = 4y$; $y = 0$; $x = 4$
7. $y^2 = 4x$; $y = x$
8. $x + 2y = 6$; $x = 2$; $x = -4$
9. $x = 2y$; $y = 3$; $x = 0$
10. $y^2 = 4x$; $x^2 = 4y$
11. $y = x^3$; $y = 0$; $x = 2$
12. $y^2 = x^3$; $x = 4$; first quadrant
13. $x = 2y$; $y^2 = -2x$
14. $y^2 + 4x = 16$; $x = 0$
15. $y = 2x - x^2$; $y = 0$
16. $x^2 - 4x + y = 0$; $y = 0$
17. $y = e^x$; $y = 0$; $x = 0$; $x = 2$
18. $y = \sin x$; $y = 0$; $x = 0$ to π
19. A right triangle has legs of 12 in. and 5 in., respectively. Find the moment of inertia and the radius of gyration with respect to (a) the 12-in. leg; (b) the 5-in. leg; (c) the hypotenuse.
20. Find the moment of inertia and the radius of gyration of an isosceles triangle whose base is 6 in. and whose altitude is 4 in., with respect to (a) altitude; (b) base.
21. Find the moment of inertia and the radius of gyration with respect to both coordinate axes of the first quadrant portion of $x^2 + y^2 = 9$.
22. Find the moment of inertia and the radius of gyration with respect to both axes of the first quadrant portion of the ellipse, $4x^2 + 9y^2 = 36$.

37.4 MOMENT OF INERTIA OF A SOLID

The moment of inertia of a solid is found in much the same way as the moment of inertia of an area. Recall that in the case of an area, we take an element of area and then multiply this element by the *square* of the arm length, or distance to an axis. In the case of a solid, we take an element of *weight* (or *mass*) and then multiply this element by the *square* of the arm length.

In connection with solids we must again take into account the *density* of the substance, that is, the weight or mass per unit volume. We shall denote the density by the constant K. Then we proceed as we did in finding the first moment. We take an element of volume and call it dV. Then the mass of this element of volume is called the *element of mass*, or $(K)(dV)$. Now, for the moment of inertia, we multiply the element of mass by the *square* of the arm length from the axis to the element of volume. Then, for the element of moment of inertia we have

$$dI = (\text{arm}^2)(K\,dV)$$

We now make use of the principle that the total moment of inertia of several masses with respect to a common point or axis is equal to the sum of the individual moments of inertia. The summation is accomplished by the definite integral, just as is done in connection with areas.

Moment of Inertia

Before we consider a problem involving calculus, let us take a specific example of this principle. Suppose we have several masses of different amounts and at varying distances from a fixed point O (Fig. 37.12). Let us assume that each mass is concentrated at a point, and all masses are in the same plane with point O as the reference point of rotation. Then the moment of inertia of each mass is its own mass multiplied by the square of its distance from O. The following table shows the several masses, their corresponding distances from point O, and the moment of inertia of each mass:

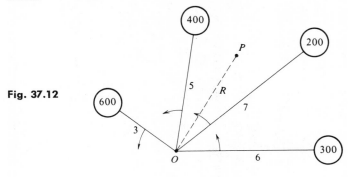

Fig. 37.12

Units of Mass	Arm Length	Square of Arm Length	Moment of Inertia
300	6	36	10,800
200	7	49	9,800
400	5	25	10,000
600	3	9	5,400
Total mass: 1500		Total moment of inertia:	36,000

The total moment of inertia is the sum of the individual moments.

Now we ask: What is the distance R from point O to a point P such that if the entire mass in the foregoing example were concentrated at the point P with a distance R from O, the total moment of inertia would still be 36,000? That is, if

$$(R^2)(\text{total mass}) = 36{,}000$$

what is the value of R? The distance R is the *radius of gyration*. To find R, we take 1500 as the total mass. Then

$$(R^2)(1500) = 36{,}000 \qquad R^2 = \frac{36{,}000}{1500} = 24$$

and

$$R = \sqrt{24} = 4.898 \text{ (approx.)}$$

Note that

$$R^2 = \frac{I(\text{total})}{\text{mass (total)}}, \qquad \text{from which} \qquad R = \sqrt{\frac{I}{m}}$$

37.4 Moment of Inertia of a Solid

In finding the moment of inertia of a solid, the element of volume may be either a disk or a shell, just as in finding the volume itself. However, the use of the disk method involves a special formula. We shall first work out an example using the shell method, and then we shall see how the result can be used as a formula in connection with the disk method.

EXAMPLE 6. Find the moment of inertia with respect to its axis of a solid right circular cylinder 20 in. long and 12 in. in diameter.

Solution. Let us place the cylinder with its axis along the x-axis and with one end at the y-axis (Fig. 37.13). Now we take a shell as an element of volume, dV. Then

$$dV = 2\pi y(20)\, dy = 40\pi y\, dy$$

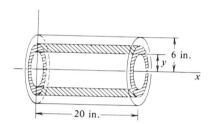

Fig. 37.13

For the element of mass, we have

$$d(\text{mass}) = (K)(40\pi y\, dy)$$

Integrating and evaluating from $y = 0$ to $y = 6$, we have

$$\text{mass} = 40K\pi \int_0^6 y\, dy = 40K\pi \left[\frac{y^2}{2}\right]_0^6 = 40K\pi(18) = 720K\pi \text{ units}$$

Of course, for the volume of a right circular cylinder, we might have used the formula

$$V = \pi r^2 h$$

Then we get

$$V = \pi(36)(20) = 720\pi$$

For the moment of inertia of the shell, we note that the entire shell is at a distance y from the axis. Then we multiply the mass of the shell by y^2 and get

$$dI_x = (y^2)(\text{element of mass}) = (y^2)(40\pi Ky\, dy) = 40\pi Ky^3\, dy$$

Integrating and evaluating between the y-limits, 0 and 6, we have

$$I_x = 40\pi K \int_0^6 y^3\, dy = 40\pi K \left[\frac{y^4}{4}\right]_0^6 = 40\pi K(324) = 12{,}960\pi K$$

For the radius of gyration we get

$$R_x^2 = \frac{I_x}{\text{mass}} = \frac{12{,}960\pi K}{720\pi K} = 18$$

$$R_x = \sqrt{18} = 4.243 \text{ (approx.)}$$

EXAMPLE 7. We shall now work the same problem as Example 5 in general terms and thereby get a formula useful in connection with the disk method.

510   Moment of Inertia

Solution. We represent the radius of the cylinder by r and the length by h. We place the cylinder in the same general position as before (Fig. 37.14). Taking a shell as an element of volume, we have

$$dV = 2\pi hy\, dy \quad \text{and} \quad d(\text{mass}) = 2\pi Khy\, dy$$

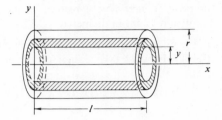

Fig. 37.14

Integrating,

$$\text{mass} = 2\pi Kh \int_0^r y\, dy = 2\pi Kh \left[\frac{y^2}{2}\right]_0^r = \pi Khr^2$$

For the element of moment of inertia, we have

$$dI_x = (y^2)(2\pi Khy\, dy) = 2\pi Khy^3\, dy$$

Integrating and evaluating again between the limits $y = 0$ and $y = r$, we get

$$I_x = 2\pi Kh \int_0^r y^3\, dy = 2\pi Kh \left[\frac{y^4}{4}\right]_0^r = \frac{1}{2}\pi Khr^4$$

For the radius of gyration, we have

$$R_x^2 = \frac{I_x}{\text{mass}} = \frac{1}{2}\frac{\pi Khr^4}{\pi Khr^2} = \frac{1}{2}r^2 \qquad R_x = \frac{r}{\sqrt{2}}$$

We state this important result in words: *The radius of gyration with respect to its axis of a solid right circular cylinder of homogeneous material of radius r is equal to $r/\sqrt{2}$.* This means also that the moment of inertia of such a cylinder with respect to its axis can be found by multiplying its mass by one half the square of the radius; that is,

$$I = (\text{mass})\left(\frac{r^2}{2}\right)$$

As we have said, to find the moment of inertia of a solid of revolution with respect to its axis, we may use either the shell or the disk method. The shell method, if it can be used, is probably most convenient. However, the disc method can be used by applying the formula we have just derived in Example 7. We shall work the next example by both methods.

EXAMPLE 8. Find the moment of inertia with respect to its axis of the solid formed by rotating about the x-axis the area bounded by the curves, $x = 2y$, $y = 0$, and $x = 8$.

Solution. Shell Method. The element of volume is a shell (Fig. 37.15); then

$$dV = 2\pi y(8 - x)\, dy$$

37.4 Moment of Inertia of a Solid

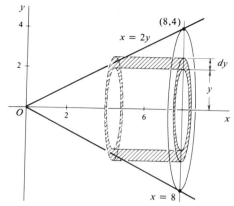

Fig. 37.15

Then the element of mass is given by

$$d(\text{mass}) = 2\pi Ky(8 - x)\, dy$$
$$= 2\pi Ky(8 - 2y)\, dy = 2\pi K(8y - 2y^2)\, dy$$

Integrating and evaluating between the limits $y = 0$ to $y = 4$, we get

$$\text{mass} = 2\pi K \int_0^4 (8y - 2y^2)\, dy = 2\pi K \left[4y^2 - \frac{2y^3}{3} \right]_0^4 = \frac{128}{3} \pi K$$

To find the moment of inertia, I_x, we multiply the element of mass by the square of the arm length. Note that every particle of the element is at a distance y from the axis. This point must be carefully observed. Then we have

$$dI_x = (y^2)(2\pi K)(8y - 2y^2)\, dy = 2\pi K(8y^3 - 2y^4)\, dy$$

Then
$$I_x = 2\pi K \int_0^4 (8y^3 - 2y^4)\, dy = 2\pi K \left[2y^4 - \frac{2y^5}{5} \right]_0^4 = \frac{1024\pi K}{5}$$

$$R^2 = \frac{I_x}{\text{mass}} = \frac{1024\pi K}{5} \div \frac{128\pi K}{3} = \frac{24}{5}$$

$$R = \sqrt{4.8} = 2.191 \text{ (approx.)}$$

Disk Method. In this case the element of volume is a disk (Fig. 37.16); then

$$dV = \pi y^2\, dx \qquad d(\text{mass}) = \pi K y^2\, dx = \pi K \frac{x^2}{4}\, dx$$

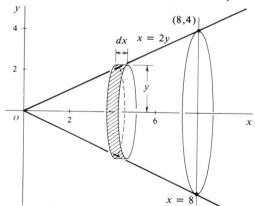

Fig. 37.16

Moment of Inertia

We have already found that the entire mass is $(128\pi K)/3$. To find the moment of inertia we cannot say that every particle of the element is at a uniform distance from the axis. However, we consider the disk itself as a cylinder (as in Example 7) whose length is dx and whose radius is y. Then we use the result of Example 7. The radius of gyration of this cylinder (the disk) is given by

$$R = \frac{y}{\sqrt{2}} \quad \text{or} \quad R^2 = \frac{y^2}{2}$$

To find the moment of inertia of this element (the disk), we multiply its mass by $y^2/2$, and get

$$dI_x = \frac{y^2}{2}(\pi K y^2 \, dx) = \frac{\pi K}{2}(y^4 \, dx) = \frac{\pi K}{2}\left(\frac{x^4}{16}\right)dx$$

Integrating and evaluating from $x = 0$ to $x = 8$, we get

$$I_x = \frac{\pi K}{2}\int_0^8 \frac{x^4}{16} dx = \frac{\pi K}{2}\left[\frac{x^5}{80}\right]_0^8 = \frac{1024\pi K}{5}$$

One point must be observed in finding the moment of inertia of a solid. For any kind of element of volume, every particle of the element must be at a uniform distance from the axis. Then this distance is taken as the arm length and is squared. To see the importance of this point, let us consider the following problem.

A rectangular solid, 5 in. long, with a cross section 3 in. by 2 in., is placed in a horizontal position with the length parallel to the x-axis so that one end is 10 in. from the y-axis, and the other end is 15 in. from the y-axis (Fig. 37.17). Now let us consider the moment of inertia with reference to the y-axis. We take a vertical slice as an element of volume; then

$$dV = 6 \, dx \quad \text{and} \quad d(\text{mass}) = 6K \, dx$$

Fig. 37.17

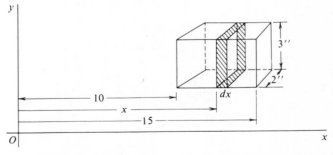

If we take the arm length as x, there is an error because every particle of the element is not at a uniform distance from the y-axis. If we consider every particle of the element at a distance x from the y-axis, we shall find that the moment of inertia becomes $4750K$, and the radius of gyration is approximately 12.588. However, the answer is slightly in error because the element is not at a uniform distance form the y-axis. If the object has a very small cross-sectional area, the error is not great.

In the next two examples, the object is a slender rod, so that the distance from the axis to the element may be considered uniform.

37.4 Moment of Inertia of a Solid

EXAMPLE 9. If a slender rod of uniform cross section and density has one end on an axis, then the radius of gyration of the rod with respect to the axis is equal to the length divided by the square root of 3. Show this is true by taking a rod 1 in. in diameter and 20 in. long (Fig. 37.18).

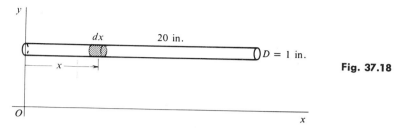

Fig. 37.18

Solution. Let us place one end on the y-axis with the rod parallel to the x-axis. We take a disk as an element of volume; then

$$dV = (\pi)\left(\frac{1}{2}\right)^2 dx$$

For the element of moment of inertia, we have

$$dI_y = (x^2)(\pi)(K)\left(\frac{1}{4}\right) dx$$

$$I_y = \frac{\pi}{4} K \int_0^{20} x^2\, dx = \frac{\pi}{4} K \left[\frac{x^3}{3}\right]_0^{20} = \frac{\pi}{4} K\left(\frac{8000}{3}\right) = \frac{2000\pi K}{3}$$

For the total mass of the rod we get $5\pi K$. Then

$$R^2 = \frac{2000\pi K}{3(5\pi K)} = \frac{400}{3} \quad \text{and} \quad R = \sqrt{\frac{400}{3}} = \frac{20}{\sqrt{3}}$$

EXAMPLE 10. Find the radius of gyration of the rod in Example 9 with respect to the axis through the rod midway between the ends (Fig. 37.19).

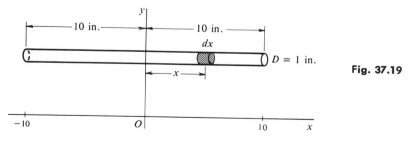

Fig. 37.19

Solution. We place the rod in a horizontal position parallel to the x-axis, and with the y-axis through the middle. The element of the moment of inertia is the same as before:

$$dI_y = (x^2)(\pi)(K)\left(\frac{1}{4}\right) dx$$

$$I_y = \frac{\pi}{4} K \int_{-10}^{10} x^2\, dx = \frac{\pi}{4} K \left[\frac{x^3}{3}\right]_{-10}^{10} = \frac{500}{3} \pi K$$

514 Moment of Inertia

Since the mass is $5\pi K$, we have

$$R^2 = \frac{100}{3} \quad \text{and} \quad R = \frac{10}{\sqrt{3}}$$

Note: In Example 9, note that the centroid is 10 units from the x-axis, whereas the radius of gyration is 11.55 units. In Example 10, if we were to consider only half the rod, the centroid would be 5 units from the y-axis, whereas the radius of gyration would be 5.77 (approximately). Now, in Example 10, with another half of the rod on the opposite side of the axis, the centroid is at the y-axis, whereas the radius of gyration is approximately 5.77 units of length. For the centroid, the entire mass may be considered to be concentrated at the y-axis. For the radius of gyration it may be considered to be concentrated at 5.77 units from the y-axis.

Another way of looking at this problem is to note that, from the standpoint of equilibrium, the portion of the rod on one side of the axis exactly balances the portion on the other side, making the rod balance and the centroid at zero. However, the two halves of the rod make the moment of inertia twice as much as the moment for one half. Since the mass is also doubled, the radius of gyration remains the same.

EXERCISE 37.2

Find the moment of inertia and the radius of gyration with respect to its axis of the solid of revolution formed by rotating the area bounded by the given curves about the given axis. (All solids are homogeneous.)

1. $x = 3y; x = 6; y = 0;$ x-axis
2. Same as No. 1; y-axis
3. $x + 2y = 6; y = 0; x = 0;$ y-axis
4. Same as No. 3; x-axis
5. $2x - 3y = 6; y = 0; x = 6;$ x-axis
6. Same as No. 5; y-axis
7. $x^2 = 6y; x = 6; y = 0;$ y-axis
8. Same as No. 7; x-axis
9. $y^2 = 2x; x = 8;$ 1st quadrant; x-axis
10. Same as No. 9; y-axis
11. $y^2 = 4x; y = x;$ y-axis
12. Same as No. 11; x-axis
13. $y^2 = 4x; x^2 = 4y;$ x-axis
14. $y = x^3; y = 0; x = 2;$ y-axis
15. $y^2 + 4x = 16;$ 1st quadrant; x-axis
16. Same as No. 14; x-axis
17. $y = 4x - x^2; y = 0;$ x-axis
18. $y = 4x - x^2; y = x;$ y-axis
19. $y = x^3; y = x^2;$ x-axis
20. Same as No. 21; y-axis
21. $y^2 = 4x; 2x - y = 4;$ first quadrant; x-axis
22. $y = e^x; y = 0; x = 0; x = 2;$ x-axis
23. $y = \sin x; y = 0; x = 0$ to $x = \pi;$ x-axis
24. $y = \tan x; y = 0; x = \pi/4;$ x-axis
25. $y = \sec x; y = 0; x = 0; x = \pi/4;$ x-axis
26. Find the moment of inertia and the radius of gyration of the frustum of a right circular cone whose bases are 6 in. and 10 in. in diameter, respectively, and whose altitude is 5 in.
27. Same as No. 26, with radii of bases, r_1 and r_2, and altitude h.
28. Find the radius of gyration of a hollow cylinder having an outer diameter of 10 in. and an inner diameter of 8 in., with respect to its axis.

chapter

38

Length of Arc and Area of Surface of Revolution

38.1 DEFINITION OF ARC LENGTH

A portion of a curve is called an *arc*. As an example, suppose we have the graph of some function of x, $y = f(x)$. (Fig. 38.1.) Any portion of the curve, such as the portion from A to B is an arc. To denote the arc from A to B we usually write $\overparen{AB}$. The straight line segment AB is a chord and is sometimes denoted by $\overline{AB}$. The arc $\overparen{AB}$ is a curved line and of course is longer than the chord AB. That is, $\overparen{AB} > \overline{AB}$.

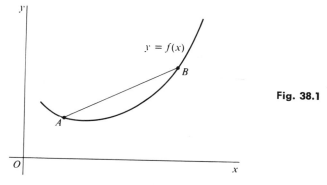

Fig. 38.1

Now we wish to find the exact length of the arc $\overparen{AB}$. This is sometimes necessary. For example, in the flight of a satellite, or a missile, the path is a curve and we may want to know the exact length of the path. The only way to determine the exact length of a curve is by calculus.

38.2 APPROXIMATION OF ARC LENGTH

Let us see first how we might find the *approximate* length of the arc. Consider the curve (Fig. 38.2) representing a general function of x, that is, $y = f(x)$. Let us take the portion of this curve from A to B, two points

Length of Arc and Area of Surface of Revolution

Fig. 38.2

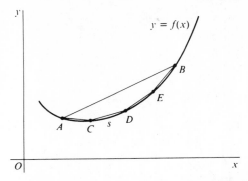

on the curve. We denote this arc by *s*. (Note that we use lower-case *s* for arc.) Now we wish to find the exact length of *s*. To get an approximation of the length, we might take a few intermediate points on the curve, such as *C*, *D*, and *E*, and draw straight-line segments connecting these five points, *A*, *C*, *D*, *E*, and *B*, in succession. Then we could measure the lengths of these straight-line segments, *AC*, *CD*, *DE*, and *EB*. The sum of these lengths would be approximately equal to the length of the arc *s*. By increasing the number of segments and making the segments shorter and shorter we could arrive at a closer approximation to the exact length of the arc. Yet the sum of even a great number of these short line segments would not equal exactly the length of the arc. The arc length will always be a trifle greater than the sum of the line segments.

38.3 THE EXACT LENGTH OF ARC

To find the exact length of arc we imagine the length of the line segments becoming smaller and smaller as the number of segments increases without limit. Then we say the exact length of arc *s* is the *limit* of the sum of the straight-line segments as the length of each segment approaches zero as a limit and as the number of segments increases without limit.

This is the same type of reasoning we have followed in finding volumes, areas, or any other quantity by use of the definite integral. Just as we have done with respect to areas and volumes, we imagine the arc *s* divided into a number of small portions of arc, one of which is called an *element of arc*. This element of arc we denote by *ds*, and call it the *differential of arc*. Then, by the definite integral, we sum up an infinite number of elements of arc between limits; that is, the total length of the arc is found by the summation of elements of arc. The summation can be done between *x*-limits or *y*-limits; that is, the length of arc *s* is

$$s = \int_{x_1}^{x_2} ds \quad \text{or} \quad s = \int_{y_1}^{y_2} ds$$

38.4 THE ELEMENT OF ARC (DIFFERENTIAL OF ARC *ds*)

Our first and most important step in finding the arc length is to state correctly the expression for the *differential of arc, ds*. After this element

38.4 The Element of Arc (Differential of Arc ds)

of arc has been correctly stated, the process of integration is almost mechanical. You recall that in finding area, volume, work, liquid pressure, moment, and so on, the first step was to write the correct expression for the *element*. Now too it is essential that we state correctly the *element of arc*. If we can do this correctly, then we find the total length of arc by integrating the element, just as we do in finding the total of any kind of quantity.

The importance of the element in all cases cannot be overemphasized. Whatever quantity Q we wish to find, *the element is simply a small bit or piece of that quantity.* To find total area we start with a bit of area called differential of area, dA. To find volume we start with dV, a bit of volume. To find total work, we take a little bit of the work done and call it dw. To find moment we start with a bit of moment and integrate to find the total. *The element is the first and most important concept.*

To state the expression for ds, the differential of arc, consider the curve of the function $y = f(x)$; see Figure 38.3. We take a point $P(x,y)$ on the curve. Now let x take on an increment $\Delta x = MN = PC$. Then y takes on an increment $\Delta y = CQ$, and the arc s changes by some increment $\Delta s = \widehat{PQ}$. Note that the chord $\overline{PQ}$ is the hypotenuse of the right triangle PCQ. Keep in mind that $\Delta s = \widehat{PQ}$ (the arc) is not the same as $\overline{PQ}$ (the chord). Note that $\overline{PQ} < \widehat{PQ}$.

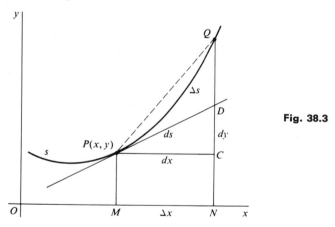

Fig. 38.3

The increments Δx and Δy are actual quantities though presumably small. Then in the right triangle PCQ, by the Pythagorean theorem we have

$$(\text{chord } PQ)^2 = (\Delta x)^2 + (\Delta y)^2$$

We have said that Δs represents $\widehat{PQ}$; then

$$(\Delta s)^2 = (PQ)^2$$

Now let us first divide and then at once multiply the quantity $(\Delta s)^2$ by the quantity $(\overline{PQ})^2$, that is, by the (chord)^2. We get

$$(\Delta s)^2 = \frac{(\Delta s)^2}{(\overline{PQ})^2} \cdot (\overline{PQ})^2$$

518   Length of Arc and Area of Surface of Revolution

Now we replace the factor $(\overline{PQ})^2$ with its equivalent $(\Delta x)^2 + (\Delta y)^2$ and get

$$(\Delta s)^2 = \frac{(\Delta s)^2}{(\overline{PQ})^2}[(\Delta x)^2 + (\Delta y)^2]$$

Dividing both sides of this equation by $(\Delta x)^2$, we get

$$\frac{(\Delta s)^2}{(\Delta x)^2} = -\frac{(\Delta s)^2}{(\overline{PQ})^2}\left[1 + \frac{(\Delta y)^2}{(\Delta x)^2}\right]$$

Now let us see what happens as Δx approaches zero as a limit. Using arrows to denote *approaches*, we have

$$\left(\frac{\Delta s}{\Delta x}\right)^2 \to \left(\frac{ds}{dx}\right)^2 \qquad \Delta s \to \overline{PQ} \qquad (\Delta s)^2 \to (\overline{PQ})^2$$

$$\frac{(\Delta s)^2}{(\overline{PQ})^2} \to 1 \qquad \frac{\Delta y}{\Delta x} \to \frac{dy}{dx}$$

The result is

$$\left(\frac{ds}{dx}\right)^2 = (1)\left[1 + \left(\frac{dy}{dx}\right)^2\right] \quad \text{or} \quad \frac{ds}{dx} = \sqrt{1 + \left(\frac{dy}{dx}\right)^2}$$

Taking ds/dx as the quotient of differentials, we multiply both sides by dx, and get, for the differential of arc,

$$ds = \sqrt{1 + \left(\frac{dy}{dx}\right)^2}\, dx$$

We have previously seen that while the line segment CQ represents Δy, the line segment CD represents dy, the differential of y. In like manner, the line segment PD represents ds, the differential of arc, whereas the arc $\widehat{PQ}$ is equal to Δs. In terms of differentials, then, we have

$$(ds)^2 = (dx)^2 + (dy)^2$$

Dividing both sides of the equation by $(dx)^2$, we get

$$\left(\frac{ds}{dx}\right)^2 = 1 + \left(\frac{dy}{dx}\right)^2 \quad \text{or} \quad \frac{ds}{dx} = \sqrt{1 + \left(\frac{dy}{dx}\right)^2}$$

For the differential of arc,

$$ds = \sqrt{1 + \left(\frac{dy}{dx}\right)^2}\, dx$$

The formula for differential of arc is often useful in an alternate form. Starting with

$$(ds)^2 = (dx)^2 + (dy)^2$$

if we divide both sides of the equation by $(dy)^2$, and solve for ds, we get

$$ds = \sqrt{1 + \left(\frac{dx}{dy}\right)^2}\, dy$$

38.5 STEPS IN FINDING ARC LENGTH

The first thing to do in determining ds, the differential of arc, is to find dy/dx, the derivative of the function. Then the derivative is squared and inserted in the formula for the differential of arc. In some problems it may be easier to find the inverted form of the derivative, dx/dy, and square this quantity. The best form to use is often discovered by "trial and error." Once we have the differential of arc for the element, we find the total length of the required arc by integrating and evaluating between limits. In some problems it may be best to use the form

$$ds = \sqrt{(dx)^2 + (dy)^2}$$

The differential of arc is especially important since it is also used in connection with finding the area of a surface of revolution. A lower case s is used for arc, because a capital S is later used to denote *surface*.

EXAMPLE 1. Find the total length of the following curve (Fig. 38.4) from the point $(-1,2)$ to the point $(1,2)$:

$$y^3 = 8x^2$$

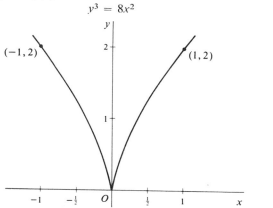

Fig. 38.4

Solution. Note that the equation remains unchanged if $(-x)$ is substituted for $(+x)$. Therefore the y-axis is an axis of symmetry. Then we simply find the length from $(0,0)$ to $(1,2)$ and double the length. We shall work the problem in two ways, using both formulas for the length of an arc.

(1) *First method.* Solving for y, we get

$$y = 2x^{2/3}$$

Differentiating,
$$\frac{dy}{dx} = \frac{4}{3}x^{-1/3}$$

Then
$$ds = \sqrt{1 + \frac{16}{9x^{2/3}}}\, dx$$

Simplifying,
$$ds = \frac{1}{3}x^{-1/3}\sqrt{9x^{2/3} + 16}\, dx$$

$$s = \frac{1}{3}\int_0^1 x^{-1/3}(9x^{2/3} + 16)^{1/2}\, dx = \frac{1}{18}\left[\frac{(9x^2 + 16)^{3/2}}{3/2}\right]_0^1$$

$$= \frac{1}{27}\left[(9x^{2/3} + 16)^{3/2}\right]_0^1 = \frac{1}{27}(125 - 64) = \frac{61}{27}$$

Now we double the answer and get the length of the curve, $122/27$.

(2) *Second method.* Solving for x, we get

$$x = \frac{1}{2\sqrt{2}} y^{3/2}$$

Now we differentiate with respect to y and get

$$\frac{dx}{dy} = \frac{3}{4\sqrt{2}} y^{1/2} \qquad \left(\frac{dx}{dy}\right)^2 = \frac{9}{32} y$$

Then
$$ds = \sqrt{1 + \frac{9y}{32}}\, dy$$

$$s = \int_0^2 \left(1 + \frac{9y}{32}\right)^{1/2} dy \qquad \text{(Note change in limits.)}$$

To apply the power rule, we need the constant $9/32$, which can be supplied. Then

$$s = \frac{32}{9} \int_0^2 \frac{9}{32}\left(1 + \frac{9y}{32}\right)^{1/2} dy = \frac{32}{9} \left.\frac{\left(1 + \frac{9y}{32}\right)^{3/2}}{3/2}\right]_0^2$$

$$= \frac{64}{27}\left(1 + \frac{9y}{32}\right)^{3/2}\bigg]_0^2 = \frac{64}{27}\left[\left(\frac{25}{16}\right)^{3/2} - 1\right] = \frac{64}{27}\left[\frac{125}{64} - 1\right] = \frac{61}{27}$$

For the entire length of the arc, we have $(2)(61/27) = 122/27 = 4.52$ (approx.).

EXAMPLE 2. A projectile is fired at an angle with the horizontal so that its path is a parabola reaching a maximum height of 4 miles and having a range of 16 miles. Write the equation of the path and find the length of the curve from the time it is fired until it strikes the ground at a distance of 16 mi horizontally from the point of firing.

Solution. Although the path is a parabola opening downward, it may be considered as upward so that values may be taken positive (Fig. 38.5). Taking the vertex of the parabola at the origin (0,0), we note that the curve passes through the point (8,4). Then the equation of the curve becomes $x^2 = 16y$. Solving for x,

$$x = 4y^{1/2}$$

$$\frac{dx}{dy} = 2y^{-1/2} \qquad \left(\frac{dx}{dy}\right)^2 = 4y^{-1}$$

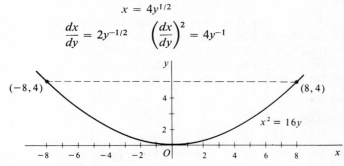

Fig. 38.5

For the differential of arc, we have

$$ds = \sqrt{1 + \frac{4}{y}}\, dy = \sqrt{\frac{y+4}{y}}\, dy = \frac{(y+4)^{1/2}}{y^{1/2}}\, dy$$

Then, for one-half of the curve, we have

$$s = \int_0^4 \frac{(y+4)^{1/2}\, dy}{y^{1/2}}$$

38.6 A Surface of Revolution

Now we shall first use algebraic substitution. Let

$$v = (y+4)^{1/2} \qquad v^2 = y+4 \qquad y = v^2 - 4 \qquad dy = 2v\, dv$$

Of course, we can also change the limits: when $y = 4$, $v = \sqrt{8}$; and when $y = 0$, $v = 2$. Making the substitutions, we get

$$s = \int_2^{\sqrt{8}} \frac{2v^2\, dv}{(v^2 - 4)^{1/2}}$$

Now we make use of trigonometric substitution. However, we leave the limits as they are because we shall convert back to v terms. Let

$$v^2 = 4\sec^2\theta \qquad v = 2\sec\theta \qquad dv = 2\sec\theta\tan\theta\, d\theta$$

Making the substitutions, we get

$$s = 8\int \sec^3\theta\, d\theta = 4[\sec\theta\tan\theta + \ln(\sec\theta + \tan\theta)]$$

Changing to v terms, we get

$$s = 4\left[\frac{v\sqrt{v^2-4}}{4} + \ln\left(\frac{v + \sqrt{v^2-4}}{2}\right)\right]_2^{\sqrt{8}}$$

$$= 4[\sqrt{2} + \ln(\sqrt{2} + 1) - 0 - 0]$$

$$= 4[1.4142 + 0.8814] = 9.1824 \text{ (approx.)}$$

Then the length of the entire path of the projectile is 18.3648 mi. The result can be checked roughly by considering the length of the chord from the origin to the point (8,4), which is approximately 8.944. Then the length of the two chords is approximately 17.888 mi.

EXERCISE 38.1

Find the length of each of the following curves between the indicated limits:
1. $2x = y^2$; (0,0) to $x=4$
2. $y^2 = x$; (1,−1) to (1,1)
3. $x^2 = 4y$; (0,0) to $y=3$
4. $y^2 = x^3$; (0,0) to (4,8)
5. $y^3 = x^2$; (0,0) to (1,1)
6. $y^3 = 8x^2$; (−8,8) to (8,8)
7. $y = 4x - x^2$; (0,0) to (4,0)
8. $y = 4 - x^2$; (−2,0) to (2,0)
9. $(y - 8)^2 = x^3$; $x = 0$ to $x = 4$
10. $3y = 2(x - 1)^{3/2}$; $x = 1$ to $x = 2$
11. $6xy = x^4 + 3$; $x = 1$ to $x = 3$
12. $y = (x + 1)^{3/2}$; $x = 1$ to $x = 4$
13. $y = \ln x$; $x = 1$ to $x = 4$
14. $y = \ln(x^2 - 1)$; $x = 2$ to $x = 3$
15. $y = \ln \sec x$; $x = 0$ to $x = 60°$
16. $x^2 + y^2 = 25$; (0,5) to (5,0)
17. $4x^2 + 9y^2 = 36$; (−3,0) to (3,0)
18. $x^{2/3} + y^{2/3} = 4$; (0,8) to (8,0)
19. $y^2 = x^2 - 4$; $x = 2$ to $x = 4$
20. $6y = x^3$; $x = 0$ to $x = 2$
21. A rifle bullet fired at an angle with the horizontal has a range of approximately 60,000 ft and attains a maximum height of 15,000 ft. Write the equation of the curve using convenient axes, and then find the distance the bullet travels from the gun until it reaches the ground.
22. A batted baseball reaches a maximum height of 75 ft and hits the ground 400 ft from the point of impact. How far does the ball travel?
23. Work Example 2 by using dy/dx.

38.6 A SURFACE OF REVOLUTION

If an arc s, a portion of a plane curve, is rotated about an axis, the arc generates a surface called a *surface of revolution*. As an example, if the arc $\widehat{AB}$ (Fig. 38.6) is rotated about the x-axis, it generates a surface. We

522 Length of Arc and Area of Surface of Revolution

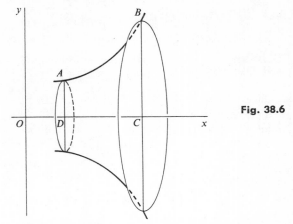

Fig. 38.6

have already seen how a solid of revolution is generated by the rotation of an area about an axis. If we consider the area $ABCD$ being rotated about the x-axis, it generates a solid of revolution at the same time the arc $\overset{\frown}{AB}$ generates a surface.

We are concerned here with the exact area of the surface. Just as we must use calculus to determine the exact length of an arc, so also must we use calculus to determine the exact area of the curved surface generated. We have used lower-case s to denote the length of the arc. Now we use a capital S for the area of the surface.

38.7 THE EXACT AREA OF A SURFACE OF REVOLUTION

If a surface is generated by a straight-line segment, the problem is simple. Then all we should need to do would be to use the formula for the frustum of a cone. (Actually this formula is derived by calculus.) In the case of an arc being rotated about an axis, the problem is not so simple. To find the exact area of a surface of revolution, we might at first glance be inclined to think that our first step would be to find the length of the arc, and then rotate this arc length about the axis for the total surface. That is, we might multiply the arc length by the circumference. However, we cannot proceed in this way because we should then have a varying radius of revolution.

Instead, our first objective is to find the correct expression for the *element of surface*, dS. Then we integrate this element and evaluate the result between limits. The total area is determined by summing up the elements of surface.

Note especially that the *element of arc*, or *differential of arc*, is distinct from the element or *differential of surface*. Since we use s to represent the length of arc, and S to represent the area of a surface, then

ds represents the element of arc s

dS represents the element of surface S

Our first step is to express correctly the element of surface dS.

38.8 THE ELEMENT OF SURFACE dS

As we have pointed out, to find the total area of a curved surface, we cannot rotate the entire arc $\widehat{AB}$ (Fig. 38.7). Instead, we take only an element of arc, ds. Now we rotate the element of arc about the axis, in this case the x-axis. As it rotates, this element ds generates a strip of surface, which we call the element of surface dS. Note especially how this element of surface dS is obtained. It might be thought of as a narrow band or belt of uniform width ds.

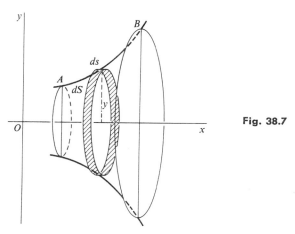

Fig. 38.7

Now we must express the area of this element of surface. This element dS (that is, the belt) has an area equal to its width times its length. The length is equal to the circumference of the belt. Since the radius of rotation is simply the radius of a circle, the circumference is equal to $(2\pi r)$. The width of the belt is the differential of arc ds. (Note that the width cannot be taken as dx or dy.) Then for dS, the element of surface area, we have the formula

$$dS = \text{(circumference)(width)} \quad \text{or} \quad dS = (2\pi r)(ds)$$

In the example shown (Fig. 38.7) the radius is equal to y, the ordinate of the curve. Then

$$dS = (2\pi y)(ds)$$

We have already seen that ds, the element of arc, is given by

$$ds = \sqrt{1 + \left(\frac{dy}{dx}\right)^2}\, dx \quad \text{or} \quad ds = \sqrt{1 + \left(\frac{dx}{dy}\right)^2}\, dy$$

If the element of surface dS has the radius y, then we might choose the formula for element of arc that contains the factor dy. Then for the element of surface we have

$$dS = 2\pi y \sqrt{1 + \left(\frac{dx}{dy}\right)^2}\, dy$$

However, in some instances it is preferable to choose the first form (containing the factor dx) and then replace y with its equivalent in terms of x.

EXAMPLE 3. Find the area of the surface generated by rotating the arc of the following curve from $x = 0$ to $x = 3$ about the x-axis: $y^2 = 4x$.

Solution. We shall solve the problem by using the two formulas in turn for the differential of arc. First we use the "dx" formula:

$$y^2 = 4x$$

Then $\qquad y = 2x^{1/2} \qquad \dfrac{dy}{dx} = x^{-1/2} \qquad \left(\dfrac{dy}{dx}\right)^2 = x^{-1} = \dfrac{1}{x}$

Therefore, $\qquad ds = \sqrt{1 + \dfrac{1}{x}}\, dx$

For the differential of surface, dS, we have

$$dS = 2\pi y \sqrt{1 + \dfrac{1}{x}}\, dx$$

$$= (2\pi)(2x^{1/2}) \sqrt{\dfrac{x+1}{x}}\, dx$$

$$= 4\pi x^{1/2} \left(\dfrac{1}{x^{1/2}}\right) \sqrt{x+1}\, dx = 4\pi \sqrt{x+1}\, dx$$

Then $\qquad S = 4\pi \displaystyle\int_0^3 (x+1)^{1/2}\, dx = \dfrac{8\pi}{3} \left[(x+1)^{3/2}\right]_0^3 = \dfrac{56\pi}{3}$

The curve and the surface generated are shown in Figure 38.8. Using the second formula we begin with $y^2 = 4x$; solving for x, we get

$$x = \dfrac{y^2}{4}$$

Now we find dx/dy:

$$\dfrac{dx}{dy} = \dfrac{y}{2} \qquad \text{and} \qquad \left(\dfrac{dx}{dy}\right)^2 = \dfrac{y^2}{4}$$

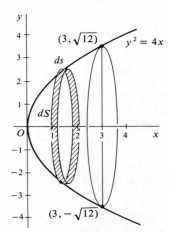

Fig. 38.8

38.8 The Element of Surface dS

Since the rotation is about the x-axis, the radius is still equal to y. Then

$$dS = 2\pi y \sqrt{1 + \frac{y^2}{4}}\, dy = 2\pi y \left(\frac{1}{2}\right)\sqrt{4 + y^2}\, dy$$

$$S = \pi \int_0^{\sqrt{12}} y(4 + y^2)^{1/2}\, dy = \frac{\pi}{3}\left[(4 + y^2)^{3/2}\right]_0^{\sqrt{12}} = \frac{56\pi}{3}$$

Note that for the second formula we must use y-limits.

EXAMPLE 4. Find the area of the surface generated by rotating the arc of the following curve from (0,2) to (2,0), first about the y-axis and then about the x-axis (Fig. 38.9):

$$y = (4 - x^2)^{1/2}$$

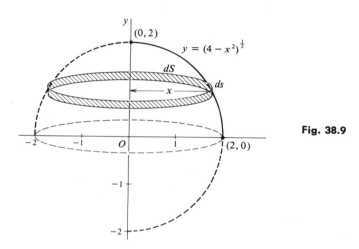

Fig. 38.9

Solution. We shall use the same formula for differential of arc for both rotations. For the differential of arc we first find the derivative dy/dx.

$$\frac{dy}{dx} = \frac{1}{2}(4 - x^2)^{-1/2}(-2x)$$

Simplifying and squaring, we get

$$\left(\frac{dy}{dx}\right)^2 = \frac{x^2}{4 - x^2}$$

Then

$$ds = \sqrt{1 + \frac{x^2}{4 - x^2}}\, dx$$

Simplifying the radical, we get, for the differential of surface,

$$dS = 2\pi x \sqrt{\frac{4}{4 - x^2}}\, dx = \frac{2\pi x}{(4 - x^2)^{1/2}}\sqrt{4}\, dx$$

$$S = 4\pi \int_0^2 x(4 - x^2)^{-1/2}\, dx = -4\pi \left[(4 - x^2)^{1/2}\right]_0^2 = 8\pi$$

Length of Arc and Area of Surface of Revolution

If the arc is rotated about the x-axis, the radius is y, and we have

$$dS = 2\pi y \sqrt{\frac{4}{4-x^2}}\, dx = 2\pi(4-x^2)^{1/2}\left[\frac{1}{(4-x^2)^{1/2}}\right]\sqrt{4}\, dx$$

$$S = 2\pi \int_0^2 dx = 4\pi\bigg[x\bigg]_0^2 = 8\pi$$

Note that in each case we are rotating a quarter of a circle about a diameter. The result is the surface of a hemisphere. By the formula for the surface of a sphere, $A = 4\pi r^2$, the surface of the entire sphere is $A = 4\pi(2)^2 = 16\pi$. This answer checks with the result in the problem for a hemisphere whose radius is 2 units.

EXERCISE 38.2

Find the area of the surface generated by revolving each of the following curves between the given limits about the indicated axis:

1. $y = x^2$; (0,0) to (4,16); about the y-axis
2. $6y = x^3$; $x = 0$ to $x = 2$; about the x-axis
3. $9y = x^3$; $y = 0$ to $y = 3$; about the x-axis
4. $27x = 8y^3$; $y = 0$ to $y = 3$; about the y-axis
5. $y = e^x$; $x = -2$ to $x = 2$; about the x-axis
6. $6xy = x^4 + 3$; $x = 2$ to $x = 5$; about the x-axis
7. $x^2 + y^2 = 25$; entire curve about the x-axis
8. $y = \frac{2}{3}(1+x^2)^{3/2}$; $x = 0$ to $x = 2$; about the y-axis
9. $4x^2 + 9y^2 = 36$; from (0,2) to (3,0); about the x-axis
10. $y^3 = x$; from (0,0) to (8,2); about both axes
11. $y^2 = x^2 - 4$; from $x = 2$ to $x = 4$; about the x-axis
12. $y^2 = 4x$; (0,0) to (4,4); about the y-axis
13. $xy = 6$; (1,6) to (6,1); about the x-axis
14. $xy = 6$; (1,6) to an infinite value of x; about the x-axis

chapter

39

Parametric Equations

39.1 DEFINITION

The relationship between two variables, such as x and y, is usually expressed by a single equation. For example, the equation

$$2x - 3y = 7$$

expresses a *direct* relation between x and y. Any value of one variable determines *directly* the value of the other.

However, a relationship between two variables is often conveniently expressed through a third variable called a *parameter*. For example, in the following equations, y is related to x, but this relation is *indirect* through the parameter t:

$$x = 3t + 5 \quad \text{and} \quad y = 2t + 1$$

For each value of t, there is some value of x and a corresponding value of y. For example, when $t = 1$, $x = 8$ and $y = 3$.

The two equations that state the relation of x and y to the parameter are called *parametric equations*, as previously mentioned in Section 26.7.

39.2 ELIMINATING THE PARAMETER

Parametric equations are useful and convenient in expressing relations between variables. However, in some instances we may wish to eliminate the parameter so that we can see the direct relation between the variables. This can often be done without much difficulty.

To eliminate the parameter in a set of two parametric equations, we usually solve one equation for the parameter and then substitute the result in the other equation. Let us take again the parametric equations

$$x = 3t + 5 \quad \text{and} \quad y = 2t + 1$$

Solving the second equation for t, we get

$$t = \frac{y - 1}{2}$$

Now we substitute this value for t in the first equation and get

$$x = 3\left(\frac{y-1}{2}\right) + 5$$

Simplifying, we get $\quad 2x - 3y = 7$

This single equation expresses the same relation between x and y as do the two parametric equations. The single equation, $2x - 3y = 7$, is called the *Cartesian* form, whereas the two separate equations involving t express the same relation in *parametric* form.

In general, if we have two parametric equations, $x = f(t)$ and $y = g(t)$, we can express the same relation between x and y directly in Cartesian form by eliminating the parameter t. However, the resulting single equation is not always identically equivalent to the parametric form, as we shall see in connection with graphing equations.

39.3 GRAPHS OF PARAMETRIC EQUATIONS

To graph a relation between x and y expressed in parametric form, we may eliminate the parameter and get the Cartesian equation showing the direct relation. However, the graph can be constructed directly from the parametric equations as shown in the following examples.

EXAMPLE 1. Sketch the graph determined by the following parametric equations:

$$x = 2t - 3 \quad \text{and} \quad y = 5 - t$$

Solution. We set up a table showing corresponding values of t, x, and y:

if $t =$	0	1	2	4	6	-1	-2
$x =$	-3	-1	1	5	9	-5	-7
$y =$	5	4	3	1	-1	6	7

Now we plot the points represented by corresponding values of x and y, and get a straight line (Fig. 39.1). If we eliminate the parameter t, we get the Cartesian equation showing the direct relation between x and y: $x + 2y = 7$. The graph of this equation is the same line.

Fig. 39.1

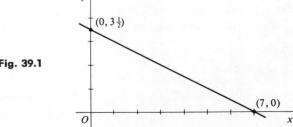

39.3 Graphs of Parametric Equations

EXAMPLE 2. Graph the curve determined by the parametric equations:
$$x = t - 2 \quad \text{and} \quad y = t^2 - 4t - 5$$

Solution. We set up a table of values:

if $t =$	0	1	2	3	5	7	-1	-2
$x =$	-2	-1	0	1	3	5	-3	-4
$y =$	-5	-8	-9	-8	0	16	0	7

When we plot the points represented by corresponding values of x and y, we get a figure that appears to be a parabola (Fig. 39.2). Now let us eliminate the parameter t and we may recognize the curve. Solving the first equation for t, we get
$$t = x + 2$$

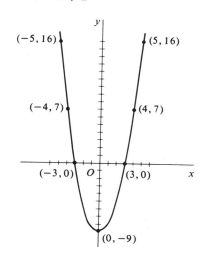

Fig. 39.2

Substituting in the second equation,
$$y = (x + 2)^2 - 4(x + 2) - 5$$
Simplifying, $\qquad y = x^2 + 4x + 4 - 4x - 8 - 5 = x^2 - 9$

Now we recognize the equation as that of a parabola with vertex at $(0, -9)$, the parabola opening upward and symmetric with respect to the y-axis. The x-intercepts are $+3$ and -3, and the y-intercept is -9.

We must be especially careful about the domain and range in connection with parametric equations. Suppose we have the following problem:

EXAMPLE 3. Graph the relation represented by the parametric equations:
$$x = 8 \cos^2 \theta \quad \text{and} \quad y = 4 \sin \theta$$

Solution. Note that whatever the value of θ, the term $\cos^2 \theta$ will be positive. This means that no part of the graph can appear on the left side of the y-axis.

However, y can be either positive or negative, depending on the value of the sine of θ. To graph the parametric equations we set a table of values of θ and corresponding values of x and y:

if $\theta =$	0	30°	45°	60°	90°	120°	135°	150°	180°	270°
$x =$	8	6	4	2	0	2	4	6	8	0
$y =$	0	2	2.83	3.46	4	−3.46	−2.83	−2	0	−4

When we plot the points represented by corresponding values of x and y, we get the curve shown by the *solid* line in Figure 39.3.

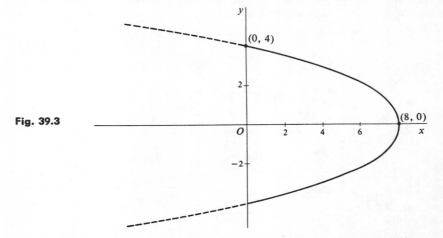

Fig. 39.3

Now, if we eliminate the parameter θ and graph the resulting equation, we get the entire parabola. To eliminate the parameter we multiply the first equation by 2 and get

$$2x = 16 \cos^2 \theta$$

Squaring both sides of the second equation we get

$$y^2 = 16 \sin^2 \theta$$

Adding the two equations,

$$y^2 + 2x = 16(\cos^2 \theta + \sin^2 \theta)$$

The result can be written,

$$y^2 = -2x + 16; \quad \text{or,} \quad y^2 = -2(x - 8)$$

Now we recognize the equation of a parabola of the form $y^2 = K(x - h)$. Since $K = -2$, the parabola opens toward the left. Since $h = 8$, the vertex of the parabola is at the point (8,0).

Graphing the Cartesian form of equation, we get the entire parabola. However, in parametric form, the domain is limited to the values $0 \leq x \leq 8$. The range is limited to the values $-4 \leq y \leq +4$.

39.3 Graphs of Parametric Equations

EXAMPLE 4. As another illustration of the limited domain and range in parametric equations, compare the graphs of each of these two sets of equations with the graph of the Cartesian equation:

(a) $x = s^2 + 3$ and $y = s^2$ (b) $x = t^2$ and $y = t^2 - 3$

Solution. In the first set of equations, s is the parameter. In these two parametric equations, x cannot be less than 3, and y cannot be less than zero. The graph is a portion of a straight line (Fig. 39.4), but only the portion extending diagonally upward to the right from the point (3,0).

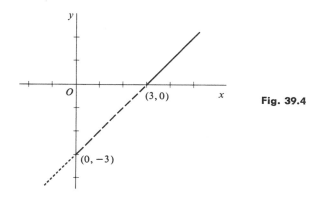

Fig. 39.4

Now let us take the second set of equations in which t is the parameter. In these equations, x cannot be less than zero, and y cannot be less than -3. The graph of these two parametric equations is the portion of the same straight line, but only the portion extending upward and to the right from the point $(0,-3)$.

If we eliminate the parameters, s in the first pair of equations and t in the second pair, we get the same Cartesian form of equation:

$$x - y = 3$$

Now the graph of this equation becomes the entire straight line shown.

EXERCISE 39.1

Graph each of the following pairs of parametric equations. Then eliminate the parameter and find the corresponding Cartesian equation. In what way, if any, do the graphs of the two forms differ?

1. $x = 5 - t;\ y = 3t + 2$
2. $x = 7 - 4t;\ y = 3t - 4$
3. $x = 3t + 7;\ y = 5t - 2$
4. $x = 4 - t;\ y = t^2 - 1$
5. $x = s - 1;\ y = s + s^3$
6. $x = s - 4;\ y = \sqrt{s^2 + 4}$
7. $x = t^2 + 1;\ y = t^3 + t$
8. $x = 2v;\ y = \sqrt{4 - v^2}$
9. $x = 2u - 1;\ y = \sqrt{u^2 - 4}$
10. $x = 2\sqrt{t + 1};\ y = \sqrt{t}$
11. $x = 3\sqrt{t} + 2;\ y = 2\sqrt{3t}$
12. $x = \sqrt{4 - t^2};\ y = 2t$
13. $x = 3\sqrt{t};\ y = \sqrt{4 - t}$
14. $x = 2t - 5;\ y = \sqrt{t - 3}$
15. $x = 4\cos\theta;\ y = 4\sin\theta$
16. $x = 5\cos\theta;\ y = 2\sin\theta$
17. $x = \cos\theta;\ y = \sin 2\theta$
18. $x = 2 - \sin\theta;\ y = \sin 2\theta$
19. $x = 4 - 4\cos^2\theta;\ y = 4\sin\theta$
20. $x = 9\cos^2\theta;\ y = 3\sin\theta$

21. $x = \dfrac{4}{1+t}; y = \dfrac{t}{1-t^2}$ 22. $x = \dfrac{4}{1+t}; y = \dfrac{t}{t^2-1}$

23. $x = \dfrac{2}{1+t^2}; y = \dfrac{2t}{1+t^2}$ 24. $x = \dfrac{3s}{s^2+1}; y = \dfrac{3s^2}{s^2+1}$

39.4 THE CIRCLE IN PARAMETRIC FORM

We have seen (Example 3) that in some parametric equations, the parameter is an angle θ. As another example, a circle can be represented by the following equations in which a is a constant:

$$x = a \cos \theta \quad \text{and} \quad y = a \sin \theta$$

The relation can be graphed from the parametric form, or the parameter can first be eliminated and the relation expressed directly in Cartesian form.

To eliminate the parameter θ in the foregoing equations, we square both sides of both equations and get

$$x^2 = a^2 \cos^2 \theta$$
$$y^2 = a^2 \sin^2 \theta$$

Adding, $\quad x^2 + y^2 = a^2 \cos^2 \theta + a^2 \sin^2 \theta$

or $\quad x^2 + y^2 = a^2 (\cos^2 \theta + \sin^2 \theta)$

Then $\quad x^2 + y^2 = a^2$

Since a^2 is a positive constant, we can now recognize the equation as that of a circle with its center at (0,0) and radius equal to a (Fig. 39.5).

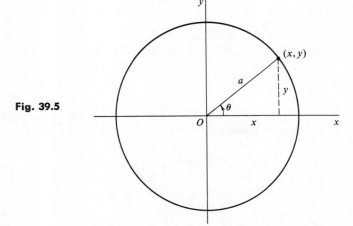

Fig. 39.5

39.5 THE ELLIPSE IN PARAMETRIC FORM

Another curve is easily expressed in the following parametric form:

$$x = a \cos \theta \quad \text{and} \quad y = b \sin \theta$$

39.5 The Ellipse in Parametric Form

In this example, *a* and *b* are constants but may be unequal. To take a specific example, suppose we have

$$x = 5 \cos \theta \quad \text{and} \quad y = 3 \sin \theta$$

We could graph the relation directly from the parametric form, but it may be easier to recognize the graph if the parameter θ is eliminated. Squaring both sides of both equations, we get

$$x^2 = 25 \cos^2 \theta$$
$$y^2 = 9 \sin^2 \theta$$

Since we know that $\cos^2 \theta + \sin^2 \theta = 1$, we attempt to equalize the coefficients of these squares. We multiply the first equation by 9 and the second equation by 25 and get

$$9x^2 = 225 \cos^2 \theta$$
$$25y^2 = 225 \sin^2 \theta$$

Adding, $\quad 9x^2 + 25y^2 = 225 (\cos^2 \theta + \sin^2 \theta) = 225$

Now we recognize the equation as that of an ellipse. If we divide both sides of the equation by 225, we get the standard form

$$\frac{x^2}{25} + \frac{y^2}{9} = 1$$

From the standard form we see that the center of the ellipse is at the origin (0,0), the semimajor axis is 5, along the *x*-axis, and the semiminor axis is 3. The graph is shown in Figure 39.6.

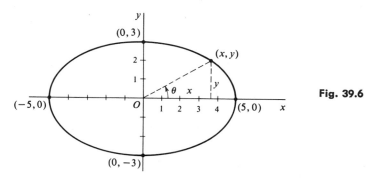

Fig. 39.6

In general, the following equations represent an ellipse with center at the origin, major axis 2*a* along the *x*-axis, and minor axis 2*b* along the *y*-axis.

$$x = a \cos \theta \quad \text{and} \quad y = b \sin \theta$$

This can be shown from Figure 39.7. The derivation is left to the student. For an ellipse with major axis along the *y*-axis, we have

$$x = b \cos \theta \quad \text{and} \quad y = a \sin \theta$$

534 Parametric Equations

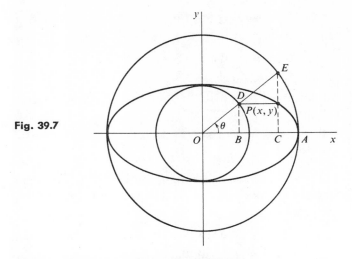

Fig. 39.7

39.6 THE PATH OF A PROJECTILE

The path of a projectile fired from a gun at an angle with the horizontal is conveniently shown in parametric form. Let us say θ represents the angle of elevation and V_0 represents initial velocity in the direction indicated by θ. Then, after the projectile leaves the gun, there is no change in horizontal velocity (neglecting air resistance). The vertical velocity is affected by the force of gravity. The horizontal distance x and the vertical distance y (Fig. 39.8) are given by the following formulas, respectively:

$$x = V_0 t \cos\theta \quad \text{and} \quad y = V_0 t \sin\theta - 16t^2$$

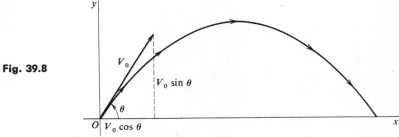

Fig. 39.8

The term $16t^2$ represents the distance downward as a result of the pull of gravity and is therefore subtracted from the upward distance indicated by $V_0 t \sin\theta$.

Suppose the projectile is fired at an angle of 30° with an initial velocity of 3000 ft/sec. Then (neglecting air resistance) the projectile follows the curve.

$$x = (3000)(t)(\cos\theta) \quad \text{and} \quad y = (3000)(t)(\sin\theta) - 16t^2$$

At the end of 5 sec, for example, the horizontal distance is

$$x = (3000)(5)(\cos 30°) = (15000)\left(\frac{\sqrt{3}}{2}\right) = 7500\sqrt{3} = 12{,}990 \text{ ft (approx.)}$$

For the height, y, above ground level, we have

$$y = (3000)(5)(\sin 30°) - 16(25) = (15000)\left(\frac{1}{2}\right) - 400 = 7100 \text{ ft (approx.)}$$

By means of the two parametric equations the position of the projectile can be spotted at any instant.

In the foregoing equations, the parameter is the time t. It can be eliminated but the resulting equation is a bit more complicated. To eliminate t, let us solve the first equation for t and substitute the result in the second equation. We have

$$x = (3000)(t)(\cos 30°) = 1500\sqrt{3}\,t; \quad t = x/1500\sqrt{3}$$

Substituting this value for t in the second equation, we get

$$y = (3000)\left(\frac{1}{2}\right)\left(\frac{x}{1500\sqrt{3}}\right) - 16\left(\frac{x}{1500\sqrt{3}}\right)^2$$

or

$$y = \frac{x}{\sqrt{3}} - 16\left(\frac{x^2}{6{,}750{,}000}\right)$$

The equation can be written in the form

$$\sqrt{3}\,x^2 + 421{,}875\sqrt{3}\,y - 421{,}875x = 0$$

Now we recognize the equation as that of a parabola opening downward and passing through the origin, but the use of the equation is probably more convenient in the parametric form.

39.7 THE CYCLOID

If a circle rolls along a straight line without slipping, a point on the rim describes a curve called a *cycloid* (Fig. 39.9). The parametric equations of the curve are

$$x = r(\theta - \sin \theta)$$
$$y = r(1 - \cos \theta)$$

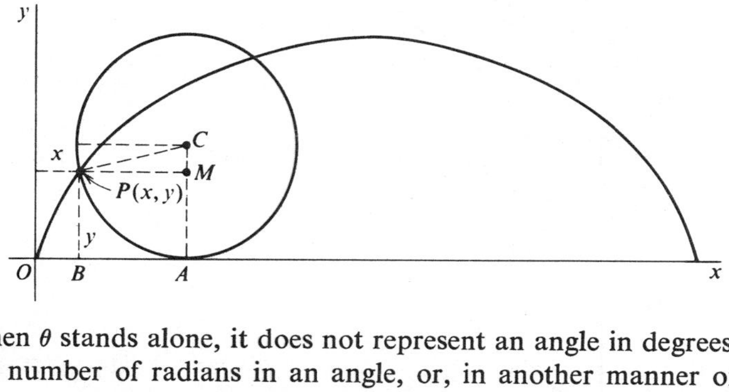

Fig. 39.9

When θ stands alone, it does not represent an angle in degrees, but instead the number of radians in an angle, or, in another manner of speaking, θ is a number. In the figure, point $P(x,y)$ is the point on the rim, which we assume began at the origin O. As the circle has turned through the angle θ, the arc $AP = OA$. Moreover, arc $AP = r\theta$, if θ is taken in radians. Then we have

$$x = OB = OA - BA = r\theta - RM = r\theta - r\sin\theta = r(\theta - \sin\theta)$$

For y, we have

$$y = BP = AM = AC - MC = r - r\cos\theta = r(1 - \cos\theta)$$

One important variation of the cycloid is the hypocycloid, the curve traced by a point on the rim of a small circle as it rolls along the inside of a larger circle as is found in the case of some mechanical clutches or gears. If the radius R of the large circle is four times the radius r of the small rolling circle, the curve is a hypocyloid of four cusps. The parametric equations of the curve are

$$x = R\cos^3\theta \quad \text{and} \quad y = R\sin^3\theta$$

The parameter can be eliminated by reducing the cubes to squares. This problem is left to the student.

39.8 CHANGING TO PARAMETRIC FORM

Eliminating a parameter is not always convenient or even desirable. Sometimes it is difficult, and the resulting direct Cartesian form may be very complicated. Moreover, the solutions of some problems are more easily handled when the relation between variables is expressed in parametric form. In fact, it sometimes happens that we change a Cartesian equation in x and y to a parametric form for convenience in a problem.

To change any Cartesian equation to a parametric form, we introduce a parameter, say t. We let x or y represent any convenient function of t, and then express the other letter in the proper terms involving t. We may sometimes let t represent the sum or product of x and y, or some other form such as the following: $y = tx$; $x = ty$; $x + y = t$; $y = x + t$. The change to a convenient parametric form is often a matter of good judgment or trial and error, but is not usually as difficult as it might seem.

The equation of a straight line is usually most convenient in the Cartesian form. However, it is sometimes desirable to find the parametric equations of a straight line. Suppose we have the equation

$$2x - 3y = 7$$

We can let x represent almost any function of t, say, $x = t - 5$. Then we substitute $(t - 5)$ for x in the equation and solve for y. We obtain

$$2(t - 5) - 3y = 7$$

from which:
$$y = \frac{2t - 17}{3}$$

Of course, in this instance the graph is most easily constructed from the Cartesian form: $2x - 3y = 7$.

39.8 Changing to Parametric Form

Probably the most useful form of parametric equations of a straight line can be written from the slope and a point on the line. Suppose the two numbers a and b are two numbers whose ratio in that order represent the slope. Suppose further that the constants c and d are, respectively, the x and y coordinates of any point satisfying the equation. Then we often use the following forms for the parametric equations of the line:

$$x = bt + c \quad \text{and} \quad y = at + d$$

EXAMPLE 5. Express the equation $2x - 3y = 7$ in parametric form using the foregoing approach.

Solution. In this equation we see that the slope is $\frac{2}{3}$. Then we use the two numbers, 2 and 3, as coefficients of t, respectively, but reversed as to association with x and y; that is, we write

$$x = 3t + c \quad \text{and} \quad y = 2t + d$$

Now we take c and d as the coordinates of any point on the line; that is, a pair of numbers satisfying the equation. One such pair is (5,1). Then the parametric equations become

$$x = 3t + 5 \quad \text{and} \quad y = 2t + 1$$

As a rule, this method avoids fractions, and the equations are in convenient forms. However, the straight line is probably best expressed in Cartesian form.

In the case of a circle or an ellipse, the parametric forms are most conveniently expressed with an angle as the parameter. With a as the radius the equation of a circle can be expressed in the forms

$$x = a \cos \theta \quad \text{and} \quad y = a \sin \theta$$

For an ellipse we have the forms

$$x = a \cos \theta \quad \text{and} \quad y = b \sin \theta$$

EXAMPLE 6. Express the circle $x^2 + y^2 = 36$ in parametric form.

Solution. We note that the radius is equal to 6. Then we have at once

$$x = 6 \cos \theta \quad \text{and} \quad y = 6 \sin \theta$$

EXAMPLE 7. Express the equation $4x^2 = 9y^2 = 36$ in parametric form.

Solution. We first write the equation in standard form. Dividing by 36, we get

$$\frac{x^2}{9} + \frac{y^2}{4} = 1$$

From this form we can identify the semimajor axis a and the semiminor axis b of the ellipse:

$$a = 3 \quad \text{and} \quad b = 2$$

Referring to the general parametric form, we have

$$x = 3 \cos \theta \quad \text{and} \quad y = 2 \sin \theta$$

EXAMPLE 8. Express the hyperbola $x^2 - 9y^2 = 36$ in parametric form.

Solution. For a hyperbola the parametric form is conveniently written in algebraic terms, taking a letter such as t for the parameter. If we let t represent y^2, then x^2 and y^2 can be stated in terms of t. Then, if $y^2 = t$,

$$9y^2 = 9t \quad \text{and} \quad x^2 = 36 + 9t$$

Solving for x and y,

$$y = \pm\sqrt{t} \quad \text{and} \quad x = \pm 3\sqrt{4 + t}$$

Both signs are used to insure getting the complete hyperbola.

EXAMPLE 9. Express the parabola $y^2 = 8x - 24$ in parametric form.

Solution. For a parabola it is convenient to take the focal length as the coefficient of the parameter t. Writing the equation in standard form,

$$y^2 = 8(x - 3)$$

Now we can identify the focal length as $8 \div 4 = 2$. Then we let $x = 2t$. To find the expression for y, we substitute $2t$ for x and get

$$y = 2\sqrt{4t - 6}$$

EXAMPLE 10. Express in parametric form:

$$(x + 3y)^3 - x(x + 3y)^2 - 2 = 0$$

Solution. Note that this equation would be difficult to graph in its Cartesian form. For the parametric form, we let t represent the binomial. If

$$t = x + 3y$$

then

$$x = t - 3y$$

Substituting,

$$t^3 - (t - 3y)t^2 - 2 = 0 \quad \text{or} \quad t^3 - t^3 + 3t^2y - 2 = 0$$

Then

$$y = \frac{2}{3t^2}$$

Since

$$x = t - 3y$$

we get

$$x = \frac{t^3 - 2}{t^2}$$

Now the graph can be constructed more easily from the parametric form.

EXERCISE 39.2

Eliminate the parameter in each of the following pairs of parametric equations. Sketch the curve using the most convenient form of equation:

1. $x = 5t - 3;\ y = 4 - 3t$
2. $x = \sqrt{17}\cos\theta;\ y = \sqrt{17}\sin\theta$
3. $2x = 3\cos\theta;\ 3y = 5\sin\theta$
4. $x = 20t;\ y = 10t - 15t^2$
5. $x = \sqrt{t};\ y = 3\sqrt{2t - 5}$
6. $x = 3t^2;\ y = 4t - 3$
7. $x = 4\sin\theta;\ y = 2\cos^2\theta$
8. $x = 8\cos^3\theta;\ y = 8\sin^3\theta$

Change the following to parametric form. Sketch the curve from the most convenient form of equation.

9. $3x - 5y = 20$
10. $4x + 3y = 17$
11. $2x - 7y = 3$
12. $x + 3y = 9$
13. $x^2 + y^2 = 16$
14. $4x^2 + 25y^2 = 100$
15. $x^2 - y^2 = 36$
16. $y^2 = 12x$
17. $x^2 + y^2 = 6x$
18. $4x^2 + y^2 = 16$
19. $4y^2 - 9x^2 = 144$
20. $3x^2 = 20y$
21. $x^2 = 15 - 12y$
22. $y^2 - 12 = 6x$
23. $xy = 16$
24. $x^2y = 12$
25. $(x - y)^2 = x$
26. $x^2 + xy + 1 = 0$
27. $(x + y)^2 - 2y = 0$
28. $xy - 6x + 2y = 8$
29. $x^2 + y^2 + 6xy = 0$
30. $(x - y)^3 - x(x - y)^2 - 12 = 0$
31. $(x + y)^3 - y(x + y)^2 = 6$
32. $(x + 2y)^3 - x(x + 2y)^2 = 4$
33. $y(x + y)^2 - (x + y)^3 + 1 = 0$
34. $x^2 + 2xy + y^2 + y + 6 = 0$
35. $x^3 + y^3 + 6xy = 0$
36. $(x + y - 3)^3 - x(x + y - 3)^2 = 1$
37. $x^2 - 3xy + y^2 - 3x + 2y = 7$

Find the position (horizontal and vertical components) of each of the following projectiles for time indicated, projected at the given angle θ.

38. $V_0 = 2500$ ft/sec; $\theta = 20°$; time, 4 sec
39. $V_0 = 3000$ ft/sec; $\theta = 60°$; time, 10 sec
40. $V_0 = 3000$ ft/sec; $\theta = 45°$; time, 10 sec

39.9 DIFFERENTIATION OF PARAMETRIC EQUATIONS

Up to this point, in finding a derivative we have always started with a direct relation between x and y (or other two variables). To find the derivative, dy/dx, from two given parametric equations, we might eliminate the parameter and then differentiate the result. However, it is possible to find the derivative without eliminating the parameter. To do so, we

(1) First, differentiate each parametric equation with respect to the parameter, and
(2) Second, find dy/dx through the relation

$$\frac{dy}{dx} = \frac{dy/dt}{dx/dt}$$

EXAMPLE 11. Given the equations $y = t^2 - 3t$ and $x = t + 2$, find the derivative, dy/dx, and then evaluate the derivative for $t = 4$.

Solution. Differentiating each equation with respect to t, we get

$$\frac{dy}{dt} = 2t - 3 \quad \text{and} \quad \frac{dx}{dt} = 1$$

Then

$$\frac{dy}{dx} = \frac{dy/dt}{dx/dt} = \frac{2t - 3}{1} = 2t - 3$$

At $t = 4$, $dy/dx = 5$.

Now let us eliminate the parameter and see whether we get the same value for the derivative. Solving the second equation for t, we get $t = x - 2$. Substituting in the other parametric equation we get

$$y = (x - 2)^2 - 3(x - 2) \quad \text{or} \quad y = x^2 - 7x + 10$$

Now we differentiate directly:

$$\frac{dy}{dx} = 2x - 7$$

To evaluate the derivative, we find that when $t = 4$, $x = 4 + 2 = 6$. Then, at $x = 6$, $dy/dx = 2(6) - 7 = 5$, the same value as before.

EXAMPLE 12. Given the equations,
$$x = 5t - 2 \quad \text{and} \quad y = (t^2 + 5)^{1/2}$$
Find dy/dx and evaluate the derivative for $t = 2$. Then eliminate the parameter, find dy/dx directly, and evaluate.

Solution. Differentiating with respect to the parameter t,
$$\frac{dx}{dt} = 5 \quad \text{and} \quad \frac{dy}{dt} = \frac{t}{(t^2 + 5)^{1/2}}$$
Then
$$\frac{dy}{dx} = \frac{t}{5(t^2 + 5)^{1/2}}$$
at $t = 2$,
$$\frac{dy}{dx} = \frac{2}{15}$$
To eliminate the parameter, we solve the first equation for t:
$$t = \frac{x + 2}{5}$$
Substituting in the second equation,
$$y = \frac{(x^2 + 4x + 129)^{1/2}}{5}$$
Differentiating,
$$\frac{dy}{dx} = \frac{x + 2}{5(x^2 + 4x + 129)^{1/2}}$$
at $t = 2$, $x = 8$,
$$\frac{dy}{dx} = \frac{2}{15}$$

EXAMPLE 13. Given the equations
$$x = 5 \cos \theta \qquad y = 2 \sin \theta$$
Find the slope of the curve at $\theta = \pi/3$. Then eliminate the parameter and find the slope from the direct relation between x and y.

Solution. Differentiating each equation with respect to θ, the parameter,
$$\frac{dx}{d\theta} = -5 \sin \theta \qquad \frac{dy}{d\theta} = 2 \cos \theta$$
Then
$$\frac{dy}{dx} = \frac{dy/d\theta}{dx/d\theta} = \frac{2 \cos \theta}{-5 \sin \theta} = -\left(\frac{2}{5}\right) \cot \theta$$
At $\theta = \pi/3$,
$$m = \frac{dy}{dx} = -\left(\frac{2}{5}\right) \cot \frac{\pi}{3} = -\left(\frac{2}{5}\right)\left(\frac{\sqrt{3}}{3}\right) = -\frac{2\sqrt{3}}{15} = -0.231$$
To eliminate the parameter θ, we square both sides of both equations and get
$$x^2 = 25 \cos^2 \theta \quad \text{and} \quad y^2 = 4 \sin^2 \theta$$
Multiplying both sides of the first equation by 4 and the second equation by 25, we get
$$4x^2 = 100 \cos^2 \theta$$
$$25y^2 = 100 \sin^2 \theta$$

Adding and simplifying, $4x^2 + 25y^2 = 100$, the equation of an ellipse. Differentiating implicitly,

$$8x + 50y\left(\frac{dy}{dx}\right) = 0$$

then

$$\frac{dy}{dx} = -\frac{4x}{25y}$$

To evaluate, we find the values of x and y when $\theta = \pi/3$. When $\theta = \pi/3$,

$$x = 5\cos\frac{\pi}{3} = \frac{5}{2} \qquad y = (2)\left(\frac{\sqrt{3}}{2}\right) = \sqrt{3}$$

Then

$$\frac{dy}{dx} = -\frac{2\sqrt{3}}{15} = -0.231,$$

as before.

39.10 THE SECOND DERIVATIVE, d^2y/dx^2

We must be especially careful in finding the second derivative in connection with parametric equations. We recall that the second derivative is the derivative of the first derivative; that is,

$$\frac{d^2y}{dx^2} = \frac{d}{dx}\left(\frac{dy}{dx}\right)$$

This is not as simple as it might first appear.

The derivative in parametric form usually contains the parameter t. For example, consider the following parametric equations:

$$x = t^2 - 5t \quad \text{and} \quad y = 2t^2 - 3t$$

Then $\dfrac{dx}{dt} = 2t - 5 \qquad \dfrac{dy}{dt} = 4t - 3 \qquad \dfrac{dy}{dx} = \dfrac{4t-3}{2t-5}$

Note that the derivative contains the parameter t. Now we must differentiate this derivative, but with respect to t; that is, we first find

$$\frac{d}{dt}\left(\frac{dy}{dx}\right) = \frac{d}{dt}\left(\frac{4t-3}{2t-5}\right)$$

Here we must use the quotient rule. We get

$$\frac{d}{dt}\left(\frac{4t-3}{2t-5}\right) = \frac{(2t-5)(4) - (4t-3)(2)}{(2t-5)^2} = \frac{-14}{(2t-5)^2}$$

Now we must divide again by dx/dt; then

$$\frac{d^2y}{dx^2} = \frac{-14}{(2t-5)^3}$$

The first and second derivative can now be evaluated for any value of t. For example, when $t = 1$,

$$\frac{dy}{dx} = \frac{4-3}{2-5} = -\frac{1}{3} \quad \text{and} \quad \frac{d^2y}{dx^2} = \frac{-14}{(2-5)^3} = \frac{14}{27}$$

If we wish to find a maximum or minimum point of the curve, we set the first derivative equal to zero and solve for t:

$$\frac{4t-3}{2t-5} = 0$$

then $\quad 4t - 3 = 0 \quad t = \dfrac{3}{4} \quad x = -\dfrac{51}{16} \quad y = -\dfrac{9}{8}$

In general it should be noted that

$$\frac{d^2y}{dx^2} = \frac{\dfrac{d}{dt}\left(\dfrac{dy}{dx}\right)}{\dfrac{dx}{dt}}$$

In the foregoing example, we found that

$$\frac{d^2y}{dx^2} = \frac{-14}{(2x-5)^3}$$

Note that the second derivative contains the denominator which is the *cube* of dx/dt; that is, $(dx/dt)^3$. Now we might ask: Is this always true? Will the second derivative always contain $(dx/dt)^3$ as the denominator? Let us see whether this is true by using more general terms.

We begin with the general parametric equations

$$x = f(t) \quad \text{and} \quad y = g(t)$$

Using primes to denote derivatives, we get

$$\frac{dx}{dt} = f'(t) \qquad \frac{dy}{dt} = g'(t)$$

then

$$\frac{dy}{dx} = \frac{g'(t)}{f'(t)}$$

Now we differentiate the first derivative with respect to t and get

$$\frac{d}{dt}\left[\frac{g'(t)}{f'(t)}\right] = \frac{[f'(t)][g''(t)] - [g'(t)][f''(t)]}{[f'(t)]^2}$$

Dividing by dx/dt,

$$\frac{d^2y}{dx^2} = \frac{[f'(t)][g''(t)] - [g'(t)][f''(t)]}{[f'(t)]^3}$$

We see then that the denominator of the second derivative is the cube of dx/dt. In most examples it is best to find the first derivative and then find the second derivative by the quotient rule.

EXAMPLE 14. Find any possible maximum or minimum points of the curve

$$x = 2t - t^3 \quad \text{and} \quad y = t^2 - 4t - 3$$

39.10 The Second Derivative, d^2y/dx^2

Solution. $\dfrac{dx}{dt} = 2 - 3t^2 \qquad \dfrac{dy}{dt} = 2t - 4$

then

$$\dfrac{dy}{dx} = \dfrac{2t - 4}{2 - 3t^2}$$

and

$$\dfrac{d^2y}{dx^2} = \dfrac{(2 - 3t^2)(2) - (2t - 4)(-6t)}{(2 - 3t^2)^3} = \dfrac{6t^2 - 24t + 4}{(2 - 3t^2)^3}$$

For $m = 0$, we have

$$\dfrac{2t - 4}{2 - 3t^2} = 0$$

from which $t = 2 \quad x = -4 \quad y = -7$

The value $t = 2$, for a possible maximum or minimum point, may be tested by the second derivative. At $t = 2$,

$$y'' = \dfrac{24 - 48 + 4}{-1000} = + \text{ (positive)}$$

The point $(-4, -7)$ is therefore a minimum.

EXAMPLE 15. Find the total area of the following ellipse (Fig. 39.10) without eliminating the parameter θ:

$$x = 6 \cos \theta \qquad y = 4 \sin \theta$$

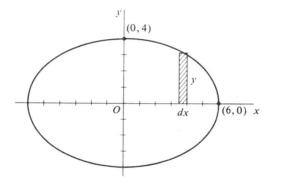

Fig. 39.10

Solution. We shall find the area in the first quadrant and then multiply the result by 4. For the element of area we have $dA = y \, dx$; then

$$A = 4 \int_0^6 y \, dx$$

Now we must replace y and dx with their equivalents in terms of θ. Since

$$x = 6 \cos \theta$$

then

$$dx = (-6 \sin \theta) \, d\theta$$

Then we have,

$$dA = (4 \sin \theta)(-6 \sin \theta) \, d\theta = -24 \sin^2 \theta \, d\theta$$

In evaluating the definite integral, we can no longer use x-limits. Instead we must use equivalent θ limits. For the new limits we have,

when $x = 6$, $\theta = 0$; when $x = 0$, $\theta = \dfrac{\pi}{2}$, or $90°$

Then
$$A = 4\int_{\pi/2}^{0} -24 \sin^2 \theta \, d\theta = -96 \int_{\pi/2}^{0} \left(\frac{1}{2} - \frac{1}{2}\cos 2\theta\right) d\theta$$
$$= -96\left[\frac{\theta}{2} - \frac{1}{4}\sin 2\theta\right]_{\pi/2}^{0} = -96\left[0 - 0 - \frac{\pi}{4} + 0\right] = 24\pi$$

EXERCISE 39.3

For each of the following parametric equations, find dy/dx and d^2y/dx^2:
1. $x = 3t$; $y = 3 - t^2$
2. $x = t^2$; $y = t^3$
3. $x = t^2 - 2$; $y = t - t^3$
4. $x = t^2$; $y = 2/t^2$
5. $x = e^t$; $y = e^{-t}$
6. $x = e^{-2t}$; $y = e^{3t}$
7. $x = e^t$; $y = t^2 - e^{2t}$
8. $x = 4 \cos \theta$; $y = 4 \sin \theta$
9. $x = 5 \cos \theta$; $y = 3 \sin \theta$
10. $x = e^t \cos t$; $y = e^{-t} \sin t$

Find the equation of the tangent line to each of the following curves at the indicated value of the parameter:
11. $x = 2t$; $y = t^2 + 3$; at $t = -1$
12. $x = t^2$; $y = t^3$; for $t = \frac{1}{2}$
13. $x = t^2$; $y = 4/t^2$; at $t = 2$
14. $x = 4t - t^3$; $y = t^2$; $t = 1$; $t = 2$
15. $x = \sqrt{t}$; $y = \dfrac{t}{9} - \dfrac{1}{\sqrt{t}}$; $t = 9$
16. $x = \sin^3 \theta$; $y = \cos^3 \theta$; $\theta = 45°$

Evaluate the first derivative, dy/dx, for the indicated value of the parameter:
17. $x = \cos 2\theta$; $y = \sin \theta$; $\theta = \pi/6$
18. $x = 3 \tan \theta$; $y = 2 \sin \theta$; $\theta = 0$
19. $x = 3 \sin^2 \theta$; $y = 5 \cos^2 \theta$; $\theta = 45°$; $\theta = 60°$. Explain the answer.
20. $x = 3 - 4 \cos 2\theta$; $y = 5 - 2 \sin \theta$; $\theta = \pi/6$

Derive the formula for each of the following by working out the problem in general terms:
21. Find the area of the circle: $x = a \cos \theta$; $y = a \sin \theta$.
22. Find the area of the ellipse: $x = a \cos \theta$; $y = b \sin \theta$.
23. Find the area of one branch of the cycloid: $x = a(\theta - \sin \theta)$; $y = a(1 - \cos \theta)$.
24. Find the centroid of one portion of the ellipse: $x = a \cos \theta$; $y = b \sin \theta$. (Take the first quadrant portion.)
25. Find the centroid of the area indicated in Problem 23.
26. Find the length of the curve: $x = a \cos^3 \theta$; $y = a \sin^3 \theta$.

chapter

40

Polar Coordinates

40.1 DEFINITION

A point in a plane may be located by stating a *direction* and a *distance*. This is the principle of *polar coordinates*, which we have already mentioned in Chapter 1. Actually we make use of this principle in much of our everyday activities. If we wish to get to a particular point or place, we often simply turn our steps in the proper direction and then proceed directly toward our objective.

If we stand at one corner of a city block, say, at A (Fig. 40.1) and wish to get to the opposite corner at B, we might get there in two different ways. We might walk, say, 240 ft east and then 180 ft north. These two numbers are the *rectangular coordinates* of the point. Another way to get from A to B, if there is no obstruction, is to turn our steps 36.9° north of east and then walk 300 ft directly to point B. The two numbers, 36.9° and 300 ft, are called the *polar coordinates* of point B, with reference to point A and to the direction *east*. In stating the pair of polar coordinates, we often state the distance first and then the direction. Then the point (300, 36.9°) in polar coordinates is the same as the point (240, 180) in rectangular coordinates.

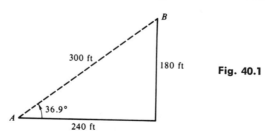

Fig. 40.1

40.2 THE POLAR COORDINATE SYSTEM

To set up the polar coordinate system, we begin with the point O called the *pole*. Then we draw a ray called the *polar axis* from O, usually toward the right. The line through the pole perpendicular to the polar axis is sometimes called the *90°-axis* (Fig. 40.2).

Polar Coordinates

A point is located in the polar system by stating its direction as measured by an angle of rotation from the polar axis. *Counterclockwise* rotation is called *positive*, and *clockwise rotation* is called *negative* rotation. The distance to the point is then measured in linear units from the pole in the direction indicated by the angle (Fig. 40.2). For example, to locate the point (5, 30°) we first lay off the angle of 30° in a positive direction of rotation from the polar axis. Then along the terminal side of the angle we measure 5 linear units from the pole O. Polar coordinate paper is convenient for this purpose since it shows the angles from the polar axis and also the various distances from the pole.

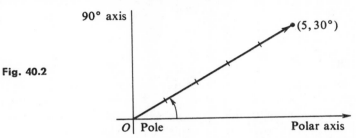

Fig. 40.2

In the notation for the general point we often use θ (theta) for the angle and r (sometimes *rho*) for the distance from O. The general point is then denoted by (r,θ). The coordinates are sometimes reversed since there can be no confusion as to the meaning in polar coordinates. The point (5, 30°) is the same as (30°, 5). In the rectangular system such a reversal of coordinates would locate a different point.

Either r or θ or both may be negative. The following points are shown in Figure 40.3:

$A(-5, 45°)$ $B(5, -45°)$ $C(-6, -60°)$

$D(6, 180°)$ $E(-6, 0°)$ $F(-150°, -6)$

A negative distance is laid off in a direction opposite to that indicated by the angle θ. To locate the point $(-5, 45°)$ we first lay off the positive angle 45°. Then for $r = -5$, we measure 5 units in a direction opposite to that indicated by the terminal side of the 45° angle. Imagine you are standing facing east. You turn your position through 45° and face northeast. Now you *back up* 5 steps. That is the meaning of a negative r.

If the angle is indicated as negative, it is laid off in a negative or clockwise direction of rotation from the polar axis. For example, to locate the point $(5, -45°)$ we first lay off the angle $-45°$. Then we measure 5 units in the direction indicated by the terminal side of the angle θ.

If the angle θ is equal to zero, then the distance r is laid off along the polar axis in a positive or negative direction depending on the sign of r. It may be that $r = 0$. For example, to locate the point (0, 30°), we lay off the 30° in a positive direction of rotation. Then, since $r = 0$, we stay right at the pole. This is something like saying you stand at the pole facing east.

40.3 Relation between the Rectangular and the Polar Systems

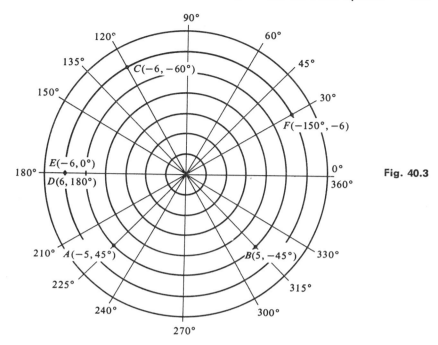

Fig. 40.3

Then you turn through an angle of 30° and *look* in that direction but do not move.

In polar coordinates a point may have more than one pair of coordinates. For example, the point (−6, 150°) is the same point as (6, −30°). In fact, the same point may have any number of pairs of polar coordinates. The point (6, −30°) may also be called (6, 330°), (−6, −210°) and others. In polar coordinates there is not the kind of one-to-one correspondence between pairs of numbers and points that we find in rectangular coordinates.

40.3 RELATION BETWEEN THE RECTANGULAR AND THE POLAR SYSTEMS

In order to express the relation between polar and rectangular coordinates of any point, we place the polar system on the rectangular system (Fig. 40.4) so that the pole coincides with the origin (0,0), and the polar axis lies along the positive direction of the *x*-axis. Then the 90° axis falls along the *y*-axis.

Now we note the relation between the polar and the rectangular coordinates. For the point (r,θ) in the polar system, we have the coordinates (x,y) in the rectangular system. Then we have the following relations:

$$\frac{x}{r} = \cos \theta \quad \text{and} \quad \frac{y}{r} = \sin \theta$$

Solving for *x* and *y*,

$$x = r \cos \theta \quad \text{and} \quad y = r \sin \theta$$

Polar Coordinates

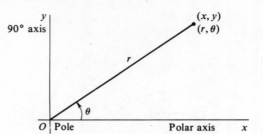

Fig. 40.4

If we have the point (x,y) in rectangular coordinates and wish to find the polar equivalents, we have the relations:

$$r = \sqrt{x^2 + y^2} \qquad \theta = \arctan \frac{y}{x} \qquad \cos \theta = \frac{x}{\sqrt{x^2 + y^2}} \qquad \sin \theta = \frac{y}{\sqrt{x^2 + y^2}}$$

EXAMPLE 1. A point has the polar coordinates $(6, 27°)$. Find the rectangular equivalents.

Solution. Using the formulas,
$$x = r \cos \theta \quad \text{and} \quad y = r \sin \theta$$
we get
$$x = 6 \cos 27° = 6(0.891) = 5.346 \qquad y = 6 \sin 27° = 6(0.454) = 2.724$$
Then the point has the rectangular coordinates $(5.346, 2.724)$.

EXAMPLE 2. Find the rectangular coordinates of the point $(-8, 150°)$.

Solution.
$$x = (-8)(\cos 150°) = (-8)\left(-\frac{\sqrt{3}}{2}\right) = 4\sqrt{3}$$
$$y = (-8)(\sin 150°) = (-8)\left(\frac{1}{2}\right) = -4$$

EXAMPLE 3. A point has the rectangular coordinates $(-6, 3)$. Find its polar coordinates.

Solution. We use the formulas,
$$r = \sqrt{x^2 + y^2} \quad \text{and} \quad \theta = \arctan \frac{y}{x}$$
Then
$$r = \sqrt{(-6)^2 + (3)^2} = \sqrt{45} = 3\sqrt{5}$$
$$\theta = \arctan\left(-\frac{3}{6}\right) = \arctan(-0.5)$$

To find the correct value of the angle θ, we note there are two positive angles less than 360° whose tangent is -0.5. They are 153.4° and 333.4°. However, since the rectangular coordinates indicate the second quadrant, we take the angle 153.4° for θ. Then the polar coordinates are $(3\sqrt{5}, 153.4°)$.

40.3 Relation between the Rectangular and the Polar Systems

EXAMPLE 4. A point has the rectangular coordinates $(-5,-13)$. Find the equivalent polar coordinates.

Solution.

$$r = \sqrt{(-5)^2 + (-13)^2} = \sqrt{25 + 169} = \sqrt{194}$$

$$\theta = \arctan \frac{-13}{-5} = \arctan (+2.6)$$

There are two positive angles less than 360° whose tangent is 2.6. They are approximately 69° and 249°. To choose the correct angle, we note that the rectangular coordinates indicate the third quadrant. Therefore, we take the angle 249°. Then the polar coordinates are $(\sqrt{194}, 249°)$. The same point could also be named by other pairs of polar coordinates, such as $(-\sqrt{194}, 69°)$; $(\sqrt{194}, -111°)$; and $(-\sqrt{194}, -291°)$.

If an equation is expressed in one form of coordinates, it can be changed to the other form by means of the relations we have mentioned.

EXAMPLE 5. Change the equation of the circle, $x^2 + y^2 = 16$, to polar form

Solution. Here we recognize that

$$x^2 + y^2 = r^2$$

Substituting in the equation, we get

$$r^2 = 16 \quad \text{or} \quad r = 4$$

The equation $r = 4$ is the polar equation of the circle with its center at the pole and radius equal to 4.

EXAMPLE 6. Change the equation $r = 6 \sin \theta$ to rectangular form, then describe the figure and sketch the graph.

Solution. Starting with the equation

$$r = 6 \sin \theta$$

we multiply both sides by r and get

$$r^2 = 6r \sin \theta$$

Now we substitute the rectangular equivalents,

$$r^2 = x^2 + y^2 \quad \text{and} \quad r \sin \theta = y$$

We get

$$x^2 + y^2 = 6y \quad \text{or} \quad x^2 + y^2 - 6y = 0$$

This is now the equation of a circle having a radius of 3 units and with the center of the circle on the y-axis at the point (0,3); see Figure 40.5.

EXAMPLE 7. Change the following equation to rectangular coordinates:

$$r = \frac{2}{1 + \cos \theta}$$

Solution. There are several ways this might be done. We might replace $\cos \theta$ immediately with its equivalent $x/\sqrt{x^2 + y^2}$. However, probably the following is the simpler method. We try to get the expression $r \cos \theta$ so that we

Fig. 40.5

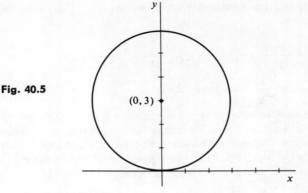

might replace it with x. Let us first multiply both sides of the equation by $(1 + \cos \theta)$, and get

$$r(1 + \cos \theta) = 2 \quad \text{or} \quad r + r \cos \theta = 2$$

Now we isolate the term $r \cos \theta$;

$$r \cos \theta = 2 - r$$

Now we substitute rectangular equivalents, and get

$$x = 2 - \sqrt{x^2 + y^2}$$

Transposing, $\sqrt{x^2 + y^2} = 2 - x$

Squaring both sides,

$$x^2 + y^2 = 4 - 4x + x^2 \quad \text{or} \quad y^2 = 4 - 4x$$

The figure is a parabola with vertex at $(1,0)$ and focus at $(0,0)$.

EXAMPLE 8. Change to rectangular form and identify:

$$r = \frac{10}{2 + 3 \sin \theta}$$

Solution. The preferable way is probably to multiply both sides of the equation by $(2 + 3 \sin \theta)$; then

$$r(2 + 3 \sin \theta) = 10$$

Removing parentheses, $2r + 3r \sin \theta = 10$

Transposing, $2r = 10 - 3r \sin \theta$

Substituting, $2\sqrt{x^2 + y^2} = 10 - 3y$

Squaring both sides, $4(x^2 + y^2) = 100 - 60y + 9y^2$

or $4x^2 + 4y^2 - 9y^2 + 60y = 100$

Combining, $4x^2 - 5y^2 + 60y = 100$

or $4x^2 - 5(y^2 - 12y) = 100$

Completing the square,

$$4x^2 - 5(y^2 - 12y + 36) = 100 - 180$$

40.3 Relation between the Rectangular and the Polar Systems

Note that in completing the square in y, we have actually *subtracted* 180. Writing as squares,

$$4x^2 - 5(y-6)^2 = -80$$

Dividing both sides by -80

$$\frac{x^2}{-20} + \frac{(y-6)^2}{16} = 1$$

or

$$\frac{(y-6)^2}{16} - \frac{x^2}{20} = 1$$

Now we recognize the equation as that of a hyperbola with its center at the point $(0,6)$ and with the transverse axis equal to 8 along the y-axis. The vertices are at the points $(0,2)$ and $(0,10)$, and the foci at $(0,0)$ and $(0,12)$. For other values we have $a = 4$; $b = \sqrt{20}$; $c = 6$; $e = 3/2$. Figure 40.6 shows the graph.

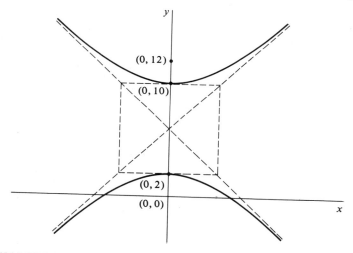

Fig. 40.6

EXAMPLE 9. Transform the following cartesian equation to polar form:

$$4x^2 + 3y^2 - 12y - 36 = 0$$

Solution. First, we recognize the equation as that of a translated ellipse. To change the equation to polar form, we might use any one of several approaches. One method is to try to get the equivalent expression for r^2; that is, $x^2 + y^2$. We can get this if we add and then subtract y^2:

$$4x^2 + 4y^2 - y^2 - 12y - 36 = 0$$

Transposing,

$$4x^2 + 4y^2 = y^2 + 12y + 36$$

or

$$4(x^2 + y^2) = y^2 + 12y + 36$$

Since the right side is a perfect square, we take the square root of both sides and get

$$2\sqrt{x^2 + y^2} = y + 6$$

Substituting polar equivalents, $2r = r \sin \theta + 6$

Transposing again, $2r - r \sin \theta = 6$

Solving for r,

$$r = \frac{6}{2 - \sin \theta}$$

EXERCISE 40.1

Change the following equations to rectangular form and identify the figure if possible.

1. $r = 4 \cos \theta$
2. $r = 5 \sin \theta$
3. $r = -6 \sin \theta$
4. $r^2 = 9 \cos^2 \theta$
5. $r = \dfrac{4}{1 + \cos \theta}$
6. $r = \dfrac{6}{1 - \sin \theta}$
7. $r = \dfrac{8}{2 + \sin \theta}$
8. $r = \dfrac{16}{5 + 3 \cos \theta}$
9. $r = \dfrac{16}{3 + 5 \sin \theta}$
10. $r = \dfrac{16}{1 - 4 \cos \theta}$
11. $r = \dfrac{12}{3 - \sin \theta}$
12. $r = \dfrac{12}{1 + 3 \cos \theta}$
13. $r = \dfrac{12}{1 + \cos \theta}$
14. $r = \dfrac{12}{3 + \cos \theta}$
15. $r = 4 - 4 \sin \theta$
16. $r = 4 + 4 \cos \theta$
17. $r = 2 - 4 \cos \theta$
18. $r = 4 + 2 \sin \theta$

Change the following equations to polar form:

19. $x^2 + y^2 = 9$
20. $x^2 - y^2 = 4$
21. $y^2 - x^2 = 9$
22. $y^2 = 6x + 9$
23. $xy = 6$
24. $x^2 + 2y^2 = 4$
25. $x^2 = 5y^2$
26. $x = 2y$
27. $x^2 + 8y = 0$
28. $x^2 = 6y + 9$
29. $5y^2 + 8x = 0$
30. $x^2 + y^2 - 6y = 0$
31. $y^2 = 8x + 16$
32. $x^2 + y^2 - 6y = 0$
33. $x^2 + 4y - 4 = 0$
34. $x^2 + y^2 + 4x - 2y = 0$
35. $9x^2 + 5y^2 - 100y - 625 = 0$.

40.4 GRAPHING EQUATIONS IN POLAR COORDINATES

In graphing an equation in polar coordinates we proceed in much the same way as with rectangular coordinates. Recall that in graphing an equation in rectangular coordinates, x and y, we find pairs of values that satisfy the equation. For example, to graph the equation, $y = x^2 - 3x - 4$, we find pairs of numbers that make the equation true. One such pair is $(2, -6)$. Then such a pair of numbers is considered as representing a point in the xy coordinate plane. Then we have the definition: *the graph of an equation is the set of all points and only those points whose coordinates satisfy the given equation.*

In polar coordinates the two related variables are the distance r and the angle θ. An equation in polar coordinates is then a statement expressing a relation between r and θ. In most instances we take θ as the independent variable. Then r is expressed as a function of θ; that is, $r = f(\theta)$. An example is the equation, $r = 8 \cos \theta$. The graph of the equation is then the set of all points (in polar coordinates) and only those points whose coordinates r and θ satisfy the equation.

To find pairs of corresponding values, we first take any value of θ and then compute the corresponding value of r. For the angle θ, we often take values from zero to 2π (that is, 0° to 360°). The angle may be stated in radians or degrees. In most cases we take angles whose trigonometric values can be computed without a table, such as 0 (0°), $\pi/6$ (30°), $\pi/4$ (45°), and so on. Then we compute the corresponding value of r as indicated by

40.4 Graphing Equations in Polar Coordinates

the equation. The pairs of values are then plotted as points on polar coordinate paper. The method is shown by examples. Note the symmetry of each figure with respect to a line or to a point.

EXAMPLE 10. Graph the equation $r = 8 \cos \theta$.

Solution. For this example we shall state the value of θ in radians. We first take values of θ, starting with zero and going to 2π. For example, if $\theta = 0$, then $r = 8 \cos 0 = 8(1) = 8$. We get the following pairs of values. Note that when θ has its terminal side in the second and third quadrants, then the cosine of θ is negative and therefore r is *negative*. The r values are approximate.

$\theta =$	0	$\frac{\pi}{6}$	$\frac{\pi}{4}$	$\frac{\pi}{3}$	$\frac{\pi}{2}$	$\frac{2\pi}{3}$	$\frac{3\pi}{4}$	$\frac{5\pi}{6}$	π	$\frac{5\pi}{4}$	$\frac{3\pi}{2}$	$\frac{7\pi}{4}$
$r =$	8	6.9	5.7	4	0	-4	-5.7	-6.9	-8	-5.7	0	5.7

The pairs of values are now plotted as points on the graph. The graph is a circle (Fig. 40.7), with its center at the point (4, 0°) and with a radius equal to 4 units. Note that the complete circle is formed by taking values of θ from zero to π. The values of θ from 180° to 360° only repeat the same circle. The figure is symmetrical with respect to the polar axis. This will always be true if the equation is unchanged when θ is replaced with $(-\theta)$.

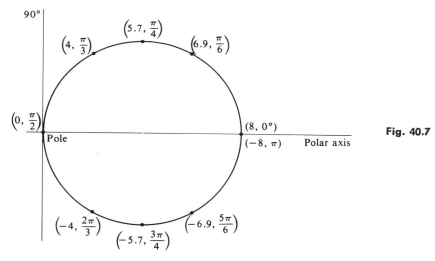

Fig. 40.7

That the equation is a circle may easily be verified if the equation is changed to rectangular form. Multiplying both sides of the original equation by r, we get

$$r^2 = 8r \cos \theta$$

Substituting rectangular equivalents, we get

$$x^2 + y^2 = 8x$$

EXAMPLE 11. Graph the equation

$$r = 4 + 4 \sin \theta$$

554   Polar Coordinates

Solution. Showing θ in degrees, we take θ from 0° to 360° and get the following pairs of values: (Note that no values of r are negative.)

$\theta =$	0	30°	45°	60°	90°	120°	135°	150°	180°
$r =$	4	6	6.83	7.46	8	7.46	6.83	6	4

$\theta =$	210°	225°	240°	270°	300°	315°	330°	360°
$r =$	2	1.17	0.54	0	0.54	1.17	2	4

The graph (Fig. 40.8) is called a *cardioid*, meaning "heart-shaped." Note that the 90°-axis is an axis of symmetry. If any equation is unchanged when θ is replaced with $(180° - \theta)$, then the figure is symmetrical with respect to the 90°-axis.

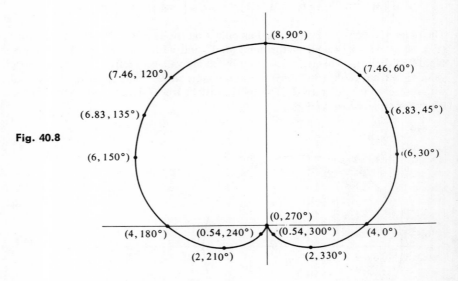

Fig. 40.8

EXAMPLE 12. Graph the equation
$$r = 3 - 6\cos\theta$$

Solution. We have the following corresponding values for θ and r:

$\theta =$	0	30°	45°	60°	90°	120°	135°	150°	180°
$r =$	−3	−2.20	−1.24	0	3	6	7.24	8.20	9

$\theta =$	210°	225°	240°	270°	300°	315°	330°	360°
$r =$	8.20	7.24	6	3	0	−1.24	−2.2	−3

The graph (Fig 40.9) is a figure called a *limaçon*. To get a more accurate graph in the neighborhood of 90° and 270°, it might be well to consider also 70°, 80°, 280°, and 290° and consult tables of values of $\cos\theta$.

40.4 Graphing Equations in Polar Coordinates

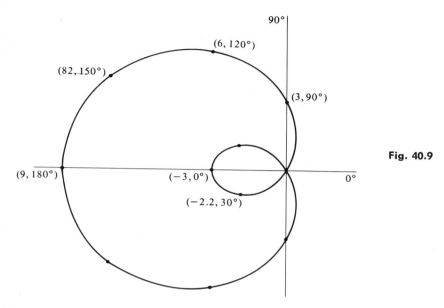

Fig. 40.9

EXAMPLE 13. Graph the equation

$$r = \frac{16}{5 - 3 \cos \theta}$$

Solution. Since $\cos(-\theta) = \cos(+\theta)$, the equation will not be altered if we replace θ with $(-\theta)$. Therefore, the polar axis is an axis of symmetry. Then we can graph the curve for values of θ from $0°$ to $180°$, and sketch the other half of the curve so that the whole is symmetrical with respect to the polar axis. However, let us list all values from $0°$ to $360°$ and compute the values of r. In an equation of this kind, one must be especially careful to compute the values of r by noting every step in the process. For example, if $\theta = 120°$, we have $\cos 120° = -0.5$. Then, $5 - 3 \cos 120° = 5 - 3(-0.5) = 5 + 1.5$, and $r = 16/6.5 = 2.46$ (approx.). For a table of values we have

$\theta =$	0	30°	45°	60°	90°	120°	135°	150°	180°
$r =$	8	6.67	5.56	4.57	3.2	2.46	2.25	2.11	2
$\theta =$		210°	225°	240°	270°	300°	315°	330°	360°
$r =$		2.11	2.25	2.46	3.2	4.57	5.56	6.67	8

If the values are carefully plotted, the graph (Fig. 40.10) appears to be an ellipse, with a major axis equal to 10, with the center at $(3, 0°)$, focus at the pole, and with one vertex at $(2, 180°)$. Let us convert the equation to rectangular coordinates. Clearing the original equation of fractions, we get

$$r(5 - 3 \cos \theta) = 16 \quad \text{or} \quad 5r - 3r \cos \theta = 16$$

Substituting rectangular equivalents and transposing, we get

$$5\sqrt{x^2 + y^2} = 3x + 16$$

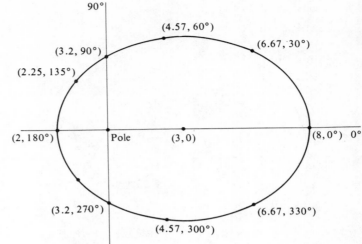

Fig. 40.10

Squaring both sides of the equation and reducing to the form of an ellipse, we get the equation

$$\frac{(x-3)^2}{25} + \frac{y^2}{16} = 1 \text{ (an ellipse)}$$

EXAMPLE 14. Graph the equation

$$r = \frac{4}{1 + \sin \theta}$$

Solution. In this example, as in Example 13, note that the denominator cannot be negative, and therefore, r cannot be negative. However, unlike Example 3, in this example, the denominator may become zero. Then r becomes infinite. The following table shows corresponding values (approx.):

$\theta =$	0	30°	45°	60°	90°	120°	135°	150°	180°
$r =$	4	2.7	2.4	2.1	2	2.1	2.4	2.7	4

$\theta =$	210°	225°	240°	270°	300°	315°	330°	360°
$r =$	8	14	30	—	30	14	8	4

The graph is a parabola (Fig. 40.11) with vertex at (2, 90°), focus at (0,0°), and symmetrical with respect to the 90°-axis.

To convert the equation to rectangular form, we first clear of fractions and get

$$r + r \sin \theta = 4 \quad \text{or} \quad r = 4 - r \sin \theta$$

Substituting rectangular equivalents,

$$\sqrt{x^2 + y^2} = 4 - y$$

Squaring both sides and simplifying,

$$x^2 = -8(y - 2)$$

40.4 Graphing Equations in Polar Coordinates

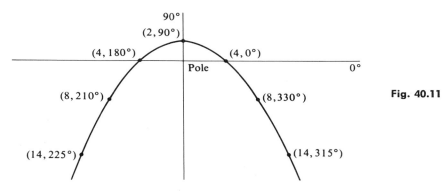

Fig. 40.11

Now we recognize the equation of a translated parabola with vertex at the point (0,2) and focus at the point (0,0), the parabola opening downward.

EXAMPLE 15. Graph the equation

$$r = 6 \sin 2\theta$$

Solution. Note that we must double θ before computing the value of r. For example, if $\theta = 15°$, then

$$r = 6 \sin 2(15°) = 6 \sin 30° = 3$$

For this reason we take any values of θ such that when they are doubled, we still get the usual special angles. The table shows some of the first values.

$\theta =$	0	15°	22.5°	30°	45°	60°	67.5°	75°	90°	105°	112.5°
$r =$	0	3	4.2	5.2	6	5.2	4.2	3	0	−3	−4.2

The graph (Fig. 40.12) is a four-leafed rose. As θ varies from 90° to 180°, $2(\theta)$ varies from 180° to 360°, and then r is negative for values of θ in the second quad-

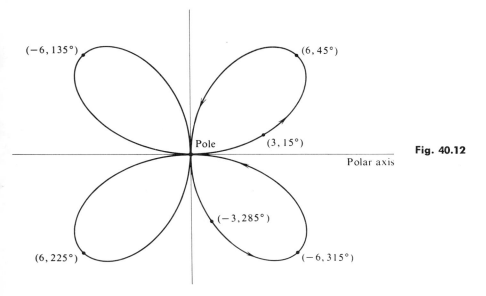

Fig. 40.12

rant. For these values we get the fourth leaf of the rose going counterclockwise. In fact, the entire figure can be drawn with one smooth continuous stroke of the pencil always going counterclockwise.

EXERCISE 40.2

Sketch the graph in polar coordinates of each of the following equations:

1. $r = 4 \cos \theta$
2. $r = 6 \sin \theta$
3. $r = -6 \cos \theta$
4. $r = -2 \sin \theta$
5. $r = 3 + 3 \sin \theta$
6. $r = 4 - 4 \cos \theta$
7. $r = 5(1 - \sin \theta)$
8. $r = 2(1 + \cos \theta)$
9. $r = 2 + 4 \cos \theta$
10. $r = 2(1 - 2 \sin \theta)$
11. $r = \sin^2 \theta$
12. $r = \cos^3 \theta$
13. $r = 4 \tan \theta$
14. $r = 4 \cos 2\theta$
15. $r = 8 \sin 2\theta$
16. $r = 6 \sin 3\theta$
17. $r = 6 \cos 4\theta$
18. $r^2 = 9 \sin 2\theta$
19. $r^2 = 16 \cos 2\theta$
20. $r = 4$
21. $\theta = 30°$
22. $r = \dfrac{18}{4 - 5 \sin \theta}$
23. $r = \dfrac{18}{5 - 4 \cos \theta}$
24. $r = \dfrac{10}{3 - 2 \sin \theta}$
25. $r = \dfrac{4}{1 - \cos \theta}$
26. $r = \dfrac{4}{\cos \theta}$
27. $r = \dfrac{2}{\cos \theta}$
28. $r = 4 \sin \theta + 4 \cos \theta$
29. $r = \theta$
30. $r = 4/\theta$
31. $r = 2^\theta$
32. $r = \sqrt{|\tan \theta|}$ $\quad (0 \leq \theta \leq \pi)$
33. $r = -\sqrt[3]{|\tan \theta|}$ $\quad (\pi \leq \theta \leq 2\pi)$

40.5 THE CONICS IN POLAR COORDINATES

When a cone is cut by a plane, the intersection is called a *conic section*. The conic sections include the circle, the ellipse, the parabola, and the hyperbola. (The Cartesian equations of these curves have been discussed in Chapters 4–7.)

The conics have special types of polar equations. We have already graphed the *circle*, $r = 8 \cos \theta$, in Example 10. For any equation of the following forms, the graph is a circle passing through the pole and with a radius equal to $a/2$, where a is a constant:

$$r = a \cos \theta \quad \text{and} \quad r = a \sin \theta$$

These are the standard forms of the polar equations of circles.

For the equation, $r = a \cos \theta$, the center of the circle is on the polar axis at the point $(a/2, 0)$. For the equation, $r = a \sin \theta$, the center is on the 90°-axis at the point $(a/2, 90°)$.

EXAMPLE 16. Graph the equation

$$r = 6 \sin \theta$$

Solution. We could graph this equation by finding several pairs of coordinates. However, the graph is easily constructed if we recognize the equation as that of a circle with its center at $(3, 90°)$ and radius of 3.

In Example 13 we sketched the graph of the equation

$$r = \dfrac{16}{5 - 3 \cos \theta}$$

40.5 The Conics in Polar Coordinates

We found that the graph is an *ellipse*. If the equation contained sin θ in place of cos θ, the graph would still be an ellipse. Note especially that the constant term, 5, in the denominator does *not* have the same absolute value as the coefficient of cos θ. An equation of this form represents either an *ellipse* or a *hyperbola*.

In general, if K, a, and b, are constants, not zero, then the following forms represent a parabola, an ellipse, or a hyperbola having one focus at the pole:

$$\text{(A)} \quad r = \frac{K}{a \pm b \sin \theta} \qquad \text{(B)} \quad r = \frac{K}{a \pm b \cos \theta}$$

The eccentricity of the curve is equal to the absolute value of b/a. In the example

$$r = \frac{16}{5 - 3 \cos \theta}$$

the eccentricity of the curve is $\tfrac{3}{5}$, which represents an ellipse.

If we change the constant term, 5, in the denominator to 1 by dividing numerator and denominator by 5, we get

$$r = \frac{16/5}{1 - \left(\frac{3}{5}\right) \cos \theta}$$

When the constant term in the denominator has been reduced to 1, then the absolute value of the coefficient of cos θ represents the eccentricity of the curve. In this example we see at once that $e = \tfrac{3}{5}$.

In the equation forms (A) and (B), if $a = b$, then $e = 1$, and the equation represents a parabola with the focus at the pole. (See Example 14.)

In the forms of the polar equations of the parabola, the ellipse, and the hyperbola, if the term in the denominator is cos θ, then the curve is symmetrical about the polar axis. The polar axis then coincides with the major axis of the ellipse, the transverse axis of the hyperbola, or the principal axis of the parabola. If the term in the denominator is sin θ, then the 90°-axis coincides with the axes mentioned. The curve is then symmetrical about the 90°-axis.

In sketching the graphs of the conics, we might first take the values 0°, 90°, 180°, and 270° for θ. From these values we get the points of intersection of the curve with the corresponding axes. In the case of the parabola, one value will lead to a zero denominator, which makes the curve infinite. For a fairly accurate shape of the curve we might take a few more values of θ. Then, taking account of the symmetry of the curve, we can get a good idea of its shape. Of course, for the hyperbola also, some values will lead to a zero denominator.

EXAMPLE 17. Describe and sketch the curve

$$r = \frac{12}{2 - 4 \sin \theta}$$

560   Polar Coordinates

Solution. Dividing the numerator and denominator by 2,

$$r = \frac{6}{1 - 2 \sin \theta}$$

The coefficient of $\sin \theta$ has an absolute value of 2. Since this represents the eccentricity, the curve is a hyperbola with one focus at the pole. The presence of $\sin \theta$ indicates that the transverse axis coincides with the 90°-axis. This means the hyperbola opens upward and downward. For two values we have the following: when $\theta = 90°$, $r = -6$; when $\theta = 270°$, $r = 2$. These two values represent the vertices. For two other points, we find that when $\theta = 0$ and $180°$, $r = 6$. From these values we can get a good idea of the shape of the curve (Fig. 40.13).

Note that the denominator becomes zero when $\sin \theta = \frac{1}{2}$; that is, when $\theta = 30°$ and $150°$. No point on the curve will be found for these values. In fact, the slope of r is then the same as the slope of the asymptotes, which is $\pm 1/\sqrt{3}$. Note that the center of the hyperbola is at (4, 270°).

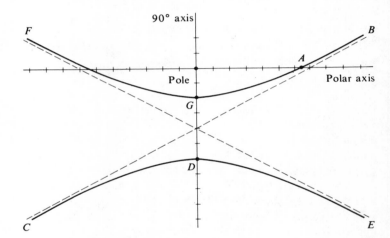

Fig. 40.13

EXAMPLE 18. Describe and sketch the graph of the equation

$$r = \frac{3}{1 + \cos \theta}$$

Solution. Since the eccentricity is 1, the curve is a parabola with the focus at the pole. The presence of $\cos \theta$ indicates that the principal axis coincides with the polar axis. The smallest absolute value that r can have occurs when $\cos \theta = 1$; that is, when $\theta = 0$. For this value, $r = \frac{3}{2}$. This point represents the vertex of the parabola. Note that when $\theta = 180°$, then r becomes infinite. The parabola therefore opens toward the left (Fig. 40.14). For other values, we have the following:

when $\theta = 90°$ and 270°, $r = 3$

These few points will show the approximate shape of the curve.

From the graph it can be seen that the Cartesian form of the equation is

$$y^2 = -6\left(x - \frac{3}{2}\right)$$

40.6 Intersection of Polar Curves

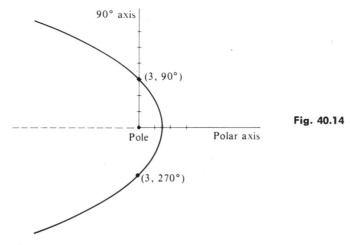

Fig. 40.14

which reduces to $y^2 + 6x - 9 = 0$. One form of the equation can be transformed into the other by use of the relations between coordinates.

EXERCISE 40.3

For each of the following equations, (a) identify the curve, (b) graph the polar equation, and (c) transform the equation to Cartesian form:

1. $r = 6 \sin \theta$
2. $r = -8 \cos \theta$
3. $2r = -5 \sin \theta$
4. $r = \dfrac{4}{1 - \sin \theta}$
5. $r = \dfrac{6}{1 - \cos \theta}$
6. $r = \dfrac{5}{2 + 2 \cos \theta}$
7. $r = \dfrac{10}{3 + 2 \cos \theta}$
8. $r = \dfrac{6}{2 - \sin \theta}$
9. $r = \dfrac{20}{5 - 3 \cos \theta}$
10. $r = \dfrac{6}{1 - 2 \cos \theta}$
11. $r = \dfrac{10}{2 - 3 \sin \theta}$
12. $r = \dfrac{14}{4 + 3 \sin \theta}$

Transform each of the following equations to polar form, and then describe and sketch the curve from the polar equation:

13. $y^2 = 8x + 16$
14. $x^2 + 4y + 4 = 0$
15. $5x^2 + 9y^2 - 20x = 25$
16. $3x^2 - y^2 + 12x + 9 = 0$
17. $3x^2 + 4y^2 - 12x - 36 = 0$

40.6 INTERSECTION OF POLAR CURVES

Points of intersection of polar curves are found by solving two equations as a system just as with rectangular coordinates.

EXAMPLE 19. Find the points of intersection of the two circles:

$$r = 6 \cos \theta \quad \text{and} \quad r = 8 \sin \theta$$

Solution. Since the two equations are both solved for r, we set

$$8 \sin \theta = 6 \cos \theta$$

Dividing both sides by (8 cos θ), we get

$$\frac{\sin \theta}{\cos \theta} = \frac{6}{8}$$

The equation reduces to $\tan \theta = \frac{3}{4}$

Solving for θ, $\theta = \arctan \frac{3}{4} = 36.9°$ (approx.); then $r = \frac{24}{5}$

Of course another possible value of θ is 218.9°, but the result is the same point since r is negative for this value of θ. The point of intersection is (4.8, 36.9°).

In solving a system of polar equations, points of intersection are not always what they seem. In the foregoing example, note that both curves pass through the pole. This surely seems to be, and, in fact, is a common point. Yet this point does not appear in the solution. Our question now might well be: why does the point at the pole not appear in the solution of the system?

The reason is that in polar coordinates, a particular point does not have a unique pair of coordinates. In graphing the circle $r = 8 \sin \theta$, we are already at the pole when $\theta = 0$. On the other hand, the circle $r = 6 \cos \theta$ does not reach the pole until $\theta = 90°$. Both curves pass through the pole *but not at the same time*. At one time the pole is represented as the point (0, 0°), and at another time as (0, 90°). These two pairs of values for the same point do not satisfy both equations.

In many instances the points of intersection of two polar equations must be determined from the graphs. As another example, consider the two equations

$$r = 4 \cos \theta \quad \text{and} \quad r = 2 + 2 \cos \theta$$

Since both equations are solved for r, we set one value equal to the other:

$$4 \cos \theta = 2 + 2 \cos \theta$$

Solving, $\quad 2 \cos \theta = 2 \quad$ or $\quad \cos \theta = 1$

The only solution is the value

$$\theta = 0°$$

If we graph the two equations, we shall find that the two curves do intersect at the point (4, 0°). However, both curves also pass through the pole. Yet the pole does not appear in the solution of the system. The reason is that in graphing the circle, $r = 4 \cos \theta$, we reach the pole when $\theta = 90°$. The cardioid, $r = 2 + 2 \cos \theta$, does not reach the pole until $\theta = 180°$. The two curves do pass through the pole, but not at the same time. In one case the pole is represented by the coordinates (0, 90°); in the other case by the pair (0, 180°).

Or take another example:

$$r = 4 \cos \theta \quad \text{and} \quad r = 4 + 4 \cos \theta$$

40.6 Intersection of Polar Curves

If we attempt to solve the system, we set

$$4 \cos \theta = 4 + 4 \cos \theta$$

Solving, we get $0 = 4$, an impossible situation. Yet both curves pass through the pole. The pair of coordinates $(0, 90°)$ satisfies the circle but not the cardioid; whereas the pair $(0, 180°)$ satisfies the cardioid but not the circle. Yet both pairs represent the pole.

EXAMPLE 20. Find the points of intersection of the curves

$$r = 9 \sin \theta \quad \text{and} \quad r = \frac{10}{3 - 2 \sin \theta}$$

Solution. We set the two values of r equal to each other and get

$$9 \sin \theta = \frac{10}{3 - 2 \sin \theta}$$

Clearing of fractions, $\quad 27 \sin \theta - 18 \sin^2 \theta = 10$

This can be written, $\quad 18 \sin^2 \theta - 27 \sin \theta + 10 = 0$

Factoring, $\quad (6 \sin \theta - 5)(3 \sin \theta - 2) = 0$

Solving, $\quad \sin \theta = \dfrac{5}{6} \quad \text{and} \quad \sin \theta = \dfrac{2}{3}$

For the value, $\sin \theta = \frac{5}{6}$, we find that $r = \frac{15}{2}$. When $\sin \theta = \frac{2}{3}$, $r = 6$. Note that there are two values of θ for each value of $\sin \theta$. For the value $\sin \theta = \frac{5}{6}$, $\theta = 56.5°$ and $123.5°$ (approx.). For the value $\sin \theta = \frac{2}{3}$, $\theta = 41.8°$ and $138.2°$ (approx.). If the two curves are graphed, the four points of intersection can be identified.

EXERCISE 40.4

Find the points of intersection of each pair of curves. Sketch the curves.

1. $r = 5 \sin \theta$
   $r = -5 \cos \theta$
2. $r = 3 \cos \theta$
   $r = 4 \sin \theta$
3. $r = 6 \cos \theta$
   $r = 2 + 2 \cos \theta$
4. $r = 2 + 4 \cos \theta$
   $r = 8 \cos \theta$
5. $r = 1 + \cos \theta$
   $r = 1 - \sin \theta$
6. $r = 1 + \cos \theta$
   $r = \sin 2\theta$
7. $r = 4 \cos 2\theta$
   $r = 4 \sin \theta$
8. $r = \dfrac{4}{\cos \theta}$
   $r = 4 \cos \theta$
9. $r = 3 \sin \theta$
   $r = 1 + \sin \theta$
10. $r = 4 - 2 \sin \theta$
    $r = 10 \cos \theta$
11. $r = 4 \sin 3\theta$
    $r = 2$
12. $r = 4 \cos \theta$
    $r = \dfrac{6}{1 - \cos \theta}$
13. $r = \dfrac{6}{1 - \sin \theta}$
    $r = \dfrac{10}{2 + \sin \theta}$
14. $r = \dfrac{12}{1 + \cos \theta}$
    $r = \dfrac{12}{3 + 2 \cos \theta}$
15. $r = \dfrac{6}{3 - 2 \sin \theta}$
    $r = \dfrac{6}{3 + 2 \cos \theta}$
16. $r = \dfrac{10}{2 - 3 \cos \theta}$
    $r = \dfrac{10}{1 - 4 \cos \theta}$
17. $r = 6 \sin \theta$
    $r = 3 + 3 \sin \theta$
18. $r = 2 \cos \theta$
    $r = 2 + 2 \cos \theta$

40.7 DIFFERENTIATION IN POLAR COORDINATES

We have already encountered the differentiation of trigonometric functions in connection with rectangular coordinates. For example, if

$$y = 4 \sin x$$

then

$$\frac{dy}{dx} = 4 \cos x$$

However, in this example, the variables x and y are understood to be *rectangular* and *not* polar coordinates.

Now suppose we have the following equation in polar coordinates:

$$r = 4 \sin \theta$$

In this example we can differentiate the function r with respect to θ and get

$$\frac{dr}{d\theta} = 4 \cos \theta \qquad \text{(often written } r' = 4 \cos \theta\text{)}$$

The derivative, $dr/d\theta$, can be evaluated for any particular value of θ. For example, when $\theta = \pi/3$, then

$$\frac{dr}{d\theta} = 4 \cos \frac{\pi}{3} = (4)\left(\frac{1}{2}\right) = 2$$

The reference in this example is to polar coordinates, the radius vector r and the angle θ. But what does this mean when we say $dr/d\theta = 2$? It does *not* mean the slope of the curve. The derivative in polar coordinates is not equivalent to the slope of the curve. The expression $dr/d\theta = 2$ simply means the rate of change of the radius vector r with respect to the angle θ. At the particular instant when $\theta = 60°$, the change in r is twice the change in the angle if the two variables are measured in the proper units. The radius vector r changes by linear units, such as inches or feet. For the correct meaning, we must measure the angle θ in radians. Then the derivative, $dr/d\theta = 2$, means that r changes instantaneously by 2 linear units for each change of one radian in the angle θ. This is the rate of change at that particular instant.

40.8 THE SLOPE OF A CURVE IN POLAR COORDINATES

The slope of a curve at a point on the curve is, of course, defined as the slope of the tangent line at that point. In rectangular coordinates, the derivative, dy/dx, is equivalent to the slope of the curve. But this just happens to be true in rectangular coordinates. In polar coordinates the derivative, $dr/d\theta$, is not equivalent to the slope of the curve. The expression for the slope is a bit more complicated.

Let us sketch the graph of the equation $r = f(\theta)$, as shown in Figure 40.15. We denote by *psi* (ψ) the angle between the radius vector r and the tangent t to the curve at a point. We read the angle ψ as a directed angle of rotation

40.8 The Slope of a Curve in Polar Coordinates

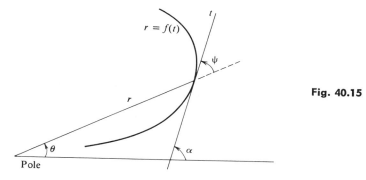

Fig. 40.15

from r to the tangent t, as indicated by the curved arrow. We shall find that this angle is involved in the expression for the slope of the curve.

We let α represent the inclination of the tangent line. From the figure we see the following relation:

$$\alpha = \psi + \theta$$

Since the slope of the curve is equivalent to the tangent of the angle α, then

$$m = \tan \alpha = \tan(\psi + \theta)$$

From a trigonometric identity, we get

$$m = \frac{\tan \psi + \tan \theta}{1 - (\tan \psi)(\tan \theta)}$$

Now we shall derive another expression for the slope and equate the two expressions. We have already used the following conversion formulas:

$$y = r \sin \theta \quad \text{and} \quad x = r \cos \theta$$

Now, if we can formulate from these two equations the correct expression for dy/dx, we shall have another form of the slope. We differentiate each equation with respect to θ. However, we must remember that r and θ are both variables. Then we must use the product rule. We get

$$\frac{dy}{d\theta} = r(\cos \theta) + (\sin \theta)\left(\frac{dr}{d\theta}\right)$$

and

$$\frac{dx}{d\theta} = r(-\sin \theta) + (\cos \theta)\left(\frac{dr}{d\theta}\right)$$

Then

$$m = \frac{dy}{dx} = \frac{dy/d\theta}{dx/d\theta} = \frac{r(\cos \theta) + (\sin \theta)(dr/d\theta)}{-r(\sin \theta) + (\cos \theta)(dr/d\theta)}$$

If we divide numerator and denominator of the fraction by the quantity $(\cos \theta)(dr/d\theta)$ and reverse the terms of the denominator, we get

$$m = \frac{dy}{dx} = \frac{(r/r') + \tan \theta}{1 - (r/r')(\tan \theta)} \quad \left(\text{where } r' = \frac{dr}{d\theta}\right)$$

Polar Coordinates

We have already derived the following formula for the slope:

$$m = \frac{\tan \psi + \tan \theta}{1 - (\tan \psi)(\tan \theta)}$$

Now we have two expressions for the slope of the curve. From the appearance of the two we see they are equivalent provided the following is true:

$$\tan \psi = \frac{r}{r'}$$

To show that this relation is true, we can equate the two expressions and solve the resulting equation for $\tan \psi$; that is, we set

$$\frac{\tan \psi + \tan \theta}{1 - (\tan \psi)(\tan \theta)} = \frac{(r/r') + \tan \theta}{1 - (r/r')(\tan \theta)}$$

If we multiply both sides of this equation by the lowest common denominator and then simplify the resulting expression, we shall find, after much algebraic manipulation, that

$$\tan \psi = \frac{r}{r'}$$

or, expressed in another form,

$$\tan \psi = (r)\left(\frac{d\theta}{dr}\right)$$

EXAMPLE 21. Given the circle, $r = 8 \cos \theta$ (Fig. 40.16), find (a) the angle between the radius vector and the tangent line when $\theta = 30°$; (b) the slope of the curve at the point; and (c) the value of θ for $m = 0$.

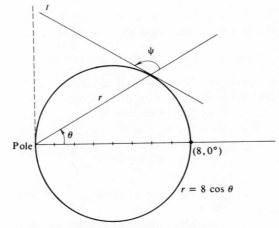

Fig. 40.16

Solution. We have $r = 8 \cos \theta$. Differentiating,

$$\frac{dr}{d\theta} = r' = -8 \sin \theta$$

$$\tan \psi = \frac{r}{r'} = \frac{8 \cos \theta}{-8 \sin \theta} = -\cot \theta$$

At $\theta = 30°$, $\quad \tan \psi = -\cot 30° = -\sqrt{3}$

Then $\quad \psi = \arctan(-\sqrt{3}) = 120°$

40.8 The Slope of a Curve in Polar Coordinates

Note that the angle ψ is read as a directed angle *from* r (the radius vector) *to* t (the tangent line).

To find the slope of the curve at the point where $\theta = 30°$, we take

$$m = \frac{dy}{dx} = \frac{\tan \psi + \tan \theta}{1 - (\tan \psi)(\tan \theta)}$$

For the value $\theta = 30°$, $\tan \psi = -\sqrt{3}$ and $\tan \theta = 1/\sqrt{3}$. Substituting values in the formula, we get

$$m = \frac{dy}{dx} = \frac{-\sqrt{3} + \dfrac{1}{\sqrt{3}}}{1 + (\sqrt{3})\left(\dfrac{1}{\sqrt{3}}\right)} = \frac{-3 + 1}{2\sqrt{3}} = \frac{-\sqrt{3}}{3} = -0.5774$$

To find a value for which $m = 0$, we set $dy/dx = 0$; that is,

$$\frac{\tan \psi + \tan \theta}{1 - (\tan \psi)(\tan \theta)} = 0$$

from which,
$$\tan \psi + \tan \theta = 0$$

Since $\tan \psi = -\cot \theta$, we have
$$-\cot \theta + \tan \theta = 0$$

Solving for θ,
$$\theta = 45°;\ 135°;\ 225°;\ 315°$$

Since the entire curve is formed as θ goes from 0 to 180°, we need consider only the values 45° and 135°.

EXERCISE 40.5

In each of the following equations, find (a) the angle, ψ, between the radius vector and the tangent line; and (b) the slope of the curve at the indicated values.

1. $r = 6 \sin \theta$; at 30°; at 60°
2. $r = -4 \cos \theta$; at 120°; at 135°
3. $r = 2 + 2 \cos \theta$; at 30°; at 60°
4. $r = 2 - 4 \sin \theta$; at 180°
5. $r = 6 - 6 \sin \theta$; at 30°; at 45°
6. $r^2 = 4 \sin 2\theta$; at 30°; at 45°
7. $r = 6 \cos 2\theta$; at 30°; at 90°
8. $r = 4 \sin 2\theta$; at 30°; at 45°
9. $r = \dfrac{2}{1 + \cos \theta}$; at 120°
10. $r = \dfrac{10}{2 + 3 \sin \theta}$; at 30°; at 150°
11. $r = 6 \sin 3\theta$; at $\pi/6$; at $\pi/4$
12. $r = 4 \sin 4\theta$; at $\pi/6$
13. $r = 4 + 4 \sin \theta$; find the values of θ for which $m = 0$.
14. $r = 8 \cos \theta$; find the values of θ for which $m = 0$.

Find the angle between the two curves at their points of intersection for each of the following pairs of equations:

15. $r = \dfrac{4}{1 + \sin \theta}$ and $r = \dfrac{6}{1 - \sin \theta}$
16. $r = 5 \sin \theta$ and $r = -5 \cos \theta$
17. $r = 6 \cos \theta$ and $r = 2 + 2 \cos \theta$
18. $r = \cos 2\theta$ and $r = \sin \theta$
19. $r = \dfrac{6}{3 - 2 \sin \theta}$ and $r = \dfrac{6}{3 + 2 \cos \theta}$
20. $r = 4 \sin 3\theta$ and $r = 2$

40.9 AREAS IN POLAR COORDINATES

Finding the area under a curve or between curves expressed in polar coordinates is somewhat similar but also different from the same type of problem in rectangular coordinates. In rectangular coordinates, we imagine the area to be found is divided into strips, each strip having a uniform width. In polar coordinates the portions of area are sectors.

Suppose we have the graph of an equation in polar coordinates: $r = f(\theta)$, as shown in Figure 40.17. Now we wish to find the area bounded by the curve and the two lines, r_1 and r_2. As the angle θ changes from θ_1 to θ_2, the radius vector r sweeps out the area we wish to find. However, the length of r is also probably changing.

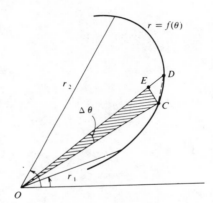

Fig. 40.17

Just as we do in rectangular coordinates, to find an area we take a small portion, a slice or strip, of area and call it an *element* of area. In connection with polar coordinates, we imagine the entire area divided into sectors. We take one sector, COD, as an element, dA. Over this element of area, the angle COD is an increment in θ, which we call $\Delta\theta$. Also, we have the corresponding increment in r, or Δr, represented in the figure by ED.

If we take the straight line segment CD, then in triangle COD we have the area

$$A = \frac{1}{2} r(r + \Delta r) \sin (\Delta\theta)$$

Dividing both sides by $\Delta\theta$, we get the ratio of increments:

$$\frac{\Delta A}{\Delta\theta} = \frac{1}{2} r(r + \Delta r) \frac{\sin (\Delta\theta)}{\Delta\theta}$$

Then, as $\Delta\theta$ approaches zero, Δr also approaches zero as a limit, and we have the derivative,

$$\frac{dA}{d\theta} = \frac{1}{2} r^2 (1) \quad \text{or} \quad dA = \frac{1}{2} r^2 \, d\theta$$

Then

$$A = \int \frac{1}{2} r^2 \, d\theta$$

40.9 Areas in Polar Coordinates

evaluated between limits of θ. Since the integrand contains the differential $d\theta$, the rest of the integrand must be expressed in terms of θ.

EXAMPLE 22. Find the area of the circle, $r = 6 \cos \theta$.

Solution. We recall that the entire circle is formed as θ goes from zero to 180° (Fig. 40.18). For the element of area we have

$$dA = \frac{1}{2} r^2 \, d\theta$$

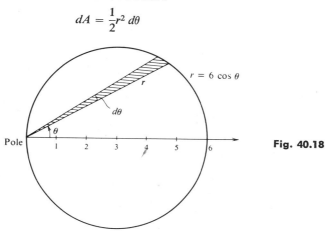

Fig. 40.18

Substituting for r its equivalent, $6 \cos \theta$, we get

$$dA = \frac{1}{2}(36 \cos^2 \theta) \, d\theta = 18 \cos^2 \theta \, d\theta$$

Then
$$A = 18 \int_0^\pi \cos^2 \theta \, d\theta$$

For $\cos^2 \theta$, we make use of the trigonometric identity,

$$\cos^2 \theta = \frac{1}{2} + \frac{1}{2} \cos 2\theta$$

Substituting,

$$A = 18 \int_0^\pi \left(\frac{1}{2} + \frac{1}{2} \cos 2\theta\right) d\theta = 9 \int_0^\pi (1 + \cos 2\theta) \, d\theta$$
$$= 9 \left[\theta + \frac{1}{2} \sin 2\theta \right]_0^\pi = 9\pi$$

The result agrees with the formula for the area of a circle, $A = \pi r^2$. In this circle the radius is 3 in.

In this example, if we mistakenly take the limits of θ as zero and 360°, we shall find that we get an area of $2\pi r^2$, or twice the correct answer. This would happen because the radius vector r sweeps over the entire area of the circle as θ goes from zero to 180°. Then as θ goes from 180° to 360°, the radius vector r sweeps over the same area again.

EXAMPLE 23. Find the total area enclosed by the three-leaf rose

$$r = 4 \sin 3\theta$$

570 Polar Coordinates

Solution. One leaf of the rose is formed as θ goes from 0 to 60°, or $\pi/3$. (Fig. 40.19.) We therefore compute the area of one leaf and then multiply the result by 3. For the element of area we have

$$dA = \frac{1}{2} r^2 \, d\theta = \frac{1}{2} (16 \sin^2 3\theta) \, d\theta$$
$$= 8 \sin^2 3\theta \, d\theta$$

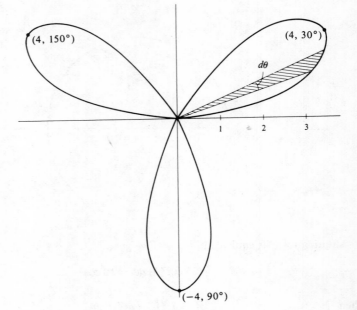

Fig. 40.19

For the area of one leaf, we have

$$A = 8 \int_0^{\pi/3} \sin^2 3\theta \, d\theta = \frac{8}{2} \int_0^{\pi/3} (1 - \cos 6\theta) \, d\theta = 4 \left[\theta - \frac{1}{6} \sin 6\theta \right]_0^{\pi/3}$$
$$= 4 \left[\frac{\pi}{3} - \frac{1}{6} \sin 2\pi \right] = \frac{4}{3} \pi$$

For the three leaves, we have the total area of 4π.

EXAMPLE 24. Find the area that is inside the circle $r = 6 \cos \theta$ but outside the cardioid, $r = 2 + 2 \cos \theta$ (Fig. 40.20).

Solution. To find the limits of integration, we solve the two equations as a system for the points of intersection. Setting the two r values equal to each other, we get

$$6 \cos \theta = 2 + 2 \cos \theta$$

Then $\cos \theta = \frac{1}{2}$; and $\theta = \pi/3$ and $-\pi/3$. Because of the symmetry of the figure, we can integrate from $\theta = 0°$ to 60° and then double the result.

To arrive at the element of the desired area, we let

$$dA_2 = \text{element of area of the circle}$$
$$dA_1 = \text{element of the area of the cardioid}$$

40.9 Areas in Polar Coordinates

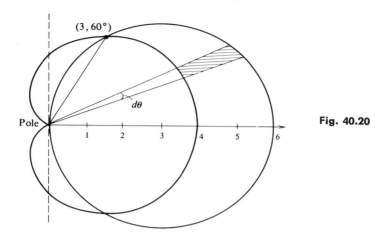

Fig. 40.20

Then for the element of the area to be found, we have the shaded portion,

$$dA = dA_2 - dA_1 = \frac{1}{2}r_2^2\,d\theta - \frac{1}{2}r_1^2\,d\theta$$

Substituting for r_2 and r_1 their equivalents, we get the element of area,

$$dA = \frac{1}{2}[(6\cos\theta)^2 - (2 + 2\cos\theta)^2]\,d\theta$$

$$= [16\cos^2\theta - 4\cos\theta - 2]\,d\theta = \left[16\left(\frac{1}{2} + \frac{1}{2}\cos 2\theta\right) - 4\cos\theta - 2\right]d\theta$$

For the total area to be found, we have

$$A = 2\int_0^{\pi/3}(6 + 8\cos 2\theta - 4\cos\theta)\,d\theta = 4\pi$$

It is extremely important in connection with polar coordinates that the sector taken as the element of area be representative over the entire area to be found. Sometimes it is necessary to use two different elements, as in the following example.

EXAMPLE 25. Find the area common to the two curves: $r = 2 + 2\cos\theta$, and $r = 6\cos\theta$ (Fig. 40.21).

Solution. We consider the area in two portions, A_1 and A_2, as shown. For the element dA_1, we take the r of the cardioid, $r = 2 + 2\cos\theta$. Note that the area A_1 is inside both curves. Yet if we integrate for this area from 0 to 60°, we shall not have included the small area indicated by A_2. The area A_2 is also inside both curves, but for this area we must use a different element. For dA_2 we take the r of the circle alone and integrate from 60° to 90°. Of course, the sum, $A_1 + A_2$, as shown will not give us the entire area called for, but since the figure is symmetrical about the polar axis, we shall double the area, $A_1 + A_2$.

For the two elements, we have

$$dA_1 = \frac{1}{2}(2 + 2\cos\theta)^2\,d\theta \quad \text{and} \quad dA_2 = \frac{1}{2}(6\cos\theta)^2\,d\theta$$

Polar Coordinates

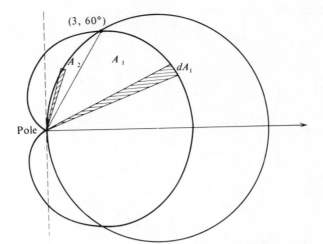

Fig. 40.21

Integrating and evaluating between the proper limits, we get the area of $A_1 + A_2$. The solution is left to the student. For answers, we should get

$$A_1 = \pi + \frac{9\sqrt{3}}{4} \quad \text{and} \quad A_2 = \frac{3\pi}{2} - \frac{9\sqrt{3}}{4}$$

The total area then inside both curves is 5π.

40.10 LENGTH OF ARC IN POLAR COORDINATES

We have already seen (Section 38.5) how to compute the length of an arc of a curve expressed in rectangular coordinates. We first express the differential of arc, given by the formula

$$ds = \sqrt{1 + (dy/dx)^2}\, dx \quad \text{or} \quad ds = \sqrt{(dx)^2 + (dy)^2}$$

The element of arc can also be expressed in polar coordinates. We have already used the relations:

$$x = r \cos \theta \quad \text{and} \quad y = r \sin \theta$$

Differentiating, we get $\quad dx = \cos \theta\, dr - r \sin \theta\, d\theta$

and $\quad dy = \sin \theta\, dr + r \cos \theta\, d\theta$

Now, if we square both sides of these two equations and substitute in the foregoing formula for ds, we shall get the formula for the differential of arc:

$$ds = \sqrt{(dr)^2 + r^2 (d\theta)^2} \quad \text{or} \quad ds = \sqrt{r^2 + (r')^2}\, d\theta$$

EXAMPLE 26. Find the total length of the cardioid, $r = 2 + 2 \cos \theta$.

Solution. Differentiating,

$$r' = -2 \sin \theta$$

Substituting in the formula for ds, we get

$$ds = \sqrt{(2 + 2 \cos \theta)^2 + (-2 \sin \theta)^2}\, d\theta$$
$$= \sqrt{4 + 8 \cos \theta + 4 \cos^2 \theta + 4 \sin^2 \theta}\, d\theta = \sqrt{8(1 + \cos \theta)}\, d\theta$$

Now we make use of the trigonometric identity,

$$1 + \cos\theta = 2\cos^2\frac{\theta}{2}$$

and get

$$ds = 4\cos\frac{\theta}{2}\,d\theta$$

For one half the total length of the curve we integrate from 0 to π and get

$$\frac{s}{2} = 4\int_0^\pi \cos\frac{\theta}{2}\,d\theta = 8\sin\frac{\theta}{2}\Big]_0^\pi = 8$$

then

$$s = 16$$

EXERCISE 40.6

Find the area indicated in each of the following problems:
1. Inside the curve $r = 6\sin\theta$
2. Inside the curve $r = 4\cos\theta$
3. Inside the curve $r = 6\cos 2\theta$
4. Inside the curve $r = 6\cos 3\theta$
5. Bounded by the curve $r = 3 - 3\cos\theta$, from $\theta = 0$ to $\theta = 90°$
6. Bounded by the curve $r = 4 + 4\sin\theta$, from $\theta = 0$ to $\theta = 90°$
7. Inside the circle $r = 6\sin\theta$ and outside the cardioid $r = 2 + 2\sin\theta$
8. Inside the cardioid $r = 2 + 2\cos\theta$ and outside the circle $r = 6\cos\theta$
9. Bounded by the curves $r = 3 + 3\cos\theta$ and $r = 3 + 3\sin\theta$
10. Between the curves $r = 4\cos\theta$ and $r = 2 + 2\cos\theta$
11. Between the curves $r = 4\cos\theta$ and $r = 4 + 4\cos\theta$
12. Total area inside the curve $r = 3 + \cos\theta$
13. The inner loop of the curve $r = 3 + 6\cos\theta$
14. Total area inside the curve $r = 3 - \sin\theta$
15. Total area inside the curve $r = 3 + 6\cos\theta$
16. Inside the curve

$$r = \frac{2}{1 + \cos\theta}$$

from $\theta = -90°$ to $\theta = +90°$
17. Inside the circle $r = 4\cos\theta$ and outside the curve

$$r = \frac{8}{3 + 2\cos\theta}$$

18. Outside the circle $r = 3$, and inside the curve, $r = 6\sin 2\theta$

Find the total length of each of the following curves:
19. $r = 6$ 20. $r = 8\cos\theta$ 21. $r = 4 + 4\sin\theta$
22. $r = 5 - 5\cos\theta$ 23. $r = 4 + 2\cos\theta$ 24. $r = 3 + 6\cos\theta$
25. Find the length of the curve given by the parametric equations $x = 8\cos^3\theta$ and $y = 8\sin^3\theta$, from $\theta = 0$ to $\theta = 90°$.

chapter
41

Hyperbolic Functions

41.1 THE CATENARY

If a chain or any perfectly flexible cord of uniform size and density hangs by its own weight alone, it takes the shape of a curve called a *catenary* (Fig. 41.1). The equation of this curve is basically

$$y = (K)\frac{e^u + e^{-u}}{2}$$

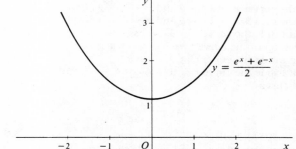

Fig. 41.1

In the equation, the variable u is some function of x. If $u = x$, then the equation in its simplest form becomes

$$y = \frac{e^x + e^{-x}}{2}$$

If x has some coefficient, say a, then the form of the equation is

$$y = \left(\frac{1}{a}\right)\frac{e^{ax} + e^{-ax}}{2}$$

Note that the constant coefficient of x in the function is a, while the coefficient of the function itself is $1/a$. These two coefficients are reciprocals of each other. Then another form of the same type of equation is

$$y = (a)\frac{e^{x/a} + e^{-x/a}}{2}$$

Here are some examples of the equations of such curves:

(a) $y = (25)\dfrac{e^{0.04x} + e^{-0.04x}}{2}$ (b) $y = (50)\dfrac{e^{0.02x} + e^{-0.02x}}{2}$

The value of the constant a will determine the amount of "spread" of the curve or the amount of "sag" in the chain between the points of support.

Note that in equation (a), the coefficient of x is 0.04, whereas in equation (b) it is much less, only 0.02. Then we shall find that for a chain fastened to two supports a given distance apart, the "sag" will be much less in the case of (b). In other words, equation (b) represents a greater tension and therefore a shorter chain between the supports (Fig. 41.2). It also means that when the equation is graphed on the coordinate system, the curve has its lowest point at a greater distance from the x-axis. Of course, the curve can be made nearer the x-axis by subtracting some constant, in which case we get some such equation as

$$y = 50\dfrac{e^{0.02x} + e^{-0.02x}}{2} - 20$$

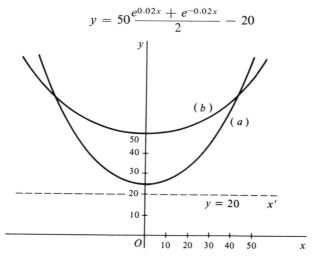

Fig. 41.2

41.2 THE HYPERBOLIC FUNCTIONS

The following expressions occur so frequently in engineering that they have been given special names:

$\dfrac{e^x - e^{-x}}{2}$ is called the *hyperbolic sine of x;* written sinh x

$\dfrac{e^x + e^{-x}}{2}$ is called the *hyperbolic cosine of x;* written cosh x

That is, by definition, in general terms, if u is a function of x,

(a) $\sinh u = \dfrac{e^u - e^{-u}}{2}$ (b) $\cosh u = \dfrac{e^u + e^{-u}}{2}$

The reason for the names *hyperbolic sine* and *hyperbolic cosine* is that the expressions bear somewhat the same relation to the equilateral hyperbola,

$x^2 - y^2 = 1$, as the circular functions, sine and cosine, bear to the circle, $x^2 + y^2 = 1$.

In finding the value of some particular hyperbolic function, such as, say, sinh 1.5, we do not look for an angle. Instead, we simply use the number, 1.5, in the formula and compute the value:

$$\sinh 1.5 = \frac{e^{1.5} - e^{-1.5}}{2} = \frac{4.4817 - 0.2231}{2} = \frac{4.2586}{2} = 2.1293$$

This is somewhat like finding the usual circular trigonometric functions of numbers rather than angles. For example, sin 1.5 = 0.9975.

There are four additional hyperbolic functions that correspond to the other four circular trigonometric functions. The definitions of these functions correspond to certain identities of the circular functions. We begin with

(1) $\sinh u = \dfrac{e^u - e^{-u}}{2}$ \qquad (2) $\cosh u = \dfrac{e^u + e^{-u}}{2}$

From these two, we define the following:

(3) $\tanh u = \dfrac{\sinh u}{\cosh u} = \dfrac{e^u - e^{-u}}{e^u + e^{-u}} =$ hyperbolic tangent u

(4) $\coth u = \dfrac{\cosh u}{\sinh u} = \dfrac{e^u + e^{-u}}{e^u - e^{-u}} =$ hyperbolic cotangent u

(5) $\operatorname{sech} u = \dfrac{1}{\cosh u} = \dfrac{2}{e^u + e^{-u}}$ \qquad (6) $\operatorname{csch} u = \dfrac{1}{\sinh u} = \dfrac{2}{e^u - e^{-u}}$

41.3 HYPERBOLIC IDENTITIES

A few identities involving hyperbolic functions are often useful and should be noted. There are some similarities to the identities involving the circular functions, but there are also some differences.

For the unit circle, we have the equation

$$x^2 + y^2 = 1$$

As a consequence we have the identity:

$$\cos^2 \theta + \sin^2 \theta = 1$$

In general terms, this identity may be stated:

$$\cos^2 u + \sin^2 u = 1.$$

For the corresponding equation of the equilateral hyperbola, we have the equation

$$x^2 - y^2 = 1$$

Then we might expect the identity:

$$\cosh^2 u - \sinh^2 u = 1$$

41.4 Graphs of the Hyperbolic Functions

Let us see whether this equation is true in general, and, consequently, an identity.

Let us recall the definitions:

$$\cosh u = \frac{e^u + e^{-u}}{2} \quad \text{and} \quad \sinh u = \frac{e^u - e^{-u}}{2}$$

Then

$$\cosh^2 u = \left(\frac{e^u + e^{-u}}{2}\right)^2 = \frac{e^{2u} + 2 + e^{-2u}}{4}$$

and

$$\sinh^2 u = \left(\frac{e^u - e^{-u}}{2}\right)^2 = \frac{e^{2u} - 2 + e^{-2u}}{4}$$

Then

$$\cosh^2 u - \sinh^2 u = \left(\frac{e^{2u} + 2 + e^{-2u}}{4}\right) - \left(\frac{e^{2u} - 2 + e^{-2u}}{4}\right) = \frac{4}{4} = 1$$

Therefore,

$$\cosh^2 u - \sinh^2 u = 1 \quad \text{or} \quad 1 + \sinh^2 u = \cosh^2 u$$

Two additional identities that are sometimes useful are the following:

(a) $\cosh x + \sinh x = e^x$ \quad (b) $\cosh x - \sinh x = e^{-x}$

These identities also can be derived from the definitions.

41.4 GRAPHS OF THE HYPERBOLIC FUNCTIONS

To graph any hyperbolic function, we first express it in exponential form. Then, as usual, we take several values of x and compute corresponding values of the function. A table of values of e^x and of e^{-x} is helpful in determining the values of the function. As x increases in positive values from zero, note especially that e^x increases and e^{-x} decreases, since $e^{-x} = 1/e^x$. For large positive values of x, e^x is large but e^{-x} is small. In fact, for very large values of x, the value of e^{-x} may even be disregarded.

As an example, let us graph the function

$$y = \sinh x$$

or, in exponential form

$$y = \frac{e^x - e^{-x}}{2}$$

Now we compute a few values of the function: if $x = 0$,

$$y = (\tfrac{1}{2})(e^0 - e^{-0}) = (\tfrac{1}{2})(1 - 1) = 0$$

If $x = 0.5$, $y = (\tfrac{1}{2})(e^{0.5} - e^{-0.5}) = (\tfrac{1}{2})(1.6487 - 0.6065) = 0.5211$

If $x = 1$, $y = (\tfrac{1}{2})(e^1 - e^{-1}) = (\tfrac{1}{2})(2.7183 - 0.3679) = 1.1752$

If $x = 1.5$, $y = (\tfrac{1}{2})(e^{1.5} - e^{-1.5}) = (\tfrac{1}{2})(4.4817 - 0.2231) = 2.1293$

If $x = 2$, $y = (\tfrac{1}{2})(e^2 - e^{-2}) = (\tfrac{1}{2})(7.3891 - 0.1353) = 3.6269$

If $x = 3$, $y = (\tfrac{1}{2})(e^3 - e^{-3}) = (\tfrac{1}{2})(20.086 - 0.0498) = 10.018$

If we take negative values of x, the function will have the same numerical values but they will all be negative. For example, if $x = -2$, then

$$y = (\tfrac{1}{2})(e^{-2} - e^{2}) = -3.6269$$

The graph is shown in Figure 41.3.

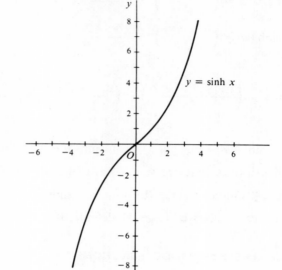

Fig. 41.3

Note especially that e^{-x}, the second term in the function, becomes very small as x increases in value from zero. When $x = 3$, $e^{-x} = 0.0498$. In fact, when x is large, the value of sinh x is close to $e^{x}/2$. In all cases, the major portion of the function value is determined by the positive power on e.

From the graph we can see that the function has no maximum or minimum value. However, it does have a point of inflection.

We have already seen that the graph of cosh x is a curve called a *catenary*. To graph the function, we write

$$y = \cosh x = \frac{e^{x} + e^{-x}}{2}$$

Now we can use the same values for e^{x} and e^{-x} as we have used for sinh x. For cosh x, the values of e^{x} and e^{-x} are added. Therefore, a negative value of x will not change the function value; that is,

$$\cosh(-x) = \cosh(+x)$$

The graph is shown in Figure 41.4. Note again that for large values of x, the term e^{-x} adds very little to the value of the function. Again the value of the function is mostly determined by the larger power on e. Moreover, if x

41.4 Graphs of the Hyperbolic Functions

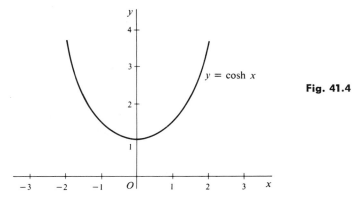

Fig. 41.4

is large positively, then the value of cosh x is very close to the value of $e^x/2$. The value of cosh x, just as with sinh x, is very little affected by a large negative power on e. For example, when $x = 3$, we have

$$y = \cosh 3 = \frac{e^3 + e^{-3}}{2} = \frac{20.086 + 0.0498}{2} = 10.069$$

The term e^{-3} affects the value of cosh 3 by only about 0.025.

Note that the function $y = \cosh x$ has a minimum point but no point of inflection.

We have pointed out that as x takes on larger and larger positive values, then the values of sinh x and cosh x are less and less affected by the term e^{-x}. It is well to note the following facts as x increases:

(a) Both sinh x and cosh x approach more closely to $e^x/2$ in value.
(b) Both functions approach each other in value.

As an example,

$$\sinh 6 \approx \cosh 6 \approx \left(\frac{1}{2}\right)e^6$$

The approximate values are as follows:

$$\sinh 6 = 201.71 \qquad \left(\frac{1}{2}\right)e^6 = 201.715 \qquad \cosh 6 = 201.72$$

The graphs of the remaining hyperbolic functions are shown in Figures 41.5–41.8. Note that the graph of $y = \tanh x$ lies entirely between the

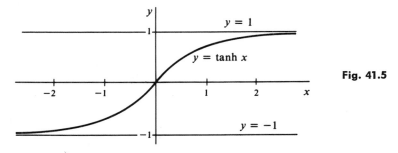

Fig. 41.5

Fig. 41.6

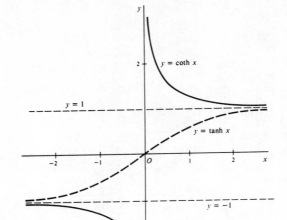

Fig. 41.7

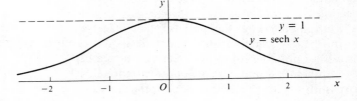

Fig. 41.8

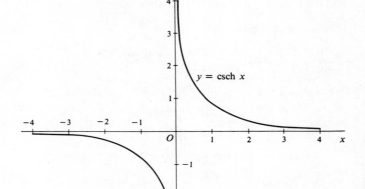

lines $y = 1$ and $y = -1$. In fact, these two lines are asymptotes. In the function

$$y = \tanh x = \frac{e^x - e^{-x}}{e^x + e^{-x}}$$

we see that the numerator is always less than the denominator in numerical value. Therefore, the value of the function will be less than 1 in absolute value. For $x = 0$, the value of the function is also zero.

In the case of the function $y = \coth x$, the numerator always has a value greater than that of the denominator. Therefore, its values are excluded between the lines $y = -1$ and $y = +1$. These lines are asymptotes. Moreover, the line $x = 0$ is also an asymptote since the denominator becomes zero when $x = 0$.

EXERCISE 41.1

Prove the following identities:

1. $\cosh x + \sinh x = e^x$
2. $\cosh x - \sinh x = e^{-x}$
3. $\operatorname{sech}^2 u = 1 - \tanh^2 u$
4. $\coth^2 u - 1 = \operatorname{csch}^2 u$
5. $\tanh u + \coth u = 2 \coth 2u$
6. $\tanh u - \operatorname{csch} u = \coth u$
7. $\sinh(-x) = -\sinh x$
8. $\cosh(-x) = \cosh x$
9. $\sinh(x + y) = \sinh x \cosh y + \cosh x \sinh y$
10. $\cosh(x + y) = \cosh x \cosh y + \sinh x \sinh y$
11. $\sinh 2x = 2 \sinh x \cosh x$
12. $\cosh 2x = \cosh^2 x + \sinh^2 x$

Graph the following equations:

13. $y = \sinh x$
14. $y = \cosh x$
15. $y = \tanh x$
16. $y = \coth x$
17. $y = 4 \cosh \dfrac{x}{4}$
18. $y = \tfrac{1}{2} \sinh 2x$

19. Sketch the curve for each of the following equations on the same graph, using x-values from $x = -4$ to $x = +4$. Compare the amount of "sag" or dip in the curves.
    (a) $y = \cosh 0.4x$ (b) $y = \cosh 0.5x$ (c) $y = \cosh 0.6x$

20. $y = 5 \cosh 0.2x - 2$
21. $y = 10 \cosh 0.05x$
22. $y = 20 \cosh 0.05 x - 10$
23. A cable hangs to form a curve defined by the equation, $y = 50 \cosh 0.02x - 20$. Find the sag in the cable if it is suspended from two points on the same level, 100 ft apart. How far is the lowest point from the ground? How high above ground are the points of suspension?

41.5 DIFFERENTIATION OF HYPERBOLIC FUNCTIONS

To find the derivative of any hyperbolic function, we first express the function in exponential form according to the definition. Then we differentiate as with any exponential function. For example, let us find the derivative of sinh u. First we have

$$y = \sinh u$$

Hyperbolic Functions

In exponential form, we have

$$y = \left(\frac{1}{2}\right)(e^u - e^{-u})$$

Now we differentiate according to the rule for any exponential function:

$$\frac{dy}{dx} = \left(\frac{1}{2}\right)(e^u + e^{-u})\left(\frac{du}{dx}\right)$$

or

$$\frac{dy}{dx} = \frac{e^u + e^{-u}}{2}\left(\frac{du}{dx}\right)$$

Note that the derivative is the expression for the hyperbolic cosine u. Then we can say

$$\frac{d}{dx}(\sinh u) = (\cosh u)\left(\frac{du}{dx}\right)$$

In the same way we shall find that

$$\frac{d}{dx}(\cosh u) = (\sinh u)\left(\frac{du}{dx}\right)$$

For the derivative of tanh u we write

$$y = \tanh u = \frac{\sinh u}{\cosh u}$$

By the quotient rule, we get

$$\frac{dy}{dx} = \text{sech}^2 u \frac{du}{dx}$$

Note that the derivatives of the three foregoing hyperbolic functions are all *positive*. The derivatives of the remaining three hyperbolic functions are *negative*. These can also be derived by the quotient rule.

$$\frac{d}{dx}(\coth u) = -\text{csch}^2 u \frac{du}{dx};$$

$$\frac{d}{dx}(\text{sech } u) = -\text{sech } u \tanh u \frac{du}{dx};$$

$$\frac{d}{dx}(\text{csch } u) = -\text{csch } u \coth u \frac{du}{dx}$$

Note the similarity to the derivatives of the circular functions with the exception of some signs.

EXAMPLE 1. Find the derivative of the function, $y = \text{sech } x^2$, and write in exponential form. Then evaluate the derivative for $x = 1$.

Solution. By the formula,

$$\frac{dy}{dx} = -2x \text{ sech } x^2 \tanh x^2$$

In exponential form,

$$\frac{dy}{dx} = (-2x)\left(\frac{2}{e^{x^2} + e^{-x^2}}\right)\left(\frac{e^{x^2} - e^{-x^2}}{e^{x^2} + e^{-x^2}}\right)$$

$$= \frac{-4x(e^{x^2} - e^{-x^2})}{e^{2x^2} + 2 + e^{-2x^2}}$$

At $x = 1$,

$$\frac{dy}{dx} = \frac{-4(e^1 - e^{-1})}{e^2 + 2 + e^{-2}} = \frac{-4(2.7183 - 0.3679)}{7.3891 + 2 + 0.1353} = -0.987$$

EXAMPLE 2. Find any possible maximum or minimum point of the curve $y = \cosh 3x$. Also find the slope of the curve where $x = 1$.

Solution. $\quad \dfrac{dy}{dx} = 3 \sinh 3x = \dfrac{3}{2}(e^{3x} - e^{-3x})$

Setting the derivative equal to zero,

$$\frac{3}{2}(e^{3x} - e^{-3x}) = 0 \quad \text{or} \quad e^{3x} = e^{-3x}$$

Multiplying both sides of the equation by e^{3x},

$$e^{6x} = 1$$

Taking the natural log of both sides,

$$6x = 0 \quad \text{or} \quad x = 0$$

A graph of $y = \cosh 3x$ shows at once that the curve has a minimum value at $x = 0$. However, if we wish to check mathematically, we take two values, one on either side of $x = 0$, say, $x = -1$ and $x = +1$. Then we have these values: at $x = -1$, $m = (3/2)(e^{-3} - e^3) = -$ (negative); at $x = +1$, $m = +$ (positive). Therefore, at $x = 1$, the curve has a minimum point. The point is (0,1).

We might also test the value $x = 0$ by the use of the second derivative. For the second derivative, we get

$$y'' = \frac{9}{2}(e^{3x} + e^{-3x}) = 9 \cosh 3x$$

At $x = 0$, $y'' = (9/2)(e^0 + e^0) = +9$ (positive). Since the second derivative is positive at $x = 0$, the value represents a minimum point.

EXERCISE 41.2

Find the derivative of each of the following functions with respect to the independent variable:

1. $y = \sinh 5x$
2. $y = \cosh 4t^2$
3. $y = \tanh (3x - 4)$
4. $y = \coth (t + 2)$
5. $s = \text{sech } 2t$
6. $x = \text{csch } (5t - 3)$
7. $y = \sinh^2 3x$
8. $y = \cosh^2 4t$
9. $y = \ln \sinh 2x$
10. $s = \cosh \sqrt{t}$
11. $s = t^2 \tanh 3t$
12. $y = \text{sech } (\sin x)$
13. $s = e^t \sinh t$
14. $i = e^{\cosh t}$
15. $q = 0.5 \cosh 0.2t$

Find any maximum and/or minimum points of the following curves:

16. $y = \sinh^2 2x$
17. $y = \tanh^2 2x$
18. $y = 50 \cosh 0.02x$

19. Show that the curve, $y = \sinh x$, has no horizontal tangent but that it has a point of inflection. Find the slope at the point of inflection.
20. Show that the curve, $y = \cosh 5x$, has a minimum point but no point of inflection. Test the minimum point by two methods.
21. The current in a certain electric circuit is given by the formula, $i = 0.015 \sinh 1.2t$. Find the rate of change of current, di/dt, when $t = 1.3$ sec.
22. The charge on a certain capacitor is given by the formula, $q = 0.3 \cosh 0.1t$. Find the rate of transport of charge, dq/dt, when $t = 1.2$ sec.
23. A hanging cable forms a curve whose equation is $y = 50 \cosh 0.02x - 20$. Find the slope of the curve where $x = 50$.
24. The amount of charge taken from a certain capacitor is given by the formula, $q = 0.1 \cosh 0.05t$ coulomb. Find the discharge current, that is the rate of transport of charge, dq/dt, when $t = 1.2$ sec.

41.6 INTEGRATION OF HYPERBOLIC FUNCTIONS

To integrate any hyperbolic function, we can first express it in exponential form. However, some integrals follow directly from a knowledge of derivatives and should be recognized at sight. We have seen that

$$\frac{d}{dx}(\sinh u) = \cosh u \frac{du}{dx}$$

Then, reversing the process, we have at once

$$\int \cosh u \, du = \sinh u + C$$

In a like manner we have the following:

$$\int \sinh u \, du = \cosh u + C$$

$\int \text{sech}^2 u \, du = \tanh u + C$ $\qquad \int \text{csch}^2 u \, du = -\coth u + C$

$\int \text{sech } u \tanh u \, du = -\text{sech } u + C$ $\qquad \int \text{csch } u \coth u \, du = -\text{csch } u + C$

Of course, in integration, we must be sure, as always, that we have the factor du present in the integrand. If only a constant is lacking, it can be supplied as usual.

EXAMPLE 3. Find the indefinite integral

$$\int \sinh 5x \, dx$$

Solution. In this example we must have the factor ($5 \, dx$). The factor 5 is supplied and its reciprocal written outside the integral sign. We get

$$\int \sinh 5x \, dx = \frac{1}{5} \int 5 \sinh 5x \, dx = \frac{1}{5} \cosh 5x + C$$

EXAMPLE 4. Find the area under the curve $y = \cosh x$ from $x = -1$ to $x = +1$.

Solution. The curve is shown in Figure 41.9. For the element of area we have

$$dA = y \, dx = \cosh x \, dx$$

41.6 Integration of Hyperbolic Functions

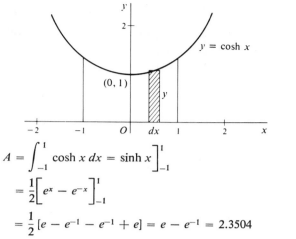

Fig. 41.9

Then
$$A = \int_{-1}^{1} \cosh x \, dx = \sinh x \Big]_{-1}^{1}$$
$$= \frac{1}{2}\Big[e^x - e^{-x}\Big]_{-1}^{1}$$
$$= \frac{1}{2}[e - e^{-1} - e^{-1} + e] = e - e^{-1} = 2.3504$$

EXAMPLE 5. Find the volume of the solid of revolution generated by rotating the first-quadrant portion of the curve $y = \cosh x$ to the line $x = 1$ about the y-axis.

Solution. The solid is indicated in Figure 41.10. Using the shell method, we find the element of volume is

$$dV = 2\pi r \, dA = 2\pi x \cosh x \, dx$$

Then
$$V = 2\pi \int_{0}^{1} x \cosh x \, dx$$

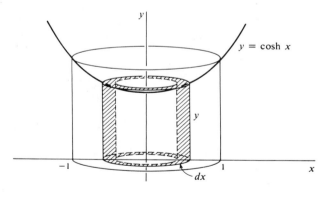

Fig. 41.10

To integrate, we use *the method of parts:*

let $u = x$	$dv = \cosh x \, dx$
$du = dx$	$v = \sinh x$

Then we get
$$V = 2\pi[x \sinh x - \int \sinh x \, dx]_0^1 = 2\pi[x \sinh x - \cosh x]_0^1$$
$$= 2\pi(\sinh 1 - \cosh 1 - 0 + \cosh 0) = 2\pi(1.1752 - 1.5431 + 1) = 3.97$$

Hyperbolic Functions

EXAMPLE 6. Find the length of the curve $y = \cosh x$ from $x = -1$ to $x = 1$.

Solution. For the element of arc, we first find the derivative:

$$\frac{dy}{dx} = \sinh x$$

then

$$\left(\frac{dy}{dx}\right)^2 = \sinh^2 x$$

For the element of arc we have

$$ds = \sqrt{1 + \sinh^2 x}\, dx$$

Now we make the substitution,

$$1 + \sinh^2 x = \cosh^2 x$$

and get

$$ds = \sqrt{\cosh^2 x}\, dx = \cosh x\, dx$$

Then

$$s = \int_{-1}^{1} \cosh x\, dx = \sinh x \Big]_{-1}^{1} = \frac{1}{2}\left[e^x - e^{-x}\right]_{-1}^{1}$$

$$= \frac{1}{2}(e - e^{-1} - e^{-1} + e) = (e - e^{-1})$$

$$= 2.7183 - 0.3679 = 2.3504$$

EXERCISE 41.3

Integrate each of the following:

1. $\int \cosh 5x\, dx$
2. $\int \sinh 2x\, dx$
3. $\int x \tanh x^2\, dx$
4. $\int t \coth^2 t\, dt$
5. $\int \sinh^3 x \cosh x\, dx$
6. $\int \tanh^3 t \operatorname{sech}^2 t\, dt$
7. $\int e^x \sinh x\, dx$
8. $\int \sinh^2 x\, dx$
9. $\int \cosh^3 2t\, dt$
10. $\int e^{-2x} \cosh 3x\, dx$
11. $\int \tanh 5x\, dx$
12. $\int \operatorname{sech} t\, dt$
13. Find the area under the curve, $y = \sinh x$, from $x = 0$ to $x = 2$.
14. If the area in Problem 13 is rotated about the x-axis, find the volume of the solid generated.
15. Find the volume generated if the area in Problem 13 is rotated about $x = 0$.
16. Find the centroid of the area in Problem 13.
17. Find the centroid of the area bounded by $y = \cosh x$; $x = 0$; $x = 1$; $y = 0$.
18. Find the moments of inertia of the area in Problem 17.
19. Find the area bounded by the curves, $y = \tanh^2 x$; $y = 1$; $x = -1$; $x = 1$.
20. Find the length of the curve, $y = 50 \cosh 0.02x$, from $x = 0$ to $x = 100$.
21. A cable 210 ft long is fastened, one end to the top of each of two posts that are 50 ft high and 200 ft apart on level ground. How far is the lowest point of the cable above the ground?

chapter

42

Partial Differentiation

42.1 A FUNCTION OF MORE THAN ONE VARIABLE

We have seen that a functional relationship exists between two variables, such as x and y, if for every value of x there exists a corresponding value of y. In other words, the value of y depends on the value assumed for x. This relationship is stated mathematically as

$$y = f(x)$$

Up to this point we have considered the dependent variable, y, to be a function of only one other single variable, x. In the case of a circle, for example, the area A is a function of the single variable, the radius r; that is,

$$A = f(r)$$

However, it often happens that the value of one variable depends on more than one other variable. In the case of a rectangle, the area is determined not only by the *length* but also by the *width*. We cannot say that the area is a function of the length only unless we assume that the width remains constant. If we assume that both length and width may vary, then we must say the area is a function of both length and width. This condition is stated mathematically by

$$A = f(l,w)$$

In the same way, the volume of a rectangular solid is a function of the length l, the width w, and the altitude h. That is,

$$V = f(l,w,h)$$

The concept of a function of more than one variable is a common notion in many everyday situations. The total cost (C) for the gasoline on an auto trip depends on several variables such as distance (s) traveled, the efficiency of the car as indicated by the number (n) of miles traveled per gallon of gasoline, the price (p) per gallon, and somewhat on the average velocity (v) of the car. Stated mathematically,

$$C = f(s,n,p,v)$$

588 Partial Differentiation

Notice that this statement does not imply any particular operation among the variables. It simply means that if s, n, p, and v are known, there exists some particular unique total cost related to these variables.

In an electric circuit involving only direct current, the amount of current I depends on the voltage E and the resistance R in the circuit; that is,

$$I = f(E,R)$$

If the circuit also contains inductance, which is always the case to some extent, then we must include inductance L as a determining factor of the current. In an alternating-current circuit, the current at any particular instant (instantaneous current i) depends on the particular time instant t; that is,

$$i = f(E,R,L,t)$$

42.2 DERIVATIVE OF A FUNCTION OF MORE THAN ONE VARIABLE

To find the derivative of a function of two or more variables, we must let one or more of the independent variables remain constant and let *only one* of them vary. For example, suppose Z is a function of x and y. Then

$$Z = f(x,y)$$

To differentiate the function, we must assume that x or y remains constant, while we differentiate with respect to the other variable. This is not so difficult as it might seem. We simply look upon one of the variables as we look upon any other constant. We illustrate the method by an example.

EXAMPLE 1. Differentiate the function indicated by

$$Z = x^2 - 3xy + 2y^2$$

Solution. If we wish to differentiate the function with respect to x as the independent variable, we simply assume that y is a constant. This may be more easily seen if we take a similar equation in which we replace y and y^2 by other letters such as a and b, which are usually considered as constants. Let us write the equation as

$$Z = x^2 - 3ax + 2b$$

Now it is probably easier to consider that a and b are constants. Then the derivative with respect to x is

$$\frac{dZ}{dx} = 2x - 3a$$

Now let us go back to the equation with which we began

$$Z = x^2 - 3xy + 2y^2$$

Now we shall assume that y is a constant. However, we know that it is really a variable which we only temporarily assume to be constant. Then the derivative of the function with respect to the variable x might be considered only a *partial derivative* of the function Z. A partial derivative is often denoted by a "curly d"

42.2 Derivative of a Function of More Than One Variable

or ∂. It indicates that there are other independent variables that are temporarily taken to be constant. The partial derivative of Z with respect to x is then denoted as follows for the above equation:

$$\frac{\partial Z}{\partial x} = 2x - 3y$$

Now suppose we wish to consider the variable x as constant and find the partial derivative of Z with respect to y as the independent variable. Then we get

$$\frac{\partial Z}{\partial y} = -3x + 4y$$

Other forms of notation are used for the partial derivatives. (Partial derivatives are often called simply *partials*.) If we denote the function of x and y by $f(x,y)$ instead of Z, then we may denote the partials of the function as follows:

If
$$f(x,y) = x^2 - 3xy + 2y^2$$

then $\qquad \dfrac{\partial f(x,y)}{\partial x} = 2x - 3y \qquad$ and $\qquad \dfrac{\partial f(x,y)}{\partial y} = -3x + 4y$

Partials are also denoted by subscripts, such as $f_x(x,y)$ and $f_y(x,y)$. The subscript denotes the independent variable which is assumed to vary while the other independent variables are temporarily held constant.

EXAMPLE 2. Find the partials of the following function with respect to x and y; then evaluate each partial for the values $x = 3$ and $y = 1$.

$$f(x,y) = 3x^2 - 5xy + 4y^2 - 7x + 2y + 11$$

Solution. Holding y constant we differentiate with respect to x and get

$$f_x(x,y) = 6x - 5y - 7$$

Holding x constant we differentiate with respect to y and get

$$f_y(x,y) = -5x + 8y + 2$$

Now we evaluate the partials for the given values: $x = 3$ and $y = 1$.

$$f_x(3,1) = 18 - 5 - 7 = 6$$
$$f_y(3,1) = -15 + 8 + 2 = -5$$

The significance of the partials may be better understood if we think of the results in Example 2 as follows: For the values $x = 3$ and $y = 1$, the rate of change of the function $f(x,y)$ with respect to x is 6. This means that at the particular instant when $x = 3$ and $y = 1$, the function is changing 6 times as fast as x, provided that y remains constant. Also at the same instant, the function is changing -5 times as fast as y, provided that x remains constant, the negative indicating opposite directions of change.

EXAMPLE 3. In the following function find the partial with respect to x, and the partial with respect to y:

$$Z = 5x^3 - 3x^2y + 4xy^2 - 2y^3$$

Solution.

$$\frac{\partial Z}{\partial x} = 15x^2 - 6xy + 4y^2 \qquad \frac{\partial Z}{\partial y} = -3x^2 + 8xy - 6y^2$$

42.3 EXAMPLES FROM ELECTRIC CIRCUIT THEORY

The electromotive force in an electric circuit is given by Ohm's law

$$E = IR$$

where E = electromotive force in volts, I = current in amperes, and R = resistance in ohms. Let us suppose we hold R constant at some value, say, 40 ohms. We differentiate E with respect to I as a variable, and get

$$\frac{\partial E}{\partial I} = R$$

If we assume that $R = 40$, the partial derivative, $\partial E/\partial I$, in this case means that the voltage varies by 40 volts for each change of one ampere in current. In general terms, the voltage E varies at a rate of R volts per ampere of change in current. For example, if I is changing at an instantaneous rate of 2 amperes per second, then E is changing by 80 volts per second. (It is not strictly correct to say that the voltage changes 40 times as much as current because voltage does not change by amperes.)

Now, if instead, the resistance R varies and we hold current I constant, then we find the partial of E with respect to the variable R, and get

$$\frac{\partial E}{\partial R} = I$$

The result means that if the current is assumed to remain constant, say, at 5 amperes, then the voltage varies instantaneously at a rate of 5 volts for each change of 1 ohm of resistance. In general terms, the voltage varies at a rate of I volts per ohm of change in resistance.

As another example involving Ohm's law, let us take the formula

$$I = \frac{E}{R}$$

To find the partial of I with respect to R, we hold E constant and get

$$\frac{\partial I}{\partial R} = -\frac{E}{R^2}$$

If we take E as a constant of 60 volts, we see that the instantaneous rate of change of current with respect to resistance is

$$-\frac{60}{R^2}$$

This means that at the instant when the variable R is equal to 10 ohms, the current is changing at a rate of

$$\frac{\partial I}{\partial R} = -\frac{60}{10^2} = -0.6$$

That is, the current is changing at the rate of -0.6 ampere per ohm of resistance change, provided E remains constant. The negative, of course, indicates that I decreases as R increases.

EXERCISE 42.1

Evaluate the partial derivative with respect to each of the independent variables for the given values in each of the following functions:

1. $Z = 2x + 3xy - 6y;\ x = 3,\ y = -2$
2. $Z = x^2 + xy + 2y^2;\ x = -2,\ y = 4$
3. $Z = x^2y - xy^2 + 2x - 3y;\ x = 1,\ y = -3$
4. $f(x,y) = 3x^2 - 4xy + 5y^2 - 7x + 6y + 4;\ x = 4,\ y = 1$
5. $f(x,y) = 3x^2 - 5xy - 4y^2 + 8x - 7y;\ x = 2,\ y = -1$
6. $f(s,t) = 4s^2 - 3st - 2t^2 - 6s + 2t;\ s = -2,\ t = -1$
7. $g(u,v) = 2u^2 + 5uv - 3v^2 - uv^2;\ u = 1,\ v = -3$
8. $f(w,h) = w^3 - 4w^2h - 3wh^2 - 3h^3;\ w = -1,\ h = -2$
9. $h(r,s) = 2r^2 + 3rs + 2s^2 - 4r - 5s + 6;\ r = 2,\ s = -1$
10. $u(x,y) = (3x + 2)^2(4y^3 - 3)^{1/2};\ x = -1,\ y = 1$

In the following functions, find $\partial Z/\partial x$ and $\partial Z/\partial y$.

11. $Z = e^{2x+3y}$
12. $Z = x^2 e^{-3y}$
13. $Z = \ln(4x - 3y)$
14. $Z = \arctan(y/x)$
15. $Z = \sec x \tan y$
16. $Z = \sin 3x \cos 2y$
17. $Z = e^{-3x} \cos y$
18. $Z = \sin(x+y) - \cos(x-y)$
19. $Z = \ln\left(x + \sqrt{x^2 + y^2}\right)$
20. $Z = \dfrac{4x^2}{y + \cos xy}$

21. In the formula for current, $I = E/R$, find the partial derivative of I with respect to R. Then, when $E = 180$ volts, find the instantaneous rate of change of current when (a) $R = 30$ ohms; (b) $R = 120$ ohms; (c) $R = 4500$ ohms.

22. In the formula for power, $P = I^2R$, find as a partial derivative the rate of change of power with respect to current I if R is held constant. Then find the particular rate of change of current when $I = 3$ amperes, for each of the following values of R: (a) 200 ohms; (b) 32 ohms; (c) 12,400 ohms.

23. In the formula for power, $P = I^2R$, find the partial of P with respect to R. Then find the instantaneous rate of change of power for each of the following values of I: (a) 3 amperes; (b) 5 amperes; (c) 15 amperes. How does a change in R affect the rate of change of power?

24. In the formula for power, $P = E^2/R$, find the rate of change of P with respect to E. Then, taking $E = 120$ volts, find the particular rate of change of power for these values of R: (a) 4 ohms; (b) 10 ohms; (c) 6000 ohms; (d) 50,000 ohms; (e) 1.8 megohms.

25. In the formula for power, $P = E^2/R$, find the rate of change of P with respect to R. Then, when $E = 60$ volts, find the instantaneous rate of change of power when R is (a) 12 ohms; (b) 60 ohms; (c) 2400 ohms.

42.4 GEOMETRIC INTERPRETATION OF PARTIAL DERIVATIVE

The meaning of a partial derivative may be illustrated geometrically. Consider the area of a rectangle (Fig. 42.1). The area is given by the formula

$$A = lw$$

In this example the area is $(14)(8)$ or 112.

Now let us suppose that the length changes by some increment, $\Delta l = 2$. Then we get the new rectangle shown in Fig. 42.2.

Fig. 42.1

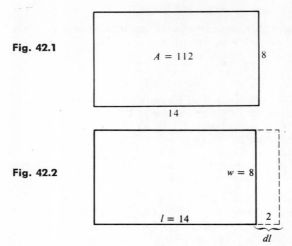

Fig. 42.2

The increment in length will result in a change in area. The increment in area is equal to

$$\Delta A = (\text{width})(\Delta l)$$

$$\Delta A = (8)\quad(2) = 16$$

This result means that the rate of change of area with respect to "l" is equal to the width. Stated mathematically, by partial differentiation, since $A = lw$, we have

$$\frac{\partial A}{\partial l} = w$$

Here we hold the width w constant and allow the length l to change.

Now, instead of holding the width constant, we may hold the length constant and allow the width to vary. Let us suppose that w changes by some increment, $\Delta w = 1.5$. We get the new rectangle shown in Fig. 42.3. The change in width will result in a change in area. This increment in area is equal to

$$\Delta A = (\text{length})(\Delta w)$$

$$\Delta A = (14)\quad(1.5) = 21$$

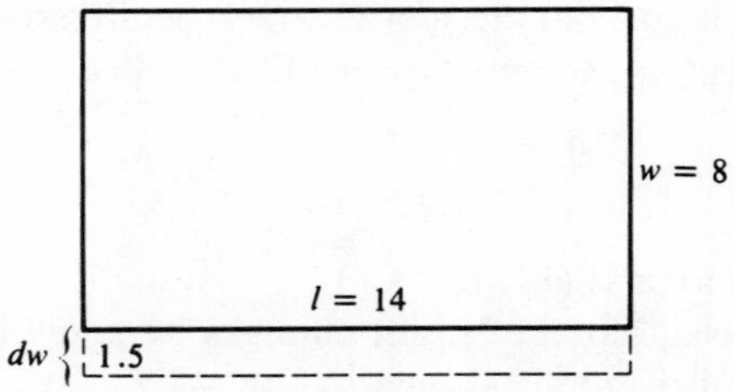

Fig. 42.3

42.5 Total Differential of a Function of Two or More Variables

In this case the rate of change of area with respect to the width is equal to the length. Stated mathematically, by partial differentiation, since $A = lw$, then we have

$$\frac{\partial A}{\partial w} = l$$

Here we hold the length l constant and allow the width w to change. In summary,

1. The area of a rectangle is a function of two variables, length and width, expressed by the formula

$$A = lw$$

2. If the width remains constant, the area will vary with the length; this rate of change is expressed by the partial derivative of A with respect to l, or

$$\frac{\partial A}{\partial l}$$

3. If the length remains constant, the area will vary with the width; this rate of change is expressed by the partial derivative of A with respect to w, or

$$\frac{\partial A}{\partial w}$$

42.5 TOTAL DIFFERENTIAL OF A FUNCTION OF TWO OR MORE VARIABLES

We have seen that the area of a rectangle is a function of two variables, length and width. We have seen that if the width remains constant, then a change in length will produce a change in area. Also, if the length remains constant, a change in width will produce a change in area.

Our question now might be: What happens to the area if changes occur in both length and width?

To understand the problem, let us begin with a particular rectangle whose length is 14 in. and width is 8 in. (Fig. 42.4). Let us assume that both length and width change by some increments, Δl and Δw, respectively, and let us

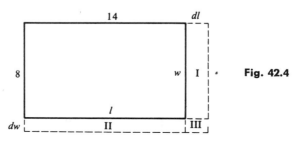

Fig. 42.4

594   Partial Differentiation

assume at this point that both increments are positive. These increments may be denoted by dl and dw. Now let these increments be as follows:

$$\text{increment in length} = dl = 2 \text{ in.}$$

$$\text{increment in width} = dw = 1.5 \text{ in.}$$

Beginning with the formula for the area of a rectangle, $A = lw$, we find the partial with respect to l:

$$\frac{\partial A}{\partial l} = w$$

That is, the rate of change of area with respect to length is equal to the width 8 in. Now if $dl = 2$ in., then the resulting increment in area is $(2)(w)$. This first increment in area, represented by the small rectangle I (Fig. 42.4), may be called $(\Delta A)_1$, so that

$$(\Delta A)_1 = (2)(w) = (2)(8) = 16 \text{ sq in. (Rectangle I)}$$

Note that this increment is equal to

$$(\Delta A)_1 = \left(\frac{\partial A}{\partial l}\right)(dl)$$

If, instead, we hold length constant, we find the partial with respect to w, and get

$$\frac{\partial A}{\partial w} = l$$

That is, the rate of change of area with respect to width is equal to the length, 14 in. Now if $dw = 1.5$ in., then the resulting increment in area is $(1.5)(l)$. This increment, represented by the small rectangle II (Fig. 42.4), may be called $(\Delta A)_2$, so that

$$(\Delta A)_2 = (1.5)(l) = (1.5)(14) = 21 \text{ sq in.}$$

Note that this increment is equal to

$$(\Delta A)_2 = \left(\frac{\partial A}{\partial w}\right)(dw)$$

Now if we add the two increments in area represented by the two small rectangles, I and II, we get the *approximate total change* in the area. This *approximate* change in area is called the *total differential of area*, and is denoted by dA. It is found as follows:

$$\text{Rectangle I} = \frac{\partial A}{\partial l}(dl) = 16 \text{ sq in.}$$

$$\text{Rectangle II} = \frac{\partial A}{\partial w}(dw) = 21 \text{ sq in.}$$

$$dA = 37 \text{ sq in.}$$

42.5 Total Differential of a Function of Two or More Variables

In general, the total differential of area, dA, is given by the formula

$$dA = \frac{\partial A}{\partial l}(dl) + \frac{\partial A}{\partial w}(dw)$$

Note especially that the *sum of the two increments* represented by the two small rectangles, I and II, *does not equal exactly the actual total increment in area*, ΔA. If we wish to find the actual increment in area, we consider the new rectangle after the changes have taken place. It has dimensions of 16 in. by 9.5 in., and a total area of 152 sq in. The actual total change in area is 40 sq in., whereas the differential of area, dA, is 37 sq in. The total differential, dA, differs from ΔA by the area of the small rectangle III.

The difference between ΔA and dA may be seen by a study of increments in general terms. We begin with the formula

$$A = lw$$

Now let us change l by Δl, and w by Δw. Then we get a change in area, ΔA.

We have $\qquad A + \Delta A = (l + \Delta l)(w + \Delta w)$

Expanding, $\qquad A + \Delta A = (l)(w) + (w)(\Delta l) + (l)(\Delta w) + (\Delta l)(\Delta w)$

Subtracting, $\qquad A \qquad\;\; = (l)(w)$

we get $\qquad\qquad \Delta A = \qquad (w)(\Delta l) + (l)(\Delta w) + (\Delta l)(\Delta w)$

$\qquad\qquad\qquad \Delta A = \qquad$ Rect. I + Rect. II + Rect. III

$\qquad\qquad\qquad dA = \qquad (w)(\Delta l) + (l)(\Delta w)$

The differential of area, dA, differs from ΔA by the product of the two small quantities, $(\Delta l)(\Delta w)$, the small rectangle III, which contains 3 sq in.

EXAMPLE 4. Let us consider an electrical problem involving Ohm's law

$$E = IR$$

Find the approximate change in voltage E when the current I changes from 5 to 5.2 amperes and the resistance R changes from 60 to 60.3 ohms.

Solution. The approximate change in E may be found by the total differential of E. This total differential, dE, while not equal exactly to the actual change, ΔE, is, for small changes in the variables, sufficiently close for practical purposes. The formula for total differential of E is

$$dE = \frac{\partial E}{\partial I}(dI) + \frac{\partial E}{\partial R}(dR)$$

In the problem as given we note that

$$dI = 0.2 \text{ ampere} \quad \text{and} \quad dR = 0.3 \text{ ohm}$$

We first find the two partial derivatives from the formula, $E = IR$:

$$\frac{\partial E}{\partial I} = R \quad \text{and} \quad \frac{\partial E}{\partial R} = I$$

596 Partial Differentiation

Substituting these values in the formula for the total differential of E, we have
$$dE = R(dI) + I(dR)$$
Substituting numerical values, we get
$$dE = (60)(0.2) + (5)(0.3)$$
$$= 12 + 1.5 = 13.5 \text{ volts}$$
The *approximate* change in voltage is, therefore, 13.5 volts. The *exact* change in voltage, ΔE, is $(5.2)(60.3) - (5)(60) = 13.56$ volts.

EXAMPLE 5. Find by the total differential the approximate error in power if the current is measured as 2.4 amperes with a possible error of ± 0.05 ampere, and the resistance is measured as 80 ohms with a possible error of ± 2 ohms.

Solution. The maximum error may be considered as a change in a variable. In this problem let us first consider errors positive, so that
$$dI = 0.05 \quad \text{and} \quad dR = 2$$
We use the formula for power,
$$P = I^2 R$$
Then the total differential of power is given by the formula
$$dP = \frac{\partial P}{\partial I}(dI) + \frac{\partial P}{\partial R}(dR)$$

The partials are $\quad \dfrac{\partial P}{\partial I} = 2IR \quad$ and $\quad \dfrac{\partial P}{\partial R} = I^2$

Therefore, $\quad dP = (2)(2.4)(80)(0.05) + (5.76)(2)$
$$= 19.2 + 11.52 = 30.72 \text{ watts}$$
The maximum error in power is therefore approximately 30.72 watts.

Of course, if one error is positive and the other negative, the approximate error in power is *less* than the 30.72 maximum error.

42.6 A FUNCTION OF MANY VARIABLES

The formula for the total differential of a function can be extended to a function of three or more variables. In such cases we hold all of the independent variables, except one, constant while we permit one to vary and find the partial with respect to the one variable. For example, if we have a function of three variables, say, x, y, and z, then we hold y and z constant while we differentiate the function with respect to x. Such a function can be represented by

$$f(x,y,z); \quad \text{or we may use another letter:} \quad V = f(x,y,z)$$

The partials may be represented as $f_x(x,y,z)$; $f_y(x,y,z)$; $f_z(x,y,z)$. If the function is represented by V, we have the partials

$$\frac{\partial V}{\partial x} \qquad \frac{\partial V}{\partial y} \qquad \frac{\partial V}{\partial z}$$

42.6 A Function of Many Variables

The total differential of the function is again given by the formula,

$$dV = \frac{\partial V}{\partial x}(dx) + \frac{\partial V}{\partial y}(dy) + \frac{\partial V}{\partial z}(dz)$$

The formula may be written in the shorter form

$$df = f_x dx + f_y dy + f_z dz$$

EXAMPLE 6. Find the approximate change in the volume of a rectangular solid when the length changes from 15 to 15.2 in., the width changes from 8 to 7.8 in., and the height changes from 12 to 12.1 in.

Solution. In this problem, $dl = 0.2$; $dw = -0.2$; $dh = 0.1$. The formula for volume is

$$V = lwh$$

Holding two of the independent variables constant and differentiating with respect to the third, we get the three partials

$$\frac{\partial V}{\partial l} = wh \qquad \frac{\partial V}{\partial w} = lh \qquad \frac{\partial V}{\partial h} = lw$$

The total differential is given by

$$dV = \frac{\partial V}{\partial l}(dl) + \frac{\partial V}{\partial w}(dw) + \frac{\partial V}{\partial h}(dh)$$

Substituting values, we have

$$dV = (wh)(dl) + (lh)(dw) + (lw)(dh)$$

Using numerical values, we get

$$dV = (8)(12)(0.2) + (15)(12)(-0.2) + (15)(8)(0.1)$$
$$dV = \quad 19.2 \quad + \quad (-36) \quad + \quad 12$$
$$= -4.8 \text{ cu in.}$$

The actual change in volume, ΔV, is equal to -5.424 cu in. The error as a result of using differentials is therefore only 0.624 cu in., which is an error of less than 0.05 percent, or less than $1/2000$.

EXAMPLE 7. Find the approximate volume of a rectangular box with the following measurements: length = 19.8 in.; width = 15.1 in.; height = 12.2 in.

Solution. We compute the volume by using the measurements, 20, 15, and 12 inches, respectively, and then find the differential of volume. We have

$$V = lwh$$

Then

$$dV = \left(\frac{\partial V}{\partial l}\right)(dl) + \left(\frac{\partial V}{\partial w}\right)(dw) + \left(\frac{\partial V}{\partial h}\right)(dh)$$

or

$$dV = (wh)(dl) + (lh)(dw) + (lw)(dh)$$
$$= (15)(12)(-0.2) + (20)(12)(0.1) + (20)(15)(0.2) = 48$$

Using the measurements, 20, 15, and 12, we get a volume of 3600 cu in. Since the differential of volume, dV, is 48, we have the approximate volume, 3648 cu in. The exact volume is 3647.556 cu in. if we use the exact measurements. The error through using differentials is less than $\frac{1}{2}$ cu in.

EXERCISE 42.2

1. A rectangle is 20 in. long and 12 in. wide. Express the rate of change of area with respect to the length. Then find the change in area when the length changes by 1.5 in.
2. A rectangular solid is 20 in. long, 15 in. wide, and 10 in. high. Express as a partial derivative the rate of change of volume with respect to height; then find the change in volume when the height changes by -0.6 in.
3. A rectangle is 18 in. long and 15 in. wide. Find the differential of area when the length changes to 18.2 in. and the width changes to 14.7 in. Find the approximate new area. Compare dA and ΔA.
4. Find the approximate change in area of a rectangle 36 in. long and 25 in. wide when the length changes to 35.82 in. and the width changes to 25.3 in. What is the exact change in area?
5. Find the approximate change in the area of a triangle having a base of 32 in. and an altitude of 20 in. when the base changes to 33.2 in. and the altitude changes to 19.6 in.
6. Using the formula, $I = E/R$, find the approximate change in current I when voltage E changes from 120 to 122.5 volts and resistance R changes from 40 to 39.4 ohms.
7. In the formula for voltage, $E = IR$, find the approximate change in voltage when current changes from 1.6 to 1.4 amperes and resistance changes from 90 to 90.15 ohms.
8. In the formula for power, $P = I^2R$, find the approximate change in power as I changes from 4.5 to 4.62 amperes and R changes from 72 to 68.6 ohms.
9. In the formula for power, $P = E^2/R$, find the approximate change in power when E changes from 150 to 152.4 volts and R changes from 90 to 93.6 ohms. What is the exact change in power?
10. In the formula for power, $P = I^2R$, the current I is measured as 6.4 amperes with a possible error of ± 0.2, and resistance R is measured as 80 ohms with a possible error of ± 5 ohms. Find the approximate maximum error in power P.
11. A rectangular solid is measured as 29.6 in. long, 19.7 in. wide, and 15.2 in. high. Find the approximate volume by taking $l = 30$, $w = 20$, and $h = 15$, and using differentials to find dV. What is the exact volume?
12. A rectangular solid is measured as 40 in. long, 30 in. wide, and 16 in. high. If the maximum error in each measurement is $\frac{1}{2}$ inch, what is the maximum error in the volume?

Evaluate the total differential in each of the following for given values:

13. $Z = 3x^2 + 4xy - 3y^2 + 1$; $x = 2$, $y = 3$, x becomes 2.06, y becomes 3.02.
14. $Z = 2x^2 + 2xy - 3y^2$; x changes from 4 to 3.96, y changes from 2 to 1.98.
15. $Z = x^3 - x^2y + 2y^3$; x changes from 4 to 3.96, y changes from 3 to 3.03.
16. $Z = 2x^2 - 3xy + 3y^2$; x changes from 4 to 3.97, y changes from 2 to 2.04.
17. Find the approximate volume of a right circular cylinder having a radius of 5.97 in. and a height of 20.2 in. (Consider the differences from $r = 6$ and $h = 20$.) $V = \pi r^2 h$.
18. Find the approximate volume of a right circular cone having a base-radius of 7.95 in. and an altitude of 16.1 in. $V = \frac{1}{3}\pi r^2 h$.
19. The total area of a right circular cylinder including lateral area and area of both ends is given by the formula, $A = 2\pi rh + 2\pi r^2$. Find the approximate total area of the cylinder in Problem 17.
20. The total resistance of two resistors, R_1 and R_2, connected in parallel is given by the formula, $R_t = R_1R_2/(R_1 + R_2)$. Find the maximum total error in the combined resistance, R_t, if R_1 and R_2 are measured as 20 ohms and 30 ohms, respectively, with a possible error of ± 0.4 ohm in each.

42.7 THE CHAIN RULE IN PARTIAL DIFFERENTIATION

In Chapter 14 we saw how to find the derivative of a function of a function by the so-called *chain rule*. For example, if $y = u^3$, then

$$\frac{dy}{du} = 3u^2$$

Now, if u is some function of x, we can also find the derivative with respect to x; that is, multiplying both sides of the equation by du/dx,

$$\left(\frac{dy}{du}\right)\left(\frac{du}{dx}\right) = 3u^2\left(\frac{du}{dx}\right), \quad \text{or} \quad \frac{dy}{dx} = 3u^2\left(\frac{du}{dx}\right)$$

By the chain rule,
$$\left(\frac{dy}{du}\right)\left(\frac{du}{dx}\right) = \frac{dy}{dx}$$

The same type of rule is applicable to functions of two or more variables. For example, suppose we have the function

$$Z = f(x,y)$$

Now, if x and y are both functions of a third variable, say t, then we can find the derivative of Z with respect to t. Let us first derive the formula. We begin with the total differential of two functions, which we have used:

$$dZ = \frac{\partial Z}{\partial x}(dx) + \frac{\partial Z}{\partial y}(dy)$$

Now, if x and y are both functions of t, to get the required formula, we divide both sides of this equation by dt, and get the formula

$$\frac{dZ}{dt} = \frac{\partial Z}{\partial x}\frac{dx}{dt} + \frac{\partial Z}{\partial y}\frac{dy}{dt}$$

EXAMPLE 8. If $Z = x^2 + 3xy - 5y^2$, $x = t^2 - 3$, and $y = 2 - t^3$, find dZ/dt.

Solution. Finding partials, we get

$$\frac{\partial Z}{\partial x} = 2x + 3y \qquad \frac{\partial Z}{\partial y} = 3x - 10y \qquad \frac{dx}{dt} = 2t \qquad \frac{dy}{dt} = -3t^2$$

Substituting values in the formula,

$$\frac{dZ}{dt} = (2x + 3y)(2t) + (3x - 10y)(-3t^2)$$

If we wish to express this derivative entirely in terms of t, we can substitute the values of x and y and get

$$\frac{dZ}{dt} = 4t^3 - 15t^4 + 87t^2 - 30t^5$$

42.8 DERIVATIVE OF IMPLICIT FUNCTIONS BY PARTIALS

In Chapter 19 we saw that it is possible to find the derivative, dy/dx, of implicit functions by implicit differentiation. We shall see that partial differentiation can be used to find the derivative of such functions in a somewhat simpler manner. For example, suppose we have the function

$$x^2 - 3xy - 4y^2 + 5x - 7y + 3 = 0$$

Now let us see how we might get the derivative by the use of partials. First we derive a formula. Again, we begin with the formula for the total differential. Let us represent the function by Z; that is,

$$Z = x^2 - 3xy - 4y^2 + 5x - 7y + 3$$

Then

$$dZ = \frac{\partial Z}{\partial x}(dx) + \frac{\partial Z}{\partial y}(dy)$$

Now, if the function Z is equal to zero, then also $dZ = 0$, and we have

$$\frac{\partial Z}{\partial x}dx + \frac{\partial Z}{\partial y}dy = 0$$

Dividing both sides by dx,

$$\frac{\partial Z}{\partial x} \cdot \frac{dx}{dx} + \frac{\partial Z}{\partial y} \cdot \frac{dy}{dx} = 0$$

Solving this equation for dy/dx, we get the formula

$$\frac{dy}{dx} = -\frac{\partial Z/\partial x}{\partial Z/\partial y}$$

EXAMPLE 9. If $x^2 - 3xy - 4y^2 + 5x - 7y + 3 = 0$ find dy/dx by partials.

Solution. If we let Z represent the function, we have

$$\frac{\partial Z}{\partial x} = 2x - 3y + 5 \quad \text{and} \quad \frac{\partial Z}{\partial y} = -3x - 8y - 7$$

Then

$$\frac{dy}{dx} = -\frac{2x - 3y + 5}{-3x - 8y - 7} = \frac{2x - 3y + 5}{3x + 8y + 7}$$

If we differentiate each term with respect to x by the method of implicit differentiation as explained in Chapter 19, we get the same result.

EXERCISE 42.3

Find dZ/dt in Problems 1–6.

1. $Z = x^2 + 4xy - 5y^2 + 2x - 3y + 5$; $x = 2 - 3t$; $y = t^2 + 1$
2. $Z = x^3 - 2x^2y - 3xy^2 + y^3 - 4xy$; $x = t^2$; $y = t^3 - 1$
3. $Z = 3x^2 + xy + 2y^2 - 5x + 3y$; $x = 4 - t^2$; $y = 3 - 2t$
4. $Z = x^2 + xy + 3y^2 - 4y$; $x = \sin 3t$; $y = \cos 2t$
5. $Z = x^2 - 3xy - 4y^2 + 5x$; $x = e^{2t}$; $y = e^{-3t}$
6. $Z = x^3 - 5x^2y + xy^2 - y^3 + x^2 - 4y$; $x = 1 - 2t$; $y = 3t + 2$

Find dF/dt in Problems 7–12.

7. $F(x,y) = x^2 + 2xy - y^2$; $x = \cos t$; $y = \sin t$
8. $F(x,y) = x^2 - xy + y^2$; $x = \cot t$; $y = \csc t$
9. $F(x,y) = \ln(x + y)$; $x = t^3$; $y = 2 - 5t$

10. $F(x,y) = e^{x^2y}$; $x = \tan t$; $y = \sec t$
11. $F(u,v) = u^2 - 2uv + 4v^2 - 3u + 5v$; $u = t^2 - 2$; $v = 4 - 5t$
12. $F(u,v) = u^3 - 3u^2v + 5uv^2$; $u = e^{3t}$; $v = e^{-2t}$

By use of partials, find dy/dx in each of the following equations, and find the slope of the curve at the given point. Identify the curve.

13. $2x^2 - 3y^2 - 4x + 5y - 6 = 0$; point $(-4,-3)$
14. $3x^2 + 2xy - 4y^2 + 5x - 3y + 22 = 0$; point $(-3,2)$
15. $2x^2 - 3xy - 4y^2 + 5x - 7y + 7 = 0$; point $(-2,1)$
16. $x^2 - 4xy + 5y^2 - 7x + 3y = 0$; point $(2,-1)$
17. $x^2 + 3xy + y^2 - 8x - 5y + 36 = 0$; point $(5,-3)$
18. $x^2 - 2xy + y^2 - 5x + 3y + 4 = 0$; point $(3,-1)$
19. $x^2 + 2xy + 4y^2 - 2x + 4y + 3 = 0$; point $(3,-1)$
20. $x^2 + 4xy + 4y^2 - 5y - 5x = 0$; point $(2,-1)$
21. $x^2 - 3xy + 4y^2 - 4x - 2y + 8 = 0$; point $(2,1)$

42.9 THE SECOND PARTIAL DERIVATIVE

It is sometimes necessary to make use of what is called the *second partial derivative*. The second partial corresponds to the second derivative of functions of only one independent variable. To find the second partial, we take the derivative of the first partial. We have seen how this is done when we have a single independent variable. For example, if $y = 7x^3$, then

$$\frac{dy}{dx} = 21x^2 \quad \text{and} \quad \frac{d^2y}{dx^2} = 42x$$

We have emphasized that in finding the derivative of a function, we must be careful to identify the *independent variable*. The expression, dy/dx, should be understood to mean "the derivative of the function y with respect to x as the independent variable." Also the expression

$$\frac{d^2y}{dx^2}$$

should be understood to mean and should be read, "the second derivative of the function y with respect to x as the independent variable *both times*."

In the case of a function of two independent variables, we hold one constant and then find the partial derivative with respect to the other. It is important that we note which variable is assumed to vary. Now, when we differentiate the function a second time (take the partial of the partial), we must again indicate which variable is assumed to vary and which is to be held constant. For the second partial, the independent variable may be the same as for the first, but it need not necessarily be the same.

Take the example

$$Z = 3x^2 - 2xy + 4y^2 - 5x + 6y$$

Let us first find the partial with respect to x:

$$\frac{\partial Z}{\partial x} = 6x - 2y - 5$$

To find the second partial, that is, the partial derivative of the first partial, we can again take x as the independent variable, and write

$$\frac{\partial}{\partial x}\left(\frac{\partial Z}{\partial x}\right) = \frac{\partial}{\partial x}(6x - 2y - 5) = 6$$

To denote the second partial derivative of Z, taking x as the independent variable for both differentiations, we use the symbol

$$\frac{\partial^2 Z}{\partial x^2}$$

This symbol means and should be read, "the second partial derivative of Z with respect to x *both times*."

However, it is sometimes necessary to change the independent variable when finding the second partial. In the foregoing example, we have the first partial

$$\frac{\partial Z}{\partial x} = 6x - 2y - 5$$

Suppose now that we wish to consider y as the independent variable for the second partial. Then we have

$$\frac{\partial}{\partial y}\left(\frac{\partial Z}{\partial x}\right) = \frac{\partial}{\partial y}(6x - 2y - 5) = -2$$

In this case the second partial with respect to y is denoted by the symbol

$$\frac{\partial^2 Z}{\partial y\, \partial x} = -2$$

One interesting and useful fact emerges from the change in the independent variable in finding the second partial derivative. Consider the function

$$Z = 3x^2 - 2xy + 4y^2 - 5x + 6y$$

If we take x as the independent variable for the first partial and then change to y as the independent variable for the second partial, we get, as we have seen,

$$\frac{\partial^2 Z}{\partial y\, \partial x} = -2$$

Now suppose we reverse the order of independent variables for the first and second partials. We take y as the independent variable for the first partial and then change to x for the second partial. We get

$$\frac{\partial Z}{\partial y} = -2x + 8y + 6$$

then

$$\frac{\partial^2 Z}{\partial x\, \partial y} = \frac{\partial}{\partial x}(-2x + 8y + 6) = -2$$

42.9 The Second Partial Derivative

The answer turns out to be the same for the reversed order of variables; that is,

$$\frac{\partial^2 Z}{\partial y\, \partial x} = \frac{\partial^2 Z}{\partial x\, \partial y}$$

The foregoing equality is true provided the function is continuous.

Various notations, in addition to the notation already shown, are used to denote the second partials of a function. The function itself may be indicated by

$$f(x,y)$$

Then the partials with respect to x and y, respectively, are indicated by

$$f_x(x,y) \quad \text{and} \quad f_y(x,y)$$

The second partials may then be indicated as follows:

$$\frac{\partial^2 f(x,y)}{\partial x^2} = f_{xx}(x,y) \qquad \frac{\partial^2 f(x,y)}{\partial y^2} = f_{yy}(x,y)$$

$$\frac{\partial^2 f(x,y)}{\partial y\, \partial x} = f_{xy}(x,y) \qquad \frac{\partial^2 f(x,y)}{\partial x\, \partial y} = f_{yx}(x,y)$$

Note especially the order of the variables in the notations for the second partials:

$$\frac{\partial^2 f(x,y)}{\partial y\, \partial x} = f_{xy}(x,y)$$

This notation means that we must differentiate first with respect to x, and then with respect to y. The meaning might be more clear if we write

$$\frac{\partial}{\partial y}\left[\frac{\partial f(x,y)}{\partial x}\right] \quad \text{or} \quad [f_x]_y(x,y)$$

Note. The foregoing symbols are often written simply as follows:

$$f_x \quad f_{xx} \quad f_y \quad f_{yy} \quad f_{xy} \quad f_{yx}$$

EXAMPLE 10. For the following function, find $f_{xx}, f_{yy}, f_{xy}, f_{yx}$.

$$f(x,y) = x^3 - 5x^2 y + 4xy^2 + 3x$$

Solution. For the first partial with respect to x, we have

$$f_x(x,y) = 3x^2 - 10xy + 4y^2 + 3$$

Now we again take x as the independent variable, and get

$$f_{xx}(x,y) = 6x - 10y$$

For the partials with respect to y, we have

$$f_y(x,y) = -5x^2 + 8xy$$

Then

$$f_{yy}(x,y) = 8x$$

Partial Differentiation

When we reverse the order of independent variables, we must observe carefully the notation. To find

$$\frac{\partial^2 f(x,y)}{\partial y\, \partial x} = f_{xy}(x,y)$$

we differentiate first with respect to x and get

$$f_x(x,y) = 3x^2 - 10xy + 4y^2 + 3$$

Differentiating next with respect to y, we get

$$f_{xy}(x,y) = -10x + 8y$$

To find $f_{yx}(x,y)$, we first find the partial with respect to y:

$$f_y(x,y) = -5x^2 + 8xy$$

Differentiating next with respect to x, we get

$$f_{yx}(x,y) = -10x + 8y$$

Here again we see that the result is the same if the order of the variables is reversed in taking successive partials; that is, in abbreviated form,

$$f_{xy} = f_{yx}$$

EXERCISE 42.4

In the following functions, find the first partial with respect to one independent variable and then the second partial with respect to the other independent variable, and show that the final result is the same if the order of the partials is reversed.

1. $Z = 5x^2 - 7xy + 4y^2 + 3x - 2y$
2. $Z = 4x^3 - 3x^2y - 7xy^2 + 5y^3$
3. $Z = x^3 + 5x^2y - 3xy^2 + 4y^3$
4. $Z = 3x^4 - 5x^3y - x^2y^2 - 7y^4$
5. $Z = 2x^3 - 3xy^2 + 4y^3 - y^4$
6. $U = x^3 - 2x^2y + 4xy^2 - 5y^3$
7. $f(x,y) = 3x^4 + 4x^3y - 6xy^2 - 7xy + 5x + 3y - 4$
8. $f(x,y) = 5x^3 - 3x^2y + 4xy^2 - 2y^3 + 7xy - 6x - 3y$
9. $f(s,t) = s^4 - 3s^3t + 2s^2t^2 - 5st^3 - st^2 - st$
10. $P(x,y) = 6x^3 - 3x^2y - 2xy^2 - 5xy + 4x - 3y + 7$
11. $Z = e^{2x+3y}$
12. $Z = x^2 e^{-y}$
13. $Z = \sin 2x \cos 3y$
14. $Z = \arctan \dfrac{x}{y}$
15. If $Z = e^x \sin y$, find $\dfrac{\partial^2 Z}{\partial x^2} + \dfrac{\partial^2 Z}{\partial y^2}$
16. Find $\dfrac{\partial^2 Z}{\partial x^2} + \dfrac{\partial^2 Z}{\partial y^2}$ if $Z = y e^x + x e^y$.

In the following functions, find f_x, f_y, f_{xx}, f_{yy}, f_{xy}, and f_{yx}; and evaluate each first and second partial for the given values of the variables:

17. $f(x,y) = x^3 + 4x^2y - 3xy^2 + 2x^2 - 5y^2 + 3x$; $x = 1, y = 2$
18. $f(x,y) = 5x^3 - 2x^2y - 4xy^2 + y^3 - 5x - 2y$; $x = -2, y = 1$
19. $f(x,y) = 3x^4 + 2x^3y - 5x^2y^2 - 3xy^3 - y^4$; $x = 1, y = -1$
20. $f(x,y) = x^4y - 3x^3y^2 - 2x^2y^3 - 3xy^4 - xy$; $x = -1, y = 2$
21. $f(x,y) = e^{3x+4y}$; $x = 2, y = -1$
22. $f(x,y) = \sin 2x \cos 3y$; $x = \pi/4; y = \pi/3$
23. $f(x,y) = \ln(x+y)$; $x = 3, y = -1$
24. $f(x,y) = e^{2x} \cos y$; $(0, \pi/3)$
25. $f(x,y) = x^2 \sin 4y$; $x = 2, y = 0$

42.10 INTEGRATION OF FUNCTIONS OF TWO VARIABLES

We have seen that when we find the partial derivative of a function of two variables, we consider one of the variables as a constant. In a similar manner we may also integrate a function of two variables by holding one variable constant. If we have more than two independent variables, we hold all but one of them constant and consider only one of them as a variable.

EXAMPLE 11. Integrate the following with respect to x as the variable, holding y as a constant:

$$\int (5 + 4x - 8xy - 6y^2)\, dx$$

Solution. To show that we wish to integrate with respect to x, we sometimes place a small circled x at the bottom of the integral sign. However, in any case, the differential dx next to the function to be integrated indicates the variable of integration. We get

$$\int_{\text{\textcircled{x}}} (5 + 4x - 8xy - 6y^2)\, dx = 5x + 2x^2 - 4x^2 y - 6xy^2 + C$$

EXAMPLE 12. Integrate the following with respect to y as the variable, holding x as a constant. Evaluate the definite integral between limits.

$$\int_1^x (5 + 4x - 8xy - 6y^2)\, dy$$

Solution. Holding x constant and integrating with respect to y, we get

$$\int_1^x (5 + 4x - 8xy - 6y^2)\, dy = 5y + 4xy - 4xy^2 - 2y^3 \big]_1^x$$

Now we evaluate the result by substituting y-limits: $y = 1$ to $y = x$; we get

$$5x + 4x^2 - 4x^3 - 2x^3 - 5 - 4x + 4x + 2 = 5x + 4x^2 - 6x^3 - 3$$

42.11 DOUBLE INTEGRATION

It is sometimes necessary or desirable to integrate a function of two variables twice, reversing the variables as we sometimes do in finding second partial derivatives. We first integrate with respect to one variable, holding the other constant. When this first integral is evaluated, the result is integrated again but with respect to the variable which was first held constant. The entire procedure is called *double integration*.

Double integration is very useful in many types of problems, such as areas, moments, volumes, and other physical problems. Triple and other multiple integration is simply an extension of double integration.

To show the method of double integration, we shall first work a problem in two steps.

Partial Differentiation

EXAMPLE 13. Integrate the following function with respect to y as the variable of integration, and evaluate between the limits, $y = 1$ to $y = x$. Then integrate the result with respect to x and evaluate between x-limits, 0 and 3:

$$\int_1^x (6y - 2xy)\, dy$$

Solution. Integrating with respect to y, and evaluating, we get

$$\int_1^x (6y - 2xy)\, dy = 3y^2 - xy^2\Big]_1^x = 3x^2 - x^3 - 3 + x$$

Now we integrate the result with respect to x and evaluate between x-limits:

$$\int_0^3 (3x^2 - x^3 - 3 + x)\, dx = x^3 - \frac{x^4}{4} - 3x - \frac{x^2}{2}\Big]_0^3 = \frac{9}{4}$$

This problem might originally have been written as

$$\int_0^3 \left[\int_1^x (6y - 2xy)\, dy\right] dx$$

The expression means that we must first integrate and evaluate the innermost integrand and then integrate and evaluate the result. The brackets are usually omitted and the expression written

$$\int_0^3 \int_1^x (6y - 2xy)\, dy\, dx$$

EXAMPLE 14. Evaluate the double integral

$$\int_1^4 \int_1^{\sqrt{y}} (6xy - 4x)\, dx\, dy$$

Solution. We first evaluate the innermost integral. Note that the innermost differential is dx. This means that for the first integration, we take x as the variable and hold y constant. We get

$$\int_1^4 \left[3x^2y - 2x^2\right]_1^{\sqrt{y}} dy = \int_1^4 (3y^2 - 2y - 3y + 2)\, dy$$

$$= \int_1^4 (3y^2 - 5y + 2)\, dy$$

Now, integrating with respect to y as the independent variable, we get

$$\int_1^4 (3y^2 - 5y + 2)\, dy = y^3 - \frac{5y^2}{2} + 2y\Big]_1^4$$

$$= 64 - 40 + 8 - 1 + \frac{5}{2} - 2 = 31\tfrac{1}{2}$$

EXAMPLE 15. Find by double integration the area between the following curves:

$$y^2 = 4x \qquad 2x - y = 4$$

Then for this area find the moment of inertia and the radius of gyration with reference to the y-axis; that is, I_y and R_y.

Solution. We first take a horizontal strip of area (Fig. 42.5). The width of this strip is dy. Next, we take a small vertical element of the horizontal strip. The height of the small element is dy, and the width is dx. Then this small element has an area of $(dx)(dy)$. We shall first integrate with respect to x to sum up the small elements over the domain, x_1 to x_2. By the second integration we add the strips over the range, y_1 to y_2. Solving the equations, we get $x_1 = y^2/4$, and $x_2 = 2 + y/2$. For y-limits we get $y_1 = -2$, and $y_2 = 4$. Now we set up the double integral:

$$A = \int_{-2}^{4}\int_{x_1}^{x_2} dx\, dy = \int_{-2}^{4}\Big[x\Big]_{x_1}^{x_2} dy = \int_{-2}^{4}\left(2 + \frac{y}{2} - \frac{y^2}{4}\right) dy$$

$$= 2y + \frac{y^2}{4} - \frac{y^3}{12}\bigg]_{-2}^{4} = 9$$

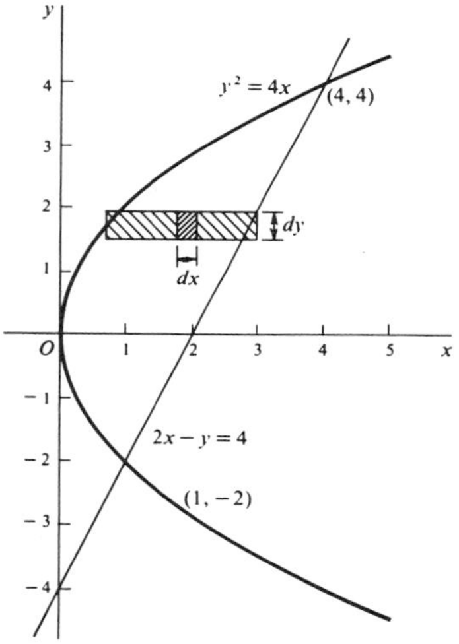

Fig. 42.5

To find I_y, we note that the arm length for the small element is x. For the moment of inertia of the small element we have

$$x^2\, dx\, dy$$

Then

$$I_y = \int_{-2}^{4}\int_{x_1}^{x_2} x^2\, dx\, dy = \int_{-2}^{4}\left[\frac{1}{3}x^3\right]_{x_1}^{x_2} dy = \frac{1}{3}\int_{-2}^{4}(x_2^3 - x_1^3)\, dy$$

$$= \frac{1}{3}\int_{-2}^{4}\left[\left(\frac{y}{2}+2\right)^3 - \left(\frac{y^2}{4}\right)^3\right] dy = \frac{1}{3}\int_{-2}^{4}\left(\frac{y^3}{8} + \frac{3y^2}{2} + 6y + 8 - \frac{y^6}{64}\right) dy$$

$$= \frac{1}{3}\left[\frac{y^4}{32} + \frac{y^3}{2} + 3y^2 + 8y - \frac{y^7}{448}\right]_{-2}^{4} = \frac{423}{14}$$

Then $\quad R_y^2 = \dfrac{423}{14} \div 9 = \dfrac{47}{14} \qquad R_y = \sqrt{\dfrac{47}{14}} = 1.832$ (approx.)

EXERCISE 42.5

In Problems 1–10 integrate with respect to the indicated variable of integration and evaluate as indicated.

1. $\int (4 + 6xy - y + 3x^2 - 6y^2)\, dx$
2. $\int (6x - 5y - 4)\, dx$
3. $\int (4 + 4xy - y + 3x^2 - 6y^2)\, dy$
4. $\int (-5x + 4y - 3)\, dy$
5. $\int (6x^2 + 10xy - 3xy^2 - 4x)\, dx$
6. $\int (5x^2 - 6xy - 8y)\, dy$
7. $\int (12x^2y^3 - 4xy^3)\, dx$
8. $\int (12x^2y^3 - 4xy^3)\, dy$
9. $\int_0^x (5 + 2xy - 3x^2 + 2y)\, dy$
10. $\int_0^y (3 - 4x + 5y + 2y^2)\, dx$

Evaluate the double integrals in Problems 11–16.

11. $\int_0^4 \int_1^x 6xy^2\, dy\, dx$
12. $\int_0^2 \int_0^{2y} 12xy^2\, dx\, dy$
13. $\int_0^1 \int_0^x e^{x+y}\, dy\, dx$
14. $\int_1^2 \int_0^{\pi} \sin y\, dy\, dx$
15. $\int_0^2 \int_1^{\sqrt{x}} 12x^2y\, dy\, dx$
16. $\int_1^2 \int_2^{x^2} (4x - 10y)\, dy\, dx$

Use double integration to solve Problems 17–21.
17. For the area bounded by the curves, $y^2 = 4x$ and $y = x$, find I_x and R_x.
18. For the area bounded by the curves, $y^2 = x$ and $x - y = 2$, find I_y and R_y.
19. For the area bounded by $y = 2x$, $x = 3y$, and $x = 3$, find I_x and R_x.
20. For the area bounded by $x = y$, $y = 2x$, and $y = 6$, find I_y and R_y.
21. For the area bounded by $x = y$ and $y = 4x - x^2$, find I_x and R_x.

chapter

43

Infinite Series

43.1 A SEQUENCE AND A SERIES

A *sequence* is a succession of terms in which each term is formed according to some fixed rule. A *series* is the indicated *sum* of the terms of a sequence. For example, the following are sequences:

$$\text{(a) } 1, 3, 5, 7, 9, 11 \qquad \text{(b) } 1, \frac{1}{2}, \frac{1}{3}, \frac{1}{4}, \frac{1}{5}$$

When we indicate the sum of the terms in a sequence, we have a series, as

$$\text{(a) } 1 + 3 + 5 + 7 + 9 + 11 \qquad \text{(b) } 1 + \frac{1}{2} + \frac{1}{3} + \frac{1}{4} + \frac{1}{5}$$

Each of the series shown has a finite number of terms. Then the sum can be found simply by adding the terms. The first series (a) has the sum 36. The second (b) has the sum 137/60. These series are *finite*.

If the number of terms of a series is understood to be unlimited, the series is said to be *infinite*. To denote that a series is infinite, we usually write a few terms of the series and then three dots to indicate that the series does not end with the terms shown. To indicate that the above series are infinite, we might write

$$\text{(a) } 1 + 3 + 5 + 7 + 9 + \cdots \qquad \text{(b) } 1 + \frac{1}{2} + \frac{1}{3} + \frac{1}{4} + \frac{1}{5} + \cdots$$

In an infinite series it is impossible to find the sum as we usually understand the meaning of the word *sum* because no matter how many terms we might add, there would always be more terms to be added. However, we shall see that a special meaning may be given to the *sum* of an infinite series.

The terms of a series can be numbered, as the first, the second, the third, and so on. That is, the terms may be put into a one-to-one correspondence with the natural numbers. When we wish to denote the general term of a series, we call it the *n*th term. Then the general term is described in terms of *n*, the *number* of the term. In general, the terms of an infinite series are often denoted by the letter u with subscripts indicating the number of the term, as

$$u_1 + u_2 + u_3 + u_4 + \cdots + u_n$$

In writing the terms of a series we often make use of the factorial symbol, $n!$. This symbol indicates the product of all positive integers from 1 to n, inclusive. It is read "n factorial" or "factorial n." For example, the symbol "4!" is read "4 factorial" and means

$$1 \cdot 2 \cdot 3 \cdot 4 = 24$$

In problems involving factorials, the meaning must be carefully observed. For example,

$$\frac{6!}{5!} = 6, \qquad (6!)(7) = 7!, \qquad \frac{(n+2)!}{n!} = (n+1)(n+2)$$

In some problems it is necessary to use the definition: $0! = 1$.

The first few terms of a series will usually show the law or rule by which the terms are formed. Sometimes it is rather easy to discover the rule. In other cases it is much more difficult. In trying to discover the rule for the general term where fractions are involved, it is best to analyze the numerator and the denominator separately. The following examples are meant to illustrate some of the approaches that may be used in determining the general term.

EXAMPLE 1. Write the general term of the infinite series:

$$2 + 5 + 8 + 11 + 14 + \cdots$$

Solution. We might first note that this is an arithmetic progression in which the first term is 2 and the common difference is 3. Then, using the rule for the nth term of an arithmetic progression, we get

$$2 + (n-1)3 \quad \text{or} \quad 3n - 1$$

The general term is then $(3n - 1)$. We find that this expression holds true for any term in the series. For example, the fifth term is $(3)(5) - 1 = 14$.

EXAMPLE 2. Find the general term of the series

$$1 + \frac{1}{2} + \frac{1}{3} + \frac{1}{4} + \frac{1}{5} + \cdots$$

Solution. We first observe that the numerator of each fraction is 1. The denominator of each fraction represents the number of the term. Then the nth or general term is $1/n$.

EXAMPLE 3. Write the notation for the general term of the series

$$\frac{1}{2} + \frac{1 \cdot 2}{6} + \frac{2 \cdot 3}{24} + \frac{3 \cdot 4}{120} + \frac{4 \cdot 5}{720} + \cdots$$

Solution. Now we wish to describe the general term using n as the number of the term. First, we might notice that the denominators can be written as factorials:

$$\frac{1}{2!} + \frac{1 \cdot 2}{3!} + \frac{2 \cdot 3}{4!} + \frac{3 \cdot 4}{5!} + \frac{4 \cdot 5}{6!} + \cdots$$

To describe the general term, let us now look at some particular term, such as the fourth term:

$$\frac{3 \cdot 4}{5!}$$

The factorial number of the denominator is 5, which is one more than the number of the term. This rule we note also holds true for the other terms. Then the denominator can be denoted in general by

$$(n + 1)!$$

The numerator in the fourth term has the factors, 3 and 4. This is the product of the number of the term and one less than the number. Then for the numerator it appears that we might write $(n - 1)(n)$. Checking this formula in the other terms, we find that it holds true for all the numerators. Now we are in a position to write the general term:

$$\frac{(n - 1)(n)}{(n + 1)!}$$

To indicate the sum of an infinite series we sometimes use the summation sign, the Greek letter *sigma* (Σ) before the general term with the limits of n written below and above the summation sign. For example, the infinite series in Example 3 can be indicated by

$$\sum_{n=1}^{\infty} \frac{(n - 1)n}{(n + 1)!}$$

If the general term is shown together with the lower limit of n, then the first and successive terms of the series can be written by using successive integral values of n in the expression for the general term.

EXAMPLE 4. Write the first five terms of the series indicated by

$$\sum_{n=1}^{} \frac{2n + 1}{n^2 + 3}$$

Solution. For the first term, we have $n = 1$. When $n = 1$, the first term becomes

$$\frac{2 + 1}{1^2 + 3} = \frac{3}{4}$$

When $n = 2$, we have, for the second term

$$\frac{2(2) + 1}{2^2 + 3} = \frac{5}{7}$$

For the first five terms we have

$$\frac{3}{4} + \frac{5}{7} + \frac{7}{10} + \frac{9}{19} + \frac{11}{28}$$

Infinite Series

EXERCISE 43.1

Write the general term in each of the following series. Then write the next two terms after those given. (In Problem 5, write $\frac{1}{3}$ as $\frac{3}{9}$.)

1. $\dfrac{1}{2} + \dfrac{2}{3} + \dfrac{3}{4} + \dfrac{4}{5} + \cdots$
2. $\dfrac{1}{3} + \dfrac{2}{5} + \dfrac{3}{7} + \dfrac{4}{9} + \dfrac{5}{11} + \cdots$
3. $1 + \dfrac{1}{2} + \dfrac{1}{4} + \dfrac{1}{8} + \dfrac{1}{16} + \cdots$
4. $2 + \dfrac{3}{4} + \dfrac{4}{9} + \dfrac{5}{16} + \dfrac{6}{25} + \cdots$
5. $\dfrac{1}{4} + \dfrac{1}{3} + \dfrac{5}{16} + \dfrac{7}{25} + \dfrac{9}{36} + \cdots$
6. $\dfrac{1}{1 \cdot 2} + \dfrac{1}{2 \cdot 3} + \dfrac{1}{3 \cdot 4} + \dfrac{1}{4 \cdot 5} + \cdots$
7. $\dfrac{3}{1 \cdot 2} + \dfrac{5}{3 \cdot 4} + \dfrac{7}{5 \cdot 6} + \dfrac{9}{7 \cdot 8} + \cdots$
8. $1 + \dfrac{2}{3} + \dfrac{4}{9} + \dfrac{8}{27} + \dfrac{16}{81} + \cdots$
9. $\dfrac{1}{4} + \dfrac{5}{8} + \dfrac{9}{16} + \dfrac{13}{32} + \dfrac{17}{64} + \cdots$
10. $\dfrac{2}{5} + \dfrac{5}{7} + \dfrac{8}{9} + 1 + \dfrac{14}{13} + \dfrac{17}{15} + \cdots$
11. $1 + 1 + \dfrac{7}{9} + \dfrac{5}{8} + \dfrac{13}{25} + \dfrac{4}{9} + \cdots$ (Hint: Write the second 1 as 4/4.)
12. $\dfrac{1}{2} + \dfrac{2}{3} + \dfrac{3}{8} + \dfrac{2}{15} + \dfrac{5}{144} + \cdots$ (Hint: Change the fractions so that denominators may be written as factorials.)

Write the first five terms of each of the following series ($n = 1$ to $n = 5$):

13. $\sum\limits_{n=1}^{} \dfrac{n}{2n+3}$
14. $\sum\limits_{n=1}^{} \dfrac{n+1}{2n+3}$
15. $\sum\limits_{n=1}^{} \dfrac{2n-1}{2^n}$
16. $\sum\limits_{n=1}^{} \dfrac{n+3}{n^2}$
17. $\sum\limits_{n=1}^{} \dfrac{3n-1}{2n+3}$
18. $\sum\limits_{n=1}^{} \dfrac{3n}{(n+1)^2}$
19. $\sum\limits_{n=1}^{} \dfrac{n^2}{n!}$
20. $\sum\limits_{n=1}^{} \dfrac{2n+1}{n^2+n}$
21. $\sum\limits_{n=1}^{} \dfrac{n^2+1}{(n+1)(n+2)}$
22. $\sum\limits_{n=1}^{} \dfrac{2n-1}{(n-1)!}$
23. $\sum\limits_{n=1}^{} \dfrac{4n-3}{\sqrt{n}}$
24. $\sum\limits_{n=1}^{} \dfrac{n^2+1}{(n+2)!}$

43.2 CONVERGENT AND DIVERGENT SERIES

An infinite series has no *sum* as we usually understand the meaning of the word *sum*. However, if the sum of the terms approaches a limit, this limit is called the *sum* of the series. Then the series is called a *convergent* series. For example, take the *geometric* series

$$1 + \dfrac{1}{2} + \dfrac{1}{4} + \dfrac{1}{8} + \dfrac{1}{16} + \cdots + \dfrac{1}{2^{n-1}}$$

Here the common ratio is 1/2. In algebra we learn that the sum of any *geometric series approaches a limit if the ratio is less than 1*. In this example the sum of the series approaches 2 as a limit as the number of terms becomes infinitely large. Then we say the sum of the series is 2, or the series converges to 2, although we should never reach 2 for any finite number of terms. We usually express this result as

$$\lim_{n \to \infty} \text{Sum}_n = 2$$

For any geometric series in which $r < 1$, we have

$$\lim_{n \to \infty} S_n = \frac{a}{1-r} \quad \text{(where a is the first term)}$$

If a series is not convergent, that is, if the sum does not approach a limit as the number of terms increases without bound, then the series is called a *divergent* series. As an example, the following series is divergent:

$$1 + 3 + 5 + 7 + \cdots + (2n - 1)$$

The sum in this case does *not* approach a limit as n becomes infinite. Note that this series is an *arithmetic progression*.

Infinite series have certain important uses in mathematics. However, if a series is to be useful, it must be convergent. A divergent series has no practical value except as a test for a given series whose character is unknown. Any given series therefore must be tested for convergence or divergence. There are several ways in which a series may be tested.

43.3 COMPARISON TESTS

A given series may often be tested by comparing it term by term with a series that is known to be convergent or divergent. If each term of the series to be tested is always *less* than the corresponding term of a series known to be *convergent*, then it is reasonable that the given series is also convergent. For example, if a series is known to converge to the sum of, say, 4, then another series whose corresponding terms are always less than this convergent series will surely have a sum not to exceed 4.

Suppose we wish to test the following series:

$$\text{(a)} \quad 1 + \frac{1}{4} + \frac{1}{12} + \frac{1}{32} + \frac{1}{80} + \cdots$$

Now let us compare this series, term by term, with the geometric series, having $r = \frac{1}{2}$:

$$\text{(b)} \quad 1 + \frac{1}{2} + \frac{1}{4} + \frac{1}{8} + \frac{1}{16} + \cdots + \frac{1}{2^{n-1}}$$

It can be seen that each term after the first in the given series (a) is less than the corresponding term of the geometric series (b) that is known to be convergent. Therefore, the given series (a) also converges. The comparison might be more clear if we write the general term of the given series (a). The series might be written:

$$\frac{1}{1 \cdot 2^0} + \frac{1}{2 \cdot 2^1} + \frac{1}{3 \cdot 2^2} + \frac{1}{4 \cdot 2^3} + \frac{1}{5 \cdot 2^4} + \frac{1}{6 \cdot 2^5} + \cdots + \frac{1}{(n)(2^{n-1})}$$

Note that the denominators of this series continue to become larger than the denominators in the geometric series. Therefore, the terms become smaller, and the series also converges.

If each term of a given series is always *greater* than the corresponding term of a series known to be *divergent*, then the given series is also divergent.

For example, suppose the following series is known to be divergent; that is, the sum has no limit:

(a) $a_1 + a_2 + a_3 + a_4 + \cdots$

Now suppose we have another series of the following terms:

(b) $b_1 + b_2 + b_3 + b_4 + \cdots$

Now, if we can show that each term in (b) is always greater than the corresponding term in (a), then surely the series (b) also has no limit for its sum and is therefore divergent.

One important type of series that is known to be divergent is the *harmonic* series. In a harmonic series the terms are the *reciprocals* of the terms of an *arithmetic* series. The following is a harmonic series:

$$1 + \frac{1}{2} + \frac{1}{3} + \frac{1}{4} + \frac{1}{5} + \frac{1}{6} + \cdots + \frac{1}{n}$$

To show that this harmonic series is divergent, we might use the following line of reasoning. The sum of the first two terms is $\frac{3}{2}$. If we add the next *two* terms, $\frac{1}{3} + \frac{1}{4}$, we find that their sum is more than $\frac{1}{2}$. That is,

$$\frac{1}{3} + \frac{1}{4} > \frac{1}{2}$$

Therefore, the sum of all the terms through $\frac{1}{4}$ is more than 2. If we combine the next *four* terms, we find their sum is more than $\frac{1}{2}$; that is,

$$\frac{1}{5} + \frac{1}{6} + \frac{1}{7} + \frac{1}{8} > \frac{1}{2}$$

For the sum of the next *eight* terms, we get

$$\frac{1}{9} + \cdots + \frac{1}{16} > \frac{1}{2}$$

We can always take enough additional terms to get more than another $\frac{1}{2}$. Therefore, there is no limit to the sum, and the series diverges. The divergence of the harmonic series can also be proved in other ways.

EXAMPLE 5. Test the following series by comparison with the harmonic series:

$$\frac{1}{2} + \frac{2}{3} + \frac{3}{4} + \frac{4}{5} + \frac{5}{6} + \cdots + \frac{n}{n+1}$$

Solution. If each numerator were 1, we should have the harmonic series:

$$\frac{1}{2} + \frac{1}{3} + \frac{1}{4} + \frac{1}{5} + \frac{1}{6} + \cdots + \frac{n}{n-1}$$

Now we note that each term after the first is greater than the corresponding term of the harmonic series, which is known to be divergent. Therefore, the given series is also divergent.

Note. It might be pointed out that the omission of one or more terms at the beginning of a series does not affect the convergence or divergence of the series. It will affect the *sum* of a convergent series, but it will not alter its convergence.

One type of series that is useful in testing is the so-called "*p-series*." This is a series of the form

$$\frac{1}{1^p} + \frac{1}{2^p} + \frac{1}{3^p} + \frac{1}{4^p} + \frac{1}{5^p} + \frac{1}{6^p} + \cdots + \frac{1}{n^p}$$

In this series, p denotes a constant power in the denominators. Whether or not the series is convergent or divergent depends on the value of p. If $p = 1$, we have the harmonic series, which we know is divergent. If $p < 1$, the denominators are smaller than the corresponding denominators of the harmonic series, and therefore the terms are greater. Therefore, the series then diverges. It can be shown that if $p > 1$, the series is convergent.

To summarize: We can often determine the convergence or divergence of a given series by comparison with a series whose convergence or divergence is already known. However, there are certain pitfalls.

If the terms of a given series are *greater* than the corresponding terms of a series known to be *convergent*, then the given series *may* or *may not* be convergent. For instance, consider the following example:

EXAMPLE 6. Test the series

(a) $\quad 1 + \frac{2}{3} + \frac{4}{9} + \frac{8}{27} + \frac{16}{81} + \cdots + \frac{2^{n-1}}{3^{n-1}}$

Solution. We test this series by comparison with a series we have seen to be convergent:

(b) $\quad 1 + \frac{1}{2} + \frac{1}{4} + \frac{1}{8} + \frac{1}{16} + \cdots + \frac{1}{2^{n-1}}$

Here we see that the terms of the given series (a) are greater than the corresponding terms of the series (b), which we have already seen to be convergent. We might be inclined to guess that series (a) is divergent, since its terms are greater than those of the known series. Yet we shall find that series (a) is also convergent, and the sum approaches the limit 3. Note that (a) is a geometric series in which $r = \frac{2}{3}$. This fact is sufficient to establish convergence.

We have seen that if the terms of a given series are greater than the corresponding terms of a series known to be divergent, then the given series is also divergent. However, if the terms of a given series are *less* than those of a *divergent* series, then the given series *may* or *may not* be convergent. Testing with a divergent series will not establish convergence of another series. For this reason a comparison with a divergent series is of little value except as a means of discarding entirely a given series.

There are other pitfalls in the use of comparison tests. One danger is in comparing only a few terms at the beginning of a series and then jumping to a conclusion concerning convergence or divergence. This danger is illustrated in the following example:

Infinite Series

EXAMPLE 7. Test the following series by comparison:

$$\text{(a)} \quad 1 + \frac{2}{3} + \frac{4}{9} + \frac{8}{27} + \frac{16}{81} + \cdots$$

Solution. Let us compare series (a) with the following harmonic series:

$$\text{(b)} \quad \frac{1}{2} + \frac{1}{3} + \frac{1}{4} + \frac{1}{5} + \frac{1}{6} + \cdots$$

Note that every term in the given series (a) is greater than the corresponding term in the harmonic series (b). Therefore, we might conclude that the given series (a) is also divergent, since the harmonic series is divergent; yet series (a) is convergent. Note that it is a geometric series with $r = \frac{2}{3}$; therefore, it is convergent. If we take a few more terms in each series, we shall find that the terms in (a) become less than the corresponding terms in (b).

EXERCISE 43.2

Use comparison tests or other criteria to determine the convergence or divergence of the following series; write the general term.

1. $1 + \frac{3}{4} + \frac{9}{16} + \frac{27}{64} + \cdots$
2. $2 + \frac{4}{3} + \frac{8}{9} + \frac{16}{27} + \frac{32}{81} + \cdots$
3. $\frac{2}{3} + \frac{2}{5} + \frac{2}{7} + \frac{2}{9} + \frac{2}{11} + \cdots$
4. $2 + 3 + \frac{9}{2} + \frac{27}{4} + \frac{81}{8} + \cdots$
5. $\frac{2}{3} + \frac{3}{5} + \frac{4}{7} + \frac{5}{9} + \frac{6}{11} + \cdots$
6. $\frac{1}{3} + \frac{1}{6} + \frac{1}{9} + \frac{1}{12} + \cdots$
7. $\frac{1}{2} + \frac{1}{4} + \frac{1}{6} + \frac{1}{8} + \frac{1}{10} + \cdots$
8. $1 + \frac{1}{1 \cdot 2} + \frac{1}{2 \cdot 3} + \frac{1}{3 \cdot 4} + \frac{1}{4 \cdot 5} + \cdots$
9. $1 + \frac{1}{1 \cdot 2} + \frac{1}{3 \cdot 4} + \frac{1}{5 \cdot 6} + \frac{1}{7 \cdot 8} + \cdots$
10. $1 + \frac{1}{4} + \frac{1}{9} + \frac{1}{16} + \frac{1}{25} + \frac{1}{36} + \cdots$
11. $1 + \frac{1}{2!} + \frac{1}{3!} + \frac{1}{4!} + \frac{1}{5!} + \cdots$
12. $\frac{1}{2 \cdot 3} + \frac{2}{3 \cdot 4} + \frac{3}{4 \cdot 5} + \frac{4}{5 \cdot 6} + \frac{5}{6 \cdot 7} + \cdots$
13. $\frac{5}{4} + \frac{5}{9} + \frac{5}{14} + \frac{5}{19} + \frac{5}{24} + \cdots$
14. $\frac{1}{3} + \frac{2}{7} + \frac{3}{11} + \frac{4}{15} + \frac{5}{19} + \frac{6}{23} + \cdots$
15. $\frac{1}{2} + \frac{1}{5} + \frac{1}{10} + \frac{1}{17} + \frac{1}{26} + \cdots$
16. $\frac{1}{\sqrt{1}} + \frac{1}{\sqrt{2}} + \frac{1}{\sqrt{3}} + \frac{1}{\sqrt{4}} + \cdots$
17. $\frac{1}{2} + \frac{1}{6} + \frac{1}{12} + \frac{1}{20} + \frac{1}{30} + \cdots$
18. $\frac{1}{\ln 2} + \frac{1}{\ln 3} + \frac{1}{\ln 4} + \frac{1}{\ln 5} + \cdots$
19. $1 + \frac{1}{3^2} + \frac{1}{5^2} + \frac{1}{7^2} + \frac{1}{9^2} \cdots$
20. $1 + \frac{2}{\sqrt{3}} + \frac{3}{\sqrt{5}} + \frac{4}{\sqrt{7}} + \frac{5}{\sqrt{9}} + \cdots$

43.4 LIMIT OF u_n, THE nth TERM

If an infinite series is to be convergent, then the nth term, u_n, must approach zero as a limit as n becomes infinite. This is a necessary condition for convergence; that is, the following must be true:

$$\lim_{n \to \infty} u_n = 0$$

If this limit is not equal to zero, then the series is not convergent.

However, if the nth term does approach zero as a limit, this condition alone does not mean the series necessarily converges. The series may still diverge. As examples, let us consider the following series:

$$\text{(a)} \quad 1 + \frac{1}{2} + \frac{1}{4} + \frac{1}{8} + \frac{1}{16} + \cdots + \frac{1}{2^{n-1}}$$

$$\text{(b)} \quad 1 + \frac{1}{2} + \frac{1}{3} + \frac{1}{4} + \frac{1}{5} + \cdots + \frac{1}{n}$$

Now let us look at the limit of the nth term in each series.

For (a), $\quad \lim_{n \to \infty} \frac{1}{2^{n-1}} = 0 \quad$ (This series is convergent.)

For (b), $\quad \lim_{n \to \infty} \frac{1}{n} = 0 \quad$ (This series is divergent.)

A *necessary* condition for convergence is that the nth term approach zero as a limit as n becomes infinite, but this condition is *not sufficient* to establish convergence.

43.5 ALTERNATING SERIES

An alternating series is one in which the terms alternate in sign, as in this example:

$$1 - \frac{1}{2} + \frac{1}{3} - \frac{1}{4} + \frac{1}{5} - \frac{1}{6} + \cdots$$

To indicate the general term of an alternating series we use the trick of writing the factor (-1) raised to some power. If (-1) is raised to an even power, the result is positive; if it is raised to an odd power, the result is negative.

To show the general term of an alternating series, then, we simply multiply the term by (-1) raised to an odd or an even power. This power can also be indicated by use of the number of the term, n. Note that in the alternating series in the example shown, the even-numbered terms are negative. If we use $(-1)^n$, we should get the wrong sign for each term. So we write $(-1)^{n+1}$ or $(-1)^{n-1}$. Then we can write

$$1 - \frac{1}{2} + \frac{1}{3} - \frac{1}{2} + \frac{1}{5} - \frac{1}{6} + \cdots (-1)^{n+1}\left(\frac{1}{n}\right)$$

One fact might be noted concerning an alternating series. In such a series, if the limit of the nth term, the general term, is zero as n becomes infinite, then this condition is *sufficient* to establish convergence. Note that the series shown is actually a harmonic series with alternating signs. If all the terms of this series had like signs, the series would be divergent. It is convergent only on the condition that the signs alternate. Then we say such a series is *conditionally* convergent. If any series is convergent regardless of the signs, then we say it is *absolutely* convergent.

618 Infinite Series

43.6 RATIO TEST

One of the most useful tests for convergence or divergence of a series, is the so-called *ratio test*. To use this test we first write the expression for the general, or nth term, that is, u_n. Then we write the expression for the following term, u_{n+1}. In writing this term, we substitute $(n+1)$ for n in the general term. We now set up the following ratio:

$$\frac{u_{n+1}}{u_n}$$

Now let us see what happens to this ratio as n becomes infinite. That is, we determine the following limit:

$$\lim_{n \to \infty} \frac{u_{n+1}}{u_n}$$

Let us denote this limit by L. Then we have three possible conditions:

If $L < 1$, the series converges.

If $L > 1$, the series diverges.

If $L = 1$, the test fails.

EXAMPLE 8. Test the following series by use of the ratio test:

$$\frac{2}{1!} + \frac{4}{2!} + \frac{8}{3!} + \frac{16}{4!} + \frac{32}{5!} + \frac{64}{6!} + \cdots + \frac{2^n}{n!}$$

Solution. With the powers of 2 in the numerators, we might suspect that the series is divergent. However, let us apply the ratio test. We have shown the nth term. The following term is found by replacing n with $n+1$. For the $(n+1)$th term, we get

$$\frac{2^{n+1}}{(n+1)!}$$

The ratio, $(u_{n+1})/u_n$ reduces to

$$\frac{2^{n+1}}{(n+1)!} \div \frac{2^n}{n!} = \frac{2}{n+1}$$

Since

$$\lim_{n \to \infty} \frac{2}{n+1} = 0$$

the series is convergent.

43.7 L'HOSPITAL'S RULE

In evaluating the limit of a fraction, it often happens that we run into a situation in which the fraction appears to have no meaning. We may get a result such as 0/0, infinity divided by infinity, a constant divided by zero, or infinity divided by a constant. These results are called *indeterminate* forms because they appear to have no meaning. Yet in many instances they do have limits, or it can be shown that the fraction has no limit.

43.7 L'Hospital's Rule

One device that is very useful in determining the limit of a so-called indeterminate form is the technique known as *L'Hospital's rule*, named for a French mathematician. We shall simply state the rule and show its use without giving the proof.

Let us suppose we have a problem involving the limit of a quotient of two functions, such as

$$\lim_{x \to a} \frac{f(x)}{g(x)}$$

When we try to evaluate the limit, we may get one of the indeterminate forms. Then, in many instances, the limit may still be found. In a previous chapter (Chapter 11) we have already evaluated some such limits. However, L'Hospital's rule is often more convenient than the method previously shown. This rule states that under certain conditions, one of which is that the functions be continuous, we may differentiate numerator and denominator separately and then find the limit of the resulting fraction. That is,

$$\lim_{x \to a} \frac{f(x)}{g(x)} = \lim_{x \to a} \frac{f'(x)}{g'(x)}$$

Warning: Do not differentiate the entire fraction by the quotient rule. Instead we differentiate numerator and denominator separately.

L'Hospital's rule may be applied to the same fraction as many times in succession as necessary until a limit is determined or until it can no longer be applied to any use. If the limit at any point can be evaluated, this should be done as soon as possible.

EXAMPLE 9. Evaluate

$$\lim_{x \to 2} \frac{x^2 - 4x + 7}{2x^2 - 8x + 8}$$

Solution. By direct substitution, we get the result: 3/0. Now we differentiate numerator and denominator, and get

$$\lim_{x \to 2} \frac{x^2 - 4x + 7}{2x^2 - 8x + 8} = \lim_{x \to 2} \frac{2x - 4}{4x - 8}$$

Direct substitution again results in the indeterminate form: 0/0. Now we differentiate numerator and denominator again and state the result:

$$\lim_{x \to 2} \frac{2}{4} = \frac{1}{2}$$

Therefore we see that the limit does exist and that it is $\frac{1}{2}$.

EXAMPLE 10. Evaluate the limit

$$\lim_{n \to \infty} \frac{n^2 - 2n + 5}{3n^2 + 4n + 3}$$

Solution. We cannot substitute infinity as we substitute a finite number in the fraction. Instead we apply the differentiating rule:

$$\lim_{n \to \infty} \frac{n^2 - 2n + 5}{3n^2 + 4n + 3} = \lim_{n \to \infty} \frac{2n - 2}{6n + 4} = \lim_{n \to \infty} \frac{2}{6} = \frac{1}{3}$$

EXAMPLE 11. Evaluate

$$\lim_{n \to \infty} \frac{n^2 + 3n + 1}{n^3 + 5n^2 + 3}$$

Solution. We apply L'Hospital's rule until a limit is determined.

$$\lim_{n \to \infty} \frac{n^2 + 3n + 1}{n^3 + 5n^2 + 3} = \lim_{n \to \infty} \frac{2n + 3}{3n^2 + 10n} = \lim_{n \to \infty} \frac{2}{6n + 10} = \lim_{n \to \infty} \frac{0}{6} = 0$$

The limit of the fraction is zero.

EXAMPLE 12. Evaluate the limit

$$\lim_{n \to \infty} \frac{n(n + 1)}{3n - 1}$$

Solution. Expanding the numerator and applying L'Hospital's rule, we get

$$\lim_{n \to \infty} \frac{n^2 + 2n}{3n - 1} = \lim_{n \to \infty} \frac{2n + 2}{3} = \lim_{n \to \infty} \frac{2}{0} = \infty$$

At the second step, we note that the numerator contains the quantity, n, which becomes infinite, while the denominator is 3. Then, at that point we can say the limit does not exist. If we go to the third step, the result shows that the limit does not exist.

EXAMPLE 13. Evaluate the limit

$$\lim_{x \to 0} \frac{\sin x + \ln \cos x}{\cos x - 1}$$

Solution. Direct substitution of zero for x results in the form $0/0$. Each time we apply the differentiating rule, we must try to evaluate the result so that we do not pass up any opportunity. Applying the rule,

$$\lim_{x \to 0} \frac{\sin x + \ln \cos x}{\cos x - 1} = \lim_{x \to 0} \frac{\cos x - \tan x}{-\sin x} = \lim_{x \to 0} \frac{-\sin x - \sec^2 x}{-\cos x}$$

We can now substitute $x = 0$ in the final fraction and get the limit: 1.

EXERCISE 43.3

Evaluate the following limits:

1. $\lim\limits_{x \to 2} \dfrac{x^2 - 9}{x^2 - 4}$

2. $\lim\limits_{x \to 2} \dfrac{3x - 5}{x^3 - 3x + 4}$

3. $\lim\limits_{x \to 2} \dfrac{4x - 5}{x^3 - 8}$

4. $\lim\limits_{n \to \infty} \dfrac{2x^2 - 5}{2x^2 - 5x + 1}$

5. $\lim\limits_{x \to \infty} \dfrac{3x^2 - 4}{x^3 - 2x^2 + x}$

6. $\lim\limits_{n \to \infty} \dfrac{n + 1}{n^2 + 1}$

7. $\lim\limits_{n \to \infty} \dfrac{n + 3}{2n + 1}$

8. $\lim\limits_{n \to \infty} \dfrac{n}{n + 1}$

9. $\lim\limits_{n \to \infty} \dfrac{2n^2 - n}{n^2 + 1}$

10. $\lim\limits_{x\to\infty} \dfrac{x^2 + 3x}{x + 2}$

11. $\lim\limits_{x\to 0} \dfrac{e^x - e^{-x}}{\sin x}$

12. $\lim\limits_{x\to 0} \dfrac{1 - \cos x}{x^2}$

13. $\lim\limits_{x\to 2} \dfrac{x - 2}{x^4 - 16}$

14. $\lim\limits_{x\to 0} \dfrac{\tan x - x}{\sin x - x}$

15. $\lim\limits_{x\to 1} \dfrac{\ln x - 1}{x^2 - 1}$

16. $\lim\limits_{x\to 0} \dfrac{\ln x - 1}{e^x}$

17. $\lim\limits_{x\to 0} \dfrac{x + \sin x}{1 - \cos x}$

18. $\lim\limits_{x\to 0} \dfrac{x + \sin x}{x}$

19. $\lim\limits_{x\to\infty} \dfrac{x^3}{e^{2x}}$

20. $\lim\limits_{x\to\infty} \dfrac{\ln(x + 1)}{\ln x}$

21. $\lim\limits_{n\to\infty} \dfrac{(n+1)(n+2)}{n(2n-1)}$

22. $\lim\limits_{n\to\infty} \dfrac{\sqrt{n}}{\sqrt{n+1}}$

23. $\lim\limits_{n\to 1} \dfrac{3n^3 - 2n^2 + 2n - 3}{4n^3 - 5n^2 - 2n + 3}$

24. $\lim\limits_{n\to\infty} \dfrac{n(n+3)}{2n(n+1)}$

Write the general term of each of the following series and then test the series for convergence or divergence by use of the ratio test.

25. $\dfrac{2}{1} + \dfrac{3}{2} + \dfrac{4}{2^2} + \dfrac{5}{2^3} + \dfrac{6}{2^4} + \cdots$

26. $\dfrac{3}{1!} + \dfrac{3}{2!} + \dfrac{3}{3!} + \dfrac{3}{4!} + \dfrac{3}{5!} + \cdots$

27. $\dfrac{3}{1\cdot 2} + \dfrac{3^2}{2\cdot 4} + \dfrac{3^3}{3\cdot 8} + \dfrac{3^4}{4\cdot 16} + \dfrac{3^5}{5\cdot 32} \cdots$

28. $\dfrac{1}{3} + \dfrac{3}{3^2} + \dfrac{5}{3^3} + \dfrac{7}{3^4} + \dfrac{9}{3^5} + \dfrac{11}{3^6} \cdots$

29. $\dfrac{1}{2} + \dfrac{1\cdot 2}{4} + \dfrac{1\cdot 2\cdot 3}{8} + \dfrac{1\cdot 2\cdot 3\cdot 4}{16} + \cdots$

30. $\dfrac{5}{3} + \dfrac{7}{6} + \dfrac{9}{12} + \dfrac{11}{24} + \dfrac{13}{48} + \cdots$

31. $\dfrac{2}{2} + \dfrac{4}{5} + \dfrac{6}{10} + \dfrac{8}{17} + \dfrac{10}{26} + \cdots$

32. $\dfrac{2^0}{4} + \dfrac{2^1}{7} + \dfrac{2^2}{10} + \dfrac{2^3}{13} + \dfrac{2^4}{16} + \dfrac{2^5}{19} \cdots$

33. $\dfrac{1}{2} + \dfrac{4}{6} + \dfrac{9}{24} + \dfrac{16}{120} + \dfrac{25}{720} + \cdots$

34. $\dfrac{1\cdot 2}{1} + \dfrac{2\cdot 3}{2} + \dfrac{3\cdot 4}{6} + \dfrac{4\cdot 5}{24} + \dfrac{5\cdot 6}{120} + \cdots$

35. $1 + 4 + \dfrac{7}{2} + \dfrac{10}{6} + \dfrac{13}{24} + \dfrac{16}{120} + \cdots$

36. $\dfrac{1}{1} + \dfrac{1\cdot 2}{2^2} + \dfrac{1\cdot 2\cdot 3}{3^3} + \dfrac{1\cdot 2\cdot 3\cdot 4}{4^4} + \cdots$

43.8 A POWER SERIES

You may wonder: What is the use of an infinite series? For one thing, infinite series are used to compute approximate numerical values of irrational numbers, such as the trigonometric functions, the logarithms of numbers, the values of π, e, and the roots of numbers. For another, they are used to solve differential equations that cannot be solved by other methods. They have other uses in the study of advanced mathematics.

One of the most important kinds of series is the so-called *power* series. One example of a power series is the geometric series:

$$a + ar + ar^2 + ar^3 + \cdots + ar^{n-1}$$

Notice that the series consists of terms in ascending positive integral powers of the variable r. In the geometric power series, the coefficients are all alike.

A power series in a variable, such as r in the geometric series, may converge or diverge depending on the value of the variable. In a *general power* series, the coefficients, although constants, may not all be the same as they are in a geometric series. Then convergence will also depend on the coefficients. The following is another example of a power series:

$$\frac{x}{1} + \frac{x^2}{2} + \frac{x^3}{3} + \frac{x^4}{4} + \frac{x^5}{5} + \frac{x^6}{6} + \cdots \frac{x^n}{n}$$

With the given coefficients in this power series, convergence will depend on the value of x. If x is less than 1, the series will converge because, even without the increasing values of the denominators, it would be a geometric series.

To evaluate a series, we take some value of the variable x and substitute this value in the series as far as we wish to go. Let us evaluate this series for, say, $x = 0.2$. Then we shall find that the value through six terms is 0.223131.

The general power series is often represented by

$$a_0 + a_1 x + a_2 x^2 + a_3 x^3 + a_4 x^4 + \cdots + a_{n-1} x^{n-1}$$

The terms contain increasing powers of the variable x. The coefficients, a_0, a_1, a_2, and so on, represent constants. If we wish to evaluate any power series, we must first know the value of each coefficient. These coefficients can be found by differentiation, provided the series is convergent. If a series converges, it can be differentiated term by term. However, the series and all its derivatives must be defined at $x = 0$.

Let us assume that a certain function can be expressed as a power series; that is,

$$f(x) = a_0 + a_1 x + a_2 x^2 + a_3 x^3 + a_4 x^4 + a_5 x^5 + \cdots$$

Now, if we know the coefficients, a_0, a_1, a_2, and so on, then we should have the particular power series for this function. These coefficients can be found by differentiating the series several times. The method involves the so-called Maclaurin's series, named after Colin Maclaurin (1698–1746), a Scottish mathematician.

43.8 A Power Series

Let us differentiate the general power series several times and then evaluate each derivative and the function itself for $x = 0$:

$f(x) = a_0 + a_1x + a_2x^2 + a_3x^3 + a_4x^4 + a_5x^5 + a_6x^6 + \cdots$

$f'(x) = a_1 + 2a_2x + 3a_3x^2 + 4a_4x^3 + 5a_5x^4 + 6a_6x^5 + \cdots$

$f''(x) = 2a_2 + 6a_3x + 12a_4x^2 + 20a_5x^3 + 30a_6x^4 + \cdots$

$f'''(x) = 6a_3 + 24a_4x + 60a_5x^2 + 120a_6x^3 + \cdots$

and so on. If we continue to differentiate and evaluate the function and its derivatives for $x = 0$, we shall find the value of each coefficient and then we shall be able to write the particular power series for the given function. Putting zero (0) for x in each, we get

$f(0) = a_0 \quad f'(0) = a_1 \quad f''(0) = 2a_2 \quad f'''(0) = 6a_3 \quad f^{iv}(0) = 24a_4$

Solving for the coefficients,

$$a_0 = f(0) \quad a_1 = f'(0) \quad a_2 = \frac{f''(0)}{2!} \quad a_3 = \frac{f'''(0)}{3!} \quad a_4 = \frac{f^{iv}(0)}{4!}$$

Now the series can be written with the coefficients:

$$f(x) = f(0) + \frac{f'(0)x}{1!} + \frac{f''(0)x^2}{2!} + \frac{f'''(0)x^3}{3!} + \frac{f^{iv}(0)x^4}{4!} + \cdots + \frac{f^n(0)x^n}{5!}$$

Notice that the order of the derivative in each term is the same as the power of x and as the factorial number in the denominator. Remember, however, that in working out the expansion of a function, it may turn out that some coefficients are zero. Then those terms will drop out in the expansion.

EXAMPLE 14. Find the coefficients, assuming that e^x can be expressed as a power series:

$$f(x) = e^x = a_0 + a_1x + a_2x^2 + a_3x^3 + a_4x^4 + a_5x^5 + \cdots$$

Solution.

$f'(x) = e^x = a_1 + 2a_2x + 3a_3x^2 + 4a_4x^3 + 5a_5x^4 + \cdots$

$f''(x) = e^x = 2a_2 + 6a_3x + 12a_4x^2 + 20a_5x^3 + \cdots$

$f'''(x) = e^x = 6a_3 + 24a_4x + 60a_5x^2 + \cdots$

Evaluating each for $x = 0$, we get

$f(0) = e^0 = a_0$, then $a_0 = 1$; $\qquad f'(0) = e^0 = a_1$, then $a_1 = 1$

$f''(0) = e^0 = 2a_2$, then $a_2 = \frac{1}{2}$; $\qquad f'''(0) = e^0 = 6a_3$, then $a_3 = \frac{1}{6}$

If we continue the differentiation and evaluation for $x = 0$, we find that e^x can be expressed as the particular series

$$e^x = 1 + x + \frac{x^2}{2!} + \frac{x^3}{3!} + \frac{x^4}{4!} + \frac{x^5}{5!} + \frac{x^6}{6!} + \frac{x^7}{7!} + \cdots$$

EXAMPLE 15. Find the value of e by using the series obtained in Example 14.

Solution. To find the value of e, we take the series for e^x in which $x = 1$. Then we put 1 for x in the power series for e^x:

$$e^1 = 1 + 1 + \frac{1}{2!} + \frac{1}{3!} + \frac{1}{4!} + \frac{1}{5!} + \frac{1}{6!} + \frac{1}{7!} + \frac{1}{8!} + \frac{1}{9!} + \cdots$$

Using ten terms, we get the value of e correct to seven places: 2.7182815.

EXAMPLE 16. If we derive the power series expansion for $\sin x$, some of the coefficients will turn out to be zero, and we shall get

$$\sin x = x - \frac{x^3}{3!} + \frac{x^5}{5!} - \frac{x^7}{7!} + \frac{x^9}{9!} - \frac{x^{11}}{11!} + \cdots$$

Using this series, find the value of $\sin 0.2$.

Solution. Note that the signs of the terms alternate. Replacing x with 0.2 in the first four terms of the series, we get

$$\sin 0.2 = 0.2 - \frac{0.008}{6} + \frac{0.00032}{120} - \frac{0.0000128}{5040}$$

If we use more terms, we get a higher degree of accuracy. Even by using only four terms, we get $\sin 0.2 = 0.19866933$. A table shows $\sin 0.2 = 0.19867$.

For large values of x, a power series converges more slowly, if at all. In all cases where we have fractional coefficients, the convergence is much faster if the numerators are small. However, even for $\sin 1$, the value can be computed fairly accurately by using only five terms.

We have said that in order to be evaluated by the Maclaurin series, the function and all its derivatives must be defined for $x = 0$. Therefore, the Maclaurin series cannot be used to evaluate such functions as $\ln x$, $\csc x$, and $\sqrt{x}$. The function, $\ln x$, does not exist at $x = 0$, since zero has no real logarithm. In the case of the function $\sqrt{x}$, the function does exist at $x = 0$, but its derivatives do not.

For such functions, we use an expansion called Taylor's series, in which we substitute $(x - a)$ for x, where a is some convenient constant in the neighborhood of x. The result is Taylor's series, which has the general form

$$f(x) = f(a) + f'(a)(x - a) + \frac{f''(a)(x - a)^2}{2!} + \frac{f'''(a)(x - a)^3}{3!} + \cdots$$

As an example, to find a convenient expansion for the evaluation of a logarithm, we might use the following form:

$$f(x) = \ln(x + 1)$$

Taylor's series is also sometimes used when convergence is very slow through the use of Maclaurin's series. For example, to find a convenient expansion for the evaluation of $(\sin 63°)$, we might use the form, in which we take $a = 60° = \pi/3$. Then we have

$$x - a = 3°$$

EXERCISE 43.4

Derive the following important series:

1. $e^x = 1 + x + \dfrac{x^2}{2!} + \dfrac{x^3}{3!} + \dfrac{x^4}{4!} + \dfrac{x^5}{5!} + \dfrac{x^6}{6!} + \cdots$

2. $\sin x = x - \dfrac{x^3}{3!} + \dfrac{x^5}{5!} - \dfrac{x^7}{7!} + \dfrac{x^9}{9!} - \cdots$

3. $\cos x = 1 - \dfrac{x^2}{2!} + \dfrac{x^4}{4!} - \dfrac{x^6}{6!} + \dfrac{x^8}{8!} - \cdots$

4. $\ln(1+x) = x - \dfrac{x^2}{2} + \dfrac{x^3}{3} - \dfrac{x^4}{4} + \dfrac{x^5}{5} - \dfrac{x^6}{6}$

5. Derive the series for e^{-x} in two ways.
6. Find the first four terms of the expansion for xe^{2x}.
7. Using the series in Problem 1, find the value of e (which is e^1) to ten places. Compare your answer with the value in a table.
8. Using the first five terms in the series for $\sin x$, find the approximate value of $\sin 0.1$, and compare your answer with a table.
9. Using the first five terms of the series for $\cos x$, find the approximate value of $\cos 0.1$ and compare your answer with a table.
10. Using the series for e^x, find the series for e^{jx}, where $j = \sqrt{-1}$. Recall the consecutive powers of j (or $i = \sqrt{-1}$). Then rearrange the terms of the series so that all the terms containing j are grouped together. Then factor out the j. How does your result compare with the series for $\cos x$ and $\sin x$? Show that the following equation is true: $e^{jx} = \cos x + j \sin x$. This equation is called Euler's formula.
11. Write out the first six terms of the expansion of Taylor's series for the function, $f(x) = \ln(x + 1)$. Then evaluate $\ln 2$.
12. Write out the first five terms of the expansion of Taylor's series for the function, $f(x) = \sin(x)$. Then find the approximate value of $\sin 63°$, by taking $a = 60° = \pi/3$, so that $(x - a) = 3° = \pi/60$.

chapter
44

Differential Equations

44.1 DEFINITION

A *differential equation* is an equation containing a derivative or differentials. To see just what this means consider the function that represents a parabola:

$$y = x^2 + 3$$

If we differentiate the function, we get the differential equation

$$\frac{dy}{dx} = 2x$$

Let us now state the result in the form of differentials:

$$dy = 2x\, dx$$

Note that we now have an *equation* containing differentials.

Suppose now that we have the reverse problem in integration, which we have already considered:

$$\int 2x\, dx$$

Integrating, we get the indefinite integral,

$$x^2 + C$$

The original equation can then be written,

$$y = x^2 + C$$

Note especially that the following expression is *not an equation* but is simply a direction to integrate:

$$\int 2x\, dx$$

On the other hand, the following form is an equation containing differentials:

$$dy = 2x\, dx$$

There is nothing really new about a differential equation. It is simply the statement of differentials in equation form. Basically, it is solved by integration. A simple differential equation has already been solved in Chapter 22. Now we shall look at the problem in somewhat more detail.

44.2 SOLUTION OF A DIFFERENTIAL EQUATION

In many problems in science it often happens that it is possible, from certain known information, to set up an equation showing the rate of change of one variable with respect to another variable. This rate of change is, of course, the derivative. Our problem then is to find the relation between the variables themselves. This is done by solving a differential equation. There are several ways in which this can be done, all of which involve integration.

To *solve* a differential equation we attempt to work back to the original equation from which the differential equation was derived. That is, we try to eliminate the derivative, or differentials, and find the original equation expressing the correct relation between the variables. This relation is called the *solution* of the differential equation.

First, let us recall that for the differential of any function, the integral of the differential is the function itself, as shown here:

$$\int dt = t + C \qquad \int dA = A + C \qquad \int dy = y + C$$

$$\int 2x\,dx = x^2 + C \qquad \int di = i + C \qquad \int i^2\,di = \frac{1}{3}i^3 + C$$

Let us take again the differential equation

$$dy = 2x\,dx$$

To solve this equation, we integrate *both sides* and get

$$\int dy = \int 2x\,dx$$

The integral of the left side is y plus some constant. The integral of the right side is x^2 plus some constant. That is, by integrating, we get

$$y + C_1 = x^2 + C_2$$

We need not, however, include a separate constant with both integrals, because we can write the result

$$y = x^2 + C_2 - C_1$$

Since C_1 and C_2 are both constants, their difference is also a constant, which we can call simply C. The solution can therefore be written

$$y = x^2 + C$$

This is the *general solution* of this given differential equation. To check the result, we differentiate and get the original differential equation. Note that in the check, the constant C is eliminated. That is,

if $\quad y = x^2 + C, \quad$ then $\quad dy = 2x\,dx$

44.3 THE GENERAL SOLUTION

In the foregoing example, we began with the differential equation

$$dy = 2x\, dx$$

We obtained the *general solution*,

$$y = x^2 + C$$

We also discovered that the solution satisfied the original differential equation.

The given differential equation would also be satisfied by many *particular solutions*, containing specific values of the constant C, such as any of the following:

$$y = x^2 + 5 \qquad y = x^2 - 3 \qquad y = x^2 - \ln 6 \qquad y = x^2$$

However, the most general solution is the one containing the arbitrary constant C. In all cases, a general solution must contain at least one arbitrary constant. It may contain several, depending on the *order* of the differential equation.

44.4 ORDER OF A DIFFERENTIAL EQUATION

The *order* of a differential equation is the order of the *highest derivative* in the equation. A differential equation of the *first order* is one containing a first but no higher derivative. A second-order equation is one containing a second but no higher derivative. A second-order differential equation may or may not contain a first derivative. A differential equation of order n is one containing a derivative of order n but no higher derivative.

The following examples illustrate orders of differential equations:

first order: $\quad dy = 2x\, dx \qquad y' + 3y = 5x$

second order: $\quad \dfrac{d^2 i}{dt^2} = 3t \qquad y'' - 5y' + 6y = 3x + \sin x$

third order: $\quad y''' + 4y' = x^2 \qquad y''' - 3y' + 2y = 3x^5$

The *degree* of a differential equation refers to the power and is the *degree of the highest order* of derivative. The foregoing equations are all of the first degree. The following is a differential equation of the second degree:

$$\left(\frac{d^2 y}{dx^2}\right)^2 + \left(\frac{dy}{dx}\right)^3 + y = x^4$$

The general solution of any differential equation will contain as many arbitrary constants as the *order* of the equation. The constants are usually denoted by $C_1, C_2, C_3, \cdots$, or by A, B, C, D, and so on.

Warning: If two arbitrary constants such as C_1 and C_2 can be combined into one, then we have essentially only one constant. For example, in solving the differential equation

$$dy = 2x\, dx$$

we first found the solution

$$y + C_1 = x^2 + C_2$$

However, since the differential equation is of the first order, it must not contain more than one such constant. This is seen to be true, since we can, by transposing, combine the two C's into a single constant.

The general solution of a second-order differential equation must contain two distinct constants which cannot be combined into one. This principle also holds for differential equations of higher order. For example, a differential equation of the fourth order must have four arbitrary constants in its most general solution.

Solving a differential equation is often a complicated process. A few differential equations are easily solved. Many are very difficult. Most are impossible. Fortunately, most of those we encounter in our practical work in physics, electronics, and other fields can be solved without too much difficulty. Several methods are used, depending on the type of equation. Of course, solving a differential equation in all cases depends on a good knowledge of integration techniques.

44.5 SEPARATION OF VARIABLES

Some differential equations can be solved by a method called *separation of variables*. In this method we first group all the y terms with dy, and all the x terms with dx. That is, we separate the variables by simple algebraic processes. Each term can then be integrated by itself.

In general terms, such an equation can be written in the form

$$M(x)\,dx + N(y)\,dy = 0$$

This form is sometimes stated in general terms as follows:

$$f(x)\,dx + g(y)\,dy = 0$$

or simply

$$M\,dx + N\,dy = 0$$

If the equation can be put into some such form in which the variables are separated, it can be solved by integrating term by term.

EXAMPLE 1. Find the equation of the curve of which the slope is always equal to the ratio of the abscissa to the ordinate.

Solution. The condition of the problem is stated by the differential equation

$$\frac{dy}{dx} = \frac{x}{y}$$

Multiplying both sides of the equation by the lowest common denominator, $y\,dx$, we get

$$y\,dy = x\,dx$$

Integrating term by term,

$$\frac{y^2}{2} = \frac{x^2}{2} + C_1$$

Multiplying by 2 and transposing, we get
$$y^2 - x^2 = 2C_1$$
Replacing the $2C_1$ with a single constant C, we have
$$y^2 - x^2 = C$$
The solution represents a family of hyperbolas. If C is positive, the hyperbolas open upward and downward with the transverse axis along the y axis. If C is negative, the transverse axis lies along the x axis. If C is zero, the solution represents the degenerate hyperbola, two intersecting straight lines.

EXAMPLE 2. Solve the following differential equation:
$$\sqrt{4 - y^2}\, dx = y(x^2 + 9)\, dy$$

Solution. Separating the variables, we get
$$\frac{dx}{x^2 + 9} = \frac{y\, dy}{\sqrt{4 - y^2}}$$

Now each term can be integrated. The left side integrates into an arctangent. For the right side we use the power rule. Then
$$\int \frac{dx}{x^2 + 9} = \int y\,(4 - y^2)^{-1/2}\, dy$$
Integrating,
$$\frac{1}{3} \arctan \frac{x}{3} = -(4 - y^2)^{1/2} + C_1$$
Multiplying both sides by 3, and transposing, we get
$$\arctan \frac{x}{3} + 3\sqrt{4 - y^2} = 3C_1$$
Using C for $3C_1$, we have
$$\arctan \frac{x}{3} + 3\sqrt{4 - y^2} = C$$

Note on the Arbitrary Constant. The constant C can be written in any form that represents a constant. It should be written in a form most convenient. In Example 2, we might have written $C/3$ instead of C_1, since we already have the fraction $\frac{1}{3}$ as the coefficient of one term. In some cases it is convenient to write the constant as a logarithm or as an exponent, or even as a power. The constant can be written, for example, in any of the following forms, as well as others, all of which represent constants:
$$C, -C, 3C, \tfrac{1}{2}C, C^4, \ln C, 10^C, e^C, \text{ and others}$$

EXAMPLE 3. Solve the equation
$$x(y - 2)\, dx = 3(x^2 + 4)\, dy$$

Solution. Separating the variables, we get
$$\frac{x\, dx}{x^2 + 4} = \frac{3\, dy}{y - 2}$$
Integrating,
$$\frac{1}{2} \ln (x^2 + 4) = 3 \ln (y - 2) + \ln C$$
or
$$\ln \sqrt{x^2 + 4} = \ln (y - 2)^3 + \ln C$$

The result can be written
$$\ln \sqrt{x^2 + 4} = \ln C(y - 2)^3$$
Then, taking antilogarithms, the simplest form of the solution becomes
$$\sqrt{x^2 + 4} = C(y - 2)^3$$
To check the solution we must differentiate and see that in the result the constant C is eliminated.

EXAMPLE 4. The acceleration of a particle down an inclined plane is given by the formula
$$a = 6 \text{ ft/sec}^2$$
Find the general equation for the velocity and for the distance at any time t.

Solution. We have seen that the following relations exist:
$$a = \frac{dv}{dt} \quad \text{and} \quad v = \frac{ds}{dt}$$

so that
$$a = \frac{dv}{dt} = \frac{d^2s}{dt^2}$$

Since $a = 6$, we first write
$$\frac{dv}{dt} = 6$$

Multiplying by dt,
$$dv = 6 \, dt$$

Integrating,
$$v = 6t + C_1$$

Since $v = ds/dt$, we write
$$\frac{ds}{dt} = 6t + C_1$$

Multiplying both sides by dt,
$$ds = (6t + C_1) \, dt$$

Integrating both sides,
$$s = 3t^2 + C_1 t + C_2$$

Note that the general solution for the distance s contains two arbitrary constants, C_1 and C_2, since we began with the second derivative of s. The two C's are independent and cannot be combined into a single constant.

EXERCISE 44.1

Find the general solution of each of the following equations:

1. $dy/dx = 6$
2. $dy/dx = 5x - 7$
3. $y' = 4 - 3x$
4. $dy/dx = -x/y$
5. $dy/dx = x/y^3$
6. $y' = -y/x$
7. $dy/dx = x^2/y^3$
8. $x^3 \, dy = y^4 \, dx$
9. $x \, dy + y \, dx = 0$
10. $dy/dx = -xy$
11. $dy/dx = x^2/y$
12. $dy/dx = -x/2y$
13. $dy/dx = 6x^2 - 4x + 3$
14. $dy/dx = x^3 y^2 - xy^2$
15. $(4 - y) \, dx = (x + 1) \, dy$
16. $y' = \sin x \cos^2 y$
17. $y' = \tan x \sec y$
18. $y' = xy^2 (1 - x^2)^{1/2}$
19. $(y^2 - 4) \, dx = (x^2 - 1) \, dy$
20. $(xy^2 + 4xy) \, dx = (x^2 y + 4y) \, dy$
21. $(x^2 + 4) \, dy = (y^2 + 9) \, dx$
22. $xy^3 \, dx + \sqrt{1 - x^2} \, dy = y^3 \, dx$
23. $\csc^2 x \, dy = \csc^2 y \, dx - dx$
24. $\cos x \cos^2 y \, dx = \sin x \, dy$
25. $y^3 \, dx = \sqrt{4 - 9x^2} \, dy$
26. $0.3(di/dt) + 12i = 6$
27. $xy^2 \, dx + 6xy \, dx + 10x \, dx = x^2 y \, dy - 4xy \, dy + 5y \, dy$
28. $x^3 \sqrt{y^2 + 4} \, dy = y \sqrt{x^2 - 1} \, dx$

Find the general solution of each of the following by direct integration until the first derivative has been obtained:

29. $d^2y/dx^2 = 8$ 30. $y'' = 6x - 7$ 31. $y''' = 12x$
32. The acceleration a due to gravity at the earth's surface is 32 ft/sec², approximately. If a ball is projected directly upward, find the general equation of the distance s."

44.6 PARTICULAR SOLUTION

In practical work, we usually require a specific answer for a particular problem. The general solution does not give us this kind of answer. It represents an entire family of specific solutions for many problems of a special type. Instead, to be useful, the solution must satisfy one particular situation.

We have seen that a general solution contains one or more arbitrary constants. Now, if we can determine the specific value of the constants for some particular problem, we shall have a *particular solution*. Then we shall have the form that can be applied to a specific situation. The values of the constants will be determined by certain known facts in the problem.

For example, consider again the differential equation

$$dy = 2x\, dx$$

The solution of the equation is

$$y = x^2 + C$$

This general solution represents a family of parabolas whose axis of symmetry lies along the y axis. A few are shown in Figure 44.1. Now if we

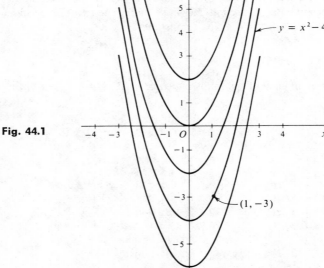

Fig. 44.1

44.6 Particular Solution

specify that we desire the equation of the particular parabola that passes through the point $(1,-3)$, we impose this condition on the equation. The constant, C, is then determined. Since the point $(1,-3)$ must satisfy the equation, we substitute the coordinates of the point in the general solution, $y = x^2 + C$, and get:

$$-3 = 1 + C \quad \text{or} \quad C = -4$$

The equation of this *particular* parabola is then

$$y = x^2 - 4$$

The imposed conditions are called *boundary* or *initial conditions*. The word *initial* is often used when one variable is time t.

EXAMPLE 5. In a certain electric circuit, an ammeter indicates a current of 0.5 ampere. Find the equation for the charge Q on a capacitor in the circuit if the initial charge is 0.2 coulomb.

Solution. Since the current is simply the rate of transport of charge, we can write

$$i = \frac{dQ}{dt} = 0.5$$

Multiplying by dt, $\qquad dQ = 0.5\, dt$

Integrating both sides, $\qquad Q = 0.5t + K$

In this example we use K for the arbitrary constant, since the letter C is often used to represent *capacitance* in an electric circuit. The specific value of K now depends on the initial charge on the capacitor. From the given information, we have $Q = 0.2$ when $t = 0$. Substituting these values, we get

$$0.2 = 0 + K \quad \text{or} \quad K = 0.2$$

The particular solution then becomes

$$Q = 0.5t + 0.2$$

This particular solution enables us to find the charge on the capacitor at any time t in this circuit.

EXAMPLE 6. The acceleration of a particle down a certain inclined plane is 4 ft/sec². If the particle is started with an initial thrust of 6 ft/sec *upward* at a distance of 10 feet from the bottom of the incline, find the particular equations for the velocity and the distance at any time t. (Disregard friction.)

Solution. If we call the upward direction positive, then we have the initial conditions: when $t = 0$, $v = 6$, $s = 10$. Since *upward* is taken as *positive*, then

$$a = \frac{dv}{dt} = -4$$

or, $\qquad dv = -4\, dt$

Integrating, $\qquad v = -4t + C_1$

From the fact that $v = 6$ when $t = 0$, we have

$$6 = 0 + C_1 \quad \text{or} \quad C_1 = 6$$

The particular solution for the velocity becomes
$$v = -4t + 6$$
Now we have
$$v = \frac{ds}{dt} = -4t + 6 \quad \text{or} \quad ds = (-4t + 6)\,dt$$
Integrating,
$$s = -2t^2 + 6t + C_2$$
From the fact that $s = 10$ when $t = 0$, we find that $C_2 = 10$. The particular equation for the distance s is then
$$s = 10 + 6t - 2t^2$$

A particular solution of a differential equation can often be found conveniently by *evaluating between corresponding limits*. We integrate both sides as usual, term by term, and then evaluate the definite integrals with their limits, just as we do in finding the definite integral. When this is done, the general solution with its arbitrary constant C will not appear. Instead, we get the particular solution directly, as shown in the following example:

EXAMPLE 7. In the following problem find the particular solutions by integrating between limits: The acceleration of a particle down an inclined plane is 6 ft/sec^2. If the particle is projected upward along the plane with an initial thrust of 10 ft/sec at a distance of 12 ft above the bottom of the plane, find the particular equations for the velocity and the distance at any time t.

Solution. Calling upward direction positive, we have, for the initial conditions: when $t = 0$, $v = 10$, $s = 12$. Acceleration is always $= -6$. Then, since, $a = dv/dt$, we have
$$\frac{dv}{dt} = -6$$
Separating variables, $\quad dv = -6\,dt$

In the integration, we make use of the following corresponding limits: when $t = 0$, $v = 10$; when $t = t$, $v = v$. Then
$$\int_{10}^{v} dv = \int_{0}^{t} -6\,dt$$
Integrating, $\qquad v\Big]_{10}^{v} = -\Big[6t\Big]_{0}^{t}$

Evaluating, $\qquad v - 10 = -6t \quad \text{or} \quad v = 10 - 6t$

Since $v = ds/dt$, we have $\qquad \dfrac{ds}{dt} = 10 - 6t$

Separating variables, $\qquad ds = (10 - 6t)\,dt$

Now we have the following limits: when $t = 0$, $s = 12$; when $t = t$, $s = s$. Then
$$\int_{12}^{s} ds = \int_{0}^{t} (10 - 6t)\,dt$$
Integrating, $\qquad s\Big]_{12}^{s} = 10t - 3t^2\Big]_{0}^{t}$

Evaluating, $\quad s - 12 = 10t - 3t^2 \quad \text{or} \quad s = 12 + 10t - 3t^2$

44.6 Particular Solution

EXAMPLE 8. Work the following problem, first, by finding the general solution including the constant C, and the particular solution from the initial conditions; and, second, by integrating between limits to find the particular solution directly. Find the particular solution of the following differential equation describing a certain electric circuit in which $L = 0.2$ henry, $R = 4$ ohms, and $E = 16$ volts; $i = 0$ when $t = 0$.

$$0.2 \frac{di}{dt} + 4i = 16$$

(a) *First Method.* Multiplying both sides of the equation by $5\, dt$, we get

$$di + 20i\, dt = 80\, dt \quad \text{or} \quad di = 80\, dt - 20\, i\, dt$$

Factoring,

$$di = 20(4 - i)\, dt$$

Separating variables,

$$\frac{di}{4 - i} = 20\, dt$$

The left side integrates directly into a natural logarithm provided the numerator contains the factor, (-1). Multiply both sides of the equation by -1, and get

$$\frac{-di}{4 - i} = -20\, dt$$

Integrating, $\quad\quad \ln(4 - i) = -20t + \ln C$

Transposing, $\quad\quad \ln(4 - i) - \ln C = -20t$

or $\quad\quad \ln \dfrac{4 - i}{C} = -20t$

In exponential form,

$$\frac{4 - i}{C} = e^{-20t} \quad \text{or} \quad 4 - i = Ce^{-20t}$$

Solving for i, $\quad\quad i = 4 - Ce^{-20t}$

which is the general solution. When $t = 0$, $i = 0$, and we have

$$0 = 4 - C$$

from which $C = 4$. The particular solution becomes

$$i = 4(1 - e^{-20t})$$

(b) *Second Method.* This method follows the first as far as the equation

$$\frac{-di}{4 - i} = -20\, dt$$

Now we integrate and evaluate between the following corresponding limits: when $t = 0$, $i = 0$; when $t = t$, $i = i$. Then

$$\int_0^i \frac{-di}{4 - i} = \int_0^t -20\, dt$$

Integrating, $\quad\quad \ln(4 - i) \Big]_0^i = -20t \Big]_0^t$

Evaluating, $\quad\quad \ln(4 - i) - \ln 4 = -20t$

or $\quad\quad \ln \dfrac{4 - i}{4} = -20t$

In exponential form, $\quad\quad \dfrac{4 - i}{4} = e^{-20t}$

Then $\quad\quad 4 - i = 4e^{-20t} \quad \text{or} \quad i = 4(1 - e^{-20t})$

Note that the second method, that of evaluating the integrals between limits, leads directly to the particular solution.

EXERCISE 44.2

Find the particular solution of each of the following equations with the given initial conditions (or boundary values); the first ten show points on the curves.

1. $dy/dx = y/x$; point $(2,-6)$
2. $y' = 4x - 3$; point $(-1,1)$
3. $dy/dx = -x/y$; point $(-3,-2)$
4. $dy/dx = x/y$; point $(0,-3)$
5. $y' = 3x^2 - 4x + 1$; point $(-1,3)$
6. $y' = 3x^2 - 6x - 9$; point $(-1,8)$
7. $x^3\,dy = y^2\,dx$; point $(2,-4)$
8. $y' = y/x + 2$; point $(1,-5)$
9. $(3-x)y' = y + 2$; point $(5,-1)$
10. $y' = xy^3\sqrt{5-x^2}$; point $(1,2)$

In each of the following examples, find the particular solution for velocity v and distance s from the initial conditions (a is acceleration).

11. $a = -6$ ft/sec^2; when $t = 0$, $v = 18$ ft/sec; $s = 5$ ft. Find each when $t = 3$.
12. $a = -5$ ft/sec^2; when $t = 0$, $v = 60$; $s = 0$. Find each when $t = 2$.
13. $a = -2$ ft/sec^2; when $t = 0$, $v = -3$; $s = 60$. Find each when $t = 1$.
14. $a = -8$ ft/sec^2; when $t = 1$, $v = 7$; $s = 15$. Find each when $t = 3$.
15. $a = 32$ ft/sec^2; when $t = 1$, $v = 32$; $s = 16$. Find each when $t = 5$.
16. $a = -10$ ft/sec^2; when $t = 2$, $v = 4$; $s = 26$. Find each when $t = 4$.
17. $a = 0$; when $t = 0$, $v = 30$ ft/sec; $s = 0$. Find each when $t = 3$.
18. A ball is projected directly upward from the top of an 80-ft building with an initial velocity of 96 ft/sec. If $a = -32$ ft/sec^2, find the particular equations for velocity and distance covered at any time t. Then find the velocity and position when $t = 4$ seconds.

Find the particular solution for each of the following electric circuit problems, taking for initial conditions, when $t = 0$, $i = 0$.

19. $(0.4)(di/dt) + 6i = 10$
20. $(0.2)(di/dt) + 4i = 8$
21. $(0.02)(di/dt) + 10i = 0.1$
22. $(0.75)(di/dt) + 4.8i = 2$
23. $(0.05)(di/dt) + 20i = 4$
24. $(0.032)(di/dt) + 0.8i = 6$
25. Solve the following equation for the particular solution if $i = 0$ when $t = 0$:

$$L\frac{di}{dt} + Ri = E$$

44.7 PHYSICAL APPLICATIONS

Many types of problems involving physical laws and relations can be solved by differential equations. We have already seen how an electric circuit problem can be solved. Let us look at a few other kinds.

One type of problem involves the laws of growth and decay. In the disintegration of many substances, the rate of loss or decay is directly proportional to the amount of the substance present at any instant. As the disintegration takes place, the amount of the substance decreases. Then the *amount of decay or loss* also *decreases*, but the *rate of decay* remains constant.

Let us suppose we begin with one pound of a substance whose rate of decay is proportional to the amount present. Let us further suppose that the substance loses half its weight in one year. At the end of one year

$\frac{1}{2}$ pound remains. During the second year the loss is half the remaining portion, or $\frac{1}{4}$ pound. If the rate of loss continues uniformly, the loss is $\frac{1}{8}$ pound the third year, $\frac{1}{16}$ pound the fourth year, and so on. Theoretically, the substance would never completely disintegrate. For this reason it is customary to speak of the *half-life* of a substance. For example, the half-life of radium is approximately 1600 years. This means it loses half its amount in that period of time.

If we let x represent the number of pounds of a substance remaining at any instant, then the rate of decay can be indicated by

$$\frac{dx}{dt}$$

Now we wish to state that this rate is proportional to the amount present. Since the amount decreases, the *rate is negative with respect to time t*: then

$$\frac{dx}{dt} = -Kx$$

The constant K is the constant of proportionality and varies with different substances. If the amount of the substance increases with time, as with the increase in the number of bacteria in a culture, then we take K positive.

Note that we have a differential equation that can be solved by separating the variables. After the equation has been solved, the value of K can be determined from certain given conditions. Then the resulting formula can be used to find the amount of the substance that will remain after any certain period. For the half-life of a substance, the portion that remains is 50 percent of the original amount.

EXAMPLE 9. A certain substance disintegrates in direct proportion to the amount present at any instant. If it loses 20 percent of its weight in 24 hours, find (a) its half-life; (b) the amount remaining after 200 hours; (c) the length of time until only 1 percent remains.

Solution. It is often convenient to begin with some specific amount. Let us say the original amount is 100 grams. Now we

let $x =$ the number of grams present at any time t

Then we write the differential equation:

$$\frac{dx}{dt} = -Kx$$

Solving, $$\frac{dx}{x} = -K\, dt$$

then $$\ln x = -Kt + \ln C$$

from which, $$\ln x - \ln C = -Kt \quad \text{or} \quad x = Ce^{-Kt}$$

Since $x = 100$ when $t = 0$, we find that $C = 100$, and we have the particular solution

$$x = 100e^{-Kt}$$

Now we must find the value of K. We have the information that when $t = 24$, $x = 80$. Substituting these values, we get
$$80 = 100e^{-24K} \quad \text{or} \quad e^{24K} = 1.25$$
Solving, $24K = \ln 1.25$

or $$K = \frac{\ln 1.25}{24} = \frac{0.223}{24} = 0.0093 \text{ (approx.)}$$

Now we have the particular solution with the specific value of K:
$$x = 100e^{-0.0093t}$$
To find the half-life of the substance, we must have $x = 50$ grams, or
$$50 = 100e^{-0.0093t}$$
Dividing by 100, $\quad 0.5 = e^{-0.0093t} \quad \text{or} \quad 2 = e^{0.0093t}$

Taking the natural logarithm of both sides,
$$\ln 2 = 0.0093t$$
Solving for t, $\quad t = \dfrac{\ln 2}{0.0093} = \dfrac{0.693}{0.0093} = 74.5 \text{ (approx.)}$

The half-life of the substance is therefore approximately 74.5 hours.

To find the amount left after 200 hours, we use the formula with $t = 200$. Then
$$x = 100e^{-0.0093(200)} = 100e^{-1.86} = 15.57 \text{ (approx.)}$$
To find the time when 1 percent is left, we take $x = 1$ in the formula and get
$$1 = 100e^{-0.0093t} \quad \text{or} \quad 100 = e^{0.0093t}$$
Solving for t, we first have
$$0.0093t = \ln 100$$
Then $\quad t = \dfrac{\ln 100}{0.0093} = \dfrac{4.605}{0.0093} = 495.2 \text{ (approx.)}$

The result means that 1 percent of the substance will remain after 495.2 hours.

It might be mentioned that the value of K can easily be determined by evaluating the integrals between corresponding limits. For example, let us begin with the differential equation
$$\frac{dx}{x} = -K\,dt$$
When we integrate, we can use the limits: when $t = 0$, $x = 100$; when $t = 24$, $x = 80$. Then
$$\ln x \Big]_{100}^{80} = -Kt \Big]_{0}^{24}$$
Evaluating,
$$\ln 80 - \ln 100 = -24K \quad \text{or} \quad 24K = \ln \frac{100}{80} = \ln 1.25$$
Solving for K, $\quad K = \dfrac{\ln 1.25}{24} = 0.0093$

Once the value of K has been determined, all the answers may be found by integrating and evaluating between corresponding limits.

Another type of physical problem that can be solved by the use of differential equations involves *Newton's law of cooling*. Consider the following problem: You have before you a cup of hot coffee without cream. Because of a sudden interruption you realize you will not be able to drink your coffee for ten minutes, yet you do not want it to cool more than necessary. Would you pour the cream into the coffee at once or wait until the end of the ten-minute interval?

Newton's law of cooling states that the rate of change of heat in a substance is proportional to the *difference* between the temperature of the substance and the temperature of the surrounding medium. A hot substance loses heat to a cooler surrounding medium at a rate proportional to the difference in temperature. In a surrounding atmosphere of 60°F, water at 150°F will lose heat faster than water at 100°F. In the first case, the difference is 90°. In the second case it is only 40°.

EXAMPLE 10. A liquid whose temperature is 180°F is exposed to the surrounding air whose temperature is 60°F. If the temperature of the liquid is 150°F at the end of 4 minutes, what is the temperature at the end of 10 minutes?

Solution. Let

T = the temperature (in degrees Fahrenheit) at any time t

Then

$T - 60$ = the difference in temperature at any time t

and $\dfrac{dT}{dt}$ = rate of change of temperature with respect to time

Again, using K as the constant of proportionality and noting that T decreases as t increases, we have

$$\frac{dT}{dt} = -K(T - 60)$$

Solving, $$\frac{dT}{T - 60} = -K\,dt$$

then $$\ln(T - 60) = -Kt + \ln C$$

We get the general solution $T = 60 + Ce^{-Kt}$

From the initial conditions, $T = 180$ when $t = 0$, we find that $C = 120$. Then we get the particular solution

$$T = 60 + 120e^{-Kt}$$

The value of K can be determined from the fact that $T = 150$ when $t = 4$; then

$$150 = 60 + 120e^{-4K}$$

from which $90 = 120e^{-4K}$ or $e^{-4K} = 0.75$

Solving for K, $-4K = \ln 0.75$

then $$K = \frac{-0.28768}{-4} = 0.0719 \text{ (approx.)}$$

Now we have the particular solution with the specific value of K:
$$T = 60 + 120e^{-0.0719t}$$
For the temperature T at the end of 10 minutes, we have
$$T = 60 + 120e^{-0.719} = 60 + 120(0.487) = 60 + 58.44 = 118.44 \text{ (approx.)}$$

EXERCISE 44.3

1. A certain substance reduces in weight from 100 grams to 90 grams in 4 hours. Find (a) its half-life; (b) the amount left after one day; (c) how long it will take until the weight has been reduced to 10 grams.
2. One hundred grams of a certain substance reduces to 98 grams in 20 days. Find (a) its half-life; (b) the amount left after 300 days; (c) how long it will take until its weight has been reduced to 2 grams.
3. Radium loses about 1.3 percent of its weight in 30 years. Find (a) its half-life; (b) the amount left after 400 years; (c) the amount left after 5000 years; (d) how long it will take to lose 99 percent of its weight.
4. One form of carbon, C^{14}, has a half-life of 5570 years. (a) What percent of it remains after 1000 years? (b) How long will it be until only 20 percent remains?
5. A certain chemical substance, Na^{24}, has a half-life of 15 hours. (a) What percent of the substance remains after 10 hours? (b) after 50 hours? (c) after 200 hours? (d) How long will it take for 99.9 percent of it to disappear?
6. The substance, U^{235}, has a half-life of about $7(10^8)$ years. What percent of it is lost (a) in 1000 years? (b) in 1,000,000 years?
7. A certain substance increases in size in proportion to the amount present. If it increases by 20 percent in 5 hours, by what percent will it increase in 32 hours?
8. If the number of bacteria in a culture increases in proportion to the number present, and if there were 1000 bacteria at noon and 1500 at 2 P.M., of the same day, how many will there be at 10 P.M. the same day? How many were there at 6 hours before noon? How many hours before noon were there 10?
9. A thermometer showing 70°F inside a house is taken outside where the temperature is 20°. In 2 minutes it shows a reading of 50°F. What will it show in 10 minutes after being taken outside?
10. A thermometer shows a reading of 80°F inside a room at 9 A.M. It is taken outside where the temperature is −10°F. At 9:02 A.M. the thermometer shows 30°F. What will it show at 9:10 A.M? at 9:15 A.M?
11. A thermometer shows an outside reading of 20°F. It is taken inside where the temperature is 70°F. Two minutes later the thermometer shows a reading of 30°F. What will it show 10 minutes after being taken inside?
12. A thermometer inside a house shows a reading of 70°F. At 9 A.M. it is brought outside where the temperature is 30°F. At 9:03 it shows a reading of 50°F. At 9:10 A.M. it is brought inside again. What will it show at 9:15?

chapter

45

Perfect Differentials

45.1 DEFINITION

We have seen that if we can separate the variables in a differential equation, then the equation can be solved. All we need to do then is to integrate each part. The solution, of course, depends on a good knowledge of integration techniques.

Now we might think that solving any differential equation is therefore a simple matter. We just separate the variables. However, sometimes this cannot be done. In fact, this is true in most practical problems. When the variables cannot be separated, then other methods must be tried.

Some differential equations can be solved by recognizing a perfect differential of some function. If we notice that a portion of the equation is a perfect differential, then we integrate the differential and get the function itself. Actually, this is what we do in a simple equation such as the following:

$$dy = 2x\,dx$$

Here we recognize dy as the differential of y, and $2x\,dx$ as the differential of x^2. Then we get, at once,

$$y = x^2 + C$$

45.2 DIFFERENTIALS OF PRODUCTS AND QUOTIENTS

In some equations a perfect differential can be recognized rather easily. However, we run into trouble especially when we have the differential of the product or the quotient of two variables. For example, take the differential of a product

$$d(xy) = x\,dy + y\,dx$$

Now, the expression, $x\,dx + y\,dx$, is the perfect differential of the product xy. But can we immediately recognize this fact? If we recognize a particular expression as the perfect differential of a product, then we can *integrate the entire perfect differential as a whole in one stroke.*

642 Perfect Differentials

Let us see how the foregoing differential may appear in a problem. Suppose we have the differential equation

$$x\,dy + y\,dx - x^2\,dx = 0$$

If we attempt to separate the variables, we find this cannot be done. However, we might recognize part of the equation as a perfect differential. Then we rewrite the equation

$$x\,dy + y\,dx = x^2\,dx$$

Now we might see at once that the left side of the equation is the perfect differential of the product xy. Then we integrate the left side *as a whole* into xy. The right side is easily integrated by itself. If we wish we might first write

$$\int (x\,dy + y\,dx) = \int x^2\,dx$$

For the solution, we get

$$xy = \frac{1}{3}x^3 + C$$

If we wish, we can solve the equation for y:

$$y = \frac{1}{3}x^2 + \frac{C}{x}$$

Note that the foregoing equation was first put into the form

$$x\,dy + y\,dx = f(x)\,dx$$

If the right side had now been zero or a constant, the equation could have been solved by separating the variables.

EXAMPLE 1. Solve the following equation in two ways and find the particular solution if $x = 5$ when $y = -2$:

$$x\,dy + y\,dx = 4\,dx$$

Solution. First, we solve the equation by noting the perfect differential of the product xy; this portion of the equation can be integrated as a whole:

$$\int (x\,dy + y\,dx) = \int 4\,dx$$

Integrating, $\qquad xy = 4x + C$

which may be written $\qquad x(y - 4) = C$

Now let us solve the same equation by separating the variables:

$$x\,dy + y\,dx = 4\,dx$$

The equation may be written

$$x\,dy = 4\,dx - y\,dx$$

or $\qquad x\,dy = (4 - y)\,dx$

Separating variables, $\qquad \dfrac{dy}{4 - y} = \dfrac{dx}{x}$

Multiplying by -1, $\qquad \dfrac{-dy}{4 - y} = -\dfrac{dx}{x}$

Integrating both sides, $\quad \ln(4-y) = -\ln x + \ln C$

Transposing, $\quad \ln(4-y) + \ln x = \ln C$

or $\quad \ln x(4-y) = \ln C$

Taking antilogarithms, $\quad x(4-y) = C$

Comparing the answer with that obtained by the first method, we see that the two results look different, but if one C is the negative of the other, then the two results are equivalent. The important fact is that the particular solution is exactly the same in both instances. From the given information, we get the particular solution

$$y = 4 - \frac{30}{x}$$

45.3 INTEGRATING FACTOR

If one side of an equation is not exactly the perfect differential of a function, it can sometimes be made so by some multiplier called an *integrating factor*, as in the following example.

EXAMPLE 2. Solve the equation

$$x\, dy + 3y\, dx = \arctan x\, dx$$

Solution. The left side of the equation looks as though it might be the differential of some product. The coefficient 3 is puzzling, but it may remind us that just such a factor appears in the derivative of the *third power*, a *cubic*. The equation therefore probably requires a multiplier as an integrating factor. This factor is probably a term in x, since the right side of the equation is a function of x. Moreover, if we use the factor x^2 we shall have the cubic x^3 on the left side and we may then have a perfect differential. So we try it, and get, multiplying by x^2,

$$x^3\, dy + 3x^2 y\, dx = x^2 \arctan x\, dx$$

By inspection we now discover that the left side is the perfect differential of the product, $x^3 y$. We can now *integrate the left side as a whole* and get

$$x^3 y = \int x^2 \arctan x\, dx$$

The right side can be integrated by parts by letting $u = \arctan x$, and $dv = x^2\, dx$. Then, on the right side we get

$$\frac{x^3}{3}(\arctan x) - \frac{1}{3}\int \frac{x^3\, dx}{x^2 + 1}$$

For the final solution we get

$$x^3 y = \frac{x^3}{3}(\arctan x) - \frac{x^2}{6} + \frac{1}{6}\ln(x^2+1) + C$$

EXAMPLE 3. Solve the equation

$$x\, dy - y\, dx = x^3 e^x\, dx$$

644 Perfect Differentials

Solution. The left side is not a perfect differential. However, the minus sign might remind us of the differential of a quotient. Then we try the differential of y/x:

$$d\left(\frac{y}{x}\right) = \frac{x\,dy - y\,dx}{x^2}$$

This is similar to the form in the differential equation except for the denominator. The left side can be put into the proper form by dividing both sides by x^2 (or multiplying by $1/x^2$). The equation then becomes

$$\frac{x\,dy - y\,dx}{x^2} = xe^x\,dx$$

Now the left side is integrated as a whole into y/x. The right side is integrated by parts. We get

$$\frac{y}{x} = xe^x - e^x + C$$

If we wish, we can solve for y, and get

$$y = e^x(x^2 - x) + Cx$$

In some instances the integrating factor can be determined by inspection. In many instances this is a matter of "trial and error." It is suggested that the student write out the differentials for the following products and quotients for reference. The similarity between one of these and a part of a differential equation may lead to the discovery of the correct integrating factor. However, later we shall see how to determine such a factor more easily.

Work the following differentials for handy reference:

(a) $d(xy)$ (e) $d(xy^3)$ (i) $d\left(\dfrac{x}{y}\right)$ (l) $d\left(\dfrac{x}{y^2}\right)$

(b) $d(x^2y)$ (f) $d(x^2y^3)$ (j) $d\left(\dfrac{y}{x}\right)$ (m) $d\left(\dfrac{1}{xy}\right)$

(c) $d(xy^2)$ (g) $d(x^3y^2)$ (k) $d\left(\dfrac{x^2}{y}\right)$ (n) $d\left(\dfrac{x^2}{y^2}\right)$

(d) $d(x^3y)$ (h) $d(x^2y^2)$

EXERCISE 45.1

Solve the following differential equations by discovering an integrating factor by inspection.

1. $x\,dy + y\,dx - \sin^3 x\,dx = 0$
2. $x\,dy + 2y\,dx - e^x\,dx = 0$
3. $3x\,dy + y\,dx - \cos y\,dy = 0$
4. $y\,dx - x\,dy - 2x\,dx = 0$
5. $x\,dy - 2y\,dx - 3x\,dx = 0$
6. $x\,dy - 3(y + x)\,dx = 0$
7. $x\,dy - (x^2 + 3x + y)\,dx = 0$
8. $x\,dy + (x^2 \sec^2 x - y)\,dx = 0$
9. $x\,dy + y\,dx = x^2y^2\,dx$
10. $x\,dy + (2y - \sqrt{4 - x^2})\,dx = 0$
11. $y\,dx + x\,dy - 5\,dy = 0$
12. $y\,dx + [3x + (y^3 + 1)^{1/3}]\,dy = 0$

45.4 FIRST-ORDER, FIRST-DEGREE DIFFERENTIAL EQUATION

We have seen that some differential equations can be solved by use of a proper integrating factor. Up to this point we have had to rely somewhat

45.4 First-Order, First-Degree Differential Equation

on guesswork to discover the correct factor. However, there is one particular type of differential equation for which the proper integrating factor can always be found directly. This is a differential equation of the first order and first degree. It is sometimes called a *linear differential equation of the first order*.

This particular type of equation contains the derivative of only the first order, dy/dx. It contains no higher derivative but it may contain the function, y, itself. However, neither y nor its derivative, dy/dx, is raised to a power higher than the first. Here is an example of the type we are now considering:

$$\frac{dy}{dx} + 2xy = 5x^3$$

This type of equation is of vital importance because it occurs often in practical problems in science. It can be represented by the general form

$$\frac{dy}{dx} + Py = Q \quad \text{or} \quad y' + Py = Q$$

In this equation, when the coefficient of (dy/dx) is 1, then P represents the coefficient of y, and Q represents the term on the right side of the equation. The coefficient P may be a constant or a function of x. The quantity Q may be zero, a constant, or a function of x. For this reason the general differential equation of this type is sometimes written

$$\frac{dy}{dx} + P(x)y = Q(x)$$

to indicate that P and Q may be functions of x.

We shall find that this form of equation can always be solved because the proper integrating factor can always be found directly. Equations of this kind are so important that as a student you should try to recognize them quickly and to identify the quantities P and Q.

Let us look at another equation of this type:

$$3\frac{dy}{dx} - 2xy = 5x^2$$

Dividing both sides by 3, we get

$$\frac{dy}{dx} - \frac{2}{3}xy = \frac{5}{3}x^2$$

Now we can recognize the form, $y' + Py = Q$, where

$$P = -\frac{2}{3}x \quad \text{and} \quad Q = \frac{5x^2}{3}$$

To identify P and Q, we must be sure that the coefficient of y' is 1.

If we cannot solve this equation by separating the variables, and if we cannot immediately recognize a proper integrating factor, then we can always *make the left side a perfect differential*. To do so, we multiply both sides of the equation by this special exponential integrating factor:

$$e^{\int P(x)\,dx}$$

You may wonder and ask: Where did this weird-looking factor come from? And why does it work? Let us try to answer both questions.

We begin with the general equation of this type:

$$\frac{dy}{dx} + P(x)y = Q(x)$$

Now let us suppose there is some function of x, as yet unknown, that is the correct integrating factor. We shall call it $F(x)$, or simply F. Let us assume that this factor, F, times y produces the integral of the left side of the equation. Now we try to discover the factor F. For the time being, F may be called just a "dummy" factor because eventually we shall discover its identity. Anyway, we proceed to multiply the equation by F and get

$$(F)\left(\frac{dy}{dx}\right) + (F)P(x)y = (F)Q(x)$$

Multiplying by dx,

$$(F)\,dy + (F)P(x)y\,dx = (F)Q(x)\,dx$$

Now we do not worry about the right side of the equation because it is some function of x alone and theoretically can be integrated.

If the left side is now to be a perfect differential, let us assume it is the differential of $(F)(y)$. In other words, the left side is the differential of the product of y and our multiplier, whatever our multiplier turns out to be. Then let us find the differential of this product. The differential must then equal the original differential on the left side of the equation. Since $(F)(y)$ is a product, its differential is

$$(F)\,dy + [y][d(F)]$$

Note that this expression is exactly like the left side of the preceding equation if we say that

$$d(F) = [F][P(x)]\,dx$$

The two expressions are equal provided F has the correct value. So we solve the equation for F. Dividing both sides by F, we get

$$\frac{d(F)}{F} = P(x)\,dx$$

Integrating, $\qquad \ln F = \int P(x)\,dx \qquad$ or $\qquad F = e^{\int P(x)\,dx}$

Now we realize that the multiplier must be

$$e^{\int P(x)\,dx}$$

Now we have discovered that this must be the multiplier if the left side is to be a perfect differential. Moreover, it is the perfect differential of the

45.4 First-Order, First-Degree Differential Equation

product of y and the multiplier. The left side therefore integrates as a whole into the product

$$[y][e^{\int P(x)\, dx}]$$

EXAMPLE 4. Solve the equation $dy = (3y + x)\, dx$, if $y = 0$ when $x = 0$.

Solution. In this equation we cannot separate the variables. Let us see whether it can be put into the form: $y' + Py = Q$. Dividing both sides by dx, we get

$$\frac{dy}{dx} = 3y + x \quad \text{or} \quad \frac{dy}{dx} - 3y = x$$

Now we can recognize the form, $y' + Py = Q$, in which $P = -3$, and $Q = x$. Then the integrating factor becomes

$$e^{\int -3\, dx} = e^{-3x}$$

Multiplying both sides by e^{-3x} and also by the factor dx, we get

$$e^{-3x}\, dy - 3ye^{-3x}\, dx = xe^{-3x}\, dx$$

Now the left side integrates as a whole into y times the integrating factor. Then

$$ye^{-3x} = \int xe^{-3x}\, dx$$

Integrating the right side by the method of parts, we have

$$ye^{-3x} = -\frac{1}{3}xe^{-3x} - \frac{1}{9}e^{-3x} + C$$

We can solve for y by multiplying both sides of the equation by e^{3x}:

$$y = -\frac{x}{3} - \frac{1}{9} + Ce^{3x}$$

For the particular solution, we have the information that $y = 0$ when $x = 0$. Substituting,

$$0 = 0 - \frac{1}{9} + C \quad \text{or} \quad C = \frac{1}{9}$$

The particular solution becomes

$$y = -\frac{x}{3} - \frac{1}{9} + \frac{1}{9}e^{3x} = \frac{1}{9}[e^{3x} - 3x - 1]$$

EXAMPLE 5. Find the general solution:

$$x\, dy - 2y\, dx = (x^3 + x)\, dx$$

Solution. Dividing by $x\, dx$, we get

$$\frac{dy}{dx} - \frac{2y}{x} = (x^2 + 1)$$

Now we can recognize the form, $y' + Py = Q$, where $P = -2/x$; $Q = x^2 + 1$. The integrating factor becomes

$$e^{\int -2/x\, dx} = e^{-2\ln x} = \frac{1}{x^2}$$

Multiplying both sides of the equation by $(1/x^2)\,dx$, we get

$$\frac{dy}{x^2} - \frac{2y\,dx}{x^3} = \left(1 + \frac{1}{x^2}\right)dx$$

Integrating the left side as a whole,

$$\frac{y}{x^2} = x - \frac{1}{x} + C \quad \text{or} \quad y = x^3 - x + Cx^2$$

EXAMPLE 6. Solve $\dfrac{di}{dt} + 3i = \sin 5t$

Solution. In the present form we recognize: $P = 3$; $Q = \sin 5t$. The integrating factor is e^{3t}. Multiplying both sides by $e^{3t}\,dt$, we get

$$e^{3t}\,di + 3ie^{3t}\,dt = e^{3t}\sin 5t\,dt$$

Integrating, $\qquad ie^{3t} = \int e^{3t}\sin 5t\,dt$

The right side can be integrated by applying the *parts* rule twice, or a table of integral forms can be used. The form on the right should be memorized because it occurs often. The result is

$$ie^{3t} = \frac{e^{3t}}{34}(3\sin 5t - 5\cos 5t) + C$$

Solving for i, $\qquad i = \dfrac{1}{34}(3\sin 5t - 5\cos 5t) + Ce^{-3t}$

EXERCISE 45.2

Solve by the use of the integrating factor $e^{\int P\,dx}$:

1. $dy - (3y + 4)\,dx = 0$
2. $dy = (e^x + 5y)\,dx$
3. $dy - (xy + x)\,dx = 0$
4. $dy + 2x(y - e^{2x^2})\,dx = 0$
5. $x\,dy + (y - \sin x)\,dx = 0$
6. $dy = (\sin x - y\cot x)\,dx$
7. $dy = (\sin x - 4y)\,dx$
8. $(x^2 + 1)\,dy = (x^3 + x + xy)\,dx$
9. $x\,dy - (x + 2 + 3y)\,dx = 0$
10. $dy = (1 + y\tan x)\,dx$
11. $di + 5i\,dt = -3t\,dt$
12. $di/dt + 4i = \cos 2t$
13. $0.2\dfrac{di}{dt} + 6i = 8$
14. $L\dfrac{di}{dt} + Ri = E$
15. $\dfrac{di}{dt} + 8i = \cos 6t$
16. $0.4\dfrac{di}{dt} + 12i = 8\sin 20t$
17. $L\dfrac{di}{dt} + Ri = E\sin \omega t$
18. $\dfrac{ds}{dt} + 5s = t^2$
19. $x\,dy = (e^x - 2y)\,dx$
20. $x\,dy = (e^x - 3y)\,dx$
21. In a certain L-R electric circuit, $L = 0.2$ henry, $R = 10$ ohms, and $E = 40$ volts. Using the form $(L)(di/dt) + Ri = E$, find the particular solution for the current i at any time t if $i = 0$ when $t = 0$.
22. In a certain L-R circuit, $L = 0.5$ henry, $R = 3$ ohms, $E = 6$ volts. Find the particular solution for the current i at any time t if $i = 0$ when $t = 0$. Then, from the particular solution, find the instantaneous current when $t = 0.1$ sec.
23. Solve for i in an electric circuit in which $L = 0.4$, $R = 6$, $E = \sin 20t$.
24. Solve for I in a circuit in which $L = 0.3$, $R = 12$, $E = \sin 30t$.
25. Solve the problems in Exercise 45.1 by use of the integrating factor.

chapter

46

Homogeneous Equations; Constant Coefficients

46.1 DEFINITIONS

We have already solved equations of the first order. Now let us consider a linear differential equation of any order; that is, an equation containing a number of derivatives of any order, such as y', y'', y''', and so on. The equation may also contain y itself.

Now, if none of the derivatives nor y itself is raised to a power higher than the first, then the equation is called a *linear differential equation in y and its derivatives*. The following are examples of this kind:

$$y'' - 5y' + 6y = 0 \qquad \frac{d^2s}{dt^2} + 16s = 0$$

$$\frac{di}{dt} + 4i = 3t^2 \qquad 3xy''' + x^2y'' - 4y' = x^3 \sin x$$

Note that all forms of y and its derivatives are of the first degree.

In general, linear differential equations may be written in the form:

(a) Second order: $\qquad y'' + f(x)y' + g(x)y = q(x)$

(b) Third order: $\qquad y''' + f(x)y'' + g(x)y' + h(x)y = q(x)$

and so on for higher derivatives. Note that the coefficients of y and its derivatives may be constants or functions of x. The right side of the equation, $q(x)$, is a function of x alone.

Now, if the right side of the equation, $q(x)$, is equal to zero, then the equation is called *homogeneous*. These equations are homogeneous:

(a) $y''' + 3xy'' + x^2y' + 2y = 0 \qquad$ (b) $y'' - 5y' + 6y = 0$

We shall first consider second-order homogeneous equations in which all coefficients of y and its derivatives are constants, such as these:

(a) $y'' - 5y' + 6y = 0 \qquad$ (b) $\dfrac{d^2i}{dt^2} + 6\dfrac{di}{dt} + 8i = 0$

649

This type of equation is important not only for itself alone but also because it is used in solving nonhomogeneous equations. In much work in science we are faced with solving equations with constant coefficients.

46.2 DERIVATIVE OPERATOR

To derive a simple method for solving the type of equation just mentioned, it is necessary that we recall what is meant by the derivative operator. We have used the expression, dy/dx, to denote the derivative of y with respect to x. We have also used the expression, d/dx, to indicate the operation of differentiation. For example, we can write

$$\frac{d}{dx}(5x^3 - 3x^2) = 15x^2 - 6x$$

The symbol, d/dx, is called the *derivative operator*. It indicates that the operation of differentiation is to be performed on any expression that follows it.

For convenience, we sometimes use the notation D_x for the operator. If there is no question as to the independent variable, we sometimes use the single letter D to denote the derivative operator, d/dx. Thus,

$$D(5x^3 - 3x^2) = 15x^2 - 6x$$

For the second derivative we have used the notation, d^2y/dx^2, or y'', and similar notations for higher derivatives. We can also use a corresponding D notation for higher derivatives, such as

$$D^2 = \frac{d^2}{dx^2} \qquad D^3 = \frac{d^3}{dx^3} \qquad D^n = \frac{d^n}{dx^n}$$

The notations, $D, D^2, D^3, \ldots D^n$, do not indicate powers but instead the successive derivatives of the quantity that follows the operator D^n.

Using this D notation for the derivative operator, we can write the successive derivatives of a function. For example, if $f(x) = x^4$,

for the first derivative, $\qquad D(x^4) = 4x^3$

for the second derivative, $\qquad D^2(x^4) = 12x^2$, which is $D[D(x^4)]$

for the third derivative, $\qquad D^3(x^4) = 24x$, and so on

The use of D, D^2, D^3, and so on for the derivative operator is not only convenient but also forms the basis for a simple method of solving many equations. Suppose we have the function x^5 and wish to find:

(a) the second derivative of the function

(b) plus 3 times the first derivative

(c) plus 2 times the function itself

That is, we wish to find

$$D^2(x^5) + 3D(x^5) + 2(x^5)$$

46.2 Derivative Operator

The expression becomes
$$20x^3 + 15x^4 + 2x^5$$
If we let y represent the function, we have
$$D^2y + 3Dy + 2y = 20x^3 + 15x^4 + 2x^5$$
Let us consider a little more carefully the expression
$$D^2(x^5) + 3D(x^5) + 2(x^5)$$
Now we might ask whether we can factor out the function x^5, making use of the distributive law and say that the expression could be written
$$(D^2 + 3D + 2)(x^5)$$
If we understand that this expression means that the derivative operators are to be applied to the function, x^5, in turn and the result added, then the distributive law is valid here. As a matter of fact, this is the meaning of the expression. Then we can say that up to this point, everything looks fine.

Now let us go one step further. We have seen that
$$(D^2 + 3D + 2)(x^5) = 20x^3 + 15x^4 + 2x^5$$
Now let us ask whether we can assume that the expression $(D^2 + 3D + 2)$ can be treated as an algebraic polynomial that can be factored into
$$(D + 1)(D + 2)$$
That is, does $(D^2 + 3D + 2)(x^5)$ give the same result as $(D + 1)(D + 2)(x^5)$? Remember, D^2 means the second derivative, *not the second power*, of D.

However, let us see if we get the same answer by using the factored form of the operators:
$$(D + 1)(D + 2)(x^5)$$
First, we operate on the function, x^5, with the second factor as operator:
$$(D + 2)(x^5) = 5x^4 + 2x^5$$
Now we operate on this result with the first factor as operator and get
$$(D + 1)(5x^4 + 2x^5) = D(5x^4 + 2x^5) + 1(5x^4 + 2x^5)$$
$$= 20x^3 + 10x^4 + 5x^4 + 2x^5 = 20x^3 + 15x^4 + 2x^5$$

The amazing fact is that we get the same answer when we factor the expression $(D^2 + 3D + 2)$ as an algebraic polynomial and then use each factor in turn as an operator!

In general, if we have a number of operators expressed in the form of a polynomial, this polynomial may be treated as any algebraic polynomial and may be factored. Then each factor may be considered operating on the function in turn. This rule is valid when the coefficients of D^n are constants, but it is not valid if such coefficients are variables.

EXERCISE 46.1

Show that the result in each of the following expressions is the same whether the given function is differentiated by the operators in the form shown or by the operators in factored form.

1. $(D^2 + 4D + 3)(x^5)$
2. $(D^2 + 2D - 3)(x^3 + 4x^2)$
3. $(D^2 + D - 6)(\sin 3x)$
4. $(D^2 - 3D - 10)(e^{5x})$
5. $(D^2 - 2D - 8)(e^{-2x} + e^{4x})$
6. $(D^2 + 2D - 15)(e^x + e^{-x})$
7. $(D^2 + 8D + 12)(e^{-6x})$
8. $(D^2 - 4D - 12)(5e^{-2x} - e^{6x})$
9. $(D^2 + D)(x^4 + 5x^3 - 3x^2)$
10. $(D^2 + 5D + 6)(\tan 3x)$
11. $(2D^2 + 7D + 3)(x^4 + 6x^3)$
12. $(3D^2 - 5D - 2)(x^3 - x)$
13. $(D^2 + 4)(5 \sin 2x)$
14. $(D^2 + 2D + 1)(xe^{-x})$
15. $D(D^2 - 3D + 2)(x^6 - x^3)$
16. $(D - 1)(D^2 - D - 6)(x^5)$
17. $(D^2 - 3D + 2)(4e^x - 5e^{2x})$
18. $(D + 1)(D^2 + 2D - 8)(\cos x)$

46.3 SOLUTIONS BY OPERATORS

The use of operators is a convenient method of solving some differential equations. Moreover, we shall find that this method leads to a simple form.

As an example to illustrate the use of operators, consider the equation

$$y'' - 5y' + 6y = 0$$

In operator form, $\quad D^2y - 5Dy + 6y = 0$

Factoring out y, $\quad (D^2 - 5D + 6)y = 0$

In factored form, $\quad (D - 3)(D - 2)y = 0$

We shall first solve the equation in steps.

We can think of this expression as meaning that the second factor $(D - 2)$ operates on y first. Then we can think of the first factor $(D - 3)$ as operating on the quantity $(D - 2)y$. Let us say that the quantity $(D - 2)y$ is some quantity we shall represent by another letter, say z. That is, we let

$$(D - 2)y = z$$

Now, instead of saying that $(D - 3)$ operates on the quantity $(D - 2)y$, we can say $(D - 3)$ operates on z. Then we can state the differential equation,

$$(D - 3)z = 0$$

where z represents $(D - 2)y$. Then we have

$$\frac{dz}{dx} - 3z = 0$$

We can solve this differential equation by separating variables and get

$$z = C_1 e^{3x}$$

Now that we have found the expression for z, we go back to the statement in which we let

$$(D - 2)y = z$$

Then we have $(D - 2)y = C_1 e^{3x}$

The expression means $\dfrac{dy}{dx} - 2y = C_1 e^{3x}$

The variables in this case cannot be separated, but we note the form

$$\frac{dy}{dx} + Py = Q$$

We recall that the integrating factor in this form is

$$e^{\int P\, dx}$$

In this example the integrating factor becomes e^{-2x}. Multiplying by this factor and dx, we get

$$e^{-2x}dy - 2ye^{-2x}dx = C_1 e^x\, dx$$

Integrating, $\qquad\qquad ye^{-2x} = C_1 e^x + C_2$

Multiplying both sides by e^{2x}, $\qquad y = C_1 e^{3x} + C_2 e^{2x}$

EXERCISE 46.2

Solve the following differential equations by operators:

1. $y'' - 3y' - 4y = 0$
2. $y'' - 8y' + 12y = 0$
3. $y'' - 7y' + 10y = 0$
4. $y'' + 6y' + 8y = 0$
5. $y'' - 4y = 0$
6. $y'' + 3y' = 0$
7. $4y'' - 9y = 0$
8. $2y'' + 5y' = 0$
9. $y'' + 5y' - 6y = 0$
10. $y'' + y' - 2y = 0$
11. $2y'' - 3y' = 0$
12. $y'' - 100y = 0$
13. $y'' - y' - 12y = 0$
14. $y'' + 2y' - 15y = 0$
15. $2y'' - 7y' + 3y = 0$
16. $4y'' - 5y' - 6y = 0$
17. $y''' + 4y'' + 3y' = 0$
18. $y'' - 6y' + 9y = 0$
19. $y'' + 4y' + 4y = 0$
20. $y'' + 8y' + 16y = 0$
21. $y'' - 4y' - 5y = 3$
22. $y'' - 2y' - 3y = x$
23. $y''' - 9y' = 4e^{-x}$
24. $y'' - 5y' + 6y = \sin 5x$

46.4 AUXILIARY EQUATION

Note that in one example we began with the homogeneous equation

$$\frac{d^2y}{dx^2} - 5\frac{dy}{dx} + 6y = 0 \quad \text{or} \quad y'' - 5y' + 6y = 0$$

For the solution, we obtained

$$y = C_1 e^{3x} + C_2 e^{2x}$$

Now let us see how we might have found the solution more easily. Note especially that the solution contains terms in e^{3x} and e^{2x}. Suppose we set up an equation in some letter, say m, resembling the operator equation $(D^2 - 5D + 6)y = 0$. We set up an equation in m, making the *power* of m the same as the *order* of the derivative. That is, we write

$$m^2 - 5m + 6 = 0$$

654 Homogeneous Equations; Constant Coefficients

The roots of this equation are $m = 2$ and $m = 3$. Note that these roots are exactly the coefficients of x as exponents on e in the solution. This is not a coincidence. *In fact, this will always be true.*

The equation in m is called the *auxiliary equation*. (Some authors use the letter r, but any letter may be used.) The first step then is to set up the auxiliary equation, in which the powers of m are the same as the *order* of the derivatives in the differential equation. The roots of this equation will be the coefficients of x in the exponents on e in the solution. For the arbitrary constants, we simply use some letter before each of the exponentials. It is common practice to use C_1, C_2, C_3, and so on, depending on the number of terms. The capital letters A, B, C, and so on are sometimes used.

EXAMPLE 1. Solve by use of the auxiliary equation: $y'' - 3y' - 10y = 0$.

Solution. The auxiliary equation is $m^2 - 3m - 10 = 0$. The roots of the auxiliary equation are $m = -2$ and $m = +5$. Then the solution is

$$y = C_1 e^{-2x} + C_2 e^{5x}$$

Note on the Constants. Consider the equation in Example 1. If this problem were solved by the use of operators, the result would be that one of the arbitrary constants would turn out to be negative. However, since the arbitrary constant can represent any constant, positive or negative, then they can both be called positive. Of course, in working with the solution, we must stick to the chosen sign of the constant. For example, if

then
$$y = Ce^{-3x}$$
$$y' = -3Ce^{-3x}$$
$$y'' = +9Ce^{-3x}$$

and
$$y''' = -27Ce^{-3x}$$

EXAMPLE 2. Solve by use of the auxiliary equation; check the solution.

$$\frac{d^2y}{dx^2} + 2\frac{dy}{dx} - 8y = 0$$

Solution. The auxiliary equation is

$$m^2 + 2m - 8 = 0$$

Solving by factoring, $\quad (m - 2)(m + 4) = 0$

Then the roots of this equation are $m = 2$ and $m = -4$. These two roots are now used as the coefficients of x for exponents on e and we get the solution:

$$y = C_1 e^{2x} + C_2 e^{-4x}$$

To check, we must find $\quad y' = 2C_1 e^{2x} - 4C_2 e^{-4x}$

The equation has y'': $\quad y'' = 4C_1 e^{2x} + 16C_2 e^{-4x}$

The equation has $2y'$: $\quad 2y' = 4C_1 e^{2x} - 8C_2 e^{-4x}$

The equation has $-8y$: $\quad -8y = -8C_1 e^{2x} - 8C_2 e^{-4x}$

Then adding, $\quad y'' + 2y' - 8y = 0 + 0$

The result checks, and the solution of the equation is $y = C_1 e^{2x} + C_2 e^{-4x}$.

46.4 Auxiliary Equation

EXAMPLE 3. Find the general solution of the equation

$$\frac{d^2y}{dt^2} + 3\frac{dy}{dt} + 2y = 0$$

Then find the particular solution from the following initial conditions: when $t = 0$, $y = 0$, and $y' = 0.4$. Finally, if t is time (in seconds), find the value of y when $t = 2$ sec.

Solution. The auxiliary equation is

$$m^2 + 3m + 2 = 0$$

Solving for the roots, m_1 and m_2, we get $m_1 = -1$ and $m_2 = -2$. Then the general solution can be written immediately:

$$y = C_1 e^{-t} + C_2 e^{-2t}$$

Since y' is given, we find

$$y' = -C_1 e^{-t} - 2C_2 e^{-2t}$$

Now we use the initial conditions to find the value of C_1 and C_2. Since we have $y = 0$, when $t = 0$,

$$0 = C_1 e^0 + C_2 e^0 \quad \text{or} \quad C_1 + C_2 = 0$$

Since $y' = 0.4$ when $t = 0$,

$$0.4 = -C_1 e^0 - 2C_2 e^0 \quad \text{or} \quad C_1 + 2C_2 = -0.4$$

Now we have the system of equations:

$$C_1 + C_2 = 0$$
$$C_1 + 2C_2 = -0.4$$

Solving the system, we get $C_1 = 0.4$ and $C_2 = -0.4$. Inserting these values for C_1 and C_2, respectively, in the general solution, we get the *particular* solution:

$$y = 0.4 e^{-t} - 0.4 e^{-2t} \quad \text{or} \quad y = 0.4(e^{-t} - e^{-2t})$$

Now that we know the constants, we can find the value of y when $t = 2$. Then

$$y = 0.4(e^{-2} - e^{-4}) = 0.4(0.13534 - 0.01832) = 0.0468 \text{ (approx.)}$$

EXAMPLE 4. Solve the following differential equation for the general solution. Then check the general solution. Finally, show that any particular solution also satisfies the equation:

$$\frac{d^2i}{dt^2} - 16i = 0$$

Solution. The auxiliary equation is $m^2 - 16 = 0$. Then $m = +4$, and -4. Using the roots $+4$ and -4, we get the general solution:

$$i = C_1 e^{4t} + C_2 e^{-4t}$$

To check the solution, we find

$$\frac{di}{dt} = 4C_1 e^{4t} - 4C_2 e^{-4t}$$

For the second derivative,

$$\frac{d^2i}{dt^2} = 16C_1 e^{4t} + 16C_2 e^{-4t}$$

The equation has $-16i$:

$$-16i = -16C_1 e^{4t} - 16C_2 e^{-4t}$$

Adding, we get the differential equation: $\dfrac{d^2 i}{dt^2} - 16i = 0 + 0$

Now let us take some particular solution, say, where $C_1 = -5$ and $C_2 = 0$. Then

$$y = -5e^{4t} \qquad y' = -20e^{4t} \qquad y'' = -80e^{4t}$$

Inserting in the differential equation, we get $-80e^{4t} - 16(-5e^{4t}) = 0$.

The differential equation could be checked with any particular solution; that is, for any values of C_1 and C_2. However, the most general solution is that in which there are two distinct arbitrary constants that cannot be combined into a single constant.

EXAMPLE 5. Solve the equation $y'' + 4y' + y = 0$.

Solution. For the auxiliary equation, we have $m^2 + 4m + 1 = 0$. Solving this quadratic equation in m by the quadratic formula, we find that the roots are irrational: $m_1 = -2 + \sqrt{3}$ and $m_2 = -2 - \sqrt{3}$. These roots are most useful when changed to decimals: then $m_1 = -0.268$ and $m_2 = -3.732$ (approx.). Then the solution is

$$y = C_1 e^{(-0.268)x} + C_2 e^{(-3.732)x}$$

EXAMPLE 6. Solve the differential equation

$$\frac{d^2 s}{dt^2} + 9s = 0$$

Solution. The auxiliary equation is $m^2 + 9 = 0$; then $m = \pm 3i$. Note that the roots of the auxiliary equation are imaginary. We can still write the general solution in the usual manner. That is, the general solution is

$$s = C_1 e^{+(3i)t} + C_2 e^{-(3i)t}$$

We shall find that when the roots of the auxiliary equation are imaginary or complex, then the solution can be written in a useful form involving sine and cosines. However, at this point, we see that the solution may be written in the same way as when the roots are real.

EXAMPLE 7. Find the general solution of the equation

$$y'' + 2y' + 10y = 0$$

Solution. For the auxiliary equation we have $m^2 + 2m + 10 = 0$. Solving the equation in m by the quadratic formula, we find that the roots are complex:

$$m_1 = -1 + 3i \qquad m_2 = -1 - 3i$$

The general solution is

$$y = C_1 e^{(-1+3i)x} + C_2 e^{(-1-3i)x}$$

EXERCISE 46.3

Solve the following equations by use of the auxiliary equation. Check Problems 1–6.

1. $y'' + 5y' + 6y = 0$
2. $y'' - 2y' - 15y = 0$
3. $y' + 5y = 0$
4. $y'' - y' - 2y = 0$
5. $2y'' + 7y' + 3y = 0$
6. $2y'' - 7y' = 0$
7. $y''' + 3y'' - 4y' = 0$
8. $y'''' - 2y'' = 3y'$
9. $3y'' + 4y' = 0$
10. $3y''' - 5y'' = 2y'$
11. $2y''' - 7y'' = 4y'$
12. $4y'''' = 25y'$
13. $4y'' - 5y' - 6y = 0$
14. $y'' + 6y' + y = 0$
15. $y'' - 5y = 0$
16. $y'' + 2y' - 4y = 0$
17. $y'' - 4y' = 20y$
18. $8y'' = 3y$
19. $y''' - y' + 2y = 2y''$
20. $y'''' - 3y' + 2y = 0$
21. $y'' + 4y = 0$
22. $4y'' + 25y = 0$
23. $y'' - 4y' + 13y = 0$
24. $s'' + 50s = 0$
25. $s'' + 2s' + 5s = 0$
26. $2i''' - 2i' + 5i = 0$
27. $y'' + 2y = 0$
28. $\dfrac{d^2i}{dt^2} + 20i = 0$
29. $\dfrac{d^2s}{dt^2} + 192s = 0$
30. $\dfrac{d^2x}{dt^2} + 72x = 0$
31. $\dfrac{d^2i}{dt^2} + 10\dfrac{di}{dt} + 125i = 0$
32. $\dfrac{d^2i}{dt^2} + 20\dfrac{di}{dt} + 500i = 0$

Find the particular solution of each of the following:
33. $y'' - 3y' - 10y = 0$; when $x = 0$, $y = 0$, and $y' = 4$.
34. $y'' - 2y' - 8y = 0$; when $x = 0$, $y = 12$, and $y' = 0$.
35. $y'' - 9y = 0$; when $x = 0$, $y = 6$, and $y' = 10$.

46.5 REPEATED ROOTS OF THE AUXILIARY EQUATION

It sometimes happens that the auxiliary equation has two or more equal roots. Then we say the root is *repeated*. We often see this condition in connection with quadratic and higher degree equations. For example, suppose we have the quadratic equation

$$x^2 - 6x + 9 = 0$$

In this example, the roots are $+3$ and $+3$; that is, the root is repeated. The root, $+3$, in this case is sometimes called a *double* root.

A similar situation may occur in connection with a cubic equation, such as

$$x^3 + 6x^2 + 12x + 8 = 0$$

Solving this equation, we find that the roots are -2, -2, and -2. Again we have a repeated root of the equation.

In solving a differential equation, when we set up and solve the auxiliary equation in m, we may also get two or more equal roots. Take the example

$$y'' - 8y' + 16y = 0$$

The auxiliary equation is $m^2 - 8m + 16 = 0$. The roots are $m_1 = 4$, and $m_2 = 4$. In this equation, the root, $+4$, is repeated.

We run into a new difficulty when the auxiliary equation has two or more equal roots. We might think that all we need to do is to use the two equal roots in the usual manner. But let us see what happens.

Suppose we have the equation

$$\dfrac{d^2y}{dx^2} - 6\dfrac{dy}{dx} + 9y = 0$$

Let us solve this equation in the usual way. We write the auxiliary equation:

$$m^2 - 6m + 9 = 0$$

Solving, we get $m_1 = 3$ and $m_2 = 3$. The roots are equal.

Homogeneous Equations; Constant Coefficients

Now let us see what happens when we write the solution in the usual way, using the two equal roots as coefficients of x for exponents on e; we get

$$y = C_1 e^{3x} + C_2 e^{3x}$$

We can factor out e^{3x}, and get

$$y = e^{3x}(C_1 + C_2)$$

In this case, note that the two C's can be combined into a single constant. Whatever the value of C_1 and C_2, their sum is also a constant, which we might denote by a single C. Then, in the result, we have essentially only one arbitrary constant. Therefore, the answer we have obtained cannot be the most general solution. Since we begin with a second-order differential equation, the solution must contain two arbitrary constants that cannot be combined into one.

The answer is not the correct one for the most general solution. How, then, can we find the correct answer?

To get the correct solution for this problem, we fall back on the use of derivative operators. If the problem is solved by operators, the result will be the correct form of the most general solution with two distinct arbitrary constants. Let us begin with the equation

$$y'' - 6y' + 9y = 0$$

For operator form, we write

$$(D^2 - 6D + 9)y = 0$$

In factored form, $\qquad (D - 3)(D - 3)y = 0$

Now we solve this equation as we have already shown by the use of operators. First, we let some other letter, say z, represent the second factor operating on y; that is,

$$(D - 3)y = z$$

Then we can say that the first factor, $D - 3$, operates on z; or

$$(D - 3)z = 0$$

This means that $\qquad \dfrac{dz}{dx} - 3z = 0$

This equation can be solved by any one of several methods we have already used: by separating the variables, by use of the exponential integrating factor, or by the use of the auxiliary equation. In any case we get

$$z = C_1 e^{3x}$$

Now that we have found the expression for z, we go back to the statement in which we let

$$(D - 3)y = z$$

Then we have $\qquad (D - 3)y = C_1 e^{3x}$

This means that $\qquad \dfrac{dy}{dx} - 3y = C_1 e^{3x}$

46.5 Repeated Roots of the Auxiliary Equation

The variables cannot here be separated, but we recall the form

$$\frac{dy}{dx} + Py = Q$$

in which the integrating factor is

$$e^{\int P\, dx}$$

In this example, the integrating factor is e^{-3x}. Multiplying both sides of the equation by this factor and dx, we get

$$e^{-3x}\, dy - 3ye^{-3x}\, dx = C_1\, dx$$

Now we integrate both sides, the left side of the equation as a *whole*. Then

$$ye^{-3x} = C_1 x + C_2$$

Multiplying both sides by e^{3x}, we get the general solution:

$$y = C_1 x e^{3x} + C_2 e^{3x}$$

The general solution may be written $y = e^{3x}(C_1 x + C_2)$. However, note that the two constants cannot be combined into a single constant.

Note especially the form of the general solution in the foregoing example. The auxiliary equation has the root $+3$, which is repeated. In the solution we do have the two terms, both of which contain the same factor e^{3x}. However, in one term this factor is multiplied by the factor x. This will always be true where we have repeated roots of the auxiliary equation. In the case of a triple root of the auxiliary equation — for example, if we have the roots 3, 3, and 3 — we should still have the terms containing the factor e^{3x}. However, in one term we should have the factor x, and in the third term we should have the factor x^2.

EXAMPLE 8. Find the general solution of the equation

$$y'' + 8y' + 16y = 0$$

Solution. The auxiliary equation is $m^2 + 8m + 16 = 0$. The roots of the auxiliary equation are the equal roots, -4 and -4. Then the solution will contain two terms, each with the factor e^{-4x}. However, since the roots are equal, we must multiply one of these by x. The general solution becomes

$$y = C_1 e^{-4x} + C_2 x e^{-4x}$$

Either factor may be multiplied by x, the independent variable.

EXAMPLE 9. Find the solution of the equation

$$4y'' - 12y' + 9y = 0$$

Solution. The auxiliary equation is $4m^2 - 12m + 9 = 0$. It has the two equal roots: $+3/2$ and $+3/2$. Then one of the factors is multiplied by the independent variable x. The general solution is

$$y = C_1 e^{(3/2)x} + C_2 x e^{(3/2)x}$$

Note. If the independent variable is t, then the extra factor must be t, not x. It is important to identify the independent variable.

EXAMPLE 10. Solve the equation
$$y''' - 12y'' + 48y' - 64y = 0$$

Solution. Since this is a differential equation of the third order, we shall have three arbitrary constants. For the auxiliary equation we have
$$m^3 - 12m^2 + 48m - 64 = 0$$
Solving by synthetic division, we get three equal roots: $m = 4, 4,$ and 4. The general solution will contain three terms, each with the factor e^{4x}. In the case of a *triple* root, one term will contain the extra factor x, and another term will contain the extra factor x^2. For the general solution we get
$$y = C_1 e^{4x} + C_2 x e^{4x} + C_3 x^2 e^{4x}$$
This can be written
$$y = e^{4x}(C_1 + C_2 x + C_3 x^2)$$

The rule in the case of repeated roots is usually stated by saying that if the auxiliary equation has repeated roots of a multiplicity n, then the successive terms of the solution must be multiplied by the extra factor x raised to a power from zero to $n - 1$, respectively.

EXAMPLE 11. Solve the equation $y''' - 4y'' + 4y' = 0$.

Solution. The auxiliary equation is $m^3 - 4m^2 + 4m = 0$. Solving, we get three roots, two of which are equal: $m = 0, 2,$ and 2. In the case of the repeated roots, 2 and 2, we observe the multiplication by the factor x, and get the general solution containing three constants:
$$y = C_1 e^{0x} + C_2 e^{2x} + C_3 x e^{2x}$$
Simplifying,
$$y = C_1 + e^{2x}(C_2 + C_3 x)$$

EXAMPLE 12. Solve the equation $y''' - 3y'' = 0$.

Solution. The auxiliary equation is $m^3 - 3m^2 = 0$; then $m^2(m - 3) = 0$. Taking $m^2 = 0$, we must consider that zero (0) is a double root. The third root is 3. Then the solution is
$$y = C_1 e^{0x} + C_2 x e^{0x} + C_3 e^{3x}$$
This simplifies to
$$y = C_1 + C_2 x + C_3 e^{3x}$$

EXAMPLE 13. Solve the equation $y^{iv} - 2y'' + y = 0$.

Solution. The auxiliary equation is $m^4 - 2m^2 + 1 = 0$. Factoring, we get
$$(m - 1)(m + 1)(m - 1)(m + 1) = 0$$
The auxiliary equation has two double roots: $1, 1, -1, -1$. The solution is
$$y = C_1 e^x + C_2 x e^x + C_3 e^{-x} + C_4 x e^{-x}$$

EXAMPLE 14. Solve the equation $y^{iv} + 4y''' + y'' - 12y' - 12y = 0$.

Solution. For the auxiliary equation we have
$$m^4 + 4m^3 + m^2 - 12m - 12 = 0$$

By synthetic division we discover that two roots are -2 and -2. When the equation is reduced by these two roots, the remainder is $m^2 - 3 = 0$. The two remaining roots are therefore $\pm\sqrt{3}$. Although these roots have the same numerical value, they are not equal because they differ in sign. Then we have the solution

$$y = C_1 e^{-2x} + C_2 x e^{-2x} + C_3 e^{\sqrt{3}x} + C_4 e^{-\sqrt{3}x}$$

EXAMPLE 15. Solve the equation $y^{iv} - 2y''' + 5y'' - 8y' + 4y = 0$.

Solution. For the auxiliary equation we have

$$m^4 - 2m^3 + 5m^2 - 8m + 4 = 0$$

By synthetic division we discover that two roots are 1 and 1. The reduced equation then becomes $m^2 + 4 = 0$. The remaining roots are $\pm 2i$. Then we have

$$y = C_1 e^x + C_2 x e^x + C_3 e^{2ix} + C_4 e^{-2ix}$$

EXERCISE 46.4

Solve the following equations:

1. $y'' - 2y' + y = 0$
2. $y''' - 4y'' + 4y' = 0$
3. $y''' + 3y'' = 0$
4. $y''' + 6y'' + 9y' = 0$
5. $y^{iv} + 4y'' = 0$
6. $y^{iv} = 4y'''$
7. $y''' + 8y'' + 16y' = 0$
8. $y^{iv} + 36y = 13y''$
9. $y^{iv} = 3y''$
10. $4y'' + 25y = 20y'$
11. $4s'' - 12s' + 9s = 0$
12. $y^{iv} + 2y'' = 0$
13. $y^{iv} - 6y''' + 4y'' = 0$
14. $y^{iv} - 27y'' = 0$
15. $y^{iv} + 32y'' = 0$
16. $y^{iv} + 2y'' + 8y' + 5y = 0$
17. $y^{iv} - 4y''' + 3y'' + 4y' - 4y = 0$
18. Work the following problem by operators: $y'' - 10y' + 25y = 0$.
19. $y'' + 2y' + y = 0$. Find the particular solution: $y = 0$, $y' = 2$, when $x = 0$.
20. Solve for x: $\dfrac{d^2x}{dt^2} + 20\dfrac{dx}{dt} + 100x = 0$.

chapter
47

Imaginary and Complex Roots

47.1 THE GENERAL SOLUTION

We have already encountered imaginary and complex roots of the auxiliary equation in m. Up to this point, in writing the solution of a differential equation we have used the same form for imaginary roots as for real roots. The general solution takes the form

$$y = C_1 e^{m_1 x} + C_2 e^{m_2 x} + \cdots + C_n e^{m_n x}$$

Let us review two examples:

(a) If $y'' + 9y = 0$; then $m = \pm 3i$; and the general solution is

$$y = C_1 e^{(3i)x} + C_2 e^{(-3i)x}$$

(b) If $y'' + 4y' + 13y = 0$; then $m = -2 \pm 3i$; the general solution is

$$y = C_1 e^{(-2+3i)x} + C_2 e^{(-2-3i)x}$$

47.2 EULER'S FORMULA

Now we shall see that when the roots of the auxiliary equation are imaginary or complex numbers, the solution can be changed to a form involving sines and cosines. This change comes about through the use of Euler's formula:

$$e^{ix} \equiv \cos x + i \sin x$$

This identity is derived through the use of the infinite series expansion for e^x, for $\sin x$, and for $\cos x$. (See pages 623–624.)

The Maclaurin's series expansion for e^x is as follows:

$$e^x = 1 + \frac{x}{1!} + \frac{x^2}{2!} + \frac{x^3}{3!} + \frac{x^4}{4!} + \cdots + \frac{x^{n-1}}{(n-1)!}$$

Since this expansion holds for all values of x, it is true for e^{ix}. When the powers of (ix) are computed correctly, noting that $i = \sqrt{-1}$, we get

$$e^{ix} = 1 + \frac{ix}{1!} - \frac{x^2}{2!} - \frac{ix^3}{3!} + \frac{x^4}{4!} + \frac{ix^5}{5!} - \frac{x^6}{6!} - \frac{ix^7}{7!} + \frac{x^8}{8!} \cdots$$

Also we have the following series for sine and cosine:

$$\sin x = x - \frac{x^3}{3!} + \frac{x^5}{5!} - \frac{x^7}{7!} + \frac{x^9}{9!} \cdots$$

$$\cos x = 1 - \frac{x^2}{2!} + \frac{x^4}{4!} - \frac{x^6}{6!} + \frac{x^8}{8!} \cdots$$

If we multiply the sine series by i and add it to the cosine series, we get

$$\cos x + i \sin x = \left(1 - \frac{x^2}{2!} + \frac{x^4}{4!} - \frac{x^6}{6!} \cdots\right) + i\left(x - \frac{x^3}{3!} + \frac{x^5}{5!} - \frac{x^7}{7!} \cdots\right)$$

The result is exactly the same as the series for e^{ix} when the terms are rearranged. Then we can say

$$e^{ix} = \cos x + i \sin x$$

Let us see now how the change in a general solution can be made.

47.3 PURE IMAGINARY ROOTS OF THE AUXILIARY EQUATION

Suppose we take again the equation

$$y'' + y = 0$$

Since $m = \pm i$, the general solution is

$$y = C_1 e^{ix} + C_2 e^{-ix}$$

Let us apply Euler's formula to each term. We get

$$C_1 e^{ix} = C_1 \cos x + iC_1 \sin x \qquad C_2 e^{-ix} = C_2 \cos(-x) + iC_2 \sin(-x)$$

Then the solution can be written:

$$y = C_1 \cos x + iC_1 \sin x + C_2 \cos(-x) + iC_2 \sin(-x)$$

From trigonometry recall that $\cos(-x) = \cos x$, but $\sin(-x) = -\sin x$. Then we can write

$$y = C_1 \cos x + C_2 \cos x + iC_1 \sin x - iC_2 \sin x$$

Combining,

$$y = (C_1 + C_2) \cos x + i(C_1 - C_2) \sin x$$

We can now use a single constant, say A, to represent $(C_1 + C_2)$; and another single constant, say B, to represent $[i(C_1 - C_2)]$. Then the solution can be written

$$y = A \cos x + B \sin x$$

It is common practice to write the cosine term first and the sine second.

EXAMPLE 1. Solve $y'' + 16y = 0$.

Solution. For the auxiliary equation, we have $m^2 + 16 = 0$; $m = \pm 4i$. Then we have the general solution:

$$y = C_1 e^{4ix} + C_2 e^{-4ix} \quad \text{or} \quad y = A \cos 4x + B \sin 4x$$

Warning: Do not change a solution to sine and cosine form if the roots of the auxiliary equation are not imaginary.

47.4 COMPLEX ROOTS OF THE AUXILIARY EQUATION

If the roots of the auxiliary equation are complex numbers of the form $(a + bi)$ and a is not zero, then the solution is a little more complicated. Let us again take the example

$$y'' + 4y' + 13y = 0$$

The roots of the auxiliary equation are $-2 + 3i$ and $-2 - 3i$. Then the solution becomes

$$y = C_1 e^{(-2+3i)x} + C_2 e^{(-2-3i)x}$$

Now we can separate the two parts of each exponent by a simple principle of exponents. We know that $(e^a)(e^b) = e^{a+b}$. Reversing the process, $e^{(-2+3i)x}$ can be written

$$(e^{-2x})(e^{3ix})$$

Then the solution can be written

$$y = C_1 e^{-2x} e^{3ix} + C_2 e^{-2x} e^{-3ix}$$

Factoring, $\quad y = e^{-2x}(C_1 e^{3ix} + C_2 e^{-3ix})$

Note that the second factor can be changed to sine and cosine form. Then

$$y = e^{-2x}(A \cos 3x + B \sin 3x)$$

In the exponent on e, note that the coefficient of x is the *real* part of the complex root of the auxiliary equation. The *coefficient* of i in the complex root becomes the *coefficient* of x in the sine and cosine form of the solution. In general, if the roots of the auxiliary equation are of the form, $a \pm bi$, then the solution has the form

$$y = e^{ax}(A \cos bx + B \sin bx)$$

EXAMPLE 2. Solve the equation $y^{iv} + 6y''' + 34y'' = 0$.

Solution. The auxiliary equation is $m^4 + 6m^3 + 34m^2 = 0$. Solving for m:

$$m = 0, 0, -3 \pm 5i$$

Then $\quad y = C_1 + C_2 x + e^{-3x}(A \cos 5x + B \sin 5x)$

47.5 Vibration of a Spring

EXAMPLE 3. If $L = 0.125$ henry, $R = 5$ ohms, and $C = 4(10^{-4})$ farad, solve the following equation for i:

$$L\frac{d^2i}{dt^2} + R\frac{di}{dt} + \frac{i}{C} = 0$$

Solution. Substituting values,

$$0.125\frac{d^2i}{dt^2} + 5\frac{di}{dt} + \frac{i}{4(10^{-4})} = 0$$

Multiplying by 8, and simplifying,

$$\frac{d^2i}{dt^2} + 40\frac{di}{dt} + 20{,}000i = 0$$

The auxiliary equation is

$$m^2 + 40m + 20{,}000 = 0 \qquad m = -20 \pm 140i$$

Then the general solution is

$$y = e^{-20t}(A\cos 140t + B\sin 140t)$$

EXERCISE 47.1

Solve the following differential equations and state the solution in the form of sines and cosines, if possible.

1. $y'' + 9y = 0$
2. $y'' + 2y = 0$
3. $3y'' + 4y = 0$
4. $\frac{d^2s}{dt^2} + 25s = 0$
5. $4\frac{d^2s}{dt^2} + 9s = 0$
6. $\frac{d^2s}{dt^2} + 128s = 0$
7. $\frac{d^2i}{dt^2} + 20i = 0$
8. $\frac{d^2i}{dt^2} + 250i = 0$
9. $\frac{d^2i}{dt^2} + 80i = 0$
10. $\frac{d^2y}{dt^2} + 300y = 0$
11. $\frac{d^2x}{dt^2} + 72x = 0$
12. $\frac{d^2Q}{dt^2} + 450Q = 0$
13. $y'' + 1600y = 0$
14. $y'' - 100y = 0$
15. $8y'' + 3y = 0$
16. $y'' + 4y' + 13y = 0$
17. $y'' - 6y' + 58y = 0$
18. $\frac{d^2s}{dt^2} + 2\frac{ds}{dt} + 5s = 0$
19. $2\frac{d^2i}{dt^2} - 2\frac{di}{dt} + 25i = 0$
20. $y'' + 64y = 0$. Find the particular solution if $y(0) = 3$; $y'(0) = 0$.
21. $s'' + 2s' + 10s = 0$. Find the particular solution if $s(0) = 0$; $s'(0) = 2$.
22. $\frac{d^2y}{dt^2} + 125y = 0$. Find the particular solution if $y(0) = \frac{1}{2}$; $y'(0) = 0$.
23. $y'' + 4y' + 40y = 0$. Find the particular solution if $y(0) = 10$; $y'(0) = 4$.
24. $\frac{d^2x}{dt^2} + 162x = 0$. Find the particular solution if $x(0) = 3$; $x'(0) = 0$.

47.5 VIBRATION OF A SPRING

Imaginary and complex roots of the auxiliary equation occur frequently in science. One such appearance is in connection with the vibration of a spring. Let us suppose a spring is suspended and a weight is attached to

666 Imaginary and Complex Roots

the lower end. Now the weight is pulled down some distance below the point of equilibrium and released. Our problem now usually consists of two questions. We wish to find the equation that represents the motion, showing the position of the object at any instant t. Also, we wish to know the frequency of vibration per second or per minute.

The vibration of the spring is caused by opposing forces. We shall here consider only a simple situation in which the two opposing forces are the force of gravity pulling downward, and the force of the spring pulling upward.

When the weight is attached and allowed to remain at rest, we call this position the position of *static equilibrium*. We use E to represent this position (Fig. 47.1). When the weight is pulled down below this position, we represent the distance of displacement by s. When the object is displaced from E, the *upward* force of the spring is equal but opposite to the *downward* force of gravity; that is,

$$\text{force}_1 \text{ (gravity)} = -\text{force}_2 \text{ (spring)}$$

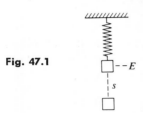

Fig. 47.1

Now let us analyze these two opposing forces. The force of gravity depends on the *mass* of the object. If we measure force in pounds, then we must take mass in *slugs*. A *slug* is defined as the weight, in pounds, *divided by* the acceleration due to gravity. In the foot-pound system, acceleration due to gravity is approximately 32 ft/sec². Then,

$$\text{mass} = \text{weight} \div (32)$$

Now we refer to Newton's second law of motion which states that

$$\text{force} = \text{mass} \cdot \text{acceleration} \quad \text{or} \quad f = ma$$

We already have an expression for mass. Now we need an expression for acceleration.

We have used s to represent the distance of displacement of the object from E at any time t. Then, as the spring vibrates, the distance s is a variable. At E, $s = 0$. Since s is a function of time t, we can write

$$s = f(t)$$

We have seen that velocity v at any time is given by the first derivative of s; that is,

$$v = \frac{ds}{dt} = f'(t)$$

Also, acceleration a is given by the second derivative of the distance s; that is,

$$a = \frac{dv}{dt} = \frac{d^2s}{dt^2}$$

Then, from Newton's second law, $f = ma$, we have, at any time t, the downward force

$$F_1 = \left(\frac{wt}{32}\right)\left(\frac{d^2s}{dt^2}\right)$$

Now let us analyze the upward pulling force of the spring. The pulling force of any spring is determined by two factors. One is the amount of distortion or stretching of the spring. According to Hooke's law, the force exerted by a stretched spring (or other elastic body) is proportional to the distorting force (if kept within elastic limits). A particular spring stretched 6 in. exerts twice as much force as it does when stretched 3 in.

The second factor affecting the force exerted by the spring is a quality inherent in the spring itself. For example, to stretch one spring a distance of 1 ft may require 8 lb of force, whereas for another spring it may require 20 lb of force. This characteristic of any particular spring is called the *spring constant*, usually denoted by K or some other letter. Let us define the spring constant K. For any particular spring, *the spring constant is the number of pounds of force required to stretch the spring one foot* (in the pound-foot system). For example, suppose a force of 3 lb will stretch a certain spring 2 in., which is $\frac{1}{6}$ of a foot. Then a force of 18 lb will be required to stretch the spring 1 ft. The spring constant for this particular spring is 18. We can always find K by the formula

$$K = \text{(force, in pounds)} \div \text{(distance, in feet)} \quad \text{or} \quad K = \frac{F}{s}$$

In the example mentioned here, $K = (3) \div (\frac{1}{6}) = 18$.

If a spring has a constant of 18, this means that for any distance s that the spring is stretched, the force exerted by the spring is given by

$$\text{force} = 18s$$

As a formula for the upward pulling force of the spring, we have

$$F_2(\text{upward}) = (K)(s)$$

Now we go back to our original statement showing the relation between forces:

$$\text{force}_1 \text{ (downward)} = -\text{force}_2 \text{ (upward)}$$

We can now write the equation relating the two forces:

$$\left(\frac{wt}{32}\right)\left(\frac{d^2s}{dt^2}\right) = -(K)(s)$$

EXAMPLE 4. A weight measured as 2.25 lb stretches a suspended spring a measured distance of 1.5 in. Now the 2.25-lb weight is removed and replaced

668 Imaginary and Complex Roots

with a 3-lb weight. This weight is now pulled down 3 in. below the point of equilibrium and released. Find the particular equation of the motion.

Solution. First we must find the spring constant. The measured distance 1.5 in. must be changed to feet by dividing by 12. This becomes 0.125 ft. Often it is more convenient to work with common fractions. In any event, we find that the spring constant is 18. Even though a different weight is now attached, the spring constant remains the same, 18. With the 3-lb weight attached, we write the basic equation form:

$$\frac{3}{32} \cdot \frac{d^2s}{dt^2} = -18s \quad \text{or} \quad \frac{3}{32} \cdot \frac{d^2s}{dt^2} + 18s = 0$$

This equation reduces to $\quad \dfrac{d^2s}{dt^2} + 192s = 0$

We write the auxiliary equation and solve:

$$m^2 + 192 = 0 \qquad m = \pm 8\sqrt{3}\,i$$

Then the general solution is

$$s = A \cos 8\sqrt{3}\,t + B \sin 8\sqrt{3}\,t$$

We shall need the derivative:

$$v = s' = (8\sqrt{3})(A)(-\sin 8\sqrt{3}\,t) + (8\sqrt{3})(B)(\cos 8\sqrt{3}\,t)$$

Now we have the initial conditions: when $t = 0$, $v = 0$, and $s = \frac{1}{4}$ ft. Substituting, for $s = \frac{1}{4}$,

$$\frac{1}{4} = A \cos 0 + B \sin 0 \qquad \text{then } A = \frac{1}{4}$$

Substituting, for $v = 0$,

$$0 = (8\sqrt{3})(A)(-\sin 0) + (8\sqrt{3})(B)(\cos 0)$$

Since $\sin 0 = 0$, the first term is zero. Then the value of B is zero. The particular solution becomes

$$s = \frac{1}{4} \cos 8\sqrt{3}\,t$$

The result agrees with the idea that the maximum distance s occurs when $t = 0$, and this maximum amplitude is $\frac{1}{4}$ ft.

47.6 FREQUENCY

The frequency refers to the number of complete vibrations or cycles per second in connection with harmonic motion. In some instances we are concerned with the frequency per minute. Frequency will depend on several factors. For one thing, the weight of the object will influence the frequency. A light weight will vibrate faster than a heavy weight, other things being constant. The spring constant is another factor that will affect the frequency. A stiff spring will cause a faster vibration, or oscillation, than a weak spring with the same weight attached.

To see how we might compute frequency, we must understand what is meant by angular velocity. Let us first review a formula involving linear velocity. We know that at any time t, the distance s is a product of time t and velocity v; that is,

$$s = vt$$

If velocity is measured in feet per second and time in seconds, then distance s is measured in feet.

Now let us see the similarity to angular measure. An angle θ is defined as the amount of rotation of a ray about its end point. Now, if the ray rotates at a steady rate, then the amount of rotation per unit of time is called *angular velocity*, often denoted by *omega* (ω). To find the amount of rotation, θ, we can multiply *angular velocity* by *time*. For example, if the ray rotates at a rate of 30 degrees per second, then in 4 seconds it will generate angle θ of 120 degrees. In general, we can say

$$\theta = \omega t$$

Note the similarity to the formula for distance in linear motion:

$$s = vt$$

In most problems in rotary motion, the angle is measured in radians. Then one complete rotation is 2π radians. If the angular velocity is one revolution per second, then $\omega = 2\pi$. If the angular velocity is six revolutions per second, then the velocity becomes

$$\omega = 6(2\pi) = 12\pi \text{ rad/sec}$$

By frequency f we mean the number of revolutions per second. If the frequency is 60 revolutions or cycles per second, then angular velocity ω is 120π rad/sec. Note that we have the following formula:

$$\omega = 2\pi f$$

Dividing both sides by 2π, we get the formula for frequency:

$$f = \frac{\omega}{2\pi}$$

As an example in the use of this last formula, suppose we have the expression,

$$\sin 60t$$

In this expression, the number 60 represents ω in radians. Then we find the frequency by

$$f = \frac{60}{2\pi} = \frac{30}{\pi} = 9.55 \text{ (approx.)}$$

We usually take π to be 3.1416 or 3.14. Instead of dividing by this number, it is often convenient to use the reciprocal of π, which is approximately 0.3183. Then we can use multiplication instead of division:

$$30\left(\frac{1}{\pi}\right) = 30(0.3183) = 9.55$$

In Example 4 in the previous section we found that

$$s = \frac{1}{4} \cos 8\sqrt{3}\, t$$

To find the frequency, we take $\omega = 8\sqrt{3}$; then

$$f = \frac{8\sqrt{3}}{2\pi} = 2.2052 \text{ cycles per second (or 132.3 per minute)}$$

47.7 THE *LRC* ELECTRIC CIRCUIT

In an electric circuit containing inductance L, resistance R, and capacitance C in series with voltage E, the total voltage drop in the circuit must be equal to the voltage impressed in the circuit. Then the basic equation is

$$L\frac{di}{dt} + Ri + \frac{1}{C}\int i\, dt = E$$

Now we consider only the simpler situation in which E is a constant, as are also L, R, and C. Differentiating both sides of the equation, we get

$$L\frac{d^2i}{dt^2} + R\frac{di}{dt} + \frac{i}{C} = 0$$

To find the instantaneous current at any time t we solve this differential equation for i. Of course, more complicated situations arise when E is not a constant. In circuit problems we use $j = \sqrt{-1}$, since i represents current.

EXAMPLE 5. Using the foregoing formula, if $L = 0.4$; $R = 8$; $C = 8(10^{-4})$; find (a) the general formula for current i and (b) the frequency.

Solution. Substituting values in the equation, we get

$$(0.4)\frac{d^2i}{dt^2} + 8\frac{di}{dt} + \frac{1}{8(10^{-4})}i = 0$$

The equation reduces to

$$\frac{d^2i}{dt^2} + 20\frac{di}{dt} + 3125i = 0$$

For the roots of the auxiliary equation we get

$$m = -10 \pm 55j$$

The general solution is

$$i = e^{-10t}(A \cos 55t + B \sin 55t)$$

From the general solution, we note that $2\pi f = 55$. Then, using the formula for frequency f, we have

$$f = \frac{55}{2\pi} = 8.753 \text{ (approx.)}$$

EXERCISE 47.2

1. A force of 3 lb will stretch a certain spring 1.5 in. The spring is suspended and a 3-lb weight attached to the lower end. The weight is then pulled down 2 in. below the point of equilibrium and released. Find the particular equation of the motion and the frequency per minute.
2. Find the particular solution and the frequency per minute in Problem 1 if the 3-lb weight is replaced with a 4-lb weight.
3. Find the particular solution and the frequency per minute in Problem 1 if the 3-lb weight is replaced with a 6-lb weight.
4. A certain spring is stretched $1\frac{1}{3}$ inches by a force of 4 lb. Find the general solution for the motion and the frequency per minute if the spring is suspended and a 4-lb weight is attached to the lower end, and then pulled down 3 in. below equilibrium and released.
5. Work Problem 4 if the 4-lb weight is replaced with a 3-lb weight.
6. Work Problem 4 if the 4-lb weight is replaced with a 6-lb weight.
7. A certain spring is stretched 1.2 in. by a force of 3.2 lb. Find the general solution and the frequency if each of the following weights are attached to the lower end, pulled down below the point of equilibrium, and released: (a) 4 lb; (b) 6 lb; (c) 8 lb.
8. A certain spring has a constant of 16 lb/ft. Set up the proper differential equation and solve for the frequency for each of the following weights attached:
   (a) 2 lb   (b) 4 lb   (c) 8 lb
   (d) How is frequency related to the weight attached?   (e) If the weight is doubled, what is the effect on frequency?
9. Using K as the spring constant, w as the weight attached, and g as the acceleration due to gravity, set up the differential equation and find the formula for frequency.
10. Using the formula derived in No. 9, find the frequency when $K = 24$, $w = 4$, and $g = 32$.

Find the general solution and the frequency for each of the following LRC electric circuits, using the formula:

$$L\frac{d^2i}{dt^2} + R\frac{di}{dt} + \frac{1}{C}i = 0$$

where L is in henrys, R in ohms, and C in farads.

11. $L = 0.2$; $R = 10$; $C = 8(10^{-4})$
12. $L = 0.5$; $R = 4$; $C = 4(10^{-3})$
13. $L = 0.2$; $R = 8$; $C = 2.5(10^{-3})$
14. $L = 0.5$; $R = 15$; $C = 3.2(10^{-3})$
15. $L = 0.5$; $R = 10$; $C = 4(10^{-3})$
16. $L = 0.2$; $R = 4$; $C = 1.6(10^{-3})$
17. $L = 0.25$; $R = 10$; $C = 3.2(10^{-4})$
18. $L = 0.25$; $R = 3.5$; $C = 1.6(10^{-3})$
19. $L = 0.4$; $R = 28$; $C = 1.6(10^{-4})$
20. $L = 0.125$; $R = 8.5$; $C = 4(10^{-4})$

chapter
48

Nonhomogeneous Differential Equations

48.1 DEFINITION

We have seen that a differential equation containing terms in only y and its derivatives is called homogeneous. If these terms are all found on the left side of the equation, then the right side is zero, such as in

$$y'' + 5y' + 6y = 0$$

Now, if the equation contains terms other than y and its derivatives, the equation is called *nonhomogeneous*. The extra terms are written at the right side of the equation. A general equation of this form may be shown as

$$a_0 y^n + a_1 y^{n-1} + a_2 y^{n-2} + \cdots + a_n y = f(x)$$

The right side represented by $f(x)$ may be any function of x or a constant.
  Here are two examples of nonhomogeneous differential equations:

(a) $\quad y'' + 5y' + 6y = 3x^2 + 2 \qquad$ (b) $\quad y'' + 9y = 3e^x + 6\cos x$

Our problem now is to determine the solution that will satisfy such a nonhomogeneous differential equation.

48.2 THE HOMOGENEOUS SOLUTION

Let us see again the meaning of a homogeneous solution. Take the equation

$$y'' + 5y' + 6y = 0$$

The general solution of this equation is

$$y = C_1 e^{-2x} + C_2 e^{-3x}$$

Let us see what this means. If we take this solution and find the first and second derivatives, and then substitute these values together with y itself into the differential equation, the result will be zero. Let us check to see whether this is true.
We have

$$y = C_1 e^{-2x} + C_2 e^{-3x}$$

then
$$y' = -2C_1 e^{-2x} - 3C_2 e^{-3x}$$

and
$$y'' = 4C_1 e^{-2x} + 9C_2 e^{-3x}$$

To check in the original differential equation, $y'' + 5y' + 6y = 0$, we take

$$y'' = 4C_1 e^{-2x} + 9C_2 e^{-3x}$$
$$5y' = -10C_1 e^{-2x} - 15C_2 e^{-3x}$$
$$6y = 6C_1 e^{-2x} + 6C_2 e^{-3x}$$

Adding, $\quad y'' + 5y' + 6y = \quad 0 \quad + \quad 0$

Let us repeat: When we solve a homogeneous differential equation, we get a solution that must check in the original equation. When y and its derivatives are inserted in the differential equation, the result must add up to zero. Otherwise the solution is not correct.

48.3 THE NONHOMOGENEOUS EQUATION

Now let us take the nonhomogeneous equation

$$y'' + 5y' + 6y = 3x^2 + 2$$

Suppose we set the right side equal to zero so that we have the homogeneous equation

$$y'' + 5y' + 6 = 0$$

We have seen that the solution of this equation is

$$y_h = C_1 e^{-2x} + C_2 e^{-3x}$$

Let us call this solution y_h to signify the solution of the homogeneous equation.

Now we know that if we check the homogeneous solution y_h in the original differential equation, the result will be zero. Therefore the complete solution must be something more, in order to make the right side equal to $3x^2 + 2$; that is, the complete solution must be

$$y_h + \text{something more}$$

In fact, the extra portion of the solution must be something that will check to the value of the right side, $3x^2 + 2$. This extra added portion is called the *particular solution*. We denote this portion of the solution by y_p. Our problem is now to determine this *particular solution*. This is not so difficult as it might seem.

48.4 THE METHOD OF UNDETERMINED COEFFICIENTS

To determine the particular solution of a differential equation we can often use a method called *undetermined coefficients*. Let us take again the equation

$$y'' + 5y' + 6y = 3x^2 + 2$$

Now we must find the particular solution that will check to $3x^2 + 2$. This means that if we take y_p and the first and second derivatives and substitute these into the differential equation, the result will be $3x^2 + 2$. That is,

$$y_p'' + 5y_p' + 6y_p = 3x^2 + 2$$

Let us see whether we can determine the kind of terms to be found in y_p. Since the right side of the equation has a term containing x^2, it is safe to assume that the particular solution, y_p, also has a term containing x^2. Moreover, for each derivative of y_p the power of x is reduced by 1. Then the particular solution may have a term in x^2, a term in x, and a constant. In fact, using the letters, A, B, and C, for undetermined coefficients, we can say that, in general terms, the particular solution has the form:

$$y_p = Ax^2 + Bx + C$$

These undetermined coefficients — A, B, C, and so on — do not have the same meaning as the arbitrary constants in the homogeneous solution. In this case, these letters in the particular solution are only "dummy" letters and their values will be found at once. Any letters may be used, since they will disappear when their values are discovered.

If the particular solution, y_p, is to satisfy the differential equation, we must find y_p' and y_p''. Since

$$y_p = Ax^2 + Bx + C$$

then $\qquad y_p' = 2Ax + B \qquad$ and $\qquad y_p'' = 2A$

Now we substitute these values in the differential equation and get

$$y_p'' = 2A$$
$$5y_p' = 10Ax + 5B$$
$$6y_p = 6Ax^2 + 6Bx + 6C$$

Adding,

$$y'' + 5y_p' + 6y_p = 6Ax^2 + 10Ax + 6Bx + 2A + 5B + 6C$$

48.4 The Method of Undetermined Coefficients

Now the right side of this equation must be equal to the quantity, $3x^2 + 2$. This means that the coefficients of like powers of x must be equal. Equating coefficients of like powers, we get

for the coefficients of x^2, $\quad 6A = 3$

for the coefficients of x, $\quad 10A + 6B = 0$

for the constant terms, $\quad 2A + 5B + 6C = 2$

Solving the system, we get

$$A = \frac{1}{2} \quad B = \frac{-5}{6} \quad C = \frac{31}{36}$$

Now that we have discovered the values of the coefficients, A, B, and C, we can write the particular solution

$$y_p = \frac{1}{2}x^2 - \frac{5}{6}x + \frac{31}{36}$$

Let us see whether this solution checks. Differentiating, we get

$$y'_p = x - \frac{5}{6} \quad \text{and} \quad y''_p = 1$$

Substituting in the differential equation, we get

$$1 + 5x - \frac{25}{6} + 3x^2 - 5x + \frac{31}{6} = 3x^2 + 2$$

Then the *complete general solution* is given by $y = y_h + y_p$, or

$$y = C_1 e^{-2x} + C_2 e^{-3x} + \frac{1}{2}x^2 - \frac{5}{6}x + \frac{31}{36}$$

Will this complete solution also check in the original differential equation? Why?

To use the method of undetermined coefficients to find the particular solution, we first consider cases in which no term in the homogeneous solution is related through differentiation to the terms on the right side of the equation. Special cases in which this relation exists will be taken up in another section.

To find the particular solution we have the following steps:

1. Set up the form of the particular solution, y_p, by writing down all the terms similar to those on the right side, as well as all forms of their derivatives.

2. Give each term an undetermined coefficient, such as A, B, C, and so on.

3. Now find all the derivatives of y_p as called for in the differential equation.

4. Write the sum of y_p and its derivatives as indicated on the left side of the equation, and equate this sum to the right side of the equation.

Nonhomogeneous Differential Equations

5. Equate coefficients of like terms and solve for each undetermined coefficient.
6. Write the particular solution using the values found for the coefficients.
7. Write the complete solution: $y = y_h + y_p$.

In most problems the right side of the equation involves only one or two terms. However, for an example here we shall use one that involves several types of terms to show the method.

EXAMPLE 1. Solve

$$y'' - 2y' - 8y = 4x^2 - 5 + 2e^{-x} + 8\cos 2x$$

Solution. For the homogeneous solution, we have

$$y_h = C_1 e^{4x} + C_2 e^{-2x}$$

First, let us note that no term in the homogeneous solution is related through differentiation to any term on the right side of the equation. Now we set up the form of the particular solution, y_p. We write the form of all the terms on the right and their derivatives, giving each term an undetermined coefficient:

$$y_p = Ax^2 + Bx + C + De^{-x} + E\cos 2x + F\sin 2x$$

Note that the derivatives of all the terms at the right can have no other form of derivatives. Moreover, none of these terms is related to any term in the homogeneous solution, y_h. Differentiating,

$$y_p' = 2Ax + B - De^{-x} - 2E\sin 2x + 2F\cos 2x$$
$$y_p'' = 2A \quad\quad + De^{-x} - 4E\cos 2x - 4F\sin 2x$$

For the differential equation as originally given, we must have

$$y_p'' = 2A + De^{-x} - 4E\cos 2x - 4F\sin 2x$$
$$-2y_p' = -4Ax - 2B + 2De^{-x} + 4E\sin 2x - 4F\cos 2x$$
$$-8y_p = -8Ax^2 - 8Bx - 8C - 8De^{-x} - 8E\cos 2x - 8F\sin 2x$$

Adding the terms on the right, we have the following (equating coefficients):

for the coefficients of x^2, $\quad -8A = 4$
for the coefficients of x, $\quad -4A - 8B = 0$
for the constant terms, $\quad 2A - 2B - 8C = -5$
for the coefficient of e^{-x}, $\quad -5D = 2$
for the coefficients of $\cos 2x$, $\quad -12E - 4F = 8$
for the coefficients of $\sin 2x$, $\quad 4E - 12F = 0$

Solving this system, we get the following values:

$$A = -\frac{1}{2} \quad B = \frac{1}{4} \quad C = \frac{7}{16} \quad D = -\frac{2}{5} \quad E = -\frac{3}{5} \quad F = -\frac{1}{5}$$

For the particular solution, we have

$$y_p = -\frac{1}{2}x^2 + \frac{1}{4}x + \frac{7}{16} + \frac{2}{5}e^{-x} - \frac{3}{5}\cos 2x - \frac{1}{5}\sin 2x$$

48.5 Modification of the Particular Solution Form

EXERCISE 48.1

Find the complete general solution of each of the following differential equations by the use of the method of undetermined coefficients.

1. $y'' + 2y' - 8y = 5x^2 - 8$
2. $y'' + 3y' - 10y = 2x^2 - 5x$
3. $y'' - y' - 6y = 3x^2 + 2$
4. $y'' - y' - 2y = 4e^{3x}$
5. $y'' + 4y' + 3y = 5xe^x$
6. $y'' + 2y' - 8y = 2e^x - 3x^2$
7. $y'' - 6y' + 9y = 5x^2 - 4e^x$
8. $y'' + 2y' + y = 6x^2 - e^{-2x}$
9. $y'' - 4y' + 4y = \sin 3x$
10. $y'' + 5y' + 6y = \cos 2x + e^{2x}$
11. $y'' + 9y = x^3 + 4x$
12. $y'' - 4y = \sin x + \cos 5x$
13. $y' + 4y = \sin 3x + \cos 2x$
14. $y' - 3y = 5e^{-2x} + 3e^{2x}$
15. $y'' + 5y' = 3e^{-2x} + \sin 4x$
16. $y'' - 3y' = 5xe^{2x}$

Find the particular solution satisfying the initial conditions:

17. $y'' - 4y = 3x^2 - 5;\ y(0) = 1;\ y'(0) = 0$
18. $y'' - 9y = 2e^x - 3xe^{-x};\ y(0) = 0;\ y'(0) = 0$
19. $y'' + 2y' + y = 5x^2 - 3e^{2x};\ y(0) = 0;\ y'(0) = -2$
20. $y'' - 3y' + 2y = 3 \sin 2x + 5 \cos 2x;\ y(0) = 0;\ y'(0) = 3$

48.5 MODIFICATION OF THE PARTICULAR SOLUTION FORM

When the form of the particular solution has first been set up showing the forms of all the terms on the right side and their derivatives, we must look back at the homogeneous solution. If any term in the homogeneous solution is related to any term in the form of the particular solution, then we must make a slight adjustment in the form of the particular solution y_p.

Let us define a *family* of terms as those terms that are alike or can be made alike through differentiation. One family consists of terms such as x^3, x^2, x, and a constant. All of these are related through differentiation. The terms, e^{-2x} and e^{-3x} belong to different families because one cannot come from the other through differentiation. However, the terms xe^{-2x} and e^{-2x} belong to the same family because the second comes from the first by differentiation. The terms $\sin 2x$ and $\cos 2x$ belong to the same family, but the term $\sin 3x$ belongs to a different family.

Now, if any term in the particular solution as it is set up belongs to the same family as any term in the homogeneous solution, then that term in the particular solution y_p must be *modified by multiplying it by a power of x*. For example, if the particular solution form shows the term e^{-2x}, and the homogeneous solution also contains a term in e^{-2x}, then that term in the *particular solution* must be multiplied by x to make it different from the term in the homogeneous solution. Briefly, any term appearing in the particular solution as it is first set up must be *different* from any term in the homogeneous solution. To make it different we multiply by a power of x.

EXAMPLE 2. Solve the equation

$$y''' - 5y'' + 6y' = 6x^2 - 4 + 3e^{2x} + 8e^x$$

Solution. For the homogeneous solution, we have

$$y_h = C_1 + C_2 e^{2x} + C_3 e^{3x}$$

For the particular solution we set up the tentative form:
$$y_p = Ax^2 + Bx + C + De^{2x} + Ee^x$$
Now we note that the homogeneous solution also has a constant, C_1. Then the three terms in y_p — Ax^2, Bx, and C — all in the same family, must each be multiplied by x, to make them different from any term in y_h. We note that y_h contains a term in e^{2x}. Therefore, in the particular solution, the term De^{2x} must be multiplied by x. As a result we have the form
$$y_p = Ax^3 + Bx^2 + Cx + Dxe^{2x} + Ee^x$$
The term e^x is not multiplied by x because it belongs to a different family. Differentiating,
$$y'_p = 3Ax^2 + 2Bx + C + 2Dxe^{2x} + De^{2x} + Ee^x$$
$$y''_p = 6Ax + 2B + 4Dxe^{2x} + 2De^{2x} + 2De^{2x} + Ee^x$$
$$y'''_p = 6A + 8Dxe^{2x} + 4De^{2x} + 4De^{2x} + 4De^{2x} + Ee^x$$
For the differential equation, we must have
$$y'''_p = 6A + 8Dxe^{2x} + 12De^{2x} + Ee^x$$
$$-5y''_p = -30Ax - 10B - 20Dxe^{2x} - 20De^{2x} - 5Ee^x$$
$$6y_p = 18Ax^2 + 12Bx + 6C + 12Dxe^{2x} + 6De^{2x} + 6Ee^x$$
Combining terms on the right and equating coefficients, we have

for the coefficients of x^2, $18A = 6$

for the coefficients of x, $-30A + 12B = 0$

for the constant terms, $6A - 10B + 6C = -4$

for the coefficients of e^{2x}, $-2D = 3$

for the coefficients of e^x, $2E = 8$

Solving,
$$A = \frac{1}{3} \quad B = \frac{5}{6} \quad C = \frac{7}{18} \quad D = -\frac{3}{2} \quad E = 4$$
Note that the coefficient of the term xe^{2x} results in zero. However, the value of the coefficient D must be taken as $-3/2$ in the form of the particular solution. Then we have
$$y_p = \frac{1}{3}x^3 + \frac{5}{6}x^2 + \frac{7}{18} - \frac{3}{2}xe^{2x} + 4e^x$$
To get the complete general solution, we combine: $y = y_h + y_p$. The particular solution, y_p, must check in the original equation. Of course, the entire general solution must also check since the check of y_h will result in zero.

EXAMPLE 3. Set up the form for the particular solution of the equation
$$y'' + 9y = 15 \sin 2x + 12 \cos 3x$$

Solution. We first find the homogeneous solution, y_h. It is
$$y_h = C_1 \cos 3x + C_2 \sin 3x$$
We have here used C's with subscripts so that these constants will not be confused

48.6 Limitations of the Method of Undetermined Coefficients

with the undetermined coefficients of the particular solution. For the particular solution we set up the tentative form:

$$y_p = A \sin 2x + B \cos 2x + C \sin 3x + D \cos 3x$$

Now we note that the homogeneous solution also contains the form, sin $3x$. Therefore, in the particular solution, the terms in this family group must be multiplied by x to make them different from the terms in y_h. Then the correct form for the particular solution is

$$y_p = A \sin 2x + B \cos 2x + Cx \sin 3x + Dx \cos 3x$$

48.6 LIMITATIONS OF THE METHOD OF UNDETERMINED COEFFICIENTS

The method of undetermined coefficients cannot be used for all types of nonhomogeneous differential equations, but it can be useful in many instances involving practical problems. Briefly, it can be used to solve nonhomogeneous equations in which the terms on the right side have a *limited number of kinds of derivatives*.

For example, if the terms on the right contain such terms as x^4, e^{3x}, sin $5x$, cos $2x$, or constants, then this method can be used. The term x^4 has a limited number of forms of derivatives, which include x^3, x^2, x, and a constant. The derivatives of e^{3x} can only be terms containing e^{3x}. The derivatives of sin $5x$ will all contain the form sin $5x$ or cos $5x$ but no others. In a term, such as x^n, if n is a positive integer, the term will have a limited number of kinds of derivatives. Even such a term as the product, $x^2 e^{3x}$, has only three forms of derivatives: $x^2 e^{3x}$; xe^{3x}; and e^{3x}.

Undetermined coefficients cannot be used when the right side contains terms such as the following:

$$x^{1/2} \qquad x^{-2} \qquad \tan x \qquad \sec x \qquad \ln x$$

and others. In equations of this kind, other methods must be used to find the solutions. One method often used in such cases is the one called *variation of parameters*. Another method is the use of power series. These methods are studied in more complete courses in differential equations.

EXERCISE 48.2

Solve the following differential equations by the method of undetermined coefficients and the modification rule.

1. $y'' - 2y' - 3y = 4e^{3x} - 3x^2$
2. $y''' + y'' - 2y' = 8x^2 + 4$
3. $y'' + 2y' = 4x^2 - 5 - 6e^{-2x}$
4. $y''' - 2y'' = 2x^2 + 3 + e^{-2x}$
5. $y'' + 4y = 3e^{2x} + 6 \cos 2x$
6. $y'' - y = 2xe^x + 3e^{-x}$
7. $y'' + y = 6\varsigma e^{x^3x} + 4 \sin x$
8. $y^{iv} + 4y'' = 8x^2 + 5 \cos 2x$
9. $y''' - 9y' = 5x^2 + 4e^{3x} + e^{2x}$
10. $y^{iv} + y'' = 2x + 3 \cos x + \sin 3x$
11. $y^{iv} - y'' = 4x^2 - 3 + 2e^x + 3xe^x$
12. $y^v + 4y''' = x^2 + 4 \sin 2x + \cos 3x$
13. $y'' + 4y = 3x^2 - 8$. Find the particular solution if $y(0) = 1$; $y'(0) = 0$.
14. $y''' - 4y' = 6e^{-2x} - 3e^x$. Find the particular solution for the following initial conditions: $y(0) = 2$; $y'(0) = -3$; $y''(0) = -7$.

Table I Natural Logarithms of Numbers

x	$\ln x$	x	$\ln x$	x	$\ln x$	x	$\ln x$
0.01	−4.6052	0.90	−0.1054	3.3	1.1939	7.8	2.0541
.02	−3.9120	.95	−0.0513	3.4	1.2238	7.9	2.0669
.03	−3.5066	1.00	0.0000	3.5	1.2528	8.0	2.0794
.04	−3.2189	1.05	0.0488	3.6	1.2809	8.1	2.0919
.05	−2.9957	1.10	0.0953	3.7	1.3083	8.2	2.1041
.06	−2.8134	1.15	0.1398	3.8	1.3350	8.3	2.1163
.07	−2.6593	1.20	0.1823	3.9	1.3610	8.4	2.1282
.08	−2.5257	1.25	0.2231	4.0	1.3863	8.5	2.1401
.09	−2.4079	1.30	0.2624	4.1	1.4110	8.6	2.1518
.10	−2.3026	1.35	0.3001	4.2	1.4351	8.7	2.1633
.11	−2.2073	1.40	0.3365	4.3	1.4586	8.8	2.1748
.12	−2.1203	1.45	0.3716	4.4	1.4816	8.9	2.1861
.13	−2.0402	1.50	0.4055	4.5	1.5041	9.0	2.1972
.14	−1.9661	1.55	0.4382	4.6	1.5261	9.1	2.2083
.15	−1.8971	1.60	0.4700	4.7	1.5476	9.2	2.2192
.16	−1.8326	1.65	0.5008	4.8	1.5686	9.3	2.2300
.17	−1.7720	1.70	0.5306	4.9	1.5892	9.4	2.2407
.18	−1.7148	1.75	0.5596	5.0	1.6094	9.5	2.2513
.19	−1.6607	1.80	0.5878	5.1	1.6292	9.6	2.2618
0.20	−1.6094	1.85	0.6152	5.2	1.6487	9.7	2.2721
.22	−1.5141	1.90	0.6419	5.3	1.6677	9.8	2.2824
.24	−1.4271	1.95	0.6678	5.4	1.6864	9.9	2.2925
.26	−1.3471	2.00	0.6931	5.5	1.7047	10	2.3026
.28	−1.2730	2.05	0.7178	5.6	1.7228	11 *	2.3979
.30	−1.2040	2.10	0.7419	5.7	1.7405	13	2.56495
.32	−1.1394	2.15	0.7655	5.8	1.7579	17	2.8332
.34	−1.0788	2.20	0.7885	5.9	1.7750	19	2.9444
.36	−1.0217	2.25	0.8109	6.0	1.7918	23	3.1355
.38	−0.9676	2.30	0.8329	6.1	1.8083	29	3.3673
0.40	−0.9163	2.35	0.8544	6.2	1.8245	31	3.4340
.42	−0.8675	2.40	0.8755	6.3	1.8405	37	3.6109
.44	−0.8210	2.45	0.8961	6.4	1.8563	41	3.7136
.46	−0.7765	2.50	0.9163	6.5	1.8718	43	3.7612
.48	−0.7340	2.55	0.9361	6.6	1.8871	47	3.85015
.50	−0.6931	2.60	0.9555	6.7	1.9021	53	3.9703
.52	−0.6539	2.65	0.9746	6.8	1.9169	59	4.0775
.54	−0.6162	2.70	0.9933	6.9	1.9315	61	4.1109
.56	−0.5798	2.75	1.0116	7.0	1.9459	67	4.2047
.58	−0.5447	2.80	1.0296	7.1	1.9601	71	4.2627
.60	−0.5108	2.85	1.0473	7.2	1.9741	73	4.2905
.65	−0.4308	2.90	1.0647	7.3	1.9879	79	4.36945
.70	−0.3567	2.95	1.0818	7.4	2.0015	83	4.4188
.75	−0.2877	3.00	1.0986	7.5	2.0149	89	4.4886
.80	−0.2231	3.1	1.1314	7.6	2.0281	97	4.5747
.85	−0.1625	3.2	1.1632	7.7	2.0412		
x	$\ln x$	x	$\ln x$	x	$\ln x$	x	$\ln x$

*Prime numbers. For composite numbers add logarithms of their factors.

Table II Values of e^{-x} and e^{-x}

x	e^x	e^{-x}	x	e^x	e^{-x}	x	e^x	e^{-x}
0.00	1.0000	1.0000	1.00	2.7183	0.3679	3.50	33.115	0.0302
0.02	1.0202	0.9802	1.05	2.8577	.3499	3.55	34.813	.0287
0.04	1.0408	.9608	1.10	3.0042	.3329	3.60	36.598	.0273
0.06	1.0618	.9418	1.15	3.1582	.3166	3.65	38.475	.0260
0.08	1.0833	.9231	1.20	3.3201	.3012	3.70	40.447	.0247
0.10	1.1052	0.9048	1.25	3.4903	0.2865	3.75	42.521	0.0235
0.12	1.1275	.8869	1.30	3.6693	.2725	3.80	44.701	.0224
0.14	1.1503	.8694	1.35	3.8574	.2592	3.85	46.993	.0213
0.16	1.1735	.8521	1.40	4.0552	.2466	3.90	49.402	.0202
0.18	1.1972	.8353	1.45	4.2631	.2346	3.95	51.935	.0193
0.20	1.2214	0.8187	1.50	4.4817	0.2231	4.00	54.598	0.0183
0.22	1.2461	.8025	1.55	4.7115	.2122	4.05	57.397	.0174
0.24	1.2712	.7866	1.60	4.9530	.2019	4.10	60.340	.0166
0.26	1.2969	.7711	1.65	5.2070	.1920	4.15	63.434	.0158
0.28	1.3231	.7558	1.70	5.4739	.1827	4.20	66.686	.0150
0.30	1.3499	0.7408	1.75	5.7546	0.1738	4.25	70.105	0.0143
0.32	1.3771	.7261	1.80	6.0496	.1653	4.30	73.700	.0136
0.34	1.4049	.7118	1.85	6.3598	.1572	4.35	77.478	.0129
0.36	1.4333	.6977	1.90	6.6859	.1496	4.40	81.451	.0123
0.38	1.4623	.6839	1.95	7.0287	.1423	4.45	85.627	.0117
0.40	1.4918	0.6703	2.00	7.3891	0.1353	4.50	90.017	0.0111
0.42	1.5220	.6570	2.05	7.7679	.1287	4.55	94.632	.0106
0.44	1.5527	.6440	2.10	8.1662	.1225	4.60	99.484	.0101
0.46	1.5841	.6313	2.15	8.5849	.1165	4.65	104.58	.0096
0.48	1.6161	.6188	2.20	9.0250	.1108	4.70	109.95	.0091
0.50	1.6487	0.6065	2.25	9.4877	0.1054	4.75	115.58	0.0087
0.52	1.6820	.5945	2.30	9.9742	.1003	4.80	121.51	.0082
0.54	1.7160	.5827	2.35	10.486	.0954	4.85	127.74	.0078
0.56	1.7507	.5712	2.40	11.023	.0907	4.90	134.29	.0074
0.58	1.7860	.5599	2.45	11.588	.0863	4.95	141.17	.0071
0.60	1.8221	0.5488	2.50	12.182	0.0821	5.00	148.41	0.0067
0.62	1.8589	.5379	2.55	12.807	.0781	5.05	156.02	.0064
0.64	1.8965	.5273	2.60	13.464	.0743	5.10	164.02	.0061
0.66	1.9348	.5169	2.65	14.154	.0707	5.15	172.43	.0058
0.68	1.9739	.5066	2.70	14.880	.0672	5.20	181.27	.0055
0.70	2.0138	0.4966	2.75	15.643	0.0639	5.25	190.57	0.0052
0.72	2.0544	.4868	2.80	16.445	.0608	5.30	200.34	.0050
0.74	2.0959	.4771	2.85	17.288	.0578	5.35	210.61	.0047
0.76	2.1383	.4677	2.90	18.174	.0550	5.40	221.41	.0045
0.78	2.1815	.4584	2.95	19.106	.0523	5.45	232.76	.0043
0.80	2.2255	0.4493	3.00	20.086	0.0498	5.5	244.69	0.0041
0.82	2.2705	.4404	3.05	21.115	.0474	6.0	403.43	.0025
0.84	2.3164	.4317	3.10	22.198	.0450	6.5	665.14	.0015
0.86	2.3632	.4232	3.15	23.336	.0429	7.0	1096.6	.0009
0.88	2.4109	.4148	3.20	24.533	.0408	7.5	1808.0	.0006
0.90	2.4596	0.4066	3.25	25.790	0.0388	8.0	2981.0	0.0003
0.92	2.5093	.3985	3.30	27.113	.0369	8.5	4914.8	.0002
0.94	2.5600	.3906	3.35	28.503	.0351	9.0	8103.1	.0001
0.96	2.6117	.3829	3.40	29.964	.0334	9.5	13360	.00007
0.98	2.6645	.3753	3.45	31.500	.0317	10	22026	.00005

| x | e^x | e^{-x} | x | e^x | e^{-x} | x | e^x | e^{-x} |

Answers to Selected Odd-Numbered Problems

Exercise 1.1, Page 5
1. $-7, -5, -4, -2, -1, 0, 1, 2, 3, 5$ **3.** (a) $-3 < 5$; (c) $2 > -7$; (e) $6 = 6$; (g) $0 > -3$; (i) $\sqrt{2} < \sqrt{3}$ **5.** (a) 10; (c) -4; (e) 5.5; (g) 0; (i) $-\frac{1}{2}$ **7.** (a) ± 4; (c) 7; -3; (e) 2; (g) $-\infty$ to -4; (i) $-\infty$ to 3; (k) 5 to $+\infty$; (m) $-\infty$ to -3; (o) -2 to $+\infty$.

Exercise 1.2, Page 14
1. (a) $\sqrt{130}$; $(-\frac{1}{2}, \frac{5}{2})$; (c) $\sqrt{74}$; $(\frac{7}{2}, -\frac{5}{2})$; (e) $\sqrt{50}$; $(-\frac{7}{2}, \frac{5}{2})$ **3.** (a) $\sqrt{160}$; $\sqrt{68}$; $\sqrt{244}$; (b) (2,4); $(-3,-2)$; (3,0); (c) $\sqrt{116}$; $\sqrt{185}$; $\sqrt{53}$; (d) $2(\sqrt{40} + \sqrt{17} + \sqrt{61})$ **5.** isosceles; (b) scalene; (c) isosceles **7.** (a) $\sqrt{80}$; (b) 3; (c) 4 **9.** A, B, D **11.** $(0, -3)$; $\sqrt{26}$ **13.** opposite sides equal **15.** sides: $\sqrt{34}$; diag.: $\sqrt{68}$ **17.** $AB + AC = BC$ **19.** sides unequal.

Exercise 1.3, Page 20
1. (a) 1; 45°; (c) -3; 108.4°; (e) 0, 0°; (g) $\frac{1}{3}$; 18.4°; (i) 2; 63.4° **3.** (a) $\frac{1}{3}$; -4; $\frac{6}{5}$; (b) $\frac{5}{2}$; $\frac{8}{11}$; $-\frac{2}{7}$; (c) -3; $\frac{1}{4}$; $-\frac{5}{6}$ **5.** two slopes: $\frac{2}{3}$; $-\frac{3}{2}$ **7.** $m = \frac{2}{3}$ **9.** (a) opposite sides have equal slopes; (b) diagonals: length $= 10$; slopes: $\frac{3}{4}$; $-\frac{4}{3}$ **11.** sides have unequal slopes **13.** (a) opposite sides have equal slopes; (b) not a square; (c) slopes of diagonals: -3; $\frac{1}{3}$.

Exercise 1.4, Page 24
1. 32.47° **3.** 0° **5.** 22.2°; 144.9°; 12.9° **7.** 40°14.2′; 34°30.5′; 105°15.3′ **9.** 38.1°; 38.3°; 103.6° **11.** $\tan D = 13/0$; $D = 90°$ **13.** angles unequal **15.** all right angles.

Exercise 1.5, Page 27
3. $\sqrt{3}$ **5.** $\sqrt{90}$ **7.** $\sqrt{101}$; $\sqrt{90}$; $\sqrt{126}$; 3; $\sqrt{122}$.

Exercise 3.1, Page 48
1. $3x - 5y = 22$ **3.** $4x - 3y = -22$ **5.** $3x - 8y = 32$ **7.** $3x + 12 = y$ **9.** $x + 3y = 9$ **11.** $3x + 4y = 7$ **13.** $7x = 2y$ **15.** $2x - 5y = 15$ **17.** $x = 4$ **19.** $x = -3$ **21.** $4y - 3x = 20$ **23.** $4x - 7y = 21$ **25.** $3x + 5y = -12$ **27.** $4x + 3y = 24$ **29.** $x - 6y = 6$ **31.** $5y - 4x = 3$ **33.** (a) $3x + 5y = 17$; (b) $2x - 3y = -16$ **35.** (a) $5x - 3y = 17$; (b) $3x + 2y = -11$ **37.** (a) $y = -3/7x + 25/7$; $m = -3/7$; (c) $y = \frac{3}{2}x - 20/3$; $m = \frac{3}{2}$.

Exercise 3.3, Page 56
1. (a) $\dfrac{3x + 7y - 25}{\sqrt{58}} = 0$; $\dfrac{25}{\sqrt{58}}$; (b) $\dfrac{9x - 6y - 40}{\sqrt{117}} = 0$; $\dfrac{40}{\sqrt{117}}$ **3.** 3; 9; 6 **5.** 2 **7.** $18/\sqrt{13}$ **9.** $16/\sqrt{10}$ **11.** 0 **13.** $5x + 4y = -21$; $56/\sqrt{41}$; 28 **15.** (a) $x + 11 = 2y$; $x + 2y = -3$; $5x + 2y = 17$; (b) $x + y = 1$; $2y - 7x = 5$; $x + 10y = 13$; (c) $2x + y = 6$; $y - 2x = 4$; $5y - 2x = 24$; (d) area $= 48$; (e) $2x + y = -2$; $y - 2x = 1$; $2x - 5y = 1$; (f) points: $(\frac{1}{2}, 5)$; $(-\frac{1}{3}, \frac{4}{3})$; $(-\frac{3}{4}, -\frac{1}{2})$.

Exercise 4.1, Page 64
1. $x^2 + y^2 - 10x - y = 10$ 3. $x^2 + y^2 + 14x + 8y = -16$ 5. $x^2 + y^2 + 10y = 6x$
7. $x^2 + y^2 + 10x + 4y = -9$ 9. $x^2 + y^2 - 6x - 14y = -49$
11. $x^2 + y^2 - 6x + 12y = -20$; $4x - 3y = 5$ 13. $x^2 + y^2 - 4x + 8y = -7$;
$2x - 3y = 3$ 15. $x^2 + y^2 - 2x + 4y = 11$; $x = -3$ 17. $C(4,-5)$; $r = 6$
19. $C(-1,2)$; $r = \sqrt{13}$ 21. $C(4, \frac{7}{2})$; $r = \frac{1}{2}\sqrt{65}$ 23. $C(-\frac{5}{3}; \frac{1}{3})$; $r = \frac{5}{3}\sqrt{2}$ 25. $C(4,-2)$;
$r = 0$ 27. $(4,2)$ 29. $(3,-4)$; $(-3,4)$ 31. imaginary 33. $(-3,\pm 4)$.

Exercise 4.2, Page 70
1. $x^2 + y^2 - 4x + 2y = 35$ 3. $x^2 + y^2 + 2x - 4y = 45$
5. $x^2 + y^2 + 3x + y = 10$ 7. $x - 3y = 7$ 9. $5x - 2y = -20$ 11. 5; $\sqrt{24}$
13. 7; 1 15. 8; 0; imag. 17. 0; 10; imag.

Exercise 5.1, Page 78
1. focus: $(\frac{7}{2},0)$ 3. focus: $(0,-\frac{3}{2})$ 5. focus: $(\frac{7}{3},0)$ 7. focus: $(0,-\frac{8}{7})$ 9. focus: $(\frac{1}{4},0)$
11. focus: $(0,-\frac{1}{12})$ 13. $y^2 = 24x$ 15. $y^2 = -98x$ 17. $y^2 = 18x$ 19. $3y^2 = 25x$;
$5x^2 = 9y$ 21. $2y^2 = -9x$; $3x^2 = -4y$ 23. $5y^2 = -x$ 25. $x^2 = 7y$ 27. $y^2 = 6x$
29. $192\sqrt{3}$.

Exercise 5.2, Page 81
1. $x^2 - 10x + 33 = 8y$ 3. $x^2 + 4x + 16y = 44$ 5. $x^2 - 4x + 8y = 4$
7. $x^2 + 2x - 4y = 7$ 9. $y^2 - 8y + 4 = 12x$ 11. $x^2 + 4x + 4 = 12y$ 13. $(4,3)$;
$(6,3)$ 15. $(-1,3)$; $(-1,4)$ 17. $(2,-3)$; $(3,-3)$ 19. $(3,0)$; $(3,-\frac{5}{2})$ 21. $(-4,-2)$;
$(-4,-1)$ 23. $y^2 + 4y + 13 = 18x$.

Exercise 5.3, Page 85
1. $(4,4)$; $(1,-2)$ 3. imaginary 5. $(3,6)$ 7. $(0,0)$; $(4,8)$ $(-4,8)$ 9. $(1,5)$; $(1,-3)$
11. $(\pm\sqrt{32},-5)$ 13. $5x^2 = 8y$ 15. $\frac{1}{24}$ in.

Exercise 6.1, Page 94
1. (a) 6; (b) 4; (c) $(\pm\sqrt{5},0)$; (d) $(\pm 3,0)$; (e) $\frac{1}{3}\sqrt{5}$; (f) $x = \pm 9/\sqrt{5}$; (g) $\frac{8}{3}$ 5. (a) $4\sqrt{6}$;
(b) $4\sqrt{2}$; (c) $(0,\pm 4)$; (d) $(0,\pm 2\sqrt{6})$; (e) $\frac{1}{3}\sqrt{6}$; (f) $y = \pm 6$; (g) $\frac{4}{3}\sqrt{6}$ 9. (a) $6\sqrt{2}$; (b) 8;
(c) $(\pm\sqrt{2},0)$; (d) $(\pm 3\sqrt{2},0)$; (e) $\frac{1}{3}$; (f) $x = \pm 9\sqrt{2}$; (g) $\frac{16}{3}\sqrt{2}$ 13. $9x^2 + 25y^2 = 225$
15. $7x^2 + 16y^2 = 112$ 17. $9x^2 + 25y^2 = 900$ 19. $3x^2 + 4y^2 = 48$
21. $9x^2 + 25y^2 = 225$ 23. $5x^2 + 9y^2 = 80$ 25. $25x^2 + 21y^2 = 100$.

Exercise 6.2, Page 98
1. $4x^2 + 9y^2 + 16x - 18y = 11$ 3. $5x^2 + 2y^2 - 10x + 12y = 17$
5. $5x^2 + 9y^2 - 10x - 54y = 94$ 7. $16x^2 + 25y^2 - 64x - 100y = 236$
9. $7x^2 + 16y^2 - 42x + 32y = 33$ 11. $9x'^2 + 16y'^2 = 144$ 13. $x'^2 + 4y'^2 = 4$
15. $25x'^2 + 16y'^2 = 400$ 17. 125 ft 19. $(3,2)$; $(4,\frac{3}{2})$ 21. $(-1,-3)$; $(13/7,19/7)$
23. $(0,-3)$; $(3,-\frac{3}{2})$ 25. $(\frac{3}{2},\pm\sqrt{3})$ 27. $(\pm 3,\pm 2)$ 29. $(0,-4)$; $(\pm\frac{4}{3}\sqrt{2},\frac{4}{3})$.

Exercise 7.1, Page 108
1. (a) 6; (b) 8; (c) $(\pm 5,0)$; (d) $(\pm 3,0)$; (e) $\frac{5}{3}$; (f) $4x \pm 3y = 0$; (g) $x = \pm\frac{9}{5}$; (h) $32/3$
5. (a) $4\sqrt{2}$; (b) $4\sqrt{3}$; (c) $(0,\pm 2\sqrt{5})$; (d) $(0,\pm 2\sqrt{2})$; (e) $\frac{1}{2}\sqrt{10}$; (f) $\sqrt{3}y \pm \sqrt{2}x = 0$;
(g) $y = \pm\frac{4}{5}\sqrt{5}$; (h) $6\sqrt{2}$ 9. (a) $4\sqrt{3}$; (b) $2\sqrt{30}$; (c) $(0,\pm\sqrt{42})$; (d) $(0,\pm 2\sqrt{3})$; (e) $\frac{1}{2}\sqrt{14}$;
(f) $\sqrt{5}y \pm \sqrt{2}x = 0$; (g) $y = \pm\frac{2}{7}\sqrt{42}$; (h) $10\sqrt{3}$ 13. $9x^2 - 16y^2 = 144$
15. $5y^2 - 4x^2 = 80$ 17. $7y^2 - 9x^2 = 63$ 19. $x^2 - 3y^2 = 16$ 21. $9y^2 - 4x^2 = 36$
23. $\dfrac{x^2}{291600} - \dfrac{y^2}{198400} = 1$.

Exercise 7.2, Page 111
1. $2x^2 - 5y^2 - 12x - 10y = 7$ 3. $y^2 - 2x^2 + 10y + 8x = -1$
5. $9x^2 - 16y^2 + 36x + 32y = 124$ 7. $3x^2 - y^2 - 6x + 4y = 13$
9. $7y^2 - 9x^2 + 14y + 36x = 92$ 11. (a) vertex: $(2,2)$; focus: $(\sqrt{8},\sqrt{8})$; (c) vertex:

Answers to Selected Odd-Numbered Problems   685

$(\sqrt{12}, \sqrt{12})$; focus: $(\sqrt{24}, \sqrt{24})$ **13.** $3x'^2 - y'^2 = 3$ **15.** $5y'^2 - 4x'^2 = 80$
17. $2x'^2 - 3y'^2 = 24$ **19.** $x = -\frac{2}{5} \pm \frac{3}{5}\sqrt{11}$; $y = -\frac{9}{5} \pm \frac{1}{5}\sqrt{11}$ **21.** $(\pm\sqrt{28}, \pm\sqrt{8})$
23. $(3,1); (-3,-1)$ **25.** $(2,3)$ **27.** $(4,2); (-4,-2)$ **29.** $(\pm 8, \pm\sqrt{45})$.
Exercise 7.3, Page 115
1. ellipse   **3.** parabola   **5.** hyperbola   **7.** ellipse   **9.** hyperbola   **11.** circle   **13.** hyperbola   **15.** hyperbola.
Exercise 9.1, Page 133
1. $-1; -7; -5; a^2 + 3a - 5; 9x^2 + 9x - 5; x^2 + 7x + 5$ **5.** $4; 5; 0; 4; 3i$ **9.** $-2;$
$3; -3; 463$ **13.** $-22; -18; -18; -18; 6i$ **17.** $16; 729; \frac{1}{2}; 27/8; 65,536$ **21.** $1; 2;$
$3; 0$ **25.** $0; 0.8415; 0.9093; -0.8415; 1; 0.4794$ **29.** $x^2 + 4x - 5; x^2 + 2x + 1;$
$x^3 - 11x + 6; x + 6 + 16/(x-3); 8; x^2 - 3x - 2; -4$ **33.** $-11; 21; 3; -1$
37. $8; 8; -27$ **39.** $-2; -2$.
Exercise 10.1, Page 143
1. 208 ft/sec   **3.** 0.5 mph/sec   **5.** -1.5 ft/sec^2 **7.** $(12°/7)$/hr   **9.** -0.035 **11.** -3.5
13. 105   **15.** 13.5   **17.** 107   **19.** $-\frac{3}{2}$ **21.** 3   **23.** 8.4   **25.** 3   **27.** 4.2   **29.** -0.4.
Exercise 11.1, Page 156
9. 11   **11.** 7   **13.** 5   **15.** 4   **17.** 2   **19.** 0.
Exercise 11.2, Page 162
1. -13 **3.** -19 **5.** -10 **7.** -9 **9.** 0   **11.** none   **13.** none   **15.** 0   **17.** 0   **19.** 0
21. $\frac{4}{3}$ **23.** $\frac{5}{4}$ **25.** 2   **27.** none   **29.** 0   **31.** 2   **33.** none   **35.** 64   **37.** 1   **39.** none
41. 1   **43.** 2.
Exercise 12.1, Page 172
1. $2x; -6$ **3.** $-6x; -4.5$ **5.** $3x^2 - 2; 10$ **7.** $6x - 3x^2; -9$ **9.** $4x^3; 32/27$
11. $4x - 3; -3$ **13.** $2t - 5; -1$ **15.** $3t^2 - 8t - 5; -2$ **17.** $-2x^{-2}; -0.02$
19. $-6(x-1)^{-2}$; no value   **21.** $-8x^{-3}; 1$ **23.** $-3/2x^{3/2}; -3/16$ **25.** $6x - 4; 2; -4$
27. $2t + 2; 2.4; 2$ **29.** ± 2 **31.** $1; 3$.
Exercise 13.1, Page 181
1. $6x - 5$ **3.** $12x^3 - 45x^2$ **5.** $10t - 7$ **7.** $6t^2 - 12t + 3$ **9.** $10t + 2$ **11.** $20 + 8v$
13. $-6x^{-3} - 3x^{-1/2}$ **15.** $\frac{7}{4}x^{-3/4}$ **17.** $\frac{2}{3}x^{-1/3}$ **19.** $8x^{-5}$
21. $3x^{-1/2} - x^{-2/3} + 6x^{1/2} - 4x^{-1/3}$ **23.** $-2x^{-5} + \frac{8}{3}x^{-5/3}$ **25.** $12x^2 - 30x + 6;$
$24x - 30$ **27.** $-8 - 12x; -12$ **29.** $\frac{3}{4}t^{-1/2} - \frac{8}{9}t^{-5/3}; -\frac{3}{8}t^{-3/2} + (40/27)t^{-8/3}$
31. $4x^3 - 15x^2 + 8x - 3; 12x^2 - 30x + 8; 24x - 30; 24$ **33.** $24x^3 - 6x^2;$
$72x^2 - 12x; 144x - 12$ **35.** 25   **37.** -3 **39.** $2t + 2; 2.4; 2$ **41.** $1; 2$ **43.** $\frac{1}{3}(2 \pm \sqrt{10})$
45. 2   **47.** 10π **49.** 72π.
Exercise 14.1, Page 187
1. $32x - 24$ **3.** $6(2x - 3)^2$ **5.** $2(x^2 - 3x + 5)(2x - 3)$ **7.** $15(3x + 2)^4$
9. $2(t^2 - t + 2)(2t - 1)$ **11.** $5(t^2 - 3t)^4(2t - 3)$ **13.** $6(x^3 - 3x^2 + 4)^5(3x^2 - 6x)$
15. $(3t - 2)^{-2/3}$ **17.** $4(6x - 5)^{-1/3}$ **19.** $-\frac{3}{2}x^2(8 - x^3)^{-1/2}$ **21.** $8(2x - 5)^3$
23. $-6t(4 - t^2)^2$ **25.** $4(x^3 - 2x^2 - 5x)^3(3x^2 - 4x - 5)$ **27.** $-36x(x^2 + 3)^{-4}$
29. $-2x(x^2 + 9)^{-3/2}$.
Exercise 14.2, Page 193
1. $12x^2 - 6x - 20$ **3.** $(x^3 + 5)^3(14x^4 + 10x)$ **5.** $(x^3 - 2x)^3(24x^2 - 14x^4 - 8)$
7. $(x^2 - 2x)^2(x + 4)^4(11x^2 + 8x - 24)$ **9.** $\frac{1}{2}t^{1/2}(7t^3 + 26t)(t^3 + 6t)^{-1/3}$
11. $(x^3 - 10x)(x^2 - 5)^{-3/2}$ **13.** $(2x^3 + 2x)(x^2 - 9)^{-3}$ **15.** $-24x(x^2 + 3)^{-3}$
17. $t^{-4}(2t^2 - 15)(5 - t^2)^{-1/2}$ **19.** $-(6t^3 + 6t)(t^2 - 1)^{-3}$
21. $(x^5 + 12x^3 - 32x)/(x^2 + 4)^2(x^2 - 4)^{1/2}$ **23.** $-3(2 - y - y^2)^{2/3}/(1 + 2y)$
25. $1/48$ **27.** $\frac{3}{2}$ **29.** $5a^5t^4; 5a^4t^4$ **31.** $(-1,1); (3,-1)$.
Exercise 15.1, Page 201
1. $s = 7; 10; v = 3; 3; a = 0; 0$ **5.** $s = 0; 4; v = 6; 2; a = -4; -4$ **9.** $s = 10;$

-10; $v = -15$; -24; $a = -12$; -6 **13.** 20 ft; 9 ft/sec; -12 ft/sec^2 **15.** -54 ft; -60 ft/sec; -24 ft/sec^2 **17.** (a) 80; (b) 160; (c) 2.5; (d) 260; (e) 129 **19.** 10,000 ft; 50 sec; 800 ft/sec **21.** 185.9 ft/sec **23.** 20 ft/sec; $\frac{5}{8}$ sec **25.** (a) -3; -8; (b) -2; 4; (c) 1; 5; (d) -2; 62.

Exercise 16.1, Page 210
1. $m = 0$ at $(2, -9)$ **3.** $m = 0$ at $(1.5, 6.25)$ **5.** $3x - y = 10$; $x + 3y = -10$ **7.** $m = 0$ at $x = -2$; 4 **9.** $m = 0$ at $(2, -1)$ **11.** $m = 0$ at $(1, -6)$ **13.** 6 **15.** -12.5 **17.** $-\frac{8}{3}$ **19.** $6x + 2 = y$ **21.** $7x + 2y = -3$ **23.** $9x + 2y = 13$ **25.** $27y - 5x = 8$ **27.** $(0,3)$ **29.** none

Exercise 17.1, Page 219
1. min., $(4, -25)$ **3.** none **5.** min., $(0,0)$ **7.** min., $(\frac{3}{2}, -\frac{27}{4})$ **9.** max., $(-1/3, 29/27)$; min., $(3/2, -45/4)$ **11.** max., $(-1/3, 68/27)$; min., $(3, -16)$ **13.** max., $(-2, 86)$; min., $(2, -42)$; max., $(3, -39)$ **15.** max., $(-2, 40)$; min., $(4, -68)$ **17.** max., $(2, 56)$; min., $(-1, -25)$; $(5, -25)$ **19.** max., $(-1, 15)$; min., $(3, -17)$ **21.** min., $(1, 2)$; $m = \frac{1}{5}\sqrt{5}$ at $x = 2$ **23.** min., $(0,0)$; max., $(\pm\sqrt{2}, 2)$; $m = \frac{3}{2}\sqrt{2}$ at $x = 1$ **25.** min., $(0,0)$; max., $(\pm 1, \frac{3}{4})$.

Exercise 17.2, Page 223
1. min., $(2, -9)$ **5.** max., $(-2, 19)$; min., $(2, -13)$; infl., $(0, 3)$ **9.** max., $(2, 34)$; min., $(-4, -74)$; infl., $(-1, -20)$ **13.** infl., $(0,0)$ **17.** max. at $t = 2 - \frac{1}{3}\sqrt{21}$; min. at $t = 2 + \frac{1}{3}\sqrt{21}$; infl., $(2,4)$ **21.** max., $(2,1)$; min., $(1,0)$; infl., $(\frac{3}{2}, \frac{1}{2})$ **25.** max., $(0, 12)$; min., $(-1, 9)$; $(4, -116)$; infl. at $x = 1 \pm \frac{1}{3}\sqrt{21}$ **29.** infl., $(2, -11)$ **33.** min., $(2, 1)$.

Exercise 18.1, Page 231
1. 18; 18 **3.** 8; 16 **5.** 22.5 by 45 **7.** 12 by 15 **9.** 4 in. **11.** 2.94 (approx.) **13.** $w = l = \frac{8}{3}\sqrt{3}$; $h = \frac{4}{3}\sqrt{3}$ **15.** $l = 12$; $w = 6$; $h = 8$; $V = 576$ **17.** $l = 12$; $w = 8$; $h = 4.8$; $V = 460.8$ **19.** 256π **21.** 96π **23.** $(32/81)\pi r^3$ **25.** $l = 4\sqrt{6}$; $w = 4\sqrt{3}$ **31.** 8000 sq ft; 188.7 ft **33.** $72\pi\sqrt{3}$ sq ft **35.** 432 **37.** 144 **39.** 73.9 **41.** $(6, 6\sqrt{2})$ **43.** $\pm\frac{1}{2}$.

Exercise 19.1, Page 240
1. $-\frac{3}{4}$ at all points **3.** $\frac{3}{2}$; $3x + 5 = 2y$ **5.** $\frac{6}{5}$; $6x - 5y = -34$ **7.** $-11/8$; $11x + 8y = -2$ **9.** -2; $2x + y = 5$ **11.** $20/23$; $23y - 20x = 11$ **13.** $x = 2$ **15.** $\frac{5}{7}$; $5x - 7y = 19$ **17.** $-\frac{1}{2}$; $x + 2y = 80$ **19.** max., $(0,1)$; min., $(4, -3)$.

Exercise 20.1, Page 249
1. $\frac{9}{8}$ ft/sec **3.** $\sqrt{143}$ ft/sec **5.** $9/4\pi$ in./min **7.** $\frac{1}{15}$ sq ft/sec **9.** $27/100\pi$ cm/min **11.** 44.74 mph (approx.) **13.** 1 sq ft/min **15.** 864 cu in./min **17.** 0.0012 in./min **19.** 0.24 watt/sec **21.** 800 ergs/sec **23.** 0.54 ohm/sec.

Exercise 21.1, Page 256
1. $dy = 5.12$; $\Delta y = 5.15852816$ **3.** $dy = -0.0256$; $\Delta y = -0.025597760064$ **5.** 400 ft; 32 ft **7.** -2.4 watts **9.** -40 watts **11.** -3.5π sq in. **13.** $25\pi/2$ cu in. **15.** 0.5π; 0.5% **17.** 0.04392 watt **19.** 128 watts/sec **21.** 25.6 **23.** -1.14 **25.** -0.58 **27.** $-1/90$ **29.** -0.11; 0.11 **31.** -0.072; 0.00522.

Exercise 22.1, Page 264
1. $4x + C$ **3.** $\frac{4}{3}x^3 + C$ **5.** $q + C$ **7.** $w + C$ **9.** $4t^2 + C$ **11.** $\frac{1}{2}x^2 - 2x + C$ **13.** $\frac{1}{3}t^3 - t^2 + C$ **15.** $-t^{-1} - \frac{3}{4}t^{4/3} + C$ **17.** $\frac{1}{3}i^3 - \frac{1}{2}i^2 + C$ **19.** $a_1 y + \frac{1}{2}a_2 y^2 + C$ **21.** $\frac{1}{3}y^3 + y^2 - 15y + C$ **23.** $-\frac{1}{2}x^{-2} - x + C$ **25.** $-x^{-1} + C$ **27.** $\frac{9}{5}x^5 - \frac{3}{2}x^4 + \frac{1}{3}x^3 + C$ **29.** $\frac{1}{5}x^5 - \frac{3}{8}x^8 + C$ **31.** $x^3 - \frac{1}{2}x^2 + 3x - 5x^{-1} - \frac{1}{2}x^{-2} + C$ **33.** $2x^{1/2} - \frac{3}{2}x^{2/3} - 3x^{5/3} - \frac{1}{2}x^2 + C$ **35.** $\frac{1}{9}x^3 + C$ **37.** $x^2 - \frac{4}{3}x^{3/2} + C$ **39.** $y = x^2 + 3x + C$; parabola **41.** $s = t^3 - t^2 + 5t + C$.

Answers to Selected Odd-Numbered Problems

Exercise 22.2, Page 270
1. $\frac{1}{8}(x^2 + 5)^4 + C$ 3. $-\frac{5}{3}(8 - y^2)^{3/2} + C$ 5. $\frac{1}{4}(x^2 - 4x + 3)^2 + C$
7. $-\frac{1}{9}(8 - t^3)^3 + C$ 9. $-\frac{1}{2}(x^2 - x - 2)^4 + C$ 11. $\frac{1}{18}(3x - 2)^6 + C$
13. $-\frac{3}{7}(2 - x)^7 + C$ 15. $-(3/10)(3 - 2x)^{5/3} + C$ 17. $10t^2 - \frac{8}{3}t^3 + C$
19. $\frac{1}{6}y^6 - \frac{8}{5}y^5 + 3y^4 - \frac{8}{3}y^3 + C$ 21. $7(x^2 - 7)^{1/2} + C$ 23. $(2 - 5x^2)^{-1/2} + C$
25. $-\frac{1}{16}(3 - x^2)^4 + C$.

Exercise 22.3, Page 273
1. $y = 3x^2 - 5x + C$ 3. $i = t^2 - t + C$ 5. $2s = 5t^2 - 6t + C$
7. $y = x^3 - x^2 + C$ 9. $6q = 4t^3 - 9t^2 + C$ 11. $4y^3 = 3x^4 + C$ 13. $x^2 + y^2 = C$
15. $5x^3 + 3y^5 = C$ 17. $2x^3 - 3x^2 - 9y^2 + 30y = C$ 19. $y = x^2 + 5x + C$
21. $y = x^3 - x^2 + x + C$ 23. $3y = 6x - 9x^2 - x^3 + C$
25. $y = 4x^2 - \frac{1}{3}x^3 - \frac{1}{4}x^4 + C$ 27. $y = 2x - \frac{7}{2}x^2 - \frac{8}{3}x^3 + C$
29. $y = 5x - \frac{1}{3}x^3 - \frac{1}{4}x^4 + C$ 31. $y = x^2 + C$.

Exercise 23.1, Page 279
1. $x^2 - 5x + 12$ 3. $5x - 3x^2 + 14$ 5. $\frac{1}{2}x^2 - 3x + 12.375$ 7. $4y - \frac{5}{2}y^2 + 4.5$
9. $3.5x^2 - 1.5$ 11. $5s + 12.5$ 13. $x^3 - 2x^2 + 5x + 4$ 15. $4s + 2.5s^2 - s^3 - 2$
17. $\frac{1}{3}x^3 - 3x + 1267/192$ 19. $\frac{1}{3}s^3 - \frac{3}{2}s^2 + 5s$ 21. $y = x^2 - 3x - 5$
23. $y = \frac{1}{2}x^2 - 3x + 2.5$ 25. $y = 17 - x - 3x^2$ 27. $y = \frac{1}{3}x^3 - 2x^2 + x + 5$
29. $y = \frac{1}{3}x^3 - \frac{1}{2}x^2 - 3x$ 31. $3y = 2x^{3/2} - 6x^{1/2} + 26$.

Exercise 23.2, Page 281
1. $y = x^2 - 6x + 7$ 3. $s = 5t - 3t^2 + 22$ 5. $r = 3t - t^2 + 20$
7. $s = t^3 - t^2 + 2t + 2.312$ 9. $y = x^3 - \frac{1}{4}x^2 - 4x + 61$ 11. $x^2 - y^2 = 12$
13. $3x^2 + 5y^2 = 120$ 15. $2x^3 - 3y^2 = 6$ 17. $x^2 = 2y^3$
19. $3x^3 - 6x^2 + 4x + 2y^2 - 5y = 20$.

Exercise 23.3, Page 285
1. 2   3. 2   5. 21   7. 21   9. 30   11. 0   13. 3   15. 10/3   17. 18   19. 8.1   21. 7625
23. 92   25. 11/3   27. 2/9   29. 24   31. 13/6   33. 28/27   35. 48   37. 9.

Exercise 24.1, Page 294
1. 32/3   3. 4   5. 64/3   7. 16/3   9. 19/3   11. $\frac{2}{3}$ inside; $\frac{1}{3}$ outside   15. 61/4.

Exercise 24.2, Page 297
1. 9   3. 16/3   5. 64/3   7. 16/3   9. 2/15   11. 9   13. 128/9   15. $\frac{9}{2}$ 17. 32   19. 8
21. $\frac{5}{3}$ 23. $\frac{8}{3}$.

Exercise 25.1, Page 302
1. (a) 32π; (b) $(256/5)\pi$ 3. (a) $(128/15)\pi$; (b) $(32/3)\pi$ 5. (a) $(128/7)\pi$; (b) $(64/5)\pi$
7. $(16384/15)\pi$ 9. 80π 11. $(412/15)\pi$ 13. (a) $(2/35)\pi$; (b) $(1/10)\pi$ 15. $(512/15)\pi$.

Exercise 25.2, Page 306
1. (a) $(256/5)\pi$; (b) 32π 3. (a) $(32/3)\pi$; (b) $(128/15)\pi$ 5. (a) $(128/7)\pi$; (b) $(64/5)\pi$
7. (a) 4π; (b) 30π 9. $(27/2)\pi$ 11. 48π 13. $(1/10)\pi$ 15. 24π.

Exercise 26.1, Page 318
1. $s = t^2 - t + C$ 3. $s = t^3 - t^2 + C$ 5. $s = 60t - 16t^2 + C$ 7. $s = 20t - 4t^2 - 6$
9. $s = \frac{1}{3}t^3 - t^2 - 3t$ 11. $s = \frac{1}{3}t^3 + \frac{1}{2}t^2 - 2t + 298/3$ 13. $v = 20t + 16$;
$s = 10t^2 + 16t + 20$ 15. $v = -10t - 24$; $s = 60 - 24t - 5t^2$ 17. $v = 12t - 6$;
$s = 6t^2 - 6t + 10$ 19. $v = 10 - 4t$; $s = 10t - 2t^2$ 21. $v = 160 - 32t$;
$s = 160t - 16t^2$ 23. $v = 160 - 32t$; $s = 640 + 160t - 16t^2$ 25. 36 ft from
bottom   27. 50.56 ft/sec (approx.).

Exercise 26.2, Page 320
1. 144 ft; 112 ft; 304 ft   3. 144 ft; 592 ft; 128 ft   5. 40 ft; 90 ft   7. 48 ft; 16 ft
9. 2656 ft; 2624 ft

Exercise 26.3, Page 325
1. 25,824.5 ft; 123,100 ft   3. max. ht: 90,000 at 90°; range: 90,000 at 45°   5. range:

(a) 27,062.5 ft; (b) 108,250 ft; (c) 243,562.5 ft; range and max. height are proportional to square of initial velocity   7. $s = \tfrac{1}{3}t^3 + C_1 t^2 + C_2 t + C_3$ 9. (a) 12.5°; (b) 19°40'; (c) 28.8°.

Exercise 27.1, Page 333
1. $2\tfrac{3}{4}$ 3. 20 (333/16)   5. 6.564   7. 2.659   9. 7.991   11. 675/64   13. 73   15. 15.01   17. 22.88   19. 2.155   21. 21.125   23. 1.4335   25. 2.468   27. 0.9943.

Exercise 28.1, Page 338
1. 2π; 1   5. 2π; 1   9. π; 2   13. 2π; 1   17. π; 2   21. 2π; 1   25. 2π; 1   29. π; 2.

Exercise 28.2, Page 347
1. $3\cos 3x$ 3. $5\sec^2 5x$ 5. $-4\csc 4x \cot 4x$ 7. $2x \cos x^2$ 9. $-\cos(6-x)$
11. $-3x^2 \csc x^3 \cot x^3$ 13. $400 \cos 8t$ 15. $300 \sec^2 50t$ 17. $1200 \sec 20t \tan 20t$
19. $dy/dt = a \cos at$; $dy/d(at) = \cos at$ 21. $de/dt = 200\pi f \cos(2\pi f)t$; $de/d(2\pi f)t = 100 \cos(2\pi f)t$ 23. $600 \cos 30t - 450 \sin 30t$ 25. $300 \cos 10t + 150 \sin 10t$ 27. $15 \sec^2 5t - 20 \csc^2 5t$ 29. $300 \cos 15t - 15 \sin 15t$
31. $5 \cos 10x$ 33. $2t \cos 3t - 3t^2 \sin 3t$ 35. $20x \cos 5x + (2 - 25x^2) \sin 5x$
37. $(x-4)\sin(3-x) + 2\cos(3-x)$ 39. 2; -2; 45°; 135°; 225°; 315°
41. 0°; 45°; 90°; 135°; 180°; 225°; 270°; 315°; 360°   43. 1; 4.

Exercise 28.3, Page 350
1. $28 \sin^3 7x \cos 7x$ 5. $-6 \sec^3(5-2t) \tan(5-2t)$ 9. $5 \tan^4(\theta - 2) \sec^2(\theta - 2)$
13. $2x(1 + \sin^2 2x + x \sin 4x)$ 17. $12 \sin 3x - 18 \sin^3 3x$ 21. $6x - 9x \cos 3x - 3 \sin 3x + 3 \sin 6x$ 25. $6x(3 - x^2)^2 \sin^2 x^2 [(3 - x^2) \cos x^2 - \sin x^2]$
29. $\sin 2x - 2 \csc^2 x \cot x$ 33. $\tfrac{1}{3}x^{-2} (\sin^3 3x)(12x \cos 3x - \sin 3x)$
37. $-50 \sin 10x$ 41. $-\cos x - 2 \sin 2x$ 45. (approximate values): max. at $x = 40.9°$; 139.1°; 270°;   min. at $x = 90°$; 220.9°; 319.1°; infl. at $x = 0°$; 57°; 123°; 180°; 237° 303°; 360°   49. min. at $x = 0$.

Exercise 29.1, Page 363
1. $5(1 - 25x^2)^{-1/2}$ 3. $3(9x^2 + 1)^{-1}$ 5. $1/x(4x^2 - 1)^{1/2}$ 7. $(9 - t^2)^{-1/2}$
9. $(t^2 + 6t + 10)^{-1}$ 11. $2(-4x^2 - 4x)^{-1/2}$ 13. $15(9x^2 + 25)^{-1}$
15. $3/x(16x^2 - 9)^{1/2}$ 17. $5(x^2 + 6x + 34)^{-1}$ 19. $\tfrac{1}{2}(x - x^2)^{-1/2}$ 21. $\dfrac{3\sqrt{x+1}}{2(x+1)^3 + 1}$
23. $\dfrac{x^2}{\sqrt{1-x^2}} + 2x \arcsin x$ 25. $\dfrac{3}{9x^3 + x} - \dfrac{\arctan 3x}{x^2}$ 27. $1 + 2x \arctan x$
29. $-2(4 + x^2)^{-1}$ 31. $(2 + 4x)(1 - 4x^2)^{-1/2}$ 33. $x = 0$; $m = 1.155$ 35. 18 ft
37. 3.103°/sec.

Exercise 30.2, Page 376
1. $(2x - 7)(x^2 - 7x)^{-1}$ 3. $-2 \tan 2t$ 5. $3 \cot \theta$ 7. $\sec t \csc t$ 9. $\tfrac{1}{2}(y^2 - 4)y^{-1}$
11. $2x^4(x^2 + 3)^{-1} + 3x^2 \ln(x^2 + 3)$ 13. $3(6x - 8)^{-1}$ 15. $(x - 1)^{-1}$
17. $(3x - 15)^{-1}$ 19. $12(3t - 5)^{-1}$ 21. $-3(\cot 3\theta + \tan 3\theta)$ 23. $\sec t$ 25. $\sec x \csc x + x^{-1}$ 27. $6x(x^2 - 1)^{-1} + (2x + 1)^{-1}$ 29. $3x^2(x^3 - 4)^{-1} \log_7 e$
31. $(13x^2 - 10x - 48)/6x(x^2 - 4)$ 33. x^{-1}; $2x^{-1}$; $3x^{-1}$; nx^{-1} 35. $-49/24$ 37. none
39. min. at $x = \pm\sqrt{2}$ 41. $-y$ 43. $x^{-2}(1 - \ln x)$.

Exercise 30.3, Page 378
1. 12   3. 11   5. $-33/2$ 7. $2052/343$ 9. $(-189/208)\sqrt{13}$ 11. $-44/3$ 13. $-\tfrac{5}{3}$
15. $5x + 8 = 4y$; $16x + 20y = -1$.

Exercise 30.4, Page 383
1. $5e^{5x}$ 3. $(2x - 4)e^{x^2 - 4x + 2}$ 5. $-4(2^{-4x}) \ln 2$ 7. $12t^3 e^{3t^4}$ 9. $3 \cos 3t\, e^{\sin 3t}$
11. $2t e^{-2t}(1 - t)$ 13. $\sec^2 t\, e^{\tan t}$ 15. $-2t \sin(t^2 - 3) e^{\cos(t^2 - 3)}$
17. $-e^{-t}(3 \sin 3t + \cos 3t)$ 19. $2(e^{2t} + e^{-2t})$ 21. $4(e^{8x} + 1)^{-1} e^{4x}$

Answers to Selected Odd-Numbered Problems   689

23. $\dfrac{e^x}{x^2+1} + e^x \arctan x$ 25. $\dfrac{3e^{(3x-2)^{1/2}}}{2(3x-2)^{1/2}}$ 27. $\dfrac{(4x^2 + 8x + 1)e^{(x+2)^2}}{2\sqrt{x}}$

29. $e^{2x}(e^{2x} + 1)^{-1} + e^x \arctan e^x$ 31. $5e^{5t}(7 \cos 30t - 5 \sin 30t)$
33. $50e^{-10t}(8 \cos 5t - \sin 5t)$ 35. $e^x(x+2)$ 37. $2e^x(5 \cos 5x - 12 \sin 5x)$
39. $(e^x - e^{3x})(e^{2x} + 1)^{-2}$ 41. $-200e^{-10t}(\sin 30t + 7 \cos 30t)$ 43. $\tfrac{1}{2}e^{-1}$
45. $0; \infty; -\tfrac{1}{2}e^2; e^{-2}$ 47. 0   49. max. at $x = 2$; min. at $x = 0$; infl. at $x = 2 \pm \sqrt{2}$.

Exercise 31.1, Page 389
1. $-\tfrac{1}{2} \ln (3 - 2x) + C$ 3. $-5 \ln x + C$ 5. $-\tfrac{5}{4}(x^2 - 4)^{-2} + C$
7. $-\tfrac{3}{2} \ln (4 - x^2) + C$ 9. $-(9 - x^2)^{1/2} + C$ 11. $-4 \ln (2 - x) + C$
13. $\tfrac{3}{2} \ln (x^2 + 3) + C$ 15. $-\tfrac{1}{2} \ln \cos 2x + C$ 17. $\ln (x - \cos x) + C$
19. $\tfrac{1}{3} \ln (e^{3x} - 2) + C$ 21. $-\tfrac{1}{2} \ln (e^{-2x} + 5) + C$ 23. $\ln \sec x + C$
25. $-\tfrac{1}{3} \ln (x^3 - 3x^2 - 3x + 9) + C$ 27. $\ln (\sec x + \tan x) + C$
29. $x - 2 \ln (x + 2) + C$ 31. $\tfrac{3}{2}x^2 + 6x + 8 \ln (2x - 3) + C$
33. $\tfrac{3}{2}x^2 - 5x - \ln (x^2 - 4x + 1) + C$ 35. 16   37. 0.29542.

Exercise 31.2, Page 396
1. $\tfrac{1}{5}e^{5x} + C$ 3. $\tfrac{1}{2}e^{x^2-4x} + C$ 5. $\tfrac{1}{8}e^{4t^2} + C$ 7. $-\tfrac{1}{5}e^{-5x}(\log_{10} e) + C$ 9. $-\tfrac{1}{2}e^{6x-x^2} + C$
11. $-\tfrac{1}{4}e^{-t^4} + C$ 13. $e^{t-3} + C$ 15. $-\tfrac{1}{2}e^{2/x} + C$ 17. $e^{\sin^2 x} + C$ 19. 4405
21. 0.85914   23. 2.3504   25. 2701   27. 12.778.

Exercise 31.3, Page 401
1. 1   3. ∞ 5. $\tfrac{1}{8}$ 7. 2   9. ∞ 11. $\tfrac{1}{4}$ 13. 1   15. 1   17. $2\sqrt{3}$ 19. $\tfrac{3}{4} + \tfrac{3}{4}(9^{1/3})$ 21. ∞
23. e 25. $V = 36\pi;\ A = \infty$.

Exercise 32.1, Page 405
1. $\tfrac{1}{7} \cos 7x + C$ 3. $\tfrac{1}{3} \ln \sec (3x + 2) + C$ 5. $\tfrac{1}{4} \ln (\sec 4x + \tan 4x) + C$ 7. 1
9. $\tfrac{1}{2} \ln 2$ 11. $\tfrac{1}{2} \ln 2$ 13. $-(1/\omega) \cos \omega t + C$ 17. 0.14383   19. 0.4142
21. $(5/\pi f)(1 - \cos 0.006\pi f)$.

Exercise 32.2, Page 412
1. $\tfrac{1}{5} \sin 5t - (1/15) \sin^3 5t + C$ 3. $(1/36) \cos^9 4t - (1/28) \cos^7 4t + C$
5. $\tfrac{3}{8}x - \tfrac{1}{4} \sin 2x + (1/32) \sin 4x + C$ 7. $\tfrac{3}{8}t + (1/4\omega) \sin 2\omega t + (1/32\omega) \sin 4\omega t + C$
9. $\tfrac{1}{8}x - (1/32) \sin 4x + C$ 11. $-(1/30) \cot 2t(3 \cot^4 2t + 10 \cot^2 2t - 15) + C$
13. $3 \tan (t/3) - t + C$ 15. $-(1/15\omega)(3 \cot^5 \omega t + 5 \cot^3 \omega t) + C$
17. $t + \tfrac{1}{3} \cot 3t - \tfrac{1}{9} \cot^3 3t + C$
19. $(1/25) \tan^5 5t - (1/15) \tan^3 5t + \tfrac{1}{5} \tan 5t - t + C$
21. $-(1/30) \cot^{10} 3x - (1/12) \cot^8 3x - (1/18) \cot^6 3x + C$
23. $\tfrac{1}{2}\omega t + \tfrac{1}{4} \sin 2\omega t + C$ 25. $\tfrac{4}{3}$ 27. 2/15   29. $\pi/4$ 31. 0; 4/15.

Exercise 32.3, Page 420
1. $\tfrac{1}{2} \ln (x^2 + 25) + C$ 3. $-(16 - x^2)^{1/2} + C$ 5. $\tfrac{5}{8} \ln (4x^2 + 9) + C$
7. $\tfrac{1}{5} \ln (e^{5x} + 4) + C$ 9. $-\tfrac{2}{3}(1 - e^{3x})^{1/2} + C$ 11. $\tfrac{1}{3} \arcsin (3 \sin x/2) + C$ 13. 0.1638
15. 0.8704   17. $(1/36)\pi\sqrt{3}$ 19. $(5/3) \arcsin (3x/2) + C$ 21. 2π.

Exercise 32.4, Page 425
1. $\tfrac{3}{2} \ln (x^2 + 16) - \tfrac{7}{4} \arctan (x/4) + C$ 3. $\tfrac{1}{6} \ln (9x^2 + 16) + \tfrac{5}{12} \arctan (3x/4) + C$
5. $\tfrac{5}{6} \ln (3x^2 + 5) + (\sqrt{15}/5) \arctan (\sqrt{15}\,x/5) + C$ 7. $\tfrac{3}{2} \arctan [(x - 5)/2] + C$
9. $\tfrac{1}{2} \ln (x^2 - 4x + 16) - (\sqrt{3}/2) \arctan [(x - 2)/2\sqrt{3}] + C$
11. $(1/16)(25 - 16x^2)^{1/2} + \tfrac{3}{4} \arcsin (4x/5) + C$ 13. $-(x + 3)^{-1} + C$
15. $(7 - 6x + x^2)^{1/2} + C$ 17. 0.8529   19. -0.3273 21. 0.247   23. -0.0094
25. $(3/\sqrt{13}) \arctan (2x - 3)/\sqrt{13} + C$
27. $\tfrac{1}{6} \ln (3x^2 - 5x + 7) + (10/3\sqrt{59}) \arctan (6x - 5)/\sqrt{59} + C$
29. $3(16 - 6x - x^2)^{1/2} + 13 \arcsin (x + 3)/5 + C$.

Exercise 33.1, Page 432
1. $5 \ln (x - 2) - 2 \ln (x + 3) + C$ 3. $\tfrac{5}{2} \ln (x + 4) - \tfrac{1}{2} \ln (x + 2) + C$

5. $\ln x^3 - \ln (x + 3)^5 + C$
7. $\ln (x - 1)^5 - \ln (x + 1)^2 + C$
9. $\frac{4}{3} \ln (x + 3) - \frac{1}{3} \ln (x - 3) + C$
11. $\frac{1}{16} \ln (x - 8)/(x + 8) + C$
13. $\frac{1}{6}\sqrt{3}[\ln (x - \sqrt{3}) - \ln (x + \sqrt{3})] + C$
15. $\ln [(x - 2)^4(x + 5)] - \ln x^3 + C$
17. $\frac{3}{2} \ln x(x + 4) - \ln (x - 3)^2 + C$
19. $\frac{1}{2} \ln x^3 - \frac{1}{2} \ln (x + 6)^5 + \ln (x - 2)^2 + C$
21. $\ln [(x - 2)^2(x + 3)^3] - \ln (x + 5)^2 + C$
23. $\ln (x - 1)^4 - \ln [(x + 1)(x + 2)^3] + C$
25. $\frac{5}{3} \ln (x + 2) - \frac{1}{3} \ln (2x - 5) + C$
27. $\ln (x - 3)^2 - \ln (2x - 1)^3 + C$
29. $x + \frac{3}{4} \ln (x - 2) - (15/4) \ln (x + 2) + C$.

Exercise 33.2, Page 435
1. $\ln (x - 2)^3 - \ln (x - 4)^2 + C$
3. $\frac{5}{2} \ln x - \frac{11}{2} \ln (x + 2) + C$
5. $(13/7) \ln (x - 3) + (8/7) \ln (x + 4) + C$
7. $- \ln (x + 1)^3 - \ln (x - 5)^4 + C$
9. $\ln C(x - 2)(3x - 1)$
11. $(25/11) \ln (3x + 2) - (1/11) \ln (x - 3) + C$
13. $\ln x^2(x + 3)^6 - \ln (x - 2)^5 + C$
15. $\ln (x + 2)^6 - \ln x(x - 3)^3 + C$
17. $\ln (x + 1)^4 - \ln [(x - 2)(x - 1)^2] + C$
19. $\ln x + \frac{7}{5} \ln (x - 3) - \ln (x - 2) - \frac{2}{5} \ln (x + 2) + C$.

Exercise 33.3, Page 438
1. $3x^{-1} + \ln (x - 2)^4 - \ln x^2 + C$
3. $4x^{-1} + \ln (x + 3)^5 - \ln x^2 + C$
5. $(x - 3)^{-1} + \ln (x - 3)^3 - \ln x^2 + C$
7. $2(x - 1)^{-1} + \ln (x - 1)^4 - \ln (x + 2)^3 + C$
9. $\frac{1}{4} \ln [(x + 2)(x - 4)^7] - \frac{1}{2}(x - 4)^{-1} + C$
11. $\frac{1}{2}(6x + 1)x^{-2} + \ln x^2(x - 2) + C$
13. $-\frac{5}{2}x^{-2} - \ln x^2(x - 2)^3 + C$
15. $(4x + 9)(x^2 + 3x)^{-1} + \ln x^2(x + 3) + C$
17. $\frac{3}{2}x^{-2} + \ln (x - 4)^5 - \ln x^2 + C$
19. $(7x + 12)/x(x - 4) - \ln (x - 4) + C$
21. $(3x - 15)(x^2 - 9)^{-1} + \ln (x - 3)^2 + C$
23. $\frac{3}{2} \ln (x + 2) - \frac{1}{2} \ln x - x^{-1} + C$
25. (a) logarithm; (b) power; (c) arctangent.

Exercise 33.4, Page 442
1. $\ln x^5 - \frac{3}{2} \ln (x^2 + 9) + \frac{2}{3} \arctan (x/3) + C$
3. $\ln (x^2 + 4) - 3x^{-1} - 4 \arctan (x/2) + C$
5. $\ln (x - 2)^5 + \frac{3}{2} \ln (x^2 + 2x + 4) - (7/\sqrt{3}) \arctan (x + 1)/\sqrt{3} + C$
7. $\ln x^4(x^2 + 1)^{1/2} - \ln [(x - 2)^2(x + 1)^3] - 5 \arctan x + C$
9. $(x^2 + 2)x^{-1} - \ln x - (7/3) \arctan (x/3) + C$
11. $(3x^2 + 5)x^{-1} + \ln x^2 - \ln (x^2 + 4)^2 - \arctan (x/2) + C$
13. $\frac{3}{2}(x^2 + 4)^{-1} + \ln x^2 - \ln (x^2 + 4) + \frac{5}{2} \arctan (x/2) + C$
15. $x^2 - 6x + \frac{3}{2} (x^2 + 4) - \ln (x + 3)^2 - \frac{5}{2} \arctan (x/2) + C$
17. $\frac{3}{4} \ln (x - 2) + (17/4) \ln (x + 2) - \frac{3}{2} \ln (x + 4) + C$
19. $\ln (x - 2)^6 - \ln (x - 1) - \frac{3}{2} \ln (x^2 + 1) + 2 \arctan x + C$
21. $\frac{1}{4} \ln (4e^{-4x} + 1) - \frac{1}{4} \ln (3e^{-4x} - 1) + C$
23. $\frac{1}{2} \ln (\sin x + 2) - \frac{1}{2} \ln (\sin x + 4) + C$.

Exercise 34.1, Page 449
1. $-\frac{1}{3}xe^{-3x} - \frac{1}{9}e^{-3x} + C$
3. $\frac{1}{4}(x^4 - 1) \arctan x - (x^3/12) + (x/4) + C$
5. $\frac{1}{6}x^6 \ln x - (x^6/36) + C$
7. $-\frac{1}{4}x^{-2}(\ln x^2 + 1) + C$
9. $x \arctan x - \frac{1}{2} \ln (x^2 + 1) + C$
11. $(1/27)(9x^2 - 2) \sin 3x + \frac{2}{9} x \cos x + C$
13. $\frac{1}{2}[\sec \theta \tan \theta + \ln (\sec \theta + \tan \theta)] + C$
15. $\sec t + \ln (\csc t - \cot t) + C$
17. $(1/216)(18t^2 - 1) \sin 6t - (1/36)(6t^3 - t) \cos 6t + C$
19. $(1/34)e^{-5x}(3 \sin 3x - 5 \cos 3x) + C$
21. $[1/(\pi^2 + 1)]e^{\pi\theta}(\pi \sin \theta - \cos \theta) + C$
23. $\frac{3}{2}[\sec x \tan x + \ln (\sec x + \tan x)] - \csc x \sec^2 x + C$
25. $[1/(25 + \omega^2)]e^{-5t}[(\omega - 5) \sin \omega t - (\omega + 5) \cos \omega t] + C$
27. $[1/(m^2 + n^2)]e^{mx}(n \sin nx + m \cos nx) + C$
29. 3.6854π
31. 2.5452.

Answers to Selected Odd-Numbered Problems   691

Exercise 34.2, Page 454
1. $\frac{1}{5}(x^2 + 6)(x^2 - 9)^{3/2} + C$ 3. $-(1/27)(16 - 9x^2)^{3/2} + C$
5. $\frac{1}{4}(x^2 - 36)(x^2 - 9)^{1/2} + 27 \arctan[(x^2 - 9)^{1/2}/3] + C$
7. $\frac{2}{5}(x - 6)(x + 9)^{3/2} + C$ 9. $(1/105)(15x^4 - 48x^2 + 128)(x^2 + 4)^{3/2} + C$
11. $(1/76545)(-2048 - 1728x^2 - 1215x^4)(16 - 9x^2)^{3/2}$
13. $-(9 - x^2)^{1/2} + C$ 15. $\frac{1}{5}\arcsin(5x/3) + C$
17. $\ln(x + 2x^{1/2} + 1) + 2/(\sqrt{x} + 1) + C$
19. $\frac{1}{3}(x^2 + 16)(4 + x^2)^{1/2} + 4[\ln(\sqrt{4 + x^2} - 2) - \ln(\sqrt{4 + x^2} + 2)] + C$
21. $\frac{1}{6}\ln\frac{v^2 - v + 1}{(v + 1)^2} - \frac{5v^2 + 6v - 2}{2(v + 1)(v^2 - v + 1)} + \frac{\sqrt{12}}{27}\arctan\frac{2v - 1}{3} + C$;
$[v = (x^2 - 1)^{1/3}]$ 25. 1412/5   27. $-96/7$ 29. 0.426   31. none   33. 0.79578.

Exercise 34.3, Page 462
1. $\frac{1}{4}x^{-1}(x^2 - 4)^{1/2} + C$ 3. $-\frac{1}{4}x^{-1}(4 - x^2)^{1/2} + C$ 5. $(1/48)x^{-3}(x^2 - 16)^{3/2} + C$
7. $\frac{1}{5}\arctan\frac{(x^2 - 25)^{1/2}}{5} + C$ 9. $\frac{1}{27}\left[\frac{3x}{(4 - 9x^2)^{1/2}} - \arcsin\frac{3x}{2}\right] + C$
11. $\frac{1}{5}(x^2 + 6)(x^2 - 9)^{3/2} + C$ 13. $(x^2 - 4)^{1/2} - 2\arctan\frac{(x^2 - 4)^{1/2}}{2} + C$
15. $\frac{1}{3}(x^2 - 16)(x^2 - 4)^{1/2} + 8\arctan\frac{(x^2 - 4)^{1/2}}{2} + C$
17. $\frac{1}{3}(x^2 + 18)(x^2 - 9)^{1/2} + C$ 19. $\frac{1}{4}\arctan\frac{(x^2 - 4)^{1/2}}{2} - \frac{1}{8}x^{-2}(x^2 - 4)^{1/2} + C$
21. $\frac{2}{27}\arcsin\frac{3x}{2} - \frac{1}{18}x(4 - 9x^2)^{1/2} + C$ 23. $\frac{1}{15}(3x^2 - 8)(4 + x^2)^{3/2} + C$
25. $\frac{1}{3}x^{-3}(4x^2 - 1)(1 - x^2)^{1/2} + \arcsin x + C$ 27. $-\frac{1}{3}(3x^2 + 8)(4 + x^2)^{-3/2} + C$
29. $\frac{1}{2}x(x^2 - 4)^{1/2} + 2\ln(x + \sqrt{x^2 - 4}) + C$
31. $-x^{-1}(x^2 + 1)^{1/2} + \ln(x + \sqrt{x^2 + 1}) + C$.
33. $\frac{1}{2}[x(x^2 + 1)^{1/2} + \ln(\sqrt{x^2 + 1} - x)] + C$ 35. none; improper integral.

Exercise 35.1, Page 470
1. 39,270   3. 11,232   5. 16,900/3   7. 7488   9. top: 4069; bottom: 5042   11. $8\sqrt{2}$ ft from top   13. 279,552   15. 3660.8 lb   17. top: 1965.6; bottom: 2246.4   19. 26.325.

Exercise 35.2, Page 474
1. 12.6 in.-lb; 1.05 ft-lb   3. (5/2) ft-lb   5. $\frac{8}{3}$ ft-lb   7. (49/72) ft-lb   9. $\frac{5}{8}$ ft-lb; (15/8) ft-lb   11. 37.5 ft-lb   13. 3   15. 3 ft-lb   17. 0.5 ft.

Exercise 35.3, Page 478
1. $40,435.2\pi$ 3. 240,240 ft-lb   5. $14,040\pi$ ft-lb   7. $51,480\pi$ ft-lb   9. 56,160 ft-lb
11. 3080 ft-lb   13. 5940 ft-lb   15. 16,150 ft-lb.

Exercise 36.1, Page 486
1. 5 ft   3. 6.46 ft   5. 9 ft   7. (2,3)   9. (7.5,1.5)   11. (2,2).

Exercise 36.2, Page 490. Some answers approximate.
1. $(\frac{8}{5},2)$ 3. $(\frac{12}{5},\frac{3}{2})$ 5. (16/5,2)   7. (46/25,8/5)   9. (9/5,9/5)   11. (3/5,12/35)
13. $(-11/5,24/75)$ 15. $(-\frac{1}{5},-\frac{3}{7})$ 17. $(-1,1.6)$ 19. (2.5,1.1)   21. (1.313,2.097)
23. $(-0.313, 0.772)$ 25. $(0.5708, 0.3927)$ 27. $(4/\pi,8/3\pi)$.

Exercise 36.3, Page 496
1. (a) $(\frac{8}{3},0)$; (b) $(0,\frac{5}{3})$ 3. $(\frac{10}{3},0)$ 5. (3.5,0)   7. $(0,-2)$ 9. (a) $(-\frac{4}{3},0)$; (b) $(0,\frac{5}{4})$ 11. $(\frac{3}{4},0)$
13. (a) (10/9,0); (b) (0,20/9)   15. (35/16,0)   17. (3a/4,0)   19. 4(32/49) in. from base   21. $(\frac{9}{8},0)$ 23. (102/49,0)   25. (ln 27,0)   27. (0,2.441).

Exercise 37.1, Page 507. R_x and R_y are shown.
1. $\frac{6}{7}\sqrt{7}$; $\frac{6}{5}\sqrt{15}$ 3. $\frac{1}{2}\sqrt{6}$; $\sqrt{6}$ 5. $\frac{1}{2}\sqrt{6}$; $\sqrt{38}$ 7. $\frac{2}{5}\sqrt{30}$; $\frac{2}{7}\sqrt{42}$ 9. $\frac{3}{2}\sqrt{2}$; $\sqrt{6}$ 11. $(8/15)\sqrt{30}$;

$\frac{2}{3}\sqrt{6}$ **13.** $\frac{2}{5}\sqrt{30}$; $\frac{4}{7}\sqrt{42}$ **15.** $(2/35)\sqrt{70}$; $\frac{1}{5}\sqrt{30}$ **17.** $I_x = \frac{1}{9}(e^6 - 1)$; $R_x = 2.65\ldots$,
$I_y = 2(e^2 - 1)$; $R_y = \sqrt{2}$ **19.** (a) $\frac{5}{8}\sqrt{6}$; (b) $2\sqrt{6}$; (c) $(10/13)\sqrt{6}$ **21.** $R_x = R_y = \frac{3}{2}$.

Exercise 37.2, Page 514. Radius of gyration shown.
1. $\frac{1}{5}\sqrt{30}$ **3.** $\frac{3}{5}\sqrt{30}$ **5.** $\frac{1}{5}\sqrt{30}$ **7.** $2\sqrt{3}$ **9.** $\frac{4}{3}\sqrt{3}$ **11.** $\frac{4}{3}\sqrt{3}$ **13.** $\frac{4}{9}\sqrt{30}$ **15.** $\frac{4}{3}\sqrt{3}$
17. $(8/21)\sqrt{42}$ **19.** $(1/39)\sqrt{455}$ **21.** $\frac{2}{5}\sqrt{35}$ **23.** $\frac{1}{4}\sqrt{6}$ **25.** $\frac{1}{3}\sqrt{6}$.

Exercise 38.1, Page 521. Approximate answers.
1. 5.124 **3.** 4.781 **5.** 1.440 **7.** 9.294 **9.** 9.0734 **11.** 14/3 **13.** 3.352 **15.** 1.3170
17. $\frac{5}{2}\sqrt{2}\pi$ **19.** 4.06 **21.** 68,868 ft.

Exercise 38.2, Page 526. Approximate answers.
1. 87.17π **3.** $(98/81)\pi$ **5.** $\pi\left[e^{-4}(e^6 - 1)(e^4 + 1)^{1/2} + 2 + \ln\dfrac{e^2 + (e^4 + 1)^{1/2}}{1 + (e^4 + 1)^{1/2}}\right]$
7. 100π **9.** $4\pi[2 + \frac{9}{5}\sqrt{5}\arcsin(\sqrt{5}/3)]$ **11.** 17.336π **13.** 29.42π.

Exercise 39.1, Page 531
1. $3x + y = 17$ **3.** $5x - 3y = 41$ **5.** $y = x^2 + 3x + 2$ **7.** $y^2 = x^3 - x^2$
9. $4y^2 = x^2 + 2x - 15$ **11.** $4x^2 - 3y^2 = 72$ **13.** $x^2 + 9y^2 = 36$ **15.** $x^2 + y^2 = 16$
17. $y^2 = 4x^2 - 4x^4$ **19.** $y^2 = 4x$ **21.** $x^2 + 8xy - 4x = 16y$ **23.** $x^2 + y^2 = 2x$.

Exercise 39.2, Page 538
1. $3x + 5y = 11$ **3.** $100x^2 + 81y^2 = 225$ **5.** $18x^2 - y^2 = 45$ **7.** $x^2 + 8y = 16$
9. $x = 5t$; $y = 3t - 4$ **11.** $x = 7t + 5$; $y = 2t + 1$ **13.** $x = 4\cos\theta$; $y = 4\sin\theta$
15. $x = \pm(36 + t)^{1/2}$; $y = \pm t^{1/2}$ **17.** $x = 3 + (9 - t)^{1/2}$; $y = (t - 3)^{1/2}$
19. $x = \frac{1}{3}(t^2 - 144)^{1/2}$; $y = \frac{1}{3}t$ **21.** $x = 3t$; $y = \frac{5}{4} - \frac{3}{4}t^2$ **23.** $x = t$; $y = 16t^{-1}$
25. $x = t^2$; $y = t^2 - t$ **27.** $x = t - t^2\frac{1}{2}$; $y = \frac{1}{2}t^2$ **29.** $x = \frac{1}{2}(t + t\sqrt{2})$; $y = \frac{1}{2}(t - t\sqrt{2})$
31. $x = 6t^{-2}$; $y = t - 6t^{-2}$ **33.** $x = t^{-2}$; $y = t - t^{-2}$.

Exercise 39.3, Page 544
1. $-\frac{2}{3}t$; $-\frac{2}{9}$ **3.** $\dfrac{1 - 3t^2}{2t}$; $\dfrac{-3t^2 - 1}{4t^3}$ **5.** $-e^{-2t}$; $2e^{-3t}$ **7.** $2te^{-t} - 2e^t$;
$2e^{-2t}(1 - t - e^{2t})$ **9.** $-\frac{3}{5}\cot\theta$; $-(3/25)\csc^3\theta$ **11.** $x + y = 2$ **13.** $x + 4y = 8$
15. $7x - 9y = 15$ **17.** $-\frac{1}{2}$ **19.** $m = -5/3$, straight line **21.** πa^2 **23.** $3\pi a^2$
25. $(\pi a, 5a/6)$.

Exercise 40.1, Page 552
1. $x^2 + y^2 = 4x$ **3.** $x^2 + y^2 = -6y$ **5.** $y^2 + 8x = 16$ **7.** $4x^2 + 3y^2 + 16y = 64$
9. $9x^2 - 16y^2 + 160y = 256$ **11.** $9x^2 + 8y^2 - 24y = 144$ **13.** $y^2 + 24x = 144$
15. $x^2 + y^2 = 4(x^2 + y^2)^{1/2} - 4y$ **17.** $x^2 + y^2 = 2(x^2 + y^2)^{1/2} - 4x$ **19.** $r = 3$
21. $r^2 = \dfrac{-9}{\cos 2\theta}$ **23.** $r^2 = \dfrac{6}{\cos\theta\sin\theta}$ **25.** $\cot^2\theta = 5$ **27.** $r = -8\tan\theta\sec\theta$
29. $r = -\frac{8}{5}\cot\theta\csc\theta$ **31.** $r = \dfrac{4}{1 - \cos\theta}$.

Exercise 40.3, Page 561
1. $x^2 + y^2 = 6y$ **3.** $2x^2 + 2y^2 = -5y$ **5.** $y^2 = 12x + 36$
7. $5x^2 + 9y^2 + 40x = 100$ **9.** $16x^2 + 25y^2 - 120x = 400$
11. $4x^2 - 5y^2 - 60y = 100$ **13.** $r = \dfrac{4}{1 - \cos\theta}$ **15.** $r = \dfrac{5}{3 - 2\cos\theta}$
17. $r = \dfrac{6}{2 - \cos\theta}$.

Exercise 40.4, Page 563. Values of θ are shown.
1. 0°; 135° **3.** origin; 60°; 300° **5.** origin; 135°; 315° **7.** origin; 30°; 90°; 150°
9. origin; 30°; 150° **11.** 10°; 50°; 130°; 170°; 250°; 290° **13.** 187.2°; 352.8°
15. 135°; 315° **17.** origin; 90°.

Exercise 40.5, Page 567
1. (a) 30°; 60°; (b) $\sqrt{3}$; $-\sqrt{3}$ 3. (a) 105°; 120°; (b) -1; 0 5. (a) 150°; 157.5°; (b) 0; $\sqrt{2}-1$ 7. (a) 163.9°; 90°; (b) $\frac{1}{7}\sqrt{3}$; ∞ 9. 30°; $-\frac{1}{3}\sqrt{3}$ 11. (a) 90; 161.6°; (b) $-\sqrt{3}$; $\frac{1}{2}$ 13. 90°; 210°; 330° 15. 90° 17. 30° 19. 83.46° at $\theta = 135°$.

Exercise 40.6, Page 573
1. 9π 3. 18π 5. $(27/8)\pi - 9$ 7. 4π 9. $(81/8)\pi - 9/4 - 9\sqrt{2}$ 11. 20π
13. $9\pi - (27/2)\sqrt{3}$ 15. $18\pi + (45/2)\sqrt{3}$ 19. 12π 21. 32 25. 12.

Exercise 41.2, Page 583
1. $5\cosh 5x$ 3. $3\operatorname{sech}^2(3x - 4)$ 5. $-2\operatorname{sech} 2t \tanh 2t$ 7. $3\sinh 6x$ 9. $2\coth 2x$
11. $3t^2 \operatorname{sech}^2 3t + 2t \tanh 3t$ 13. e^{2t} 15. $0.1 \sinh 0.2t$ 17. min., (0,0) 19. infl., (0,0); $m=1$ 21. 0.0447 23. 1.175.

Exercise 41.3, Page 586
1. $\frac{1}{5}\sinh 5x + C$ 3. $\frac{1}{2}\ln\cosh x^2 + C$ 5. $\frac{1}{4}\sinh 4x$ 7. $x \cosh x - \sinh x + C$
9. $\frac{1}{2}\sinh 2t + \frac{1}{6}\sinh^3 2t + C$ 11. $\frac{1}{5}\ln\cosh 5x + C$ 13. 2.762 15. 7.795π
17. (0.538, 0.599) 19. 1.5232.

Exercise 42.1, Page 591
1. -4; 3 3. -13; 4 5. 25; -9 7. -20; 29 9. 1; -3 11. $2e^{2x+3y}$; $3e^{2x+3y}$
13. $4(4x-3y)^{-1}$; $-3(4x-3y)^{-1}$ 15. $(\tan y)(\sec x)(\tan x)$; $(\sec x)(\sec^2 y)$
17. $-3e^{-3x}\cos y$; $-e^{-3x}\sin y$ 19. $\dfrac{(x^2+y^2)^{1/2} + x}{x(x^2+y^2)^{1/2} + x^2 + y^2}$; $\dfrac{y}{x(x^2+y^2)^{1/2}+x^2+y^2}$ 21. $-1/5$; $-1/80$; $-1/112500$ 23. 9; 25; 225 25. (a) -25; (b) -1; (c) $-1/1600$.

Exercise 42.2, Page 598
1. 18 3. $dA = -2.4$; $\Delta A = -2.46$ 5. 5.6 7. -17.76 9. -2; $-121/65$ 11. 8865; 8863.424 13. 1.24 15. 0.18 17. 720π 19. 312.48π.

Exercise 42.3, Page 600
1. $-3(2x+4y+2) + 2t(4x-10y-3)$ 3. $-2t(6x+y-5) - 2(x+4y+3)$
5. $4x^2 + 3xy + 10x + 24y^2$ 7. $(-\sin t)(2x+2y) + (\cos t)(2x-2y)$
9. $(3t^2-5)(x+y)^{-1}$ 11. $4tu - 4tv - 6t + 10u - 40v - 25$ 13. 20/23
15. $-\frac{2}{3}$; hyperbola 17. $\frac{7}{4}$; hyperbola 19. -1; ellipse 21. ∞; ellipse.

Exercise 42.4, Page 604
1. -7 3. $10x - 6y$ 5. $-6y$ 7. $12x^2 - 12y - 7$ 11. $6e^{2x+3y}$ 13. $-6\cos 2x \sin 3y$
15. 0 17. 14; -28; 26; -16; -4 19. -1; 7; 14; -4; 17 21. $3e^2$; $4e^2$; $9e^2$; $16e^2$; $12e^2$ 23. $\frac{1}{2}$; $\frac{1}{2}$; $-\frac{1}{4}$; $-\frac{1}{4}$; $-\frac{1}{4}$ 25. 0; 16; 0; 0; 16.

Exercise 42.5, Page 607
1. $4x + 3x^2y - xy + x^3 - 6xy^2 + C$ 3. $4y + 2xy^2 - \frac{1}{2}y^2 + 3x^2y - 2y^3 + C$
5. $2x^3 + 5x^2y - \frac{3}{2}x^2y^2 - 2x^2 + C$ 7. $4x^3y^3 - 2x^2y^3 + C$ 9. $5x + x^2 - 3x^3$
11. 1968/5 13. $\frac{1}{2}e^2 - e + \frac{1}{2}$ 15. 8 17. 64/5; $\frac{2}{5}\sqrt{30}$ 19. 53.75; 2.677 21. 4131/140; $(3/70)\sqrt{3570}$.

Exercise 43.1, Page 612
1. $\dfrac{n}{n+1}$; $\dfrac{5}{6} + \dfrac{6}{7}$ 3. $1/2^{n-1}$; $\dfrac{1}{32} + \dfrac{1}{64}$ 5. $\dfrac{2n-1}{(n+1)^2}$; $\dfrac{9}{36} + \dfrac{11}{49}$ 7. $\dfrac{2n+1}{2n(2n-1)}$; $\dfrac{11}{9 \cdot 10} + \dfrac{13}{11 \cdot 12}$ 9. $\dfrac{4n-3}{2^{n+1}}$; $\dfrac{21}{128} + \dfrac{25}{256}$ 11. $\dfrac{3n-2}{n^2}$; $\dfrac{19}{49} + \dfrac{22}{64}$
13. $\dfrac{1}{5} + \dfrac{2}{7} + \dfrac{3}{9} + \dfrac{4}{11} + \dfrac{5}{13}$ 15. $\dfrac{1}{2} + \dfrac{3}{4} + \dfrac{5}{8} + \dfrac{7}{16} + \dfrac{9}{32}$ 17. $\dfrac{2}{5} + \dfrac{5}{7} + \dfrac{8}{9} + \dfrac{11}{11} + \dfrac{14}{13}$

19. $\dfrac{1}{1} + \dfrac{4}{2} + \dfrac{9}{6} + \dfrac{16}{24} + \dfrac{25}{120}$ 21. $\dfrac{2}{6} + \dfrac{5}{12} + \dfrac{10}{20} + \dfrac{17}{30} + \dfrac{26}{42}$

23. $\dfrac{1}{1} + \dfrac{5}{\sqrt{2}} + \dfrac{9}{\sqrt{3}} + \dfrac{13}{\sqrt{4}} + \dfrac{17}{\sqrt{5}}$.

Exercise 43.2, Page 616

1. convergent; $(\tfrac{3}{4})^{n-1}$ 3. divergent; $\dfrac{2}{2n+1}$ 5. divergent;   7. divergent; $\dfrac{n+1}{2n+1}$
$1/2n$ 9. convergent; $\dfrac{1}{4n^2 - 10n + 6}$ 11. convergent; $1/n!$ 13. divergent; $\dfrac{5}{5n-1}$
15. convergent; $\dfrac{1}{n^2+1}$ 17. convergent; $\dfrac{1}{n^2+n}$ 19. convergent; $\dfrac{1}{(2n-1)^2}$.

Exercise 43.3, Page 620

1. 1   3. $\tfrac{1}{3}$ 5. 0   7. $\tfrac{1}{2}$ 9. 2   11. 2   13. 1/32   15. $\tfrac{1}{2}$ 17. 2   19. 0   21. $\tfrac{1}{2}$ 23. 1   25. $\dfrac{n+1}{2^{n-1}}$; convergent   27. $\dfrac{3^n}{n(2^n)}$; divergent   29. $n!/2^n$; divergent   31. $\dfrac{2n}{n^2+1}$; convergent   33. $\dfrac{n^2}{(n+1)!}$; convergent   35. $\dfrac{3n-2}{(n-1)!}$; convergent.

Exercise 44.1, Page 631

1. $y = 6x + C$ 3. $2y = 8x - 3x^2 + C$ 5. $y^4 = 2x^2 + C$ 7. $3y^4 = 4x^3 + C$
9. $xy = C$ 11. $3y^2 = 2x^3 + C$ 13. $y = 2x^3 - 2x^2 + 3x + C$
15. $(x+1)(y-4) = C$ 17. $\sin y = \ln \sec x + C$ 19. $\left(\dfrac{y+2}{y-2}\right)\left(\dfrac{x-1}{x+1}\right)^2 = C$
21. $2 \arctan(y/3) - 3 \arctan(x/2) = C$ 23. $4(\tan y - y) = 2 - \sin 2x + C$
25. $2y^2 \arcsin(3x/2) + 3 = Cy^2$ 27. $6 \arctan(y+3) + 4 \arctan(x-2) = \ln \dfrac{C(y^2 + 6y + 10)}{x^2 - 4x + 5}$ 29. $y = 4x^2 + C_1 x + C_2$ 31. $2y = x^4 + C_1 x^2 + C_2 x + C_3$.

Exercise 44.2, Page 636

1. $y = -3x$ 3. $x^2 + y^2 = 13$ 5. $y = x^3 - 2x^2 + x + 7$ 7. $8x^2 + 3x^2 y = 4y$
9. $(x-3)(y+2) = 2$ 11. 0; 32   13. -5; 56   15. 160; 400   17. 30; 90
19. $i = \tfrac{5}{3}(1 - e^{-15t})$ 21. $i = 0.01(1 - e^{-500t})$ 23. $i = \tfrac{1}{5}(1 - e^{-400t})$
25. $i = (E/R)(1 - e^{-Rt/L})$.

Exercise 44.3, Page 640

1. (a) 26.3 hr; (b) 53.144 gm; (c) 87.42 hr   3. (a) 1589 yr; (b) 84.0%; (c) 11.3%; (d) 10,560 yr   5. (a) 63.0%; (b) 9.92%; (c) 0.01%; (d) 149.3 hr   7. 221.2%   9. 23.89°   11. 53.55°.

Exercise 45.1, Page 644

1. $3xy = \cos^3 x - 3 \cos x + C$ 3. $xy^3 = y^2 \sin y + 2y \cos y + \sin y + C$
5. $y = Cx^2 - 3x$ 7. $y = x^2 + x \ln Cx$ 9. $x^2 y + 1 = Cxy$ 11. $xy = 5y + C$.

Exercise 45.2, Page 648

1. $3y = Ce^{3x} - 4$ 3. $y + 1 = Ce^{x^2/2}$ 5. $xy + \cos x = C$
7. $y = (1/17)(4 \sin x - \cos x) + Ce^{-4x}$ 9. $6y + 3x + 4 = Cx^3$
11. $i = 3/25 - 3t/5 + Ce^{-5t}$ 13. $i = \tfrac{4}{3} + Ce^{-30t}$
15. $i = 1/100)(6 \sin 6t + 8 \cos 6t) + Ce^{-8t}$
17. $i = \dfrac{L^2}{R^2 + \omega^2 L^2}\left(\dfrac{R}{L} \sin \omega t - \omega \cos \omega t\right) + Ce^{-Rt/L}$ 19. $x^2 y = xe^x - e^x + C$
21. $i = 4(1 - e^{-50t})$ 23. $i = (1/50)(3 \sin 20t - 4 \cos 20t) + Ce^{-15t}$.

Answers to Selected Odd-Numbered Problems   695

Exercise 46.2, Page 653
1. $y = C_1 e^{-x} + C_2 e^{4x}$ 3. $y = C_1 e^{2x} + C_2 e^{5x}$ 5. $y = C_1 e^{-2x} + C_2 e^{2x}$
7. $y = C_1 e^{-1.5x} + C_2 e^{1.5x}$ 9. $y = C_1 e^{-6x} + C_2 e^x$ 11. $y = C_1 + C_2 e^{1.5x}$
13. $y = C_1 e^{-3x} + C_2 e^{4x}$ 15. $y = C_1 e^{0.5x} + C_2 e^{3x}$ 17. $y = C_1 + C_2 e^{-x} + C_3 e^{-3x}$
19. $y = C_1 e^{-2x} + C_2 x e^{-2x}$ 21. $y = C_1 e^{-x} + C_2 e^{5x} - \frac{3}{5}$
23. $y = C_1 + C_2 e^{-3x} + C_3 e^{3x} - \frac{4}{9}x$.

Exercise 46.3, Page 656
1. $y = C_1 e^{-2x} + C_2 e^{-3x}$ 3. $y = C e^{-5x}$ 5. $y = C_1 e^{-0.5x} + C_2 e^{-3x}$
7. $y = C_1 + C_2 e^{-4x} + C_3 e^x$ 9. $y = C_1 + C_2 e^{-4x/3}$ 11. $y = C_1 = C_2 e^{-0.5x} + C_3 e^{4x}$
13. $y = C_1 e^{-3x/4} + C_2 e^{2x}$ 15. $y = C_1 e^{-\sqrt{5}x} + C_2 e^{\sqrt{5}x}$
17. $y = C_1 e^{-2.899x} + C_2 e^{6.899x}$ 19. $C_1 e^{-x} + C_2 e^x + C_3 e^{2x}$
21. $y = C_1 e^{-2ix} + C_2 e^{2ix}$ 23. $y = C_1 e^{(2-3i)x} + C_2 e^{(2+3i)x}$
25. $y = C_1 e^{(-1-2i)x} + C_2 e^{(-1+2i)x}$ 27. $y = C_1 e^{-\sqrt{2}ix} + C_2 e^{\sqrt{2}ix}$
29. $s = C_1 e^{-13.86jt} + C_2 e^{13.86jt}$ 31. $i = C_1 e^{(-5-10j)t} + C_2 e^{(-5+10j)t}$
33. $y = \frac{4}{7}e^{5x} - \frac{4}{7}e^{-2x}$ 35. $y = \frac{4}{3}e^{-3x} + \frac{14}{3}e^{3x}$.

Exercise 46.4, Page 661
1. $y = C_1 e^x + C_2 x e^x$ 3. $y = C_1 + C_2 x + C_3 e^{-3x}$
5. $y = C_1 + C_2 x + C_3 e^{2ix} + C_4 e^{-2ix}$ 7. $y = C_1 + C_2 e^{-4x} + C_3 x e^{-4x}$
9. $y = C_1 + C_2 x + C_3 e^{\sqrt{3}x} + C_4 e^{-\sqrt{3}x}$ 11. $s = C_1 e^{1.5x} + C_2 x e^{1.5x}$
13. $y = C_1 + C_2 x + C_3 e^{-0.764x} + C_4 e^{-5.236x}$
15. $y = C_1 + C_2 x + C_3 e^{-\sqrt{32}ix} + C_4 e^{\sqrt{32} \cdot x}$
17. $y = C_1 e^x + C_2 e^{-x} + C_3 e^{2x} + C_4 x e^{2x}$ 19. $y = 2x e^{-x}$.

Exercise 47.1, Page 665
1. $y = A \cos 3x + B \sin 3x$ 3. $y = A \cos 1.155x + B \sin 1.155x$
5. $s = A \cos 1.5t + B \sin 1.5t$ 7. $i = A \cos 4.472t + B \sin 4.472t$
9. $i = A \cos 8.944t + B \sin 8.944t$ 11. $x = A \cos 8.485t + B \sin 8.485t$
13. $y = A \cos 40x + B \sin 40x$ 15. $y = A \cos 0.6124x + B \sin 0.6124x$
17. $y = e^{3x}(A \cos 7x + B \sin 7x)$ 19. $i = e^{0.5t}(A \cos 3.5t + B \sin 3.5t)$
21. $s = \frac{2}{3}e^{-t} \sin 3t$ 23. $y = e^{-2x}(10 \cos 6x + 4 \sin 6x)$.

Exercise 47.2, Page 671
1. $s = \frac{1}{6} \cos 16t$; 152.8   3. $s = \frac{1}{6} \cos \sqrt{128}\, t$; 108.0   5. $s = \frac{1}{4} \cos \sqrt{384}\, t$; 187.1
7. frequencies: (a) 152.8; (b) 124.75; (c) 108.0   9. $f/\min. = (30/\pi)\sqrt{Kg/w}$
11. $i = e^{-25t}(A \cos 75t + B \sin 75t)$; $f = 11.94/\sec$
13. $i = e^{-20t}(A \cos 40t + B \sin 40t)$; $f = 6.366/\sec$
15. $i = e^{-10t}(A \cos 20t + B \sin 20t)$; $f = 3.183/\sec$
17. $i = e^{-20t}(A \cos 110t + B \sin 110t)$; $f = 17.507/\sec$
19. $i = e^{-35t}(A \cos 120t + B \sin 120t)$; $f = 19.10/\sec.$

Exercise 48.1, Page 677
1. $y = C_1 e^{-4x} + C_2 e^{2x} - \frac{5}{8}x^2 - (5/16)x + 49/64$
3. $y = C_1 e^{-2x} + C_2 e^{3x} - \frac{1}{2}x^2 + \frac{1}{6}x - 19/36$
5. $y = C_1 e^{-3x} + C e^{-x} + \frac{5}{8}x e^x - (15/32)e^x$
7. $y = C_1 e^{3x} + C_2 x e^{3x} + \frac{5}{9}x^2 + (20/27)x + 10/27 - e^x$
9. $y = C_1 e^{2x} + C_2 x e^{2x} - (5/169) \sin 3x + (12/169) \cos 3x$
11. $y = A \cos 3x + B \sin 3x + \frac{1}{9}x^3 + (10/27)x$
13. $y = C_1 e^{-4x} + (4/25) \sin 3x - (3/25) \cos 3x + \frac{1}{10} \sin 2x + \frac{1}{5} \cos 2x$
15. $y = C_1 + C_2 e^{-5x} - \frac{1}{2}e^{-2x} - (1/41) \sin 4x - (5/164) \cos 4x$
17. $y = \frac{1}{16}e^{-2x} + \frac{1}{16}e^{2x} - \frac{3}{4}x^2 + \frac{7}{8}$
19. $y = -(89/3)e^{-x} - 11x e^{-x} + 5x^2 - 20x + 30 - \frac{1}{3}e^{2x}.$

Exercise 48.2, Page 679
1. $y = C_1 e^{-x} + C_2 e^{3x} - xe^{3x} + x^2 - \frac{4}{3}x + \frac{14}{9}$
3. $y = C_1 + C_2 e^{-2x} + \frac{2}{3}x^3 - x^2 - \frac{3}{2}x + 3xe^{-2x}$
5. $y = A \cos 2x + B \sin 2x + \frac{3}{8}e^{2x} + \frac{3}{2}x \sin 2x$
7. $y = A \cos x + B \sin x + \frac{3}{5}e^{-3x} - 2x \cos 2x$
9. $y = C_1 + C_{e_2}^{-3x} + C_{e_3}^{3x} - (5/27)x^3 - (10/81)x + \frac{2}{9}xe^{3x} - \frac{1}{10}e^{2x}$
11. $y = C_1 + C_2 x + C_3 e^{-x} + C_4 e^x - \frac{1}{3}x^4 - \frac{5}{2}x^2 - \frac{11}{4}xe^x + \frac{3}{4}x^2 e^x$
13. $y = (27/8) \cos 2x + \frac{3}{4}x^2 - 19/8$.

Index

Abscissa, 7
Absolute convergence, 617
Absolute value, 2
　graphs of, 33
Acceleration, average, 199
　instantaneous, 199
　positive and negative, 201
Algebraic curves, 122
Algebraic substitution, 449
　necessary conditions, 452
Alternating series, 617
Amplitude, 337
Analytic geometry, problems of, 42
Angle, of incidence, 83
　of inclination, 18
　between lines, 21
　rate of change of, 357
　of reflection, 83
Antiderivative, 258
Apex of a cone, 113
Approximate integration, 326
Arc, of a circle, 339
　of a curve, 515
　differential of, 516
　length of, 515, 519, 572
Area, centroid of, 484
　under a curve, 288
　between curves, 295
　by double integration, 606
　element of, 290
　first moment of, 484
　moment of inertia of, 498
　in polar coordinates, 568
　second moment of, 498
　of a sector, 339
　separate portions of, 292
　of a surface of revolution, 521
Arithmetic progression, 613
Asymptotes, 31, 38
Auxiliary equation, 653
　complex roots of, 664
　imaginary roots of, 662
　repeated roots of, 657
Average rate of change, 138
Axis, conjugate of hyperbola, 106
　major, to ellipse, 89
　minor, of ellipse, 89
　of parabola, 73
　of symmetry, 36
　transverse, of hyperbola, 101
　x-axis, 1, 6, 25
　y-axis, 6, 25
　z-axis, 25

Base, in exponential functions, 365, 393
　of natural logarithms, 368, 370
Boundary conditions, 276, 633

Cardioid, 554
Catenary, 574
Cartesian coordinates, 6
Catesian form of equation, 323, 528
Center, of gravity, 480
　of mass, 481, 490
Centroid, definition, 480
　disc method, 493
　of plane area, 484
　shell method, 495
　of solid, 490
　x-bar, 485, 486
　y-bar, 485, 486
Chain problem, 477
Chain rule, 184, 599
　in partials, 599
Choosing proper coefficient, 456
Circle, arc of, 339
　determining conditions, 65
　equation of, 57–59

graphing, 60
imaginary, 59
intercepts, 62
intersection with lines, 63
parametric equations of, 532
tangent, length of, 69
tangent to, 68
Comparison tests, series, 613
Complex roots of auxiliary equation, 664
Components of velocity, 321
Concavity, 213
Conditional convergence, 617
Cone, apex of, 113
  definition, 113
  nappes, 113
  right circular, 113
Conic sections, 112
  degenerate forms, 113–114
  equations of, 114, 558
  identification of, 114
  in polar coordinates, 558
Constant, 126
Constant (or modulus), spring, 472
Constant factor in integration, 262, 392
Constant of integration, 387
Constants, in differential equations, 628, 630
Convergence, absolute, 617
  conditional, 617
  conditions for, 617
  of infinite series, 612
  tests for, 613, 618
Cooling, Newton's law of, 639
Coordinate systems, 1, 6, 25, 545
Coordinates, polar, 6, 545
  rectangular, 6
  space, 25
Cosine, derivative of, 343
Cosine series, 624
Curvature, 116
  reversal of, 120
Curve, discontinuous, 38
  slope of, 204
Cycloid, 535

Decreasing function, 211
Definite integral, 282, 283, 318
Degree of differential equation, 628

Delta method for derivative, 169
Delta notation, 4, 16, 165–166
Density, 491
Dependent variable, 29, 127
Derivative, 164
  application, 225, 230
  chain rule, 184, 599
  definition, 167, 169
  by delta method, 167, 169
  by formula, 174
  geometric meaning, 205
  higher order, 179
  as instantaneous rate of change, 165
  by inversion, 192
  notations for, 170
  as quotient of differentials, 253
  as slope of a curve, 204, 344
Derivative, of a constant, 178
  of exponential function, 378
  of hyperbolic functions, 574
  of implicit functions, 237, 599
  of inverse trigonometric functions, 358–361
  of logarithmic functions, 369, 373
  of parametric equations, 541
  partial, 587
  power, rule, 174, 183, 185, 348
  product rule, 188
  quotient rule, 190
  second partial, 601
  of trigonometric functions, 341–348
Derivative operator, 176, 670
Descartes, René (1596–1650),, 6, 30
Differential, 251
  as an approximation, 254
  of arc, 516
  of product and quotient, 641
  of surface, 522–523
  total, 593
  of volume, 299
Differential equations, 270, 279, 309, 626
  applications, 309, 636, 670
  auxiliary equation, 653
  constants in solution of, 628
  degree of, 628
  electric circuit problem, 635, 670
  evaluating between limits, 310, 634
  first order, linear, 644

general solution of, 271, 310, 628
homogeneous, 649
integrating factor, 643
modification rule, 677
nonhomogeneous, 672
order of, 628
particular solution of, 279, 310, 632, 674
perfect differential, 641
solution, by operators, 652
  by separating variables, 629
  trigonometric form of, 663
  undetermined coefficients, 674
Differentiation, 169, 183, 334, 352, 564, 581
  implicit, 237
  logarithmic, 376
  of parametric equations, 539
  partial, 587
  of polar equations, 564
Directed distance, 3
Directrix, 72, 87, 100
Disc method, 300, 493
Discontinuous curve, 38
Discriminant of a quadratic denominator, 443
Distance, from a line to a point, 52
  from origin to a point, 26
  between points, 3, 10, 27
Divergent series, 613
Domain, 33
Double integration, 605

e, natural base, 368
Eccentricity, 73
  of ellipse, 87
  of hyperbola, 100
  of parabola, 73
Electric circuit problems, 590, 633–635, 655, 670
Element, 291
  of arc, 516
  of area, 290, 297
  of cone, 113
  of force, 468
  of mass, 493
  of moment, 487, 499
  of surface of revolution, 522–523
  of volume, 299
  of work, 473

Ellipse, 87, 88
  as a conic section, 113
  directrix, 87, 91
  eccentricity, 87
  equation of, 89, 93
  focus, 87, 91
  graphing, 92
  major and minor axes, 89
  parametric equations of, 532
  reflecting principle of, 98
  standard position of, 95
  translation of, 94
  vertices of, 89
Equation, Cartesian form of, 323, 528
  of circle, 57, 58, 59
  differential, 279, 309, 627
  of ellipse, 89
  first degree, 31
  fourth degree, 120
  graph of, 30
  higher degree, 116
  of hyperbola, 102
  linear, 31
  of parabola, 73
  second degree, 35
  solution of, 29
  straight line forms, 42–50
  third degree, 117
Equations, parametric, 527
Equations and graphs, 29
Euler's formula, 662
e^x, infinite series, 623
Excluded values, 374
Exponential curve, slope of, 381
Exponential functions, 365
  base of, 365
  derivatives of, 378
  general base of, 393
  graph of, 367
  integration of, 390
Exponential integrating factor, 646
Explicit functions, 234
Extent of a curve, 33

Factor, integrating, 266, 643, 646
Factorial, 610
Family of curves, 275
Family of terms, 677
Finite series, 609
First-degree equation, 31

700 Index

Focal chords of parabola, 73
Focal length of parabola, 73
Focus, of ellipse, 87, 91
   of hyperbola, 100–101
   of parabola, 72
Foot-pound, 472
Four-step rule for derivative, 169
Fourth-degree equation, 120
Fraction, improper, 441
   irrational, 427
   proper, 441
   rational, 427
Fractions, partial, 427
Function, 124
   algebraic, 183
   combinations of, 131
   decreasing, 211
   derivative of, 167, 168
   explicit, 234
   of a function, 131, 184
   hyperbolic, 574
   implicit, 234
   increasing, 211
   increment of, 135
   limit of, 149
   as a machine, 132
   multiple-valued, 127
   periodic, 336
   rate of change of, 165
   of several variables, 587, 596, 605
   single-valued, 127
   transcendental, 334
   trigonometric, 334

   value of, 130
   of x, 128
Function value, 130
Fundamental theorem of calculus, 290

General form of equation, of circle, 59
   of ellipse, 92
   of hyperbola, 104, 108
   of parabola, 79
   of second degree, 114
   of straight line, 43
General point, 8
General term of series, 609
Geometric series, 613
Graphs of equations, 30

algebraic, 122
exponential, 367
logarithmic, 367
hyperbolic, 577
parametric, 520
polar, 552
Gravity, force on planets, 318
Greater than, symbol, 2
Growth and decay, rate of, 636
Gyration, radius of, 502, 508

Half-life, 637
Harmonic series, 614
Higher degree curves, 116
Higher derivatives, 179
Homogeneous differential equation, 649
Hooke's Law, 472
Horizontal line, 19, 44
Hybrid point, 9
Hyperbola, 100, 101
   application, 110
   center, 101
   as a conic section, 114
   conjugate, 106
   conjugate axis, 106
   eccentricity, 100
   equation of, 102, 106
   equilateral, 111
   graph of, 104
   translation, 108
   transverse axis of, 101
Hyperbolic functions, 574
   derivatives of, 581
   graphs of, 577
   identities, 576
   integration of, 584
Hypocycloid, 536

Identities, hyperbolic, 576
Identification of conics, 114
Imaginary circle, 59
Imaginary roots of auxiliary equation, 662
Implicit differentiation, 237
Implicit functions, 234
   derivative of, 237, 598
Improper fractions, integration, 441
Improper integral, 396
Inch-pound, 473

Incidence, angle of, 83
Incident ray, 83
Inclination of a line, 18
Increasing function, 211
Increments, 4, 16, 134, 135
    relation between, 137
Indefinite integral, 261, 308
Independent variable, 29, 127
Indeterminate forms, 618
Inertia, moment of, 498, 509
    radius of gyration of, 502, 508
Infinite series, 609
    alternating, 617
    arithmetic, 613
    convergent, 612
    for cosine, 624
    divergent, 612
    for e^x, 623
    general term of, 609
    geometric, 613
    limit of nth term, 616
    p series, 615
    power series, 627
    for sine, 624
    sum of, 609
    tests, comparison, 613
        ratio test, 618
    use of, 621
Infinitesimal, 148
Infinity, 149
Inflection point, 117, 222
Initial conditions, 276, 633
Instantaneous rate of change, 164
Instantaneous velocity, 139, 196
Intercept form of equation, 44
Intercepts, 32, 61
Integral, as antiderivative, 258
    definite, 282, 318
    double, 605
    improper, 396
    indefinite, 261, 308
    particular, 274, 308
Integrand, 259
Integrating factor, 266, 643, 646
Integration, 258
    of algebraic fractions, 386, 413, 417, 420
    by algebraic substitution, 449
    approximate, 327
    constant of, 260, 387
    double, 605

    of exponential functions, 390
    of functions of two variables, 605
    of hyperbolic functions, 584
    leading to arcsine, 413
        to arctangent, 417
        in linear motion, 307
        to logarithms, 386
        to powers, 384
    by parts, 445
    of polar functions, 568
    power rule, 265, 384
    special rules, 263
    special techniques, 445
    supplying constant factor, 268
    trapezoidal rule, 327
    of trigonometric functions, 402–411
    by trigonometric substitution, 455
Intercepts, 32
Inverse functions, 352
    derivatives of, 358
    exponentials and logarithms, 366
    trigonometric, 358
Irrational fraction, 427

Latus rectum, of ellipse, 92
    of hyperbola, 104
    of parabola, 73
Length of arc, 515, 519, 572
Less than, symbol, 2
L'Hospital's rule, 618
Limaçon, 554
Limit, mathematical, 146–147
    of a function, 149
    $(\sin x)/x$, 340
    special cases, 157
    theorems on, 153, 206
Limits of integration, 283
Line, horizontal, 19, 44
    slope of, 15
    vertical, 19, 44
Linear equation, 31
Linear motion, 194
    differentiation, 196
    integration, 307, 318
Liquid pressure, 464
Logarithmic curve, slope of, 373
Logarithmic differentiation, 376
Logarithmic function, 365
    base of, 365, 370

702   Index

derivative of, 369
graph of, 367
as an integral, 386
natural base, 368, 369
slope of, 373

Maclaurin series, 622
Mass, center of, 481, 490
   moment of, 493, 507
   second moment of, 499, 510
Maxima and minima, 211–213
   point, 212
   tests for, 222
   value, 212, 213, 218
Midpoint, 12
Modification rule, 677
Modulus, 472
Moment, of area, 484
   first, 481, 486
   of force, 481
   of inertia, 498, 509
   of mass, 493, 507
   total, 482
Moment of inertia, of area, 498
   of cylinder, 509
   of solid, 491, 507
Motion, of a projectile, 321
   rectilinear, 319
Multiple integral, 605
Multiple-valued function, 127

Nappes of a cone, 113
Natural base e, 368
Natural logarithm, 368
Newton's law of cooling, 639
Nonhomogeneous differential equations, 672, 673
Normal form, straight line equation, 51
Normal intercept, 51
Number line, 1

Octant, 26
One-dimensional system, 1
One-to-one correspondence, 7
Operator, derivative, 176, 650
   solution by, 652, 658
Order of differential equation, 628
Ordered pair, 7, 127
Ordinate, 7

Origin, 1, 6, 25

p series, 615
Parabola, 36, 72
   as a conic section, 114
   directrix of, 72
   equation of, 73, 76
   focus of, 72
   graph of, 75
   intersection with other curves, 82
   principal axis, 73
   reflecting principle, 83
   translation of, 79
Parallel lines, slope, 20
Parameter, 323, 527
   elimination of, 527
Parametric equations, 323, 527
   of circle, 532
   of cycloid, 535
   differentiation of, 537
   of ellipse, 532
   graphing, 528
   of projectile, 323, 534
   second derivative of, 541
Partial derivative, 587
   chain rule, 599
   geometric meaning, 591
   in implicit differentiation, 599
   notations for, 603
   second partial, 601
Partial fractions, 427
   methods of findings, 429–440
Particular integral, 274, 276, 308
Particular solution, 279, 310, 632, 674
Parts, integration by, 445
Perfect differential, 641
Period, 336
Periodic curves, 336
Perpendicular lines, 20
Phase angle, 336
Point, general, 8
   hybrid, 9
   specific, 2, 8
Point-circle, 59
Point-slope form of equation, 42
Points, distance between, 3, 10, 27
   in a plane, 6
   in space, 25
Polar axis, 6, 545
Polar coordinates, 6, 545

arc length in, 572
area in, 568
conics, 558
differentiation, 564
graphing, 552
relation to rectangular, 547
Polar curves, intersection, 561
  slope of, 564
Pole, 6
Power rule, in differentiation, 174, 183, 348, 349
  in integration, 265, 266, 406
  limitations of, 384
Power series, 621
Pressure, liquid, 464
Principal axis, of parabola, 73
Principal values, 355
Problem solving by derivative, 225, 230
Problems of analytic geometry, 42
Product rule, differentiation, 188
Projectile, 321, 534
Projection, 9

Quadrants, 7
Quadratic factors in partial fractions, 439
Quotient rule, differentiation, 190

Radian, 334
Radius of gyration, 502, 508
  of a cylinder, 510
  of a slender rod, 513
  of a solid, 508
Range, of a function, 33
  of a projectile, 322, 534
Rate of change, of angle, 357
  average, 138
  of a function, 165
  instantaneous, 139, 164
Rates, related, 241, 243
  time-rates, 242
Ratio-test of infinite series, 618
Rational fraction, 427
Rectangular coordinates, 6
  relation to polar, 547
Reflected ray, 83
Reflection, angle of, 83
Reflecting principle, of ellipse, 98
  of parabola, 83
Related rates, 241, 243

Related time-rates, 243
Relation, 127
Repeated factors in partial fractions, 436
Repeated roots of auxiliary equation, 657
Rotation of axes, 114

Scalar components of a vector, 17
Secant, derivative of, 347
Second-degree equation, 35, 114
Second partial derivative, 601
Sector of a circle, area, 339
Separation of variables, 629
Sequence, 609
Series, 609
  convergent, 612
  divergent, 613
  finite, 609
  infinite, 609
  tests for convergence, 613, 618
Shell method, for first moment, 494
  for second moment, 509
  for volume, 303
Simpson's rule, 333
Sine, derivative of, 341
Sine and cosine solution of differential equation, 663
Sine series, 624
Single-valued functions, 127, 354
Slope of a curve, 203, 212
  of exponential curve, 381
  of logarithmic curve, 373
  of parallel lines, 20
  of perpendicular lines, 20
  of polar curves, 564
  rate of change of, 219
  of straight line, 15
  of trigonometric curves, 344
Slope-intercept form of equation, 46
Slug, 666
Solid, first moment of, 491
  moment of inertia of, 507
  of revolution, 299
Solution, of a differential equation, 627
  of an equation, 29
  general solution, 271, 310, 627
  particular solution, 279, 310, 632, 674
Space coordinates, 25

Specific point, 2, 8
Spherical aberration, 84
Spring, constant or modulus, 472, 667
Spring problem, 472
   frequency of, 668
   vibration, 665
Straight line, 42
   equation of, 43
   general form, 48
   horizontal, 47
   intercept form, 45
   normal form, 51
   point-slope, 42, 44
   slope-intercept, 46
   through two points, 44
   vertical, 47
Substitution, algebraic, 449
   trigonometric, 455, 461
Supplying a constant factor, 268, 392
Surface of revolution, area, 521
   element of, 523
Symbols, greater or less than, 2
Symmetry, 36
System of coordinates, one-dimensional, 1
   polar, 545
   rectangular, 6
   three-dimensional, 25
   two-dimensional, 5

Tangent, derivative of, 346
Tangents to a circle, 68
   length of, 69
Tank problem, 475
Taylor's series, 624
Theorems, on limits, 153, 206
Third-degree equation, graph, 117
Three-step rule, 349
Time-rates of change, 242
Torque, 481
Total differential, 593
Transcendental functions, 334
Transformation triangle, 458
Translation of axes, 78, 95, 108
Trapezoidal rule, 327
Trigonometric curves, 335
   slopes, 334
Trigonometric functions, derivatives, slopes, 344
   graphs, 335
   integration, 402
   three-step rule, 349
Trigonometric substitutions, 455, 461
   transformation triangle for, 458
Two-dimensional coordinate system, 5

Undetermined coefficients, 674
Upper limit of definite integral, 283

Variable, dependent, independent, 29, 126–127
Variable height, liquid pressure, 466
Variation of parameters, 679
Vector, scalar components, 17
Velocity, average, 125, 195
   components, 321
   instantaneous, 125, 196
   reference to time, 242
Vertical line, equation of, 44
Volume, by disc method, 300
   element of, 299
   by shell method, 303
   of solid of revolution, 299

Weight versus mass, 491
Work, chain problem, 477
   foot-pound, 471, 472
   inch-pound, 473
   spring problem, 472
   tank problem, 475

x-axis, 6
x-bar, 485
x-intercept, 32

y-axis, 6, 25
y-bar, 485
y-intercept, 32

z-axis, 25
Zero factorial, 610